APPLIED CALCULUS

APPLIED CALCULUS

JOHN C. HEGARTY
BENTLEY COLLEGE

JOHN WILEY & SONS
NEW YORK • CHICHESTER • BRISBANE • TORONTO • SINGAPORE

Cover and text designed by Ann Marie Renzi
Cover photograph by Photographic Associates

Printed in Singapore
10 9 8 7 6 5 4 3 2 1

To Marie

PREFACE

Applied Calculus has been written for use in an introductory calculus course for students pursuing professional careers in the management, life, or social sciences. The goal of the book is to present, in an intuitive and clear manner, the fundamental concepts of differential and integral calculus to a student who is primarily interested in learning how calculus can be used. For this reason numerous examples and applications, where appropriate, are given in each section to illustrate and enhance new concepts. In addition, attention has been paid to making the explanations and examples as complete and detailed as possible to increase the readability of the book.

Although the presentation is not mathematically rigorous, basic concepts such as the limit, continuity, the derivative, and definite and indefinite integrals are developed carefully and systematically. Because students assimilate mathematical concepts more quickly and easily when they can visualize their meaning, there is a heavy reliance and emphasis on graphing not only in the presentation of new material, but also in the examples and in the problems. For example, graphs are used in Exercises 3.1, 3.2, and 3.3 to test and enhance a student's understanding of the derivative and the rules of differentiation. They are also used in Exercises 4.1 and 4.3 to test and enhance a student's understanding of the meaning and significance of the signs of the first and second derivatives. Exercises of this type have the advantage that no lengthy algebraic analysis is required; as a result a student can attain a considerable understanding of the basic concepts in a short period of time.

Basic curve sketching, whenever appropriate, is emphasized. Although curve sketching can, in many cases, be carried out more quickly with a computer or a calculator, critical points and inflection points are of such fundamental importance that a student should know how to determine and, simultaneously, to *visualize* the behavior of a function in the vicinity of each type. The ability to visualize the results of an analysis should be an important objective of any calculus course; curve sketching provides such a vehicle. For this reason the development of the curve-sketching process is carried out in detail for each example in Section 4.4. The same attention to detail is used in the curve-sketching examples for exponential and logarithmic functions in Sections 5.3 and 5.4.

The number and scope of the applications have been kept at a reasonable level. Students should realize that calculus is an important analytical tool in many disciplines and should gain some experience with its applications. On the other hand, there may be a price to pay when students are confronted with an overwhelming number of applications. Students often spend an inordinate amount of time struggling to understand the dynamics of an application without gaining a commensurate increase in their understanding of the fundamental concepts of calculus. Since the majority of students enrolled in applied calculus courses are in the fields of business and economics, the majority of applications are in these areas. The book contains two new elementary business applications developed by the author. Both have recently been published.*

Through problems in the exercises, I have attempted, in a small way, to focus each student's attention on the relationship between differential and integral calculus and on the types of problems that each addresses. The problems range from simple true–false questions in Exercise 6.1 to water supply and consumers' surplus problems in Exercises 6.3 and 6.4; the latter require a student to use optimization techniques to find the upper limit for a definite integral which is to be evaluated.

The text also makes extensive use of tables as visual aids or summaries. Tables are used extensively in Chapter 2 in developing the concept of the limit. Tables that highlight the signs of the first and second derivatives are used throughout Chapter 4. Table 1 of Section 3.3 describes the relationship between the last operation in an algebraic function and the appropriate rule used to differentiate the function. Table 1 in Section 6.1 lists the names of derivatives and their antiderivatives in some basic applications.

Supplements

An Instructor's Manual which contains

- complete solutions to all exercises and review problems,
- a syllabus for one- and two-semester courses,
- a listing of business applications by course,
- a review of algebra, which an instructor can reproduce for students who may be weak in the mathematical prerequisites.

A Student Study Guide and Solutions Manual which contains

- terms, concepts, and formulas,
- over 80 worked-out problems with solutions tips,
- solutions to most problems with emphasis on word problems,
- a review of algebra.

*1. Hegarty, J. C. "A Depreciation Model for Calculus Classes," *The College Mathematics Journal,* May 1987, Vol. 18, No. 3, pp. 219–221.

2. Hegarty, J. C. "Calculus Model of Sum-of-the-Years-Digits Depreciation," *Mathematics and Computer Education,* Fall 1987, Vol. 21, No. 3, pp. 159–161.

Three Test System options averaging 50 questions per chapter

- Printed
- IBM PC compatible
- Macintosh

John C. Hegarty

ACKNOWLEDGMENTS

I want to take this opportunity to express publicly my appreciation to the people who have contributed to this book.

First, I want to thank the following friends and colleagues at Bentley College who were extremely generous and gracious with their comments and suggestions: Amir Aczel, Claire Archambault, Yvonne Aghassi, Brad Allen, Dermott Breault, Mary Briggs, David Carhart, Larry Dolinsky, Michael Epelman, James Fullmer, Alfred Galante, Larry House, Norman Josephy, Joseph Kane, Nikolaos Kondylis, Jerry Kuliopolos, Philip Laurens, Harvey Leboff, Kenneth McGourty, Carroll McMahon, Barbara Nevils, Thalia Papageorgiou, Harold Perkins, John Powers, Michael Saxe, Karen Schroeder, Richard Swanson, Erl Sorensen, Stanley Tannenholtz, and Jason Taylor.

I am grateful to the following reviewers for their insightful comments, opinions and advice.

Marcia Bain, Florida Community College at Jacksonville
Jeff Brown, University of North Carolina at Wilmington
Lyle Dixon, Kansas State University
James Gauthier, Louisiana State University at Alexandria
Neil Gretsky, University of California at Riverside
Larry Griffey, Florida Community College at Jacksonville
Kempton Huehn, California State Polytechnic Institute
Gary Itzkowitz, Glassboro State College
Joe Jenkins, State University of New York at Albany
Giles Wilson Maloof, Boise State University
Ed Mealy, University of Wisconsin at River Falls
Tom Miles, Mount Pleasant, Michigan
Edward Miranda, Saint Johns University
Elgin Schilhab, Austin Community College, Reagan Campus
Edith Silver, Mercer County Community College
Phil Smith, American River College
Sheng Seung-So, Central Missouri State University
Alan Sultan, Queens College
Richard Weimer, Frostburg State University
Earl Zwick, Indiana State University

Finally, I wish to thank the following people at John Wiley for their untiring efforts and assistance in this endeavor: Elizabeth Austin, Donald Ford, Page Mead, Stacey Memminger, Ann Renzi, and Priscilla Todd.

J. H.

CONTENTS

1

PRELIMINARIES

INTRODUCTION

Calculus is the branch of mathematics that enables us to analyze or determine the effects of change. For example, the president of a fast-food chain would like to know what effect a 25¢ increase in the price of the company's popular cheeseburger will have on sales. Similarly, an ecologist would want to know how much the acidity of a lake will be changed over the next ten years by the acid content of the rainfall during this period.

In order to analyze the effect that a change in one variable has on another, it is necessary to know the mathematical relationship between the two variables. This relationship is generally expressed as a function. This chapter is devoted to studying some simple functions, their graphs, and applications.

1.1 FUNCTIONS AND THEIR GRAPHS

When the value of one variable depends on the value of a second variable, we say that the first is a function of the second and attempt to express the relationship mathematically. For example, A, the area of a square, depends on x, the length of each side, and the relationship between the two variables can be written as

$$A = x^2$$

The variable A is also said to be a *function* of the variable x. For each value that is assigned to x, the equation enables us to find the corresponding value of A; for example, if $x = 5$, then $A = (5)^2 = 25$. The variable x to which values are assigned is called the **independent** variable; A, the variable whose value is found from the equation, is called the **dependent** variable. The functional relationship between two variables is expressed in the following definition.

Definition A **function** f is a rule that matches or pairs each value of an independent variable, say, x, with exactly one value of a second variable, say, y. Each matching is written as an **ordered pair** (x, y) and y is said to be a function of x.

For example, the function f defined by the equation

$$y = x^2 + 1$$

is a rule that tells us to square each value of x and then to add 1 to get the corresponding value of y. Thus the number $x = 2$ is paired with the number

$$y = (2)^2 + 1 = 5$$

written as $(2, 5)$.

In applications, variables other than x and y are often used in representing functions, as illustrated in Example 1.

Example 1 The Speedy Rent-a-Car Agency charges \$15 per day plus 25¢ per mile for its compact autos. Find an equation that gives D, the charge in dollars, as a function of M, the daily mileage.

Solution The equation for D contains two terms: (1) the \$15 that is paid to take the car onto the road and (2) the mileage charge, in dollars, which is calculated by multiplying M by 0.25. Adding these two terms gives us the equation

$$D = 15 + 0.25M$$

If we drive 200 miles during the day, the charge becomes

$$D = 15 + 0.25(200) = \$65$$

■

f(x) Notation

When we define a function, we often write the variable y as $f(x)$ (read "f of x"). This notation is very useful in calculus because any value assigned to x is retained in the parentheses while we determine the corresponding value of y. For example, suppose the function f is defined as

$$y = f(x) = x^2 + 2x - 1$$

If x is assigned the value 3, the corresponding value of y, written $f(3)$, becomes

$$\begin{aligned} f(3) &= (3)^2 + 2(3) - 1 = 9 + 6 - 1 \\ &= 14 \end{aligned}$$

This tells us that $x = 3$ is paired with $y = 14$.

Example 2 Suppose the function f is defined by the equation

$$f(x) = 3x^2 - 5x + 7$$

Find $f(4)$ and $f(-1)$.

Solution We find $f(4)$ by substituting 4 for x everywhere in the equation that defines $f(x)$:

$$\begin{aligned} f(4) &= 3(4)^2 - 5(4) + 7 \\ &= 35 \end{aligned}$$

Similarly, we get for $f(-1)$

$$f(-1) = 3(-1)^2 - 5(-1) + 7 = 15$$

■

In addition to $f(x)$, functional notation can be described by other letters such as $g(x)$ and $A(x)$, which may be better descriptors of the relationship between the independent and dependent variables. This is particularly true in applications.

Example 3 A pipe on an oil rig has ruptured and crude oil is spilling into the sea in all directions, forming a circular oil slick. The radius of the oil slick is a function of t, the time, in hours, from the rupture; the relationship is given by the equation

$$R(t) = 1000\sqrt{t}$$

where $R(t)$ is the radius, in feet. An additional crew has been dispatched to cap the well and to make the necessary repairs; however, it will take 36 hours to stop the leaking.

(a) How large will the radius of the oil slick be when the pipe is repaired?
(b) How large an area will the oil slick cover?

Solution (a) The radius of the oil slick is found by substituting 36 for t into the equation, yielding

$$\begin{aligned} R(36) &= 1000\sqrt{36} \\ &= 1000(6) = 6000 \text{ ft} \end{aligned}$$

(b) Recall that the area of a circle is given by the formula

$$A = \pi R^2$$

We get for the area

$$A = \pi(6000)^2 = 36{,}000{,}000\pi \text{ ft}^2$$

This information is useful to the owners of the rig because it enables them to estimate the cost of cleaning up the oil spill. ■

The $f(x)$ notation is particularly useful when x is replaced by an algebraic expression instead of a number. This occurs often when analyzing a function in calculus.

Example 4 If the function f is defined as

$$f(x) = x^2 - 3x + 2$$

find $f(1 + h)$.

Solution We can find $f(1 + h)$ by substituting $1 + h$ for x everywhere in the equation

$$f(1 + h) = (1 + h)^2 - 3(1 + h) + 2$$

The expression on the right-hand side can be simplified by expanding and grouping like terms together:

$$\begin{aligned} f(1 + h) &= (1 + 2h + h^2) - 3 - 3h + 2 \\ &= 1 + 2h + h^2 - 3 - 3h + 2 \\ &= h^2 - h \end{aligned}$$

This result tells us that when x is equal to $1 + h$, the corresponding value of y is $h^2 - h$. ■

Domain

When we work with a function, it is necessary to know what values can be assigned to the independent variable; these values constitute the **domain** of the function.

In Example 1 (Speedy Rent-a-Car), the variable M has meaning only for nonnegative values. Therefore, we say that the domain consists of the values of M for which $M \geq 0$.

When the domain is not given or cannot be inferred from the context of the problem, the domain consists of all real values of x for which the function is defined.

Example 5 Find the domain of each of the following functions:

(a) $f(x) = x^2 - 2$ (b) $f(x) = \dfrac{3}{2 - x}$ (c) $f(x) = \sqrt{x - 1}$

Solution (a) The expression $x^2 - 2$ produces a real number for every value of x that is substituted into it, so we conclude that the domain consists of all real numbers.

(b) The quantity $\dfrac{3}{2 - x}$ produces a real number for every value of x except $x = 2$, where division by zero occurs; so the domain consists of all real numbers except $x = 2$.

(c) The quantity $\sqrt{x - 1}$ generates real numbers only when $x - 1 \geq 0$, so the domain consists of values of x that are greater than or equal to 1, that is, $x \geq 1$. ■

Graph of a Function

When we work with a function, it is helpful to visualize the relationship between the independent variable x and the dependent variable $y = f(x)$. We accomplish this by sketching the *graph* of the function.

Recall that an ordered pair (x, y) or $(x, f(x))$ can be represented geometrically as a point in a Cartesian or xy-coordinate system; the point representing the ordered pair $(3, 2)$ is shown in Figure 1. The **graph** of a function f is the set of all points corresponding to the ordered pairs $(x, f(x))$ that satisfy the equation defining the function.

When a function is defined by a simple equation, its graph can usually be sketched by finding a representative group of points and then connecting them by means of a smooth curve, as the next two examples illustrate.

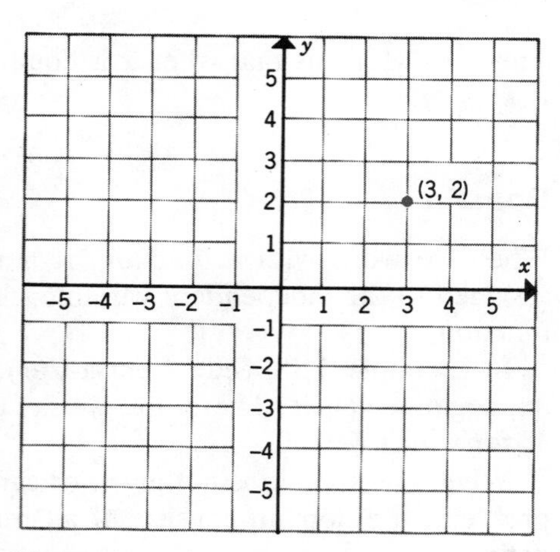

Figure 1

Example 6 Sketch the graph of the function $f(x) = x^2 - 2$.

Solution A table containing representative values of x and y is used to generate points to sketch the graph in Figure 2.

x	$y = x^2 - 2$
−3.0	7.0
−2.0	2.0
−1.0	−1.0
0.0	−2.0
1.0	−1.0
2.0	2.0
3.0	7.0

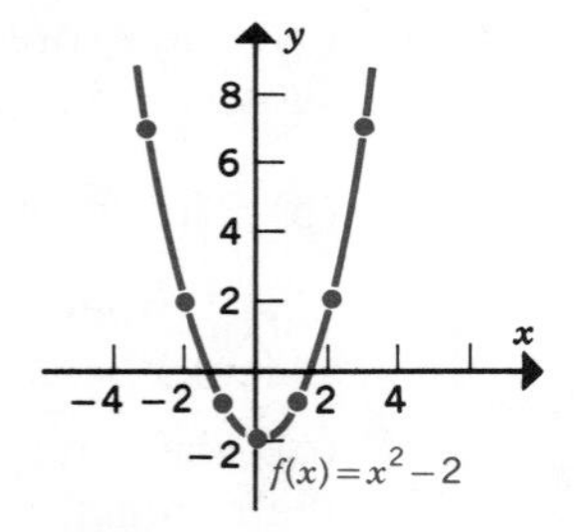

Figure 2 ■

When we graph a function, it is helpful to indicate the values of x for which the function is not defined. This procedure highlights regions of the xy-plane through which the graph does not pass. In addition, we can usually improve our sketch if we select values of x close to those that do not belong to the domain.

Example 7 Sketch the graph of the function $f(x) = \dfrac{3}{2 - x}$.

Solution First, note that the function is not defined when $x = 2$; this means that the graph will not contain a point whose x coordinate equals 2. The graph can be sketched by selecting values of x, taking care to include *nonintegral* values close to 2, and calculating the corresponding values of y. The resulting graph is shown in Figure 3.

x	$y = \dfrac{3}{2 - x}$
−4.00	0.50
−3.00	0.60
−2.00	0.75
−1.00	1.00
0.00	1.50
1.00	3.00
1.50	6.00
2.00	undefined
2.50	−6.00
3.00	−3.00
4.00	−1.50
5.00	−1.00

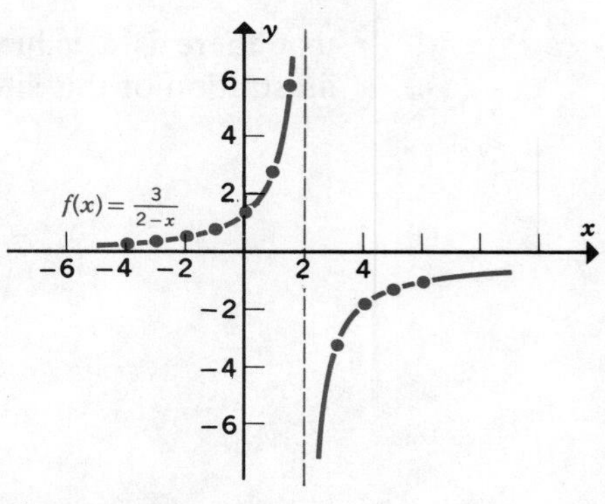

Figure 3

■

Not all equations in two variables represent functions. If the equation has the property that, for some x, there are two or more corresponding values of y, the equation does not represent a function. A simple illustration of this situation is the equation

$$y^2 = x - 1$$

Note: Except for $x = 1$, two values of y are paired with each value of x, for which a solution can be found. For example, when $x = 5$, two solutions for y are generated from the resulting equation $y^2 = 4$,

$$y_1 = +2 \qquad y_2 = -2$$

We can sketch the graph of the equation by first rewriting the equation as $y = \pm\sqrt{x - 1}$; because $x - 1 \geq 0$, or $x \geq 1$, the graph will not pass through points for which $x < 1$, as Figure 4 illustrates.

x	$y = \pm\sqrt{x - 1}$
1	0
2	±1
5	±2
10	±3

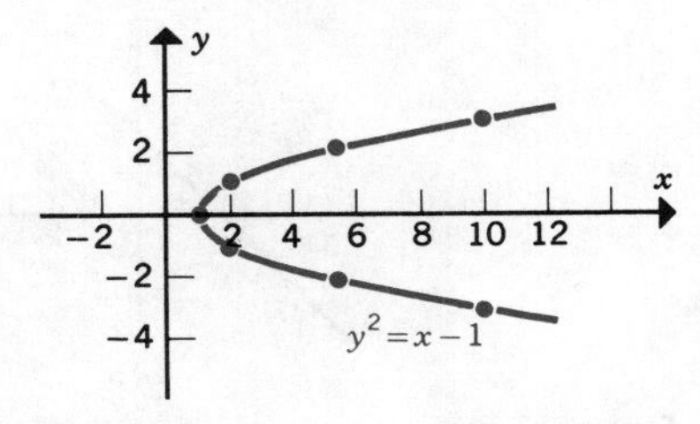

Figure 4

■

Vertical-Line Test

The graph of an equation in two variables can often indicate whether or not the equation represents a function. If a vertical line drawn on the coordinate system intersects the curve at two or more points, the curve cannot represent a function; the intersection of a vertical line with two or more points on the curve shows that there is a value of x for which there are two or more values of y. An illustration of this situation is shown in Figure 5.

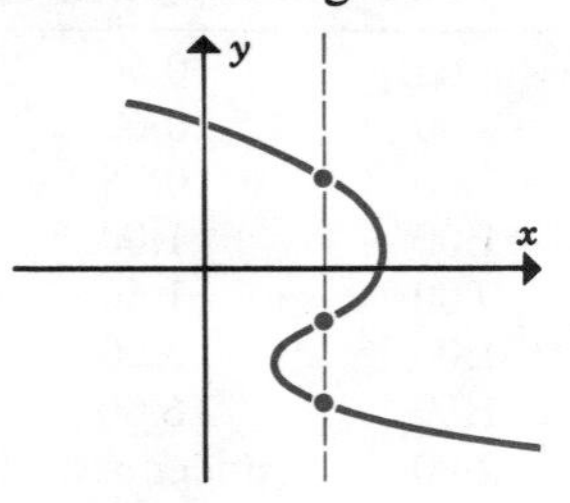

Figure 5

Piecewise Functions

Sometimes, two or more equations may be needed to define a function. In such situations, the domain consists of "pieces," one for each equation, as shown in the following example.

Example 8 Sketch the graph of the function

$$f(x) = \begin{cases} x + 2, & x < 1 \\ 3 - x, & x \geq 1 \end{cases}.$$

Solution The graph is found by first plotting the equation $y = x + 2$, restricting the variable x to values less than 1. The graph is shown in Figure 6*a*, where the open circle indicates that (1, 3) is not a point on this part of the graph. Next, the graph of the equation $y = 3 - x$ is plotted using only values of x greater than or equal to 1; the graph is shown in Figure 6*b*. Superimposing the two segments or pieces results in the graph of the function shown in Figure 6*c*.

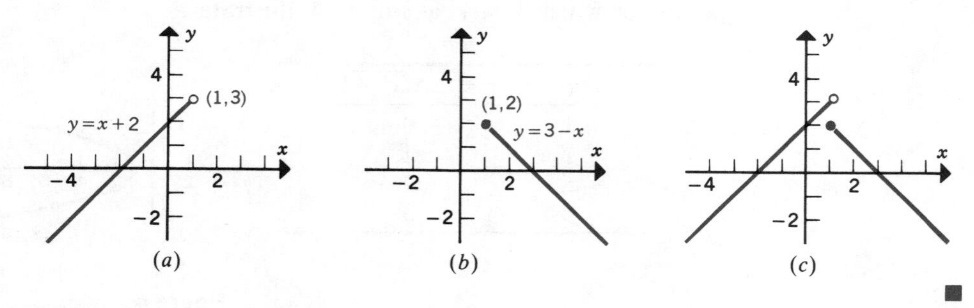

Figure 6

■

Application

The hourly salary of many workers increases when they work overtime. In these situations, one equation is not sufficient to write a worker's weekly salary as a function of the number of hours he or she works, as the next example illustrates.

Example 9 A medical technician is paid \$10 per hour for a 35-hour week. She is paid \$15 per hour (time and a half) for each hour over 35 she works. Find $S(h)$, her weekly salary, as a function of h, the number of hours she works each week.

Solution When $h \le 35$, her salary is given by the equation $S(h) = 10h$. When $h > 35$, the equation becomes $S(h) = 350 + 15h$. Now we can summarize these results as

$$S(h) = \begin{cases} 10h, & h \le 35 \\ 350 + 15h, & h > 35 \end{cases}$$

■

1.1 EXERCISES

1. If the function f is defined by the equation $f(x) = x^2 - x$, find
(a) $f(2)$ (b) $f(-1)$ (c) $f(0)$ (d) $f(8)$

2. If the function g is defined by the equation $g(x) = x^3 - x^2 + 1$, find
(a) $g(1)$ (b) $g(-1)$ (c) $g(5)$ (d) $g(-10)$

3. If the function h is defined by the equation $h(x) = x^4 + x^2 - 3$, find
(a) $h(2)$ (b) $h(-2)$ (c) $h(4)$ (d) $h(-3)$

4. If $f(x) = 3x^2 - 2x + 1$, find
(a) $f(1)$ (b) $f(-2)$ (c) $f(\frac{1}{2})$ (d) $f(\frac{2}{3})$

5. If $f(x) = x^3 + 1$, find
(a) $f(2)$ (b) $f(-1)$ (c) $f(-\frac{1}{2})$ (d) $f(b)$

6. If $g(x) = x^2 + x - 1$, find
(a) $g(3)$ (b) $g(-10)$ (c) $g(c)$ (d) $g(c + 1)$ (e) $g(c + 1) - g(c)$

7. If $f(t) = t^2 - t + 2$, find
(a) $f(d + 1)$ (b) $f(d - 1)$ (c) $f(d^2)$ (d) $f(d + 1) - f(1)$

8. If $h(x) = \dfrac{1}{x + 1}$, find
(a) $h(0)$ (b) $h(\frac{1}{2})$ (c) $h(-\frac{1}{2})$ (d) $h(c - 1)$

9. If $f(s) = \sqrt{s - 1}$, find
(a) $f(1)$ (b) $f(\frac{5}{4})$ (c) $f(a + 1)$ (d) $f(a + 1) - f(1)$

10. If $f(x) = \dfrac{x}{x^2 + 1}$, find
(a) $f(0)$ (b) $f(1)$ (c) $f(-1)$ (d) $f(\frac{1}{3})$

In Exercises 11 through 16, indicate which of the graphs represent functions and which do not.

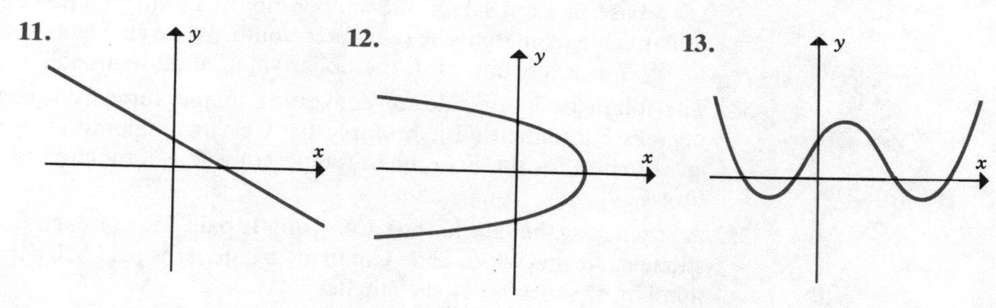

14.

15.

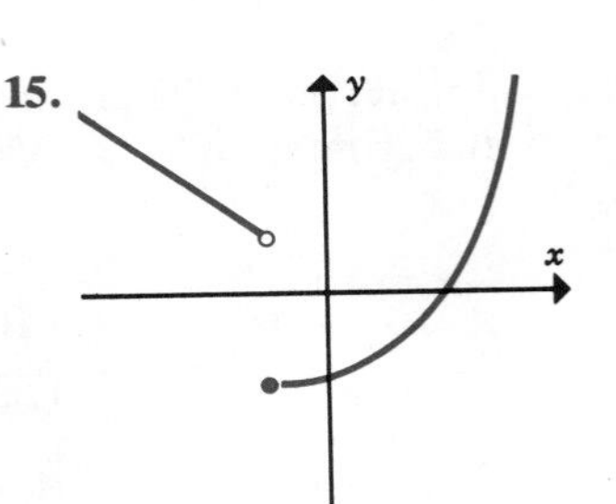

16.

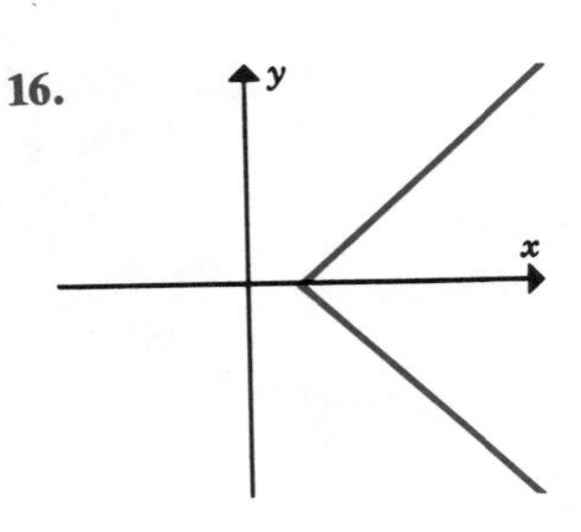

Find the domain of the functions in Exercises 17 through 22.

17. $f(x) = x^2 - 2x + 5$

18. $f(x) = \dfrac{3}{x^2}$

19. $f(x) = \dfrac{2}{x^2 - x + 2}$

20. $f(x) = \sqrt{1 - x}$

21. $g(x) = \dfrac{x - 1}{x + 1}$

22. $f(\mathrm{x}) = \sqrt{x^2 + 1}$

Plot the graph of the functions in Exercises 23 through 28.

23. $f(x) = x^3$

24. $f(x) = \dfrac{2}{x^2}$

25. $g(x) = \dfrac{-1}{x - 1}$

26. $f(x) = 1 - \sqrt{x}$

27. $f(x) = x^2 - x$

28. $f(x) = \dfrac{1}{x^2 + 1}$

29. If $f(x) = \begin{cases} x, & x < 0 \\ x^2, & x \geq 0 \end{cases}$

find (a) $f(-2)$ (b) $f(-1)$ (c) $f(1)$ (d) $f(3)$

30. Sketch the graph of the piecewise function

$$f(x) = \begin{cases} x, & x < 0 \\ x^2, & x \geq 0 \end{cases}$$

31. Sketch the graph of the piecewise function

$$f(x) = \begin{cases} x + 1, & x \leq 1 \\ -x^2, & x > 1 \end{cases}$$

32. A part-time cashier in a supermarket earns \$4 per hour. Write an equation that describes her weekly salary, S, as a function of h, the number of hours worked each week.

33. The monthly salary of a salesclerk in a large department store consists of two terms: (1) a base or fixed salary of \$400 per month and (2) a 3 percent commission on the dollar value of all items he sells each month. Write an equation that gives his monthly salary S as a function of d, the dollar value of all items sold.

34. The following is the rule for converting temperature from degrees Celsius (°C) to degrees Fahrenheit (°F): Multiply the Celsius temperature by $\frac{9}{5}$ and add 32. Write an equation for the Fahrenheit temperature F as a function of the Celsius temperature C.

35. A stitcher at the Black Shoe Company is paid 25¢ for each pair of shoes he or she stitches. At the White Shoe Company a stitcher is paid \$50 per day regardless of the number of shoes he or she stitches.

(a) Find an equation that describes the daily salary of a stitcher at the Black Shoe Company as a function of the number of pairs of shoes stitched each day. Plot the graph of this function.
(b) Repeat part (a) for a stitcher at the White Shoe Company, plotting the function on the same coordinate system as that used in part (a).
(c) Under what conditions would a stitcher be better off financially working for the Black Shoe Company? The White Shoe Company?

36. Piecewise functions are often used in applications involving price discounts. Dip Printing has the price schedule in Table 1 for photocopying a document.

Table 1

Number of Copies	Price per Copy
1–10	10¢
Over 10	8¢

(a) Write p, the price per copy, as a function of n, the number of copies to be made.
(b) Find a function that describes the cost to get n copies of a document made.
(c) Sketch the graphs of each of the functions in parts (a) and (b).

37. Exercise physiologists use the equation $M(A) = 220 - A$ to describe the relationship between $M(A)$, the maximum heart rate, in beats per minute, and the age A, in years, of a person who is exercising vigorously.
(a) Sketch the graph of this equation.
(b) What is the maximum heart rate that a 25-year-old woman should attain during an aerobic workout?

38. Accident investigators use the equation $S(L) = 5.5\sqrt{L}$ to estimate the speed $S(L)$ of a car, in miles per hour, from the length L, in feet, of the skid marks on a dry road.
(a) Sketch the graph of this equation.
(b) How fast was a car moving when the brakes were applied if the length of the skid marks is 81 ft?

1.2 LINEAR FUNCTIONS

Certain kinds of functions appear so often in applications that they deserve special attention. Linear functions fall into this category.

Definition 1 A function f is a **linear function** if it can be written as

$$y = f(x) = mx + b \tag{1}$$

where m and b are constants.

Example 1 The weekly salary for sales people at the D to D Vacuum Cleaner Company consists of two parts:

(a) A base salary of \$100.
(b) A commission of \$20 for each vacuum cleaner sold.

Find an equation that gives y, the weekly salary (in dollars), as a function of x, the number of vacuum cleaners sold each week. In addition, sketch the graph of the equation.

Solution The equation defining y contains two terms:

$$y = \text{base salary} + \text{commission}$$

where

$$\begin{aligned} \text{Base salary} &= \$100 \\ \text{Commission} &= 20 \times \text{number of vacuum cleaners sold} \\ &= 20x \end{aligned}$$

The equation takes the form

$$y = 20x + 100$$

Its graph is shown in Figure 1.

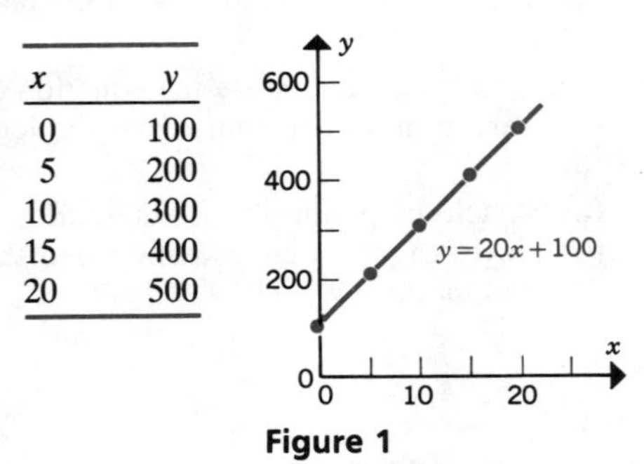

x	y
0	100
5	200
10	300
15	400
20	500

Figure 1

■

Example 1 illustrates some important properties of linear functions:

1. Their graphs are straight **lines.**
2. The coefficient of x, 20, represents a property of the line known as the **slope;** it tells us that increasing x by one unit causes y to increase by \$20.
3. The constant term, 100, represents the y coordinate of the point where the line intersects the y axis, known as the ***y* intercept.**

Before proceeding further, we need to define what is meant by the slope of the line passing through any two distinct points (x_1, y_1) and (x_2, y_2), as shown in Figure 2.

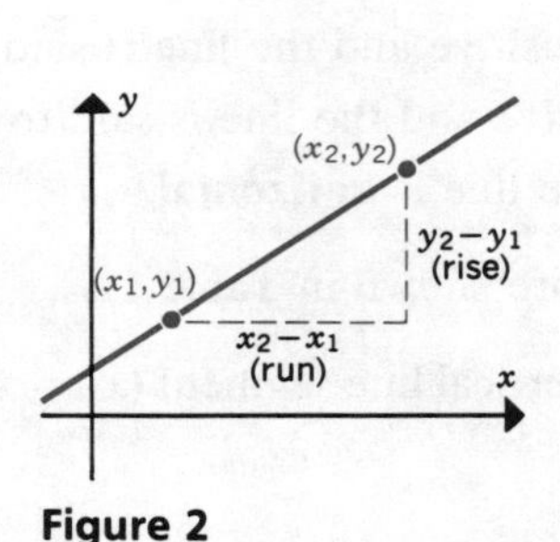

Figure 2

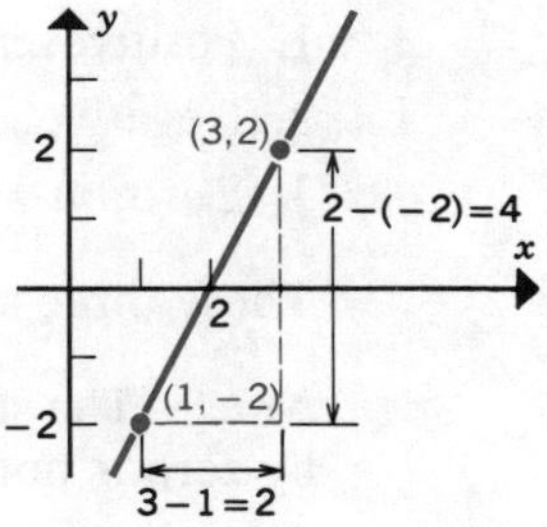

Figure 3

Definition 2 The **slope** m of the line which passes through the points (x_1, y_1) and (x_2, y_2) is defined as

$$m = \frac{y_2 - y_1}{x_2 - x_1} \quad \text{provided } x_1 \neq x_2 \tag{2}$$

The vertical change $y_2 - y_1$ is also called the *rise*, and the horizontal change $x_2 - x_1$ is called the *run*, so that the slope is often written as

$$m = \frac{\text{rise}}{\text{run}} = \frac{\text{change in } y}{\text{change in } x}$$

Example 2 Find the slope of the line which passes through $(1, -2)$ and $(3, 2)$.

Solution With $(x_1, y_1) = (1, -2)$ and $(x_2, y_2) = (3, 2)$, Equation 2 yields

$$m = \frac{y_2 - y_1}{x_2 - x_1} = \frac{2 - (-2)}{3 - 1} = \frac{4}{2} = 2$$

The same result is obtained if we let $(x_1, y_1) = (3, 2)$ and $(x_2, y_2) = (1, -2)$. This change only reverses the signs of both the numerator and denominator, so that the ratio itself is not affected. The graph is given in Figure 3. ■

Example 3 Find the slope of the line connecting $(0, 100)$ and $(15, 400)$ in Example 1.

Solution With $(x_1, y_1) = (0, 100)$ and $(x_2, y_2) = (15, 400)$, equation (2) gives

$$m = \frac{y_2 - y_1}{x_2 - x_1} = \frac{400 - 100}{15 - 0} = \frac{300}{15} = 20$$

■

The slope m tells us two things about a line: (1) its **direction** and (2) its **steepness.** Before we consider direction, let us adopt the convention that movement along a line or curve occurs from left to right and that $x_2 - x_1$, the change in x, is always a positive number. If $y_2 - y_1$, the corresponding change in y, is

1. Positive, m is also **positive** and the line is said to be **rising.**
2. Negative, m is **negative** and the line is said to be **falling.**
3. Zero, $m = 0$ and the line is **horizontal.**

These three situations are shown in Table 1.

Note: The slope of a vertical line segment ($x_1 = x_2$) is not defined since division by zero is not defined.

Table 1

Line	m	Graph
Rising	+	/
Falling	−	\
Horizontal	**0**	—
Vertical	Undefined	• (x_2, y_2), • (x_1, y_1) ; $x_2 = x_1$

The slope m is also an indicator of the steepness of a line; the greater the magnitude or size of the slope, the steeper the line. Figures 4 and 5 indicate that the lines become steeper as the magnitude of the slope gets larger.

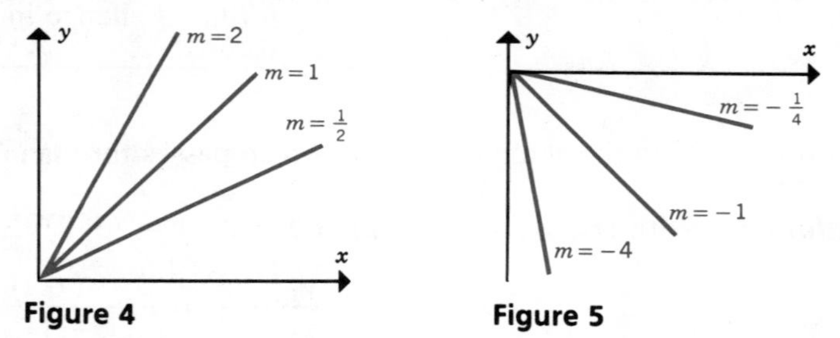

Figure 4 Figure 5

We now consider what roles the direction and steepness of a line can play in decision making. Figure 6 shows graphs of the profits as a function of time for three companies, A, B, and C. If you were going to make a long-term investment in one of the three companies, which one would you select, assuming that the trends shown on the graphs were to continue?

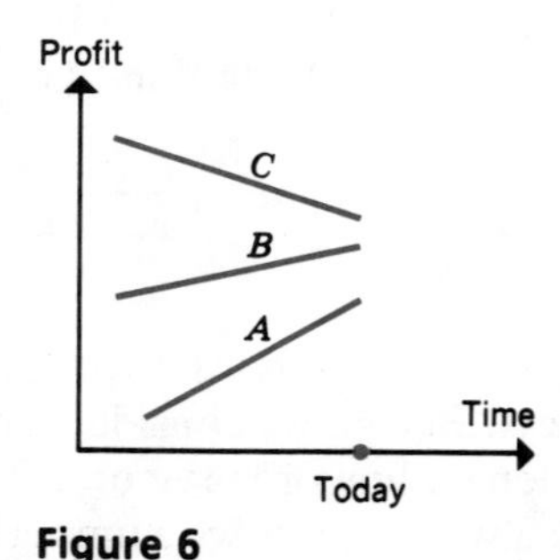

Figure 6

Although it has the highest profits, company C does not look like a desirable investment because its profits are falling ($m_C < 0$). The profits of B are larger than those of A and are rising ($m_B > 0$); however, the profit line for A is steeper than that of B ($m_A > m_B$) so that A's profits should overtake and surpass those of B if current trends continue. Therefore, your best long-term investment would be company A, the least profitable of the three companies.

Graphing a Linear Function

A linear function of the form $y = f(x) = mx + b$ is encountered often, so it is important to understand and remember its properties.

1. Its graph is a straight line whose **slope** is m.
2. The **y intercept,** denoted by b, is the y coordinate of the point where the line intersects the y axis.
3. The **x intercept,** denoted as a, is the x coordinate of the point where the line intersects the x axis:

$$a = \frac{-b}{m} \quad \text{provided } m \neq 0$$

Sketching the graph of a linear function is easy because only two points are needed to draw a line. It is recommended that the two points you select not be close to each other; otherwise a small error in plotting one or both points can cause a large deviation between the true line and the one you sketch.

Example 4 Sketch the graph of the linear function $y = f(x) = -2x + 4$.

Solution If we arbitrarily select $x_1 = -1$ and $x_2 = 3$ as the x coordinates of the two points, the corresponding y coordinates are

$$y_1 = f(-1) = -2(-1) + 4 = 6 \qquad y_2 = f(3) = -2(3) + 4 = -2$$

Using $(-1, 6)$ and $(3, -2)$ as reference points, we can sketch the graph as shown in Figure 7. In addition, the following features should be noted.

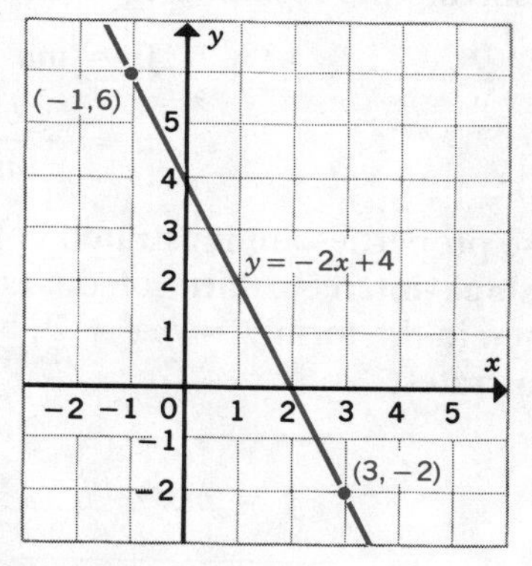

Figure 7

1. The slope of the line connecting $(-1, 6)$ and $(3, -2)$ is -2.
2. The line intersects the y axis at $(0, 4)$, so the y intercept is 4.
3. Since $-\frac{b}{m} = \frac{-4}{-2} = 2$, the line intersects the x axis at $(2, 0)$, so the x intercept is 2. ■

Properties 1 through 3 of a linear function $y = f(x) = mx + b$ can now be demonstrated.

I. The graph is a line whose slope is m. Let (x_1, y_1) and (x_2, y_2) be any two distinct points that satisfy the equation $y = mx + b$. Because (x_1, y_1) and (x_2, y_2) satisfy the equation $y = mx + b$, each of the following equations holds:

$$y_1 = mx_1 + b \tag{3}$$

$$y_2 = mx_2 + b \tag{4}$$

Subtracting the left- and right-hand sides of Equation (3) from the left- and right-hand sides, respectively, of Equation (4) and setting the resulting expressions equal to one another yields

$$y_2 - y_1 = (mx_2 + b) - (mx_1 + b)$$
$$y_2 - y_1 = m(x_2 - x_1)$$

Dividing both sides by $x_2 - x_1$ gives

$$m = \frac{y_2 - y_1}{x_2 - x_1}$$

II. The y intercept is b; it is found by setting $x = 0$ and solving for y, yielding $y = b$.

III. The x intercept a equals $-\frac{b}{m}$; it is found by setting $y = 0$ and $x = a$ and solving the resulting equation for a:

$$0 = ma + b$$

$$a = -\frac{b}{m} \quad \text{provided } m \neq 0$$

These properties are illustrated in Figure 8. The equation $y = mx + b$ is called the **slope–intercept form** of the equation of a line. If a linear equation is not written in the form $y = mx + b$, it can be put into that form easily, as Example 5 illustrates.

$$\boxed{y = mx + b} \quad \text{(slope–intercept form)} \tag{5}$$

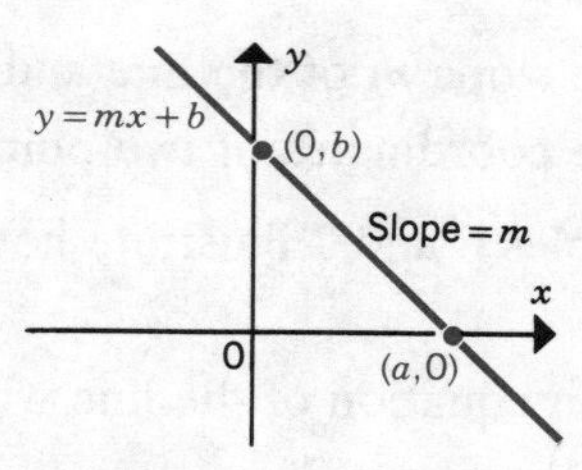

Figure 8

Example 5 Write the linear equation $4x - 3y = 7$ in the slope–intercept form; determine the slope and the x and y intercepts.

Solution The equation can be put into the slope–intercept form by subtracting $4x$ from both sides of the equation and then dividing every term by -3, the coefficient of y. The result is

$$y = \frac{4x}{3} - \frac{7}{3}$$

from which the following values can be derived:

$$m = \frac{4}{3}, \qquad b = -\frac{7}{3}, \qquad a = \frac{7}{4}$$

The graph of this equation is shown in Figure 9.

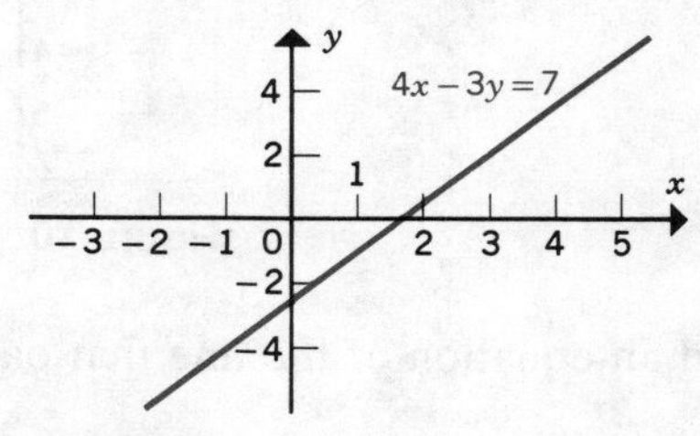

Figure 9 ■

Finding an Equation of a Straight Line

So far, it has been assumed that we have been given a linear equation from which certain characteristics of the line such as the slope and the intercepts, together with the graph, can be deduced. Now we want to look at the problem in reverse. Given certain characteristics of a line or its graph, find an equation of the line.

Just as the graph of a line can be drawn if you are given either the coordinates of two points on the line or the slope together with the coordinates of one point, so too can an equation of the line be found if either of the following is given or can be found.

1. The slope m of the line and the coordinates of one point (x_1, y_1).
2. The coordinates of two points $(x_1\ y_1)$ and (x_2, y_2).

Examples 6 and 7 illustrate how this information is used.

Example 6 Find an equation of the line whose slope is 2 and that passes through the point $(1, -3)$.

Solution Using Equation (5) with $m = 2$, we get

$$y = 2x + b$$

To find b, substitute $x = 1$ and $y = -3$ into the equation to get

$$-3 = 2(1) + b$$
$$b = -5$$

So the equation is $y = 2x - 5$. The graph is shown in Figure 10.

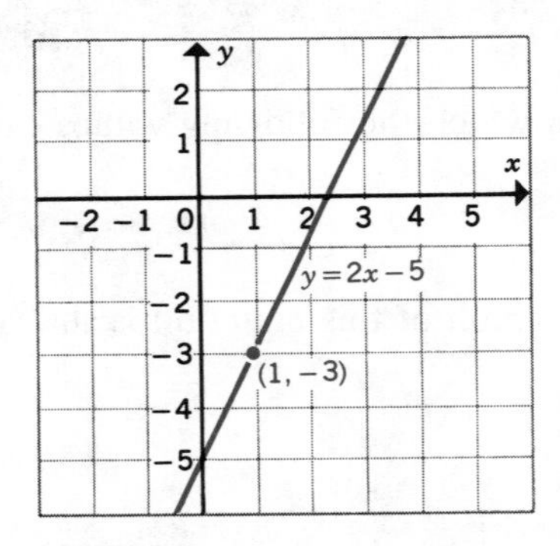

Figure 10 ■

Example 7 Find an equation of the line that passes through $(1, 3)$ and $(3, -1)$.

Solution Knowing the coordinates of two points allows us to find m from Equation (2):

$$m = \frac{y_2 - y_1}{x_2 - x_1} = \frac{-1 - 3}{3 - 1} = -2$$

The equation of the line has the form $y = -2x + b$; to find b, substitute the coordinates of either point into this equation and solve:

$$\begin{array}{cc} (1, 3) & (3, -1) \\ 3 = -2(1) + b & -1 = -2(3) + b \\ \multicolumn{2}{c}{b = 5} \end{array}$$

So the equation is $y = -2x + 5$ (Figure 11). ■

If m is the slope and (x_1, y_1) is a given point on a line, another useful form, the **point-slope form,** can be developed from the definition of m. In Figure 12

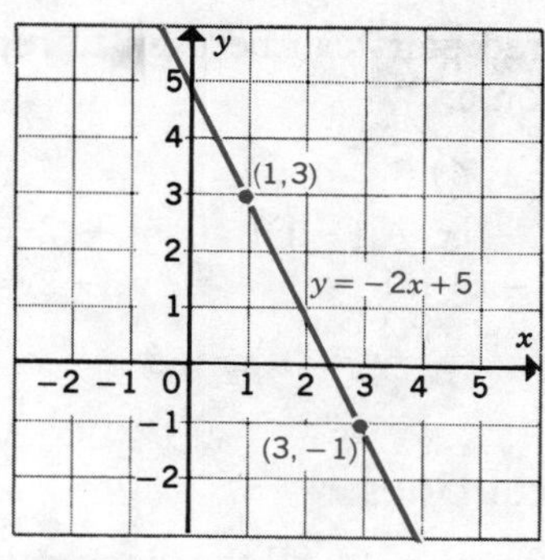

Figure 11

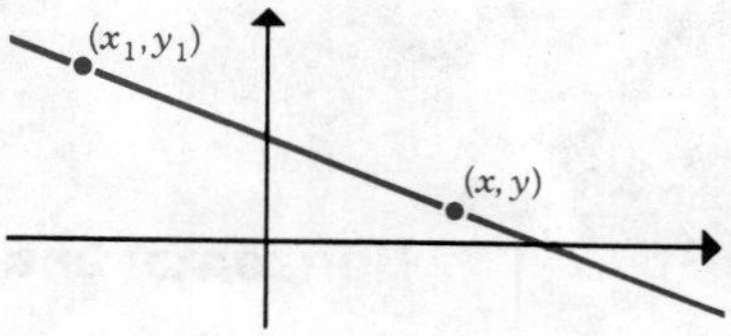

Figure 12

let (x, y) represent any point other than (x_1, y_1). The slope m of the line can be written

$$m = \frac{y - y_1}{x - x_1}$$

Multiplying both sides by $x - x_1$ gives

$$\boxed{y - y_1 = m(x - x_1)} \qquad \text{(point–slope form)} \qquad (6)$$

Note: In Equation (6), x and y are variables; x_1 and y_1 are the coordinates of a fixed point.

Example 8 Find an equation of the line whose slope is -3 and that passes through $(-1, 2)$.

Solution Let $m = -3$, $x_1 = -1$, and $y_1 = 2$ in Equation 6; this yields

$$\begin{aligned} y - 2 &= -3[x - (-1)] \\ &= -3x - 3 \\ y &= -3x - 1 \end{aligned}$$

■

The point–slope form can also be used when the coordinates of two points are given.

Example 9 Find an equation of the line that passes through $(-1, 4)$ and $(1, 2)$.

Solution The slope m can be found first:

$$m = \frac{y_2 - y_1}{x_2 - x_1} = \frac{2 - 4}{1 - (-1)} = \frac{-2}{2} = -1$$

Next, either of the ordered pairs can be used to represent the coordinates of the fixed point in Equation 6:

$$\begin{array}{ll} (-1, 4) & (1, 2) \\ y - (4) = -1[x - (-1)] & y - 2 = -1(x - 1) \\ y - 4 = -x - 1 & y - 2 = -x + 1 \\ \downarrow & \downarrow \end{array}$$

$$y = -x + 3$$

■

Horizontal and Vertical Lines

If a line is horizontal, then $m = 0$ and the slope–intercept form becomes

$$y = 0x + b$$

$$\boxed{y = b} \qquad \text{(horizontal line)}$$

The y coordinate is the same for all points regardless of the value of x (Figure 13).

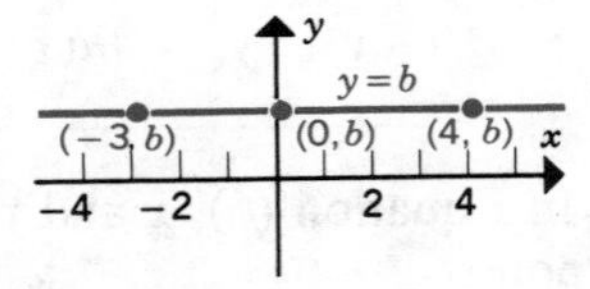

Figure 13

If a line is vertical, m is undefined; this means that we cannot use the point–slope or slope–intercept equations in this situation. However, a vertical line has the property that the x coordinates of all its points are constant and equal to the x intercept a (Figure 14).

$$\boxed{x = a} \qquad \text{(vertical line)}$$

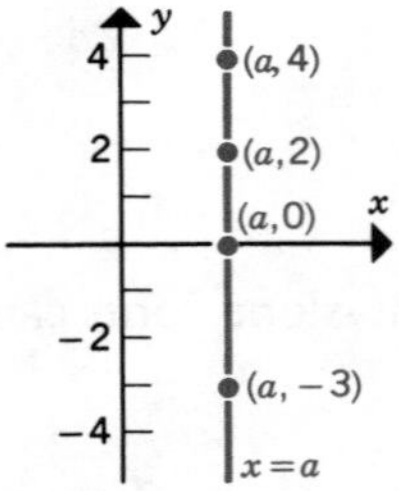

Figure 14

Example 10 Find the equations of the horizontal and vertical lines through $(-3, 2)$.

Solution The equation of the horizontal line is $y = 2$; the equation of the vertical line is $x = -3$ (Figure 15).

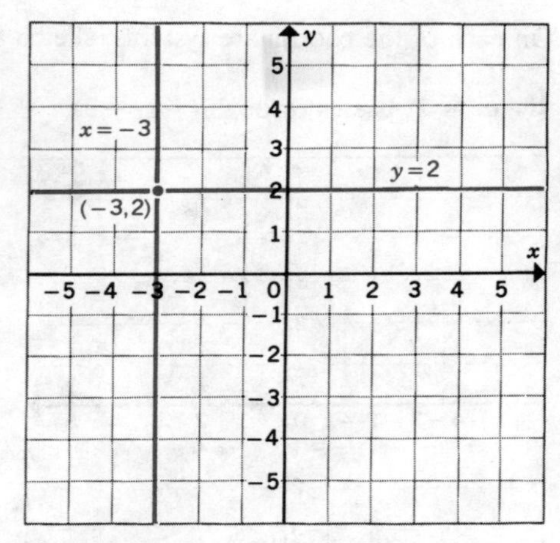

Figure 15

■

1.2 EXERCISES

Find the slope of each of the lines described in exercises 1 through 15.

1. The line passes through $(-1, -2)$ and $(3, 2)$.

2. The line passes through $(2, 6)$ and $(5, -3)$.

3. The line passes through $(0, 4)$ and $(2, 4)$.

4. The line passes through $(-5, -2)$ and $(3, 2)$.

5. The line passes through $(3, 5)$ and $(5, 13)$.

6. The line passes through $(1, 2)$ and $(4, 4)$.

7. The line passes through $(-3, 2)$ and $(-3, 6)$.

8. The line passes through $(-3, 0)$ and $(0, 2)$.

9. The line passes through $(1, -2)$ and $(1 + h, -2 + 3h)$.

10. The line passes through $(-2, 4)$ and $(-2 + h, 4 - 5h)$.

11. The line passes through $(1, 1)$ and $(1 + h, 1 + 2h + h^2)$.

12. The line passes through $(-3, -2)$ and $(-3 + h, -2 + 5h + 2h^2)$.

13. The line passes through $(2, -1)$ and $(2 + h, -1 + 2h + h^2 - h^3)$.

14. The x intercept equals 2 and the y intercept equals 4.

15. The x intercept equals the y intercept; each is nonzero.

On each of the coordinate systems, sketch the graph of the indicated line.

16. $m = 2$; passes through $(-1, -3)$

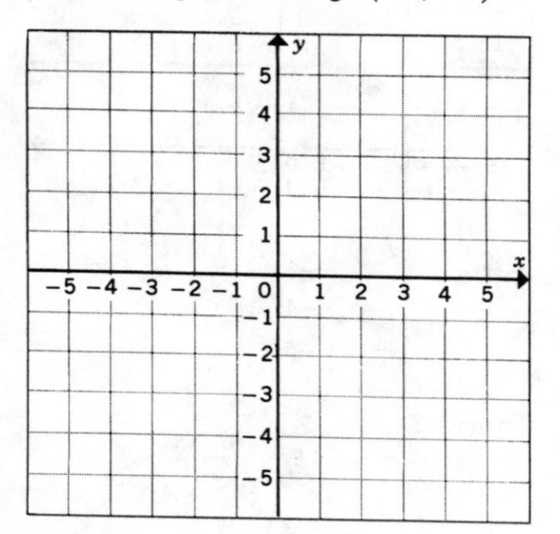

17. $m = \frac{-3}{4}$; passes through $(-2, 1)$

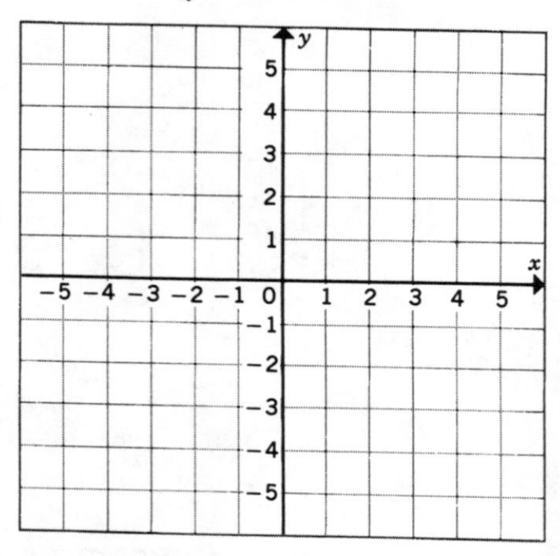

18. The slopes of the five lines shown in Figure 16 are given in the accompanying table. Match each slope with one of the lines.

m	Line
$\frac{1}{2}$	
-1	
$\frac{3}{2}$	
2	
$-\frac{5}{2}$	

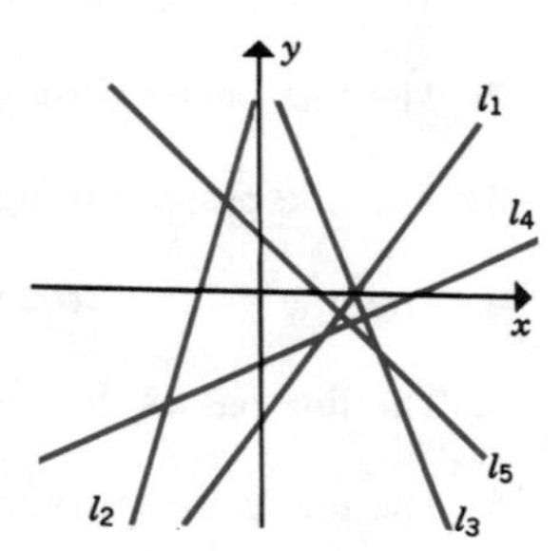

Figure 16

In Exercises 19 through 28, find (a) the slope and (b) the x and y intercepts, when they exist, of the line represented by each linear equation. In addition, sketch the graph of each equation.

19. $x + y = 1$

20. $2y = 4x - 3$

21. $2x - 3y = 6$

22. $5x + 2y = 10$

23. $x - y = 2$

24. $3y + 2x = 12$

25. $x + 3y = 6$

26. $4y + 5x = 10$

27. $x = 3$

28. $y = -4$

Find the equation of each line described in Exercises 29 through 51.

29. Slope is 1 and the line passes through $(-2, 3)$.

30. Slope is -2 and the line passes through $(1, 0)$.

31. Slope is 3 and the line passes through $(-4, -5)$.

32. Slope is 4 and the line passes through $(-2, 1)$.
33. Slope is -1 and the line passes through $(2, 2)$.
34. Slope is $\frac{1}{2}$ and the line passes through $(3, 1)$.
35. Slope is $-\frac{2}{3}$ and the line passes through $(0, 4)$.
36. Slope is $-\frac{3}{2}$ and the line passes through $(2, -1)$.
37. Slope is $\sqrt{2}$ and the line passes through $(\sqrt{2}, 3)$.
38. Slope is $-\sqrt{3}$ and the line passes through $(4\sqrt{3}, -2)$.
39. The line passes through $(-1, 3)$ and $(2, 6)$.
40. The line passes through $(-2, 4)$ and $(2, -4)$.
41. The line passes through $(-3, 0)$ and $(6, 9)$.
42. The line passes through $(1, -1)$ and $(3, 5)$.
43. The line passes through $(1, -3)$ and $(4, 3)$.
44. The line passes through $(0, 2)$ and $(3, -4)$.
45. The line passes through $(3, -1)$ and $(3, 5)$.
46. The line passes through $(\sqrt{2}, 2)$ and $(2\sqrt{2}, 5)$.
47. The line passes through $(\sqrt{2}, 3)$ and $(\sqrt{2}, 5)$.
48. The line passes through $(\sqrt{3}, 5)$ and $(2\sqrt{3}, 5)$.
49. The line passes through $(c, 2d)$ and $(2c, 4d)$, where c and d are constants $(c \neq 0)$.
50. The x intercept is -2: the y intercept is 4.
51. The x intercept is 3: the y intercept is 1.

Parallel lines have equal slopes. Use this property to find an equation of each of the lines described in Exercises 52 through 55.

52. Line passes through $(-2, 1)$ and is parallel to the line $6x + 3y = 7$.
53. Line passes through $(3, 3)$ and is parallel to the line $2x - 3y = 5$.
54. Line passes through $(1, -4)$ and is parallel to the line $3x + 6y = 2$.
55. Line passes through $(\frac{3}{2}, -\frac{1}{2})$ and is parallel to the line $4x - 2y = 3$.
56. (a) Pete the painter can paint 300 ft^2 in an hour. Find an equation that gives T, the time in hours to paint a room, as a function of A, the surface area in square feet of the room.
    (b) If Pete earns \$10 per hour, find an equation that describes C, the cost in dollars to paint a room, as a function of A.
57. A stock analyst estimates that, for each dollar drop in the price of a barrel of oil, annual profits of Standard Airlines will increase by \$5 million. If the price per barrel of oil is \$25 and annual profits of Standard are \$40 million, find an equation that gives P, annual profits, as a function of p, the price per barrel of oil.
58. Sammy Superstar earns an annual salary of \$2 million. In addition, he receives a bonus of \$50,000 for each grand-slam home run that he hits during a season. Find an equation that gives S, his annual salary, as a function of h, the number of grand-slam home runs he hits.
59. (a) A car dealer has 300 cars on his lot. If the cars are sold at the rate of 15 per week, find an equation that gives I, inventory of cars, as a function of t, the time in weeks. What is his inventory of cars after eight weeks? How long will it take him to sell all 300 cars?
    (b) Suppose the factory ships cars to him at the rate of ten cars per week. Find an equation that gives I as a function of t.

60. (a) The Speedy Rent-a-Car Agency charges \$12 per day plus 20¢ per mile for renting their compact autos. Find an equation that gives the dollar charge D as a function of the mileage M driven each day. Graph the function on the coordinate system of Figure 17.

(b) The Drive-Away Rent-a-Car Agency charges \$5 per day plus 25¢ per mile for renting their compact autos. Again, find an equation that gives the dollar charge D as a function of the mileage M driven each day, and graph the function on the coordinate system of Figure 17.

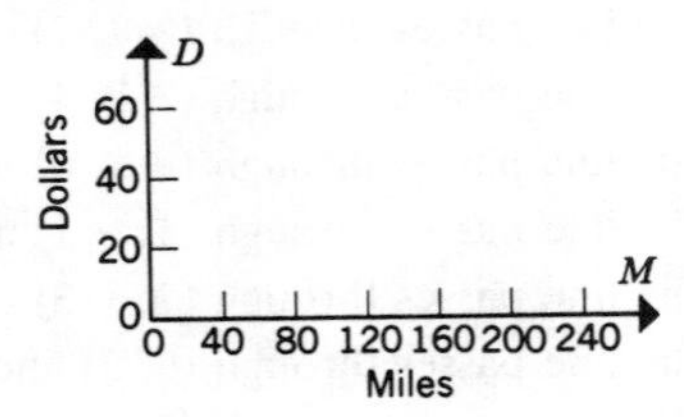

Figure 17

(c) How many miles per day would a person have to drive in order to make the Speedy offer more attractive than that of the Drive-Away Agency.

61. When money is borrowed or lent for a short period of time, usually less than one year, the interest I paid on the principal P (the amount borrowed or loaned) is often calculated according to the simple interest equation

$$I = Prt$$

where r is the annual interest rate and t is the time expressed in years. The amount or value of the loan A grows with time according to the equation

$$A = P + I = P(1 + rt)$$

(a) If a person takes out a six-month note for \$1000 at 18 percent interest per year from a finance company, find an equation that gives A as a function of t, where t is expressed in years.

(b) If the variable t is expressed in units other than years, say months, the interest rate r must be expressed as a monthly rate. For the problem in part a, express A as a function of t, where t is expressed in months.

62. The weekly salary of a union plumber, S, is a function of n, the number of hours he works. If $n \leq 40$, the hourly rate of pay is \$14; if $n > 40$, the hourly rate increases to \$22 for all hours over 40. Find a piecewise function that gives S as a function of n.

63. Because of a severe drought the amount of water in a municipal reservoir has been dropping at a rate of 5 million gallons a month.

(a) If the reservoir contains 500 million gallons today and conditions do not change, find an equation that describes the relationship between $A(t)$, the amount of water in the reservoir, in millions of gallons, and t, the time in months, from today.

(b) The commission overseeing the operation of the reservoir has an established policy of instituting water rationing as soon as the amount of water drops below 420 million gallons. When will water rationing begin if the drought persists?

64. (a) Medication is administered to a patient intravenously from a 250 milliliter (mL) bag. A saline solution containing the medication flows from the bag at a rate of

50 mL/hr. Find an equation that describes the relationship between the amount of saline solution $A(t)$, in milliliters, and the time t, in hours, from the moment the solution flows from a full bag.

(b) The nurses replace a bag with a new one when the amount of solution in the bag drops to 25 mL. How often is the bag replaced?

65. (a) The shoreline in many states is eroding, owing in part to winter storms. The width of a strip of land between the Atlantic Ocean and Cape Cod Bay is decreasing at a rate of 5 ft per year. If the strip of land is one mile wide at the present time, find an equation that describes the width W, in feet, as a function of the time t, in years, from today.

(b) How long will it be before the strip is under water?

66. In general, aviation pilots use the following rule to estimate the air temperature at any altitude. For each 1000-ft increase in altitude, there is a decrease of 2°C in the air temperature. If the temperature at the surface of the earth is 20°C, find an equation that describes the air temperature T, in °C, as a function of the height h, in feet, above the earth's surface.

1.3 APPLICATIONS OF LINEAR FUNCTIONS

Linear functions can be applied to the analysis of many elementary situations commonly encountered in the fields of business and economics; some of these applications are illustrated in this section.

Straight-Line Depreciation

When a firm purchases a piece of equipment with a useful life of many years, the total cost of the item cannot be charged as an expense at the time of purchase; instead, this expense is spread out over the useful life of the asset. As this expense is applied periodically, it is recorded on the income statement as a cost known as **depreciation**. In addition to "spreading out" the expense, depreciation also indicates how rapidly the value of the asset, recorded on the balance sheet, is decreasing.

One of the more common depreciation techniques is the **straight-line method** in which the decline in value is uniform from one period to the next. To illustrate this technique, suppose we use the following notation:

C = original cost of the item (in dollars)
T = useful life of the item (in years)
S = salvage or resale value of the item (in dollars) at the end of T years

The annual depreciation D, expressed in dollars per year, is given by the equation

$$D = \frac{C - S}{T} \tag{1}$$

The quantities C, S, and T are constants for any item. To develop an equation describing the worth or value of any item in terms of the length of time from the date of purchase, let us introduce two variables.

1. The independent variable t, which expresses the length of time, usually in years, from the date of purchase ($t = 0$).
2. The dependent variable $V(t)$, which gives the value of the item at any time t.

Assuming that the item decreases in value at a uniform rate, the relationship between $V(t)$ and t is given by the equation

$$V(t) = -Dt + C, \qquad 0 \leq t \leq T \tag{2}$$

Example 1 Amalgamated Products purchased an executive jet for \$1,500,000. The company plans to use the plane for 10 years, at which time they expect to sell it for \$500,000. What is the equation expressing the relationship between the book value $V(t)$ and t, the time from the date of purchase? Draw a graph of the function.

Solution The annual depreciation D is given by

$$D = \frac{1{,}500{,}000 - 500{,}000}{10} = \$100{,}000/\text{year}$$

so

$$V(t) = -100{,}000t + 1{,}500{,}000, \qquad 0 \leq t \leq 10$$

when

$$t = 2, \qquad V(2) = -100{,}000(2) + 1{,}500{,}000 = \$1{,}300{,}000$$
$$t = 7, \qquad V(7) = -100{,}000(7) + 1{,}500{,}000 = \$800{,}000$$

The graph of the function is shown in Figure 1.

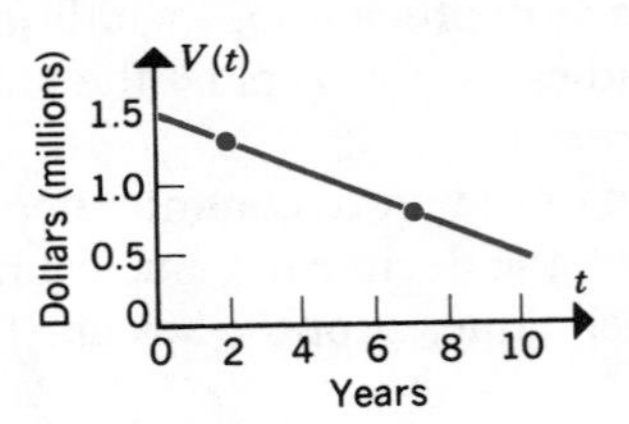

Figure 1

■

Break-even Analysis

In the operation of any enterprise, a manager is always concerned about the revenue generated through sales and the costs associated with operating the enterprise. The operation generates a profit when revenue exceeds costs, whereas it is operating at a loss when the reverse is true. The profit $P(x)$ can be written in terms of the revenue $R(x)$ and total cost $C(x)$ as

$$\text{Profit} = \text{revenue} - \text{total cost}$$

or

$$P(x) = R(x) - C(x) \tag{3}$$

In most applications, the goal is to express $R(x)$ and $C(x)$ and thus $P(x)$ as functions of x, the number of items sold. The revenue $R(x)$ can be written as

$$R(x) = px$$

where p is the unit price. In many situations the total cost $C(x)$ can be written as the sum of two terms, that is,

$$C(x) = F + V(x)$$

where F represents what are known as fixed costs, costs such as rent, office equipment, and insurance; which are independent of the number of units sold, $V(x)$ represents the variable costs, such as labor, raw material, transportation, and machinery, which are functions of the number of units produced. Like the revenue function, $V(x)$ can often be written as

$$V(x) = vx$$

where v is a constant, called the unit variable cost. The profit can then be written in terms of x as

$$\begin{aligned} P(x) &= px - (F + vx) \\ &= (p - v)x - F \end{aligned} \tag{4}$$

Example 2 A company manufacturing specialty T-shirts sells the shirts to a wholesaler for \$2 each. The company's annual fixed costs are \$70,000, and the unit variable cost is 60¢. Find the profit $P(x)$ as a function of x, the number of T-shirts produced and sold.

Solution With $F = 70{,}000$, $p = 2.00$, and $v = 0.60$, Equation 4 becomes

$$P(x) = (2.00 - 0.60)x - 70{,}000 = 1.40x - 70{,}000$$

Figure 2 contains the graph of the profit function. The x intercept on the graph is called the **break-even** point, that is, the point where $P(x) = 0$; the corre-

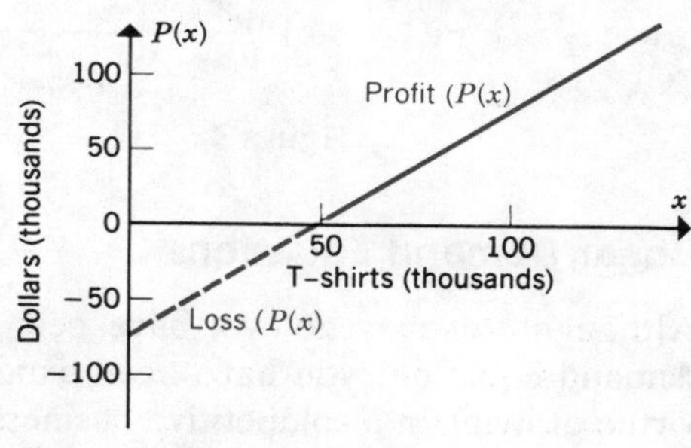

Figure 2

sponding value of x represents the number of items that must be produced and sold to cover costs. The break-even point is found by setting $P(x)$ equal to zero in Equation 4 and solving the resulting equation for x. Denoting the solution as x_{BE}, we get the following,

$$0 = (p - v)x_{BE} - F$$
$$F = (p - v)x_{BE}$$

or

$$\boxed{\quad p - v \quad} \qquad (5)$$

■

Example 3 Find the break-even point for the company producing specialty T-shirts (Example 4).

Solution For the break-even value of x, we get

$$x_{BE} = \frac{70{,}000}{2.00 - 0.60} = 50{,}000$$

Equation 3 indicates that the break-even point ($P(x) = 0$) occurs when revenue $R(x)$ equals total costs $C(x)$. This is shown for the T-shirt manufacturer in Figure 3, which contains the graphs of the revenue equation $R(x) = 2x$ and the total cost equation $C(x) = 0.60x + 70{,}000$. Break-even occurs when $x = 50{,}000$ units and $R(x) = C(x) = \$100{,}000$.

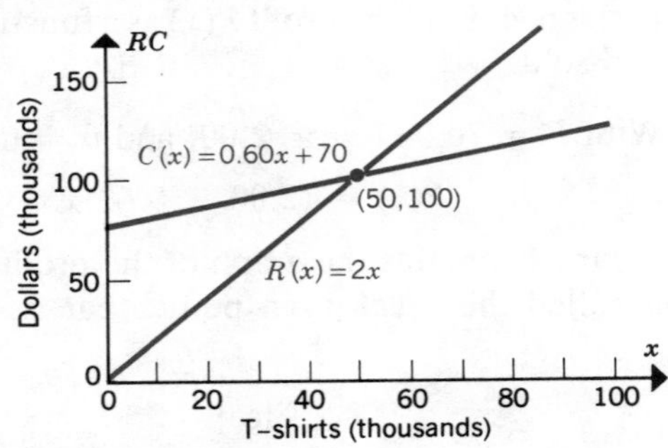

Figure 3

■

Linear Demand Equations

Although you may not yet have been formally exposed to the concept of a demand equation, you have been exposed to the basic ideas underlying it by virtue of living in a competitive business environment. You know that the price of an item plays an important part in your decision to purchase or not to purchase the item; the increased attractiveness of an item as its purchase price decreases forms the personal aspect of the demand phenomenon. As an example, the recent large price drops in personal computers and VCRs have resulted in a tremendous surge in the numbers purchased by the general public.

The **demand equation** represents an attempt to write p, the unit price of an item, as a function of q (quantity), the maximum number of items that can be sold at the price p.* If a demand equation is linear, we would expect its slope to be negative because high prices are generally associated with low sales and low prices with high sales (see Figure 4).

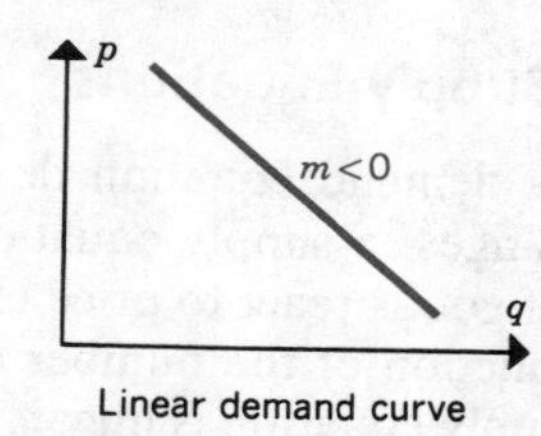

Linear demand curve

Figure 4

Example 4 The accompanying table contains the results of a market research study conducted by the MacDougall Hamburger Company to determine the relationship between unit price and weekly sales for its new Gluttonburger (1 pound of meat on a 6-inch roll); the results are shown graphically in Figure 5. Find a linear equation that describes the relationship between p and q.

q (thousands)	p ($)
50	1.00
40	1.25
30	1.50
20	1.75

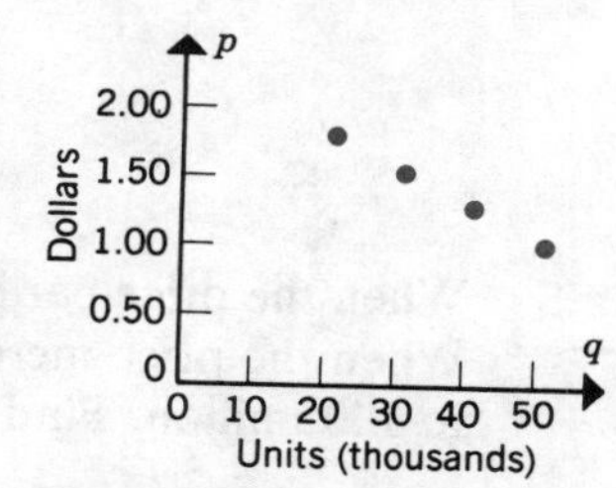

Figure 5

Solution A linear equation that gives p in terms of q can be developed using the methods described in Section 1.2. Using the extreme values given in the table, we get the slope of the line:

$$m = \frac{p_2 - p_1}{q_2 - q_1} = \frac{1.00 - 1.75}{50 - 20} = -0.025$$

Next, using the point–slope formula (Equation 6, Section 1.2) in the form

$$p - p_1 = m(q - q_1)$$

*Economists generally draw demand curves with q on the horizontal axis and p on the vertical axis. For this reason we treat q as the independent variable and p as the dependent variable, even though it is more logical to reverse their role.

we get, using (50, 1.00) for (q_1, p_1),

$$p - 1 = -0.025(q - 50)$$
$$p = -0.025q + 2.25, \qquad 20 \le q \le 50$$

Caution would have to be exercised if you were to attempt to extend the validity of the equation to values of q beyond the domain. For example, the equation implies that the company could charge nothing for the burger ($p = 0$) and yet only 90,000 would be "sold" in this unusual situation. ■

Linear Supply Equations

Just as a demand equation describes mathematically how consumers react to price changes, a supply equation describes mathematically how sellers or producers of goods react to price changes. The **supply equation** gives the unit price p as a function of the number of items q produced or made available for sale. If the supply equation is linear, we would expect its slope to be positive because producers are more inclined to step up output as unit prices rise in order to increase profits (see Figure 6).

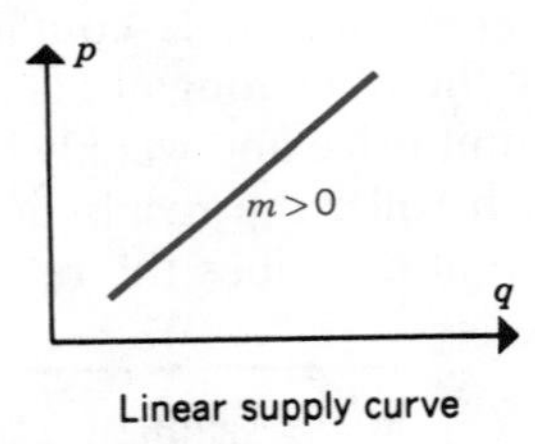

Figure 6

Example 5 When the price per bushel of wheat was \$2.50, 75 million bushels were planted. When the price increased to \$3 per bushel, the number of bushels planted rose to 100 million. Find the supply equation, assuming that it is linear.

Solution The supply curve is shown in Figure 7. Using the methods developed in Section 1.2, we get an equation by first finding the slope:

$$m = \frac{3.00 - 2.50}{100 - 75} = 0.02$$

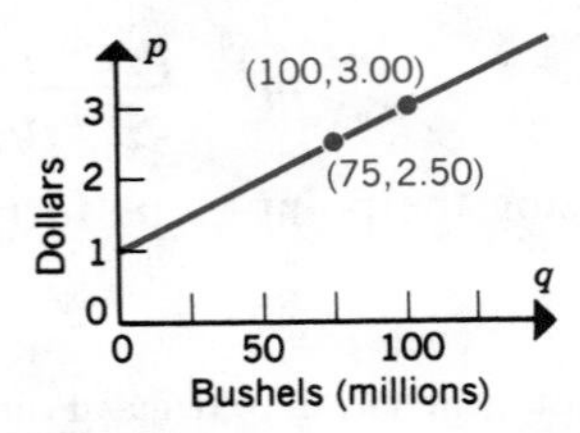

Figure 7

Again using the point–slope formula (Equation 6, Section 1.2) in the form

$$p - p_1 = m(q - q_1)$$

we get, using (100, 3) for (q_1, p_1),

$$p - 3 = 0.02(q - 100)$$
$$p = 0.02q + 1$$

■

Equilibrium

In a supply–demand situation, market **equilibrium** is achieved when the quantity offered for sale equals the quantity consumers plan to buy. Graphically, the point where the supply and demand curves intersect gives the equilibrium quantity q_E and price p_E illustrated in Figure 8.

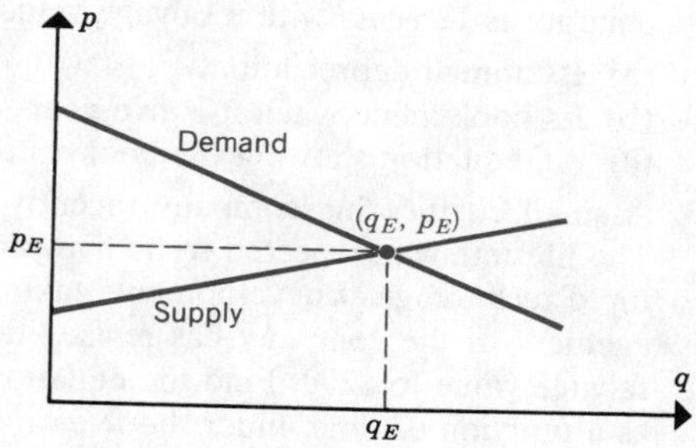

Figure 8

Example 6 Find the equilibrium quantity and price for a commodity whose supply and demand equations are

$$p = 0.5q + 3 \quad \text{(supply)} \qquad p = -q + 6 \quad \text{(demand)}$$

Solution Because the ordered pair (q_E, p_E) satisfies both the supply and the demand equation, we have at equilibrium

$$p_E = 0.5q_E + 3, \qquad p_E = -q_E + 6$$

or, equivalently,

$$0.5q_E + 3 = -q_E + 6$$

Solving this equation for q_E gives

$$q_E = 2, \qquad p_E = 4$$

as shown in Figure 9.

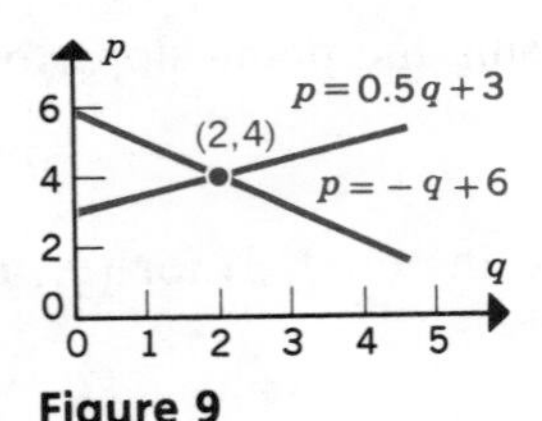

Figure 9

■

1.3 EXERCISES

1. A taxi company purchases a fleet of ten cabs for \$35,000. If the useful life of the taxis is three years and the salvage value of each taxi is estimated to be \$500 at the end of the three-year period, find an equation which gives the value of the fleet as a function of the time t, in years.
2. The owner of a photo studio purchased an enlarger for \$1500. If the lifetime of the enlarger is 12 years with a salvage value of \$300, find
   (a) Its annual depreciation.
   (b) Its book value when it is five years old.
   (c) An equation showing the book value $V(t)$ as a function of the time t, in years.
3. A small engineering company recently purchased a portable computer for \$4400. The lifetime was expected to be five years with a salvage value of \$400. Because of rapid technological developments taking place in the computer industry, the management of the company has revised downward the lifetime to four years and the salvage value to \$200. Find the equations that give the book value of the computer as a function of time under the five- and four-year lifetime assumptions.
4. The owner of a small restaurant has a new refrigeration system that costs \$6000 installed. If the salvage value at the end of ten years is \$0, find an equation that describes $V(t)$, the book value, as a function of t, the time (in years) from the date of acquisition.
5. A small company manufactures an indoor–outdoor thermometer that it sells to wholesalers for \$6 each. Labor and material costs for each thermometer are \$4, and fixed annual costs are \$25,000.
   (a) How many thermometers must be sold for the company to break even?
   (b) Find an equation that describes the company's annual profit as a function of x, the number of thermometers sold *each* year.
6. A retired carpenter rents a small shop where he makes and sells dollhouses. The houses sell for \$75 each, and material for each costs \$20. If rent and utilities cost \$220 per month, find an equation that describes his monthly profit $P(x)$ in terms of x, the number of dollhouses sold each month. How many dollhouses must be sold to break even?
7. The operator of a small self-service gasoline station pays 90¢ per gallon for gasoline; he sells it for \$1.20 at the pump. If his monthly overhead (rent, utilities, insurance) is \$600, find an equation that describes $P(x)$, his monthly profit, as a function of x, the number of gallons sold each month. How many gallons does he have to sell to break even? If he wants to earn a profit of at least \$450 per month, what is the minimum number of gallons of gasoline that he has to sell?

8. A nationwide manufacturer of financial calculators finds that its profits are \$4 million when it sells 300,000 units. If the company's fixed costs are \$800,000 and unit variable costs are \$20 per calculator, find
   (a) The unit price of each calculator.
   (b) The profit $P(x)$ as a function of x, the number of calculators sold.
   (c) The break-even point.
9. A company that manufactures lawn furniture finds it can sell 200,000 units if it charges \$10 for its best-selling chaise lounge. The company expects to sell 250,000 units when the price is decreased to \$9 per unit. Find a demand equation and sketch the graph.
10. A group sponsoring a rock concert is trying to determine the ticket price it should charge. On the basis of past experience, it estimates that an audience of 10,000 people can be drawn if the ticket price is \$15. For each dollar increase in the price of a ticket, 500 fewer people will attend the concert.
    (a) Find the demand equation, assuming that it is linear.
    (b) If the concert hall holds 8750 people, what is the maximum ticket price that will ensure a full house?
11. Because of a rapid increase in the price of silver, a film manufacturing company raised the price on its 24-exposure roll from \$1.75 to \$2.00. Immediately, monthly sales of the film decreased from 2 million to 1.8 million rolls. Find a linear equation that describes the price p, in dollars, in terms of monthly sales q expressed in millions of rolls.
12. The manufacturer of a new tennis racket produces 500 units per week, all of which are sold to a wholesaler for \$20 each. When the racket proves to be very popular, the wholesaler increases the order to 900. The manufacturer is unwilling to allocate additional resources to producing tennis rackets but agrees to do so when the wholesaler indicates a willingness to pay \$25 for each racket. Find a linear supply equation for this situation.
13. In planning production for the coming year, the operations research group in a large office equipment company recommends manufacturing 30,000 electric typewriters if the selling price remains constant at \$150 per unit. They also recommend manufacturing 4000 additional typewriters for each \$10 increase in the selling price. Find a supply equation for this product.
14. When the price per barrel of oil was \$30, Premier Oil Company produced 50,000 barrels each month. When the price dropped to \$25 per barrel, the management decided to reduce production to 40,000 barrels per month in order to divert its financial resources to other more profitable ventures.
    (a) Assuming that the relationship between price and production is linear, find a supply equation.
    (b) According to the equation developed in part (a), at what price would Premier stop producing oil?
15. Alaska Electronics produces computer chips. When the price of a chip was \$1.00, the company manufactured 40,000 chips per month. As foreign competition drove the price of the chip down, the company responded by reducing output. When the price reached \$0.75, the management of Alaska decided that the profit margin at this price was too low and terminated production of the chip. Find a supply equation for this product.

Find the values of q and p at which market equilibrium is achieved in Exercises 16 through 21.

16. $p = 0.5q + 3$ (supply)
$p = -0.3q + 7$ (demand)

17. $p = 0.02q + 5$ (supply)
$p = -0.04q + 8$ (demand)

18. $p = 0.35q + 1.3$ (supply)
$p = -0.15q + 5.5$ (demand)

19. $p = 0.6q + 2.3$ (supply)
$p = -0.8q + 6.7$ (demand)

20. $p = 0.2q + 2.8$ (supply)
$p = -q + 10$ (demand)

21. $p = 0.3q + 1.0$ (supply)
$p = -0.5q + 5$ (demand)

22. Playtime Company has developed a robotic toy, Zaptron, which the company sells for \$20. Zaptron turns out to be very popular; sales for the first year were 100,000 units. In addition, the company had back orders of 50,000 units. To handle the excess demand, management decided to increase production capacity by 50,000 units. In order to finance this expansion, the price of Zaptron was increased to \$30. During the second year, Playtime was not able to sell all 150,000 units that it produced; its year-end inventory was 25,000 units. Find a supply and a demand equation and determine what price would bring supply and demand into equilibrium?

23. A sales representative for Data Base, an internationally known manufacturer of computer equipment, has been offered two salary plans:

Plan A: base salary of \$500 per month plus a commission of 3 percent on all sales.
Plan B: base salary of \$620 per month plus a commission of 5 percent on all sales in excess of \$20,000 per month.

(a) Write a function for the monthly salary under each plan in terms of x, the dollar amount of sales. *Note:* A linear piecewise function is needed to describe Plan B.
(b) Sketch the graphs of each function found in part (a).
(c) For what range of sales is plan A preferable to plan B?

24. Sally Senior recently sold her home for \$100,000 and moved into a one-bedroom apartment in an elderly citizens' housing complex. She has been advised to invest the \$100,000 and use the income from the investments to supplement her Social Security payments. After some study she decided to invest some of the \$100,000 in AAA corporate bonds yielding 9 percent interest per year and the remaining amount in bank certificates of deposit yielding 7.5 percent interest per year.

(a) Find an equation that describes $I(x)$, her annual investment income, as a function of x, the amount invested in corporate bonds.
(b) Sketch the graph of this function.
(c) Suppose that she needs an annual income of at least \$8100 from her investments. What is the minimum amount that she should invest in corporate bonds?

1.4 QUADRATIC FUNCTIONS AND APPLICATIONS

Although linear functions are useful for analyzing and describing many simple situations in science, business, and economics, they are not adequate to handle more complex phenomena. Nonlinear functions are needed for such cases. *Quadratic functions,* which are nonlinear, are encountered in many applications, as the following example illustrates.

Example 1 A retail store finds that the relationship between p, the unit price of its best-selling VCR, and q, the maximum number it can sell per month, is described by the equation

$$p = 400 - 10q$$

Find $R(q)$, the monthly revenue, as a function of q.

Solution The relationship between $R(q)$, the revenue, and q, the maximum number that can be sold, is given by the equation

$$R(q) = pq = (400 - 10q)q = 400q - 10q^2$$

The presence of the quadratic term $-10q^2$ indicates that $R(q)$, the revenue, is not a linear function of q. The nonlinearity can be seen by plotting the graph of the equation; the results are shown in Figure 1.

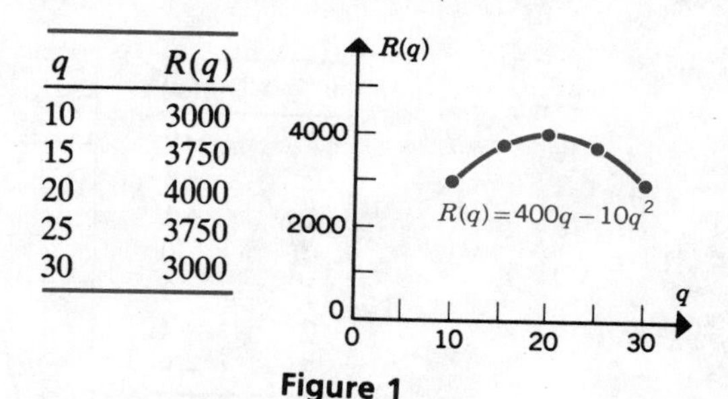

q	$R(q)$
10	3000
15	3750
20	4000
25	3750
30	3000

Figure 1

Inspection of the graph shows that revenue increases as sales increase when $q < 20$ but declines as q increases when $q > 20$. This indicates that monthly revenue is maximized when 20 VCRs are sold; to attain this goal, the manager of the store should set the unit price of each VCR at

$$p = 400 - 10(20) = \$200$$ ■

A **quadratic function** is defined as one having the form

$$\boxed{f(x) = ax^2 + bx + c} \qquad (1)$$

where a, b, and c are constants and $a \neq 0$. The graph of a quadratic function is a curve called a **parabola.**

The graphs of some simple quadratic functions illustrate some of their important characteristics.

Example 2 Sketch the graphs of the functions

$$f(x) = x^2 \quad \text{and} \quad g(x) = -(x^2) = -x^2$$

Solution The graph of each function can be drawn by using the points contained in the following tables.

x	$f(x)$
-3	9
-2	4
-1	1
0	0
1	1
2	4
3	9

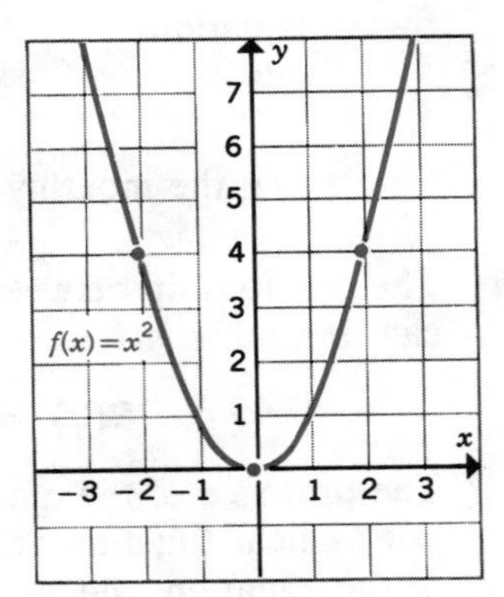

Figure 2

x	$g(x)$
-3	-9
-2	-4
-1	-1
0	0
1	-1
2	-4
3	-9

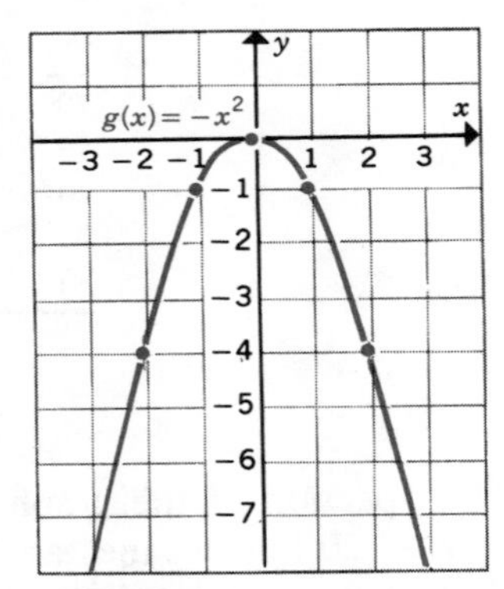

Figure 3 ■

The graphs in Figures 2 and 3 illustrate the following features.

1. (a) If a, the coefficient of x^2, is **positive,** the parabola opens **upward,** as in Figure 2.
   (b) If a is **negative,** the parabola opens **downward,** as in Figure 3.
2. The "high" or "low" point on a parabola is called the **vertex.**
3. The vertical line that passes through the vertex is called the **axis of symmetry.***

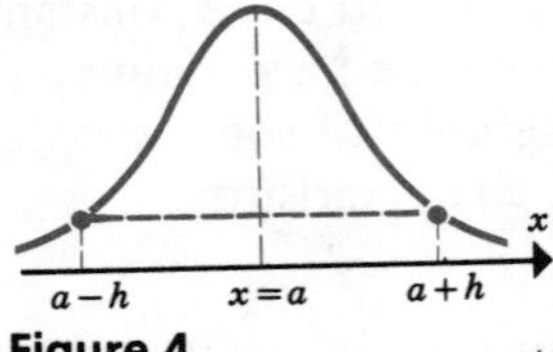

Figure 4

*A vertical line $x = a$ is an axis of symmetry of a graph if $(a - h, y)$ is a point on the graph whenever $(a + h, y)$ is a point on the graph (Figure 4).

Graphing Quadratic Functions

The graph of a quadratic function can be sketched quickly by following the procedure illustrated in Examples 3 and 4.

Example 3 Sketch the graph of the function $f(x) = x^2 - 4x + 2$.

Solution 1. Ignore temporarily the constant term; sketch the graph of the function $F(x) = x^2 - 4x$ after finding the following points:

(a) The x intercepts. Setting $F(x)$ equal to zero and solving yields

$$0 = x^2 - 4x = x(x - 4)$$

Solutions: $x_1 = 0, \quad x_2 = 4$

(b) The vertex. Since the axis of symmetry passes through the vertex and bisects the line segment connecting the x intercepts found in part a, x_V, the x coordinate of the vertex, becomes

$$x_V = \frac{0 + 4}{2} = 2$$

and y_V, the y coordinate of the vertex, becomes

$$y_V = F(2) = (2)^2 - 4(2) = -4$$

The three points found in parts (a) and (b) can be used to sketch the graph shown in Figure 5.

2. Add 2, the constant term, to the y coordinates of the three points found in part 1. Since $f(x) = F(x) + 2$, the three new points can be used to sketch the graph of the function $f(x) = x^2 - 4x + 2$. The results are shown in Figure 6.

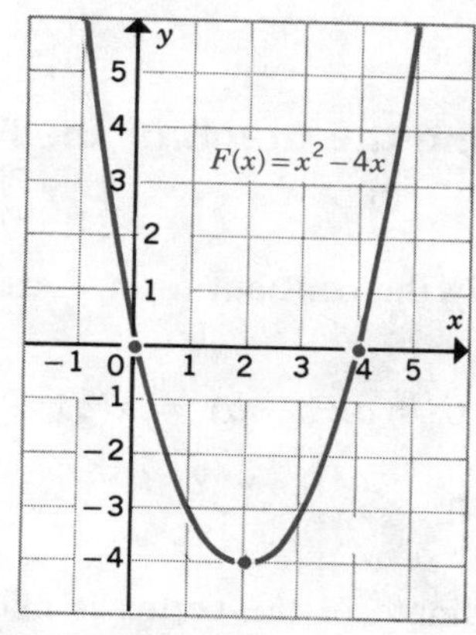

Figure 5

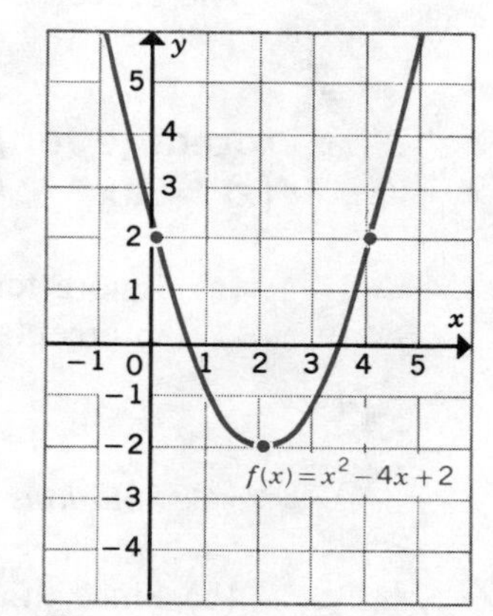

Figure 6

■

Example 4 Sketch the graph of the function

$$f(x) = -2x^2 + 6x - 5$$

Solution 1. (a) Find the x intercepts of the function

$$F(x) = f(x) + 5 = -2x^2 + 6x$$
$$0 = -2x^2 + 6x$$
$$0 = -2x(x - 3)$$

Solutions: $x_1 = 0, \quad x_2 = 3$

(b) The x coordinate of the vertex, x_V, is midway between $x_1 = 0$ and $x_2 = 3$, so

$$x_V = \frac{3}{2}, \qquad y_V = F\left(\frac{3}{2}\right) = -2\left(\frac{3}{2}\right)^2 + 6\left(\frac{3}{2}\right) = \frac{9}{2}$$

2. Since $f(x) = F(x) - 5$, its graph can be sketched by subtracting 5 from the y coordinates of each of the three points found in part 1, as shown in Figure 7.

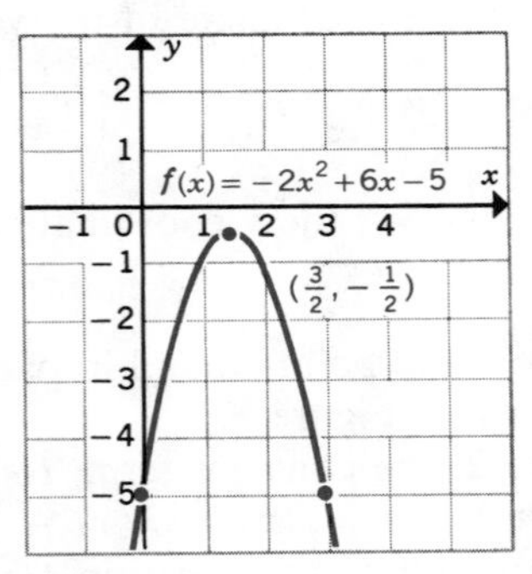

Figure 7

■

The procedure used in Examples 3 and 4 can be stated formally.

Procedure for Sketching the Graph of the Function $f(x) = ax^2 + bx + c$

1. (a) Ignore temporarily the constant term. Let $F(x) = ax^2 + bx$; find its x intercepts:

$$0 = ax^2 + bx = x(ax + b)$$

Solutions: $x_1 = 0, \quad x_2 = -\dfrac{b}{a}$

(b) Since the x coordinate of the vertex is midway between x_1 and x_2, we get for x_V,

$$x_V = -\frac{b}{2a}, \qquad y_V = F\left(-\frac{b}{2a}\right) = a\left(-\frac{b}{2a}\right)^2 + b\left(-\frac{b}{2a}\right) = -\frac{b^2}{4a}$$

2. Since $f(x) = F(x) + c$, the graph of $f(x)$ can be sketched by adding c to the y coordinates of the three points found in part 1. The resulting points are

$$(0, c), \quad \left(-\frac{b}{a}, c\right), \quad \left(\frac{-b}{2a}, \frac{4ac - b^2}{4a}\right) \text{ — vertex}$$

A possible graph is shown in Figure 8.

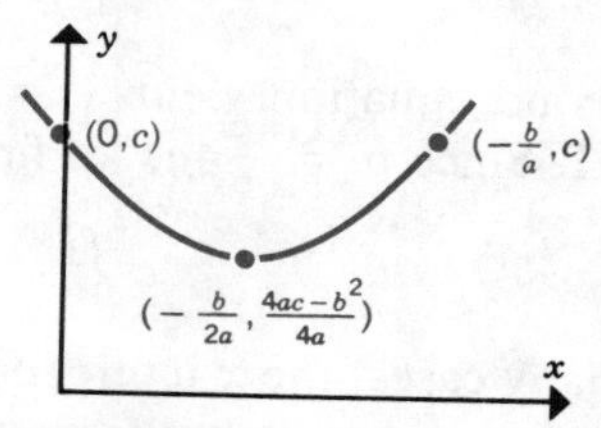

Figure 8

■

Example 5 Sketch the graph of the quadratic function $f(x) = 2x^2 - 8x + 3$.

Solution For this function $a = 2$, $b = -8$, and $c = 3$; the procedure provides the three points:

$$(0, c) = (0, 3)$$

$$\left(\frac{-b}{a}, c\right) = \left(-\frac{-8}{2}, 3\right) = (4, 3)$$

$$\left(\frac{-b}{2a}, \frac{4ac - b^2}{2a}\right) = \left(2, \frac{24 - 64}{8}\right) = (2, -5)$$

The graph is shown in Figure 9.

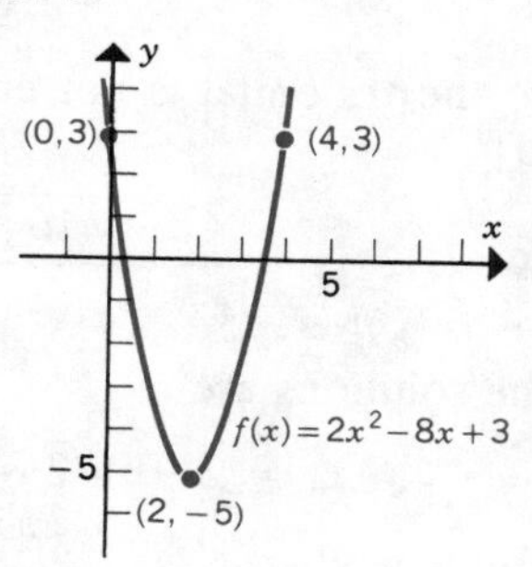

Figure 9

■

Example 6 Find an equation of the parabola whose vertex is at (2, 5) and that passes through the point (4, 1).

Solution The formula $x_V = -\dfrac{b}{2a}$ can be used to establish the relationship between the constants a and b:

$$x_V = 2 = -\frac{b}{2a} \quad \text{or} \quad b = -4a$$

This result enables us to write the equation $f(x) = ax^2 + bx + c$ as $f(x) = ax^2 - 4ax + c$. Since each of the ordered pairs (2, 5) and (4, 1) satisfies this equation, we can write the two equations

$$5 = a(2)^2 - 4a(2) + c$$
$$1 = a(4)^2 - 4a(4) + c$$

The second equation yields $c = 1$. Substituting this result into the first yields $a = -1$. Since $b = -4a$, we find that $b = 4$, and the equation for $f(x)$ is

$$f(x) = -x^2 + 4x + 1$$ ■

In many cases, the x intercepts, if they exist, are needed. They are found by setting $f(x) = ax^2 + bx + c$ equal to zero and solving the quadratic equation

$$ax^2 + bx + c = 0 \tag{2}$$

The solution(s) of Equation 2 can be found by factoring (whenever possible) or using the quadratic formula

$$x = \frac{-b \pm \sqrt{b^2 - 4ac}}{2a} \tag{3}$$

Example 7 Find the x intercepts for each of the functions in Examples 3 and 4.

Solution 1. Example 3

$$f(x) = x^2 - 4x + 2$$
$$0 = x^2 - 4x + 2$$

Since the trinomial is not easily factored, the quadratic formula is used to yield

$$x = \frac{4 \pm \sqrt{16 - 4(2)}}{2} = \frac{4 \pm \sqrt{8}}{2} = \frac{4 \pm 2\sqrt{2}}{2}$$

so the solutions are

$$x_1 = 2 + \sqrt{2} \approx 3.41, \qquad x_2 = 2 - \sqrt{2} \approx 0.59$$

2. Example 4

$$f(x) = -2x^2 + 6x - 5$$
$$0 = -2x^2 + 6x - 5$$

Once again, the quadratic formula is used:

$$x = \frac{-6 \pm \sqrt{36 - 4(-2)(-5)}}{-4} = \frac{-6 \pm \sqrt{-4}}{-4}$$

Since $\sqrt{-4}$ is not a real number, the quadratic equation has no real solutions, which means that there are no x intercepts. There are no x intercepts whenever the quantity $b^2 - 4ac$, called the *discriminant,* is negative. ■

Application: Quantity or Group Discounts

Piecewise functions in which one equation is quadratic can arise in situations involving quantity discounts, as the following example shows.

Example 8 A charter airline charges a flat rate of \$200 per passenger for groups of 50 or fewer people. For each additional passenger over 50, the airline reduces the ticket price by \$1 for all passengers in the group. Find the airline's revenue $R(x)$ as a function of x, the number of passengers in a group. In addition, find the group size that maximizes $R(x)$.

Solution First, the ticket price p can be written as

$$p = \begin{cases} 200, & x \leq 50 \\ 200 - 1(x - 50) = 250 - x, & x > 50 \end{cases}$$

Next, the revenue function $R(x) = px$ can be written as

$$R(x) = \begin{cases} 200x, & x \leq 50 \\ 250x - x^2, & x > 50 \end{cases}$$

The graph of this function is shown in Figure 10. The maximum revenue is attained when the group size is 125. The corresponding ticket price is also \$125.

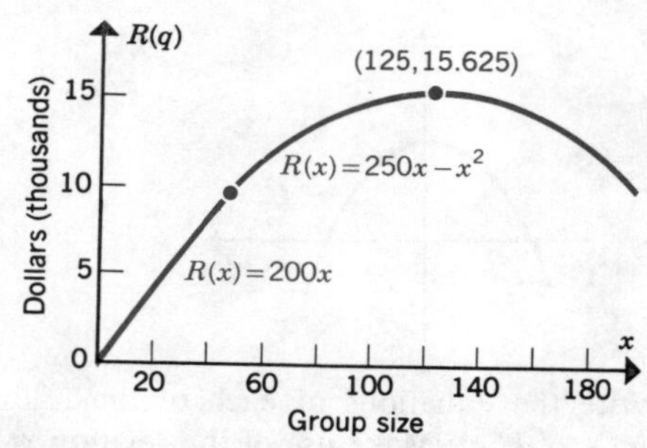

Figure 10

■

1.4 EXERCISES

Sketch the graph of each of the following functions. In addition, find (a) the vertex and (b) the x and y intercepts.

1 $f(x) = x^2 + 1$

2. $f(x) = x^2 - 2$

3. $g(x) = 1 - x^2$

4. $h(x) = (x - 1)^2$

5. $f(x) = 2x^2$

6. $g(x) = -\dfrac{x^2}{2}$

7. $f(x) = -(x + 2)^2$

8. $h(x) = x^2 - 4x + 3$

9. $g(x) = -x^2 + 2x + 8$

10. $f(x) = 2x^2 + 4x - 6$

11. $f(x) = x^2 + 6x + 9$

12. $h(x) = -3x^2 + 12x$

13. $f(x) = 9 - x^2$

14. $f(x) = x^2 - 3x + 2$

15. $g(x) = 2x^2 + 6x$

16. $f(x) = 5x^2 - 4x + 2$

17. $f(x) = -3x^2 + 8x + 2$

18. $h(x) = 2x^2 - 5x + 1$

19. $f(x) = \sqrt{2}x^2 + 4x$

20. $f(x) = x^2 - \sqrt{2}x$

For each of the quadratic functions whose graphs are shown in Exercises 21–26, determine whether the constants a, b, and c in the equation $f(x) = ax^2 + bx + c$ are positive, negative or zero.

21.

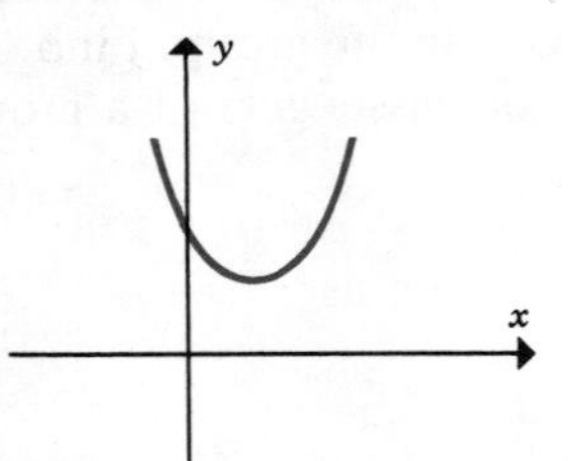

22.

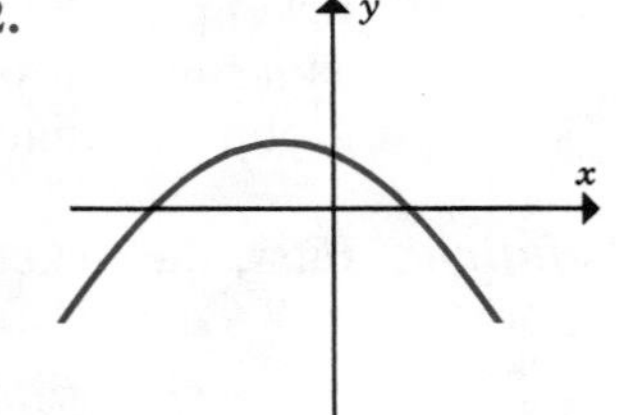

23.

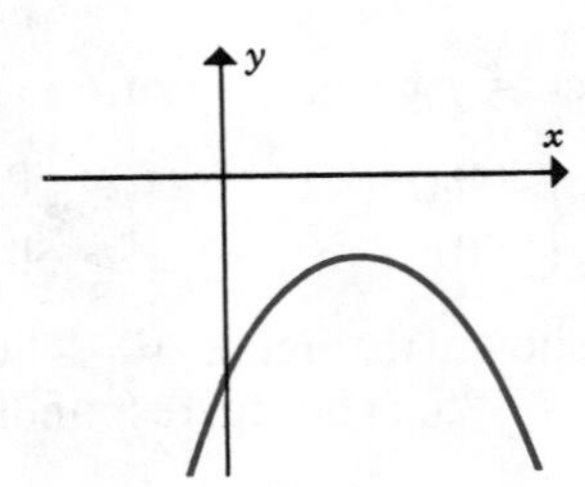

24.

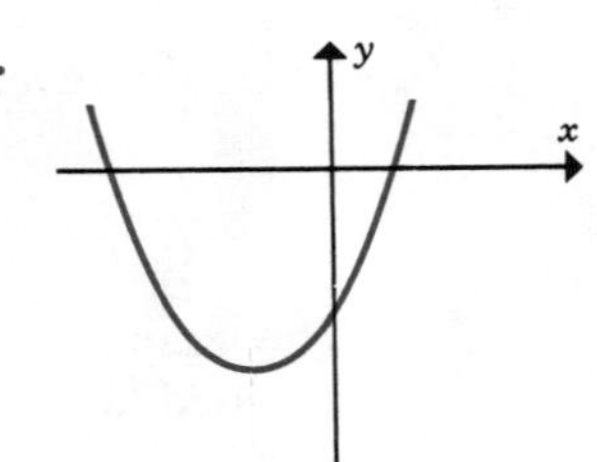

25.

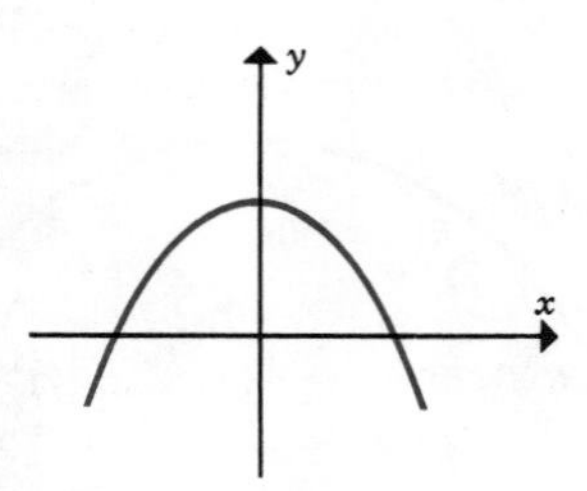

26.

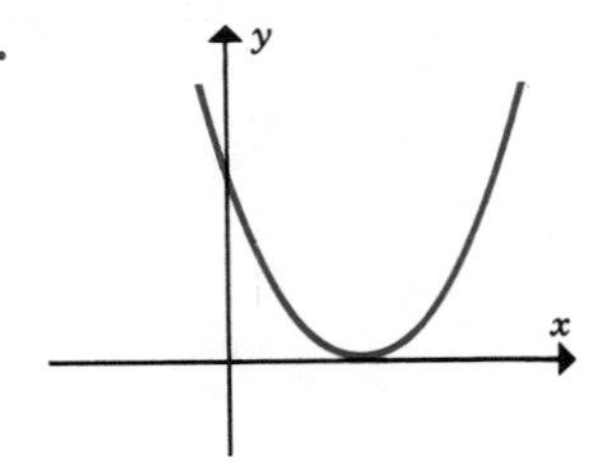

Write the equations of each of the following parabolas in the form $f(x) = ax^2 + bx + c$. *Hint:* Make use of the relation $x_V = -(b/2a)$.

27. Vertex at (0, 4); passes through (1, 0).
28. Vertex at (1, 2); passes through (0, 0).
29. Vertex at (4, 0); passes through (0, 6).
30. Vertex at (1, −2); passes through (3, 2).
31. Vertex at (−2, 3); passes through (0, −5).
32. Vertex at (−3, −4); passes through (−1, 8).
33. Vertex at (3, 0); passes through (1, 1).
34. Vertex at (6, 6); passes through (5, 5).

35. (a) Material for a poster costs \$2 per square foot. If a poster is cut in the form of a square, find an equation that describes C, the cost (in dollars) of the poster, in terms of x, the length (in feet) of one side.
(b) If the length of each side is doubled, what happens to the cost?
(c) Write an equation for C in terms of y, the length in yards.

36. For the retail store in Example 1, find an equation for R, the revenue, as a function of p, the unit price. Verify that the maximum revenue is attained when $p = \$200$.

37. For the retail store in Example 1, suppose that the owner pays \$100 for each VCR he purchases. Find an equation for P, the monthly profit, as a function of q, the number of units that can be sold. Find the selling price that maximizes P. Is it equal to the price that maximizes monthly revenue?

38. (a) (See exercise 10, Section 1.3) Suppose that the group sponsoring the rock concert wants to maximize the revenue from ticket sales. What should the ticket price be?
(b) Suppose that their costs are \$5000 to rent the concert hall and \$20,000 for the group. Find P, the profit, as a function of p, the ticket price. What value of p maximizes P?

39. A resort hotel that has 100 rooms can rent all its rooms if it charges \$50 per day for each room. For each dollar increase in the daily rate, the number of vacant rooms increases by 2. If the cost of cleaning and maintaining an occupied room is \$8 per day, what room rate will maximize the daily profit for the hotel?

40. What change occurs in the answer to exercise 39 if in addition the cost of cleaning and maintaining an unoccupied room is \$2 per day?

41. A small theater offers group discounts on tickets according to the following plan. If the group size is 20 or fewer people, the ticket price is \$6 per person: For each additional person over 20, the theater reduces the ticket price by 10¢ for everyone in the group. Find R, the revenue, as a function of x, the number of people in a group. Plot R as a function of x and determine the size of the group that maximizes R.

42. The volume V, in cubic centimeters, of a cylindrical blood vessel 8 cm long is a function of the radius r, in centimeters; the relationship is given by the equation $V = 8\pi r^2$.
(a) Sketch the graph of this function.
(b) How much blood can an artery whose normal radius is 0.20 cm hold? If the radius is reduced to 0.10 cm by the buildup of fatty deposits on the inside wall of the artery, how much blood can the artery hold?
(c) In general, what happens to the volume that a blood vessel can hold if the radius is reduced to half its normal size because fatty deposits have accumulated on the interior wall of the vessel?

43. The concentration of bacteria in a public water system has become too large and has forced public health officials to treat the water with an antibacterial agent. The biochemist in charge of the project estimates that $N(t)$, the number of bacteria per cubic centimeter, can be described by the equation

$$N(t) = 40t^2 - 320t + 1000$$

where t is the time, in days, from the treatment.
(a) When will the concentration reach its minimum level?
(b) The water is considered unsafe for drinking when the concentration exceeds 720 bacteria per cubic centimeter. How long after the treatment will it be before the water is considered safe for drinking?

(c) If no further treatments are planned, how long does it take for the water to again become unsafe to drink?

44. According to a model developed by a public health group, the number of people $N(t)$, in hundreds, who will be ill with the Manchurian flu at any time t, in months, next winter is described by the equation $N(t) = 100 + 30t - 10t^2$ where $t = 0$ corresponds to the beginning of December.

(a) When will the flu be at its peak? How many people will be afflicted at this time?
(b) Sketch the graph of this equation.
(c) When will the flu be over?

1.5 OTHER TYPES OF FUNCTIONS

Linear and quadratic functions are special cases of functions called polynomial functions. A **polynomial function** has the form

$$f(x) = a_0 + a_1x + a_2x^2 + \cdots + a_nx^n \tag{1}$$

where n is a nonnegative integer and $a_0, a_1, a_2, \ldots, a_n$ are real numbers. The following are examples of polynomial functions.

1. $f(x) = 2x^3 + 5x^2 - 3x - 1$
2. $g(x) = x^4 + 2x^3 + 3x^2 - 4x + 5$
3. $F(x) = 3x^5 - 4x^3 + 2x^2 + 6$

Except for linear and quadratic functions, polynomial functions are usually difficult to graph. The methods of differential calculus, presented in Chapter 4, often can be used to graph such functions. However, the graphs of some simple polynomial functions of the type

$$f(x) = ax^n \tag{2}$$

can be sketched directly, as the next example illustrates.

Example 1 Plot a graph of the function $f(x) = x^3$.

Solution The graph can be plotted by using the points shown in the following table to yield the curve shown in Figure 1.

x	$y = x^3$
−3	−27
−2	−8
−1	−1
0	0
1	1
2	8
3	27

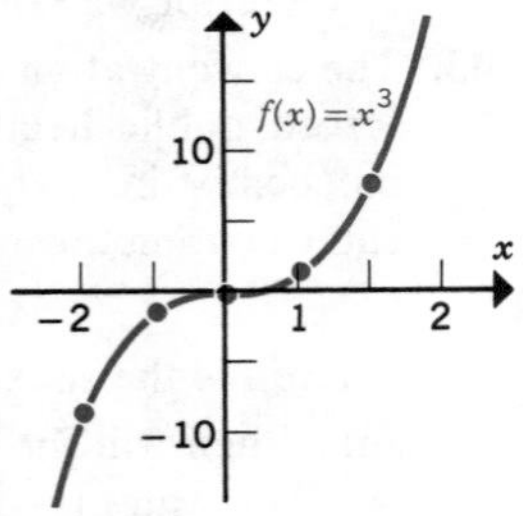

Figure 1

■

Rational Functions

Polynomial functions are special cases of yet another type of function, a rational function. A **rational function** is a function that can be written as a ratio of two polynomials; the following are examples of rational functions:

$$f(x) = \frac{x}{x+1}$$

$$g(x) = \frac{x^2-1}{x+2}$$

$$h(x) = \frac{x-3}{x^4+x^2+1}$$

Graphing rational functions generally requires the use of calculus and will be described in detail later (Section 4.4). However, it is possible to graph simple rational functions of the type

$$f(x) = \frac{a}{(x-b)^n} \tag{3}$$

where a, b are real numbers.

Example 2 Plot the graph of the function $f(x) = \dfrac{1}{x+1}$.

Solution The graph can be plotted with the aid of the following table and is shown in Figure 2.

x	$y = \dfrac{1}{x+1}$
-4	$-\frac{1}{3}$
-3	$-\frac{1}{2}$
-2	-1
$-\frac{3}{2}$	-2
$-\frac{5}{4}$	-4
-1	Undefined
$-\frac{3}{4}$	4
$-\frac{1}{2}$	2
0	1
1	$\frac{1}{2}$
2	$\frac{1}{3}$

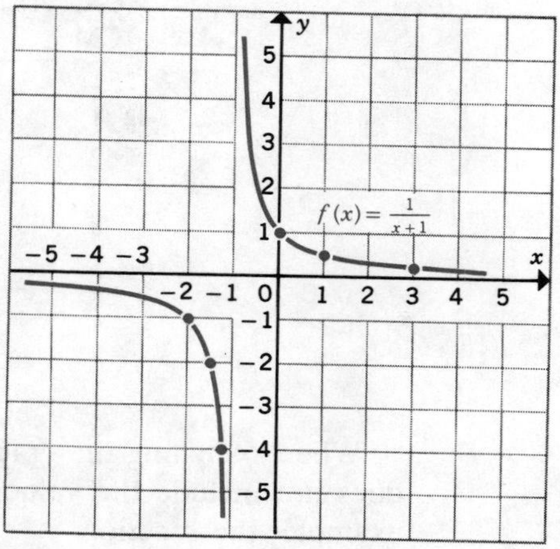

Figure 2

Note: When you are sketching the graph of a rational function, it is good practice to select values of x close to those where the function is not defined because small changes in x produce large changes in y. Since $f(x) = \dfrac{1}{x+1}$ is not defined when $x = -1$, values of x close to $x = -1$, such as $x = -\frac{3}{2}$, $-\frac{5}{4}$, . . . , were used to sketch the graph. ■

Application

In break-even analysis (Section 1.3), the quantity $p - v$ is called the *contribution margin* (CM). The relationship between x_{BE}, the break-even volume, and CM is given by the equation

$$x_{BE} = \frac{F}{CM} \tag{4}$$

Example 3 Acne Company has developed a new skin cream to treat the heartbreak of psoriasis. The cream is sold in 6-oz tubes. Annual fixed costs are \$60,000. Sketch the graph of x_{BE}, the break-even volume, as a function of CM.

Solution If F is expressed in thousands of dollars and CM in dollars per tube, x_{BE} in Equation 4 has units of thousands of tubes. The relationship between x_{BE} and CM becomes

$$x_{BE} = \frac{60}{CM}$$

The graph of this equation is shown in Figure 3.

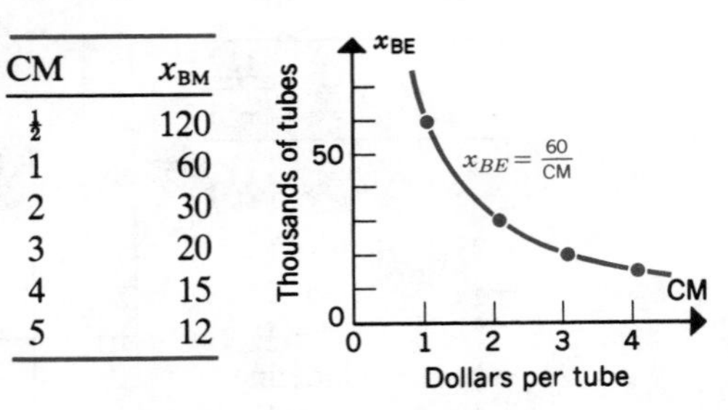

CM	x_{BM}
$\frac{1}{2}$	120
1	60
2	30
3	20
4	15
5	12

Figure 3 ■

When sketching the graph of a rational function, remember that its domain does not include the values of x for which the denominator equals zero. For example, the function

$$f(x) = \frac{2x+1}{x-5}$$

has as its domain all real values of x except $x = 5$. Similarly, the domain of the function

$$g(x) = \frac{3}{x^2 - 4} = \frac{3}{(x + 2)(x - 2)}$$

contains all real values of x except $x = 2$ and $x = -2$.

Power Functions

A *power function* has the form

$$f(x) = C[u(x)]^r$$

where $u(x)$ is a function of x and C and r are real numbers.

Example 4 Sketch the graph of the function $f(x) = (x + 2)^{1/2} = \sqrt{x + 2}$.

Solution Since $\sqrt{x + 2}$ produces real numbers only if $x + 2 \geq 0$, the domain consists of the real numbers satisfying the condition $x \geq -2$. The graph in Figure 4 is sketched using the points in the following table.

x	$y = \sqrt{x + 2}$
-2	0
-1	1
2	2
7	3
14	4

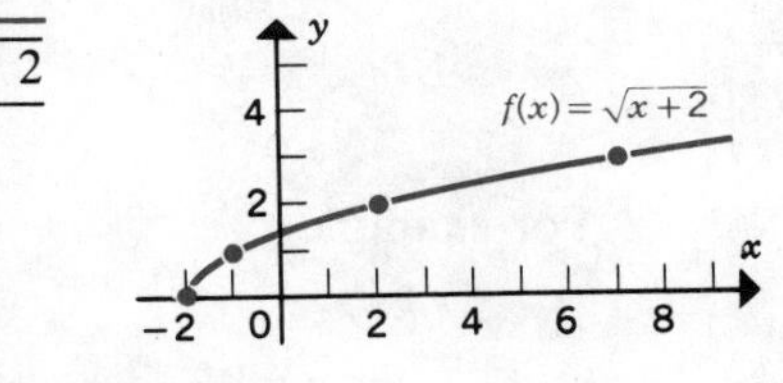

Figure 4

Note: In the table, values of x were selected for which $x + 2$ is a perfect square; this strategy allowed us to find the corresponding values of y quickly. ■

Application

Simple power functions have been used to describe how learning or experience affects the length of time required to assemble an item or to carry out an operation. The graphs of these functions are called **learning** or **experience** curves.

Example 5 ABC Manufacturing has just received a contract from the Department of State for 100 units of their new laser copier. The industrial engineering group has found that the relationship between n, the nth unit produced, and $T(n)$, the time (in hours) required to produce that unit, is given by the equation

$$T(n) = \frac{60}{\sqrt{n}} \qquad 1 \leq n \leq 100$$

Plot the graph of this function.

Solution The points in the following table can be used to sketch the graph (see Figure 5). The curve shows the effect of learning or experience in the production process; the time to produce an additional unit decreases as the number of units produced gets larger. Once developed, learning curves can be used to estimate the total cost of labor for a given level of production (exercise 40).

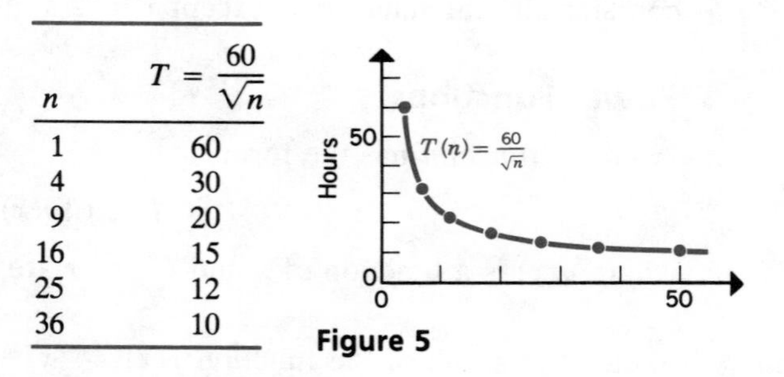

n	$T = \dfrac{60}{\sqrt{n}}$
1	60
4	30
9	20
16	15
25	12
36	10

Figure 5

■

Absolute-Value Function

The *absolute value* of any real number x, designated as $|x|$, is defined as

$$|x| = \begin{cases} x & \text{when } x \geqslant 0 \\ -x & \text{when } x < 0 \end{cases}$$

For example,

$$|5| = 5, \qquad |-8| = -(-8) = 8$$
$$|4 - 11| = |-7| = -(-7) = 7$$

The **absolute-value** function, $f(x) = |x|$, is defined as

$$f(x) = |x| = \begin{cases} x & \text{when } x \geqslant 0 \\ -x & \text{when } x < 0 \end{cases}$$

Using the following table, we can sketch the graph of the function, as shown in Figure 6. The domain of the absolute-value function is the set of all real numbers.

x	$y = \|x\|$
−3	3
−2	2
−1	1
0	0
1	1
2	2
3	3

Figure 6

Holes in Graphs

Two or more functions can be added, subtracted, multiplied, or divided to generate new functions. Holes may appear in the graphs of a new function, particularly when the operation of division is used to generate the new function.

Example 6 Plot the graph of the function $F(x) = \dfrac{f(x)}{g(x)}$, where $f(x) = x^2 - 1$ and $g(x) = x - 1$.

Solution The function $F(x)$ is defined by the equation

$$F(x) = \frac{x^2 - 1}{x - 1}$$

When $x = 1$, the denominator equals zero; therefore, $F(1)$ is not defined. The graph of $F(x)$ can be plotted quickly after the right side is simplified by reducing the fraction,

$$\frac{x^2 - 1}{x - 1} = \frac{(x + 1)(x - 1)}{x - 1} = x + 1, \qquad x \neq 1$$

so $F(x)$ can be written as

$$F(x) = x + 1, \qquad x \neq 1$$

The graph of $F(x)$ is a straight line; the restriction $x \neq 1$ means that the point (1, 2) must be removed from the graph. This is indicated by the open circle in Figure 7.

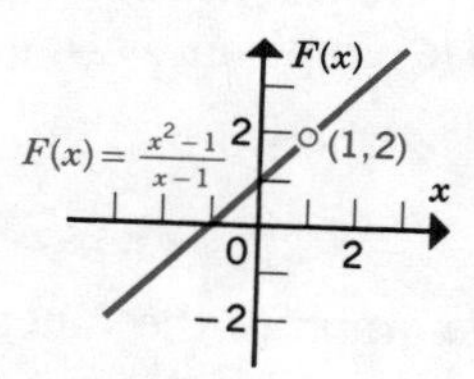

Figure 7

■

Example 7 The Public Relations Department of Wahoo College has published the following data on their star running back.

Game	Carries	Yards Gained
1	20	100
2	25	125
3	15	75

Find an equation that describes the relationship between x, total number of carries, and y, total number of yards gained. Next, find $\bar{y}$, the average number of yards per carry, as a function of x; plot the graph of this function.

Solution Figure 8 shows the data from which the equation

$$y = 5x$$

can be found. The average number of yards per carry, $\bar{y}$, is found by means of the equation

$$\bar{y} = \frac{y}{x} = \frac{5x}{x}$$

Noting that this ratio is not defined when $x = 0$, we get

$$\bar{y} = 5, \qquad x \neq 0$$

The graph is a horizontal line with the point (0, 5) removed (Figure 9).

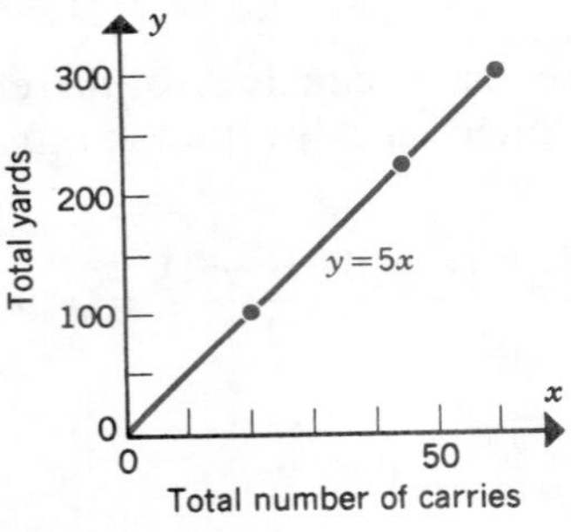

Figure 8

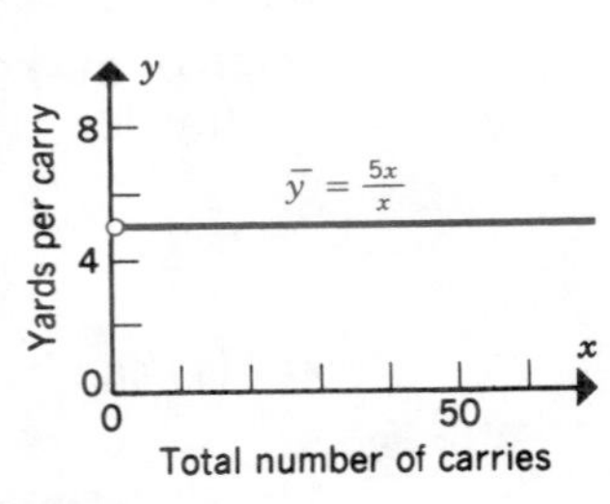

Figure 9 ■

The holes in the graphs in Figures 7 and 9 were due to the fact that the functions were not defined when $x = 1$ and $x = 0$, respectively. A hole can also appear in a graph at $x = a$ when $f(a)$ is defined, as illustrated in the next example.

Example 8 Sketch the graph of the function

$$f(x) = \begin{cases} \dfrac{x^2}{x}, & x \neq 0 \\ 2, & x = 0 \end{cases}$$

Solution When $x = 0$, $f(0) = 2$. For all other values of x, $f(x)$ is found by means of the equation

$$f(x) = \frac{x^2}{x} = x, \qquad x \neq 0$$

whose graph is a straight line with the point (0, 0) removed. Figure 10 shows the entire graph of the function.

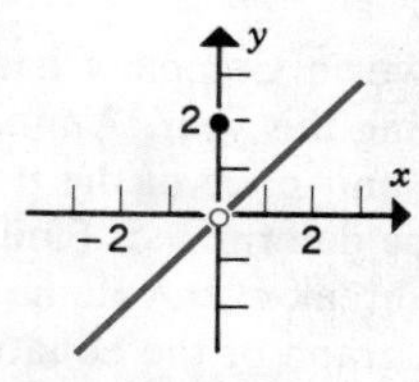

Figure 10

■

1.5 EXERCISES

Which of the following are polynomial functions?

1. $F(x) = 8x^3 + 5x^2 - 2x + 1$
2. $F(x) = 6 - x$
3. $g(x) = \dfrac{3}{x^2}$
4. $f(x) = \sqrt{2 - x}$
5. $F(x) = 2x^2 + \dfrac{3}{x}$
6. $g(x) = x^2 + 3x - 4$
7. $f(x) = 4x^3 + \sqrt{x}$
8. $f(x) = \dfrac{1 - \sqrt{x}}{x}$

Find the domain for each of the following functions.

9. $f(x) = \dfrac{2x}{x + 1}$
10. $f(x) = \dfrac{x}{x^2 - 4}$
11. $f(x) = \dfrac{2x - 1}{x^2 - 3x + 2}$
12. $g(x) = \dfrac{x}{x^2 + 1}$
13. $F(x) = \sqrt{x - 1}$
14. $f(x) = \sqrt{2x + 4}$

Sketch the graph of each of the following functions.

15. $f(x) = -x^3$
16. $g(x) = 2x^3$
17. $f(x) = x^4$
18. $F(x) = t^3 - 2$
19. $g(x) = \dfrac{-1}{x}$
20. $f(s) = \dfrac{1}{s^2}$
21. $f(x) = \dfrac{1}{x - 1}$
22. $F(x) = \dfrac{2}{x - 1}$
23. $f(x) = \sqrt{x - 1}$
24. $g(x) = \sqrt{x + 1}$
25. $f(x) = -\sqrt{x}$
26. $F(t) = \sqrt{2t}$
27. $f(x) = \sqrt{x^2}$ Caution: $\sqrt{x^2} \neq x$
28. $f(x) = \dfrac{1}{\sqrt{x}}$
29. $f(x) = \sqrt{1 - x^2}$
30. $f(x) = \sqrt{x + 3}$
31. $g(x) = -|x|$
32. $f(x) = |x - 1|$
33. $F(x) = |2x|$
34. $f(x) = x + |x|$

Find the holes in the graphs of each of the following functions.

35. $f(x) = \dfrac{2x^2 - 2}{x - 1}$
36. $F(x) = \dfrac{4x^2 - 2x^3}{x^2 - 2x}$

37. (a) Newtron Company has developed a new laser jet printer, which it will begin selling this year. Annual fixed costs are $200,000. The company has not yet set the unit price of the printer, in part because some of the variable costs remain to be determined. Find an equation that describes the relationship between x_{BE}, the break-even volume, and CM, the contribution margin of each printer. Sketch the graph of the equation.

(b) The variable cost of each printer is $400. If management does not want the break-even volume to exceed 2000 units, what is the minimum selling price?

38. (a) TOYS ARE $, Inc. has developed a new doll. Annual fixed costs for this product are $50,000. Plot x_{BE} as a function of CM.

(b) If the variable costs per doll are $20, and the break-even volume is not to exceed 10,000 units, what is the minimum price that management should set on each doll?

39. (a) The management of TOYS ARE $ has adopted the following pricing policy: The selling price of any item equals 150 percent of the variable cost. If annual fixed costs for a given toy are $20,000, find an equation that gives x_{BE} as a function of v, the variable cost.

(b) What is the minimum value of v for the toy in part (a) if management does not want the break-even volume to exceed 5000 units?

40. Dynamic Aircraft Company has received an order for 64 of its new supersonic aircraft. The relationship between $T(n)$, the time (in weeks) to produce the nth plane, and n is given by the equation

$$T(n) = \frac{36}{\sqrt[3]{n}}$$

Sketch the graph of this equation. How long will it take the company to produce the twenty-seventh aircraft?

Complete the following statement. The effect of learning in this case can be described as follows: The time required to build an aircraft decreases by 50 percent as the number of aircraft built increases by ________ percent.

41. The average amount of time that a new claim reviewer spends on a health claim is described by the equation

$$T(n) = \frac{12}{\sqrt[4]{n}}, \qquad 1 \le n \le 256$$

where $T(n)$ is the length of time, in minutes, spent on the nth claim.

(a) Sketch the graph of this equation.

(b) How much time is spent reviewing the first claim? The sixteenth?

42. Suppose that a learning curve is described by the function

$$T(n) = \frac{20}{\sqrt{n}}$$

where $T(n)$ is the time (in hours) to produce the nth unit.

(a) How much time is required to produce the fourth unit?

(b) How much time is required to produce the first four units?

(c) If the cost of labor is $10 per hour, how much money does it cost to produce the first four units?

43. Animal physiologists use the formula $S = kw^{2/3}$ to calculate S, the surface area of an animal in square meters, from its weight w, in kilograms; the value of the constant k depends on the type of animal under study.

(a) Sketch the graph of this equation for the case $k = 0.1$.
(b) Suppose the weight of an animal increases from 1 to 8 kg. What is the ratio of the animal's new surface area to the original surface area?

44. The students in a junior high school were given a list of 50 items and asked to memorize the list. Each day a group of students was selected and asked to write down as many of the items on the list as they could remember. The average number of items remembered $A(t)$ can be described by the equation

$$A(t) = 5 + \frac{18}{t}, \qquad t \geq 1$$

where t is the time, in days, from the moment the list was memorized.
(a) Sketch the graph of this equation.
(b) What happens to $A(t)$ as t becomes large?

KEY TERMS

function
independent variable
dependent variable
domain
ordered pair
solution
graph
vertical-line test
piecewise function
slope
linear function
***x* intercept**
***y* intercept**
point–slope formula
slope–intercept formula
straight-line depreciation
break-even analysis
demand equation
supply equation
equilibrium price
quadratic function
parabola
vertex
polynomial function
rational function
absolute-value function

REVIEW PROBLEMS

1. If $f(x) = x^2 + x$, find

(a) $f(1)$ (b) $f(-2)$ (c) $f(\frac{1}{2})$ (d) $f(1 + h)$ (e) $f(1 + h) - f(1)$
(f) $\dfrac{f(1 + h) - f(1)}{h}$

2. If $g(x) = \dfrac{2}{x}$, find

(a) $g(1)$ (b) $g(4)$ (c) $g(-\frac{1}{2})$ (d) $g(1 + h)$ (e) $g(1 + h) - g(1)$
(f) $\dfrac{g(1 + h) - g(1)}{h}$

3. If $f(t) = 1 - \sqrt{2t}$, find

(a) $f(0)$ (b) $f(2)$ (c) $f(\frac{1}{2})$ (d) $f(8)$ (e) $f(2a^2)$

4. If $F(x) = \dfrac{2}{x^2 + 1}$, find

(a) $F(0)$ (b) $F(1)$ (c) $F(-1)$ (d) $F(h)$ (e) $F(h) - F(0)$

(f) $\dfrac{F(h) - F(0)}{h}$

5. Indicate which of the curves shown in Figure 1 represent graphs of functions.

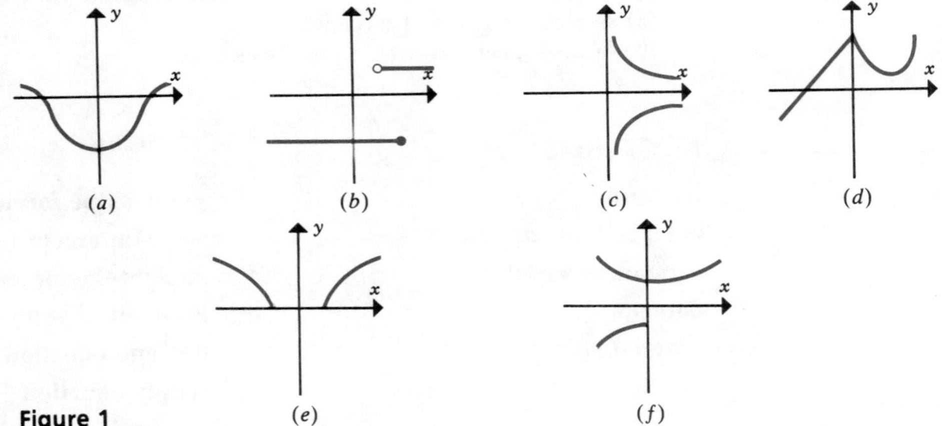

Figure 1

Sketch the graph of each of the functions in Exercises 6 through 10.

6. $f(x) = 5 - 3x$ **7.** $F(x) = x^2 + 2x$ **8.** $f(t) = \dfrac{2}{t + 2}$

9. $f(x) = \begin{cases} 3, & x \leq 0 \\ 2 + x, & x > 0 \end{cases}$ **10.** $F(x) = \begin{cases} 1 - x, & x \leq 1 \\ \sqrt{x}, & x > 1 \end{cases}$

11. The equations of the four lines shown in Figure 2 are

$$3y - 4x = 6, \quad 2y + 4x = 8, \quad 6y - 2x = 12, \quad 3y + 5x = 10$$

Match each equation with its graph.

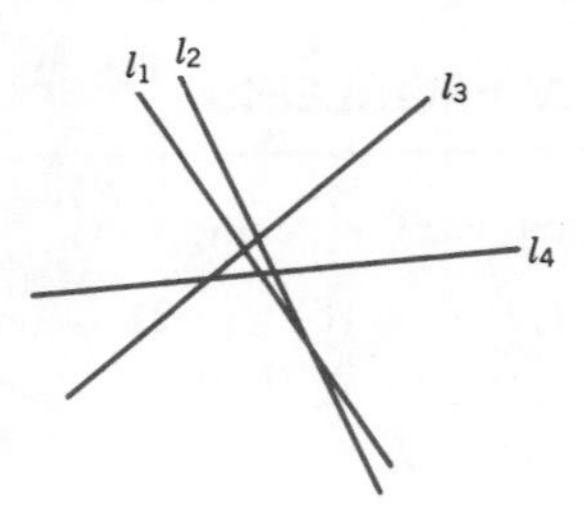

Figure 2

Find the equation of each of the lines described in problems 12 through 18.

12. The slope equals 3 and the line passes through $(1, -2)$.

13. The slope equals -2 and the line passes through $(1, 3)$.

14. The line passes through $(2, -3)$ and $(1, -3)$.
15. The line passes through $(k, 2k)$ and $(2k, -k)$; $k \neq 0$.
16. The line passes through $(5, 2)$ and $(5, -2)$.
17. The line passes through $(2, -1)$ and has a nonzero y intercept which is three times the x intercept.
18. The line is parallel to the line whose equation is $3y + 6x = 8$ and passes through the point $(1, 1)$.

Sketch the graph of each of the quadratic functions in problems 19 through 24. In addition, find the coordinates of the vertex, the y intercept, and the x intercepts when they exist.

19. $f(x) = 4x - x^2$ **20.** $f(x) = 2x^2 - 6x + 2$ **21.** $f(x) = 5 - 2x - x^2$
22. $g(t) = t^2 - 1$ **23.** $f(x) = x^2 - 4x + 4$ **24.** $f(x) = 6 - 5x - x^2$
25. Find an equation of the parabola whose vertex is at $(2, 4)$ and which passes through the point $(0, -1)$.
26. Find an equation of the parabola whose vertex is $(-1, -3)$ and which passes through $(2, 5)$.
27. The graphs of the functions $f(x) = a_1x^2 + b_1x + c_1$ and $g(x) = a_2x^2 + b_2x + c_2$ are shown in Figure 3. Determine whether

i. a_1 is greater than, less than, or equal to a_2;
ii. b_1 is greater than, less than, or equal to b_2; and
iii. c_1 is greater than, less than, or equal to c_2.

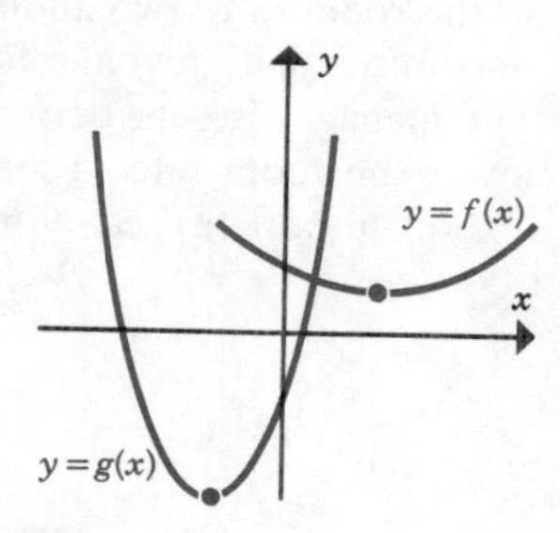

Figure 3

Sketch the graph of each of the following functions.

28. $f(x) = \dfrac{4}{x^2}$ **29.** $f(x) = \dfrac{4}{x^2 + 1}$ **30.** $g(x) = -2x^3$
31. $f(x) = -\sqrt{1 - x}$ **32.** $f(x) = 1 - |x|$ **33.** $g(x) = x^{3/2}$
34. Old McDonald has a farm. On his farm he grows soybeans. When the price of a bushel was \$3, he planted 500 acres of soybeans. When the price of a bushel dropped to \$2.50, he allocated only 400 acres of his farm to the planting of soybeans.

(a) Assuming a linear relationship between p, the price of a bushel of soybeans, and x, the number of acres of soybeans planted, find a supply equation.
(b) At what price would McDonald stop planting soybeans?

35. Regal Publishing Company purchases a Fax machine for \$1200. The company uses the straight-line method of depreciation. If the lifetime of the machine is five years and if it has zero salvage value, find an equation that describes the book value $V(t)$, in dollars, as a function of the time t, in years, from the date of purchase.
36. A blood test revealed that a woman has a high level of blood cholesterol. Her

physician prescribed medication to correct the problem. The relationship between her cholesterol level $C(t)$, in milligrams per deciliter, and the time t, in weeks, from the beginning of the treatment is described by the equation

$$C(t) = 175 + \frac{125}{1 + \sqrt{t}}$$

(a) What was her cholesterol level when the treatment began?
(b) What was her cholesterol level four weeks later?
(c) Her doctor wants her cholesterol level below 200 mg/dL. How long does it take for her to attain this level?

37. The cost of making copies on a self-service copier at Sire Speedy Printing is governed by the following schedule.

Number of Copies	Cost of Each Copy
1–100	10¢
Over 100	7¢

(a) Find a piecewise function that gives the cost $C(x)$ to make x copies at Sire Speedy.
(b) Sketch the graph of the function obtained in part (a).
(c) For what values of x would it be economically advantageous to lie and overstate the number of copies that you made on the self-service copier?

38. (a) The Bates Motel has 80 rooms, all of which can be rented each night at the current rate of \$50 per night. The manager estimates that for each \$1 increase in the room rate, two additional rooms become vacant. Find the room rate that maximizes daily revenue for the motel.
(b) The manager is considering a proposal to increase the number of rooms and to lower the room rate to maintain 100 percent occupancy. On the basis of your answer to part (a), can you make a recommendation regarding this proposal?

2

THE DERIVATIVE

INTRODUCTION

Calculus is used to study the behavior of functions. When we examine the behavior of a given function, the derivative of the function enables us to determine the rate at which the function is changing at a point. An understanding of the derivative is based upon a fundamental concept of calculus, the limit of a function. The limit is also needed to describe a characteristic of a function called continuity, which tells us whether or not the graph of the function contains any breaks or gaps.

2.1 LIMITS

A study of calculus requires an understanding of two important concepts: **limit** and **continuity.** Of the two, continuity is the easier to describe and to visualize. Let's look at the three functions whose graphs are shown in Figures 1 through 3. The function whose graph is shown in Figure 1 can be plotted on the interval $a \leq x \leq b$ without lifting your pencil off the paper; this function is said to be continuous for all values of x on the interval. On the other hand, the graphs shown in Figures 2 and 3 cannot be drawn without lifting your pencil off the page at $x = 1$; the corresponding functions are said to be discontinuous at $x = 1$.

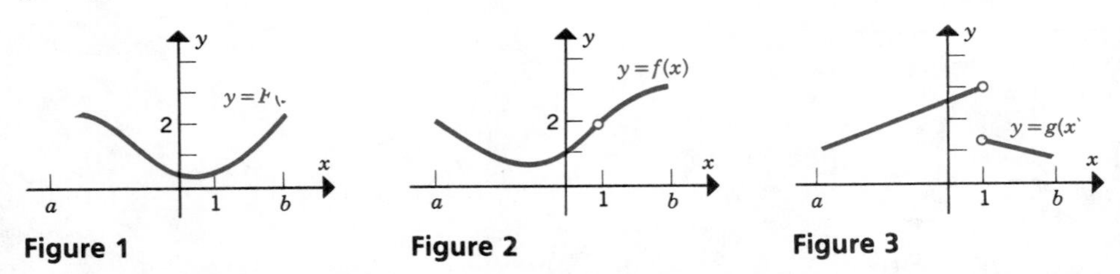

Figure 1 Figure 2 Figure 3

Although both functions shown in Figures 2 and 3 are discontinuous at $x = 1$, the graphs indicate that the functions behave differently as the independent variable x is assigned values that are close to $x = 1$. In Figure 2, as x is assigned values that get closer and closer to $x = 1$, the corresponding values of $f(x)$ get closer and closer to $y = 2$, or in mathematical terms, the limit of $f(x)$, as x approaches 1, equals 2. On the other hand, in Figure 3, as x is assigned values that get closer and closer to $x = 1$, the corresponding values of $g(x)$ do not get closer and closer to a unique number; in this instance the limit of $g(x)$, as x approaches 1, does not exist.

It is now time to describe the concepts of limit and continuity (Section 2.2) in more detail. First, when we say that the variable x is approaching the number 1, it is understood that x is assigned values that get closer and closer to 1 but do not equal 1. An example of the process is shown in Table 1 and Figure 4. In the first column of the table, arbitrary values are assigned to x subject to the following conditions.

1. Each value is less than 1.
2. From top to bottom, each value is greater than the one preceding it.

Table 1

From the Left x	From the Right x
0.00000	2.00000
0.50000	1.50000
0.90000	1.10000
0.99000	1.01000
0.99900	1.00100
0.99990	1.00010
⋮	⋮

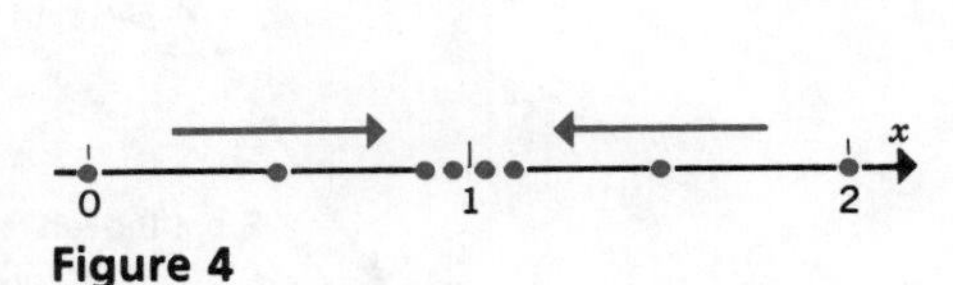

Figure 4

The reverse is true in the second column. Note that x approaches 1 from both the left and the right. This approach is symbolized as

$$x \longrightarrow 1 \quad \text{means "}x\text{ approaches 1"}$$

In general,

$$x \longrightarrow a \quad \text{means "}x\text{ approaches }a\text{"}$$

that is, the values assigned to x in the domain of a function get closer and closer to a given number a.

Next, for a given function, we want to determine whether $f(x)$, the dependent variable, approaches a unique value as "x approaches a." For example, suppose that we are given the function

$$f(x) = 3x - 1$$

As the variable x approaches 1, what value (if any) does $f(x)$ approach? Table 2 illustrates the process numerically and Figure 5 illustrates it graphically. The numbers in the first and third columns of Table 2 represent the arbitrary values of x that are getting closer and closer to 1; the numbers in the second and fourth columns are the corresponding values of $f(x)$. The numbers in columns 2 and 4 together with the graph indicate that $f(x)$ approaches 2 as x approaches 1; this result is written as

$$f(x) \longrightarrow 2 \quad \text{as } x \longrightarrow 1$$

Table 2

From the Left		From the Right	
x	$y = 3x - 1$	x	$y = 3x - 1$
0.00000	−1.00000	2.00000	5.00000
0.50000	0.50000	1.50000	3.50000
0.90000	1.70000	1.10000	2.30000
0.99000	1.97000	1.01000	2.03000
0.99900	1.99700	1.00100	2.00300
⋮	⋮	⋮	⋮

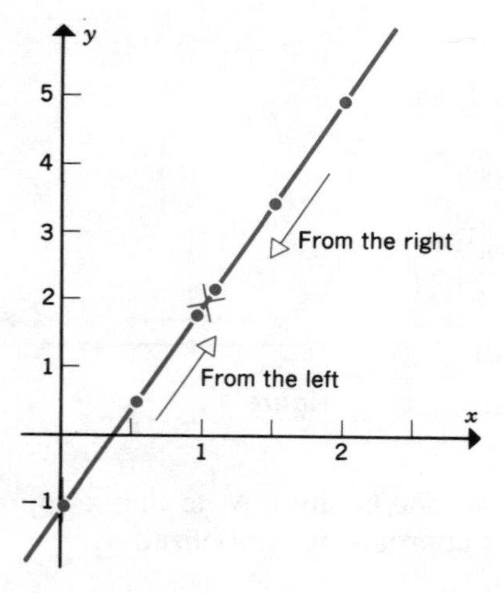

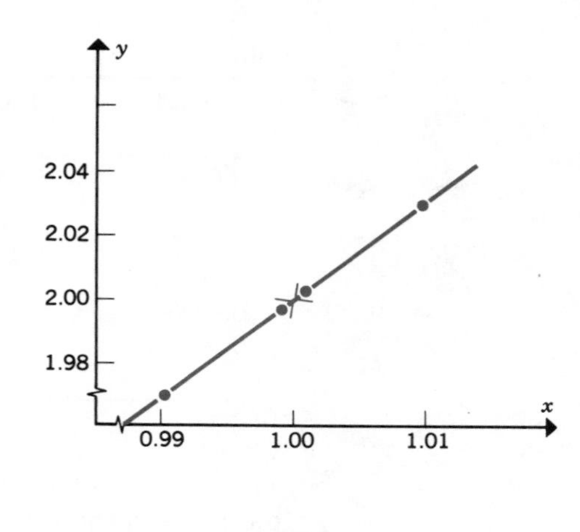

Figure 5

The number 2 is called the **limit** of the function $f(x) = 3x - 1$ as x approaches 1, written as

$$\lim_{x \to 1} f(x) = \lim_{x \to 1}(3x - 1) = 2$$

The process just described leads to the following informal definition of the limit of a function as x approaches a.

Definition The limit of a function $y = f(x)$, as x approaches a, equals L, written

$$\lim_{x \to a} f(x) = L \tag{1}$$

if $f(x)$ approaches L as x approaches a.

The graph in Figure 6 illustrates the limit process.

The graph of the function $f(x) = 3x - 1$ indicates that it can be drawn without lifting your pen from the page; therefore, it is continuous for all values of x. Because the function is continuous at $x = 1$, the limit of the function as x

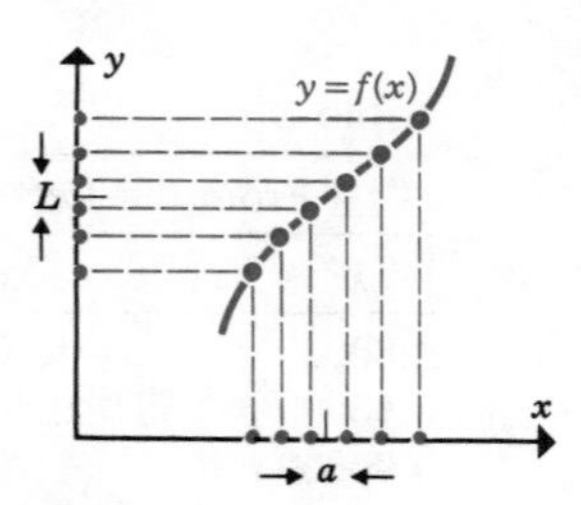

Figure 6

approaches 1 could have been found simply by substituting 1 for x in the equation defining the function, giving

$$\lim_{x \to 1}(3x - 1) = 3(1) - 1 = 2$$

The substitution process allows us to find $\lim_{x \to a} f(x)$ quickly for any function that is continuous at $x = a$.

Example 1 Find $\lim_{x \to 2} 3x^2$.

Solution The function $f(x) = 3x^2$ is a simple quadratic function whose graph can be sketched without lifting your pen from the page, so it is continuous for all x. For this reason the limit of the function $f(x) = 3x^2$ as x approaches 2 can be found by substituting 2 for x, yielding

$$\lim_{x \to 2} 3x^2 = 3(2)^2 = 12$$

The numbers in columns 2 and 4 in Table 3 and the graph (Figure 7) support this result.

Table 3

From the Left		From the Right	
x	$y = 3x^2$	x	$y = 3x^2$
1.000000	3.000000	3.000000	27.000000
1.500000	6.750000	2.500000	18.750000
1.900000	10.830000	2.100000	13.230000
1.990000	11.880300	2.010000	12.120300
1.999000	11.988003	2.001000	12.012003
1.999900	11.998800	2.000100	12.001200
1.999990	11.999880	2.000010	12.000120
⋮	⋮	⋮	⋮

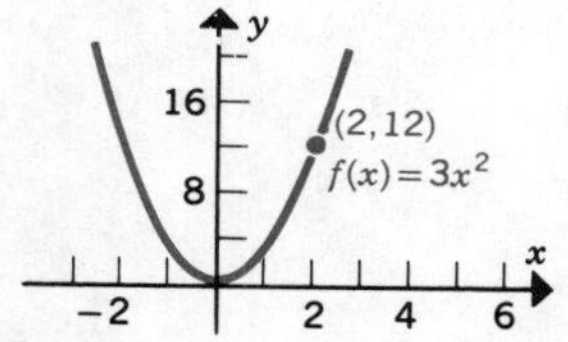

Figure 7

■

On the other hand, if the graph cannot be drawn without lifting your pencil from the paper at $x = a$, the behavior of the function in the vicinity of $x = a$ must be examined carefully to determine whether or not the limit exists. The following examples illustrate some of the possibilities.

Example 2 If $f(x) = \dfrac{3x^2 - 3}{2x - 2}$, find $\lim_{x \to 1} f(x)$.

Solution First, if we attempt to find the limit by substituting $x = 1$ into the equation that defines $f(x)$, we get

$$\frac{3(1)^2 - 3}{2(1) - 2} = \frac{0}{0}$$

which is not defined. This result only tells us that $f(x)$ is not defined when $x = 1$; it does not tell us anything about

$$\lim_{x \to 1} \frac{3x^2 - 3}{2x - 2}$$

Table 4 tells us more about the behavior of the function as x approaches 1. The numbers appearing in columns 2 and 4 indicate that $f(x)$ is approaching some number as $x \to 1$ and that

$$\lim_{x \to 1} \frac{3x^2 - 3}{2x - 2} = 3$$

Table 4

From the Left		From the Right	
x	$y = \frac{3x^2 - 3}{2x - 2}$	x	$y = \frac{3x^2 - 3}{2x - 2}$
0.00000	1.50000	2.00000	4.50000
0.50000	2.25000	1.50000	3.75000
0.90000	2.85000	1.10000	3.15000
0.99000	2.98500	1.01000	3.01500
0.99900	2.99850	1.00100	3.00150
0.99990	2.99985	1.00010	3.00015
⋮	⋮	⋮	⋮

The graph of the function can provide more information. Before attempting to graph this function, note that the fraction

$$\frac{3x^2 - 3}{2x - 2}$$

is not in its simplest form. Factoring the numerator and denominator and dividing out common factors gives

$$\frac{3x^2 - 3}{2x - 2} = \frac{3(x + 1)(x - 1)}{2(x - 1)} = \frac{3(x + 1)}{2}, \qquad x \neq 1$$

So $f(x)$ can be written

$$f(x) = \frac{3(x + 1)}{2}, \qquad x \neq 1 \tag{2}$$

The graph is shown in Figure 8. The "hole" at (1, 3) indicates why $\lim_{x \to 1} f(x)$ exists even though $f(1)$ is not defined. In searching for the limit the variable x *approaches* 1 but *never equals* 1. If we substitute the limiting value of x into Equation 2, we get

$$\lim_{x \to 1} f(x) = \lim_{x \to 1} \frac{3(x + 1)}{2} = 3$$

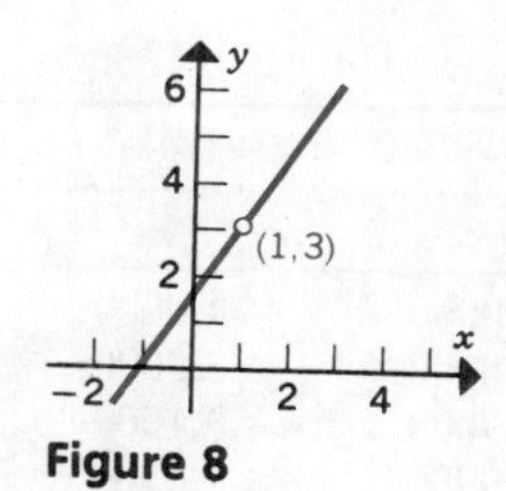

Figure 8

The next example illustrates an instance in which $\lim_{x \to a} f(x)$ does not exist.

Example 3 Find $\lim_{x \to 1} \dfrac{1}{x-1}$.

Solution The values in columns 2 and 4 of Table 5 along with the graph of the function in Figure 9 clearly show that as x approaches 1, $f(x)$ does not approach a unique value. Therefore, we can conclude that $\lim_{x \to 1} \dfrac{1}{x-1}$ does not exist.

Table 5

From the Left		From the Right	
x	$f(x) = \dfrac{1}{x-1}$	x	$f(x) = \dfrac{1}{x-1}$
0.00000	−1	2.00000	1
0.50000	−2	1.50000	2
0.90000	−10	1.10000	10
0.99000	−100	1.01000	100
0.99900	−1,000	1.00100	1,000
0.99990	−10,000	1.00010	10,000
0.99999	−100,000	1.00001	100,000

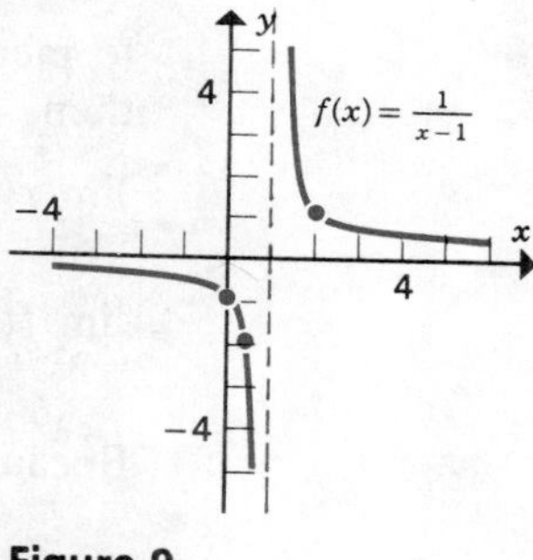

Figure 9

When we deal with a piecewise function, it is instructive to search for the limit of the function as x approaches a value for which the expression defining the function changes, as illustrated in the next two examples.

Example 4 If $f(x)$ is defined as

$$f(x) = \begin{cases} x - 2, & x \leq 3 \\ 7 - 2x, & x > 3 \end{cases}$$

find $\lim_{x \to 3} f(x)$.

Solution The numbers in columns 2 and 4 of Table 6 together with the graph in Figure 10 indicate that $f(x) \to 1$ as $x \to 3$, so we can write

$$\lim_{x \to 3} f(x) = 1$$

Table 6

From the Left		From the Right	
x	$y = x - 2$	x	$y = 7 - 2x$
2.0000	0.0000	4.0000	−1.0000
2.5000	0.5000	3.5000	0.0000
2.9000	0.9000	3.1000	0.8000
2.9900	0.9900	3.0100	0.9800
2.9990	0.9990	3.0010	0.9980
2.9999	0.9999	3.0001	0.9998
⋮	⋮	⋮	⋮

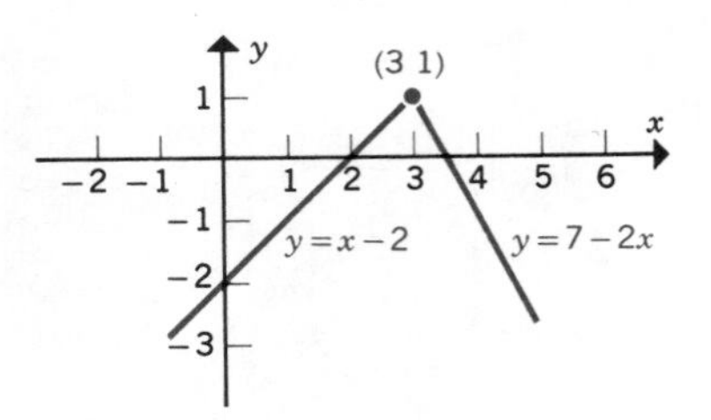

Figure 10

This result can be obtained formally using the mathematical expressions that define $f(x)$. If we use the notation*

$$x \to 3^-$$

to mean "x approaches 3 from the left" (numbers in column 1 of table 6) and

$$x \to 3^+$$

to mean "x approaches 1 from the right" (numbers in column 3 of Table 6), then

$\lim_{x\to 3^-} f(x) = \lim_{x\to 3^-}(x - 2) = 3 - 2 = 1$ [1 is called the left-hand limit of $f(x)$ as x approaches 3]

$\lim_{x\to 3^+} f(x) = \lim_{x\to 3^+}(7 - 2x) = 7 - 2(3) = 1$ [1 is called the right-hand limit of $f(x)$ as x approaches 3]

Because $\lim_{x\to 3^-} f(x) = \lim_{x\to 3^+} f(x) = 1$, we get $\lim_{x\to 3} f(x) = 1$. ■

Example 5 If $f(x)$ is defined as

$$f(x) = \begin{cases} 3 - 4x, & x \le 1 \\ x + 1, & x > 1 \end{cases}$$

find $\lim_{x\to 1} f(x)$.

Solution Using the values of x in columns 1 and 3 of Table 7 to generate the corresponding values of y in columns 2 and 4, we see that the values of y in the second column appear to be approaching -1, but those in column 4 appear to be approaching 2. Because the numbers in columns 2 and 4 are not approaching a unique value, we say that $\lim_{x\to 1} f(x)$ does not exist. This conclusion is supported by the graph in Figure 11, which shows a jump at $x = 1$. If in addition we look at the right-

*The minus and plus superscripts have nothing to do with the signs of the numbers assigned to x; 3^- means that all the values assigned to x can be written as $3 - \delta$, where δ is an arbitrary (usually small) positive number. In the same manner, 3^+ means that all the values assigned to x can be written as $3 + \delta$.

Table 7

From the Left		From the Right	
x	$y = 3 - 4x$	x	$y = x + 1$
0.0000	3.0000	2.0000	3.0000
0.5000	1.0000	1.5000	2.5000
0.9000	−0.6000	1.1000	2.1000
0.9900	−0.9600	1.0100	2.0100
0.9990	−0.9960	1.0010	2.0010
0.9999	−0.9996	1.0001	2.0001
⋮	⋮	⋮	⋮

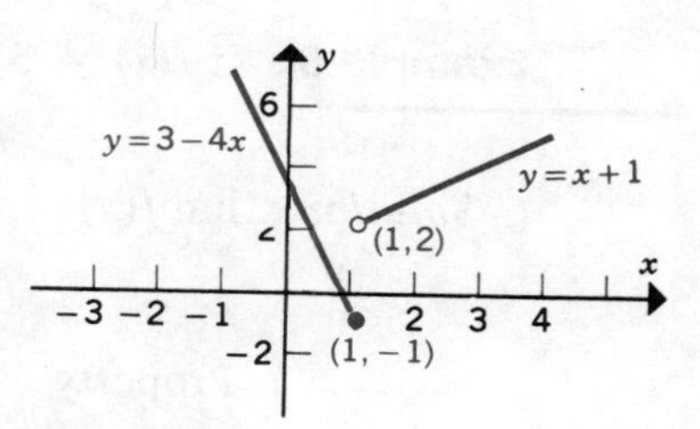

Figure 11

and left-hand limits, we get

$$\lim_{x\to 1^-} f(x) = \lim_{x\to 1^-}(3 - 4x) = 3 - 4(1) = -1 \quad \text{(left-hand limit)}$$

$$\lim_{x\to 1^+} f(x) = \lim_{x\to 1^+}(x + 1) = 1 + 1 = 2 \quad \text{(right-hand limit)}$$

Since the right-hand limit does not equal the left-hand limit, $\lim_{x\to 1} f(x)$ does not exist. ■

Before we proceed further, it is important to call to your attention two important characteristics of limits that were illustrated in the preceding examples.

1. The existence of $\lim_{x\to a} f(x)$ does not depend on
   (a) whether or not $f(a)$ is defined or
   (b) the value of $f(a)$ if $f(a)$ is defined.
2. If $f(x)$ is defined on both sides of $x = a$, then $\lim_{x\to a} f(x)$ exists if and only if the limits from both the left and the right exist and are equal.

Although the tables and graphs used in Examples 1 through 5 are useful to demonstrate the basic concepts underlying the limit, they are, in general, awkward and inefficient when searching for a limit. Fortunately there are more direct methods, using the properties of limits, for finding the limit of a function when it exists. The properties are stated and illustrated; no proofs are given because the proofs rely on a formal definition of the limit that has not been given.

Properties of Limits

If $\lim_{x\to a} f(x) = L$ and $\lim_{x\to a} g(x) = M$, where L and M are real numbers, properties I through IX hold.

Property I. $\boxed{\lim_{x\to a} c = c}$, where c is a constant

Example 6 If $f(x) = 3$, find $\lim_{x\to 5} f(x)$.

Solution $\lim_{x\to 5} f(x) = \lim_{x\to 5} 3 = 3.$ ■

Property II. $\boxed{\lim_{x\to a} x^n = a^n}$, where n is a postive integer.

Example 7 If $f(x) = x^4$, find $\lim_{x\to 2} f(x)$.

Solution $\lim_{x\to 2} x^4 = (2)^4 = 16.$ ■

Property III. $\boxed{\lim_{x\to a} \sqrt[n]{f(x)} = \sqrt[n]{\lim_{x\to a} f(x)} = \sqrt[n]{L}}$ provided $L \geq 0$ when n is even

Example 8 Find $\lim_{x\to 4} \sqrt{x^3}$.

Solution $\lim_{x\to 4} \sqrt{x^3} = \sqrt{\lim_{x\to 4} x^3} = \sqrt{64} = 8.$ ■

Property IV. $\boxed{\lim_{x\to a} [cf(x)] = c \lim_{x\to a} f(x) = cL}$, where c is a constant

Example 9 Find $\lim_{x\to 2} 4x^3$.

Solution $\lim_{x\to 2} 4x^3 = 4 \lim_{x\to 2} x^3 = 4(2^3) = 32.$ ■

Property V. $\boxed{\lim_{x\to a} [f(x) + g(x)] = \lim_{x\to a} f(x) + \lim_{x\to a} g(x) = L + M}$

The **limit of a sum** equals the **sum of the limits.**

Example 10 Find $\lim_{x\to 2} (5x + 3x^2)$.

Solution $\lim_{x\to 2} (5x + 3x^2) = \lim_{x\to 2} 5x + \lim_{x\to 2} 3x^2 = 10 + 12 = 22.$ ■

Property VI. $$\lim_{x \to a}[f(x) \cdot g(x)] = \lim_{x \to a} f(x) \cdot \lim_{x \to a} g(x) = L \cdot M$$

The **limit of a product** equals the **product of the limits.**

Example 11 Find $\lim_{x \to 3}(2x + 1)(3x^2 + 4)$.

Solution Since $\lim_{x \to 3}(2x + 1) = 7$ and $\lim_{x \to 3}(3x^2 + 4) = 31$, we can write $\lim_{x \to 3}(2x + 1)(3x^2 + 4) = 7 \cdot 31 = 217$ ■

Property VII. $$\lim_{x \to a} \frac{f(x)}{g(x)} = \frac{\lim_{x \to a} f(x)}{\lim_{x \to a} g(x)} = \frac{L}{M} \quad \text{provided } M \neq 0$$

The **limit of a ratio** equals the **ratio of the limits.**

Example 12 Find $\lim_{x \to 2} \frac{3x + 2}{1 - x^2}$.

Solution Because $\lim_{x \to 2}(3x + 2) = 8$ and $\lim_{x \to 2}(1 - x^2) = -3$, we can use property VII because $M = -3 \neq 0$, giving

$$\lim_{x \to 2} \frac{3x + 2}{1 - x^2} = \frac{\lim_{x \to 2}(3x + 2)}{\lim_{x \to 2}(1 - x^2)} = -\frac{8}{3}$$ ■

Property VII cannot be used when M, the limit of the denominator, equals zero. In this situation, the existence or nonexistence of $\lim_{x \to a} \frac{f(x)}{g(x)}$ depends on L, the limit of the numerator, as described in properties VIII and IX.

Property VIII. If $M = 0$ and $L \neq 0$, $\lim_{x \to a} \frac{f(x)}{g(x)}$ does not exist

Example 13 Find $\lim_{x \to 2} \frac{x}{x - 2}$.

Solution Since $\lim_{x \to 2} x = 2$ and $\lim_{x \to 2}(x - 2) = 0$, $\lim_{x \to 2} \frac{x}{x - 2}$ does not exist. The graph in Figure 12 indicates why the limit does not exist. As x approaches 2 from the

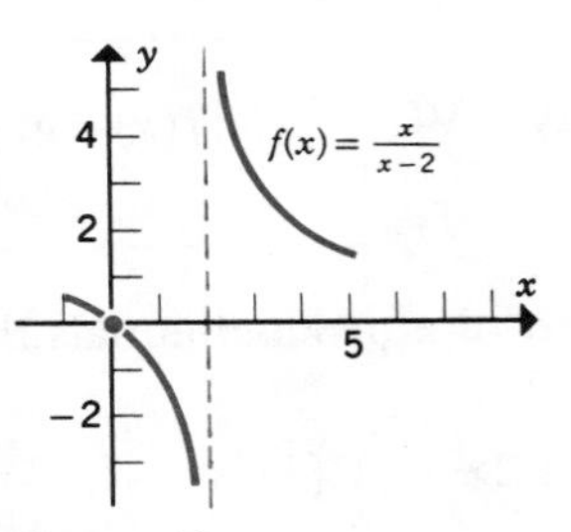

Figure 12

left, the corresponding values of y decrease more and more rapidly; the reverse is true as x approaches 2 from the right. ■

Property IX. If $M = 0$ and $L = 0$, $\lim_{x \to a} \frac{f(x)}{g(x)}$ may or may not exist

When $L = 0$ and $M = 0$, more analysis is required.* Often the ratio $f(x)/g(x)$ can be transformed into an equivalent algebraic form for which one of the other properties (I through VIII) can be applied, as illustrated in Examples 14 and 15.

Example 14 Find $\lim_{x \to 2} \frac{3x - 6}{x^2 + x - 6}$.

Solution Because $\lim_{x \to 2}(3x - 6) = 0$ and $\lim_{x \to 2}(x^2 + x - 6) = 0$, the fraction must be altered to determine whether or not the limit exists. Factoring both numerator and denominator and then reducing gives

$$\frac{3x - 6}{x^2 + x - 6} = \frac{3(x - 2)}{(x + 3)(x - 2)} = \frac{3}{x + 3}, \qquad x \neq 2$$

Now we can write

$$\lim_{x \to 2} \frac{3x - 6}{x^2 + x - 6} = \lim_{x \to 2} \frac{3}{x + 3} = \frac{3}{5}$$

The replacement of $\frac{3x - 6}{x^2 + x - 6}$ by $\frac{3}{x + 3}$ in the limit process is valid because x approaches 2 but does not equal 2. A partial graph is shown in Figure 13 supporting this result.

*When $L = 0$ and $M = 0$, the meaningless result 0/0 is called an **indeterminate** form. It means that you cannot tell whether or not $\lim_{x \to a} \frac{f(x)}{g(x)}$ exists and that you have to investigate the function in more detail as $x \to a$.

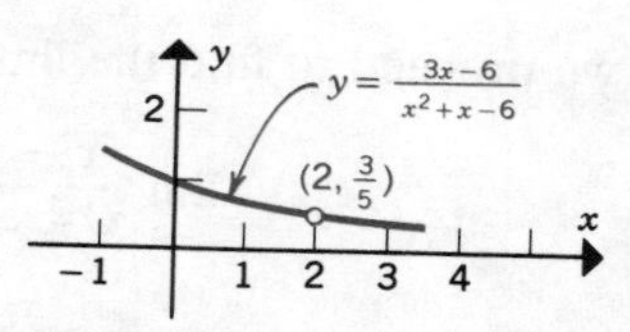

Figure 13

■

Example 15 If $f(x) = \dfrac{3x - 3}{x^2 - 2x + 1}$, find $\lim_{x\to 1} f(x)$.

Solution Because $\lim_{x\to 1}(3x - 3) = 0$ and $\lim_{x\to 1}(x^2 - 2x + 1) = 0$, we are again at an impasse; we cannot tell whether or not the limit exists. As in the previous example, the numerator and denominator are factorable, so that the ratio can be expressed as

$$\frac{3x - 3}{x^2 - 2x + 1} = \frac{3(x - 1)}{(x - 1)^2} = \frac{3}{x - 1}, \qquad x \neq 1$$

In this instance we see that

$$\lim_{x\to 1}(x - 1) = 0 \quad \text{but} \quad \lim_{x\to 1} 3 = 3 \neq 0$$

According to property VIII, we find that $\lim_{x\to 1} \dfrac{3x - 3}{x^2 - 2x + 1}$ does not exist. ■

Simplifying $\dfrac{f(x)}{g(x)}$ by factoring both numerator and denominator is only one of many techniques that can be used to determine whether or not $\lim_{x\to a} \dfrac{f(x)}{g(x)}$ exists when $\lim_{x\to a} f(x) = 0$ and $\lim_{x\to a} g(x) = 0$. It is not our intent to study a large number of these techniques; we point out only that other approaches are available. For example, when either the numerator or the denominator contains radicals, it is sometimes possible, by rationalizing, to determine whether or not the limit exists, as the following example illustrates.

Example 16 If $y = f(x) = \dfrac{x - 4}{\sqrt{x} - 2}$, find $\lim_{x\to 4} f(x)$.

Solution Once more, we have $\lim_{x\to 4}(x - 4) = 0$ and $\lim_{x\to 4}(\sqrt{x} - 2) = 0$. To resolve this situation, we can write the ratio in an equivalent form by rationalizing the denominator as follows:

$$\frac{x - 4}{\sqrt{x} - 2} = \frac{(x - 4)(\sqrt{x} + 2)}{(\sqrt{x} - 2)(\sqrt{x} + 2)} = \frac{(x - 4)(\sqrt{x} + 2)}{x - 4}$$
$$= \sqrt{x} + 2, \qquad x \neq 4$$

Next we proceed to find the limit:

$$\lim_{x \to 4} \frac{x - 4}{\sqrt{x} - 2} = \lim_{x \to 4}(\sqrt{x} + 2) = 4 \qquad \blacksquare$$

Caution When rules such as

$$\lim_{x \to a}[f(x)g(x)] = [\lim_{x \to a} f(x)][\lim_{x \to a} g(x)]$$

are used, it is important to remember that both $\lim_{x \to a} f(x)$ and $\lim_{x \to a} g(x)$ must exist. For example, the result

$$\lim_{x \to 0}[xg(x)] = (\lim_{x \to 0} x)[\lim_{x \to 0} g(x)] = 0 \cdot \lim_{x \to 0} g(x) = 0$$

is valid only if $\lim_{x \to 0} g(x)$ exists. For example this result is not valid if $g(x) = \frac{1}{x^2}$; in this instance we get

$$\lim_{x \to 0} xg(x) = \lim_{x \to 0}\left(x \frac{1}{x^2}\right) = \lim_{x \to 0}\left(\frac{1}{x}\right)$$

which does not exist.

2.1 EXERCISES

Use the graphs of each of the following functions to determine (a) $f(2)$ and (b) $\lim_{x \to 2} f(x)$.

1.

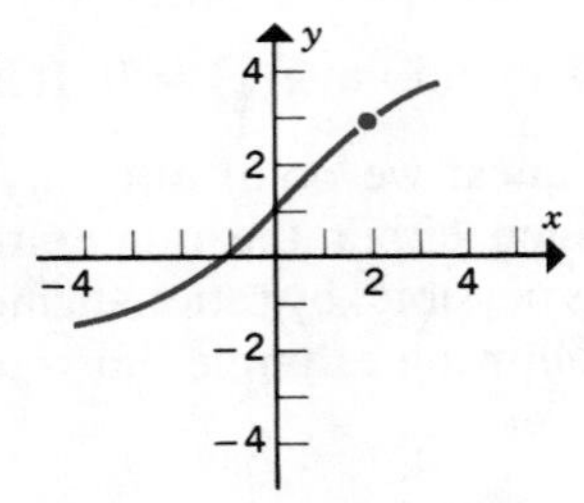

2.

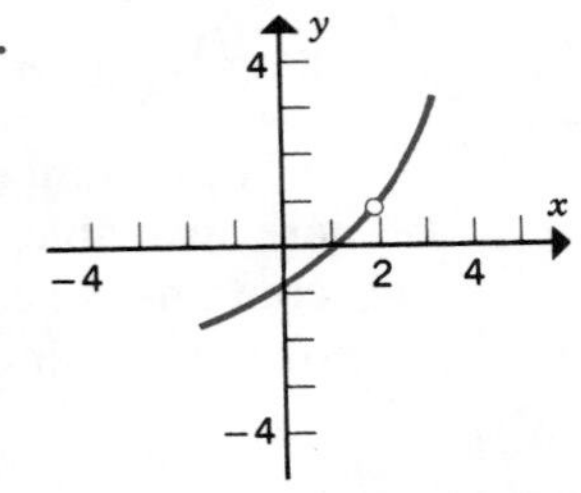

3.

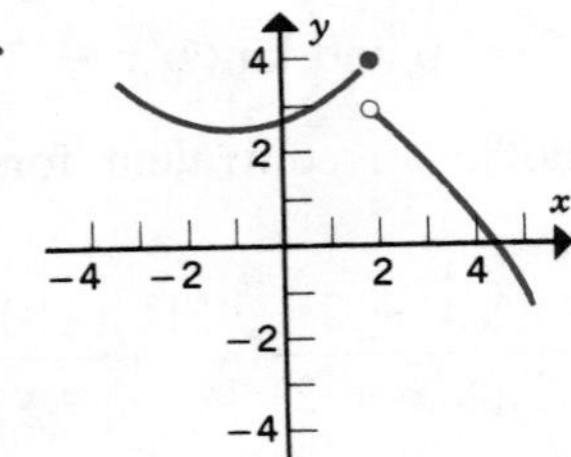

4.

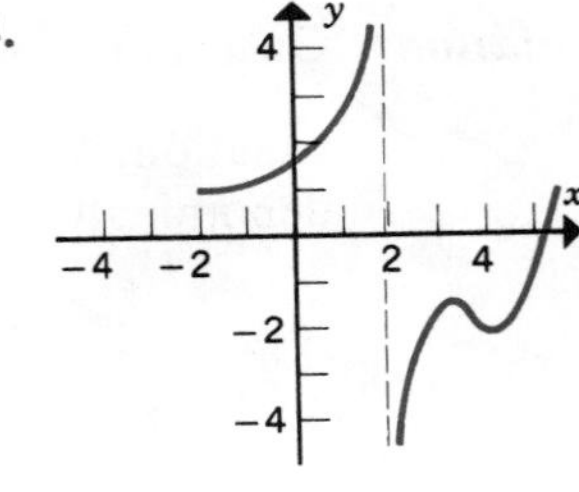

5.

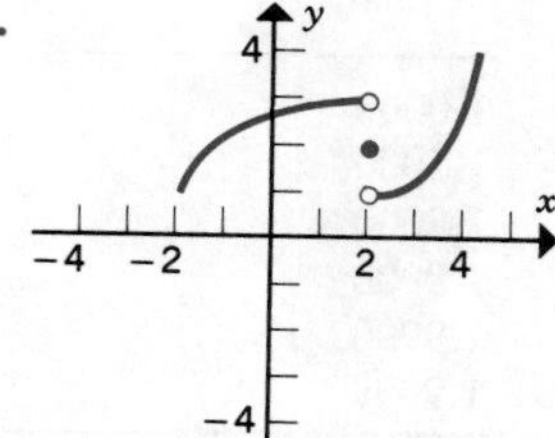

6.

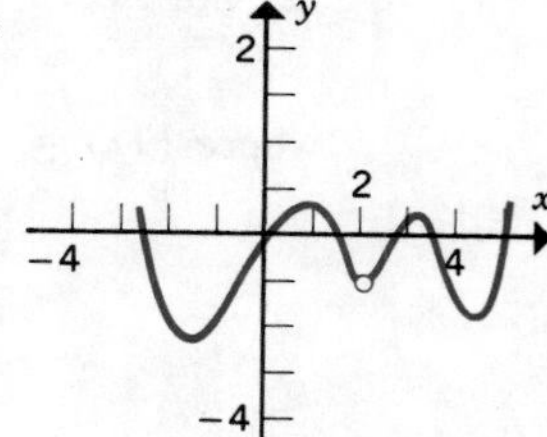

7.

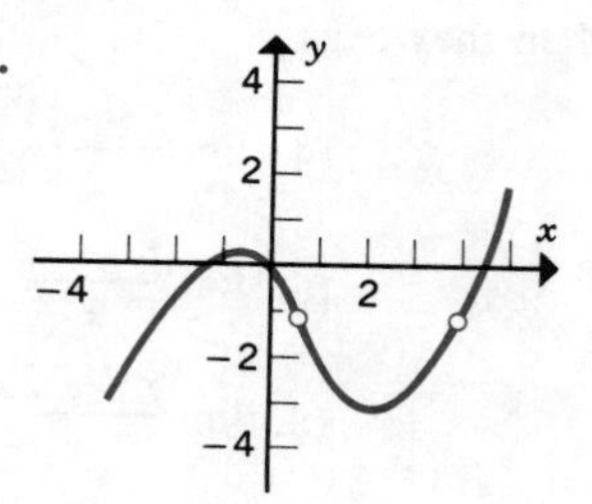

8.

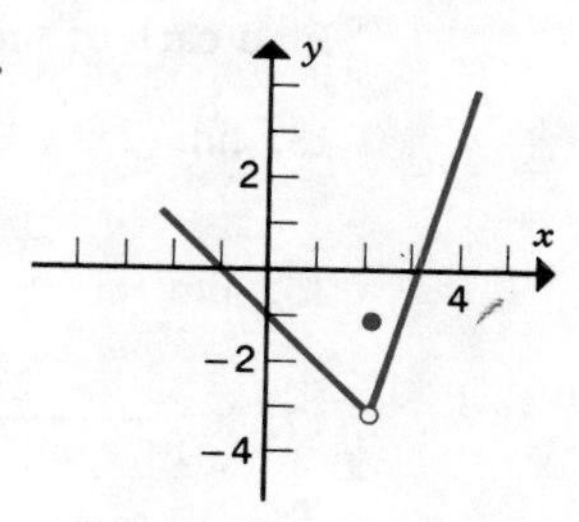

For Exercises 9 through 12, complete the tables, expressing y to four decimal places, and determine (a) whether the indicated limit exists and (b) the value of the limit when it exists. Finally, sketch the graph of the function in the vicinity of the limiting value of x.

9. $\lim_{x \to 1}(1 - 2x)$

x	y	x	y
0.0000		1.9000	
0.5000		1.5000	
0.7500		1.2500	
0.9500		1.0500	
0.9950		1.0050	
0.9995		1.0005	

10. $\lim_{x \to 3} \dfrac{x^2 - 3x}{x^2 - 9}$

x	y	x	y
2.0000		4.0000	
2.5000		3.5000	
2.9000		3.1000	
2.9900		3.0100	
2.9990		3.0010	
2.9999		3.0001	

11. $\lim_{x \to 2} \dfrac{3}{x - 2}$

x	y	x	y
1.0000		3.0000	
1.5000		2.5000	
1.9000		2.1000	
1.9900		2.0100	
1.9990		2.0010	
1.9999		2.0001	

12. $\lim_{x \to 2} F(x)$

where $F(x) = \begin{cases} 1 - x, & x < 2 \\ x^2 - 5, & x \geq 2 \end{cases}$

x	y	x	y
1.0000		3.0000	
1.5000		2.5000	
1.9000		2.1000	
1.9900		2.0100	
1.9990		2.0010	
1.9999		2.0001	

Find each of the following limits when they exist.

13. $\lim_{x \to 1} (2x + 1)$

14. $\lim_{x \to 2} \dfrac{x^2 - 4}{2}$

15. $\lim_{x \to -1} \dfrac{x - 3}{x + 1}$

16. $\lim_{x \to 0} \dfrac{x^2 + x}{x}$

17. $\lim_{x \to 4} \dfrac{x^2 - 2x - 8}{x - 4}$

18. $\lim_{x \to 2} \dfrac{x^2 - 2x + 8}{x - 2}$

19. $\lim_{t \to 1} \dfrac{t^2 - 1}{t - 1}$

20. $\lim_{x \to -1} \dfrac{2 + x - x^2}{3x + 3}$

21. $\lim_{x \to 1} \dfrac{x^2 + 1}{x - 1}$

22. $\lim_{x \to 5} \dfrac{x^2 - 4x - 5}{2x - 10}$

23. $\lim_{x \to 0} \dfrac{2x^2 + x}{x^2 - x}$

24. $\lim_{x \to 3} \dfrac{x^2 - 2x - 3}{x^2 - 9}$

25. $\lim_{x \to -3} \dfrac{x^2 - 2x - 3}{x^2 - 9}$

26. $\lim_{x \to 9} \dfrac{x - 9}{\sqrt{x} - 3}$

27. $\lim_{x \to 1} \dfrac{x^2 - 1}{x^2 - 3x + 2}$

28. $\lim_{x \to 4} \dfrac{2 - \sqrt{x}}{8 - 2x}$

29. $\lim_{x \to 0} \dfrac{x}{(2 + x)^2 - 4}$

30. $\lim_{x \to 1} \dfrac{x - 1}{x^2 - 2x + 1}$

31. $\lim_{x \to 0} \dfrac{x}{(2 + x)^3 - 8}$

32. $\lim_{x \to 0} \dfrac{x}{(2 + x)^4 - 16}$

In Exercises 33 through 36, find each of the limits when they exist. In addition, sketch the graph of the function in the vicinity of the limiting value of x.

33. $\lim_{x \to 0} f(x)$, where $f(x) = \begin{cases} x + 1, & x \leq 0 \\ 1 - x, & x > 0 \end{cases}$

34. $\lim_{x \to 1} f(x)$, where $f(x) = \begin{cases} x^2, & x \leq 1 \\ 1 + x, & x > 1 \end{cases}$

35. $\lim_{x \to 2} f(x)$, where $f(x) = \begin{cases} x, & x < 2 \\ 2 - x, & x \geq 2 \end{cases}$

36. $\lim_{x \to 1} g(x)$, where $g(x) = \begin{cases} x + 1, & x < 1 \\ 0, & x = 1 \\ 2 - x, & x > 1 \end{cases}$

37. If $f(x) = x^2$, find $\lim_{h \to 0} \dfrac{f(1 + h) - f(1)}{h}$

38. If $f(x) = 3x^2$, find $\lim_{h\to0} \dfrac{f(-1 + h) - f(-1)}{h}$

39. If $f(x) = x^3$, find $\lim_{h\to0} \dfrac{f(1 + h) - f(1)}{h}$

40. If $F(x) = \sqrt{x}$, find $\lim_{h\to0} \dfrac{F(4 + h) - F(4)}{h}$

41. Find $\lim_{x\to a} \dfrac{x^2 - ax - x + a}{x^2 - a^2}$

42. Find $\lim_{x\to1} \dfrac{x^2 - ax - x + a}{x^2 + 2ax - x - 2a}$

43. The graph in Figure 14 shows the number $N(t)$ of residents in an apartment complex as a function of the time t, in months. Use the information shown in the graph to find each of the following, when it exists.

(a) $N(2)$ (b) $\lim_{t\to2} N(t)$ (c) $N(1)$ (d) $\lim_{t\to1} N(t)$

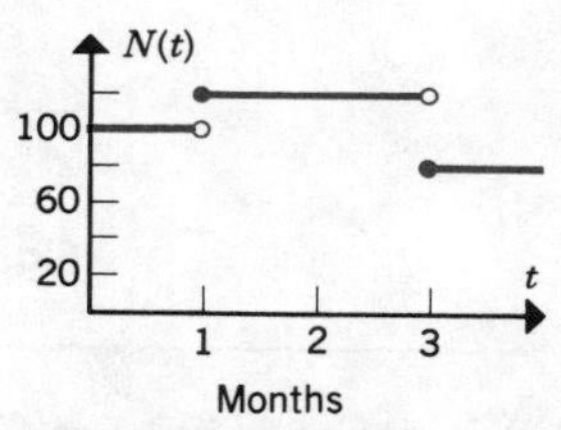

Figure 14

44. The graph in Figure 15 shows the cost $C(x)$, in dollars, to make x copies at Harry's Print Shop. Use the information in the graph to find each of the following.

(a) $C(5)$ (b) $\lim_{x\to5} C(x)$ (c) $C(10)$ (d) $\lim_{x\to10} C(x)$

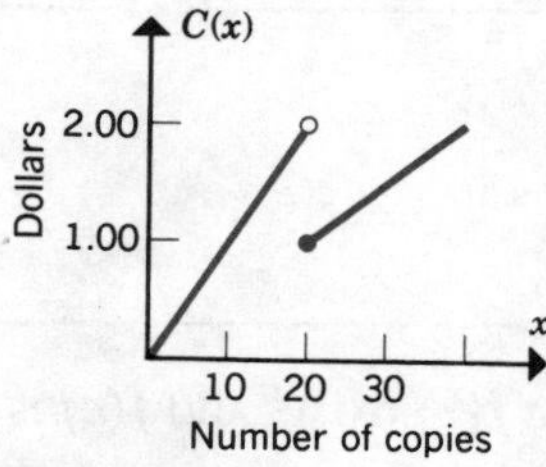

Figure 15

2.2 CONTINUITY

Intuitively a function is said to be **continuous** if its graph contains no breaks such as holes, gaps, or jumps. The resulting curve is unbroken and can be sketched without lifting your pen from the paper as illustrated in Figure 1.

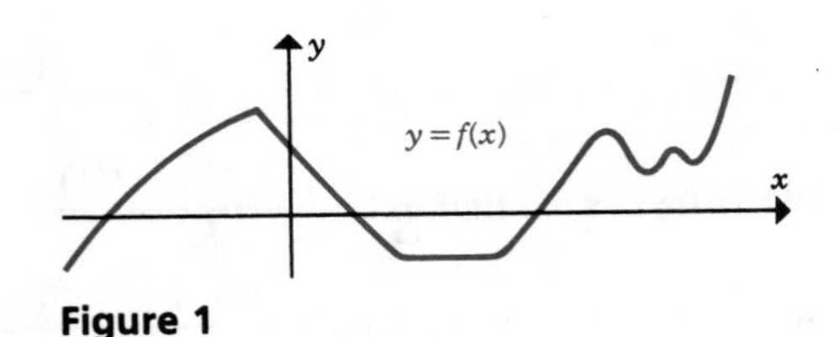

Figure 1

When a hole or gap appears in a graph at $x = a$, the function is said to be **discontinuous** at $x = a$. Conditions A through D describe the ways in which discontinuities can occur and Figures 2 through 5 illustrate some of the graphs that can appear.

A. $\lim_{x \to a} f(\mathrm{x})$ exists, but $f(a)$ is not defined.

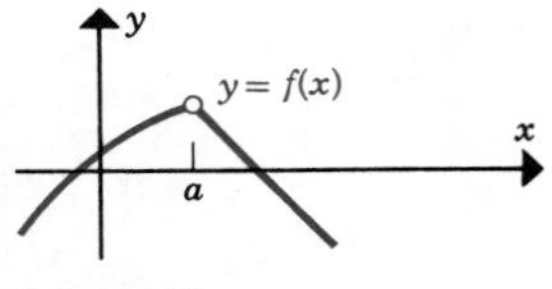

Figure 2

B. $f(a)$ is defined, but $\lim_{x \to a} f(x)$ does *not* exist.

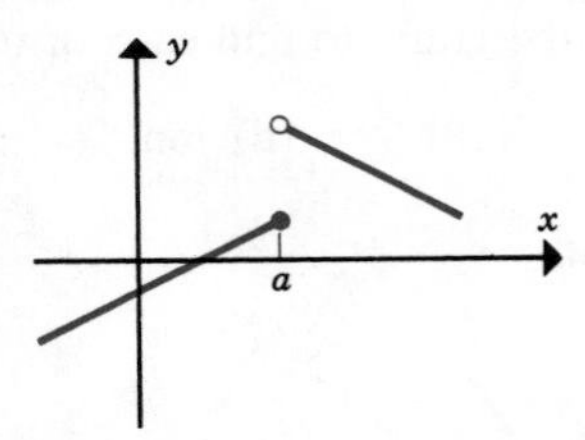

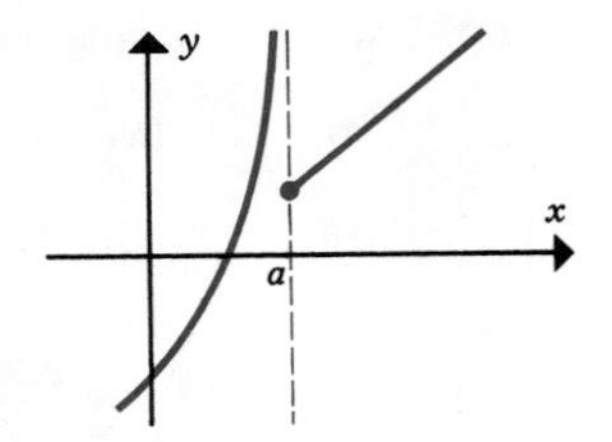

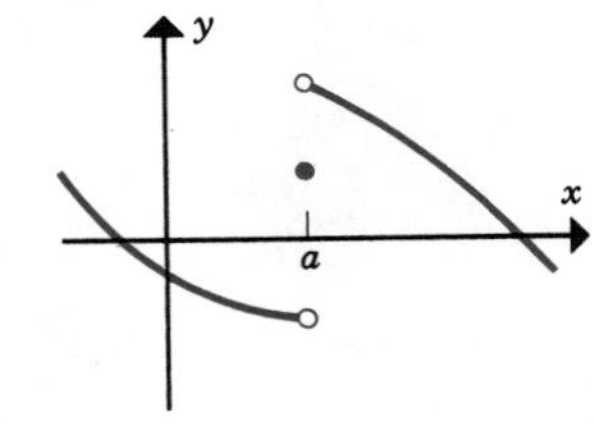

Figure 3

C. $\lim_{x \to a} f(x)$ exists and $f(a)$ is defined, but $f(a) \neq \lim_{x \to a} f(x)$.

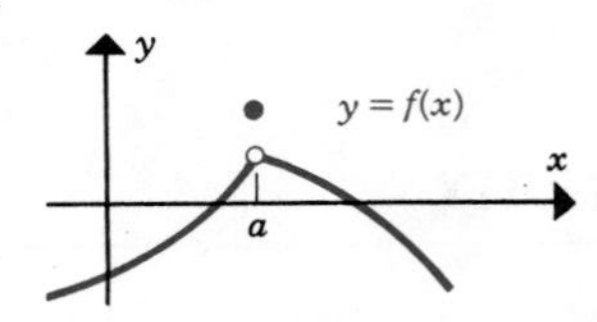

Figure 4

D. $\lim_{x \to a} f(x)$ does not exist, and $f(a)$ is not defined.

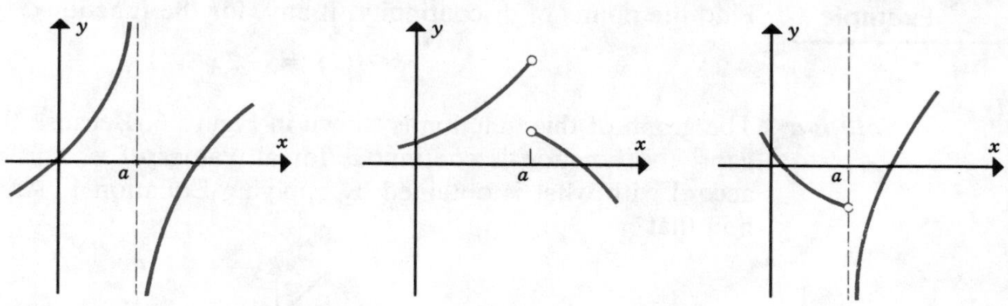

Figure 5

When any of the conditions A through D prevails at $x = a$, the function is said to be discontinuous at $x = a$. Because the behavior of the function at points such as these is somewhat unusual, additional time and effort is devoted to locating and analyzing the behavior of a function at points of discontinuity.

When a function is not exhibiting one of the unusual features described by conditions A through D at $x = a$, the function is said to be continuous at $x = a$.

Definition A function $y = f(x)$ is said to be **continuous** at $x = a$ if

$$\lim_{x \to a} f(x) = f(a) \tag{1}$$

That is, the limit of $f(x)$, as x approaches a, equals $f(a)$, the value of the function at $x = a$.

Equation 1 implies that three conditions must be satisfied in order for a function to be continuous at $x = a$:

1. $f(a)$ must be defined.
2. $\lim_{x \to a} f(x)$ must exist.
3. $\lim_{x \to a} f(x) = f(a)$.

If one or more of these conditions is violated, the function is discontinuous at $x = a$.

If a function is continuous at every point of an interval, it is said to be continuous over the interval; graphically, this means that the curve is unbroken in the sense that it can be drawn without holes or jumps, as shown earlier in Figure 1. Almost all the functions we shall study will be continuous everywhere except for one or two values of x. Our attention in the remainder of this section will be centered on locating values of x, if any, where a function is discontinuous.

Example 1 Find the points of discontinuity, if any, for the function

$$f(x) = -2x + 1$$

Solution The graph of this function is shown in Figure 6. Because there are no holes or gaps, the function is continuous for all values of x. This intuitive result is in accord with what is obtained by applying Equation 1. As x approaches a, we find that

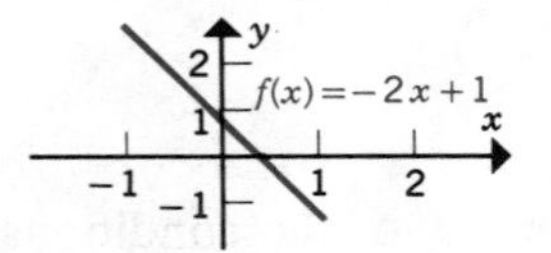

Figure 6

$$\lim_{x \to a} f(x) = \lim_{x \to a} (-2x + 1) = -2a + 1$$

The value of $f(x)$ at $x = a$ is given by

$$f(a) = -2a + 1$$

From this we see that

$$\lim_{x \to a} f(x) = f(a)$$

and we can conclude that the function is continuous for all values of x. ■

When they occur, discontinuities are often due to

1. A function having the form

$$y = f(x) = \frac{g(x)}{h(x)}$$

   with the discontinuities appearing at values of x for which $h(x) = 0$, as illustrated in Example 2.
2. A piecewise function whose graph has breaks at values of x where the equations defining the function change, as in Example 3.

Example 2 Find the values of x, if any, for which the function

$$f(x) = \frac{2}{x - 1}$$

is discontinuous.

Solution An examination of the function indicates that it is not defined at $x = 1$. In addition, the limit of the function does not exist as $x \to 1$. So we can say that the function is not continuous at $x = 1$ for two reasons.

1. $\lim_{x\to 1} f(x) = \lim_{x\to 1} \dfrac{2}{x-1}$ does not exist.
2. $f(1) = 2/0$ is not defined.

So condition D best describes the discontinuity. These conclusions are reinforced by the graph of the function shown in Figure 7.

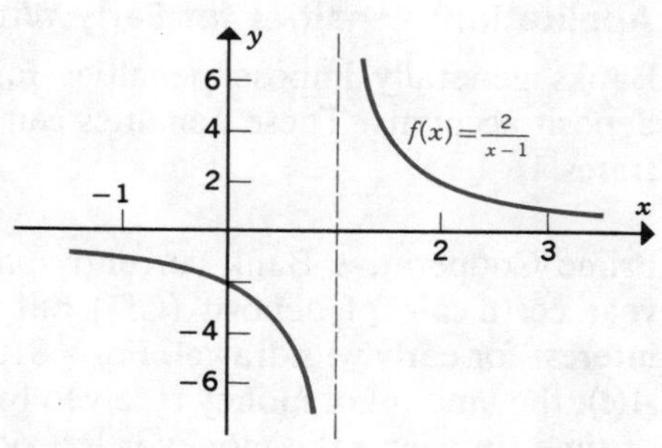

Figure 7

■

Example 3 Find the points of discontinuity, if any, for the function

$$f(x) = \begin{cases} x + 1, & x \le 1 \\ x^2, & x > 1 \end{cases}$$

Solution The graph in Figure 8 shows a jump at $x = 1$. The function is discontinuous at $x = 1$ because $\lim_{x\to 1} f(x)$ does not exist. As x approaches 1 from the left $(x \to 1^-)$,

$$\lim_{x\to 1^-} (x + 1) = 2$$

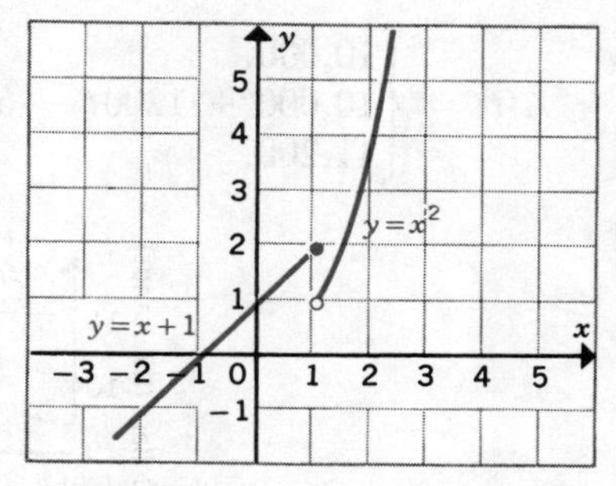

Figure 8

As x approaches 1 from the right ($x \to 1^+$),

$$\lim_{x\to 1^+} (x^2) = 1$$

Since $2 \neq 1$, the right- and left-hand limits are not equal, which means that $\lim_{x\to 1} f(x)$ does not exist. The function itself is defined at $x = 1$:

$$f(1) = 1 + 1 = 2$$

From this analysis, we conclude that condition B holds at $x = 1$. ■

Application: Penalties for Early Withdrawal

Banks generally impose penalties for early withdrawal of money from time deposit accounts. These penalties can cause discontinuities, as Example 4 illustrates.

Example 4 Prime Cooperative Bank currently pays 12 percent simple interest on its one-year certificate of deposit (CD) but imposes a penalty equal to six months' interest for early withdrawal. For a $10,000 CD, find the equation that describes $A(t)$, the amount of money received by a depositor, as a function of t, the length of time (in years) the money is left on deposit.

Solution First, if no penalty were imposed, $A(t)$ would be described by the simple-interest equation (Section 1.2, exercise 61):

$$\begin{aligned} A(t) &= 10{,}000 + (10{,}000)(0.12)t \\ &= 10{,}000 + 1200t, \qquad 0 \le t \le 1 \end{aligned}$$

The graph of this equation is represented by the broken line in Figure 9. If the money is withdrawn before maturity ($t = 1$), a penalty equal to six months' interest is imposed. The penalty P can be written

$$P = 10{,}000(0.12)(\tfrac{1}{2}) = \$600, \qquad 0 < t < 1$$

Note that the penalty is not imposed when $t = 0$ or $t = 1$. Taking the penalty into account, we can write the relationship between $A(t)$ and t as the piecewise function

$$A(t) = \begin{cases} 10{,}000, & t = 0 \\ 10{,}000 + 1200t - 600 = 9400 + 1200t, & 0 < t < 1 \\ 11{,}200, & t = 1 \end{cases}$$

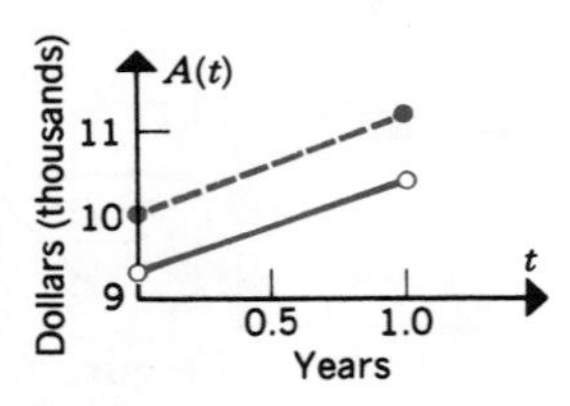

Figure 9

The graph of this function is shown in Figure 9. Because of the early withdrawal penalty, the function is discontinuous at $t = 0$ and $t = 1$. ■

2.2 EXERCISES

Use the graphs in Exercises 1 through 8 to determine whether each function is continuous or discontinuous at $x = 2$; if a function is discontinuous, state what condition, A through D, best describes the discontinuity.

A. $\lim_{x\to a} f(x)$ exists, but $f(a)$ is not defined.

B. $f(a)$ is defined, but $\lim_{x\to a} f(x)$ does not exist.

C. $f(a)$ is defined and $\lim_{x\to a} f(x)$ exists, but $f(a) \neq \lim_{x\to a} f(x)$.

D. $f(a)$ is not defined and $\lim_{x\to a} f(x)$ does not exist.

1.

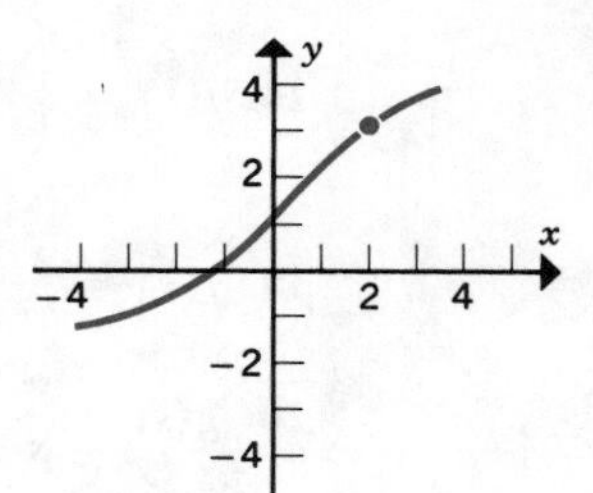

2.

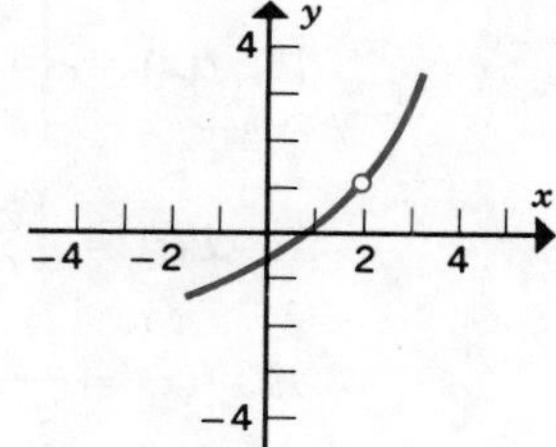

3.

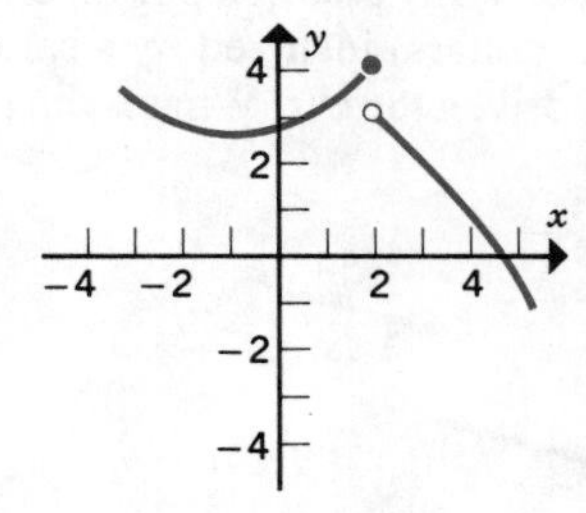

4.

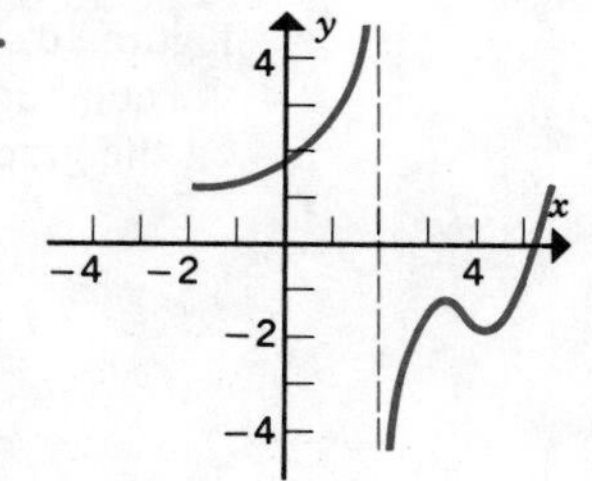

5.

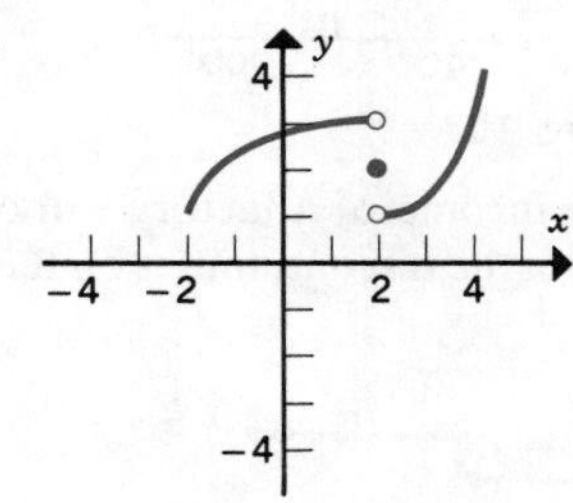

6.

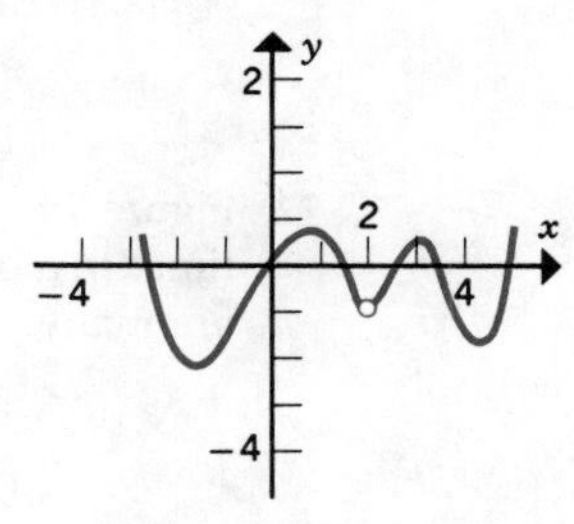

7.

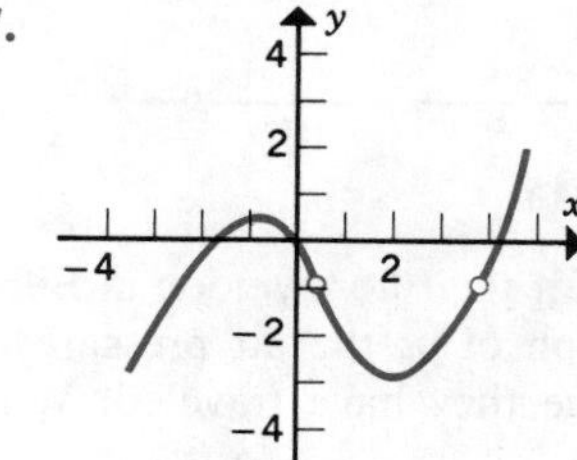

8.

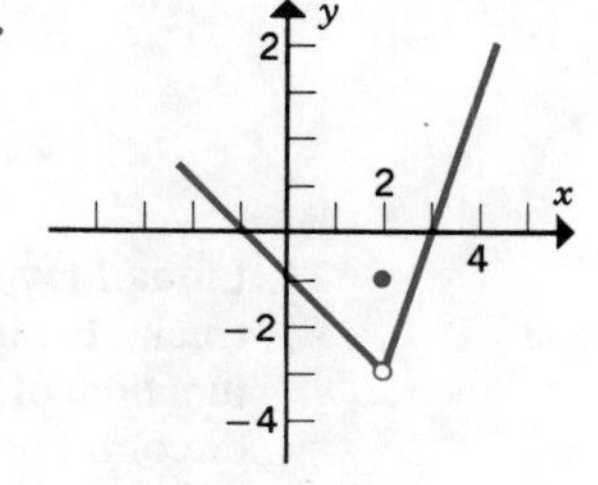

Sketch the graph of each of the functions in Exercises 9 through 20. Determine the value or values, if any, of the independent variable for which the function is discontinuous. Indicate what condition, A through D (see the preceding), best describes each discontinuity.

9. $f(x) = \dfrac{x-1}{x-1}$

10. $f(x) = \begin{cases} \dfrac{x^2+2x+1}{x+1}, & x \neq -1 \\ 2, & x = -1 \end{cases}$

11. $f(x) = \dfrac{1}{x^2}$

12. $f(x) = \begin{cases} \dfrac{1}{x^2}, & x \neq 0 \\ 1, & x = 0 \end{cases}$

13. $f(x) = \begin{cases} 3x-2, & x \leq 1 \\ x-1, & x > 1 \end{cases}$

14. $f(x) = \begin{cases} \dfrac{x^2-9}{x-3}, & x \neq 3 \\ 5, & x = 3 \end{cases}$

15. $f(x) = \begin{cases} 2-x, & x < 0 \\ 1, & x = 0 \\ x+2, & x > 0 \end{cases}$

16. $f(x) = \begin{cases} x^2, & x \leq 0 \\ x+1, & x > 0 \end{cases}$

17. $F(t) = \begin{cases} 2t, & t \leq 1 \\ t, & t > 1 \end{cases}$

18. $f(x) = \dfrac{x-1}{x^2-x}$

19. $f(x) = \dfrac{3}{x^2-1}$

20. $g(s) = \dfrac{s^2}{s}$

21. Speedy Rent-a-Car charges $20 per day plus 10¢ per mile for one of its compact cars. Figure 10 shows C, the cost (in dollars) incurred by a salesman, as a function of M, the number of miles that he has driven the car. What is the reason for the discontinuity on the graph?

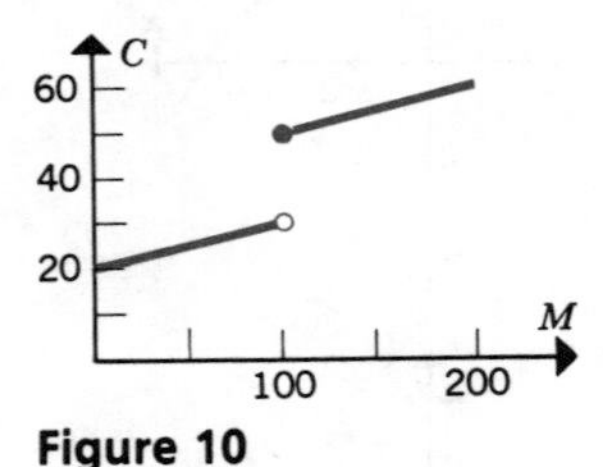

Figure 10

22. Figure 11 shows I, the monthly income of a factory worker, as a function of t, the time (in months). Can you explain the discontinuities on the graph at t_1 and t_2? (There are many.)

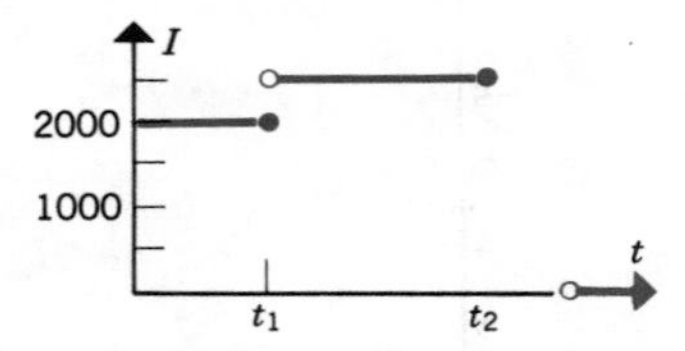

Figure 11

23. Linda May and Billy Joe leave for their honeymoon in Billy Joe's 1976 Ford pickup truck. Figure 12 shows the graph of p, the air pressure in one of the tires, as a function of t, the length of time they have traveled. What is the reason for the discontinuity of t_1?

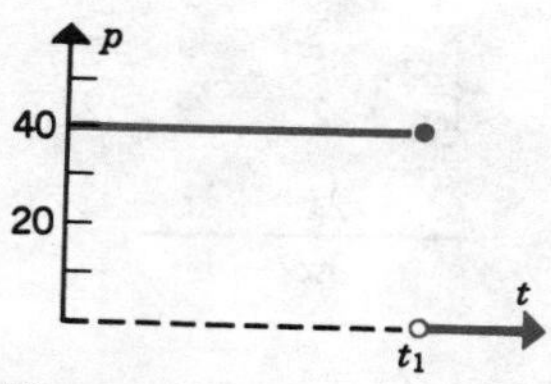

Figure 12

24. National Supermarkets is selling ground beef this week at the following prices:

 \$2.00/lb (packages less than 3 lb)
 \$1.50/lb (packages of 3 lb or more)

 (a) Find the function that describes C, the cost of a package, as a function of x, the number of pounds of ground beef in the package.
 (b) Sketch the graph of this function.
 (c) How much money does a shopper save if he buys one 3-lb package instead of three 1-lb packages? Where does this savings appear on your graph?

25. Many catalog retailers give price discounts based on the dollar value of an order. Outdoors, which sells camping equipment, gives a discount of 10 percent on all orders totaling \$100 or more for its food products. Outdoors sells 6-oz packages of freeze-dried meat for \$4 each.

 (a) Find the relationship between C, the total cost of an order, as a function of x, the number of packages ordered. Where is the function discontinuous?
 (b) Suppose that Outdoors increases the price of each package to \$5. How does this affect the results obtained in part (a)?
 (c) Outdoors imposes a shipping charge on each order as listed in Table 1.

Table 1

Cost of Order	Shipping Charge
Under \$100	\$8
\$100 or over	\$10

 How are the answers to parts a and b affected?

In Exercises 26 through 30, how should $f(2)$ be defined to make each function continuous at $x = 2$?

26.

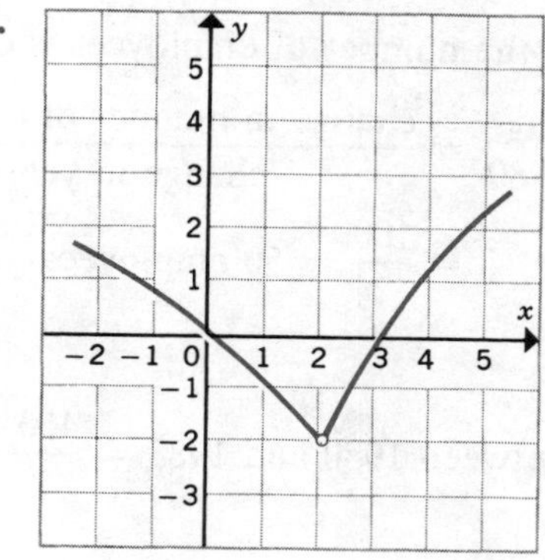

27. $f(x) = \dfrac{x^2 + x - 6}{x^2 - 2x}$

28. $f(x) = \begin{cases} x^2, & x < 2 \\ x + 2, & x > 2 \end{cases}$

29. $f(x) = \dfrac{\sqrt{2x} - 2}{2x - 4}$

30. $f(x) = \dfrac{x^3 - 2x^2 + x - 2}{3x - 6}$

Hint: Use either synthetic division or the "factoring-by-grouping" method to find the factors of the numerator.

2.3 AVERAGE RATE OF CHANGE

Calculus is the branch of mathematics that deals with change. A function called the derivative allows us to determine instantaneous rates of change for many applications. For example, the derivative enables us to determine how fast a company's profits are growing or how rapidly an oil slick is spreading out from an oil spill.

Before the derivative is defined, it will be helpful to introduce a preliminary concept, the **average rate of change** of a function. The phrase "rate of change" describes the changes occurring in a dependent variable in terms of changes taking place in an accompanying independent variable; rate of change describes the dynamics of change.

To put the preceding remarks into perspective, consider the following situations.

1. A 20°F change in temperature (dependent variable) during a change of 1 hour (independent variable) is far more dramatic and noticeable than a 20°F change over a 12-hour period.
2. A 4 percent increase in the consumer price index (dependent variable) would be a cause for concern if the change occurs over a time (independent variable) interval of 1 month; the same increase over a 12-month period would not be so worrisome.

Figure 1 shows the number of employees in a small company from 1977 to 1986. From the graph, we can find the change in the number of employees from 1978 to 1980:

$$\text{Change in number of employees} = 250 - 150 = 100$$
$$\text{Change in years} = 1980 - 1978 = 2$$

The average rate of change in the number of employees is defined as

$$\begin{aligned}\text{Average rate of change between 1978 and 1980} &= \frac{\text{change in number of employees}}{\text{change in years}} \\ &= \frac{100}{2} = 50 \text{ employees/year}\end{aligned}$$

Similarly,

$$\begin{aligned}\text{Average rate of change between 1980 and 1983} &= \frac{460 - 250}{3} \\ &= 70 \text{ employees/year}\end{aligned}$$

Geometrically, the average rate of change between 1980 and 1983 equals the slope of the line connecting the points (3, 250) and (6, 460). The line connecting any two points on a curve is called a **secant** line.

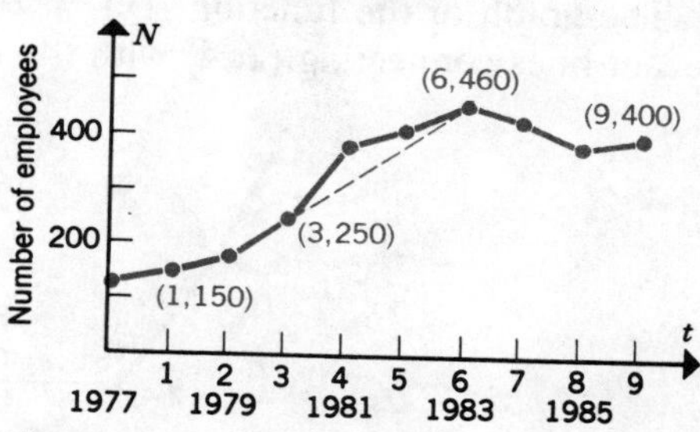

Figure 1

The method just described can be used to find the average rate of change for any function. In Figure 2, as x changes from x_1 to x_2, the average rate of change of the function $y = f(x)$ with respect to x is defined as

$$\text{Average rate of change} = \frac{\text{change in } y}{\text{change in } x} = \frac{y_2 - y_1}{x_2 - x_1}$$

As noted previously, the **average rate of change** can be interpreted geometrically as the **slope of the secant line** connecting the points P and Q in Figure 2, so we can write

$$\text{Average rate of change} = m = \frac{y_2 - y_1}{x_2 - x_1} = \frac{f(x_2) - f(x_1)}{x_2 - x_1} \tag{1}$$

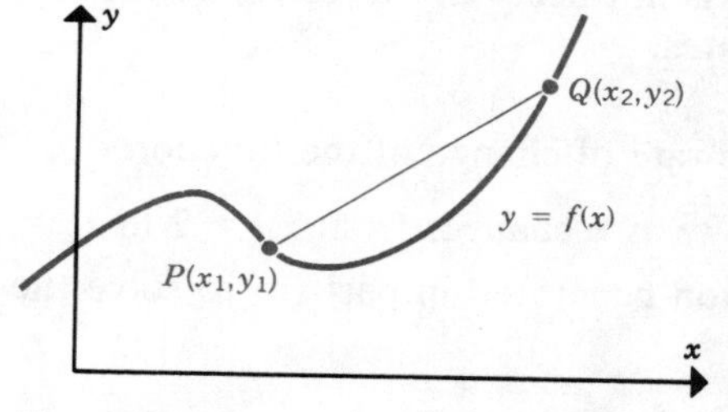

Figure 2

Example 1 Find the average rate of change of the function $f(x) = x^2$ as x changes (a) from 2 to 4, (b) from 2 to 0, (c) from 2 to -2.

Solution Using Equation 1, we get the following results.

(a) $\text{Average rate of change} = \dfrac{f(4) - f(2)}{4 - 2} = \dfrac{16 - 4}{2} = 6$

(b) $\text{Average rate of change} = \dfrac{f(0) - f(2)}{0 - 2} = \dfrac{0 - 4}{-2} = 2$

(c) Average rate of change $= \dfrac{f(-2) - f(2)}{(-2) - 2} = \dfrac{4 - 4}{-4} = 0$

The graph of the function $f(x) = x^2$ is shown in Figure 3 together with the secant lines connecting (2, 4) with (4, 16), (0, 0), and (−2, 4).

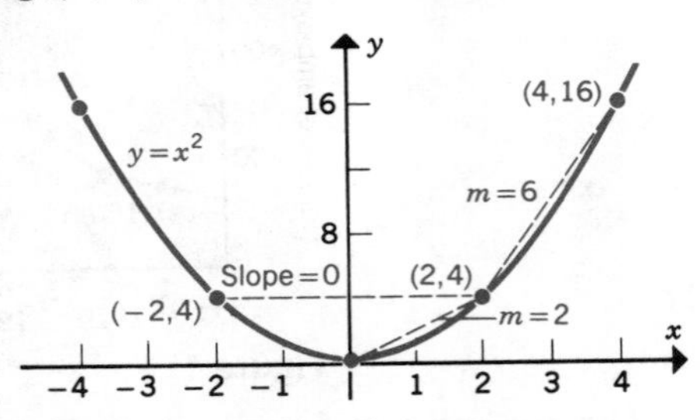

Figure 3

■

Equation 1 can be written in a more useful form. First, we write x_2 as

$$x_2 = x_1 + h$$

where h represents the **change in *x*;** for example, if $x_1 = 1$ and $x_2 = 4 = 1 + 3$, then $h = 3$. The equation for the average rate of change becomes

$$\text{Average rate of change} = m = \frac{f(x_1 + h) - f(x_1)}{(x_1 + h) - x_1}$$

$$= \frac{f(x_1 + h) - f(x_1)}{h} \tag{2}$$

This form of the equation allows you to write the average rate of change in terms of x_1 and h; it is not necessary to calculate the y coordinates separately, as Example 2 illustrates.

Example 2 (a) Find the average rate of change of the function

$f(x) = x^2$ as x changes from $x_1 = 2$ to $x_2 = 2 + h$.

(b) Use the expression generated in part (a) to solve the problem in Example 1.

Solution (a) First, it is necessary to find $f(2 + h)$:

$$f(2 + h) = (2 + h)^2 = 4 + 4h + h^2$$

The average rate of change can now be found using Equation 2:

$$m = \frac{f(2 + h) - f(2)}{h}$$

$$= \frac{(4 + 4h + h^2) - 4}{h} = \frac{4h + h^2}{h} = \frac{\not{h}(4 + h)}{\not{h}}$$

$$= 4 + h, \quad h \neq 0 \tag{3}$$

(b) When x changes from $x_1 = 2$ to $x_2 = 4 = 2 + 2$, $h = 2$, so Equation 3 yields

$$m = 4 + 2 = 6$$

when x changes from $x_1 = 2$ to $x_2 = 0$, $h = -2$, and

$$m = 4 + (-2) = 2$$

Finally, when x changes from $x_1 = 2$ to $x_2 = -2$, $h = -4$, and

$$m = 4 + (-4) = 0$$

Note that not only do the answers in part (b) agree with those obtained in Example 1, but also it was not necessary to find the y coordinates of any of the points in making the calculations. ■

The method for finding the average rate of change can be generalized as follows. If the subscript 1 is dropped in Equation 2, x represents the x coordinate of an arbitrary point, and the average rate of change in going from $(x, f(x))$ to $(x + h, f(x + h))$ is given by the equation

$$\text{Average rate of change} = m = \frac{f(x + h) - f(x)}{h} \tag{4}$$

The quantity $[f(x + h) - f(x)]/h$ is called the **difference quotient.** It represents the slope of the secant line that connects the points $[x, f(x)]$ and $[x + h, f(x + h)]$, as shown in Figure 4. The difference quotient is a very important quantity in calculus because it is used not only to define the derivative (Section 2.4) but also to find the derivatives of many simple types of functions. The use of the difference quotient to find an equation for the slope of a secant line is illustrated in the next two examples.

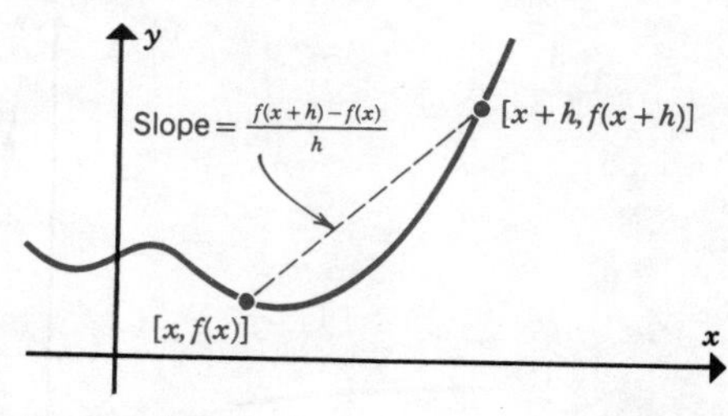

Figure 4

Example 3 Find the difference quotient for the function $f(x) = 3x^2$.

Solution 1. $f(x + h) = 3(x + h)^2 = 3x^2 + 6xh + 3h^2$

2. Find the difference $f(x + h) - f(x)$:

$$f(x + h) - f(x) = (3x^2 + 6xh + 3h^2) - 3x^2 = 6xh + 3h^2$$

3. Divide the result found in step 2 by h, yielding

$$\frac{f(x + h) - f(x)}{h} = \frac{6xh + 3h^2}{h} = 6x + 3h, \qquad h \neq 0$$

This result is shown geometrically in Figure 5.

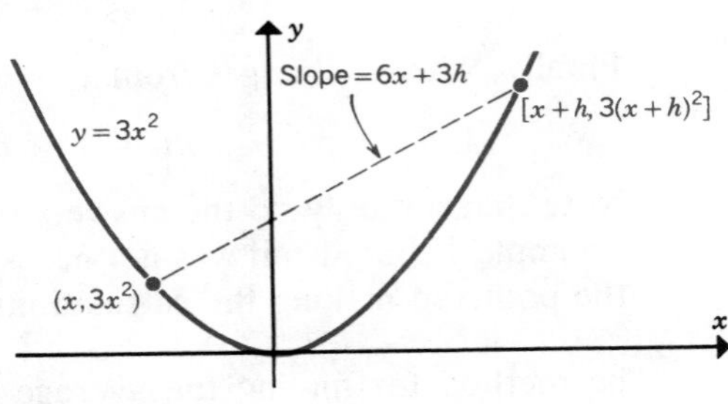

Figure 5

■

Example 4 Find the difference quotient for the function $f(x) = \frac{1}{x}$.

Solution 1. $f(x + h) = \frac{1}{x + h}$.

2. $f(x + h) - f(x) = \frac{1}{x + h} - \frac{1}{x} = \frac{x - (x + h)}{x(x + h)} = \frac{-h}{x(x + h)}$

Note: It is good practice to simplify $f(x + h) - f(x)$ as much as possible.

3. $$\frac{f(x + h) - f(x)}{h} = \frac{\frac{-h}{x(x + h)}}{h} = -\frac{h}{x(x + h)} \cdot \frac{1}{h} = \frac{-1}{x(x + h)}$$

Again, the result is shown graphically in Figure 6.

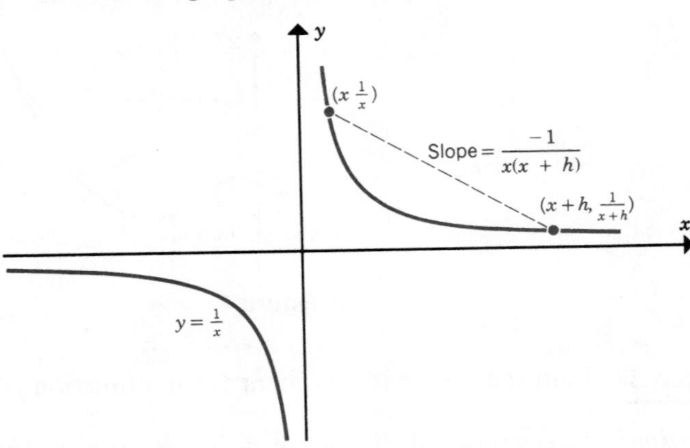

Figure 6

■

The difference quotient $\frac{f(x + h) - f(x)}{h}$ depends on both x and h. This means that values for both x and h must be provided in order to calculate the average rate of change or the slope of the secant line connecting two points. The difference quotient, once it is found, is a more efficient way of calculating the average rate of change, particularly if the average rates of change for a large number of pairs of points are to be tabulated.

Example 5 For the function $f(x) = \frac{1}{x}$, find the slope of the secant line connecting the points whose x coordinates are $\frac{1}{2}$ and 3.

Solution We use the result from Example 4

$$m = \frac{-1}{x(x + h)}$$

Letting $x = \frac{1}{2}$ and $h = 3 - \frac{1}{2} = \frac{5}{2}$, we get

$$m = \frac{-1}{(\frac{1}{2})(\frac{1}{2} + \frac{5}{2})} = -\frac{2}{3}$$

If we had set $x = 3$ and $h = -\frac{5}{2}$, we would have obtained the same result. The curve and the secant line are shown in Figure 7. ■

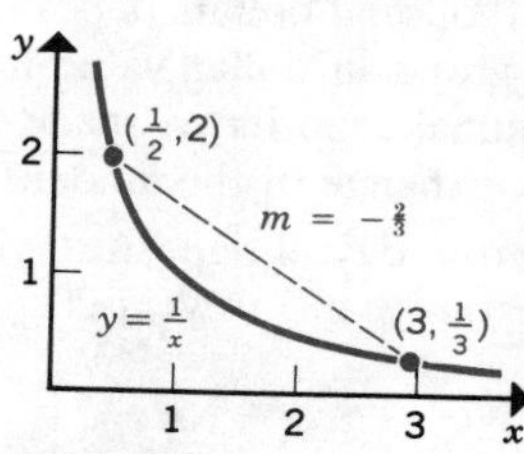

Figure 7

Average Velocity

When an object is in motion, the rate at which distance is traveled over a given interval of time yields the average velocity of the object. For example, if you traveled 160 miles in a period of 4 hours, the average velocity with which you were moving during this period is 40 miles/hour. When the distance $s(t)$ from some reference point is a known function of the time t, the **average velocity** between times t and $t + h$ is defined as

$$\text{Average velocity} = \frac{s(t + h) - s(t)}{(t + h) - t} = \frac{s(t + h) - s(t)}{h}$$

Example 6 Suppose an object is moving along a straight line and the equation describing the distance $s(t)$, in feet, of the object from some reference point is

$$s(t) = 5t^2 + 8$$

(a) Find an expression for the average velocity between t and $t + h$.
(b) Using the results from part (a), find the average velocity between $t_1 = 2$ sec and $t_2 = 5$ sec.

Solution (a) $s(t + h) = 5(t + h)^2 + 8 = 5t^2 + 10th + 5h^2 + 8$

$$s(t + h) - s(t) = (5t^2 + 10th + 5h^2 + 8) - (5t^2 + 8) = 10th + 5h^2$$

$$\text{Average velocity} = \frac{10th + 5h^2}{h} = 10t + 5h$$

(b) Letting $t = 2$ and $h = 5 - 2 = 3$, we find the average velocity from the expression $10t + 5h$:

$$\text{Average velocity} = 10(2) + 5(3) = 35 \text{ ft/sec}$$

The same result can be obtained by setting $t = 5$ sec and $h = 2 - 5 = -3$ sec, yielding,

$$\text{Average velocity} = 10(5) + 5(-3) = 35 \text{ ft/sec}$$ ■

2.3 EXERCISES

1. One thousand dollars deposited into a savings account paying 6 percent interest per year grows in dollar value in the manner shown in Figure 8 if no withdrawals or additional deposits are made. If $t = 0$ designates the day of deposit, find the average rate of change of the amount in the account.
   (a) From day of deposit to year 10. (b) From year 5 to year 15.
   (c) From year 5 to year 20.

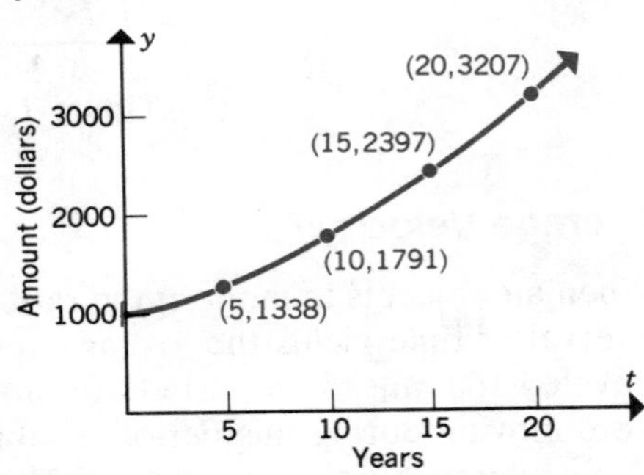

Figure 8

2. In addition to depreciating the value of an asset on a straight-line basis as described in Section 1.3, there are other methods for which the depreciation is larger in the early years than the corresponding depreciation calculated by the straight-line method. One of these methods, double-declining balance, will give the results shown on the graph in Figure 9 for an asset that costs $20,000, has a useful life of ten years, and has no salvage value at the end of the ten-year period. Using Figure 9, determine the average rate of change of the value.
   (a) Over the first five years. (b) From year 2 to year 7.
   (c) From year 2 to year 10.

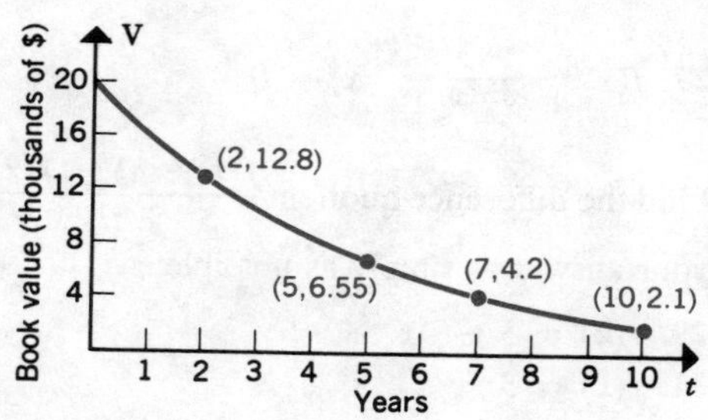

Figure 9

Find the average rate of change of each of the following functions as x changes from x_1 to x_2.

3. $f(x) = 2x^2;\quad x_1 = -1, x_2 = 2$
4. $f(x) = x^2;\quad x_1 = \frac{1}{2}, x_2 = \frac{3}{2}$
5. $f(x) = x^3;\quad x_1 = -1, x_2 = 2$
6. $F(x) = x^2 - 1;\quad x_1 = -2, x_2 = 2$
7. $g(x) = x^4;\quad x_1 = 0, x_2 = -2$
8. $f(x) = x^2 + x - 1;\quad x_1 = -2, x_2 = 3$
9. $f(x) = 3\sqrt{x};\quad x_1 = 1, x_2 = 9$
10. $F(x) = \dfrac{2}{x};\quad x_1 = -2, x_2 = -1$
11. $f(x) = \dfrac{2}{x-1};\quad x_1 = -2, x_2 = 0$
12. $f(x) = \dfrac{1}{\sqrt{x}};\quad x_1 = 4, x_2 = 36$

Find the slope of the secant line connecting the points whose x coordinates are given. In addition, sketch the graph of each function and the secant line.

13. $f(x) = x^2;\quad x_1 = -1, x_2 = 2$
14. $F(x) = x^2 - 2x;\quad x_1 = 0, x_2 = 3$
15. $g(x) = 4 - x^2;\quad x_1 = -2, x_2 = 0$
16. $f(x) = \dfrac{2}{x};\quad x_1 = -2, x_2 = -1$
17. $F(x) = 2\sqrt{x};\quad x_1 = 0, x_2 = 4$
18. $g(x) = x^3;\quad x_1 = -1, x_2 = 1$

Find the average rate of change of each of the following functions as x changes from x_1 to $x_1 + h$; express your answers in terms of h.

19. $f(x) = 3x^2;\quad x_1 = 1$
20. $f(x) = x^2 + 2x;\quad x_1 = 0$
21. $F(x) = x^2 - 3x + 2;\quad x_1 = -1$
22. $g(x) = x^3;\quad x_1 = -1$
23. $f(x) = -3x^2 - 2;\quad x_1 = -2$
24. $g(x) = -\dfrac{1}{x};\quad x_1 = -1$
25. $F(x) = 2\sqrt{x};\quad x_1 = 1$
26. $f(x) = \dfrac{1}{x+1};\quad x_1 = 0$

27. $f(x) = \frac{3}{x^2 + 1}$; $x_1 = 0$

28. $g(x) = \frac{2}{\sqrt{x}}$; $x_1 = 4$

Find the difference quotient $\frac{f(x + h) - f(x)}{h}$ for each of the following functions. Express your answer as simply as possible.

29. $f(x) = 5 - 3x$

30. $F(x) = 2x + 6$

31. $f(x) = 3x^2$

32. $f(x) = 1 - x^2$

33. $g(x) = x^2 + x$

34. $f(x) = x^2 + 2x - 1$

35. $f(x) = -\frac{2}{x}$

36. $F(x) = 2x^3$

37. $f(x) = 3\sqrt{x}$

38. An object moves according to the equation

$$s(t) = 4t^2 - 12t$$

where $s(t)$ is the distance in feet and t is the time in seconds.

(a) Find an expression for the average velocity $\frac{s(t + h) - s(t)}{h}$.

(b) Using the result of part (a), find:
  i. The average velocity between $t = 3$ and $t = 4$ sec and
  ii. The average velocity between $t = 1$ and $t = 5$ sec.

39. An object is thrown vertically upward from a tall building. Its height y (in feet) as a function of t, the time (in seconds), from the moment of release is given by the equation

$$y = f(t) = 400 + 128t - 16t^2$$

(a) Find an expression for the average velocity $\frac{f(t + h) - f(t)}{h}$.

(b) Use the expression from part (a) to find
  i. The average velocity between $t = 0$ and $t = 3$ sec.
  ii. The average velocity between $t = 2$ and $t = 8$ sec.

40. A company has developed a new self-adjusting carburetor. Annual fixed costs are estimated to be \$300,000 for this product. The relationship between x_{BE}, the break-even volume, and CM, the contribution margin (Section 1.5), is given by the equation

$$x_{BE} = f(\text{CM}) = \frac{300{,}000}{\text{CM}}$$

Find the difference quotient $\frac{f(\text{CM} + h) - f(\text{CM})}{h}$.

41. The supply of blood $S(t)$, in pints, at a local hospital is given by the equation

$$S(t) = 24 - 8t + t^2, \qquad 0 \le t \le 10$$

where t is the time, in days, from today ($t = 0$).

(a) Find the average rate of change in the blood supply from $t = 0$ to $t = 4$.

(b) Find the average rate of change in the blood supply from $t = 2$ to $t = 8$.

(c) Find the difference quotient $\frac{S(t + h) - S(t)}{h}$.

42. The time for a pendulum to make a complete swing and return to its original position is called its period and is a function of its length L. The functional relationship is

described by the equation

$$T(L) = 2\pi\sqrt{\frac{L}{g}}$$

where g is the acceleration due to gravity ($g = 32$ ft/sec²).

(a) Find the difference quotient $\dfrac{T(L + h) - T(L)}{h}$.

(b) Use the result obtained in part a to find the average rate of change in the period of a pendulum when L is changed from 2 to 8 feet.

2.4 DERIVATIVE OF A FUNCTION

In determining the average rate of change of a function or, equivalently, the slope of a secant line in the previous section, we found that the result depends on the coordinates of the two points selected. We would now like to turn our attention to the question of finding a way to determine the *instantaneous rate of change* of a function, or equivalently, the *slope of the tangent line* at any point on a curve. In the process, we shall develop and define a very important concept, the **derivative** of a function.

First, what do we mean when we use the phrase "tangent line at a point"? Initially let's think of the tangent line as a line that grazes or touches a curve in such a way that the line and the curve almost coincide in the vicinity of the point. As a result, it is difficult to distinguish between the line and the curve in the vicinity of a given point. In Figure 1, the lines at P_1, P_3, and P_5 are tangent lines, but those at P_2 and P_4 are not.

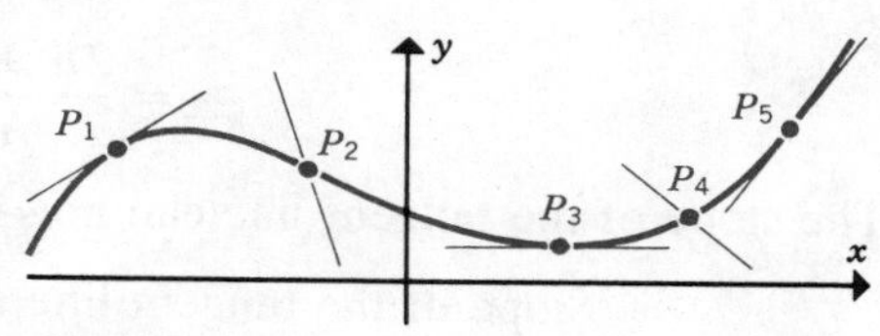

Figure 1

Next, how do we determine the slope of a tangent line? For example, how do we find the slope of the line tangent to the curve $f(x) = x^2$ at (1, 1)? Although we can make a crude attempt to sketch the tangent line at (1, 1), as shown in Figure 2, we need an analytical method in order to assign a value to the slope of the tangent line. It is possible to approximate the slope by finding the differ-

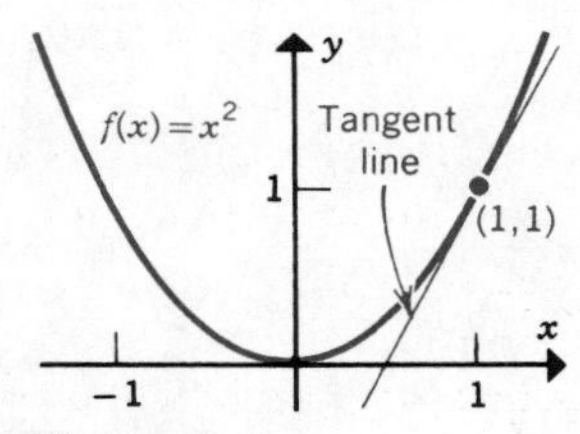

Figure 2

ence quotient $\dfrac{f(1+h)-f(1)}{h}$ for $f(x) = x^2$ and then assigning smaller and smaller values to h. A scenario of this process is shown in Figure 3. As $h \to 0$, the secant lines and the tangent line become nearly parallel, so that their slopes are almost equal. On the basis of this observation, we define the slope

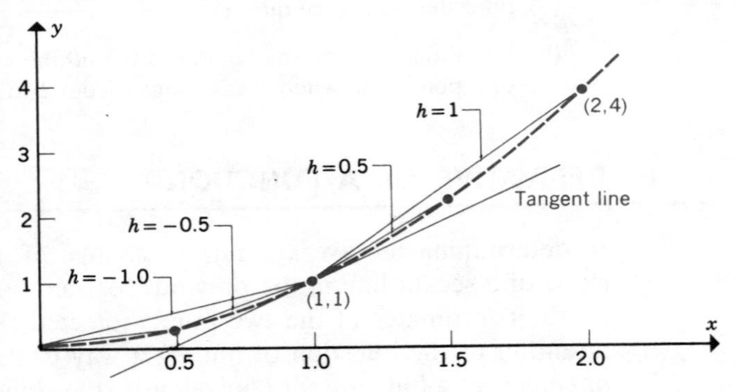

Figure 3

of the tangent line at $x = 1$ as the limit of the slopes of the secant lines as $h \to 0$. First,

$$\text{Slope of a secant line} = \frac{f(1+h)-f(1)}{h}$$

$$= \frac{(1+h)^2 - 1^2}{h} = \frac{1 + 2h + h^2 - 1}{h}$$

$$= \frac{2h + h^2}{h} = 2 + h, \qquad h \neq 0$$

The slope of the tangent line can now be found by letting $h \to 0$:

$$\text{Slope of the tangent line at } (1,1) = \lim_{h \to 0} (2 + h) = 2$$

This method can be used to find the slope of the line tangent to the curve $f(x) = x^2$ at any point; for example, if $x = -2$, we get

$$\text{Slope of the tangent line at } (-2,4) = \lim_{h \to 0} \frac{f(-2+h) - f(-2)}{h}$$

$$= \lim_{h \to 0} \frac{(-2+h)^2 - (-2)^2}{h}$$

$$= \lim_{h \to 0} \frac{4 - 4h + h^2 - 4}{h}$$

$$= \lim_{h \to 0} \frac{-4h + h^2}{h}$$

$$= \lim_{h \to 0} [-4 + h] = -4$$

This procedure can be extended to determine the slope of the tangent line at an arbitrary point on the curve $f(x) = x^2$:

$$\begin{aligned}
\text{Slope of the tangent line at } (x, x^2) &= \lim_{h\to 0} \frac{f(x+h) - f(x)}{h} \\
&= \lim_{h\to 0} \frac{(x+h)^2 - x^2}{h} \\
&= \lim_{h\to 0} \frac{x^2 + 2hx + h^2 - x^2}{h} \\
&= \lim_{h\to 0} \frac{\cancel{h}(2x+h)}{\cancel{h}} \\
&= \lim_{h\to 0} (2x + h) = 2x
\end{aligned}$$

Note: In the preceding limit process, h approached 0 while x, although arbitrary, remained constant.

Thus, for the function $f(x) = x^2$, the slope of a line tangent to the curve at any point (x, x^2) equals $2x$.

Example 1 Find the slope of the line tangent to the curve $f(x) = x^2$ at (3, 9).

Solution The slope of the tangent line equals $2x$, so

$$m = 2(3) = 6$$

The curve and the tangent line are shown in Figure 4.

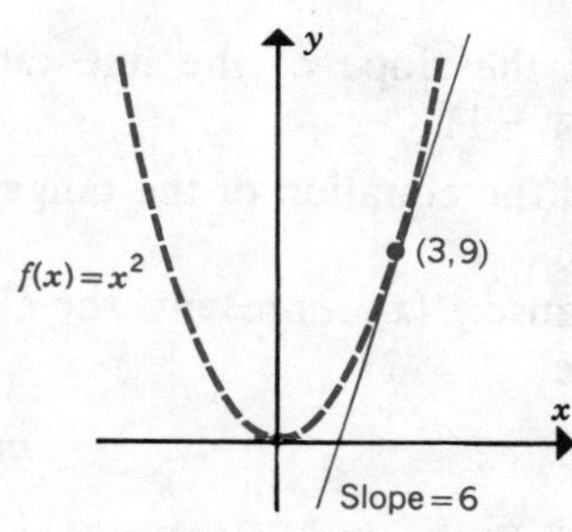

Figure 4

■

The quantity $\lim_{h\to 0} \dfrac{f(x+h) - f(x)}{h}$, if it exists, is called the derivative of the function $y = f(x)$ and is denoted $f'(x)$.

Definition 1 The **derivative** of the function $y = f(x)$, written $f'(x)$, is defined as

$$f'(x) = \lim_{h \to 0} \frac{f(x + h) - f(x)}{h} \tag{1}$$

provided that the limit exists. The function is said to be differentiable at the values of x for which the limit exists.

The derivative is itself a function that represents the slope of the line tangent to the curve $y = f(x)$ at those points where $f'(x)$ is defined. The following examples illustrate the steps used to find the derivative of some simple functions.

Example 2 Find the derivative of the function $f(x) = x^3$:

Solution

1. $f(x + h) = (x + h)^3 = x^3 + 3x^2h + 3xh^2 + h^3$
2. $f(x + h) - f(x) = x^3 + 3x^2h + 3xh^2 + h^3 - x^3$
   $= 3x^2h + 3xh^2 + h^3$
3. $\dfrac{f(x + h) - f(x)}{h} = \dfrac{3x^2h + 3xh^2 + h^3}{h} = \dfrac{\cancel{h}(3x^2 + 3xh + h^2)}{\cancel{h}}$
   $= 3x^2 + 3xh + h^2$
4. $f'(x) = \lim_{h \to 0}(3x^2 + 3xh + h^2) = 3x^2$ ■

Once the derivative has been found, not only can the slope of the tangent line at a point be found, but also its equation can be written, as Example 3 demonstrates.

Example 3

(a) Find the slope of the line tangent to the curve $f(x) = x^3$ at the point $(-1, -1)$.

(b) Find the equation of the tangent line at $(-1, -1)$.

Solution

(a) Because $f'(x)$ represents the slope of the tangent line at any point, we can write

$$m = f'(-1)$$

From Example 2, $f'(x) = 3x^2$, so

$$m = f'(-1) = 3(-1)^2 = 3$$

(b) The equation of the tangent line can be written because we know (1) the slope ($m = 3$) and (2) the coordinates of one point $(-1, -1)$ on the line (the tangent line and curve both pass through this point). If we use the point–slope formula (Section 1.2, Equation 6),

$$y - y_1 = m(x - x_1) \tag{2}$$

and substitute $m = 3$, $x_1 = -1$, $y_1 = -1$, we get

$$y + 1 = 3(x + 1)$$

Simplifying yields the equation

$$y = 3x + 2$$

for the tangent line. The curve and the tangent line are shown in Figure 5.

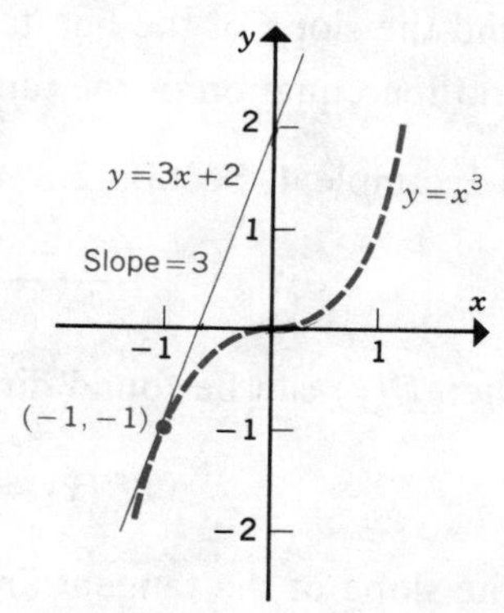

Figure 5

Caution

The variables x and y in the point–slope formula (2) refer to coordinates on the tangent line; except for the point of tangency, they do not, in general, represent the coordinates of points on the curve itself. ■

Equation of a Tangent Line

Given a function $y = f(x)$ that is differentiable at $x = a$, the equation of the tangent line at $x = a$ can be found from the point–slope formula (Equation 2) by substituting $x_1 = a$, $y_1 = f(a)$, and $m = f'(a)$ to yield

$$\boxed{y - f(a) = f'(a)(x - a)} \tag{3}$$

The relationship between the equation $y = f(x)$, which describes the curve, and Equation 3, the equation of the tangent line at $x = a$, is shown graphically in Figure 6.

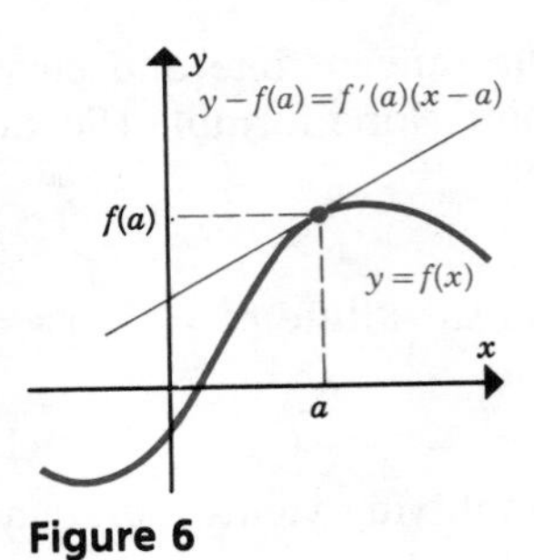

Figure 6

Example 4 (a) Find the derivative of the function $f(x) = \dfrac{1}{x}$.

(b) Find the slope of the line tangent to the curve at $(\frac{1}{2}, 2)$.

(c) Find the equation of the tangent line at $(\frac{1}{2}, 2)$.

Solution (a) In Example 4, Section 2.3, the difference quotient was found:

$$\frac{f(x + h) - f(x)}{h} = \frac{-1}{x(x + h)}$$

Then $f'(x)$ can be found directly by letting $h \to 0$:

$$f'(x) = \lim_{h \to 0} \frac{-1}{x(x + h)} = -\frac{1}{x^2}$$

(b) The slope of the tangent line at $x = \frac{1}{2}$ equals $f'(\frac{1}{2})$:

$$\begin{aligned} m &= f'(\tfrac{1}{2}) \\ &= \frac{-1}{(-\frac{1}{2})^2} = -4 \end{aligned}$$

(c) The equation of the tangent line can now be found by using Equation 3 with $a = \frac{1}{2}$, $f(a) = 2$, and $f'(a) = -4$, yielding

$$y - 2 = 4(x - \tfrac{1}{2})$$

Simplifying gives

$$y = -4x + 4$$

Figure 7 shows the curve and the tangent line.

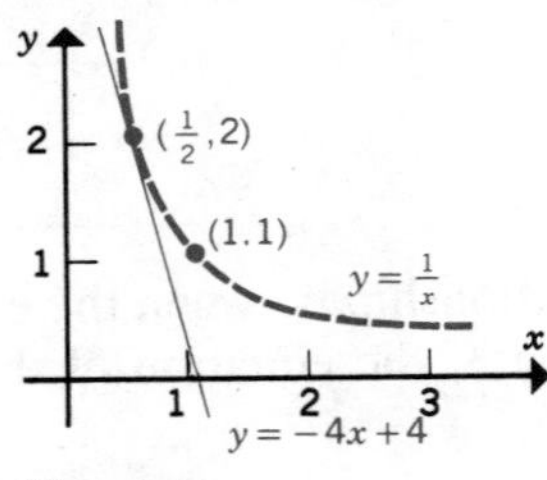

Figure 7

■

Example 5 (a) Find the derivative of the function $f(x) = \sqrt{x}$.

(b) Find the slope and the equation of the tangent line at (1, 1).

Solution (a) The difference quotient must be found first:

$$f(x + h) = \sqrt{x + h}$$

$$\frac{f(x + h) - f(x)}{h} = \frac{\sqrt{x + h} - \sqrt{x}}{h}$$

The next step is to evaluate

$$\lim_{h \to 0} \frac{\sqrt{x + h} - \sqrt{x}}{h}$$

If we attempt to substitute $h = 0$ into this expression, the result is 0/0, which is indeterminate. A situation similar to this came up in Section 2.1, Example 16; in this instance we rationalize the numerator to produce an equivalent algebraic expression:

$$\frac{\sqrt{x + h} - \sqrt{x}}{h} = \frac{(\sqrt{x + h} - \sqrt{x})}{h} \frac{(\sqrt{x + h} + \sqrt{x})}{(\sqrt{x + h} + \sqrt{x})}$$

$$= \frac{x + h - x}{h(\sqrt{x + h} + \sqrt{x})} = \frac{1}{\sqrt{x + h} + \sqrt{x}}$$

Now, if $h \to 0$, we get

$$f'(x) = \lim_{h \to 0} \frac{1}{\sqrt{x + h} + \sqrt{x}} = \frac{1}{2\sqrt{x}}$$

(b) The slope of the tangent line at $x = 1$ can be found:

$$m = f'(1) = \frac{1}{2\sqrt{1}} = \frac{1}{2}$$

Equation 3 can be used to find the equation of the tangent line; letting $a = 1, f(a) = f(1) = \sqrt{1} = 1, f'(1) = \frac{1}{2}$, we get

$$y - 1 = \tfrac{1}{2}(x - 1)$$

Multiplying both sides of the equation by 2 and combining like terms gives

$$2y = x + 1$$

The tangent line and the curve are shown in Figure 8.

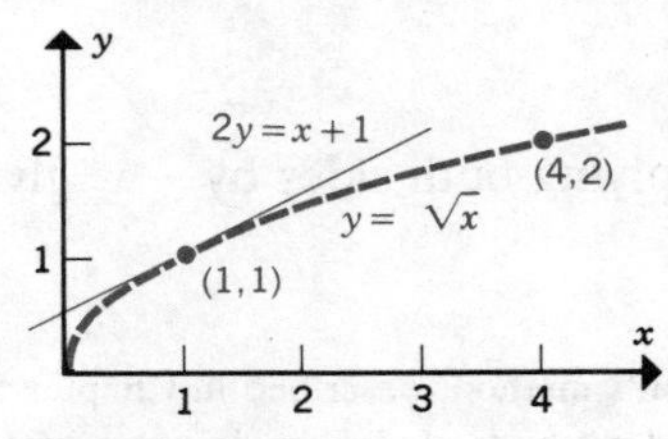

Figure 8

■

Up to this point, our attention has been focused on finding the derivative and using it to find the slope and equation of the tangent line at a given point. There are many situations in which the slope of a tangent line is given and we want to find the coordinates of the point or points where the derivative equals the given slope. This type of problem is usually more difficult to solve than the problem of finding the slope because the resulting equation may be algebraically complex and difficult to solve.* The next two examples illustrate problems of this type.

Example 6 Find the point or points on the curve $f(x) = x^2$ where the slope of the tangent line equals 3.

Solution The problem is depicted graphically in Figure 9. We are attempting to find a point on the curve $f(x) = x^2$ where the tangent line has a slope equal to 3. In Example 1 we found that $f'(x) = 2x$; in addition, we are given that $f'(x) = 3$. So we want to find values of x that satisfy the equation

$$2x = 3$$

Solving yields

$$x = \tfrac{3}{2} \quad \text{and} \quad f(\tfrac{3}{2}) = (\tfrac{3}{2})^2 = \tfrac{9}{4}$$

as the coordinates of the point.

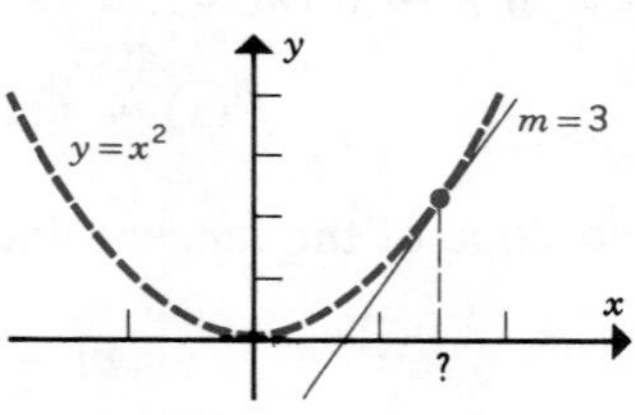

Figure 9

■

Example 7 Find the point(s) on the curve $f(x) = \dfrac{1}{x}$ where the slope of the tangent line equals -4.

Solution In Example 4, we found that $f'(x) = \dfrac{-1}{x^2}\cdot$ Since we are looking for points at which $f'(x) = -4$, we can write

$$-4 = \frac{-1}{x^2}$$

Multiplying both sides by $-x^2$ gives

$$4x^2 = +1$$

*Newton's method, described in Chapter 5, enables us to obtain approximate solutions when the standard methods of algebra do not work.

Subtracting 1 from both sides and then factoring yields the quadratic equation

$$4x^2 - 1 = (2x + 1)(2x - 1) = 0$$

Two solutions emerge:

$$x_1 = \tfrac{1}{2}, \qquad f(\tfrac{1}{2}) = 2$$
$$x_2 = -\tfrac{1}{2}, \qquad f(-\tfrac{1}{2}) = -2$$

Figure 10 shows the curve and the tangent lines at the points $(\frac{1}{2}, 2)$ and $(-\frac{1}{2}, -2)$.

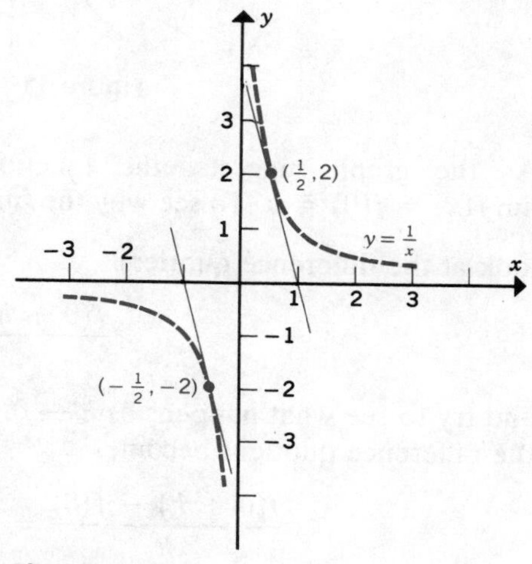

Figure 10

■

Nondifferentiable Conditions

A given function $y = f(x)$ may not be differentiable at $x = a$ for a number of reasons.

1. The function $y = f(x)$ is not continuous at $x = a$. In Example 4, where $f(x) = \dfrac{1}{x}$, we found that $f'(x) = -\dfrac{1}{x^2}$. The derivative $f'(x)$ is not defined at $x = 0$, where $f(x)$ is discontinuous.
2. The function $y = f(x)$ is continuous at $x = a$, but
   (a) The tangent line to the curve at $x = a$ is *vertical,* and therefore its slope is not defined. The function $f(x) = \sqrt{x}$, for which

   $$f'(x) = \frac{1}{2\sqrt{x}}$$

   has a vertical tangent line at $x = 0$ (Figure 8).
   (b) The direction of the curve changes abruptly, as indicated by a sharp corner or wedge-shaped appearance, as shown in Example 8.

Example 8 Show that the function

$$f(x) = \begin{cases} 2x, & x \le 0 \\ -x, & x > 0 \end{cases}$$

whose graph is shown in Figure 11 is not differentiable at $x = 0$.

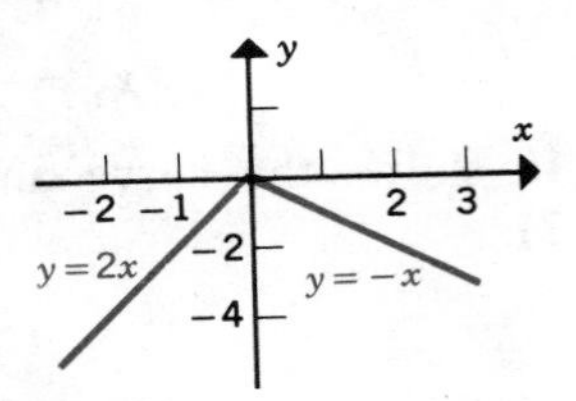

Figure 11

Solution As the graph suggests, the function is continuous at $x = 0$ because $\lim_{x \to 0} f(x) = f(0) = 0$. To see why the function is not differentiable at $x = 0$, let's look at the difference quotient.

$$\frac{f(0 + h) - f(0)}{h}$$

and try to see what happens as $h \to 0$. As $h \to 0$ from the *left*, $f(x) = 2x$, and the difference quotient becomes

$$\frac{f(0 + h) - f(0)}{h} = \frac{2(0 + h) - 2(0)}{h} = 2$$

As $h \to 0$ from the *right*, $f(x) = -x$, and the difference quotient becomes

$$\frac{f(0 + h) - f(0)}{h} = \frac{-(0 + h) - 0}{h} = -1$$

Since $2 \neq -1$,

$$\lim_{h \to 0} \frac{f(0 + h) - f(0)}{h}$$

does not exist and thus $f(x)$ is not differentiable at $x = 0$. ■

2.4 EXERCISES

Use the definition

$$f'(x) = \lim_{h \to 0} \frac{f(x + h) - f(x)}{h}$$

to find the derivative of each of the following functions.

1. $f(x) = 2x$
2. $f(x) = -3x$
3. $f(x) = x + 1$
4. $f(x) = 5 - 4x$
5. $F(x) = 3x^2$
6. $g(x) = x^2 + 3$

7. $f(x) = x^2 + x$
8. $f(x) = 2x^2 - 3x$
9. $F(x) = x^2 - x + 1$
10. $f(x) = x^2 + 4x - 2$
11. $f(x) = -x^2 + 3x - 4$
12. $g(x) = 3x^2 - 4x + 7$
13. $g(x) = 2x^3$
14. $F(x) = x^3 + x$
15. $f(x) = x^3 - x^2$
16. $g(x) = x^4$
17. $F(x) = \frac{2}{x}$
18. $f(x) = \frac{1}{2x}$
19. $f(x) = \frac{1}{x^2}$
20. $F(x) = \frac{2}{x^2}$
21. $F(x) = \sqrt{2x}$
22. $f(x) = 2\sqrt{x}$
23. $f(x) = \frac{1}{x - 1}$
24. $F(x) = \frac{1}{\sqrt{x}}$

The derivatives required for Exercises 25 through 35 have been found in Exercises 1 through 24; refer to them as needed.

25. Find the slope and equation of the line tangent to the curve $F(x) = 3x^2$ at
    (a) $x = -2$ (b) $x = 3$
    Sketch the graphs of the curve and tangent lines.
26. Find the slope and equation of the line tangent to the curve $g(x) = 2x^3$ at
    (a) $x = -2$ (b) $x = 1$
    Sketch the graphs of the curve and tangent lines.
27. Find the slope and equation of the line tangent to the curve $f(x) = \frac{1}{x - 1}$ at
    (a) $x = 3$ (b) $x = 0$
    Sketch the graphs of the curve and tangent lines.
28. Find the slope and equation of the line tangent to the curve $F(x) = 2\sqrt{x}$ at
    (a) $x = 1$ (b) $x = 4$
    Sketch the graphs of the curve and tangent lines.
29. Find the slope and equation of the line tangent to the curve $g(x) = x^4$ at
    (a) $x = -1$ (b) $x = 0$ (c) $x = 2$
    Sketch the graphs of the curve and tangent lines.
30. Find the slope and equation of the line tangent to the curve $f(x) = \frac{1}{x^2}$ at
    (a) $x = -1$ (b) $x = 1$
    Sketch the graphs of the curve and tangent lines.
31. Find the points on the curve $f(x) = x^2 + x$ at which $f'(x) = 3$.
32. Find the points on the curve $f(x) = x^3 - x^2$ at which the slope of the tangent line equals 1.
33. Find the points on the curve $g(x) = 3x^2 - 4x + 7$ at which the slope of the tangent line equals 5.
34. Find the points on the curve $F(x) = \frac{2}{x^2}$ at which the slope of the tangent line equals $-\frac{1}{2}$.

35. Find the points on the curve $F(x) = x^3 + x$ at which the slope of the tangent line equals 13.
36. Find the values of x on the graph in Figure 12 where the function is not differentiable.

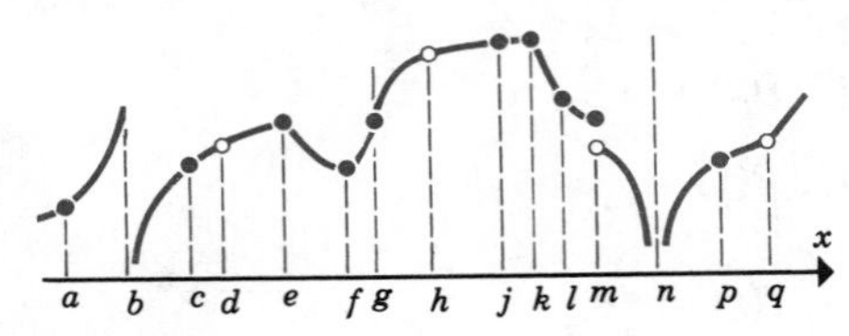

Figure 12

37. The graphs of the function $y = f(x)$ and the line tangent to the curve at $x = 3$, are shown in Figure 13. Find $f(3)$ and $f'(3)$.
38. The graphs of the function $y = F(x)$ and the line tangent to the curve at $x = -1$ are shown in Figure 14. Find $F(-1)$ and $F'(-1)$.
39. The graphs of the function $y = g(x)$ and the line tangent to the curve at $x = 1$ are shown in Figure 15. Find $g(1)$ and $g'(1)$.

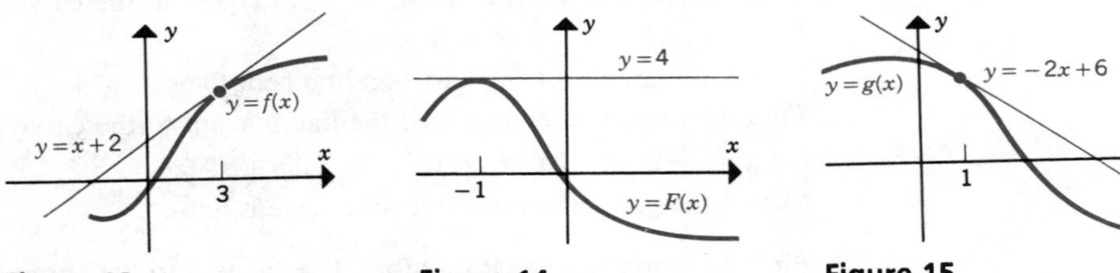

Figure 13 Figure 14 Figure 15

40. The equation $y = 3x + 4$ is an equation of the line tangent to the curve $y = f(x)$ at $x = 2$. Find $f(2)$ and $f'(2)$.
41. The equation $2y + 5x = 4$ is an equation of the line tangent to the curve $y = F(x)$ at $x = -3$. Find $F(-3)$ and $F'(-3)$.
42. The equation $x + y = 2$ is an equation of the line tangent to the curve $y = f(x)$ at $x = 5$. Find $f(5)$ and $f'(5)$.
43. Use the definition to find the derivative of the function $f(x) = \sqrt{x^3}$.
44. Use the definition to find the derivative of the function $f(x) = mx + b$, where m and b are constants.
45. Use the definition to find the derivative of the function $f(x) = \sqrt{ax}$ where a is a constant.
46. Use the definition to find the derivative of the function $f(x) = ax^2 + bx + c$, where a, b, and c are constants.

KEY TERMS

limit
indeterminate form
continuous
difference quotient
derivative
slope of tangent line

discontinuous
average rate of change
secant line
velocity
differentiability

REVIEW PROBLEMS

Use the graphs in problems 1 through 6 to determine (a) $f(1)$ and *b* $\lim_{x \to 1} f(x)$.

1.

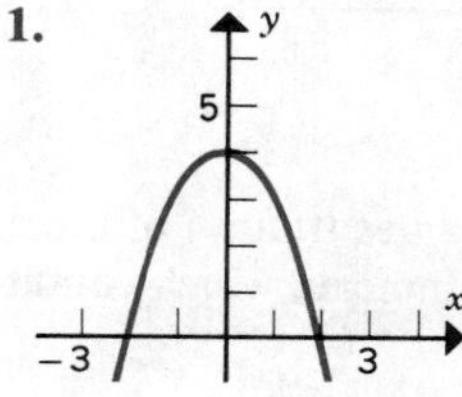

2.

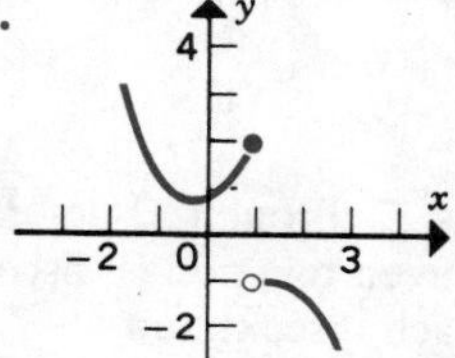

3.

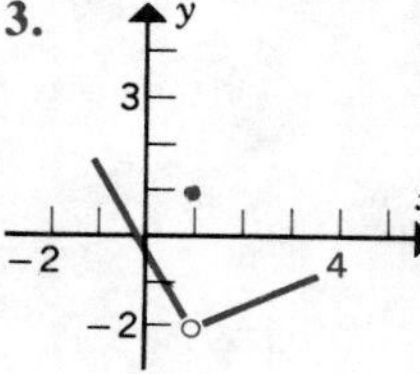

4.

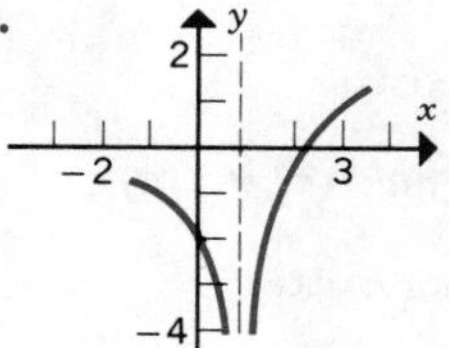

5.

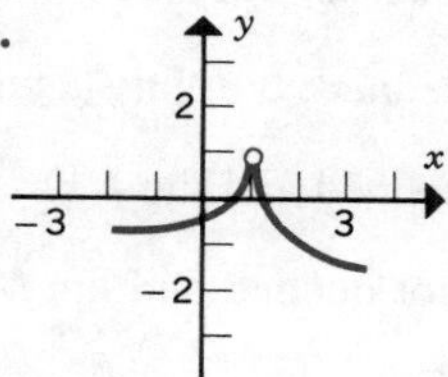

6.

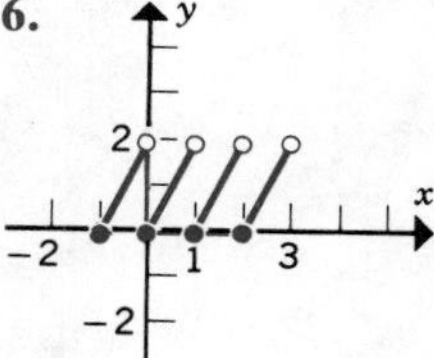

Evaluate each of the following limits when they exist.

7. $\lim_{x \to 2} (x^2 + 3)$

8. $\lim_{x \to 3} \dfrac{2x^2 - 5x - 3}{2x - 6}$

9. $\lim_{x \to 0} \dfrac{x^4 - x^3}{x^3 + x^2}$

10. $\lim_{h \to 0} \dfrac{h^2 + h}{h}$

11. $\lim_{x \to 1} \sqrt{5 - x^2}$

12. $\lim_{x \to -2} \dfrac{x^2 + 5x + 6}{2x + 4}$

13. $\lim_{x \to 1} \dfrac{x^2 - x}{x^2 - 2x - 1}$

14. $\lim_{x \to 1} \dfrac{2\sqrt{x} - 2}{x^2 - 1}$

15. $\lim_{h \to 0} \dfrac{\sqrt{h + 1} - 1}{h}$

16. $\lim_{x \to 2} \dfrac{x^3 - 8}{2x^2 - x - 6}$

17. $\lim_{h \to 0} \dfrac{(x + h) - x}{h}$

18. $\lim_{x \to a} \dfrac{x^2 - a^2}{2x - 2a}$

19. $\lim_{x \to a} \dfrac{x^2 - a^2}{2x + 2a}$

20. $\lim_{x \to 0} \dfrac{(x + 1)^4 - 1}{x}$

In problems 21 through 26 use the graphs to determine whether each of the underlying functions is continuous or discontinuous at $x = 3$. If a function is discontinuous at $x = 3$, determine which of the following conditions, A through D, best describes the discontinuity.

A. $\lim_{x \to 3} f(x)$ exists, but $f(3)$ is not defined.

B. $f(3)$ is defined, but $\lim_{x \to 3} f(x)$ does not exist.

C. $f(3)$ is defined and $\lim_{x \to 3} f(x)$ exists, but $\lim_{x \to 3} f(x) \neq f(3)$.

D. $f(3)$ is not defined and $\lim_{x \to 3} f(x)$ does not exist.

21.

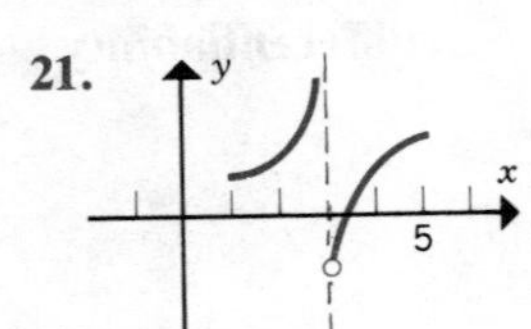

22.

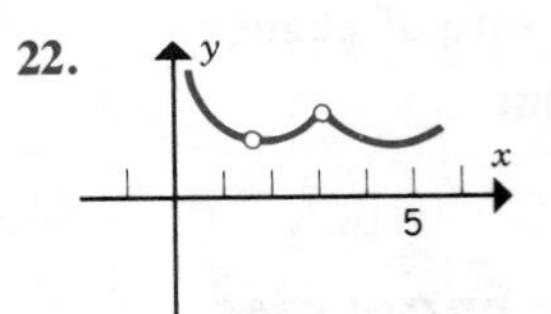

23.

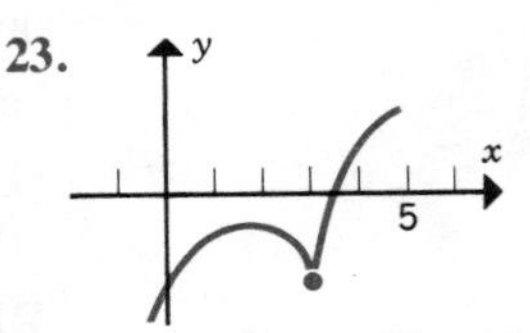

24.

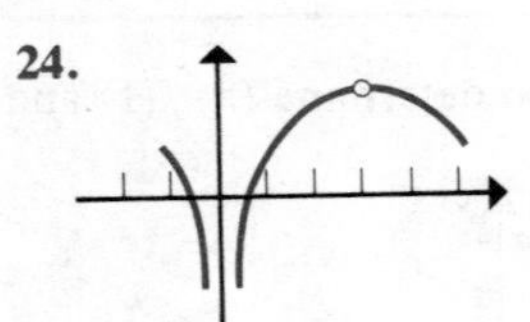

25.

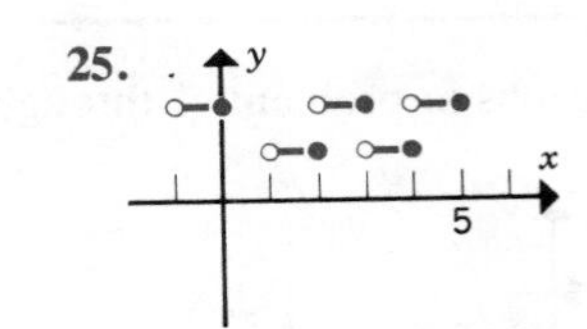

26.

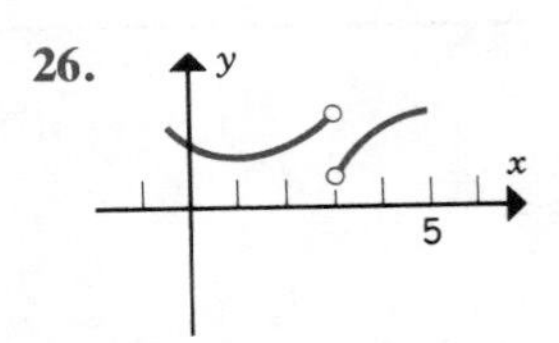

In problems 27 through 32 determine the values (if any) of the independent variable for which the given function is discontinuous. Indicate which condition, A through D, best describes each discontinuity.

A. $\lim_{x \to a} f(x)$ exists, but $f(a)$ is not defined.

B. $f(a)$ is defined, but $\lim_{x \to a} f(x)$ does not exist.

C. $f(a)$ is defined and $\lim_{x \to a} f(x)$ exists, but $\lim_{x \to a} f(x) \neq f(a)$.

D. $f(a)$ is not defined and $\lim_{x \to a} f(x)$ does not exist.

27. $f(x) = \dfrac{x^2 + 4x + 4}{2x + 4}$

28. $f(x) = \dfrac{2 - x}{x - 2}$

29. $g(t) = \begin{cases} t + 2, & t \leq 1 \\ \sqrt{t}, & t > 1 \end{cases}$

30. $F(x) = \begin{cases} \dfrac{1}{x}, & x \leq 2 \\ x + 2 & x > 2 \end{cases}$

31. $f(x) = \dfrac{2x + 2}{x^2 + x}$

32. $f(x) = \begin{cases} 1 - x, & x < 0 \\ 2, & x = 0 \\ x + 1, & x > 0 \end{cases}$

33. Sarah Jane is blowing up a balloon for a birthday party. The volume V of the balloon as function of the time t is shown in Figure 1. What could have caused the discontinuity at $t = t_1$?

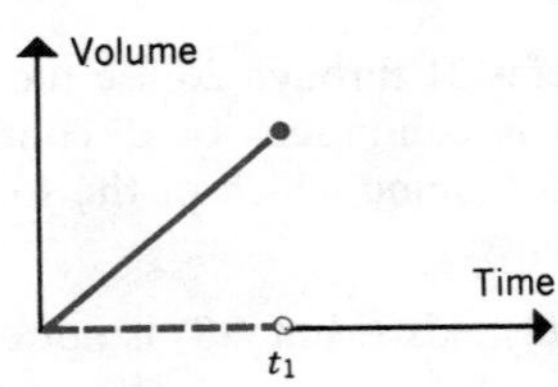

Figure 1

34. The intensity I of light in a room as a function of the time t is shown in Figure 2. Can you give a reason for the discontinuity at $t = t_1$?

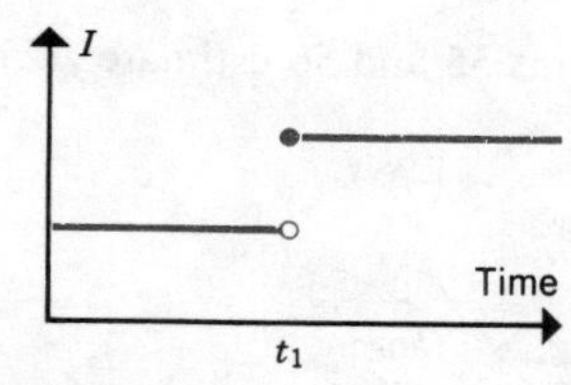

Figure 2

How should $f(1)$ be defined in order to make each of the following functions continuous at $x = 1$?

35. $f(x) = \dfrac{2x^2 - x - 1}{3x - 3}$

36. $f(x) = \begin{cases} 1 - x^2, & x < 1 \\ x^2 - 1, & x > 1 \end{cases}$

In problems 37 through 40, find the average rate of change of each function for the given changes in x.

37. $f(x) = x^2 + 2$, 0 to 4

38. $F(x) = 1 - x^3$, 0 to 1

39. $g(x) = 2\sqrt{x}$, 1 to 4

40. $f(x) = \dfrac{4}{x}$, 1 to 2

In problems 41 through 50, use the definition $f(x) = \lim_{h\to 0} \dfrac{f(x + h) - f(x)}{h}$ to find the derivative of each function. In addition, use the derivative to find the slope and the equation of the line tangent to the graph at the given point.

41. $f(x) = 3x$, $(2, 6)$

42. $f(x) = 1 + x^2$, $(1, 2)$

43. $f(x) = x^2 + 2x$, $(-1, -1)$

44. $f(x) = 3x^2 - 2x$, $(1, 1)$

45. $f(x) = x^3 + 1$, $(1, 2)$

46. $f(x) = x^3 - x$, $(-1, 0)$

47. $f(x) = \dfrac{2}{x}$, $(-1, -2)$

48. $f(x) = 2\sqrt{x}$, $(4, 4)$

49. $f(x) = x^2 + 3x - 2$, $(0, -2)$

50. $f(x) = \sqrt{x + 1}$, $(0, 1)$

In problems 51 through 54, use the graphs to determine the values of x for which the underlying functions are not differentiable.

51.

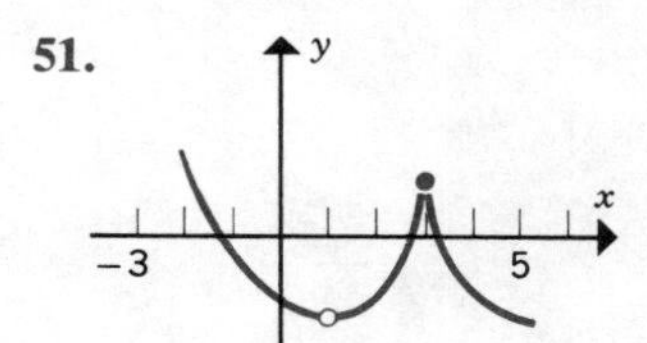

52.

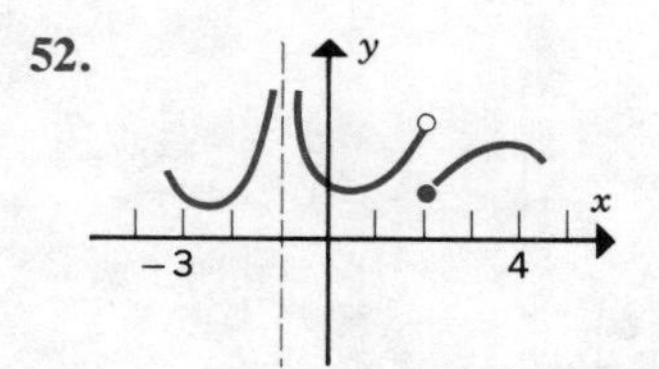

53.

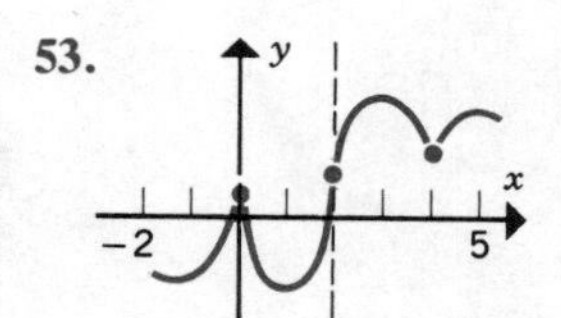

54.

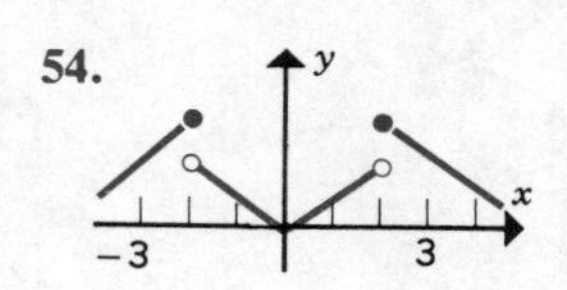

In problems 55 and 56 estimate $f(2)$ and $f'(2)$ from the graphs.

55.

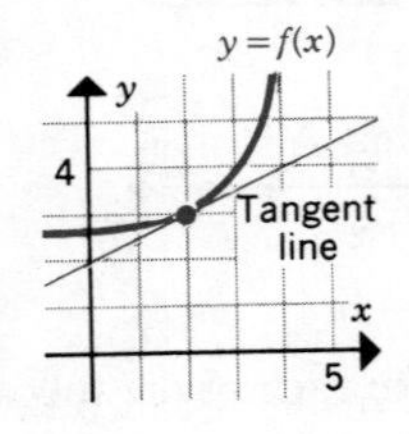

56.

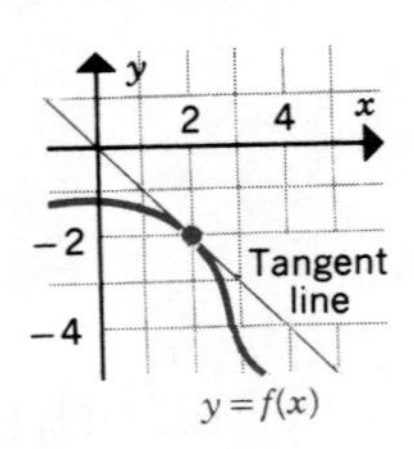

3

DIFFERENTIATION AND APPLICATIONS

INTRODUCTION

Direct use of the definition

$$f'(x) = \lim_{h \to 0} \frac{f(x + h) - f(x)}{h}$$

is generally not the most efficient way to find the derivative of a function. Except for very simple functions, the definition does not produce the derivative quickly and easily. However, the definition can be used to prove general rules that enable you to find the derivatives of most functions in a reasonable length of time. This chapter is devoted to studying these rules together with some basic applications of the derivative.

3.1 POWER RULE; SUM RULE

Before studying the power rule and the sum rule, we should point out that $f'(x)$ is not the only way to represent the derivative. Notations that are frequently used include

$$f'(x), \quad y', \quad \frac{dy}{dx}, \quad D_x y$$

For example, in Section 2.4 we found that the derivative of the function $f(x) = x^2$ equals $2x$. This result can be stated in many ways:

$$f'(x) = 2x, \qquad \frac{dy}{dx} = 2x, \qquad \frac{d(x^2)}{dx} = 2x, \qquad D_x y = 2x$$

Each function that appears in the rules of this chapter is assumed to be differentiable.

Rule 1. The Derivative of a Constant Equals Zero

If $f(x) = c$, where c is a real number, then

$$f'(x) = \frac{d(c)}{dx} = 0$$

This result should come as no surprise. The graph in Figure 1 of the function $f(x) = c$ is a horizontal line whose slope equals zero for all values of x. The rule for the constant function can be obtained from the definition.

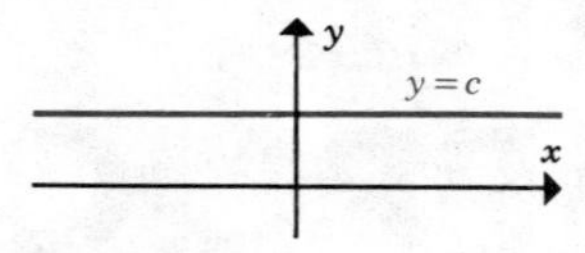

Figure 1

$$f(x) = c, \qquad f(x + h) = c$$

$$f'(x) = \lim_{h \to 0} \frac{f(x + h) - f(x)}{h} = \lim_{h \to 0} \frac{c - c}{h}$$

$$= \lim_{h \to 0} \frac{0}{h} = 0$$

Power Rule

In Section 2.4, the derivatives listed in Table 1 were found for some simple power functions.

Table 1

Function	Derivative
$f(x) = x^2$	$f'(x) = 2x$
$f(x) = x^3$	$f'(x) = 3x^2$
$f(x) = x^4$	$f'(x) = 4x^3$

These results are examples of the **power rule**. It is used for functions of the type $f(x) = x^n$, where n is a real number.

Rule 2. (Simple) Power Rule

If $f(x) = x^n$, where n is a real number, then

$$f'(x) = nx^{n-1}$$

A derivation of this rule is presented at the end of this section for the case in which n is a positive integer.

Example 1 Find the slope of the line tangent to the curve $f(x) = x^6$ at (1, 1).

Solution According to the power rule,

$$f'(x) = 6x^5$$

Then m, the slope of the tangent line at (1, 1), is

$$m = f'(1) = 6(1)^5 = 6$$

■

Simple rational or radical functions can be differentiated directly after they have been written as power functions, as shown in parts (b) and (c) of Example 2.

Example 2 Find the derivative of each of the following functions.

(a) $f(x) = x^8$ (b) $f(x) = \dfrac{1}{x^3}$ (c) $f(x) = \sqrt[4]{x}$

Solution (a) $f'(x) = 8x^7$

(b) Since $\frac{1}{x^3} = x^{-3}$, the power rule gives

$$\frac{d}{dx}(x^{-3}) = -3x^{-4} = \frac{-3}{x^4}$$

Caution $\frac{d}{dx}\left(\frac{1}{x^3}\right) \neq \frac{1}{3x^2}$. You cannot apply the power rule to the denominator of a rational expression.

(c) First we write $\sqrt[4]{x} = x^{1/4}$; applying the power rule gives

$$f'(x) = \frac{1}{4}x^{-3/4} = \frac{1}{4x^{3/4}} = \frac{1}{4\sqrt[4]{x^3}}$$

■

Rule 3. Constant Factor Rule

If $f(x) = cF(x)$, where c is a constant, then

$$f'(x) = cF'(x)$$

The constant factor rule says that the **derivative of a constant times a function equals the constant times the derivative of the function.** The rule is illustrated in the next example.

Example 3 Find the derivative of each of the following functions.

(a) $f(x) = 3x^5$ (b) $f(x) = \frac{2}{3x}$

Solution (a) According to the constant factor rule,

$$\frac{d}{dx}(3x^5) = 3\frac{d(x^5)}{dx}$$

Applying the power rule gives

$$f'(x) = 3(5x^4) = 15x^4$$

(b) $f(x) = \frac{2}{3}\cdot\frac{1}{x} = \frac{2}{3}x^{-1}$

$$f'(x) = \frac{2}{3}\frac{d}{dx}(x^{-1}) \quad \text{(constant factor rule)}$$

$$= \frac{2}{3}(-1)x^{-2} \quad \text{(power rule)}$$

$$= -\frac{2}{3x^2}$$

■

The constant factor rule can be obtained directly from the definition of the derivative:

$$f(x) = cF(x), \qquad f(x + h) = cF(x + h)$$

$$\begin{aligned} f'(x) &= \lim_{h \to 0} \frac{cF(x + h) - cF(x)}{h} \\ &= c \lim_{h \to 0} \frac{F(x + h) - F(x)}{h} \\ &= cF'(x) \end{aligned}$$

The next rule tells us how to find the derivative of a function that contains more than one term.

> **Rule 4. Sum Rule**
>
> If $f(x) = F(x) + G(x)$, then
>
> $$f'(x) = F'(x) + G'(x)$$

The sum rule says that the **derivative of a sum equals the sum of the derivatives.** Operationally, it tells us that we can differentiate a function term by term.

Example 4 Find the derivative of $f(x) = 6x^5 + 2x^4$.

Solution

$$\begin{aligned} f'(x) &= \frac{d}{dx}(6x^5) + \frac{d}{dx}(2x^4) && \text{(sum rule)} \\ &= 6\frac{d}{dx}(x^5) + 2\frac{d}{dx}(x^4) && \text{(constant factor rule)} \\ &= 6(5x^4) + 2(4x^3) && \text{(power rule)} \\ &= 30x^4 + 8x^3 \end{aligned}$$

■

Although the statement of the sum rule contains only two terms, the rule holds regardless of the number of terms to be differentiated.

Example 5 Find the derivative of each of the following functions.

(a) $f(x) = 7x^4 + 5x^3 - 2x^2 + 1$ (b) $f(x) = 4\sqrt{x} + \dfrac{3}{x} + 5$

Solution (a) Each of the four terms is differentiable, so the sum rule gives

$$\begin{aligned} f'(x) &= \frac{d}{dx}(7x^4) + \frac{d}{dx}(5x^3) + \frac{d}{dx}(-2x^2) + \frac{d}{dx}(1) \\ &= 28x^3 + 15x^2 - 4x \end{aligned}$$

(b) Anticipating the application of the power rule, we write $f(x)$ as

$$f(x) = 4x^{1/2} + 3x^{-1} + 5$$

Differentiating term by term yields

$$\begin{aligned} f'(x) &= 4(\tfrac{1}{2})x^{-1/2} + 3(-1)x^{-2} + 0 \\ &= 2x^{-1/2} - 3x^{-2} \\ &= \frac{2}{x^{1/2}} - \frac{3}{x^2} = \frac{2}{\sqrt{x}} - \frac{3}{x^2} \end{aligned}$$

■

The sum rule can be derived from the definition (Section 2.4) of the derivative:

$$\begin{aligned} f(x) &= F(x) + G(x), \qquad f(x+h) = F(x+h) + G(x+h) \\ f'(x) &= \lim_{h\to 0} \frac{F(x+h) + G(x+h) - F(x) - G(x)}{h} \\ &= \lim_{h\to 0} \left[\frac{F(x+h) - F(x)}{h} + \frac{G(x+h) - G(x)}{h} \right] \\ &= \lim_{h\to 0} \frac{F(x+h) - F(x)}{h} + \lim_{h\to 0} \frac{G(x+h) - G(x)}{h} \end{aligned}$$

The last step follows from property V (Section 2.1) of limits, namely the limit of a sum equals the sum of the limits. Noting that the first term represents $F'(x)$ and the second $G'(x)$, we can write, finally,

$$f'(x) = F'(x) + G'(x)$$

Applications

So far, the geometric interpretation of the derivative as the slope of a tangent line has been emphasized. However, in applications, the interpretation of the derivative as the **instantaneous rate of change** of the dependent variable with respect to the independent variable is emphasized. In addition, each field has developed its own terminology for the derivative; this multiplicity of meanings can be confusing when first encountered. For these reasons applications will be restricted to those that can be understood without having to possess an extensive knowledge of the field under study.

Velocity

The initial developments in calculus were motivated by a desire to describe mathematically the characteristics of moving objects, such as velocity and acceleration. In Section 2.3 the average velocity of a moving object between $t = t_1$ and $t = t_2$ was defined as

$$\text{Average velocity} = \frac{s(t_2) - s(t_1)}{t_2 - t_1}$$

where $s(t)$ represents the distance of the object from some reference point. The average velocity between two arbitrary instants of time, t and $t + h$, is given by the difference quotient:

$$\text{Average velocity} = \frac{s(t + h) - s(t)}{h}$$

The **velocity,** $v(t)$, is defined to be the limit of the average velocity as h approaches zero:

$$v(t) = s'(t) = \lim_{h \to 0} \frac{s(t + h) - s(t)}{h}$$

Example 6 An object is dropped from the roof of a 400-ft building. The object's distance above the ground, $s(t)$, in feet, as a function of t, the time in seconds, is given by the equation

$$s(t) = -16t^2 + 400$$

The position of the object for various values of t is shown in Figure 2.

(a) Find the velocity as a function of t.

(b) How fast is the object moving when $t = 3$ sec?

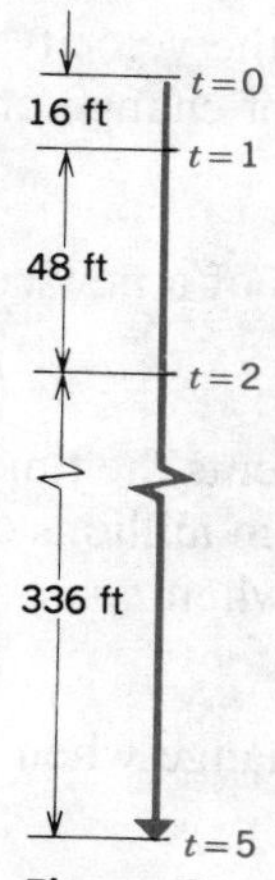

Figure 2

Solution (a) The velocity $v(t)$ at any time is found by differentiating $s(t)$ with respect to t. The result is

$$v(t) = \frac{ds}{dt} = -32t$$

(b) The velocity when $t = 3$ sec is

$$v(3) = -32(3) = -96 \text{ ft/sec}$$

The negative sign for the velocity indicates that the object is falling; a positive sign would indicate that the object is rising. ■

The velocity $v(t)$ is rarely constant. The rate of change of the velocity is called the **acceleration** $a(t)$ and is defined as

$$a(t) = v'(t) = \frac{dv}{dt} \tag{1}$$

For Example 6, the acceleration $a(t)$ is

$$a(t) = \frac{dv}{dt} = -32 \text{ ft/sec}^2$$

Time Rate of Change

If the variable t represents the time, the *time rate of change* of the function $y = f(t)$ with respect to t is described by the derivative of y with respect to t:

$$\text{Time rate of change} = \frac{dy}{dt} = f'(t)$$

For example, the velocity is the time rate of change of distance; acceleration is the time rate of change of velocity.

Example 7 Annual profits of a new computer software company are given by the equation

$$P(t) = -1 + 0.5t - 0.01t^2$$

where t represents the time (years) since the company began and $P(t)$ represents annual profits in millions of dollars. Find the rate at which the company's profits are changing when $t = 5$.

Solution The rate of change when $t = 5$ is given by $P'(5)$. First, we find $P'(t)$:

$$P'(t) = 0.5 - 0.02t$$

Evaluating $P'(t)$ when $t = 5$ yields

$$P'(5) = 0.5 - 0.02(5) = \$0.4 \text{ million/year}$$
$$= \$400{,}000\text{/year}$$

The fact that $P'(5)$ is a positive number means that the company's profits are increasing at a rate of \$400,000 per year when $t = 5$; a negative value for $P'(t)$ would mean that the company's profits are decreasing.

Derivation of the Power Rule

If the function to be differentiated has the form $f(x) = x^n$ where n is a positive integer, its derivative $f'(x)$ is

$$f'(x) = nx^{n-1}$$

This result can be demonstrated by applying the definition

$$f'(x) = \lim_{h\to 0} \frac{f(x+h) - f(x)}{h}$$

to the function $f(x) = x^n$. We get

$$f'(x) = \lim_{h\to 0} \frac{(x+h)^n - x^n}{h} \tag{2}$$

According to the binomial theorem, the expression $(x + h)^n$ can be expanded to give

$$(x+h)^n = x^n + nx^{n-1}h + \frac{n(n-1)}{2}x^{n-2}h^2 + \cdots + nxh^{n-1} + h^n$$

When the expanded version is substituted into Equation 2, we get

$$\begin{aligned} f'(x) &= \lim_{h\to 0} \frac{x^n + nx^{n-1}h + [n(n-1)/2]x^{n-2}h^2 + \cdots + h^n - x^n}{h} \\ &= \lim_{h\to 0}\left[nx^{n-1} + \frac{n(n-1)}{2}x^{n-2}h + \cdots + h^{n-1}\right] \end{aligned} \tag{3}$$

Since every term on the right-hand side of Equation 3 except the first contains h as a factor, the only term that survives as $h \to 0$ is the first, yielding

$$f'(x) = nx^{n-1}$$

3.1 EXERCISES

Find the derivative of each of the following functions.

1. $f(x) = 7$
2. $f(x) = x^2 + x + 1$
3. $f(x) = x^4 - 3x^2 + 5$
4. $f(x) = 6x^5 - 8x^4 + 7$
5. $f(x) = \sqrt{8}$
6. $f(x) = x^7 + 6x^4 - 9x^3 - 1$
7. $f(x) = \dfrac{5}{x^2}$
8. $f(x) = \dfrac{-2}{x^4}$
9. $f(x) = \dfrac{4}{3x^2}$
10. $f(x) = 2\sqrt{x} + 1$
11. $F(x) = 7x^3 - \dfrac{4}{x^3}$
12. $g(x) = \dfrac{4}{\sqrt{x}}$
13. $f(x) = \sqrt[3]{x^4}$
14. $F(x) = \sqrt{3x}$
15. $f(x) = 5x - \dfrac{4}{x^2}$
16. $g(x) = 6\sqrt{x} + \dfrac{2}{x} + 1$

17. $f(x) = \dfrac{x^3 - x^2 + x - 1}{x}$

18. $g(x) = \dfrac{x^4 + x^2 + 1}{x^2}$

19. $F(x) = \dfrac{\sqrt{x^3} + \sqrt{x}}{x}$

20. $f(x) = \dfrac{8}{x} - \dfrac{2}{\sqrt{x^3}}$

21. $f(x) = \dfrac{x^{n+1}}{n + 1} - \dfrac{x^n}{n}$

22. $g(x) = n\sqrt[n]{x}$

Find the slope and an equation of the line tangent to each of the following curves at the indicated points.

23. $f(x) = x^3 - 6x + 1$ at $(-1, 6)$

24. $g(x) = x^2 - 4x + 2$ at $(2, -2)$

25. $f(x) = x^4 + x^3 + x^2$ at $(1, 3)$

26. $F(x) = 4\sqrt{x} - x$ at $(1, 3)$

27. $f(x) = x - \dfrac{2}{x}$ at $(-2, -1)$

28. $F(x) = 3\sqrt[3]{x} - \dfrac{8}{x}$ at $(-1, 5)$

29. $f(x) = \sqrt{x^3} + \dfrac{4}{x}$ at $(4, 9)$

30. $f(x) = x^2 - \dfrac{1}{x^2}$ at $(1, 0)$

On each of the following curves, find the points where the slope of the tangent line has the value given.

31. $f(x) = x^3 + 3x^2 - x + 1,\quad m = -1$

32. $f(x) = 7x^2 - 8x + 3,\quad m = 6$

33. $F(x) = x^3 + 3x^2 + 4x + 5,\quad m = 1$

34. $F(x) = x + \dfrac{1}{x},\quad m = 0$

35. $g(x) = x - \dfrac{1}{x},\quad m = 2$

36. $f(x) = 2\sqrt{x} - \dfrac{x}{3} + 1,\quad m = 2/3$

37. $f(x) = 2\sqrt{x} + \dfrac{2}{\sqrt{x}},\quad m = 0$

38. $f(x) = \dfrac{x^n}{n} + \dfrac{x^{n-1}}{n - 1},\quad m = 0$

39. The graphs of the function $y = f(x)$ and tangent line at $x = 3$ are shown in Figure 3. If $F(x) = 2f(x)$, find $F(3)$ and $F'(3)$.

40. The graphs of the function $y = f(x)$ and tangent line at $x = -1$ are shown in Figure 4. If $g(x) = 3f(x)$, find $g(-1)$ and $g'(-1)$.

41. The graphs of the function $y = f(x)$ and tangent line at $x = 1$ are shown in Figure 5. If $F(x) = x^2 + f(x)$, find $F(1)$ and $F'(1)$.

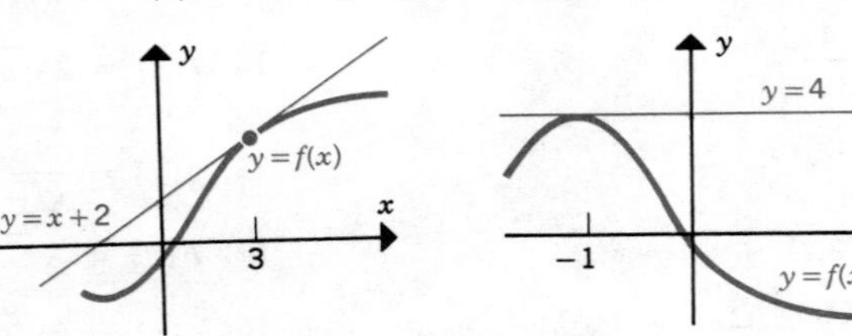

Figure 3

Figure 4

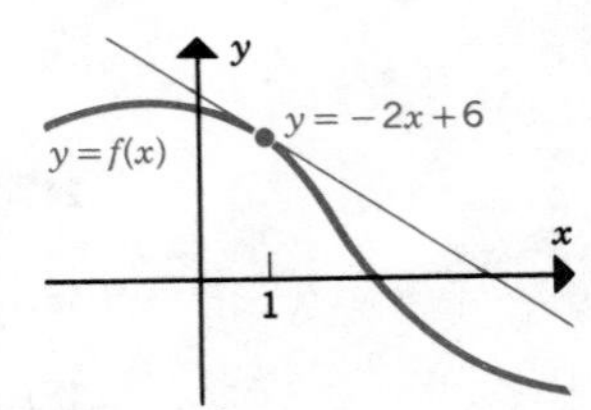

Figure 5

42. The graphs of the functions $y = f(x)$ and $y = g(x)$ together with the tangent lines at $x = 1$ are shown in Figure 6. If $F(x) = f(x) + g(x)$, find $F(1)$ and $F'(1)$.

43. The graphs of the functions $y = f(x)$ and $y = g(x)$ together with the tangent lines at $x = 2$ are shown in Figure 7. If $F(x) = f(x) - g(x)$, find $F(2)$ and $F'(2)$.

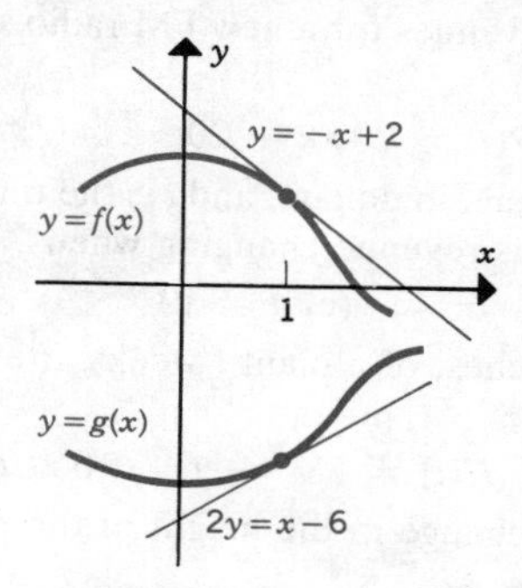

Figure 6

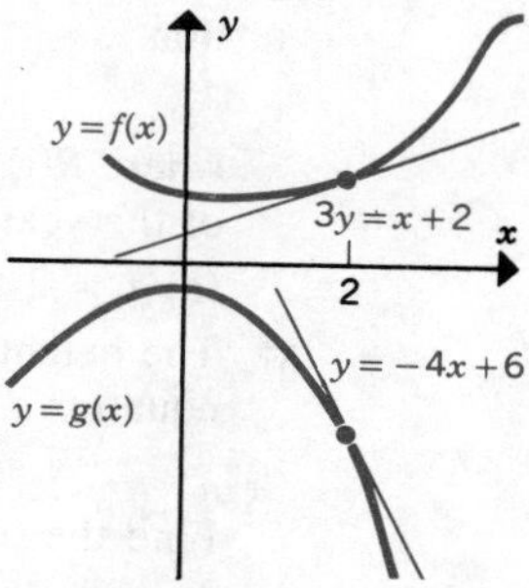

Figure 7

44. The equation $y + 2x = 3$ is an equation of the line tangent to the curve $y = f(x)$ at $x = -1$. If $F(x) = x^3 + f(x)$, find $F(-1)$ and $F'(-1)$.

45. The equation $y = x + 4$ is an equation of the line tangent to the curve $y = f(x)$ at $x = 2$. If $F(x) = 3x - f(x)$, find $F(2)$ and $F'(2)$.

Find $v(t)$, the velocity, and $a(t)$, the acceleration, in Exercises 46 through 52, where $s(t)$ is the distance, in feet, of an object from a reference point and t is the time, in seconds. In addition, evaluate the velocity and acceleration at the times indicated.

46. $s(t) = -16t^2 + 720t$, $\quad t = 10, 20, 30$

47. $s(t) = t^2 - 2t + 30$, $\quad t = 0, 1, 2$

48. $s(t) = t^3 - 12t^2 + 9t$, $\quad t = 0, 1, 2$

49. $s(t) = 100 + 320t - 16t^2$, $\quad t = 5, 10, 20$

50. $s(t) = \sqrt{t^5}$, $\quad t = 1, 4, 16$

51. $s(t) = 24t^2 - t^3$, $\quad t = 4, 8, 12$

52. $s(t) = 20t + 40$, $\quad t = 5, 10, 15$

53. A small rocket is launched from ground level. The height $h(t)$, in feet, of the rocket above the ground is given by the equation $h(t) = 960t - 16t^2$, where t is the time, in seconds, from the moment of launch ($t = 0$).

(a) Find the velocity and acceleration functions.
(b) Determine how long it takes for the rocket to reach its maximum height.
(c) How long is the rocket airborne?

54. A soft-drink company begins a ten-week campaign for its new caffeine-free diet cola. Weekly sales $S(t)$ of the cola, in thousands of dollars, t weeks after the campaign begins are described by the equation

$$S(t) = 100 + 40t - t^2, \qquad 0 \le t \le 30$$

Find the time rate of change in weekly sales when

(a) $t = 5$ (b) $t = 20$ (c) $t = 25$

55. The size of a bacteria culture, $N(t)$, as a function of t, the time in hours, is given by the equation

$$N(t) = 1000 + 5t + 2t^{3/2}$$

How fast is the culture growing when $t = 4$?

56. Monthly advertising revenues for a new FM radio station are described by the equation

$$R(t) = 2000t + 20t^2, \qquad 0 \leq t \leq 12$$

where $R(t)$ is the revenue, in dollars, and t is the time, in months, from the beginning of the year. How fast is revenue changing when
(a) $t = 2$ (b) $t = 6$ (c) $t = 10$

57. The height $H(t)$, in inches, of a plant t weeks after it breaks ground is given by the equation

$$H(t) = 8\sqrt{t} - t, \qquad 0 \leq t \leq 16$$

Find the time rate of change in the height of the plant when
(a) $t = 1$ (b) $t = 4$

58. The supply of blood $S(t)$, in pints, at a local hospital is given by the equation

$$S(t) = 24 - 8t + t^2, \quad 0 \leq t \leq 10$$

where t is the time, in days, from today ($t = 0$). Find the time rate of change in the hospital's blood supply when
(a) $t = 0$ (b) $t = 2$ (c) $t = 4$ (d) $t = 7$

3.2 PRODUCT AND QUOTIENT RULES

Product Rule

Often, a function is written as the product of two other functions; for example,

$$f(x) = (x^6 + 3x^2 - 9)(4x^5 - 8x^3 + 6x)$$

can be expressed as the product

$$f(x) = u(x) \cdot v(x)$$

where $u(x) = x^6 + 3x^2 - 9$ and $v(x) = 4x^5 - 8x^3 + 6x$. In such cases, the derivative can be found using the **product rule.**

Rule 5. Product Rule

If $y = f(x) = u(x)v(x)$, then

$$f'(x) = u(x)v'(x) + v(x)u'(x)$$

or

$$\frac{dy}{dx} = u\frac{dv}{dx} + v\frac{du}{dx}$$

In words, the rule says the following: **The derivative of a product of two functions equals the first function times the derivative of the second plus the second function times the derivative of the first.** A derivation of the product rule is given at the end of this section.

Example 1 Use the product rule to find the derivative of the function

$$f(x) = (x^3)(x^4)$$

Solution Letting

$$u(x) = x^3 \qquad v(x) = x^4$$

$u'(x)$ and $v'(x)$ are

$$u'(x) = 3x^2 \qquad v'(x) = 4x^3$$

According to the product rule, the expressions connected by the arrows are multiplied and the results added to give

$$f'(x) = (x^3)(4x^3) + (x^4)(3x^2) = 7x^6$$

The same result is obtained if we multiply before differentiating; writing

$$f(x) = (x^3)(x^4) = x^7$$

and applying the power rule gives

$$f'(x) = 7x^6$$

■

Caution *The derivative of a product does not equal the product of the derivatives.* If $f(x) = u(x)v(x)$, then

$$f'(x) \neq u'(x)v'(x)$$

In Example 1, where $f(x) = (x^3)(x^4)$,

$$f'(x) \neq (3x^2)(4x^3) = 12x^5$$

Example 2 Find the derivative of the function

$$f(x) = (x^2 + 5x - 3)(x^3 - 2x^2 + 7x)$$

Solution Let

$$u(x) = x^2 + 5x - 3 \qquad v(x) = x^3 - 2x^2 + 7x$$

Next differentiate each function

$$u'(x) = 2x + 5 \qquad v'(x) = 3x^2 - 4x + 7$$

Multiplying the expressions connected by the arrows and then adding gives

$$f'(x) = (x^2 + 5x - 3)(3x^2 - 4x + 7) + (x^3 - 2x^2 + 7x)(2x + 5)$$

After simplifying, the derivative becomes

$$f'(x) = 5x^4 + 12x^3 - 18x^2 + 82x - 21$$

■

In Example 2, you may have observed that it would have been just as easy to multiply the two polynomials first and then differentiate term by term. Doing this is not only an effective way to check your work but should also make you aware that expressions can be modified to take advantage of other derivative rules. On the other hand, at this stage, we do not advise you to multiply first and differentiate term by term simply to avoid using the product rule. You will encounter problems for which the product rule will have to be used; therefore, it is in your best interest to master it while the functions to be differentiated are easy to handle.

Example 3 Find the derivative of the function

$$f(x) = \sqrt{x}(2 + x^2)$$

Solution The product rule can be used with

$$u(x) = \sqrt{x} \qquad v(x) = 2 + x^2$$

$$u'(x) = \frac{1}{2\sqrt{x}} \qquad v'(x) = 2x$$

After combining, the product rule yields

$$f'(x) = (\sqrt{x})(2x) + (2 + x^2)\left(\frac{1}{2\sqrt{x}}\right) = 2x\sqrt{x} + \frac{2 + x^2}{2\sqrt{x}}$$

The derivative $f'(x)$ can also be written as one term:

$$f'(x) = \frac{(2x\sqrt{x})(2\sqrt{x})}{2\sqrt{x}} + \frac{2 + x^2}{2\sqrt{x}} = \frac{4x^2 + 2 + x^2}{2\sqrt{x}}$$

$$= \frac{5x^2 + 2}{2\sqrt{x}}$$

If the denominator is rationalized, $f'(x)$ can also be written as

$$f'(x) = \frac{5x^2 + 2}{2\sqrt{x}} \cdot \frac{\sqrt{x}}{\sqrt{x}} = \frac{(5x^2 + 2)\sqrt{x}}{2x}$$

■

Example 3 illustrates a common occurrence: ***f′(x)* can often be written in many equivalent forms.** Keep this in mind when checking your results. The way in which $f'(x)$ is expressed depends on the way it is to be used. If it is not stated how $f'(x)$ is to be used, then it can be written in more than one acceptable form.

Quotient Rule

In many instances a function is written as the ratio of two functions; for example,

$$f(x) = \frac{x^2}{x + 1}$$

can be written as the ratio

$$f(x) = \frac{u(x)}{v(x)}$$

where $u(x) = x^2$ and $v(x) = x + 1$. In such instances, $f'(x)$ can be found by using the **quotient rule.**

Rule 6. Quotient Rule

If $y = f(x) = \dfrac{u(x)}{v(x)}$, then

$$f'(x) = \frac{v(x)u'(x) - u(x)v'(x)}{[v(x)]^2}$$

or

$$\frac{d}{dx}\left(\frac{u}{v}\right) = \frac{v(du/dx) - u(dv/dx)}{[v(x)]^2}$$

The quotient rule can also be expressed as

$$\text{Derivative of a quotient} = \frac{(\text{denominator})(\text{numerator})' - (\text{numerator})(\text{denominator})'}{(\text{denominator})^2}$$

Notice the minus sign in the numerator; this means the order of the terms is important. A derivation of the quotient rule is given at the end of this section.

Example 4 Use the quotient rule to find the derivative of the function

$$f(x) = \frac{x^7}{x^4}$$

Solution Let

$$u(x) = x^7 \qquad v(x) = x^4$$
$$u'(x) = 7x^6 \qquad v'(x) = 4x^3$$

Applying the quotient rule gives

$$f'(x) = \frac{(x^4)(7x^6) - (x^7)(4x^3)}{(x^4)^2} = \frac{7x^{10} - 4x^{10}}{x^8}$$

$$= \frac{3x^{10}}{x^8} = 3x^2, \qquad x \neq 0$$

The same result is obtained if we divide before differentiating:

$$f(x) = \frac{x^7}{x^4} = x^3, \qquad x \neq 0$$

Applying the power rule gives

$$f'(x) = 3x^2, \qquad x \neq 0$$ ■

Caution *The derivative of a quotient does not equal the quotient of the derivatives.* If $f(x) = \dfrac{u(x)}{v(x)}$, then

$$f'(x) \neq \frac{u'(x)}{v'(x)}$$

In Example 4, where $f(x) = \dfrac{x^7}{x^4}$,

$$f'(x) \neq \frac{7x^6}{4x^3} = \frac{7x^3}{4}$$

Example 5 Find the derivative of the function

$$f(x) = \frac{x}{2x + 1}$$

Solution Because $f(x)$ is written as the ratio of two functions, the quotient rule is applied. We let

$$\begin{aligned} u(x) &= x & v(x) &= 2x + 1 \\ u'(x) &= 1 & v'(x) &= 2 \end{aligned}$$

Applying the quotient rule gives

$$f'(x) = \frac{(2x + 1)(1) - (x)(2)}{(2x + 1)^2} = \frac{1}{(2x + 1)^2}$$ ■

Application

The revenue that a company earns from the sale of a product can be written as

$$\text{Revenue} = p \cdot q$$

where p is the unit price of the product and q is the quantity or number of units sold. If both p and q change with t, the time, the revenue $R(t)$ can be written

$$R(t) = p(t) \cdot q(t)$$

The derivative $R'(t)$ describes the rate at which revenue is changing with time. Applying the product rule, we obtain the following result for $R'(t)$:

$$R'(t) = p(t)q'(t) + q(t)p'(t)$$

The first term describes the rate at which $R(t)$ is changing as the number of units sold changes; the second term describes the rate at which $R(t)$ is changing as the unit price changes.

Example 6 Over the next five years, $p(t)$, the price of a barrel of oil, is expected to vary according to the equation

$$p(t) = 20 - 2t + 0.5t^2$$

and $q(t)$, the number of barrels of oil sold annually by the Midget Oil Company, is expected to follow the equation

$$q(t) = 2000 + 60t - 10t^2$$

What is the rate at which $R(t)$ is changing

(a) when $t = 1$? (b) when $t = 4$?

Solution (a) The derivative $R'(t)$ can be found by using the product rule:

$$p(t) = 20 - 2t + 0.5t^2, \qquad q(t) = 2000 + 60t - 10t^2$$
$$p'(t) = -2 + t, \qquad q'(t) = 60 - 20t$$

Now $R'(t)$ can be written as

$$R'(t) = (20 - 2t + 0.5t^2)(60 - 20t) + (2000 + 60t - 10t^2)(-2 + t)$$

The expression for $R'(t)$ will not be simplified in order to see, separately, the effects on $R'(t)$ caused by changing prices and sales.

(a) When $t = 1$,

$$\begin{aligned} R'(1) &= (18.5)(40) + (2050)(-1) \\ &= 740 - 2050 \\ &= -\$1310/\text{year} \end{aligned}$$

The negative sign indicates that revenue is decreasing at the rate of $1310 per year; it tells us that the effect of falling prices ($q(1)p'(1) = -2050$) is greater than the effect of rising sales ($p(1)q'(1) = 740$).

(b) When $t = 4$,

$$\begin{aligned} R'(4) &= (20)(-20) + (2080)(2) \\ &= -400 + 4160 \\ &= \$3760/\text{year} \end{aligned}$$

The positive sign indicates that revenue is increasing at the rate of $3760 per year when $t = 4$; the effect of rising prices (4160) is greater than the effect of falling sales (−740).

Derivation of the Product Rule

If the function to be differentiated has the form

$$f(x) = u(x)v(x)$$

the first derivative $f'(x)$ is

$$f'(x) = u(x)v'(x) + v(x)u'(x)$$

This result can be derived from the definition. Noting that

$$f(x + h) = u(x + h)v(x + h)$$

we can write $f'(x)$ as

$$f'(x) = \lim_{h \to 0} \frac{u(x + h)v(x + h) - u(x)v(x)}{h}$$

The next step in the development is one that initially has an aura of "black magic" about it; however, it does enable us to find the limit. The expression $u(x + h)v(x)$ is added to and subtracted from the numerator in the following manner:

$$f'(x) = \lim_{h \to 0} \frac{u(x + h)v(x + h) - u(x + h)v(x) + u(x + h)v(x) - u(x)v(x)}{h}$$

Next, the right-hand side is rewritten as

$$\begin{aligned} f'(x) &= \lim_{h \to 0} \left[u(x + h)\frac{v(x + h) - v(x)}{h} + v(x)\frac{u(x + h) - u(x)}{h} \right] \\ &= \lim_{h \to 0} \left[u(x + h)\frac{v(x + h) - v(x)}{h} \right] + \lim_{h \to 0} \left[v(x)\frac{u(x + h) - u(x)}{h} \right] \end{aligned}$$

Since

$$\lim_{h \to 0} u(x + h) = u(x), \qquad \lim_{h \to 0} \frac{v(x + h) - v(x)}{h} = v'(x)$$

$$\lim_{h \to 0} \frac{u(x + h) - u(x)}{h} = u'(x)$$

it is possible to write

$$\lim_{h \to 0} \left[u(x + h)\frac{v(x + h) - v(x)}{h} \right] = \lim_{h \to 0} u(x + h) \lim_{h \to 0} \frac{v(x + h) - v(x)}{h} = u(x)v'(x)$$

$$\lim_{h \to 0} \left[v(x)\frac{u(x + h) - u(x)}{h} \right] = \lim_{h \to 0} v(x) \lim_{h \to 0} \frac{u(x + h) - u(x)}{h} = v(x)u'(x)$$

from which we get

$$f'(x) = u(x)v'(x) + v(x)u'(x)$$

Derivation of the Quotient Rule

If the function to be differentiated has the form

$$f(x) = \frac{u(x)}{v(x)}$$

its first derivative can be written as

$$f'(x) = \frac{v(x)u'(x) - u(x)v'(x)}{[v(x)]^2}$$

Like the product rule, an operation that looks initially like a "sleight of hand" is needed to generate $f'(x)$. Working from the definition, we get

$$f'(x) = \lim_{h \to 0} \frac{1}{h}\left[\frac{u(x+h)}{v(x+h)} - \frac{u(x)}{v(x)}\right]$$

$$= \lim_{h \to 0} \frac{v(x)u(x+h) - u(x)v(x+h)}{hv(x)v(x+h)}$$

Next, the expression $v(x)u(x)$ is added to and subtracted from the numerator:

$$f'(x) = \lim_{h \to 0} \frac{v(x)u(x+h) - v(x)u(x) + v(x)u(x) - u(x)v(x+h)}{hv(x)v(x+h)}$$

Proceeding as we did with the product rule, we can write

$$f'(x) = \lim_{h \to 0} \frac{\left[v(x) \cdot \dfrac{u(x+h) - u(x)}{h}\right] - \left[u(x) \cdot \dfrac{v(x+h) - v(x)}{h}\right]}{v(x) \cdot v(x+h)}$$

$$= \frac{v(x) \cdot \lim\limits_{h \to 0} \dfrac{u(x+h) - u(x)}{h} - u(x) \cdot \lim\limits_{h \to 0} \dfrac{v(x+h) - v(x)}{h}}{v(x) \cdot \lim\limits_{h \to 0} v(x+h)}$$

$$= \frac{v(x)u'(x) - u(x)v'(x)}{[v(x)]^2}$$

3.2 EXERCISES

Find the derivative of each of the following functions.

1. $f(x) = (x + 2)(3x + 4)$

2. $f(x) = (x^2 + 1)(x^2 - 1)$

3. $F(x) = (x^2 + 2x - 1)(2x + 5)$
4. $f(x) = (x + 3)(x^3 + x^2 + 1)$
5. $f(x) = (x^4 + 2x^2 + 3)(x - 1)$
6. $f(x) = (8x^2 - 3x + 6)(5x + 4)$
7. $g(x) = \sqrt{x}(x + 1)$
8. $f(x) = \frac{1}{x}(x^2 + 1)$
9. $F(x) = (\sqrt{x} + 1)(\sqrt{x} - 1)$
10. $f(x) = \left(\frac{1}{x} + 2\right)\left(\frac{1}{x} - 2\right)$
11. $f(x) = \frac{x^3 + x^2}{x}$
12. $g(x) = \frac{x}{x + 1}$
13. $f(x) = \frac{x + 2}{x - 1}$
14. $f(x) = \frac{x^2}{x^2 + 3}$
15. $f(x) = \frac{x^2 - 1}{x + 2}$
16. $F(x) = \frac{\sqrt{x}}{x + 1}$
17. $f(x) = x^n(x^n + 1)$
18. $f(x) = \frac{x^n}{x + 1}$
19. $F(x) = \frac{x}{\sqrt{x} - 1}$
20. $f(x) = \frac{(x + 1)(x - 1)}{(x + 2)}$

Find the slope and an equation of the line tangent to each of the following curves at the indicated point.

21. $f(x) = (x^3 + x - 1)(x + 2)$ at $(1, 3)$
22. $f(x) = (x^2 - 2x + 3)(x^2 - x + 2)$ at $(0, 6)$
23. $f(x) = \left(\frac{2}{x} + 1\right)(x - 2)$ at $(1, -3)$
24. $F(x) = \sqrt{x}\,(x^2 + 1)$ at $(1, 2)$
25. $f(x) = \frac{x}{x + 2}$ at $(0, 0)$
26. $f(x) = \frac{x + 5}{x + 1}$ at $(3, 2)$
27. $g(x) = \frac{x^2 + 2}{x^3 - 1}$ at $(0, -2)$
28. $f(x) = \frac{\sqrt{x}}{x - 3}$ at $(4, 2)$

Find the points on each of the following curves where the slope of the tangent line has the given value.

29. $f(x) = (x + 2)(x - 1)$, $m = 5$
30. $f(x) = (x^2 + 1)(x - 2)$, $m = 5$
31. $F(x) = x^2(3x - 4)$, $m = +1$
32. $f(x) = \frac{x^2}{x - 1}$, $m = 0$
33. $f(x) = \frac{x - 2}{x - 3}$, $m = -1$
34. $f(x) = \frac{x^2 + 1}{x - 1}$, $m = \frac{1}{2}$

35. $g(x) = 2\sqrt{x}(x + 3), \qquad m = 6$

36. $f(x) = \dfrac{2\sqrt{x}}{x + 1}, \qquad m = 0$

37. The graphs of the function $y = f(x)$ and tangent line at $x = 3$ are shown in Figure 1. If $F(x) = xf(x)$, find $F(3)$ and $F'(3)$.

38. The graphs of the function $y = f(x)$ and tangent line at $x = -1$ are shown in Figure 2. If $g(x) = \dfrac{f(x)}{x}$, find $g(-1)$ and $g'(-1)$.

39. The graphs of the function $y = f(x)$ and tangent line at $x = 1$ are shown in Figure 3. If $F(x) = (x^2 + 2)f(x)$, find $F(1)$ and $F'(1)$.

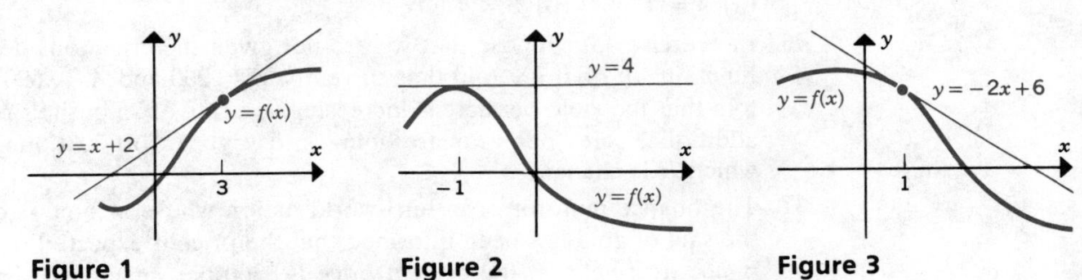

Figure 1 Figure 2 Figure 3

40. The graphs of the functions $y = f(x)$ and $y = g(x)$ together with the tangent lines at $x = 1$ are shown in Figure 4. If $F(x) = f(x) \cdot g(x)$, find $F(1)$ and $F'(1)$.

41. The graphs of the functions $y = f(x)$ and $y = g(x)$ together with the tangent lines at $x = 2$ are shown in Figure 5. If $F(x) = \dfrac{f(x)}{g(x)}$, find $F(2)$ and $F'(2)$.

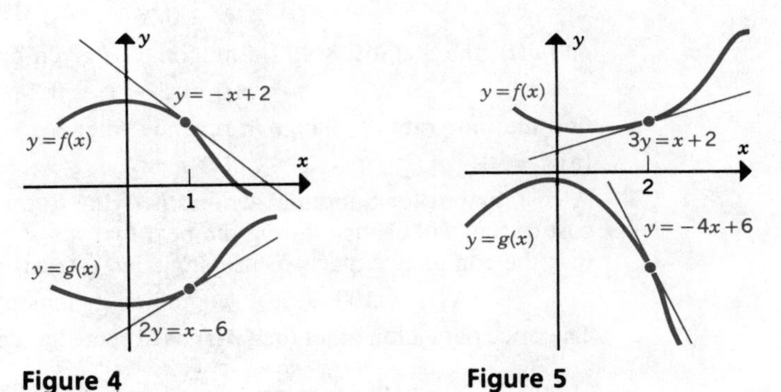

Figure 4 Figure 5

42. The equation $y + 2x = 1$ is an equation of the line tangent to the curve $y = f(x)$ at $x = 3$. If $F(x) = x^2f(x)$, find $F(3)$ and $F'(3)$.

43. The equation $y = 3x + 5$ is an equation of the line tangent to the curve $y = f(x)$ at $x = -2$. If $F(x) = \dfrac{f(x)}{x}$, find $F(-2)$ and $F'(-2)$.

44. The concentration of a drug in the bloodstream, $C(t)$, as a function of t, the time in hours after injection, is described by the equation

$$C(t) = \frac{36}{t^2 + 2}, \qquad t \geq 0$$

where the units of $C(t)$ are parts per million (ppm). How fast is the concentration changing when

(a) $t = 1$? (b) $t = 4$?

45. The owners of a large farm have decided to begin converting the land to soybean production. The annual soybean harvest, H, in bushels, is a product of Y, the yield per acre (bushels per acre), times A, the number of acres of soybeans planted. If Y and A as functions of t, the time in years from the beginning of the conversion, are described by the equations

$$Y = 100 + 10\sqrt{t}, \qquad A = 200 + 100t$$

find the rate at which H is changing when

(a) $t = 1$ (b) $t = 4$

46. In exercise 45, suppose that we are not given the equations describing Y and A as functions of t but are told that currently $Y = 200$ and $A = 600$; in addition, we are told that the yield per acre is increasing at a rate of 25 bushels per year and that 100 additional acres per year are going into soybean production. What is the rate at which H is changing?

47. The finance minister of a third-world nation whose income is derived mainly from the sale of gold has been informed that the price is expected to decrease by \$20 per ounce next year. If the current price is \$300 per ounce and the nation sells 12,000 ounces per year, by how much will production have to increase so that the time rate of change in annual revenue will equal zero? The country is heavily in debt and cannot afford to have its revenue from the sale of gold decrease.

48. Because of severe foreign competition, the price of computer chips is expected to fall during the next several years; at the same time, demand is expected to increase. If $p(t)$, the unit price of a chip in dollars, is given by the equation

$$p(t) = 2 - 0.25t^2, \qquad 0 \le t \le 4$$

and $q(t)$, the quantity sold (in millions), is given by the equation

$$q(t) = 5 + t + 0.2t^2$$

find the time rate of change in revenue when

(a) $t = 0$ (b) $t = 1$ (c) $t = 4$

49. A cost analyst for Supreme Airlines is trying to estimate the rate at which the total cost of fuel will change during the next five years. The number of gallons of jet fuel that the company expects to use annually, $N(t)$, is given by the equation

$$N(t) = 100 + 20t, \qquad \text{million gallons per year } 0 \le t \le 5$$

The price per gallon of jet fuel, $p(t)$, is expected to decrease according to the equation

$$p(t) = \frac{50}{50 + t^2}, \qquad \text{dollars per gallon } 0 \le t \le 5$$

Find the rate at which the company's annual fuel costs are changing when

(a) $t = 0$ (b) $t = 2$ (c) $t = 4$

50. The number of people $N(t)$ who vote at any time t is equal to the number of people eligible to vote $E(t)$ multiplied by $\frac{p(t)}{100}$, where $p(t)$ is the percentage of the eligibles who vote. Find the time rate of change of $N(t)$ when the number of eligibles equals 20,000 and is increasing at a rate of 500 voters per year, and when the percentage who vote is 60 percent and is decreasing at a rate of 2 percent per year.

3.3 GENERAL POWER RULE: CHAIN RULE

General Power Rule

The general power rule is used to find the derivative of functions having the form

$$f(x) = [u(x)]^n \tag{1}$$

where $u(x)$ is a function of x and n is a real number. The right-hand side of Equation 1 is a power expression whose base is a function of x. Examples of functions of this type are

1. $f(x) = (x^2 + 3x - 1)^4$, where $u(x) = x^2 + 3x - 1$ and $n = 4$
2. $f(x) = \sqrt{5 - x} = (5 - x)^{1/2}$, where $u = 5 - x$ and $n = \frac{1}{2}$

To develop a feeling for the structure of the general power rule, suppose that we attempt to find the derivative of the function

$$f(x) = [u(x)]^2$$

The right-hand side can be written as $u(x) \cdot u(x)$; applying the product rule gives

$$f'(x) = u(x)u'(x) + u(x)u'(x) = 2u(x) \cdot u'(x) \tag{2}$$

Except for the additional factor $u'(x)$, this result looks like something you would expect from the power rule (Section 3.1). The additional factor is the basic difference between the two rules.

Rule 7. General Power Rule

If $f(x) = [u(x)]^n$, where n is a real number,

$$f'(x) = n[u(x)]^{n-1}u'(x)$$

or

$$\frac{dy}{dx} = n[u(x)]^{n-1}\frac{du}{dx}$$

Note: When $u(x) = x$, the general power and the power rules give the same result for the derivative, that is, $f'(x) = nx^{n-1}$.

Although a derivation of the general power rule is beyond the scope of this book, a proof of the rule when $n = 3$ and $n = 4$ is left for the exercises.

Example 1 Find the derivative of the function

$$f(x) = (3x^2 + 8x - 1)^4$$

Solution Applying the general power rule with

$$u(x) = 3x^2 + 8x - 1, \qquad n = 4, \qquad u'(x) = 6x + 8$$

we get

$$f'(x) = \underset{\substack{\uparrow \\ n}}{4}\,(\underset{\substack{\uparrow \\ u(x)}}{3x^2 + 8x - 1})^{\overset{}{3}}(6x + 8)$$

(with arrows marking n, $u(x)$, $n-1$, and $u'(x)$)

Simplifying yields

$$f'(x) = (24x + 32)(3x^2 + 8x - 1)^3$$ ■

Example 2 Find the derivative of the function

$$f(x) = \sqrt{3x^4 + 10x} = (3x^4 + 10x)^{1/2}$$

Solution Again the general power rule can be applied. Noting that

$$u(x) = 3x^4 + 10x, \qquad n = \tfrac{1}{2}, \qquad u'(x) = 12x^3 + 10$$

we can write

$$f'(x) = \tfrac{1}{2}(3x^4 + 10x)^{-1/2}(12x^3 + 10)$$

(with arrows marking n, $u(x)$, $n-1$, and $u'(x)$)

Simplifying yields

$$f'(x) = \frac{6x^3 + 5}{\sqrt{3x^4 + 10x}}$$ ■

If two or more rules must be used to find a derivative, it is important that the rules be applied in the correct order. To decide which rule to apply, determine which algebraic operation is carried out *last* in the expression to be differentiated. Table 1 shows the relation between the last operation and the differentiation rule.

Table 1

Last Operation	Differentiation Rule
Addition–subtraction	Sum rule
Multiplication	Product rule
Division	Quotient rule
Raising to a power	Power rule

Although these procedures may appear self-evident, they are helpful in deciding where to begin when finding the derivative of a seemingly complex function. Examples 3 and 4 illustrate functions to which two or three techniques must be applied to find the derivative.

Example 3 Find the derivative of the function

$$f(x) = \sqrt{x^2 + 6x} + 8x$$

Solution The last operation is addition; therefore, the sum rule is applied first:

$$f'(x) = \frac{d}{dx}(x^2 + 6x)^{1/2} + \frac{d}{dx}(8x)$$

In differentiating the first term, we apply the general power rule, which yields

$$\frac{d}{dx}(x^2 + 6x)^{1/2} = \tfrac{1}{2}(x^2 + 6x)^{-1/2}(2x + 6) = \frac{x + 3}{\sqrt{x^2 + 6x}}$$

The derivative of the second term is found by applying the constant factor and simple power rules:

$$\frac{d}{dx}(8x) = 8\frac{d}{dx}(x) = 8$$

Adding the derivatives gives

$$\frac{d}{dx}(\sqrt{x^2 + 6x} + 8x) = \frac{x + 3}{\sqrt{x^2 + 6x}} + 8$$ ■

In order to master the differentiation rules, it is important that you avoid making algebraic mistakes. Errors are bound to occur as you work to become familiar with the rules; detecting and analyzing these errors will be easier if the algebra is carried out flawlessly. There is another reason for exercising care in carrying out the algebraic operations. In many instances successful application of the differentiation rules represents only a small fraction of the time and energy required to reach the final form of the derivative; the bulk of the effort consists of transforming the algebraic expressions to achieve a simple form for the derivative.

Example 4 Find the derivative of the function

$$f(x) = \frac{(x^2 + 1)^5}{x^3 + 7x}$$

Solution Because the last operation is *division*, the *quotient rule* is applied first. Letting

$$u(x) = (x^2 + 1)^5, \qquad v(x) = x^3 + 7x$$

Each of these functions must be differentiated in order to apply the quotient rule. The general power rule must be used to find $u'(x)$ and the sum rule to find $v'(x)$. Carrying out these operations gives

$$u'(x) = 10x(x^2 + 1)^4, \qquad v'(x) = 3x^2 + 7$$

Putting these pieces together gives

$$f'(x) = \frac{(x^3 + 7x)10x(x^2 + 1)^4 - (x^2 + 1)^5(3x^2 + 7)}{(x^3 + 7x)^2}$$

After simplification, we get

$$f'(x) = \frac{(x^2 + 1)^4 (7x^4 + 60x^2 - 7)}{(x^3 + 7x)^2}$$ ■

The general power rule is a special case of a more powerful technique known as the chain rule. The chain rule is used when the function to be differentiated can be written as a **composition** of two functions.

Composition of Two Functions

Often it is useful to create new functions by combining two given functions $g(x)$ and $u(x)$. If the variable x in the expression defining $g(x)$ is replaced by $u(x)$, the resulting function is called the **composite** of $g(x)$ with $u(x)$ and is written $g(u(x))$.

Example 5 (a) If $g(x) = 3x^2$ and $u(x) = x + 1$, find $g(u(x))$.
(b) Find $g(u(1))$.

Solution (a) Since

$$g(\quad) = 3(\quad)^2$$

we can write

$$g(u(x)) = 3(u(x))^2$$

Replacing $u(x)$ on the right-hand side by $x + 1$, we get

$$g(u(x)) = 3(x + 1)^2 = 3x^2 + 6x + 3 \tag{3}$$

Note that the composite function $g(u(x))$ is a new function. The relationship between $g(x)$, $u(x)$, and $g(u(x))$ is illustrated in part b.

(b) Substituting 1 for x everywhere in Equation 3 gives $g(u(1))$.

$$g(u(1)) = 3(1)^2 + 6(1) + 3 = 12$$

This result can also be obtained in a stepwise fashion by using the original functions $u(x)$ and $g(x)$.

1. $u(1) = 1 + 1 = 2$
2. $g(2) = 3(2)^2 = 12$

The mechanism behind the composite function $y = g(u(x))$ is illustrated in Figure 1. Pairings created by the given functions $u(x)$ and $g(x)$ are indicated by the two short segments; the pairing created by the composite function $g(u(x))$ is indicated by the long segment.

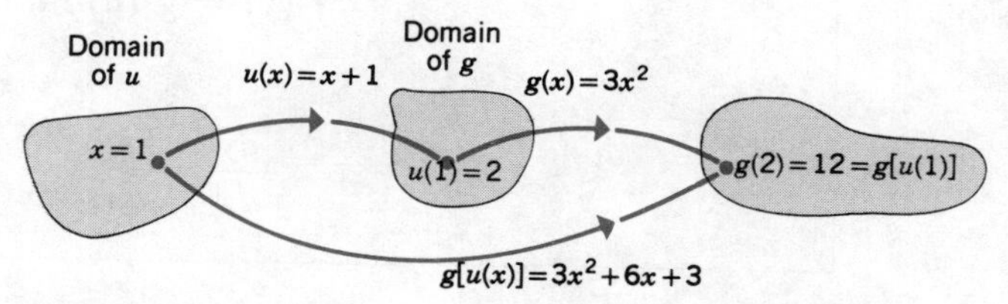

Figure 1

■

Example 6 (a) If $g(x) = \sqrt{x + 2}$ and $u(x) = x^3 - 1$, find $g(u(x))$.
(b) Evaluate $g(u(2))$.

Solution (a) Since

$$g(\quad) = \sqrt{(\quad) + 2}$$

we can write

$$g(u(x)) = \sqrt{u(x) + 2} = \sqrt{x^3 - 1 + 2}$$
$$= \sqrt{x^3 + 1}$$

(b) $g(u(2)) = \sqrt{(2)^3 + 1} = \sqrt{9} = 3$

The same result can be obtained in stepwise fashion.

1. $u(2) = 2^3 - 1 = 7$
2. $g(7) = \sqrt{7 + 2} = \sqrt{9} = 3$

■

Chain Rule

The chain rule is used to differentiate a function that can be written as a composite function. For example, the function

$$f(x) = (x^2 - 3x + 5)^4$$

can be written as a composite function. If $u(x)$ is defined as

$$u(x) = x^2 - 3x + 5$$

and $g(u)$ as

$$g(u) = u^4$$

then $f(x)$ can be written as the composite function

$$f(x) = g(u(x))$$

The chain rule tells us how to differentiate functions of this kind.

Rule 8. Chain Rule

If $y = f(x) = g(u(x))$, then

$$f'(x) = g'(u) \cdot u'(x)$$

or

$$\frac{dy}{dx} = \frac{dg}{du} \cdot \frac{du}{dx}$$

Example 7 Use the chain rule to find the derivative of the function

$$f(x) = (x^2 - 3x + 5)^4$$

Solution Let $u(x) = x^2 - 3x + 5$ and $g(u) = u^4$. Differentiating gives

$$u'(x) = 2x - 3, \qquad g'(u) = 4u^3$$

According to the chain rule,

$$f'(x) = 4u^3(2x - 3)$$

Substituting $x^2 - 3x + 5$ for u gives

$$\begin{aligned} f'(x) &= 4(x^2 - 3x + 5)^3(2x - 3) \\ &= (8x - 12)(x^2 - 3x + 5)^3 \end{aligned}$$

■

When $g(u) = u^n$, $g'(u) = nu^{n-1}$, and the chain rule yields the general power rule,

$$\begin{aligned} f'(x) &= g'(u)u'(x) \\ &= nu^{n-1}u'(x) \end{aligned}$$

The relationship between $f'(x)$ and the factors $g'(u)$ and $u'(x)$ is illustrated graphically in the next example.

Example 8 (a) Use the chain rule to find the derivative of the function $f(x) = \sqrt{5 - x^2}$.
(b) Use the chain rule to find $f'(1)$.

Solution (a) Let $u(x) = 5 - x^2$ and $g(u) = \sqrt{u} = u^{1/2}$. The derivatives are

$$u'(x) = -2x, \qquad g'(u) = \frac{1}{2\sqrt{u}}$$

According to the chain rule,

$$f'(x) = g'(u) \cdot u'(x) = \frac{1}{2\sqrt{u}}(-2x) = \frac{-x}{\sqrt{u}}$$

Substituting $5 - x^2$ for u gives

$$f'(x) = \frac{-x}{\sqrt{5 - x^2}} \tag{4}$$

(b) Before applying the chain rule to find $f'(1)$, note that $u(1) = 5 - 1 = 4$, $g(4) = \sqrt{4} = 2$. According to the chain rule, $f'(1)$ is a product of the two factors:

$$f'(1) = g'(4) \cdot u'(1)$$

The expressions for $g'(u)$ and $u'(x)$ from part (a) yield

$$g'(4) = \frac{1}{2\sqrt{4}} = \frac{1}{4}, \qquad u'(1) = -2(1) = -2$$

so that

$$f'(1) = 1/4(-2) = -\tfrac{1}{2}$$

The same result is obtained if we substitute 1 for x in Equation 4 in part (a).

$$f'(1) = \frac{-1}{\sqrt{5 - 1}} = \frac{-1}{\sqrt{4}} = -\tfrac{1}{2}$$

Figure 2 illustrates geometrically the relationship between $f'(1)$ and the factors $u'(x = 1)$ and $g'(u = 4)$.

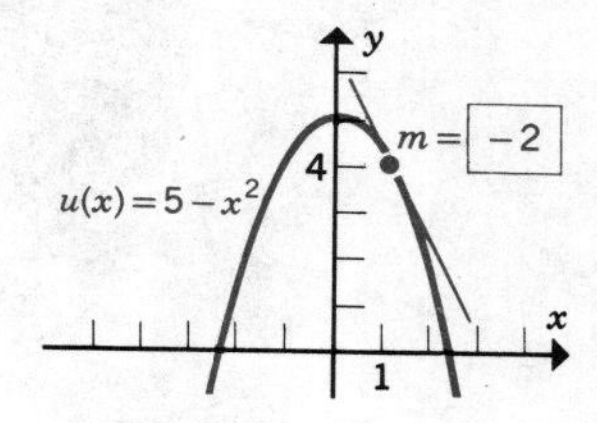

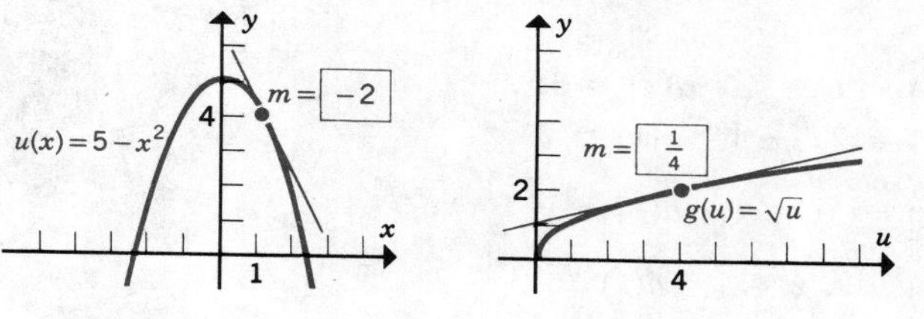

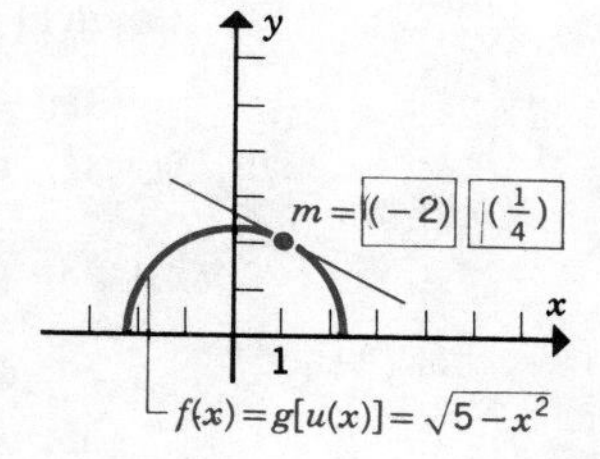

Figure 2

■

3.3 EXERCISES

Find the derivative of each of the following functions.

1. $f(x) = (x^2 + 3)^4$
2. $F(x) = 8(x^3 + 6)^5$
3. $f(x) = 3(x^2 - 7x + 1)^6$
4. $f(x) = 6(x^{10} + 4)^{12}$
5. $f(x) = \left(2 + \dfrac{1}{x}\right)^8$
6. $g(x) = (\sqrt{x} + 1)^3$
7. $F(t) = \sqrt{t^4 + 2t - 1}$
8. $f(x) = \sqrt[3]{6x - 10}$
9. $f(x) = \dfrac{1}{(x^2 + 1)^2}$
10. $f(x) = \dfrac{4}{(3 - 2x - x^2)^3}$
11. $f(x) = \dfrac{2}{\sqrt{x^2 + 1}}$
12. $f(x) = x(x^2 - 2)^3$
13. $F(x) = (x + 2)(x^4 + 5)^6$
14. $f(s) = (x^2 + 1)^4(x^3 + 2)^6$

15. $f(x) = \dfrac{x}{(2x + 1)^2}$

16. $f(x) = \dfrac{6x + 4}{(x - 3)^2}$

17. $f(x) = \left[\dfrac{x + 1}{x - 1}\right]^3$

18. $f(x) = \sqrt{1 + \sqrt{x}}$

19. $g(x) = \sqrt{x}(2x^2 + 3)^4$

20. $f(x) = (\sqrt{x} + 1)^4(\sqrt{x} - 1)^4$

Find the slope and an equation of the tangent line at the indicated point in Exercises 21 through 26.

21. $f(x) = (2 - x)^4$ at $(1, 1)$
22. $f(x) = \sqrt{2x + 4}$ at $(0, 2)$
23. $f(x) = (x^2 - 1)^3$ at $(1, 0)$
24. $f(x) = (x^3 - 4x + 2)^4$ at $(2, 16)$
25. $f(x) = (x^2 - 3)^4(x^2 + 1)$ at $(-2, 5)$
26. $f(x) = \sqrt{1 + \sqrt{x}}$ at $(9, 2)$

In Exercises 27 through 30, find the points on the given curves where the slope of the tangent line has the value given.

27. $f(x) = (x + 2)^3$, $m = 3$
28. $f(x) = (x + 3)^4$, $m = 4$
29. $f(x) = \sqrt{3 + 6x}$, $m = 2$
30. $f(x) = (x^2 - 6x + 8)^4$, $m = 0$

In Exercises 31 through 38, find $g(u(x))$ and $u(g(x))$. In addition, evaluate $g(u(1))$ and $u(g(1))$.

31. $g(x) = x^2 + 1$, $u(x) = 3x$
32. $g(x) = 2x + 4$, $u(x) = 2\sqrt{x}$
33. $g(x) = 3x + 1$, $u(x) = \dfrac{1}{x^2}$
34. $g(x) = \sqrt{x + 1}$, $u(x) = x^2 - 2$
35. $g(x) = x^2 + 2x + 3$, $u(x) = 2x + 4$
36. $g(x) = x^3 + 5$, $u(x) = x + 2$
37. $g(x) = \dfrac{1}{x^2 + 1}$, $u(x) = x - 1$
38. $g(x) = \sqrt{x^2 + 1}$, $u(x) = x^2 - 1$
39. The graphs of the functions $y = u(x)$ and $y = g(x)$ are shown in Figure 3. If $f(x) = g(u(x))$, find $f(2)$; if $F(x) = u(g(x))$, find $F(-2)$.

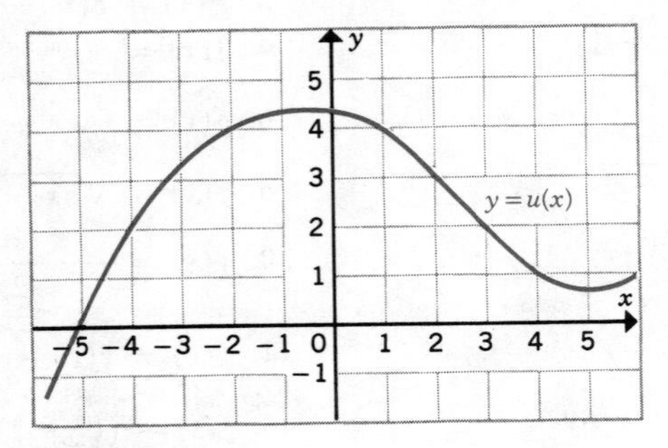

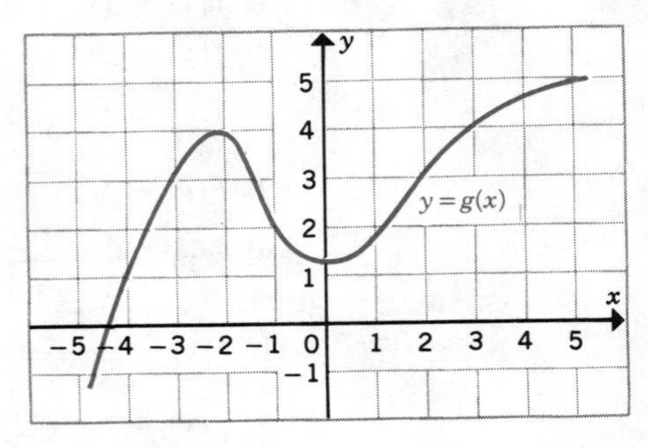

Figure 3

Use the chain rule to find the derivative of each of the following functions.

40. $f(x) = (x^2 + 1)^3$

41. $f(x) = (x^2 - 4x + 5)^6$

42. $f(x) = \sqrt{2x + 3}$

43. $f(x) = \dfrac{1}{x^3 + x^2 + x}$

44. $f(x) = \left(\dfrac{x}{x - 1}\right)^2$

45. $f(x) = \sqrt{1 - 3x^2}$

46. The graphs of the functions $y = u(x)$ and $y = g(x)$ together with the tangent lines at selected points are shown in Figure 4. If $f(x) = g(u(x))$, use the information provided on the graphs to find each of the following.

(a) $u(2)$ (b) $g(3)$ (c) $f(2)$ (d) $u'(2)$ (e) $g'(3)$ (f) $f'(2)$

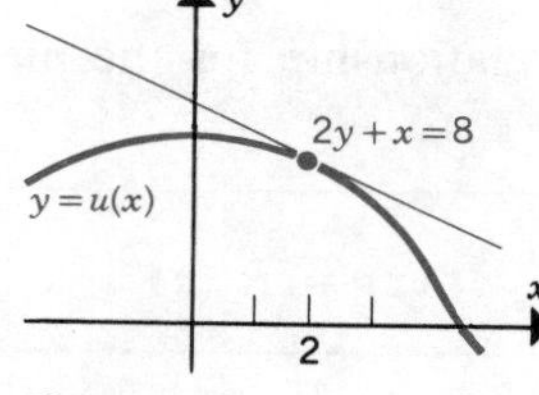

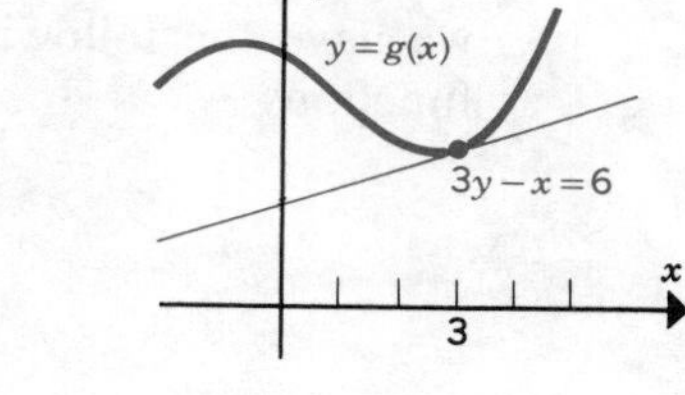

Figure 4

47. If $f(x) = [u(x)]^3$, show that $f'(x) = 3[u(x)]^2u'(x)$. *Hint:* Write $f(x) = [u(x)]^2[u(x)]$ and use the product rule together with the result stated in Equation 2 on page 129.

48. If $f(x) = [u(x)]^4$, show that $f'(x) = 4[u(x)]^3u'(x)$. *Hint:* Write $f(x) = [u(x)]^3[u(x)]$ and use the product rule together with the result stated in exercise 47.

3.4 MARGINAL ANALYSIS

When a company produces or sells a product, the revenue, cost, and ultimately the profit functions will depend on the number of items produced or sold. Managers and accountants are interested not only in determining the value of these functions at a given level of production or sales, but also in determining how sensitive such functions are to changes in production or sales. The word **marginal** is used to describe the rate of change of any of these functions with respect to the number of items produced or sold. If we denote the number of items sold or produced as x, marginal revenue, cost, and profit are defined as follows.

1. Given a revenue function $R(x)$,

$$\boxed{\text{Marginal revenue} = R'(x) = \frac{dR}{dx}} \qquad (1)$$

2. Given a cost function $C(x)$,

$$\text{Marginal cost} = C'(x) = \frac{dC}{dx} \tag{2}$$

3. Given a profit function $P(x)$,

$$\text{Marginal profit} = P'(x) = \frac{dP}{dx} \tag{3}$$

Because $R(x)$, $C(x)$, and $P(x)$ are related, that is,

$$P(x) = R(x) - C(x)$$

we have the following relationship for the marginal profit, revenue, and cost functions:

$$P'(x) = R'(x) - C'(x) \tag{4}$$

Example 1 The weekly revenue and cost functions for a firm selling x units of its most popular motorcycle, the Macho, are given by the equations

$$R(x) = 2000x - 4x^2, \qquad C(x) = 1000 + 1300x + x^2$$

Find the marginal revenue, marginal cost, and marginal profit.

Solution The marginal revenue is found by differentiating $R(x)$, yielding

$$R'(x) = 2000 - 8x$$

Similarly, the marginal cost becomes

$$C'(x) = 1300 + 2x$$

The marginal profit $P'(x)$ equals $R'(x) - C'(x)$, so we get

$$P'(x) = 700 - 10x$$

■

The marginal concept is used to determine the change in a function when an additional item is produced or sold. In Example 1, suppose the firm selling motorcycles sells five motorcycles per week. The weekly revenue $R(5)$ is

$$R(5) = 2000(5) - 4(5)^2 = \$9900$$

If sales were to increase to six motorcycles per week, weekly revenue would increase to

$$R(6) = 2000(6) - 4(6)^2 = \$11{,}856$$

The additional unit sold produces additional revenue equal to

$$R(6) - R(5) = \$1956$$

The marginal revenue at $x = 5$ equals

$$R'(5) = 2000 - 8(5) = \$1960$$

Although $R(6) - R(5)$ and $R'(5)$ are not equal, generally they differ by such a small amount that the difference can be ignored, and we write

$$R(6) - R(5) \approx R'(5)$$

where the symbol $\approx$ means "approximately equal."

Generalizing this approach, we can write

$$\boxed{R(x + 1) - R(x) \approx R'(x)} \tag{5}$$

The rationale behind this result can be seen by referring to Figure 1, which contains (1) the graph of an arbitrary revenue function, (2) the secant line connecting $[x, R(x)]$ and $[x + 1, R(x + 1)]$, and (3) the line tangent to the graph at $[x, R(x)]$. Because the points $[x, R(x)]$ and $[x + 1, R(x + 1)]$ are close together, the slope of the secant line

$$\frac{R(x + 1) - R(x)}{(x + 1) - x} = R(x + 1) - R(x)$$

will not differ significantly from the slope $R'(x)$ of the line tangent to the graph at $[x, R(x)]$. More will be said about this technique when the differential is presented in Section 4.5.

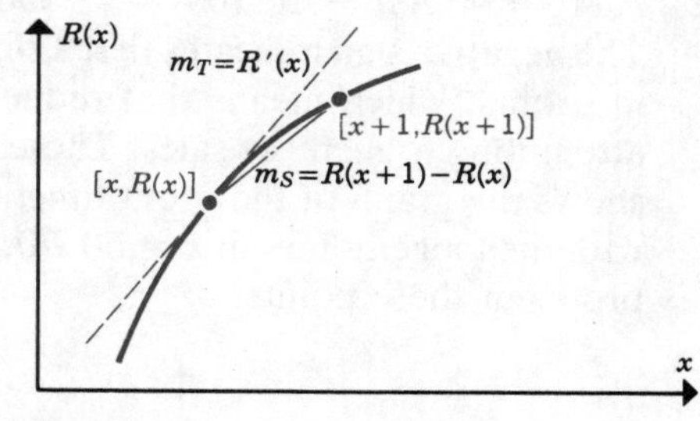

Figure 1

Example 2 For the cost function

$$C(x) = 1000 + 1300x - 20x^2$$

find the change in cost when x changes from 5 to 6,

(a) Using the first derivative $C'(x)$ at $x = 5$.

(b) Using the relation $C(x + 1) - C(x)$ with $x = 5$.

Solution (a) The marginal cost $C'(x)$ can be written as

$$C'(x) = 1300 - 40x$$

$$C'(5) = \$1100/\text{unit}$$

(b) $C(x + 1) - C(x) = C(6) - C(5)$

$$= \$1080/\text{unit}$$

■

A knowledge of the marginal profit, revenue, or cost is important in making business decisions. If $P'(x)$, the marginal profit, is positive, increasing sales or production will yield higher profits; on the other hand, if $P'(x)$ is negative, increasing sales or production will yield lower profits. The next example illustrates this.

Example 3 Use the information provided in Example 1 to find the marginal profit when

(a) $x = 50$ (b) $x = 70$ (c) $x = 100$

Solution The marginal profit $P'(x)$ was found to be

$$P'(x) = 700 - 10x$$

(a) The marginal profit when $x = 50$ is $P'(50)$

$$P'(50) = 700 - 10(50) = \$200/\text{unit}$$

This result tells us that a slight increase in sales will yield additional profit of approximately $200 for each additional unit sold.

(b) $P'(70) = 700 - 10(70) = \$0/\text{unit}$
This result tells us that a slight increase in sales will have little or no effect on profits.

(c) $P'(100) = 700 - 10(100) = -\$300/\text{unit}$
The negative sign indicates that a slight increase in sales will cause a decrease in profits, which means that reducing sales may be a better strategy than attempting to increase sales. These results are illustrated in Figure 2, which shows the graph of the profit function $P(x) = 700x - 5x^2 - 1000$ together with the tangent lines at $x = 50, 70, 100$ whose slopes represent the marginal profits at these points. ■

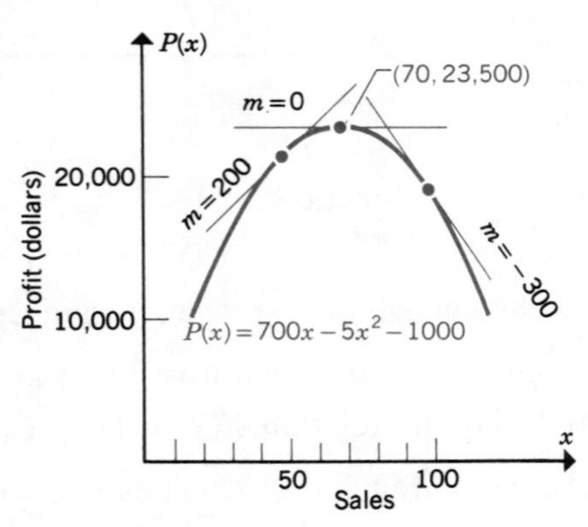

Figure 2

Equation 4 also illustrates an important concept in economics and business. In most operations $R'(x) > 0$; that is, increasing sales or production generates additional revenue. From Equation 4, we see that

1. If $R'(x) > C'(x)$, $P'(x) > 0$. If the marginal revenue exceeds the marginal cost, the marginal profit is positive, which implies that a small increase in sales produces an increase in profits.
2. If $R'(x) < C'(x)$, $P'(x) < 0$. If the marginal revenue is less than the marginal cost, the marginal profit is negative; this means that a small increase in sales results in a decrease in profits.
3. If $R'(x) = C'(x)$, $P'(x) = 0$. If the marginal revenue equals the marginal cost, the marginal profit is zero; this means that a slight increase in sales has little or no effect on profits.

These results are illustrated graphically in Example 4.

Example 4 Plot the graphs of the marginal revenue, marginal cost, and marginal profit equations for the firm described in Example 1.

Solution The equations for the marginal revenue, marginal cost, and marginal profit are

$$R'(x) = 2000 - 8x$$
$$C'(x) = 1300 + 2x$$
$$P'(x) = 700 - 10x$$

Their graphs are shown in Figure 3.

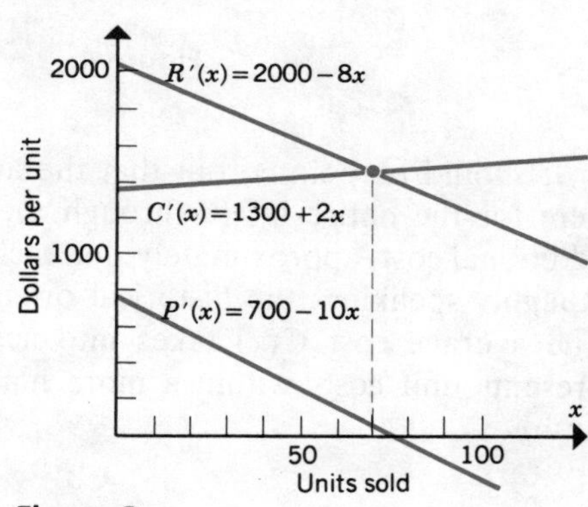

Figure 3 ■

The concept of marginal should not be confused with that of average. When given a total cost function $C(x)$, the **average** cost associated with producing x items is denoted by $\overline{C}(x)$ and is defined as

$$\overline{C}(x) = \frac{C(x)}{x}, \qquad x > 0$$

Example 5 A company has found that the relationship between the total cost $C(x)$ and the number of items produced is given by the equation

$$C(x) = 2000 + 200x + x^2, \qquad 0 \le x \le 100$$

Find the average and marginal costs when 25 items are produced.

Solution First, find the average and marginal cost functions $\overline{C}(x)$ and $C'(x)$:

$$\overline{C}(x) = \frac{2000}{x} + x + 200$$

$$C'(x) = 200 + 2x$$

Evaluating these functions at $x = 25$ gives

$$\overline{C}(25) = \$305/\text{unit}$$

$$C'(25) = \$250/\text{unit}$$

The graphs of the average and marginal cost functions are shown in Figure 4.

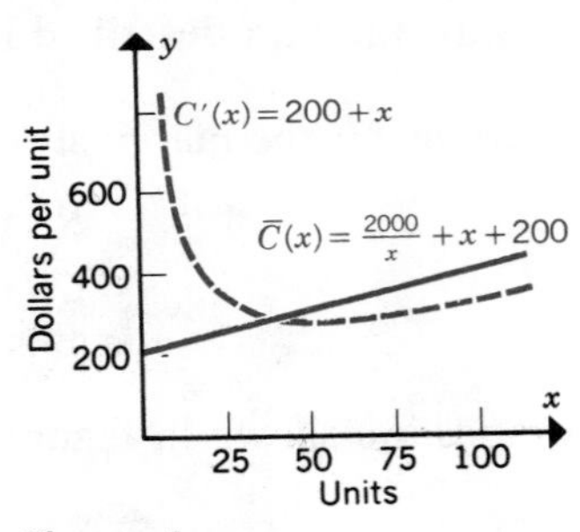

Figure 4 ■

It should be pointed out that the average cost $\overline{C}(x)$ represents the cost per item for the entire lot (0 through x), whereas the marginal cost $C'(x)$ is the additional cost, approximately, for each additional item that is produced or sold. Roughly speaking, the historical orientations of $\overline{C}(x)$ and $C'(x)$ are different. The average cost $\overline{C}(x)$ takes into account all past cost data, whereas $C'(x)$ presents unit costs within a more narrow time frame, that is, the immediate future.

3.4 EXERCISES

1. The annual revenue $R(x)$ from the sales of portable television sets is given by the equation

$$R(x) = 200x - x^2$$

where x equals the number of units sold.

(a) Find $R'(x)$, the marginal revenue.
(b) Find $R'(30)$, the marginal revenue when $x = 30$.

(c) Find $R(31) - R(30)$, the change in revenue when the number of units sold increases from 30 to 31. Compare with the answer from part (b).

2. The total monthly cost for a small manufacturer of swimwear is given by the equation

$$C(x) = 10{,}000 + 4x - 0.001x^2$$

where $C(x)$ is the cost, in dollars, and x represents the number of items produced.
(a) Find $C'(x)$, the marginal cost function.
(b) Find $C'(300)$, the marginal cost when $x = 300$.
(c) Find $C(301) - C(300)$, the cost to produce the 301st item. Compare with the answer to part (b).
(d) Find $\overline{C}(x)$, the average cost function.
(e) Find $\overline{C}(300)$, the average cost when $x = 300$.

3. Funtime has developed a new robotic toy, which it sells to wholesalers for \$20 each. The cost function is given by the equation

$$C(q) = 5000 + 4q + 0.001q^2$$

where $C(q)$ is the cost, in dollars, and q is the quantity produced.
(a) Find $P(q)$, the profit function.
(b) Find $P'(q)$, the marginal profit function.
(c) Find $P'(1000)$, the marginal profit when $q = 1000$.
(d) The company produces and sells 1000 units. Use the answer to part (c) to determine whether or not the company should increase production slightly.

4. The demand equation for a product is $p = 20 - q$, where p is the unit price, in dollars, and q is the number of units that can be sold.
(a) Find $R(q)$, the revenue, as a function of q.
(b) Find $R'(q)$, the marginal revenue function.
(c) Sketch the graph of $R'(q)$.
(d) For what values of q is $R'(q) > 0$?
(e) Find $\overline{R}(q)$, the average revenue function.

5. The demand equation for a new compact-disk player is $p = 800 - 2q$, where q is the number of units sold and p is the unit price, in dollars. The cost to produce q units is given by the equation

$$C(q) = 10{,}000 + 300q + 0.01q^2$$

(a) Find $P(q)$, the profit function.
(b) Find $P'(q)$, the marginal profit function.
(c) Find $P'(100)$, the marginal profit when $x = 100$.
(d) If the company sells 100 units, use the result in part (c) to determine whether or not the company should decrease the unit price slightly to generate more demand for the product.

6. The marginal profit for a company selling word processors is

$$P'(x) = 2000 - 4x$$

where x is the number of units sold. The company currently sells 100 units and earns a profit of \$15,000. By how much (approximately) would profits increase if an additional unit is sold?

7. The marginal cost of producing x cartons of dried food is

$$C'(x) = 8 - 0.01x$$

When $x = 200$, the company's total costs equal \$2500. By how much (approximately) does the total cost increase if x is increased by one unit to 201?

8. Demand for a new luxury car is described by the equation

$$p = 30{,}000 - 20q$$

where p is the unit price, in dollars, and q is the number of units produced and sold.

(a) Find $R(q)$, the revenue as a function of q.
(b) Find $R'(q)$, the marginal revenue function.
(c) The company wants to sell 100 cars. What price should it set on each car? What is the corresponding revenue?
(d) Find $R'(100)$.
(e) By how much (approximately) is revenue increased if the company lowers the price slightly so that 101 cars are sold?

9. A book publisher finds that inventory costs are described by the equation

$$C(x) = 200 + x + \frac{400}{x}$$

where x is the number of cartons printed and stored in the company's warehouse.

(a) Find $C'(x)$, the marginal cost function.
(b) Find $C'(10)$.
(c) Currently the company prints and stores ten cartons of books at a time. On the basis of the information provided in part (b), by how much (approximately) would the company's costs decline if they printed and stored the books eleven cartons at a time?
(d) Calculate $C(11) - C(10)$ and compare with the answer to part c.

10. For the book publisher in Exercise 9, find

(a) The average cost $\overline{C}(x)$.
(b) The marginal average cost $\frac{d}{dx}\overline{C}(x)$.
(c) The marginal average cost when $x = 10$, denoted as $\left.\frac{d}{dx}\overline{C}(x)\right|_{x=10}$
(d) The difference $\overline{C}(11) - \overline{C}(10)$. Compare with the result from part (c).

11. Using the definition $\overline{C}(x) = \frac{C(x)}{x}$, show that the marginal average cost function $\frac{d}{dx}\overline{C}(x)$ can be written

$$\frac{d}{dx}\overline{C}(x) = \frac{C'(x) - \overline{C}(x)}{x}$$

Since $x \geq 0$, this result implies that as x increases

(a) $\overline{C}(x)$ increases when $C'(x) > \overline{C}(x)$.
(b) $\overline{C}(x)$ decreases when $C'(x) < \overline{C}(x)$.
(c) $\overline{C}(x)$ changes only slightly, if at all, when $C'(x) = \overline{C}(x)$.

3.5 HIGHER-ORDER DERIVATIVES

Since the derivative $f'(x)$ is a function, we can attempt to find its derivative. The result is called the **second derivative** of $f(x)$, generally written as

$$f''(x) \quad \text{or} \quad \frac{d^2y}{dx^2}$$

That is,

$$\frac{d}{dx}[f'(x)] = f''(x), \qquad \frac{d}{dx}\left(\frac{dy}{dx}\right) = \frac{d^2y}{dx^2}$$

Example 1 Find the second derivative of the function

$$f(x) = 9x^4 - 3x^3 + x^2 + 5x - 1$$

Solution The derivative $f'(x)$ is found first; the result is

$$f'(x) = 36x^3 - 9x^2 + 2x + 5$$

The second derivative $f''(x)$ is found by differentiating $f'(x)$, yielding

$$f''(x) = 108x^2 - 18x + 2$$

■

Example 2 Find the second derivative of the function

$$y = \frac{2}{x} = 2x^{-1}$$

Solution As in Example 1, we differentiate twice:

$$\frac{dy}{dx} = -2x^{-2}$$

$$\frac{d^2y}{dx^2} = 4x^{-3} = \frac{4}{x^3}$$

■

The processing of differentiating can be continued. That is, $f''(x)$ can be differentiated to produce $f'''(x)$, the third derivative; $f'''(x)$ can be differentiated to produce $f^{(4)}(x)$, the fourth derivative; and so on. The third, fourth, and higher-order derivatives are written

$$f'''(x),\ f^{(4)}(x),\ f^{(5)}(x),\ \ldots,\ f^{(n)}(x)$$

or

$$\frac{d^3y}{dx^3}, \frac{d^4y}{dx^4}, \frac{d^5y}{dx^5}, \ldots, \frac{d^n(x)}{dx^n}$$

Example 3 Find $f'''(x)$ and $f^{(4)}(x)$ for the function given in Example 1.

Solution Since $f''(x)$ is known, $f'''(x)$ can be found immediately:

$$f'''(x) = 216x - 18$$

Differentiating again gives

$$f^{(4)}(x) = 216$$

■

Example 4 Find $\frac{d^3y}{dx^3}$ and $\frac{d^4y}{dx^4}$ for the function in Example 2.

Solution In Example 2

$$\frac{d^2y}{dx^2} = 4x^{-3}$$

Differentiating gives

$$\frac{d^3y}{dx^3} = -12x^{-4}$$

and

$$\frac{d^4y}{dx^4} = 48x^{-5} = \frac{48}{x^5}$$ ■

The time and effort you spend in finding the second derivative and higher-order derivatives will be reduced considerably if you simplify all expressions as much as possible before attempting to differentiate; in addition, the chances of making an error will be simultaneously reduced. All the rules presented in Sections 3.1 through 3.3 are applicable in finding higher-order derivatives.

Example 5 Find the second derivative of the function

$$f(x) = (x^2 + 5)^4$$

Solution The general power rule (Section 3.3) is used, giving

$$f'(x) = 4(x^2 + 5)^3(2x) = 8x(x^2 + 5)^3$$

Next the product rule (Section 3.2) is used to find $f''(x)$. Letting $u(x) = 8x$ and $v(x) = (x^2 + 5)^3$, we get

$$\begin{aligned} f''(x) &= 8x[3(x^2 + 5)^2(2x)] + (x^2 + 5)^3(8) \\ &= 48x^2(x^2 + 5)^2 + 8(x^2 + 5)^3 \\ &= (x^2 + 5)^2(48x^2 + 8x^2 + 40) = 8(7x^2 + 5)(x^2 + 5)^2 \end{aligned}$$ ■

As we shall see in Chapter 4, the second derivative plays an important role in determining the behavior of a function and in sketching its graph.

3.5 EXERCISES

Find the second derivative of each of the following functions.

1. $f(x) = 4x + 7$
2. $y = 3x^2 - 5x + 2$
3. $f(x) = 6x^3 - 9x^2 + 11$
4. $y = x^7 - 5x^4 + 3x^2 - 9$

5. $f(x) = \dfrac{8}{x^2}$

6. $y = \dfrac{-3}{x^4}$

7. $f(x) = 5\sqrt{x}$

8. $F(x) = \sqrt{5x}$

9. $y = (1 - 6x)^3$

10. $f(x) = 4(x^2 - 3)^5$

11. $f(t) = \sqrt{1 - t}$

12. $f(x) = \sqrt[3]{1 + x}$

13. $y = \dfrac{2x}{1 + x}$

14. $y = \dfrac{x^2}{2 + x}$

15. $f(x) = (x^4 + 1)^3 - \dfrac{1}{x}$

16. $f(x) = x(4 - x)^3$

17. $f(x) = \sqrt{x^2 + 1}$

18. $y = \dfrac{x + 1}{\sqrt{x}}$

19. $f(x) = x + \dfrac{1}{\sqrt{x}}$

20. $f(x) = x^n$, n is a real number

21. If $f(x) = mx + b$, where m and b are constants, show that $f''(x) = 0$.

22. If $y = ax^2 + bx + c$, where a, b, and c are constants, show that $\dfrac{d^2y}{dx^2} = 2a$.

23. If $f(x) = u(x) \cdot v(x)$, show that $f''(x) = u(x)v''(x) + 2u'(x)v'(x) + vu''(x)$.

24. Use the formula for $f''(x)$ in exercise 23 to find the second derivative of the function $f(x) = \left(x - \dfrac{1}{x}\right)(x^2 + 1)^2$.

25. If $f(x) = [u(x)]^n$, show that $f''(x) = n[u(x)]^{n-2}[u(x)u''(x) + (n - 1)(u'(x))^2]$.

26. Use the formula for $f''(x)$ in exercise 25 to find the second derivative of the function
$$f(x) = (x^2 + 1)^4$$

3.6 IMPLICIT DIFFERENTIATION: RELATED RATES

Most of the functions studied so far have been written in the form $y = f(x)$; for example, the equations

$$f(x) = 2x + 3 \quad \text{and} \quad y = 2x + 3$$

are equivalent. The dependent variable, y, or equivalently $f(x)$, appears alone on one side of the equation, and the mathematical expression that generates y or $f(x)$ for a given value of x appears on the other side. For equations of this type, y is called an **explicit** function of x.

However, in many instances the equation defining the relationship between x and y does not assume the form $y = f(x)$; in addition, the equation may represent more than one function of x. When this occurs, the function or functions are defined **implicitly**.

The following are examples of equations in which y is defined implicitly in terms of x:

$$x^3 - y + 2 = 0 \tag{1}$$

$$x - y^2 = 1 \tag{2}$$

$$2x^3y^4 + y^3x^4 = 5x^2y \tag{3}$$

Equation 1 represents a case where y can be written explicitly as a function of x. The equation can be rewritten so that all terms containing y are on one side of the equation, yielding

$$y = x^3 + 2$$

Equation 2 can also be solved for y by writing

$$y^2 = x - 1 \tag{4}$$

Notice that this equation does not represent a function; for example, when $x = 2$, $y = \pm 1$. However, taking the square root of both sides of Equation 4 reveals the two functions

$$y = f_1(x) = +\sqrt{x - 1} \quad \text{and} \quad y = f_2(x) = -\sqrt{x - 1}$$

whose graphs are shown in Figure 1.

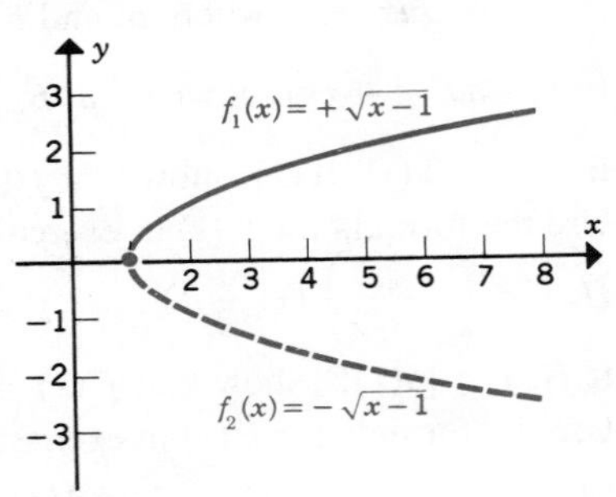

Figure 1

In many instances, however, rewriting the equation to express y explicitly in terms of one or more functions of x is either extremely difficult or impossible; Equation 3 is an example of this situation. The failure or inability to express y explicitly as a function of x does not prevent us from finding the derivative when y is differentiable. A technique called **implicit differentiation** can be used. To see how the method works, let's try to find $\frac{dy}{dx}$ for the function defined in Equation 1 by differentiating both sides with respect to x,

$$x^3 - y + 2 = 0$$

$$\frac{d}{dx}(x^3 - y + 2) = \frac{d}{dx}(0)$$

Differentiating term by term gives

$$\frac{d}{dx}(x^3) - \frac{d}{dx}(y) + \frac{d}{dx}(2) = 0$$

$$3x^2 - \frac{d}{dx}(y) + 0 = 0$$

The second term on the left is $\frac{dy}{dx}$, the derivative we are seeking, so we can write

$$3x^2 - \frac{dy}{dx} = 0$$

Solving for $\frac{dy}{dx}$ gives

$$\frac{dy}{dx} = 3x^2$$

If y had been written explicitly as a function of x, namely

$$y = x^3 - 2$$

differentiating would have yielded

$$\frac{dy}{dx} = 3x^2$$

After the implicit differentiation has been completed, all terms containing the factor $\frac{dy}{dx}$ are gathered on one side of the equation, and the resulting equation is then solved for $\frac{dy}{dx}$ in terms of x and y.

Example 1 Find the derivative of the function defined implicitly by the equation

$$3xy - x = 2 \tag{5}$$

(a) By implicit differentiation.
(b) By first writing y as an explicit function of x and then differentiating.

Solution (a) Differentiating term by term gives

$$\frac{d}{dx}(3xy) - \frac{d}{dx}(x) = \frac{d}{dx}(2)$$

The product rule must be applied to the first term on the left-hand side, yielding

$$\frac{d}{dx}(3xy) = 3x\frac{dy}{dx} + y\frac{d}{dx}(3x) = 3x\frac{dy}{dx} + 3y$$

The remaining two terms can be differentiated quickly:

$$\frac{d}{dx}(x) = 1, \qquad \frac{d}{dx}(2) = 0$$

so now we have the equation

$$3x \frac{dy}{dx} + 3y - 1 = 0$$

Solving for $\frac{dy}{dx}$ gives

$$\frac{dy}{dx} = \frac{1 - 3y}{3x}$$

(b) Because Equation 5 has a simple algebraic structure, y can be expressed directly in terms of x; the result is

$$y = \frac{2}{3x} + \frac{1}{3} = \frac{2}{3}x^{-1} + \frac{1}{3}$$

Differentiating term by term yields

$$\frac{dy}{dx} = -\frac{2}{3}x^{-2} = \frac{-2}{3x^2}$$

The results obtained by the two approaches do not appear to be identical. The differences are merely superficial; the answer to part (a) can be written in terms of x alone as follows:

$$\frac{1 - 3y}{3x} = \frac{1 - 3[(2 + x)/3x]}{3x} = \frac{1 - [(2 + x)/x]}{3x} = \frac{x - (2 + x)}{3x^2} = -\frac{2}{3x^2}$$

Although the results obtained by implicit and explicit differentiation are equivalent, their algebraic forms may be different. When this happens, some algebraic manipulation will bring the two forms into agreement. ■

The general power rule, in the form

$$\boxed{\frac{d}{dx}(y^n) = ny^{n-1}\frac{dy}{dx}} \tag{6}$$

must be used when an equation contains factors that are powers of y, as the next example illustrates.

Example 2 Find $\frac{dy}{dx}$ for the functions defined implicitly by the equation

$$x^2 + y^3 = 2xy^2$$

Solution

$$\frac{d}{dx}(x^2) + \frac{d}{dx}(y^3) = \frac{d}{dx}(2xy^2)$$

The derivatives of the terms on the left-hand side of the equation are

$$\frac{d}{dx}(x^2) = 2x, \qquad \frac{d}{dx}(y^3) = 3y^2\frac{dy}{dx}$$

The product rule must be used to find the derivative on the right-hand side:

$$\frac{d}{dx}(2xy^2) = 2x\frac{d}{dx}(y^2) + y^2\frac{d}{dx}(2x)$$

$$= 4xy\frac{dy}{dx} + 2y^2$$

Setting the left- and right-hand sides equal to one another gives

$$2x + 3y^2\frac{dy}{dx} = 4xy\frac{dy}{dx} + 2y^2$$

Putting all terms containing the factor $\frac{dy}{dx}$ on one side of the equation and the remaining terms on the other gives

$$3y^2\frac{dy}{dx} - 4xy\frac{dy}{dx} = 2y^2 - 2x$$

Finally, solving for $\frac{dy}{dx}$ gives

$$\frac{dy}{dx} = \frac{2y^2 - 2x}{3y^2 - 4xy}$$

■

When you are given an equation containing the variables x and y, the procedure for finding $\frac{dy}{dx}$ by implicit differentiation can be summarized as follows.

1. Treat y as a function of x and differentiate each term of the equation with respect to x.
2. Put all terms containing the factor $\frac{dy}{dx}$ on one side of the equation and the rest of the terms on the other side.
3. Factor out $\frac{dy}{dx}$ from all terms that contain it.
4. Solve the equation for $\frac{dy}{dx}$.

When we are finding the slope of a tangent line, it is often necessary to specify both the x and y coordinates of the point on the curve because the equation may not represent a function; each value of x may be paired with two or more values of y. This situation is illustrated in the next example.

Example 3 Find the equation of the line tangent to the curve

$$y^2 - 4x = 0$$

at the point $(1, -2)$.

Solution Recalling that $\frac{dy}{dx}$ represents the slope of the line tangent to the curve, we first find the derivative by implicit differentiation, obtaining

$$2y\frac{dy}{dx} - 4 = 0$$

Solving for $\frac{dy}{dx}$ gives

$$\frac{dy}{dx} = \frac{2}{y},$$

so that the slope of the tangent line at $(1, -2)$ becomes

$$\frac{dy}{dx} = \frac{2}{-2} = -1$$

The equation of the tangent line can now be found from the point–slope form $y - y_1 = m(x - x_1)$. The result is

$$y + 2 = -1(x - 1)$$

or

$$y = -x - 1$$

The curve and the tangent line are shown in Figure 2.

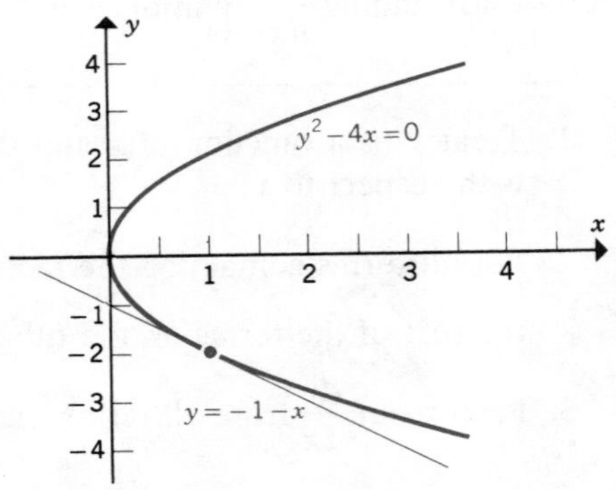

Figure 2 ■

Related Rates

Suppose that the relationship between R, the revenue, and Q, the number of units produced and sold each week, is given by the equation

$$R = 40Q - Q^2 \tag{6}$$

In addition, suppose that Q is a function of t, the time. Can we write an expression for $\frac{dR}{dt}$ even if we do not know how Q, and therefore R, depends upon t. We can, but $\frac{dR}{dt}$ will be written in terms of both Q and $\frac{dQ}{dt}$. If we differentiate both sides with respect to t, we can use the chain rule to yield

$$\begin{aligned}\frac{dR}{dt} &= \frac{dR}{dQ} \cdot \frac{dQ}{dt} \\ &= (40 - 2Q)\frac{dQ}{dt} \\ &= 40\frac{dQ}{dt} - 2Q\frac{dQ}{dt} \qquad (7)\end{aligned}$$

The time rate of change of R, which is $\frac{dR}{dt}$, is expressed in terms of $\frac{dQ}{dt}$, the time rate of change of Q. Applications in which the time rate of change of one quantity is written in terms of the time rate of change of a second variable are called **related-rate** problems.

Of what use is a result such as Equation 7? Suppose that $Q = 30$ and that management is planning to increase production by five units per week that is $\frac{dQ}{dt} = 5$; Equation 7 enables us to predict the corresponding time rate of change in R, that is, $\frac{dR}{dt}$. In this case, we get

$$\begin{aligned}\frac{dR}{dt} &= 40(5) - 20(30)(5) = 200 - 300 \\ &= -\$100/\text{week}\end{aligned}$$

Thus management's decision to increase production by five units per week when $Q = 30$ results in a decrease in weekly revenue by \$100.

Example 4 Oil is leaking from an oil tanker at a rate of 300 m^3/hr. The oil forms an expanding circular pool 1 cm deep around the tanker (Figure 3). How fast is r, the radius of the circle, changing when $r = 100$ m?

Figure 3

Solution Since the pool of oil forms a circular disk 1 cm thick, we can use the formula $V = \pi r^2 h$, with $h = 1 \text{ cm} = \frac{1}{100}$ m, to describe the relationship between V, the volume of oil that has leaked from the tanker, and r, the radius of the circular disk (Figure 4):

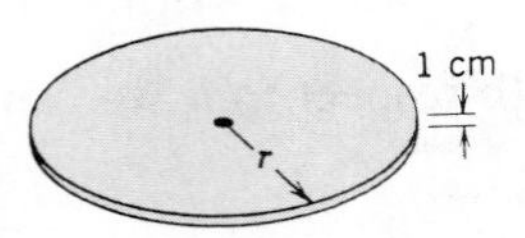

Figure 4

$$V(r) = (\pi r^2)(\tfrac{1}{100}) \tag{8}$$

The rate at which r is changing, $\frac{dr}{dt}$, can be found by differentiating both sides of Equation 8 with respect to t and using the chain rule to produce

$$\frac{dV}{dt} = \frac{dV}{dr}\frac{dr}{dt}$$

$$\frac{dV}{dt} = \frac{2\pi r}{100}\frac{dr}{dt} \tag{9}$$

Since $\frac{dV}{dt} = 300 \text{ m}^3/\text{hr}$, we can solve Equation 9 for $\frac{dr}{dt}$ to give

$$\frac{dr}{dt} = \frac{30{,}000}{2\pi r} \text{ m/hr}$$

When $r = 100$ m, we get

$$\frac{dr}{dt} = \frac{150}{\pi} \approx 47.8 \text{ m/hr}$$

■

The functional relationship between the independent and dependent variables was given explicitly in Equations 6 and 8. However, there are many applications in which the relationship is stated implicitly; this situation does not present any difficulties, as the next example illustrates.

Example 5 Shortly after an ice storm, Slow Sid props a 20-ft ladder against a wall as shown in Figure 5. While Sid is gone to get his tools, the bottom of the ladder begins

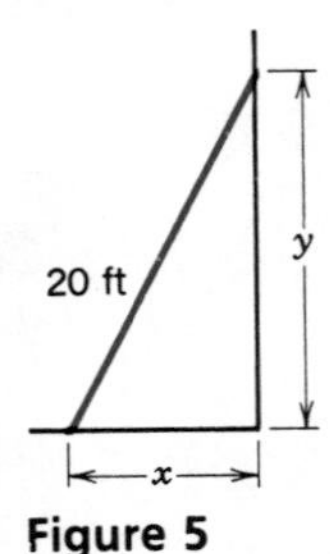

Figure 5

to slide away from the wall at a speed of 5 ft/min. Find the speed at which the top of the ladder is moving down the wall when the bottom of the ladder is 10 ft from the wall.

Solution Let x be the distance of the bottom of the ladder from the wall and y the distance of the top of the ladder from the ground. Since the triangle formed by the ladder, the ground, and the wall is a right triangle, the variables x and y are related by the equation

$$x^2 + y^2 = 20^2 \tag{10}$$

It is not necessary to establish the functional relationship between x and y; since each can be considered a function of t, we can differentiate both sides of Equation 10 with respect to t to yield

$$2x\frac{dx}{dt} + 2y\frac{dy}{dt} = 0 \tag{11}$$

Equation 11 can be solved for $\frac{dy}{dt}$, the vertical speed at which the top of the ladder is moving, giving

$$\frac{dy}{dt} = \frac{-x}{y}\frac{dx}{dt} \tag{12}$$

Two of the three quantities on the right-hand side of Equation 12 are known, that is, $x = 10$ and $\frac{dx}{dt} = 5$. The third, y, can be found by substituting $x = 10$ into Equation 10 and solving:

$$(10)^2 + (y)^2 = 20^2$$
$$y = \sqrt{300} \approx 17.32 \text{ ft}$$

This piece of information can then be substituted into Equation 12 to find $\frac{dy}{dt}$.

$$\frac{dy}{dt} \approx \frac{-10(5)}{17.32} = -2.89 \text{ ft/sec}$$

What is the significance of the negative sign in the answer? ■

3.6 EXERCISES

In Exercises 1 through 6, find the derivative (a) by solving for y explicitly in terms of x and then differentiating, and (b) by implicit differentiation.

1. $5x + 2y = 3$

2. $xy + 5 = x^4$

3. $x^2 + 3y = x^3 - 2xy$

4. $x^2 - 2y^3 = 7$

5. $x^2 + y^2 = 4$

6. $\frac{1}{x} + \frac{1}{y} = 2$

Find $\frac{dy}{dx}$ by implicit differentiation in Exercises 7 through 16.

7. $3xy + x^2 = y^2$
8. $xy^2 - 5x = 4y$
9. $y^5 + 3x^2y^2 = 10$
10. $(3 + y)^2 = 3x^4 - 5$
11. $(x + y)^2 - y^2 = 7x^2$
12. $\sqrt{x} + \sqrt{y} = 2$
13. $\frac{x + y}{y + 1} = 6x$
14. $(2x - y^2)^3 = 4y + 3$
15. $\sqrt{x + \sqrt{y}} = 2x$
16. $x^n + y^n = 2$, n is a real number

In Exercises 17 through 22, find the slope and an equation of the tangent line at the indicated point.

17. $x^2 - y^2 = 5$, $(3, 2)$
18. $x^3 + y^3 = 9$, $(1, 2)$
19. $x^2 + y^2 = 25$, $(3, 4)$
20. $y^3 + xy^2 = 1$, $(2, -1)$
21. $x^2y^2 + xy = 6$, $(1, 2)$
22. $x^n + y^n = 2$, $(1, 1)$; n is a real number

Related Rates

In Exercises 23 through 34, assume that x and y are both functions of t. Find $\frac{dy}{dt}$, expressing your answer in terms of $\frac{dx}{dt}$.

23. $y = x^3 + 2$
24. $y = \frac{1}{x}$
25. $y = \sqrt{x^2 + 1}$
26. $y = \frac{x}{x^2 - 1}$
27. $xy^2 + x^2y = 2$
28. $x^2 - y^2 = 10$
29. $x\sqrt{y} + x^2 = 4$
30. $x^3 + y^3 = xy$
31. $x^2y^2 = 1$
32. $\sqrt{x + y} = y$
33. $\frac{x}{x^2 + y^2} = 3$
34. $x^2 + y^4 = 5$

35. Because of a close pennant race, demand for Green Sox baseball tickets is growing at a rate of 200 tickets per week. If each ticket costs \$5, find the rate at which R, weekly revenue, is growing.

36. The demand for compact-disk players is given by the equation

$$p = -10q + 600$$

where p is the unit price in dollars and q is the number of units sold (in thousands). If the unit price is decreasing at a rate of \$5 per month, find the rate at which R, the monthly revenue, is changing when $p = \$300$.

37. If the temperature is kept constant, the relationship between p, the pressure, and V, the volume, of gas contained in a balloon can be described by the equation

$$pV = K$$

where K is a constant. If $p = 40$ lb/in.^{2} and is changing at a rate of 2 lb/in.^{2} per minute, find the rate at which V is changing when $V = 100$ in.^{3}

38. A rectangular swimming pool 50 ft long by 20 ft wide is being filled with water at a rate of 4 ft^3/min. How fast is the water level changing when the water is 2 ft deep? When it is 4 ft deep?

39. Water is being pumped into a swimming pool that is 50 ft long by 20 ft wide at a rate of 4 ft^3/min. The depth of the pool varies from 3 ft at the shallow end to 8 ft at the deep end, as shown in Figure 6. How fast is the water level rising when the water level at the deep end is 2 ft? When it is 4 ft?

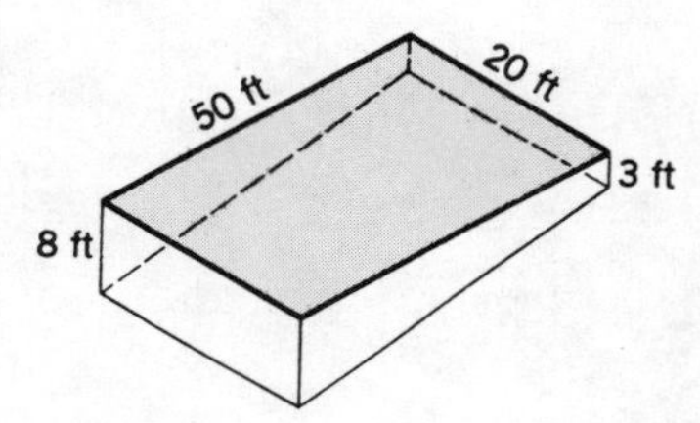

Figure 6

40. The yield Y, in bushels per acre, for an apple orchard is related to N, the number of trees planted per acre, by the equation

$$Y = 200N - N^2$$

If the manager is adding trees to each acre at a rate of four per year, find the rate at which Y is changing when $N = 50$.

41. A pebble thrown into a still lake causes a circular ripple whose radius grows at a rate of 2 cm/sec. How fast is the area of the circle increasing when $r = 20$ cm?

42. The number of fish in a pond, N, is related to x, the level (in ppm) of PCBs in the pond through the equation

$$N = \frac{200}{1 + x}$$

If the level of PCBs is increasing at a rate of 4 ppm/year, find the rate at which N is changing when $x = 10$.

43. A meltdown has occurred at a nuclear reactor. The intensity of the radiation, I, is related to d, the distance, in miles, from the reactor according to the equation

$$I = \frac{100}{d^2}$$

If a woman, upon hearing about the disaster, gets in her car and drives at 40 mph away from the reactor, how fast is I changing when $d = 10$ miles? When $d = 20$ miles?

44. The power E, in watts, radiated from a very hot object, is related to T, the object's temperature, in degrees Kelvin, by the equation

$$E = kT^4$$

where k is a constant. Write an equation for the rate at which E is changing in terms of T and $\dfrac{dT}{dt}$.

45. The total weekly cost C for the Speedy Printer Company to produce x printers is given by the equation

$$C = 5000 + 80x + 20x^{0.5}$$

If production of the units is increasing at a rate of 16 units per week, how fast are costs changing when $x = 225$?

46. A 32-ft-long ladder is leaning against a wall as shown in Figure 7. The ladder is being pulled up the wall at a rate of 2 ft/sec by workers on the roof. Assuming that the ladder maintains contact with the ground, determine how fast the base of the ladder is approaching the wall when $x = 6$ ft.

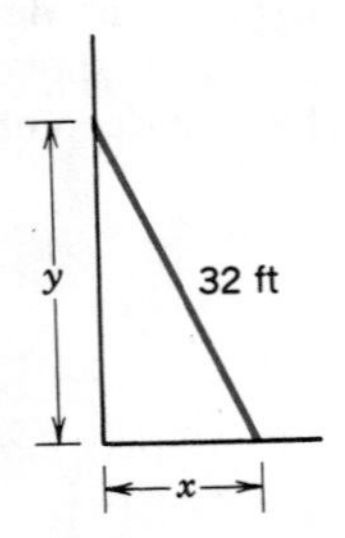

Figure 7

47. (a) The focal length f, in centimeters, of the human eye can be controlled so that an object at almost any distance d, in centimeters, in front of the lens can be focused on the retina located 2.5 cm behind the lens. The relationship between f and d is given by the simple lens formula

$$\frac{1}{f} = \frac{1}{d} + \frac{1}{2.5}$$

Find the rate at which the focal length is changing when an object moving toward a person at a constant speed of 50 cm/sec is 300 cm away.

(b) How is the answer to part (a) affected if the object is moving away from the eye instead of toward it?

48. As plaque builds up on the interior walls of an artery, its cross-sectional area A decreases. For an artery whose unobstructed radius is 40 mm, the relationship between A and the thickness of the plaque x, in millimeters, is given by the formula $A = \pi(40 - x)^2$. Find the rate at which A is changing when x equals 2 mm and is increasing at a rate of 1 mm per year.

KEY TERMS

power rule
sum rule
instantaneous rate of change
acceleration
product rule
quotient rule
general power rule
composite function
chain rule
marginal revenue
marginal cost
marginal profit
average cost
average profit
second derivative
implicit differentiation
related rates

REVIEW PROBLEMS

Find the derivative of each of the following functions.

1. $f(x) = 3$
2. $f(x) = 4x^2$
3. $F(x) = \dfrac{2}{x}$
4. $g(x) = \sqrt{5}$
5. $f(x) = x^3 + x^2 + 5x$
6. $f(x) = 2\sqrt{x}$
7. $F(x) = \sqrt{2x}$
8. $f(x) = \sqrt{3x + 2}$
9. $g(x) = \sqrt{3x} + 2$
10. $f(x) = \dfrac{2x}{x^2 + 1}$
11. $f(x) = (x^3 + x^2 + 1)(x^4 + 1)$
12. $F(x) = (x^2 + 1)^5$
13. $g(x) = x^2(x^2 + 1)^3$
14. $F(x) = \dfrac{x^2 - 1}{x^2 + 1}$
15. $f(x) = \dfrac{x^3 - x^2 + 1}{x^2}$
16. $g(t) = \dfrac{t^2}{(t + 2)^2}$
17. $f(x) = \sqrt[3]{2x} + \dfrac{4}{x^2}$
18. $F(x) = (x^3 - 1)(x^2 + 1)^4$
19. $f(x) = \frac{2}{5}(x + 1)^{5/2} - \frac{2}{3}(x + 1)^{3/2} + 1$
20. $g(t) = (t + 2)\sqrt{t + 1}$

Find the second derivative of each of the following functions.

21. $f(x) = 4x^3 - 6x^2 + 7x - 1$
22. $f(x) = \dfrac{3}{x - 1}$
23. $g(t) = \dfrac{3}{t} + \sqrt{2t}$
24. $F(x) = (x^2 + 1)^4$
25. $f(x) = \dfrac{x^4 + x^3 + x^2}{x^3}$
26. $g(x) = \dfrac{x^2 - 1}{x^2 + 1}$

In problems 27 through 30, find the slope and an equation of the tangent line at the indicated point.

27. $f(x) = x^4 - 2x^2 - 3x + 1,\ (2, 3)$
28. $F(x) = \dfrac{2x^2}{x + 1},\ (1, 1)$
29. $g(t) = 2\sqrt{x} + \dfrac{4}{x};\ (4, 5)$
30. $f(x) = \sqrt[3]{x^2 + 4};\ (2, 2)$

In problems 31 through 34, find the points, if any, where the slope of the tangent line has the value shown.

31. $f(x) = x^2 + 3x - 2,\ m = 2$
32. $F(x) = \dfrac{x}{x + 1},\ m = \frac{1}{4}$
33. $g(x) = 3\sqrt{x^2 + 5},\ m = 4$
34. $f(x) = (x^2 + 3)^{3/2},\ m = 6$

The graphs of the functions defined by the equations $y = f(x)$ and $y = g(x)$ are shown in Figures 1 and 2 together with the tangent lines at $x = 1$. Use the graphs to answer problems 35 through 40.

35. Find each of the following: $f(1)$, $f'(1)$, $g(1)$, $g'(1)$.

36. If $F(x) = f(x) + g(x)$, find $F(1)$ and $F'(1)$.

37. If $G(x) = f(x)g(x)$, find $G(1)$ and $G'(1)$.

38. If $H(x) = \dfrac{f(x)}{g(x)}$, find $H(1)$ and $H'(1)$.

39. If $K(x) = [f(x) + g(x)]^2$, find $K(1)$ and $K'(1)$.

40. If $J(x) = xf(x) + \dfrac{x}{g(x)}$, find $J(1)$ and $J'(1)$.

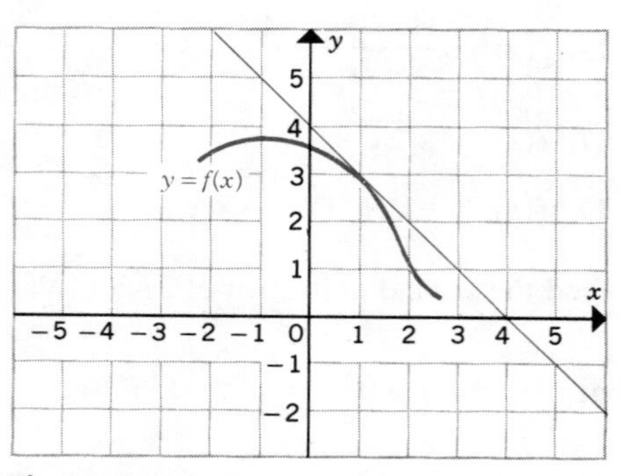

Figure 1

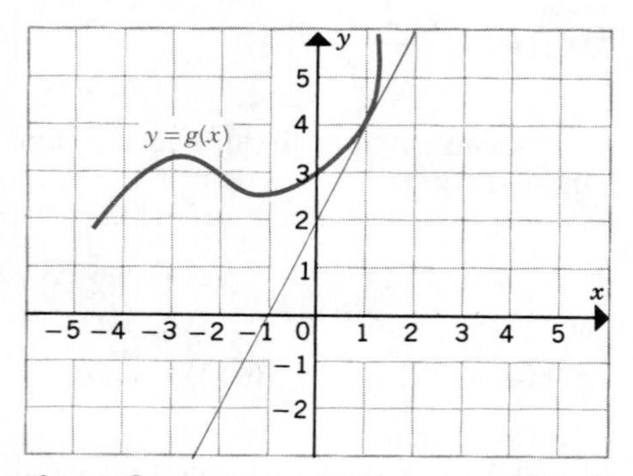

Figure 2

Use implicit differentiation to find $\dfrac{dy}{dx}$ in problems 41 through 44.

41. $x^2y + y^2x = 5$

42. $(x + y)^2 - x^2 = y$

43. $\sqrt{x} - \sqrt{y} = 3x$

44. $x^3y - (x^2 + 1)^3 = y^2$

45. The Acme Furniture Company has purchased a computer-controlled lathe. The book value $V(t)$, in dollars, at any time t, in years, can be described by the equation

$$V(t) = 4000(5 - t)^2, \qquad 0 \le t \le 5$$

(a) Find the time rate of change of $V(t)$.
(b) How fast is the book value changing when $t = 0$? $t = 2$?

46. An object is dropped from the top of a 900-foot building. The height $h(t)$, in feet, of the object above the ground at any time t, in seconds, is given by the equation

$$h(t) = 900 - 16t^2$$

where $t = 0$ corresponds to the moment when the object is released.
(a) Find the velocity of the object as a function of t.
(b) How fast is the object moving three seconds after it has been released?
(c) How fast is the object moving when it hits the ground ($h = 0$)?

47. According to a model developed by a public health group, the number of people $N(t)$, in hundreds, who will be ill with the Manchurian flu at any time t, in months, next winter is described by the equation

$$N(t) = 100 + 30t - 10t^2, \qquad 0 \le t \le 4$$

where $t = 0$ corresponds to the beginning of December.
(a) Find the time rate of change of $N(t)$.
(b) How rapidly is the flu spreading when $t = 1$?
(c) Find $N'(3)$. What does the negative sign in your answer signify?

48. A blood test reveals that a woman has a high level of blood cholesterol. Her physician prescribes medication to correct the problem. The relationship between her cholesterol level $C(t)$, in milligrams per deciliter, and the time t, in weeks, from the beginning of the treatment is described by the equation

$$C(t) = 175 + \frac{125}{1 + \sqrt{t}}$$

(a) Find the time rate of change of $C(t)$.
(b) How fast is her cholesterol level changing when $t = 0$? When $t = 4$?

49. The demand equation for a new Fax machine is

$$p = \frac{10{,}000}{4 + q^2}$$

where p is the unit price, in dollars, and q is the number of units sold each day.
(a) Find the marginal revenue $R'(q)$.
(b) Find $R'(1)$.
(c) If each machine costs \$500 to manufacture and sell, find an equation that describes the profit $P(q)$ as a function of q.

50. The relationship between the weekly cost $C(x)$ and x, the number of keyboards produced each week by the Musac Manufacturing Company, is given by the equation

$$C(x) = 2000 + 75x + \frac{50}{\sqrt{x}}, \qquad x \ge 1$$

(a) Find an equation for the marginal cost $C'(x)$.
(b) Use the answer to part (a) to estimate the cost of increasing production from 25 to 26 units.
(c) Calculate $C(26) - C(25)$, and compare the result with the answer to part (b).

51. In problem 50 suppose that keyboard production is increasing at a rate of five units per week to keep pace with demand. How fast are weekly costs changing when $x = 100$? *Hint:* You now have to think in terms of related rates.

52. The area A of a skin lesion is related to its diameter by the equation

$$A = \frac{\pi d^2}{4}$$

If the diameter is growing at a rate of 10 mm per week, find the time rate of change of A when $d = 6$ mm.

4

OPTIMIZATION AND CURVE SKETCHING

INTRODUCTION

When the graph of a function is plotted point by point, each point is considered equally important. However, when the graph is used as a decision-making tool, some points are more important than others. For example, suppose a financial analyst has developed a mathematical model that predicts P, the price of a stock, as a function of t, the time; Figure 1 represents a typical graph generated by the model. If the graph is used to make investment decisions, points such as A, B, and C indicate when to sell and points such as D, E, and F when to buy the stock. Points such as these, where the direction of a curve changes, belong to a class known as **critical** points.

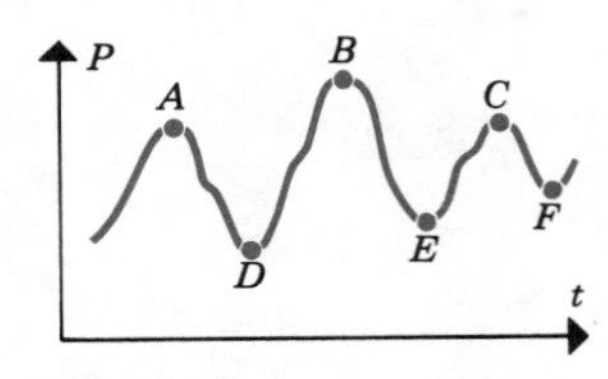

Figure 1

The graph of a function can be sketched more quickly if the shape of the curve can be determined, not only in the vicinity of the critical points, but also in the vicinity of another class of points, **inflection** points. Much of this chapter is devoted to studying and illustrating how these points are found and determining the shape of the curve in the vicinity of each one.

The ability to locate critical points enables you to **optimize,** that is, to maximize or minimize, some function. For example, farmers would want to know how much fertilizer they should buy to maximize (optimize) the yield from their crops. On the other hand, the owner of a fleet of trucks would want to know how many miles each truck should be driven between tune-ups to minimize (optimize) operating costs of his or her fleet. If the trucks are serviced too frequently, service costs plus downtime costs increase; if they are serviced infrequently, fuel costs increase because the engines operate poorly.

4.1 MAXIMA AND MINIMA

Increasing and Decreasing Functions

In Section 1.2, we found the following relationships between m, the slope, and the direction of a straight line.

1. If $m > 0$, the line is rising.
2. If $m < 0$, the line is falling.
3. If $m = 0$, the line is horizontal.

These properties are summarized in Table 1.

Table 1

m	Line	Graph
+	Rising	/
−	Falling	\
0	Horizontal	—

Similar relationships exist between the sign of $f'(x)$ at a point and the behavior of the function in the vicinity of the point. To see what these relationships are, let's look at Figure 2, which contains the graph of the function $f(x) = x^3 - 12x + 4$. In addition, the sign of the derivative,

$$f'(x) = 3x^2 - 12 = 3(x - 2)(x + 2)$$

is given on the x axis for various segments of the curve; the graph shows the following.

1. When $x < -2$, $f'(x)$ is positive, and the curve is rising, or, equivalently, $f(x)$ is increasing.
2. When $-2 < x < 2$, $f'(x)$ is negative, and the curve is falling, or, equivalently, $f(x)$ is decreasing.
3. When $x > 2$, $f'(x)$ is again positive, and again the curve is rising.

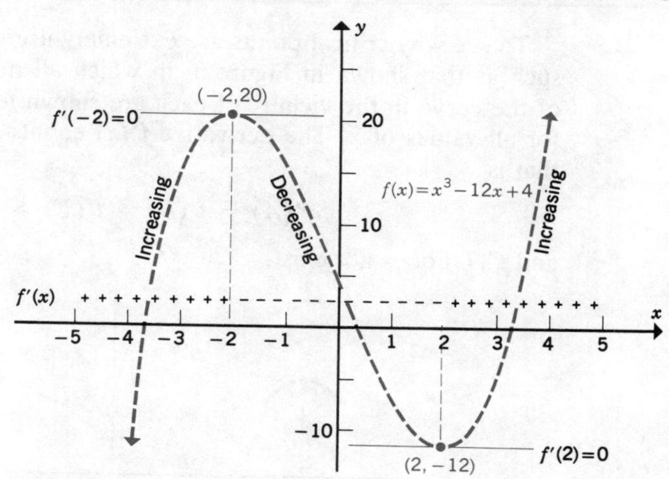

Figure 2

The results are summarized in Table 2.

Table 2

Interval	$f'(x)$	$f(x)$	Curve
$x < -2$	+	Increasing	Rising
$-2 < x < 2$	−	Decreasing	Falling
$x > 2$	+	Increasing	Rising

The relationships between $f'(x)$ and the behavior of the function $f(x) = x^3 - 12x + 4$ hold for functions in general.

Increasing and Decreasing Functions

1. If $f'(x) > 0$ on the interval $c < x < d$, the **curve is rising** or the **function is increasing** on the interval.
2. If $f'(x) < 0$ on the interval $c < x < d$, the **curve is falling** or the **function is decreasing** on the interval.

At points such as $(-2, 20)$ and $(2, -12)$, where $f'(x) = 0$, the tangent lines are horizontal ($m = 0$). Points such as $(-2, 20)$ where the curve "peaks" and $(2, -12)$ where the curve "bottoms out" are important in curve sketching because they indicate where the direction of the curve changes from rising to falling and vice versa. They belong to a class of points known as **critical points.**

Definition 1 A point $(a, f(a))$ is a **critical point**

1. If $f'(a) = 0$ or
2. If $f'(a)$ does not exist.

To see why critical points are extremely useful, suppose we have a situation such as that shown in Figure 3, in which all the critical points and the shape of the curve in the vicinity of each are shown for a function that is continuous for all values of x. The derivative $f'(x)$ equals zero at four of the five points, that is

$$f'(A) = f'(B) = f'(C) = f'(E) = 0$$

and $f'(D)$ does not exist.

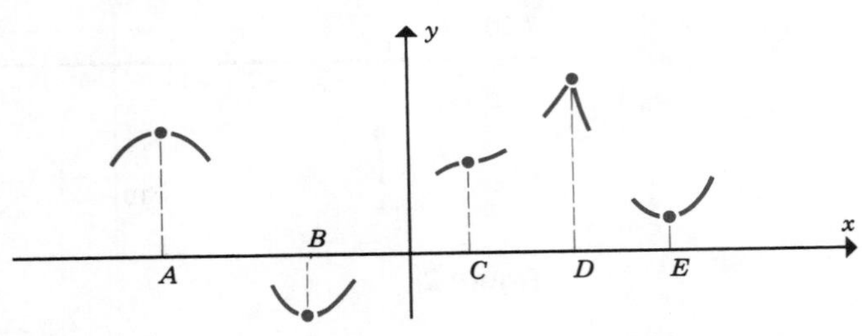

Figure 3

Suppose now that you are asked to complete the graph by making a rough sketch of the curve. Since all the critical points are shown, it is not possible for peaks or valleys other than those displayed in Figure 3 to exist. By connecting adjacent tails with smooth segments and extending the outer tails at A and E down and up, respectively, you might obtain a curve such as that shown in Figure 4. In most cases, the sketch would be a fairly accurate representation of the graph. Some features, such as the shape of the curve as x becomes very large or small, cannot be determined from a knowledge of the critical points; this is the reason for drawing multiple tails on both ends of the curve. The subject of curve sketching will be discussed in detail in Section 4.4, where we will find that locating all the critical points and determining the shape of the curve in the vicinity of each one is an essential step in obtaining the graph.

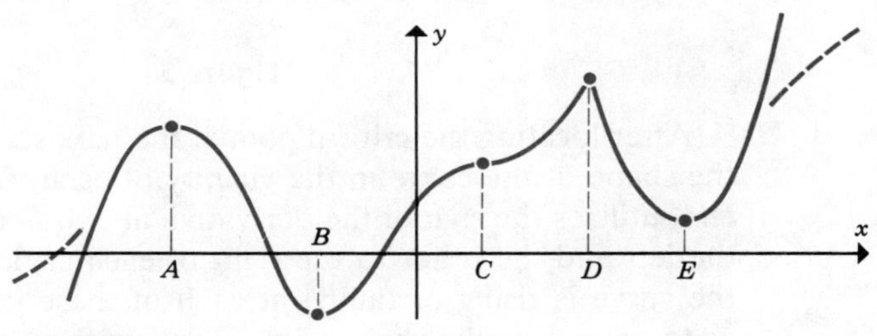

Figure 4

The peaks that occur on the curve at $x = A$ and $x = D$ are called **relative maxima.** The word *relative* signifies that the y coordinate associated with such a point is greater than the y coordinate of any other point in its vicinity. In the same way, the valleys or troughs that appear at $x = B$ and $x = E$ are called **relative minima.** A critical point such as the one that appears at $x = C$ is neither a maximum nor a minimum.

It is now time to learn how critical points are found and how the shape of the curve in the vicinity of each is determined. The critical points are found, according to Definition 1, by locating the points at which $f'(x) = 0$ or $f'(x)$ does not exist. Example 1 illustrates this process.

Example 1 Find the critical points of the function

$$f(x) = x^3 - 3x^2 - 9x + 10$$

Solution First, we find $f'(x)$, getting $f'(x) = 3x^2 - 6x - 9$. Next, we set $f'(x)$ equal to zero and solve for x:

$$0 = 3x^2 - 6x - 9$$
$$0 = 3(x - 3)(x + 1)$$

We get the solutions

$$x_1 = -1, \qquad f(-1) = 15$$
$$x_2 = 3, \qquad f(3) = -17$$

The points $(-1, 15)$ and $(3, -17)$ are the only critical points on the curve because $f'(x)$ is defined for all values of x. Therefore, condition 2 in Definition 1 does not apply. The critical points are shown in Figure 5 together with the horizontal tangent line at each point.

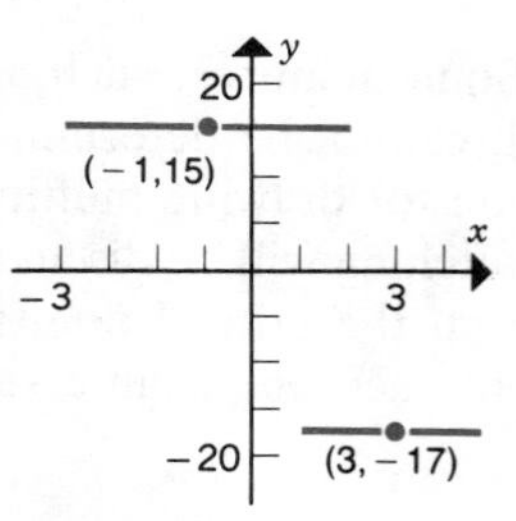

Figure 5

After locating the critical points, the next step in the analysis is to determine the shape of the curve in the vicinity of each. One method, the **first-derivative test,** utilizes the sign of the derivative at two neighboring points, one located to the left and the other to the right of each critical point, to determine whether the curve is rising or falling at each of these points. This information enables us to determine the shape of the curve in the vicinity of each critical point. The first-derivative test together with its graphical interpretation is described next.

First-Derivative Test

Suppose that a function $y = f(x)$ is continuous everywhere on the interval $c < x < d$ and that $(a, f(a))$ is the only critical point in the interval. Let x_L and x_R be values of x in the interval to the left and right, respectively, of the critical value $x = a$. The shape of the curve in the vicinity of $(a, f(a))$ can be determined from the **algebraic signs of $f'(x_L)$ and $f'(x_R)$,** as shown in Table 3.

Table 3

$f'(x_L)$	$f'(x_R)$	$(a, f(a))$	Graph ($f'(a) = 0$)
+	−	Relative maximum	$f'(x_L)>0$, $f'(x_R)<0$; x_L a x_R
−	+	Relative minimum	$f'(x_L)<0$, $f'(x_R)>0$; x_L a x_R
+	+	Neither maximum nor minimum	$f'(x_L)>0$, $f'(x_R)>0$; x_L a x_R
−	−	Neither maximum nor minimum	$f'(x_L)<0$, $f'(x_R)<0$; x_L a x_R

The first-derivative test is illustrated in the following examples.

Example 2 Find the critical points of the function $f(x) = x^2 - 6x + 10$ and then apply the first-derivative test to determine the shape of the curve in the vicinity of each critical point.

Solution 1. The critical points are found by setting $f'(x)$ equal to zero and solving for x:

$$f'(x) = 2x - 6$$
$$0 = 2(x - 3)$$

Solution: $x = 3, f(3) = 1$.

2. Apply the first-derivative test.

Since (3, 1) is the only critical point and $f'(x)$ is defined for all x, we test for the sign of $f'(x)$ in the intervals $x < 3$ and $x > 3$, as shown in Table 4 and Figure 6.

Table 4

	Interval	Value of x	$f'(x)$	Curve
Left	$x < 3$	2	-2	Falling
Right	$x > 3$	4	$+2$	Rising

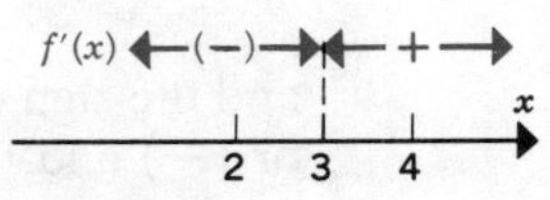

Figure 6

We can conclude, on the basis of these results, that (3, 1) is a relative minimum. A rough sketch of the curve in the vicinity of the critical point is shown in Figure 7.

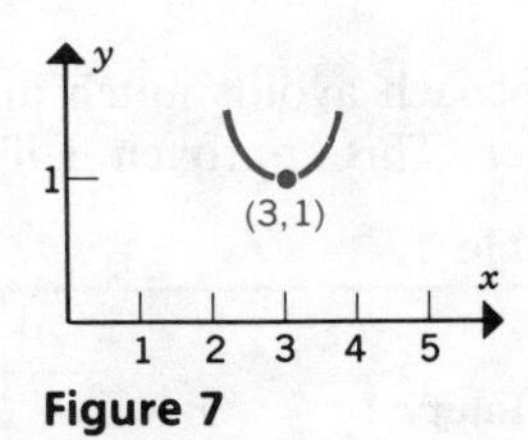

Figure 7

■

The first-derivative test can be simplified for functions that have two or more critical points and that are continuous for all values of x. Noting that changes in the sign of the derivative can occur only at critical points, we can use the x coordinates of the critical points as reference points in order to subdivide the x axis into intervals on which the function is either increasing or decreasing. Finding the sign of the derivative at an arbitrary point on each interval will provide us with enough information to determine the shape of the curve in the vicinity of each critical point. The procedure is illustrated in the next example.

Example 3 Use the first-derivative test to determine the shape of the function $f(x) = x^3 - 3x^2 - 9x + 10$ (see Example 1) in the vicinity of its critical points.

Solution We found that the function has two critical points $(-1, 15)$ and $(3, -17)$. Using $x = -1$ and $x = 3$ as reference points, we subdivide the x axis into the three intervals $x < -1$, $-1 < x < 3$, and $x > 3$, shown in Figure 8. Next, the sign of the derivative

$$f'(x) = 3x^2 - 6x - 9 = 3(x - 3)(x + 1)$$

is found at an arbitrary point on each interval to determine whether the curve is rising or falling on the interval.

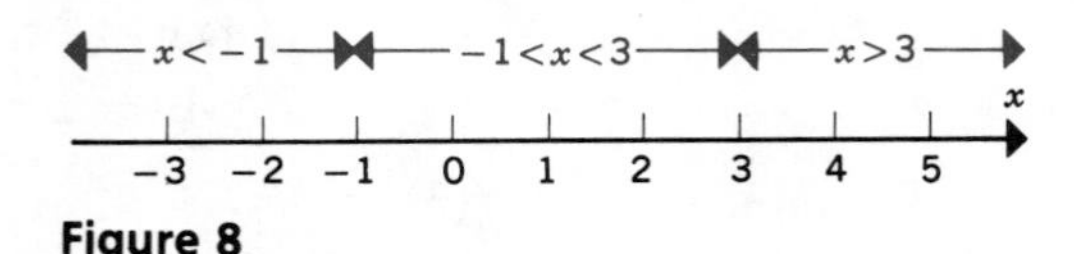

Figure 8

When $f'(x)$ is factorable, as it is here, its sign can be determined from the signs of its factors for a given value of x. For example, when $x = 2$,

$$f'(2) = 3(2 - 3)(2 + 1)$$

and the sign of $f'(2)$ can be found by representing the sign of each factor as $(+)$ or $(-)$ and then finding the sign of the product. In this case

$$\text{Sign of } f'(2) = (+)(-)(+)$$

Since the product of two positive numbers and one negative number yields a negative number, we get

$$\text{Sign of } f'(2) = (-)$$

This approach avoids much unnecessary arithmetic, particularly when x is not an integer. This approach is illustrated in Table 5 and Figure 9.

Table 5

		Signs of			
Interval	x	$x - 3$	$x + 1$	$f'(x) = 3(x-3)(x+1)$	Curve
$x < -1$	-2	$-$	$-$	$(+)(-)(-) = +$	Rising
$-1 < x < 3$	0	$-$	$+$	$(+)(-)(+) = -$	Falling
$x > 3$	4	$+$	$+$	$(+)(+)(+) = +$	Rising

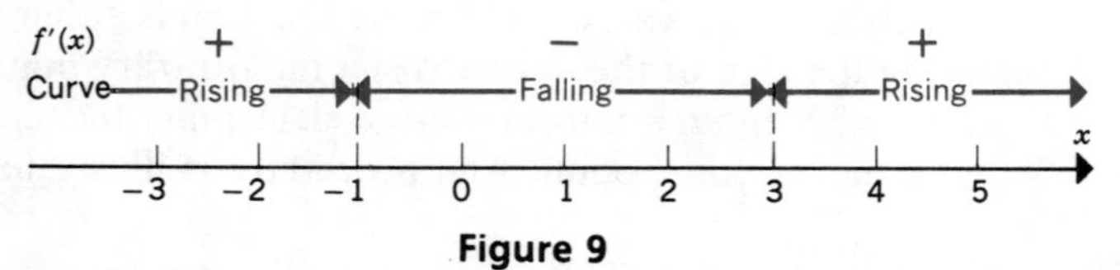

Figure 9

On the basis of this analysis, we can conclude that $(-1, 15)$ is a relative maximum and $(3, -17)$ is a relative minimum. A rough sketch of the curve in the vicinity of each point is shown in Figure 10.

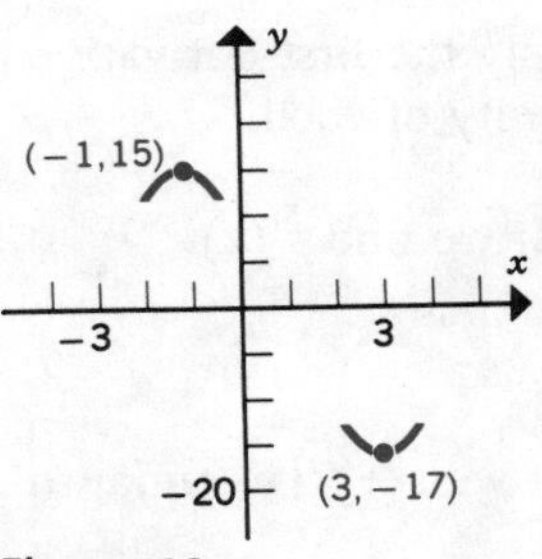

Figure 10

■

The next example shows that some critical points are neither maxima nor minima.

Example 4 Find the critical points of the function

$$f(x) = 2x^3 - 6x^2 + 6x + 1$$

and apply the first-derivative test to determine the behavior of the function in the vicinity of each critical point.

Solution First we find the critical points

$$f'(x) = 6x^2 - 12x + 6$$
$$0 = 6(x^2 - 2x + 1) = 6(x - 1)^2$$

Solving for x gives

$$x = 1, \qquad f(1) = 3$$

Next, we apply the first-derivative test. Because $f'(x) = 6(x - 1)^2 > 0$ for all values of x except $x = 1$, the curve is rising on both sides of (1, 3). From this we can conclude that (1, 3) is neither a maximum nor a minimum. A rough sketch of the curve in the vicinity of (1, 3) is shown in Figure 11. ■

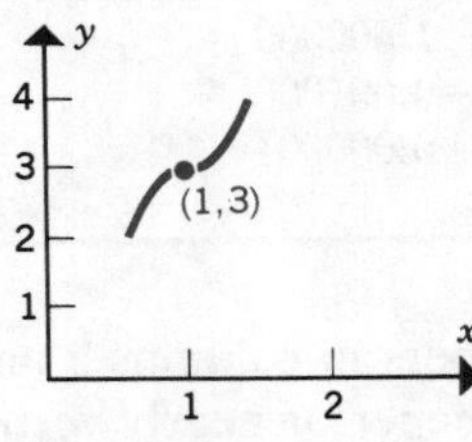

Figure 11

The next example illustrates a critical point at which $f'(x)$ does not exist.

Example 5 Find the critical points of the function

$$f(x) = \sqrt[3]{x^2} = x^{2/3}$$

and apply the first-derivative test to determine the behavior of the function in the vicinity of each.

Solution 1. First, we find $f'(x)$:

$$f'(x) = \frac{2}{3}x^{-1/3} = \frac{2}{3\sqrt[3]{x}}$$

Next we set $f'(x)$ equal to zero and solve for x if possible:

$$0 = \frac{2}{3\sqrt[3]{x}}$$

There are no values of x that satisfy this equation, so there are no critical points where $f'(x)$ equals zero. However, there is a point on the curve, namely (0, 0), where $f'(x)$ is not defined.

2. Since (0, 0) is the only critical point and $f'(x)$ is defined for all x except $x = 0$, we test for the sign of $f'(x)$ in the intervals $x < 0$ and $x > 0$, as shown in Table 6.

Table 6

Interval	Value of x	$f'(x)$	Curve
$x < 0$	-1	$-\frac{2}{3}$	Falling
$x > 0$	1	$+\frac{2}{3}$	Rising

These results indicate that the point (0, 0) is a minimum. To get a feeling for what is happening to the slope in the vicinity of (0, 0), let us look at Table 7, which is similar to those used in Section 2.1 where limits were introduced.

Table 7

From the Left		From the Right	
x	$f'(x) = \frac{2}{3\sqrt[3]{x}}$	x	$f'(x) = \frac{2}{3\sqrt[3]{x}}$
-1.000000	$-\frac{2}{3}$	1.000000	$\frac{2}{3}$
-0.001000	$-\frac{20}{3}$	0.001000	$\frac{20}{3}$
-0.000001	$-\frac{200}{3}$	0.000001	$\frac{200}{3}$
⋮	⋮	⋮	⋮

The numbers in columns 2 and 4 of Table 7 indicate that the tangent lines are getting steeper or nearly vertical as $x \to 0$. On the basis of this, a rough sketch of the curve in the vicinity of (0, 0) can be made, as shown in Figure 12.

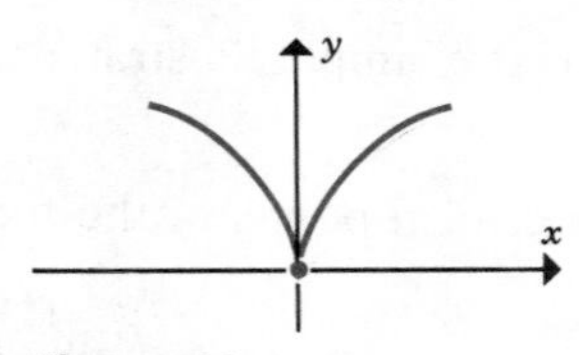

Figure 12

■

Rational Functions

When the first-derivative test is applied to a rational function, it is also important to find those values of x where the function is not defined because the sign of the derivative may change at such values. For example, the function $f(x) = \frac{1}{x^2}$ is not defined at $x = 0$; its derivative $f'(x) = -\frac{2}{x^3}$ is positive when $x < 0$ and negative when $x > 0$. The graph of the function is shown in Figure 13, which also displays the signs of $f'(x)$ on each interval.

Therefore, any analysis based on the first-derivative test should include not only the x coordinates of the critical points but also any values of x for which the function is not defined. These values together with the x coordinates of the critical points are then used to subdivide the x axis into distinct intervals over which the function is either increasing or decreasing. This approach is illustrated in the next example.

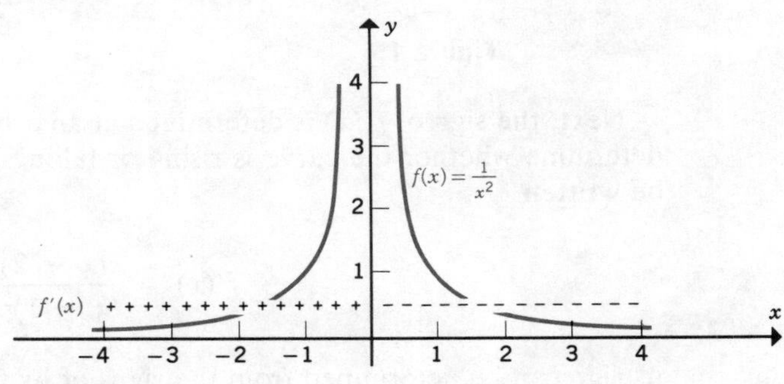

Figure 13

Example 6 Find the critical points of the function

$$f(x) = \frac{x^2}{x - 1}$$

and apply the first-derivative test to determine the shape of the curve in the vicinity of each critical point.

Solution First we find $f'(x)$ using the quotient rule. We get

$$f'(x) = \frac{2x(x - 1) - x^2(1)}{(x - 1)^2} = \frac{x(x - 2)}{(x - 1)^2}$$

Next, we set $f'(x)$ equal to zero and solve the equation for x:

$$0 = \frac{x(x - 2)}{(x - 1)^2}$$

Solutions: $x_1 = 0, \quad f(0) = 0$

$x_2 = 2, \quad f(2) = 4$

The points are shown in Figure 14. Before applying the first-derivative test, we note that the function is not defined at $x = 1$. With $x = 0, 1, 2$ as reference

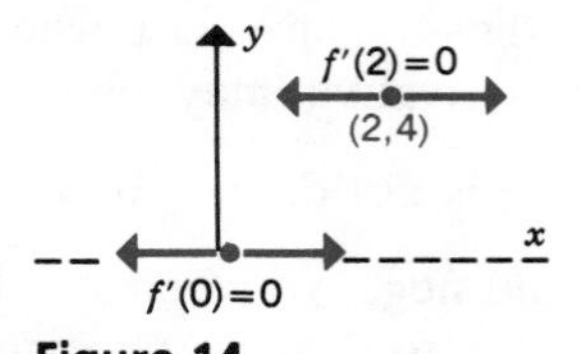

Figure 14

points, the x axis is subdivided into four separate intervals, as shown in Figure 15. The sign of $f'(x)$ remains the same for all x in each interval.

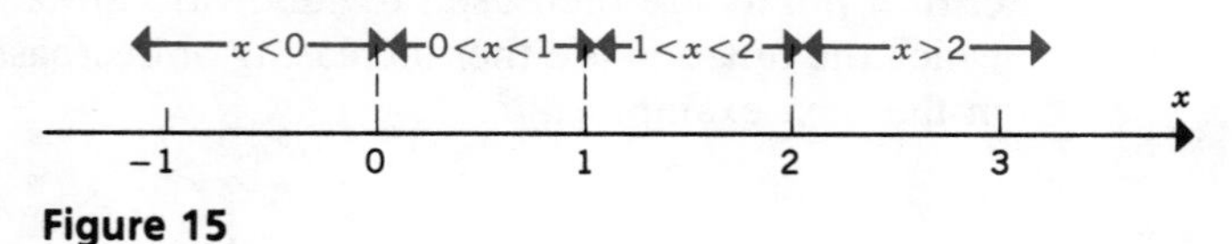

Figure 15

Next, the sign of $f'(x)$ is determined at an arbitrary point on each interval to determine whether the curve is rising or falling on the interval. Since $f'(x)$ can be written

$$f'(x) = \frac{x(x - 2)}{(x - 1)^2}$$

its sign can be determined from the signs of its factors, as illustrated in Table 8 and Figure 16. Based on the results of this analysis, we can conclude that (0, 0) is a relative maximum and (2, 4) a relative minimum. A rough sketch of the curve in the vicinity of each critical point is shown in Figure 17.

Table 8

Interval	x	Sign of $f'(x)$	$f(x)$
$x < 0$	-1	$\frac{(-)(-)}{(+)} = +$	Increasing
$0 < x < 1$	$\frac{1}{2}$	$\frac{(+)(-)}{(+)} = -$	Decreasing
$1 < x < 2$	$\frac{3}{2}$	$\frac{(+)(-)}{(+)} = -$	Decreasing
$x > 2$	3	$\frac{(+)(+)}{(+)} = +$	Increasing

■

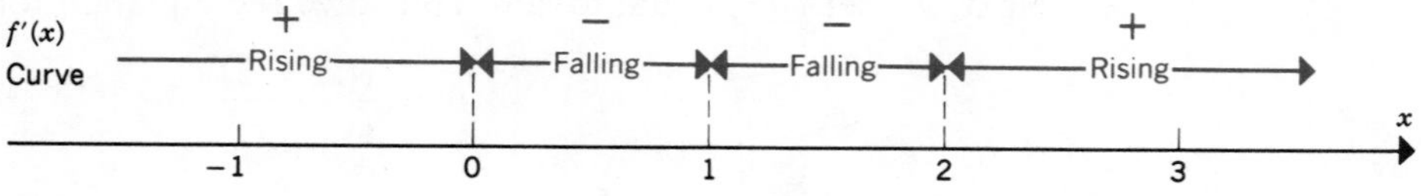

Figure 16

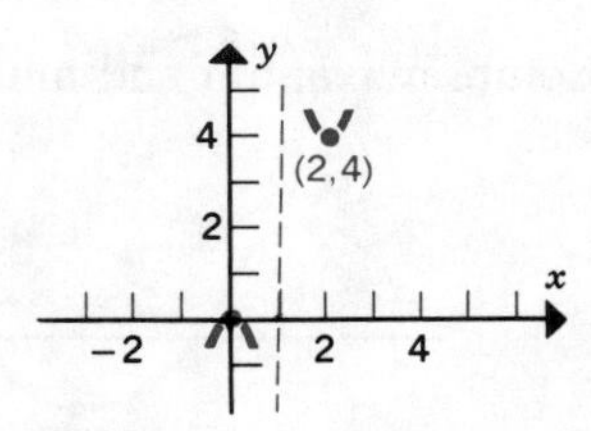

Figure 17

Absolute Maxima and Minima

The largest value that a function attains on its domain is called the **absolute maximum** of the function; similarly, the smallest value it attains is called its **absolute minimum**. A function may or may not have an absolute maximum or minimum. The graphs shown in Figure 18 illustrate the possibilities for some functions that are defined for all values of x.

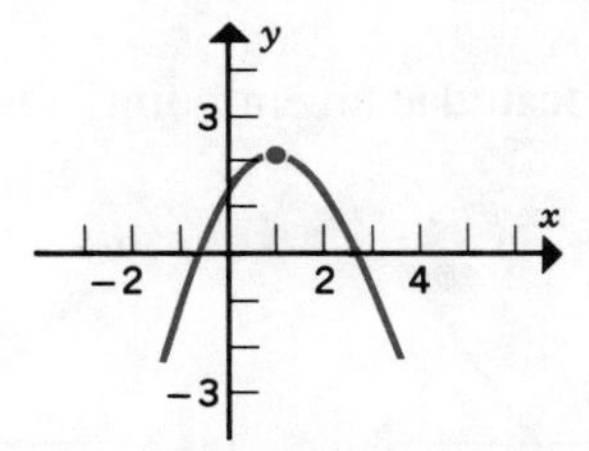

Absolute maximum at $x = 1$
No absolute minimum

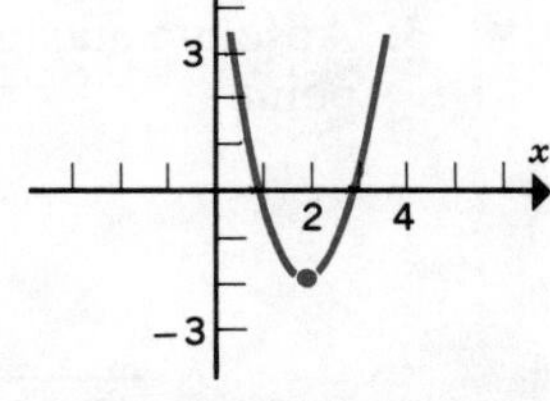

Absolute minimum at $x = 2$
No absolute maximum

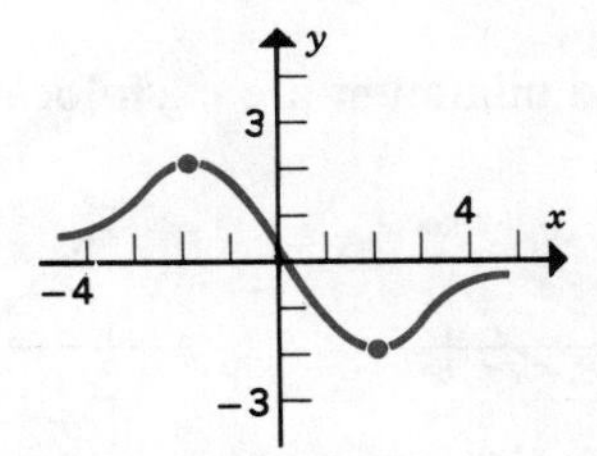

Absolute maximum at $x = -2$
Absolute minimum at $x = 2$

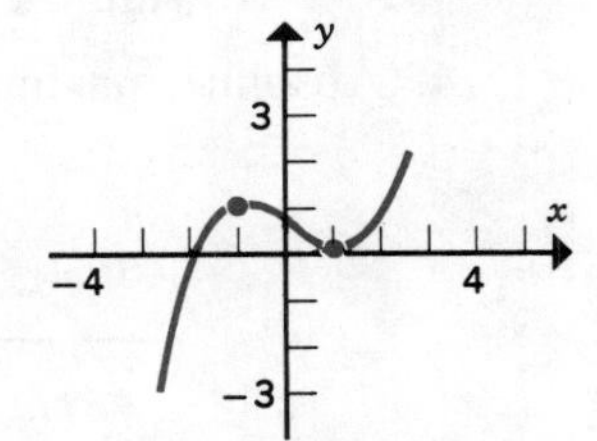

No absolute maximum
No absolute minimum
Relative maximum at $x = -1$
Relative minimum at $x = 1$

Figure 18

However, if the domain of the function is a closed interval, $a \leq x \leq b$, and if the function is continuous everywhere on the interval, the function will have an absolute maximum and an absolute minimum on the interval. They are located at either

1. The endpoints or
2. Interior points that are critical points.

Some of the possibilities are illustrated in Figures 19 through 22.

1. Absolute maximum and minimum occur at a critical point.

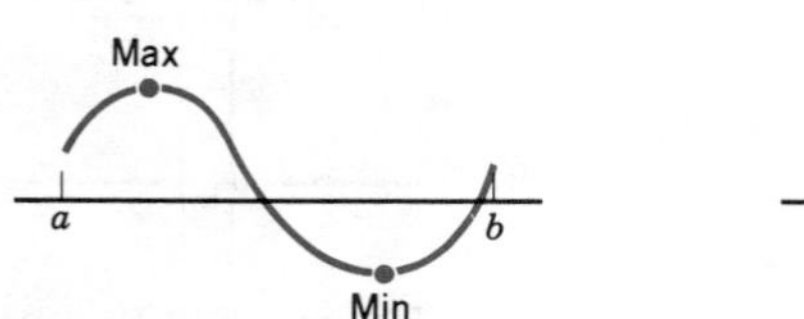

Figure 19

2. Absolute maximum is located at a critical point; absolute minimum is at an endpoint.

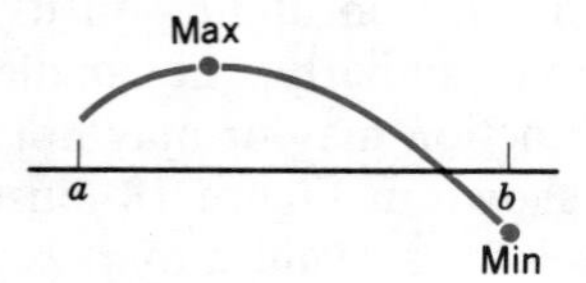

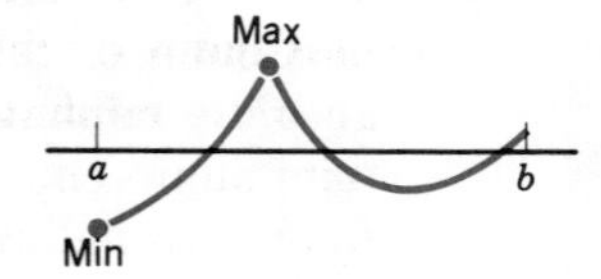

Figure 20

3. Absolute maximum is located at an endpoint; absolute minimum is at a critical point.

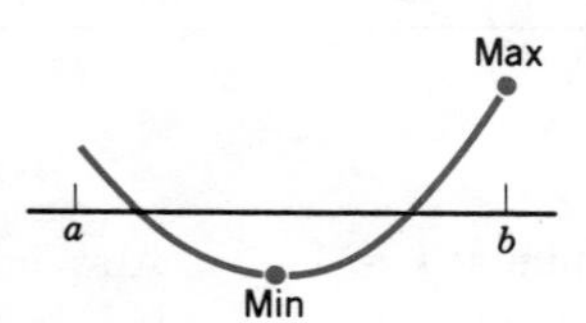

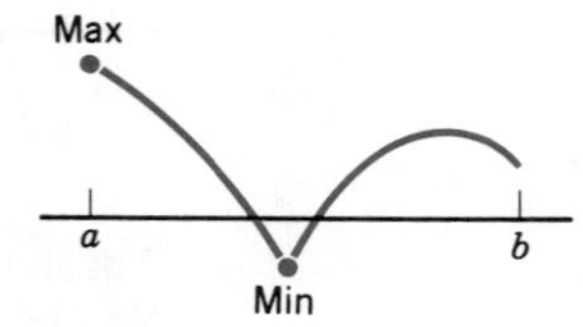

Figure 21

4. Absolute maximum and minimum are each located at an endpoint.

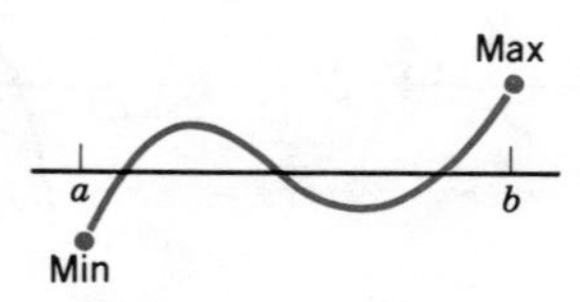

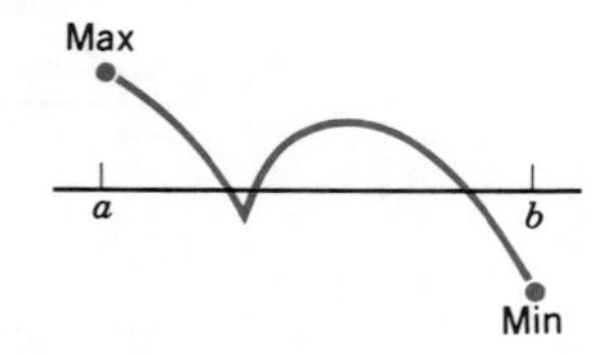

Figure 22

The following procedure is used to find the absolute maximum and minimum values of a continuous function on a closed interval $a \le x \le b$.

1. Find the y coordinates of the critical points on the closed interval.
2. Calculate $f(a)$ and $f(b)$, the y coordinates of the endpoints.
3. The largest and smallest values obtained from steps 1 and 2 represent the absolute maximum and minimum values of the function on the interval.

Example 7 Find the absolute maximum and minimum of the function $f(x) = x^2 - 6x + 2$ on the interval $0 \le x \le 5$.

Solution 1. The critical points are found by solving the equation

$$f'(x) = 2x - 6$$

for x, yielding

$$x = 3, \qquad f(3) = -7$$

2. The y coordinates of the endpoints are

$$f(0) = 2, \qquad f(5) = -3$$

3. Of the three y coordinates, 2 is the largest and -7 the smallest; therefore, the absolute maximum is 2 and the absolute minimum is -7. ■

Example 8 Find the absolute maximum and minimum of the function $f(x) = x^3 + 3x^2 + 4$ on the interval $-4 \le x \le 2$.

Solution 1. The critical points are found by solving the equation

$$f'(x) = 3x^2 + 6x = 3x(x + 2) = 0$$

The solutions are

$$x_1 = -2, \qquad f(-2) = 8$$
$$x_2 = 0, \qquad f(0) = 4$$

2. The y coordinates of the endpoints are obtained next:

$$f(-4) = -12, \qquad f(2) = 24$$

3. A comparison of the four values obtained in steps 1 and 2 indicates that the absolute maximum is 24, located at $x = 2$, and the absolute minimum is -12, located at $x = -4$. ■

Example 9 Find the absolute maximum and minimum of the function

$$f(x) = 3x^4 + 4x^3 - 36x^2 + 1$$

on the closed interval $-1 \le x \le 1$.

Solution 1. The critical points are found by solving the equation

$$f'(x) = 12x^3 + 12x^2 - 72x = 0$$
$$12x(x + 3)(x - 2) = 0$$

The solutions are

$$x_1 = 0, \qquad x_2 = -3, \qquad x_3 = 2$$

Only one of the critical values, $x_1 = 0$, is in the interval $-1 \le x \le 1$: the corresponding value of y is

$$f(0) = 1$$

2. The y coordinates of the endpoints are

$$f(-1) = -36, \qquad f(1) = -28$$

3. Of the three values found in steps 1 and 2, the largest is 1, located at $x = 0$, and the smallest is -36, located at $x = -1$. ■

4.1 EXERCISES

Find the critical points for each of the following functions. Use the first-derivative test to determine the shape of the curve in the vicinity of each critical point. Finally, make a rough sketch of the curve in the vicinity of each critical point.

1. $f(x) = x^2 + 2x$
2. $f(x) = x^2 - 4x + 3$
3. $f(x) = 2x^2 - 4x - 3$
4. $F(x) = 8x - 2x^2$
5. $f(x) = x^2 + 5x - 1$
6. $g(x) = x^3 - 6x^2$
7. $f(x) = x^3 + 3x^2 - 1$
8. $f(x) = x^4 - 4x + 2$
9. $f(x) = 2x^3 - 3x^2 - 36x + 50$
10. $f(x) = x^3 + 6x^2 - 4$
11. $F(x) = x^4 - 2x^2 + 1$
12. $f(x) = x^3 + 2$
13. $f(x) = (x - 1)^2(x - 4)$
14. $g(x) = (x - 1)^2(x + 2)^2$
15. $f(x) = (x^2 + 1)^4$
16. $f(x) = x(x^2 + 1)^4$
17. $f(x) = \dfrac{x^2}{x - 2}$
18. $f(x) = \dfrac{2}{x^2 + 1}$
19. $g(x) = \dfrac{x}{x^2 + 1}$
20. $f(x) = x + \dfrac{1}{x}$
21. $f(x) = 3x^5 - 5x^3 + 1$
22. $f(x) = \sqrt[3]{x^4}$

Find the absolute maximum and minimum for each of the following functions on the given interval.

23. $f(x) = x^2 - x, \quad -1 \le x \le 3$
24. $f(x) = 7 - 4x - x^2, \quad -1 \le x \le 3$
25. $f(x) = x^3 + 1, \quad -1 \le x \le 1$
26. $f(x) = 2x^3 - 9x^2 - 24x, \quad -2 \le x \le 10$
27. $f(x) = \dfrac{1}{x}, \quad 1 \le x \le 3$
28. $f(x) = x^3 - 3x^2 + 5, \quad 1 \le x \le 4$
29. $f(x) = 4\sqrt{x} - x, \quad 1 \le x \le 16$
30. $f(x) = 4x^5 - 5x^3, \quad -2 \le x \le 2$
31. $f(x) = 2x - \dfrac{1}{x}, \quad 1 \le x \le 5$
32. $f(x) = 3x^4 - 4x^3 - 1, \quad -1 \le x \le 2$
33. Find the values of the constants A and B if the function $f(x) = Ax^2 + Bx$ has a critical point at (2, 4). In addition, determine the shape of the curve in the vicinity of (2, 4).
34. Find the values of the constants A and B if the function $f(x) = Ax^3 + Bx^2$ has a critical point at (2, -6). In addition, determine the shape of the curve in the vicinity of (2, -6).

35. Find the values of the constants A and B if the function

$$f(x) = Ax + \frac{B}{x}$$

has a critical point at (2, 8). In addition, determine the shape of the curve in the vicinity of (2, 8).

36. Find the values of the constants A, B, and C if the function $f(x) = Ax^2 + Bx + C$ has a critical point at $(1, -3)$ and a y intercept equal to -5. In addition, determine the shape of the curve in the vicinity of $(1, -3)$.

Use the first-derivative test to match each of the functions whose derivatives are given in Exercises 37 through 40 with one of the graphs shown in Figure 23.

37. $f'(x) = x(x + 2)(x - 2)$

38. $f'(x) = x^2(x + 2)(x - 2)$

39. $f'(x) = x^2(x + 2)(2 - x)$

40. $f'(x) = x(x + 2)(x^2 - 4x + 4)$

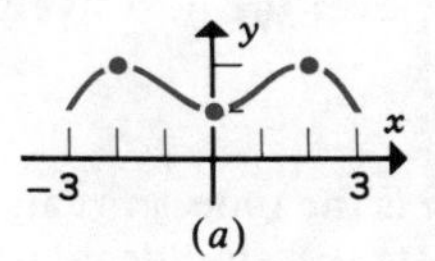

(a)

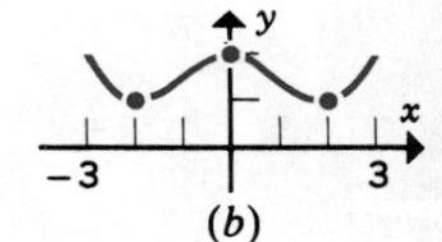

(b)

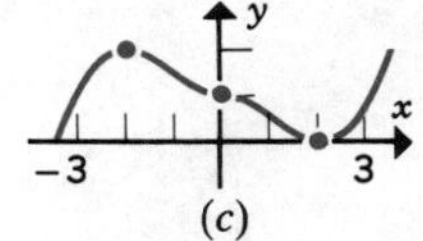

(c)

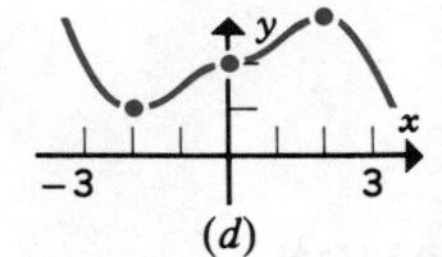

(d)

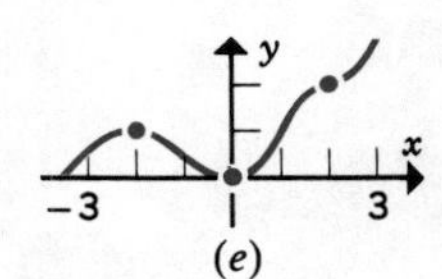

(e)

Figure 23

Use the first-derivative test to match each of the functions whose derivatives are given in Exercises 41 through 44 with one of the graphs shown in Figure 24.

41. $f'(x) = \dfrac{x(x - 4)}{x - 2}$

42. $f'(x) = \dfrac{x^2(x - 4)}{x - 2}$

43. $f'(x) = \dfrac{x(x - 4)}{(x - 2)^2}$

44. $f'(x) = \dfrac{x(x^2 - 8x + 16)}{2 - x}$

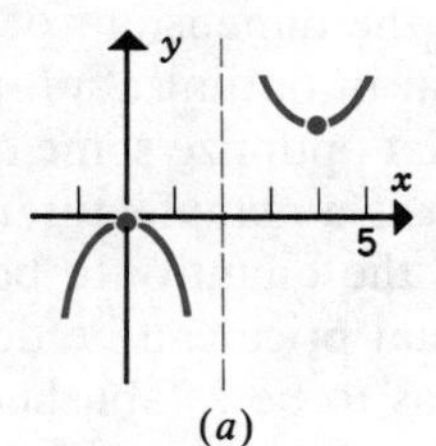

(a)

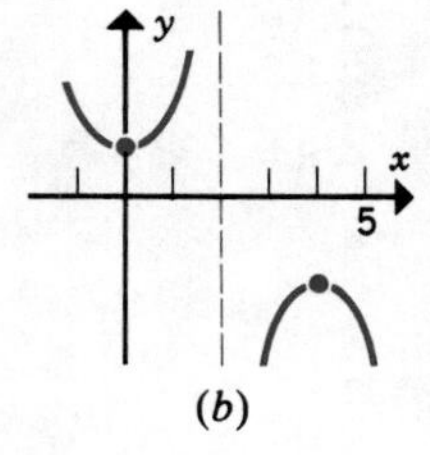

(b)

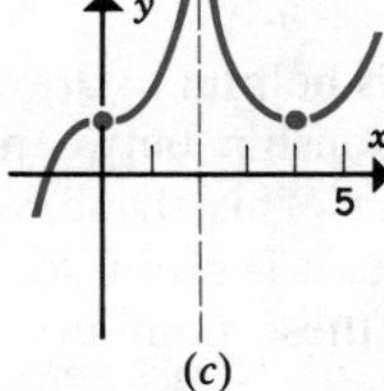

(c)

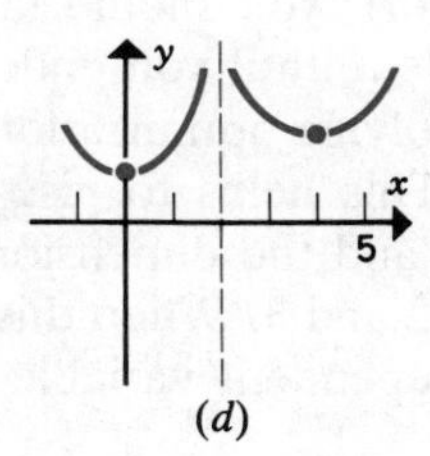

(d)

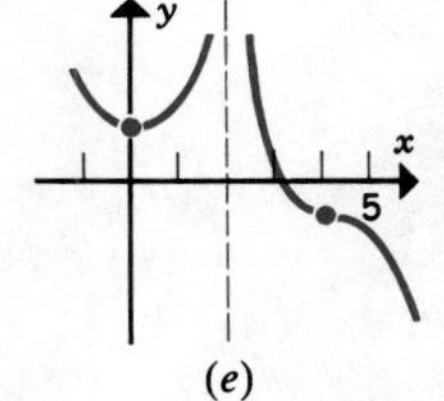

(e)

Figure 24

45. The height $H(t)$, in inches, of a plant t weeks after it breaks ground is given by the equation

$$H(t) = 8\sqrt{t} - t, \qquad 0 \le t \le 20$$

When does the plant reach its maximum height?

46. A soft-drink company begins a ten-week campaign for its new caffeine-free diet cola. Weekly sales $S(t)$ of the cola, in thousands of dollars, t weeks after the campaign begins are described by the equation

$$S(t) = 100 + 40t - t^2, \qquad 0 \le t \le 30$$

How long after the campaign begins do weekly sales reach their maximum level?

47. The supply of blood $S(t)$, in pints, at a local hospital is given by the equation

$$S(t) = 24 - 8t + t^2, \qquad 0 \le t \le 10$$

where t is the time, in days, from today ($t = 0$). Determine when the supply of blood reaches its minimum level.

48. An analyst for a major airline estimates that annual fuel costs $C(t)$, in millions of dollars, over the next five years can be described by the equation

$$C(t) = \frac{96 + t^2}{10 + t}$$

where t is the time, in years, from today ($t = 0$). Determine the time at which annual fuel costs will stop decreasing and begin to rise.

4.2 OPTIMIZATION

In many situations it is desirable to maximize or minimize some quantity; problems of this kind are called **optimization** problems. For example, a manufacturer will try to set the selling price of an item so as to **maximize** profits; this price is called the **optimum** price. On the other hand, a design engineer may want to determine the dimensions of the item that **minimizes** the cost of material.

The goal in optimization problems is to find the values of the independent variable that optimize some quantity that is the dependent variable. If the problem concerns a manufacturer, the independent variable might be p, the selling price, and the quantity to be optimized might be P, profits. However, before the optimum price can be determined, the mathematical relationship between p and P has to be established; that is, the problem must be "translated" from English to mathematics. Since most of your time and effort will be concentrated on this effort, you should take enough time to read and, if necessary, reread each problem until you understand it.

When solving optimization problems, it is helpful to draw a picture whenever possible. This helps to visualize the relationship between the quantity to be optimized and the dimensions of the object. This strategy is illustrated in Examples 1, 2 and 3. When this is not possible, it is useful to assign specific values to the independent variable and then use these numbers to calculate the corresponding values of the dependent variable, as illustrated in Example 4. This arithmetic can often help in establishing the algebraic relationship between the independent and dependent variables.

Example 1 The recreation department in a small town has been authorized to set up a rectangular playground whose area is 10,000 ft². The playground is to be enclosed on all four sides by a fence. What are the dimensions of the rectangle that minimize the number of feet of fencing required to enclose the playground?

Solution Before you try to solve a problem mathematically, it is sometimes preferable to visualize it by drawing a diagram and considering a number of different configurations. Some extreme cases might even be quite illuminating; one of the configurations that will produce a 10,000-ft² plot, shown in Figure 1, is a field measuring 1 by 10,000 ft. Aside from the fact that it is highly unsuitable for a

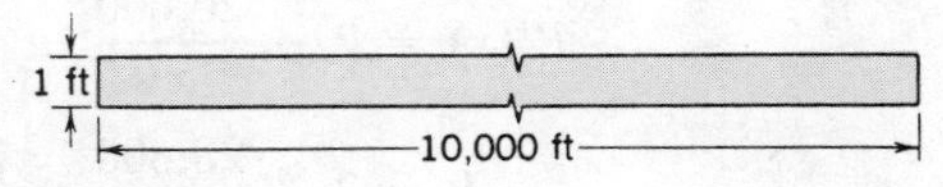

Figure 1

playground, it has the dubious distinction of requiring 20,002 ft (almost 4 miles) of fencing to enclose it, a situation that would undoubtedly please the fencing supplier. The configuration can be improved slightly by doubling the width to 2 ft and reducing the length to 5000 ft (Figure 2); the number of linear feet

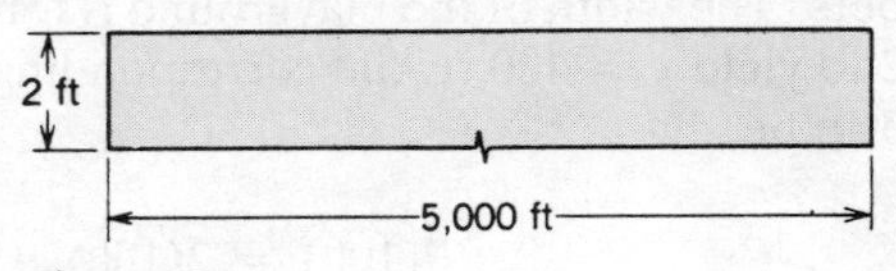

Figure 2

needed to enclose the field has been reduced to 10,004 ft. Notice that the area is the same in both cases. The optimum configuration can be determined once the mathematical relationship between P, the perimeter, and the dimensions of the rectangle has been established. Let x be the length and y the width of the rectangle (Figure 3); P can be written as

$$P = 2x + 2y, \qquad x, y > 0 \tag{1}$$

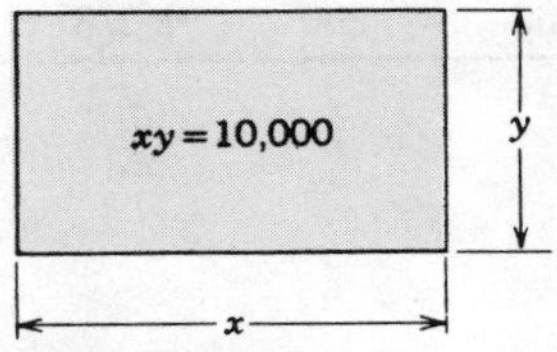

Figure 3

The expression for P contains two variables, x and y. However, they are not independent variables because A, the area, must be kept constant at 10,000 ft^2, so $A = xy = 10{,}000$. This relationship allows us to write P in terms of only one variable. Noting that y can be written in terms of x as

$$y = \frac{10{,}000}{x} \tag{2}$$

we can substitute this result in Equation 1, and P can be written as a function of x alone:

$$P(x) = 2x + 2\left(\frac{10{,}000}{x}\right) = 2x + \frac{20{,}000}{x}, \qquad x > 0 \tag{3}$$

The next step is to determine the value of x that minimizes $P(x)$. This is done by finding $P'(x)$, setting it equal to zero, and solving the resulting equation for x:

$$P'(x) = 2 - \frac{20{,}000}{x^2}$$

$$0 = 2 - \frac{20{,}000}{x^2} = \frac{2x^2 - 20{,}000}{x^2}$$

$$0 = 2x^2 - 20{,}000 = 2(x + 100)(x - 100)$$

Solutions: $x_1 = 100$ ft, $\quad x_2 = -100$ ft

The negative solution is discarded because the dimensions are positive numbers. The width of the playground is found by substituting $x = 100$ into Equation 2 to yield $y = 100$ ft; the corresponding value of $P(100)$ is found from Equation 3 to be

$$P(100) = 2(100) + 2\left(\frac{10{,}000}{100}\right) = 400 \text{ ft}$$

The first-derivative test can be applied to make sure that $P(100)$ represents a minimum (Table 1). From this, we can conclude that (100, 400) is a minimum. The graph of P is shown in Figure 4.

Table 1

Interval	x	$P'(x)$	$P(x)$
$0 < x < 100$	10	-198	Decreasing
$x > 100$	200	$+2/3$	Increasing

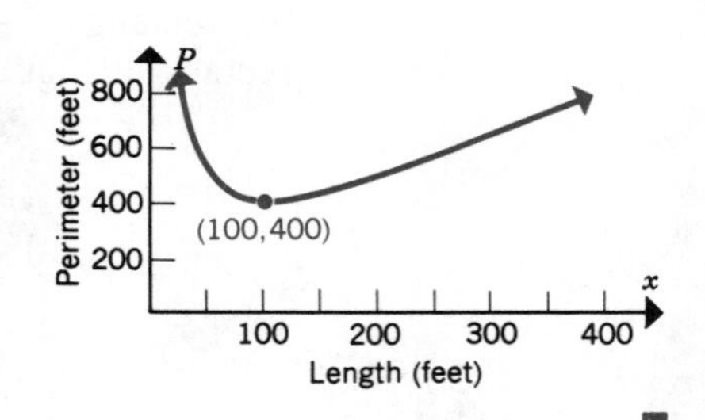

■

Example 2 Suppose that the recreation department decides to add a wading pool to the playground and to separate it from the rest of the equipment with a fence running across the width of the field as shown in Figure 5. What are the dimensions of the rectangle that will minimize the length of fencing required if the total area enclosed remains 10,000 ft^2?

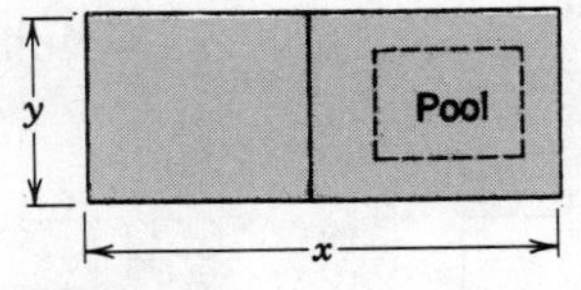

Figure 5

Solution As before, let x designate the length and y the width of the playground; they are related via the equation

$$xy = 10{,}000$$

The number of feet of fencing, L, can be written as

$$L = 2x + 3y$$

Proceeding as in Example 1, we express L in terms of x alone, getting

$$L(x) = 2x + \frac{30{,}000}{x}$$

Next, we differentiate and set the derivative equal to zero

$$L'(x) = 2 - \frac{30{,}000}{x^2} = \frac{2x^2 - 30{,}000}{x^2}$$

$$0 = 2x^2 - 30{,}000$$

$$x^2 = 15{,}000$$

which yields

$$x = \sqrt{15{,}000} \approx 122.47 \text{ ft}, \qquad y = \frac{10{,}000}{122.47} \approx 81.65 \text{ ft}$$

as the values of x and y that minimize the amount of fencing required. The minimum amount of fencing can be calculated, giving

$$L(\sqrt{15{,}000}) \approx 489.9 \text{ ft}$$

■

Example 3 A flat piece of sheet metal measuring 30 by 30 in. is to be used for making an open box to hold rivets that did not pass inspection in a quality control process. The box is to be made by cutting identical square pieces from the four corners and folding up the flaps to complete the box, as shown in Figure 6. The length and width of the cutout pieces are denoted by x. For what value of x is the volume of the box maximized?

Solution 1 To see how the volume $V(x)$ of the box depends on the size of the cutout pieces, we will consider a specific example or two before generalizing. Integral values of x will be used because the arithmetic is easier.

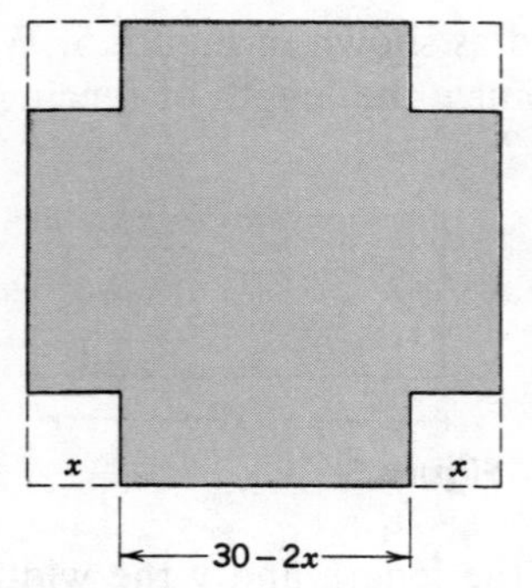

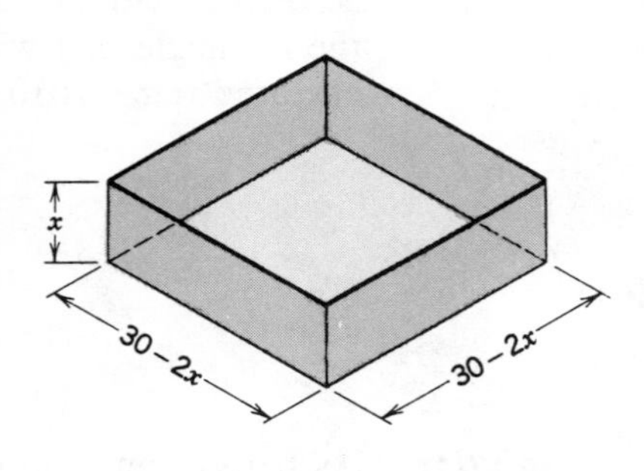

Figure 6

$x = 1$ Base of box measures 28 by 28 in., height = 1 in.
$V(1) = $ area of base × height = $28 \times 28 \times 1 = 784$ in.^{3}
$x = 2$ Base of box measures 26 by 26 in., height = 2 in.
$V(2) = 26 \times 26 \times 2 = 1354$ in.^{3}
⋮
$x = 14$ Base measures 2 by 2 in., height = 14 in.
$V(14) = 2 \times 2 \times 14 = 56$ in.^{3}

If x is arbitrary, the base measures $(30 - 2x) \times (30 - 2x)$ inches, the height is x, so $V(x)$ can be written as

$$V(x) = x(30 - 2x)^2, \qquad 0 \le x \le 15$$

where the inequality indicates the range of values that can be assigned to the variable x. This is an important part of the problem because the equation $V'(x) = 0$ may have solutions outside this range. Explicitly denoting the domain will reduce the likelihood of accepting as a bona fide solution a value of x that is not realistic.

We can use the product rule to find $V'(x)$, and we get

$$V'(x) = -4x(30 - 2x) + (30 - 2x)^2 = (30 - 2x)(30 - 6x)$$

Setting $V'(x)$ equal to zero and solving gives

$$0 = 30 - 2x \quad \text{and} \quad 0 = 30 - 6x$$

Solutions: $x_1 = 15$ in., $x_2 = 5$ in.

The corresponding values of the volume are $V(15) = 0$ and $V(5) = 2000$ in.^{3}. Because the solution $x = 5$ is the only critical value for which $V(x)$ is positive and because $V(x)$ equals zero at the endpoints, $x = 0$ and $x = 15$, we can conclude that the volume is maximized when the dimensions of each cutout piece are 5 by 5 in.

Solution 2 *Another Approach.* The flexibility available in solving problems of this type can be demonstrated by letting x equal the length and width of the box, as shown in Figure 7. The volume $V(x)$ can be written as

$$V(x) = x^2 \frac{30 - x}{2} = 15x^2 - \frac{x^3}{2}, \qquad 0 \le x \le 30$$

$$V'(x) = 30x - \frac{3x^2}{2}$$

Setting $V'(x)$ equal to zero yields

$$0 = 30x - \frac{3x^2}{2}$$

$$= \frac{60x - 3x^2}{2}$$

$$0 = 60x - 3x^2 = 3x(20 - x)$$

Solutions: $x_1 = 0, \quad V_1(0) = 0$
$x_2 = 20, \quad V_2(20) = 2000 \text{ in.}^3$

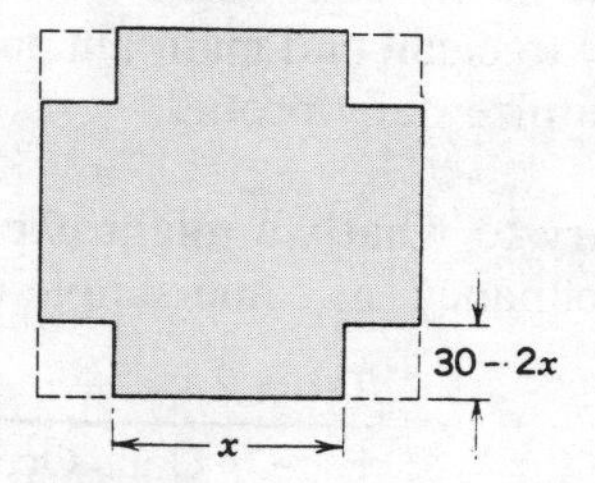

Figure 7

■

Since the process of "translating" a word problem into an equivalent mathematical statement is the most challenging part of optimization, we have listed some steps to assist you in solving optimization problems.

Steps for Solving an Optimization Problem

1. Read the problem carefully. If necessary, read it repeatedly until you understand it.
2. If possible, sketch a picture and label the dimensions.
3. Determine the quantity that is to be maximized or minimized; designate it with an appropriate letter such as L for length, P for profit, C for cost, etc.
4. Label all other variables in the problem. Write down the equations that express the relationships between these variables (these equations are called "constraint equations").
5. Write an equation that describes the functional relationship between the quantity to be maximized or minimized and the variables defined in step four.

6. **Use the constraint equations from step four to express the quantity to be maximized or minimized as a function of a single variable.**
7. **Find the critical points for the function found in step 6. Evaluate the function at each critical point together with any endpoints to determine the optimal value.**

Note: Before you attempt to carry out these steps, it is sometimes helpful to carry out some sample calculations to develop a "feel" for the mathematical relationship between the quantity to be optimized and the other variables.

Profit Maximization

Example 4 The owners of the Economy Motel have observed that all 100 rooms can be rented if they charge \$40 per day for each unit. Whenever they attempt to increase the daily rate, occupancy declines. They estimate that for each dollar increase in the daily rate, two additional units go unoccupied. It costs them \$2 per day to clean and maintain each occupied unit. What rate should they charge to maximize daily profit?

Solution Contrary to what we might expect, maximum profit is not always achieved at full occupancy, as some simple calculations will reveal (Table 2).

Table 2

Rate	Units Occupied	Revenue	Cost	Profit
40	100	4000	200	3800
41	98	4018	196	3822
42	96	4032	192	3840
⋮	⋮	⋮	⋮	⋮

Let x represent the daily charge. The number of rooms occupied depends on the daily charge:

$$\begin{aligned}\text{Number of rooms occupied} &= 100 - 2(x - 40)\\ &= 180 - 2x, \qquad x \geq 40\end{aligned}$$

The daily revenue $R(x)$ can be written as

$$\begin{aligned}R(x) &= (\text{daily rate})(\text{number of rooms occupied})\\ &= x(180 - 2x) = 180x - 2x^2\end{aligned}$$

The cost of cleaning and maintenance, $C(x)$, can be written

$$C(x) = 2(180 - 2x) = 360 - 4x$$

The daily profit $P(x)$ can be found next:

$$\begin{aligned}P(x) &= R(x) - C(x)\\ &= 184x - 2x^2 - 360\end{aligned}$$

The value of x that maximizes $P(x)$ can be found as follows:

$$P'(x) = 184 - 4x \tag{4}$$
$$0 = 184 - 4x$$

Solution: $x = \$46, \quad P(46) = \$3872$

Number of rooms occupied $= 180 - 2(46) = 88$

The first-derivative test can be used to show that $P(46)$ is a maximum. Writing $P'(x) = 4(46 - x)$, we see that $P'(x) > 0$ when $x < 46$ and $P'(x) < 0$ when $x > 46$; therefore, we conclude that $P(46)$ is a maximum. ■

Inventory Model

These optimization techniques can also be applied to the problem of minimizing the total costs associated with producing and storing a product, as illustrated in the next example.

Example 5 The Rain-and-Snow Tire Company has a factory in New Hampshire that produces and distributes, among other items, steel-belted radial tires for the New England area. Demand for these tires is consistent throughout the year and amounts to 2,700,000 tires annually. Because of the numerous items produced at this plant, production of the tires is not a continuous process; they are produced in large batches and then stored in a warehouse nearby, from which they are then shipped as orders come in. When the inventory becomes very low, another batch is produced and brought to the warehouse. Then no more are produced until the number runs low again.

Management wants to minimize the annual cost of producing and storing the tires. If they decide to produce the tires all at once, the **storage** costs, for warehouse space and handling, are very high, but the **setup** costs, for preparing the machinery for production, are low. On the other hand, if the tires are produced in small batches many times a year, the reverse is true. Management wants to find a production schedule for which the two costs will be balanced.

To visualize the inventory as a function of time, suppose that the tires are produced in batches of 900,000. This indicates that three production runs per year are required and that the inventory level over the 12-month period looks like the one shown in Figure 8. The vertical segments on the graph represent the replenishment of the inventory during production, and the declining segments represent depletion of the inventory as orders from wholesalers are filled. Management wants to know how many tires should be produced in each batch to minimize annual costs, based on the following estimates.

Setup costs per run are \$18,000.

Material and labor costs are \$5 per tire.

Storage cost is \$3 per tire per year.

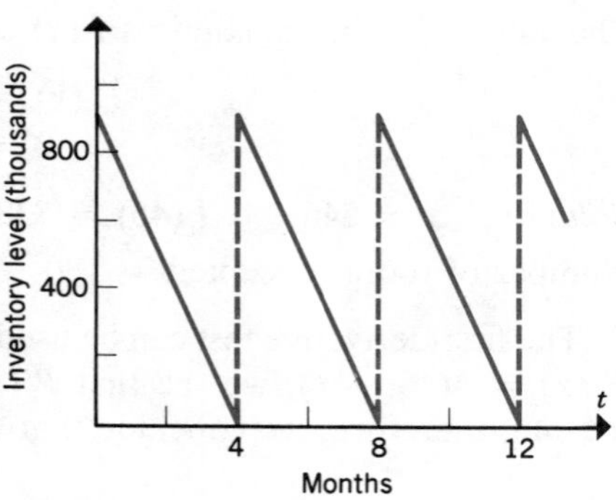

Figure 8

Solution The total annual cost of this operation consists of three terms

Total annual cost = annual set up costs
+ annual material and labor costs + annual storage costs

The number of items produced during each production run is called the **batch size.** If we let x be the batch size, the three types of costs are found as follows.

1. Annual setup costs equal the cost per setup times the number of setups per year. The number of setups equals annual demand divided by batch size. For example, if 900,000 tires are produced in each batch, the number of setups is $\frac{2{,}700{,}000}{900{,}000} = 3$. So

$$\text{Annual setup costs} = 18{,}000\left(\frac{2{,}700{,}000}{x}\right) = \frac{(486)(10^8)}{x}$$

2. $$\begin{aligned}\text{Annual material and labor cost} &= \text{number of tires produced annually} \times \text{material and labor cost per tire}\\ &= (2{,}700{,}000)(5) = 13{,}500{,}000\end{aligned}$$

3. Annual storage or inventory costs are probably one of the more difficult items to get a grip on because the inventory does not remain constant. However, if the average number of items in storage during the year were known, the annual inventory costs could be found by determining the product of the annual storage costs per tire and the average number of tires in inventory. Because the inventory level declines uniformly from a maximum of x units to zero units during each cycle, the average number of items in inventory during the year equals $\frac{x}{2}$, as shown in Figure 9. Therefore, annual storage costs are $\frac{3x}{2}$.

Putting all this together gives us the following equation for $C(x)$, the annual cost function:

$$C(x) = \frac{(486)(10^8)}{x} + \frac{3x}{2} + 13{,}500{,}000 \tag{4}$$

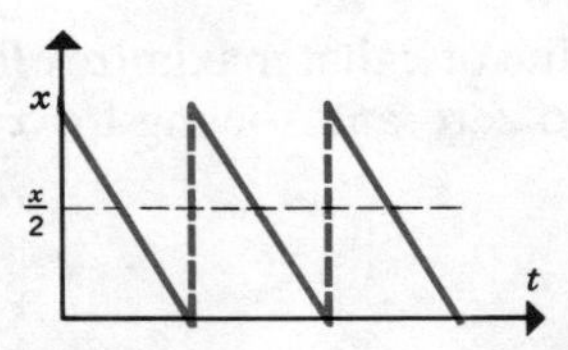

Figure 9

The value of x that minimizes $C(x)$ is found in the usual manner:

$$C'(x) = -\frac{(486)(10^8)}{x^2} + \frac{3}{2}$$

$$0 = -\frac{(486)(10^8)}{x^2} + \frac{3}{2}$$

Multiplying both sides by $2x^2$ gives

$$0 = -(972)(10^8) + 3x^2$$

Rewriting gives

$$0 = -324(10^8) + x^2$$

Solutions: $x_1 = 180{,}000, \quad x_2 = -180{,}000$

The negative solution is ignored; so the tires should be produced in batches of 180,000 tires. The number of batches or production runs equals $\dfrac{2{,}700{,}000}{180{,}000} = 15.$

The first-derivative test can be used to show that $C(x)$ is a minimum when $x = 180{,}000$, as demonstrated in Table 3.

Table 3

Interval	x	$C'(x)$	Curve
$x < 180{,}000$	100,000	-3.66	Falling
$x > 180{,}000$	200,000	$+0.29$	Rising

The cost for this production schedule is found from Equation 4 to be

$$C(180{,}000) = \$14{,}040{,}000$$

■

Example 6 The monthly revenue and cost functions for a firm selling x units of large screen television sets are given by the equations

$$R(x) = 1500x - 4x^2, \qquad C(x) = x^2 + 500x + 20{,}000$$

Find the number of units that maximizes the company's profits.

Solution The profit function $P(x)$ is found from the equation

$$P(x) = R(x) - C(x)$$
$$= 1000x - 5x^2 - 20{,}000$$

The value of x that maximizes $P(x)$ is found by setting $P'(x)$, the marginal profit, equal to zero and solving for x:

$$P'(x) = 1000 - 10x$$
$$0 = 1000 - 10x$$

Solution: $x = 100, \quad P(100) = \$30{,}000$

There is another way to find the optimum value of x. Since $P(x) = R(x) - C(x)$, it follows that

$$P'(x) = R'(x) - C'(x) \tag{5}$$

The profit function $P(x)$ is optimized when $P'(x) = 0$, so we get

$$0 = R'(x) - C'(x),$$

or

$$\boxed{R'(x) = C'(x)}$$

That is, the maximum profit is attained when the marginal revenue equals the marginal cost. This means that a solution can be obtained by differentiating $R(x)$ and $C(x)$ and setting them equal to one another, yielding

$$1500 - 8x = 500 + 2x$$

Solution: $x = 100$

Figure 10 shows the graphs of $R'(x)$, $C'(x)$, and $P(x)$ to illustrate the principle that profit is maximized when $R'(x) = C'(x)$.

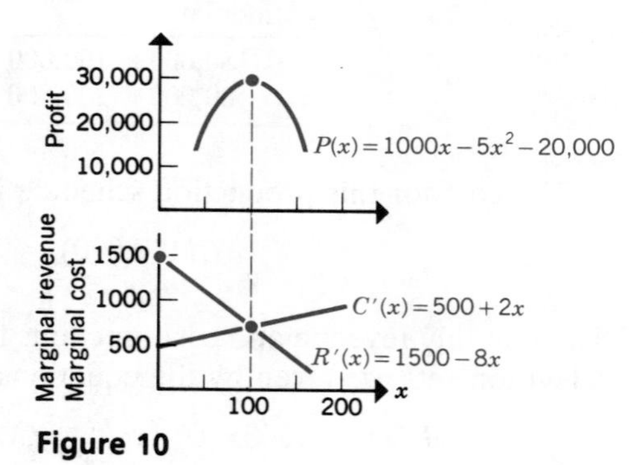

Figure 10

■

4.2 EXERCISES

1. A piece of wire 12 ft long is to be bent into a rectangle. What is the largest possible area that can be enclosed by the rectangle?

2. Let x and y be two positive numbers whose sum is 80. Which pair has the largest product?
3. Find two positive numbers whose sum is 78 and whose product is a maximum.
4. Find two numbers whose difference is 14 and whose product is a minimum.
5. Find two positive numbers whose product is 256 and whose sum is a minimum.
6. One hundred feet of fencing are available for enclosing a rectangular plot to be used for a small vegetable garden. What are the dimensions of the plot that will maximize the area of the garden?
7. The Supreme Boat Rental Company wants to install a fence around a rectangular plot of land where its canoes and rowboats can be stored. If 1800 ft^2 of land are to be enclosed and if fencing is needed on only three sides, what are the dimensions of the plot that will require the minimum length of fence (Figure 11)?

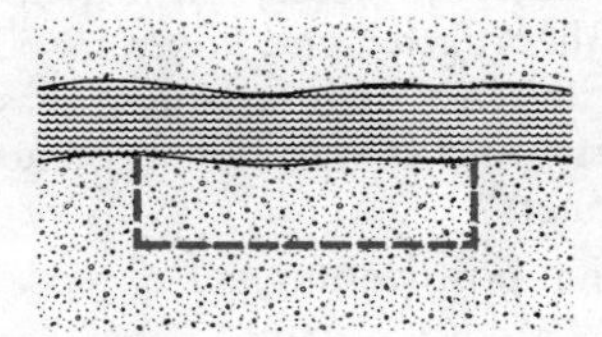

Figure 11

8. The manager of a building wants to build a 1250-ft^2 rectangular garden, which is to be divided into three equal rectangular segments (Figure 12). Find the dimensions of the plot that will minimize the length of fence that encloses the garden.

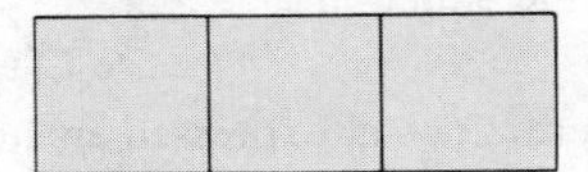

Figure 12

9. A field located next to a main highway is to be fenced in for storage of the state highway department's heavy equipment. The field is to have an area of 20,000 ft^2. What are the dimensions of the rectangular field that will minimize the cost of fencing if material for the side running along the highway costs \$3 per linear foot and material for the other three sides costs \$1 per linear foot?
10. A closed rectangular box is to be made with a square base and to have a volume of 324 $in.^3$. If the material for the sides and top cost 1¢ per square inch, and that for the base costs 2¢ per square inch, what are the dimensions of the box that minimize the cost of materials?
11. An open rectangular box is to be made with a square base and is to have a volume of 6000 $in.^3$. If the material for the sides costs 2¢ per square inch, and that for the base costs 3¢ per square inch, what are the dimensions of the box that minimize the cost of materials?
12. Identical square pieces are cut out from the four corners of a thin piece of cardboard measuring 25 by 25 in.; the flaps are then turned up to make an open rectangular box. What are the dimensions of the cutout pieces that maximize the volume?
13. Identical square pieces are cut out from the four corners of a thin piece of cardboard measuring 8 by 15 in.; the flaps are then turned up to make an open rectangular

box. What are the dimensions of the cutout pieces that maximize the volume? What is the maximum volume?

14. A 210-room hotel is filled when the room rate is \$50 per day. For each \$1 increase in the rate, three fewer rooms are rented. Find the room rate that maximizes daily revenue.

15. For the hotel in exercise 14, it costs \$4 per day to clean and maintain each room occupied. What room rate maximizes daily profit?

16. A retailer who sells television recorders finds that she can sell 50 recorders per month when the price is \$400 per unit. For each \$5 decrease in price, one additional recorder can be sold. If she purchases each unit for \$100 and pays \$20 to a technician to adjust and fine-tune each unit when it is delivered, what level of sales will maximize monthly profits?

17. A restaurant finds that if it charges \$10.75 for its prime rib dinner, it will serve 100 dinners each night. For every 50¢ increase in the price, five fewer dinners are ordered. If the cost of preparing and serving each dinner is \$2.75, what price should the restaurant charge if it wants to maximize its daily profit from the sale of prime rib dinners?

18. What positive number exceeds its square by the largest amount?

19. The publisher of *Skin*, a professional magazine for dermatologists, has learned that the demand equation for the magazine is

$$p = 3 - 0.01x$$

where x represents the number of magazines (in hundreds) and p is the price (in dollars) of each magazine. The cost of printing, distributing, and advertising is given by the equation

$$C(x) = 10 + x + 0.03x^2$$

where $C(x)$ is expressed in hundreds of dollars. At what level of sales will the company's profit be maximized? At what price should the magazine be selling in order to maximize profits?

20. The daily output per station in a small machine shop is constant at 500 units when the number of stations is 18 or fewer. For each additional station over 18, the daily output per station declines by 25 units because of overcrowding and the inability to service breakdowns quickly. At what number of stations is daily output for the entire shop maximized?

21. The owner of an apple orchard finds that the annual yield per tree is constant at 320 lb when the number of trees per acre is 50 or fewer. For each additional tree over 50, the annual yield per tree decreases by 4 lb because of overcrowding. How many trees should be planted on each acre to maximize the annual yield from an acre?

22. The state legislature is considering a bill that would impose a meals tax of t cents per dollar spent on food in restaurants, cafeterias, and the like. Annual spending in restaurants is currently \$216 million. Economists estimate that the tax will cause a decline in $A(t)$, the amount of money spent by the public on food. The relationship between $A(t)$ and t is described by the equation

$$A(t) = 216 - 2t^2, \quad 0 \leq t \leq 8$$

Find the equation that describes the government's revenue as a function of t. What value of t maximizes the government's revenue?

23. The publishers of a popular paperback book find that demand is steady at 400,000 copies per year. Setup costs for printing a batch of these books are \$2000. The cost of labor and material is \$0.75 per book, and the cost of holding one book in inventory for one year is \$0.25. How many production runs should the company schedule each year to minimize total costs?

24. Counterfeit Charlie is the West Coast printer and distributor of bogus \$10 bills for organized crime. Charlie has a contract to print 50,000 bills annually. Setup costs for each printing run are \$1250, material and labor costs are \$0.25 for each bill, and storage costs for each bill amount to \$0.25 per year. The cost of storage is high because the bills cannot be stored at one location. How many bills should be printed during each run if Charlie wants to minimize the cost of printing and storing the bogus money? Assume the demand is uniform during the year.

25. A consumer electronics store sells 1000 compact-disk players each year. The players cost \$150 each. It costs the company \$200 to place an order and \$40 to store a disk player for a year. Find the order size that minimizes total annual costs.

26. The moonshine produced and bottled in a still hidden in a wooded area is to be transported for storage across a river to a barn located 3 miles down the river from point A (Figure 13). The moonshiners want to transport the liquor to the barn in the shortest possible time to avoid the legal authorities. Travel by boat is at the rate of 6 mph, whereas travel over land is at the rate of 10 mph. How far from point A should the boat land in order to minimize transport time?

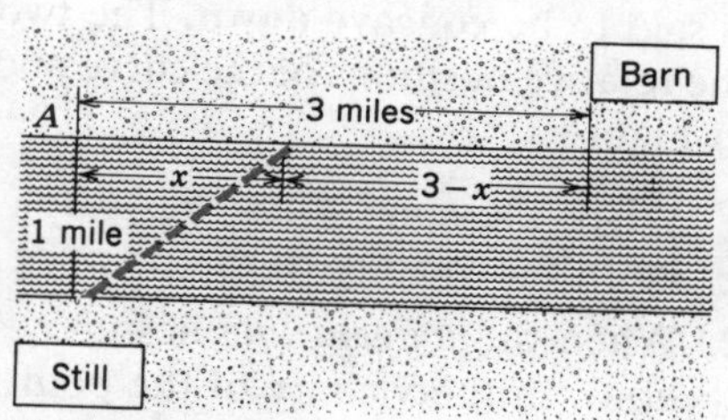

Figure 13

27. A loan company is restricted by state law to charge no more than 24 percent interest on loans to customers. Even at this exorbitant rate of interest, the company can lend out all the money it has available. To raise money for loans, the company offers savings accounts to individuals with surplus cash. The amount of money placed by depositors is proportional to the interest rate paid on savings accounts by the company. What interest rate should be paid on savings accounts to maximize the profit earned by the company?

28. For the loan company described in exercise 27, suppose that the amount of money deposited in the bank is proportional to the square of the interest rate that is paid on savings accounts. What interest rate maximizes the company's profit?

29. Suppose banking regulations forbid the loan company in exercise 27 from lending out all the money deposited with it. If the company must keep 10 percent on reserve, find the interest rate that maximizes the company's profit.

30. A closed rectangular box is to be made with a square base and to have a volume of 375 cm^3. If the material for the top and bottom costs three times as much as the material for the four sides, find the dimensions of the box that minimize the cost of materials.

31. A 36-ft piece of wire is to be cut into two parts. One part is to be bent into a square and the second into a circle. At what point should the wire be cut so that the sum of the areas is a minimum?

32. In Exercise 26, suppose that the amount of energy needed to travel one mile by water is twice the amount needed to travel one mile by land. How far from point A should the moonshiners land their boat if they want to minimize the total amount of energy expended in going from the still to the barn?

4.3 SECOND-DERIVATIVE TEST; INFLECTION POINTS

Although the second derivative $f''(x)$ has received little attention so far, it would be a mistake to regard it as unimportant. The second derivative is very useful in determining the behavior of a function because it describes the rate and direction of bending on a curve.

Before we study the second derivative in more detail, it is necessary to introduce you to the concept of **concavity;** the concept is illustrated in the two curves shown in Figure 1. When a curve opens *upward,* such as that shown in Figure 1*a*, it is said to be **concave up;** a curve like the one shown in Figure 1*b* is said to be **concave down.** The two types of concavity can be distinguished by the relative positions of the curve and its tangent lines in the two examples shown in Figure 1.

1. When the *tangent line* lies *below* the curve, the curve is said to be *concave up* in the vicinity of the point of tangency.
2. When the *tangent line* lies *above* the curve, the curve is said to be *concave down* in the vicinity of the point of tangency.

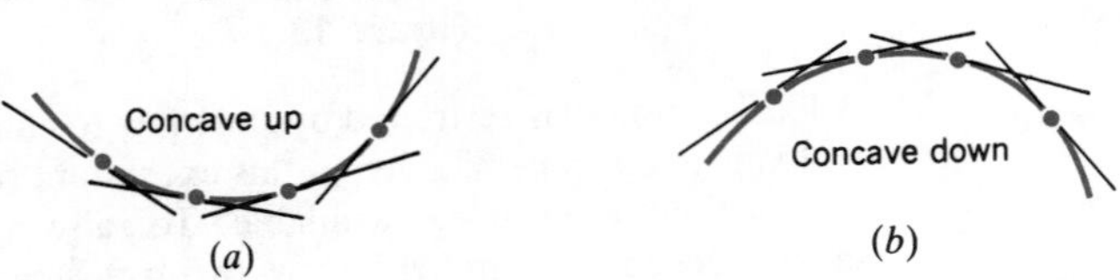

Figure 1

An understanding of the geometric meaning of $f''(x)$ requires a knowledge of the relationship between $f''(x)$ and $f'(x)$. Noting that the second derivative $f''(x)$ bears the same relationship to the derivative $f'(x)$ that $f'(x)$ bears to the function $f(x)$, we can state the relationship as follows:

1. If $f''(x)$ is **positive** everywhere on the interval $c < x < d$, the derivative $f'(x)$ is **increasing** on the interval.
2. If $f''(x)$ is **negative** everywhere on the interval $c < x < d$, the derivative $f'(x)$ is **decreasing** on the interval.

To visualize what an increasing or decreasing derivative means, let us examine the curves in Figure 2. As we move from left to right along the curve shown in Figure 2*a*, the derivative becomes larger or is increasing; on the other hand, the

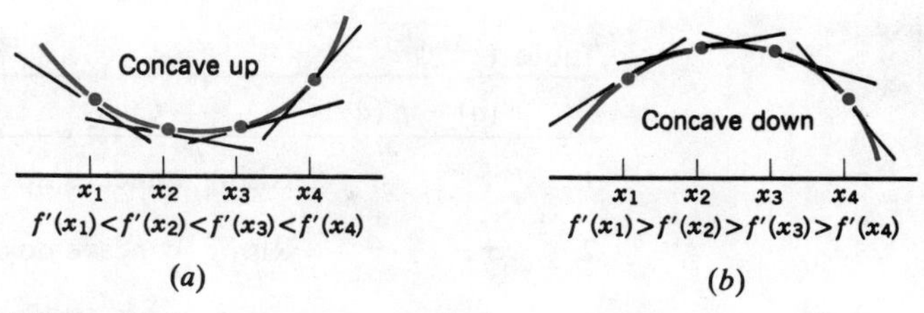

Figure 2

derivative is decreasing for the curve in Figure 2*b*. These curves indicate that an increasing derivative causes the curve to bend upward, whereas a decreasing derivative causes the curve to bend down. From this analysis, we can conclude the following important relation between the second derivative $f''(x)$ and the shape of a curve.

1. If $f''(x) > 0$ everywhere on the interval $c < x < d$, the curve is **concave up** on the interval.
2. If $f''(x) < 0$ everywhere on the interval $c < x < d$, the curve is **concave down** on the interval.

The relationship between $f''(x)$ and the shape of a curve is illustrated in Figure 3, which contains the graph of the function $f(x) = x^4 - 6x^2 + 12$; in addition, the sign of $f''(x) = 12x^2 - 12$ is displayed on the x axis. The figure indicates that when $f''(x) > 0$, the curve is concave up and when $f''(x) < 0$, the curve is concave down.

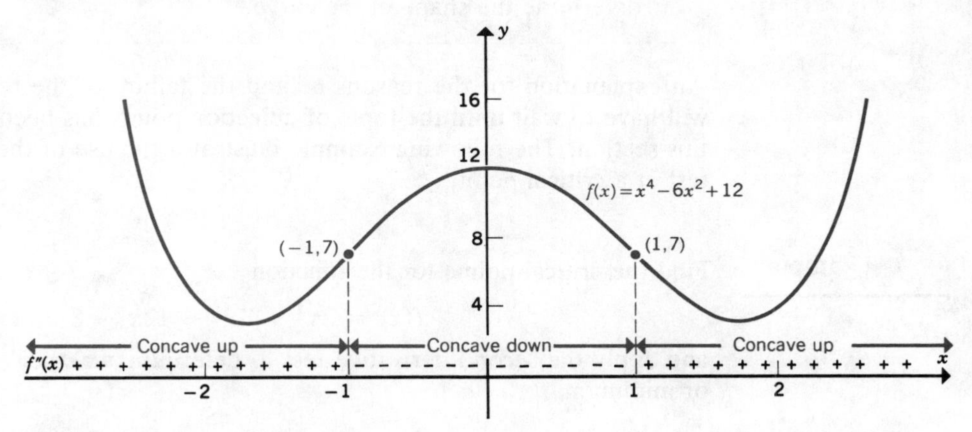

Figure 3

It is now possible to combine our knowledge of the signs of $f'(x)$ and $f''(x)$ at a point to determine the shape of a curve in the vicinity of the point. The results are shown in rows 1 through 4 of Table 1. Rows 5 and 6 in Table 1 form the basis of the second-derivative test to determine whether a critical point represents a maximum or a minimum.

Table 1

	$f'(a)$	$f''(a)$	Curve	Graph
1.	+	+	Rising, concave up	
2.	+	−	Rising, concave down	
3.	−	+	Falling, concave up	
4.	−	−	Falling, concave down	
5.	0	+	Horizontal tangent line, concave up	
6.	0	−	Horizontal tangent line, concave down	

Second-Derivative Test

If $f'(a)$ equals zero, then

1. The point $(a, f(a))$ is a relative maximum when $f''(a) < 0$.
2. The point $(a, f(a))$ is a relative minimum when $f''(a) > 0$.

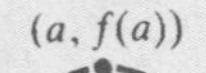

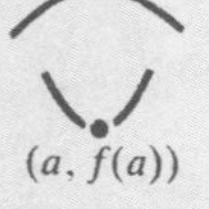

3. The test fails when $f''(a) = 0$; the critical point can be a relative maximum, relative minimum, or neither. Use the first-derivative test to determine the shape of the curve.

An explanation for the reasons behind the failure of the test when $f''(a) = 0$ will have to wait until the topic of inflection points has been presented later in this section. The following example illustrates the use of the second-derivative test at a critical point.

Example 1 Find the critical points for the function

$$f(x) = 2x^3 + 3x^2 - 12x - 8$$

and apply the second-derivative test to determine whether each is a maximum or minimum.

Solution First, we find $f'(x)$, set it equal to zero, and solve for x:

$$f'(x) = 6x^2 + 6x - 12$$

$$0 = 6(x^2 + x - 2) = 6(x + 2)(x - 1)$$

Solutions: $x_1 = -2, \quad f(-2) = 12$

$x_2 = 1, \quad f(1) = -15$

Next, we evaluate the second derivative $f''(x)$ at each critical point:

$$f''(x) = 12x + 6$$
$$f''(-2) = -18 < 0 \quad \text{(concave down)}$$

Therefore, $(-2, 12)$ is a relative maximum:

$$f''(1) = +18 > 0, \quad \text{(concave up)}$$

So $(1, -15)$ is a relative minimum. The two points and the curve in the vicinity of each are shown in Figure 4.

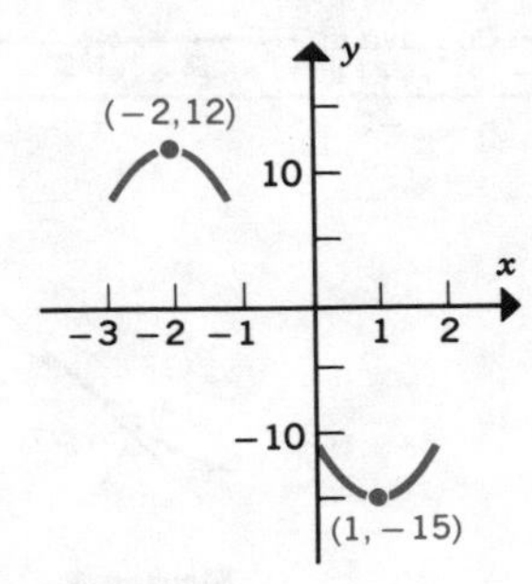

Figure 4

Although easy to apply, the second-derivative test is sometimes less desirable than an alternate test such as the first-derivative test for the following reasons.

1. If the function is complicated, finding the second derivative may require much more time and effort than that required by the first-derivative test.
2. The second-derivative test fails when both $f'(a)$ and $f''(a)$ equal zero, thus forcing us to fall back on the first-derivative test to determine the shape of the curve in the vicinity of the critical point. ■

Inflection Points

Our study thus far has avoided discussing any situation in which $f''(x) = 0$. Before investigating this case, let us again look at the graph of the function $f(x) = x^4 - 6x^2 + 12$ shown in Figure 5. Points such as $(-1, 7)$ and $(1, 7)$, which separate a segment of the curve that is concave down from one that is concave up, are called **inflection points.** Closer inspection of the curve in the vicinity of these points (Figure 6) indicates that the tangent line at an inflection point intersects the curve so that it is above the curve on one side of the inflection point and below the curve on the other.

Definition ***Inflection Point.*** A point $(a, f(a))$ is an **inflection point** if the graph of the function $y = f(x)$ changes from concave up to concave down, or vice versa, at the point.

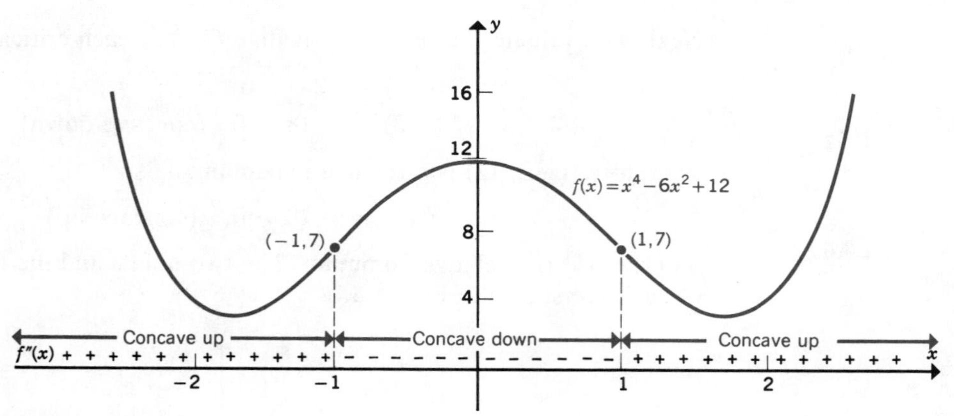

Figure 5

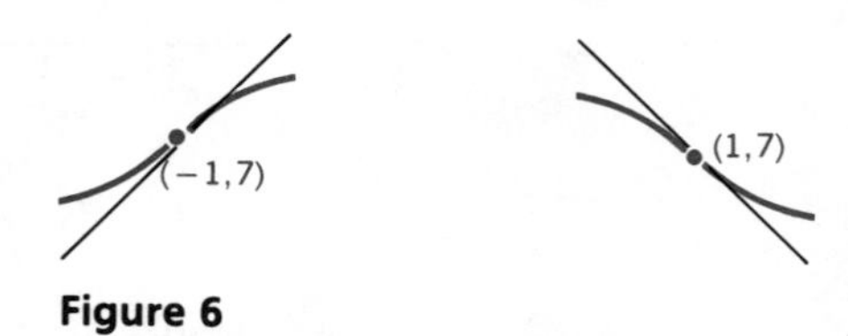

Figure 6

Having described the graphical characteristics of an inflection point, we now have to address the following question: How do we analytically locate inflection points on a curve? Finding the inflection points on a curve requires two steps. First, because an inflection point separates a segment on which the graph is concave up from one on which it is concave down, the second derivative $f''(x)$ can be neither positive nor negative at such a point. Therefore, if $(a, f(a))$ is an inflection point, we can conclude that either (1) $f''(a)$ is zero or (2) $f''(a)$ does not exist. This is similar to the method used to find critical points (Section 4.1).

Finding Inflection Points

If $(a, f(a))$ is an inflection point, either (1) $f''(a) = 0$ or (2) $f''(a)$ does not exist.

After we find those points, if any, where $f''(x)$ is zero or $f''(x)$ does not exist, the next step is a test to determine whether the points found are indeed inflection points. This step is required because the second derivative can equal zero at points that are *not* inflection points; such a situation will be illustrated in Example 3. The test itself consists of determining whether the concavity to the left of each point where $f''(x)$ is zero differs from the concavity to the right. If the concavity is different, the point is an inflection point; otherwise it is not. The test is described next.

Test for Inflection Points

Suppose that a function $y = f(x)$ is differentiable everywhere in the interval $c < x < d$ and that $(a, f(a))$ is the only point where $f''(x)$ is zero or $f''(x)$ does not exist. Let x_L be a value of x to the left of $(a, f(a))$ and x_R a value to the right. If the sign of $f''(x_L)$ is different from that of $f''(x_R)$, then $(a, f(a))$ is an inflection point; otherwise it is not. The test is summarized in Table 2.

Table 2 $f''(a) = 0$ or $f''(a)$ does not exist

$f''(x_L)$	$f''(x_R)$	$(a, f(a))$
+	−	Inflection point
−	+	Inflection point
+	+	Not an inflection point
−	−	Not an inflection point

The following example is intended not only to demonstrate the process of locating and testing for inflection points, but also to show how a rough sketch of the curve in the vicinity of an inflection point can be drawn.

Example 2 Find the points, if any, on the curve

$$f(x) = x^3 - 3x^2 + 4$$

where $f''(x)$ is zero and determine whether or not the points found are inflection points. In addition, make a rough sketch of the curve in the vicinity of each point.

Solution 1. Since $f'(x) = 3x^2 - 6x$, $f''(x)$, the second derivative, equals

$$f''(x) = 6x - 6$$

Setting $f''(x)$ equal to zero and solving for x yields

$$0 = 6x - 6$$

Solution: $x = 1, \quad f(1) = 2.$

2. Test to see whether (1, 2) is an inflection point. Since (1, 2) is the only point where $f''(x)$ is zero, we determine the sign of $f''(x)$ in the intervals $x < 1$ and $x > 1$, as shown in Table 3 and Figure 7. Since the curve is concave down to the left and concave up to the right of (1, 2), we can conclude that (1, 2) is an inflection point.

Table 3

Interval	Value of x	$f''(x)$	Concavity
$x < 1$	0	−6	Down
$x > 1$	2	+6	Up

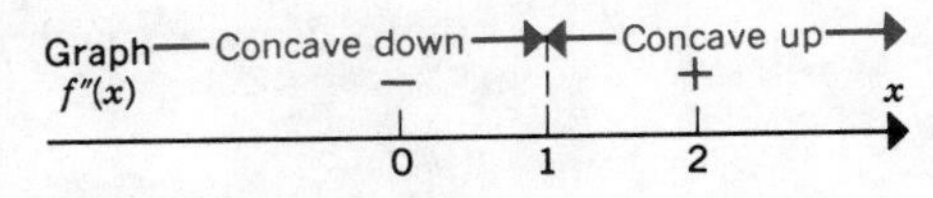

Figure 7

■

3. A rough sketch of the curve in the vicinity of (1, 2) can be drawn once we determine the sign of $f'(1)$, the derivative at (1, 2):

$$f'(1) = 3(1)^2 - 6(1) = -3, \quad \text{curve is falling}$$

A falling tangent line is drawn at (1, 2); the results from step 2 regarding the concavity are then used to sketch the curve shown in Figure 8.

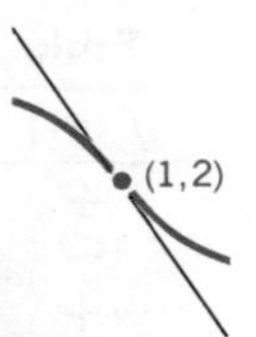

Figure 8

The next example illustrates a situation in which $f''(a)$ is zero, but the corresponding point $(a, f(a))$ is not an inflection point.

Example 3 Find the points on the curve

$$f(x) = x^4 - 4x^3 + 6x^2 + 2$$

where $f''(x)$ is zero. Determine whether or not each point is an inflection point and make a rough sketch of the curve in the vicinity of each.

Solution 1. The points where $f''(x) = 0$ are found

$$f'(x) = 4x^3 - 12x^2 + 12x$$
$$f''(x) = 12x^2 - 24x + 12$$
$$0 = 12x^2 - 24x + 12 = 12(x^2 - 2x + 1)$$
$$0 = 12(x - 1)^2$$

Solution: $x = 1, f(1) = 5$

2. Test to determine whether or not (1, 5) is an inflection point. Because $f''(x) = 12(x - 1)^2 > 0$ for all values of x except $x = 1$, the curve is concave up on both sides of (1, 5). Therefore, (1, 5) is **not** an inflection point.
3. A rough sketch of the curve in the vicinity of (1, 5) can be made once we know the sign of $f'(1)$:

$$f'(1) = 4(1)^3 - 12(1)^2 + 12(1)$$
$$= +4, \quad \text{curve is rising}$$

Knowing that the curve is rising and is concave up in the vicinity of (1, 5) enables us to draw a rough sketch such as that shown in Figure 9.

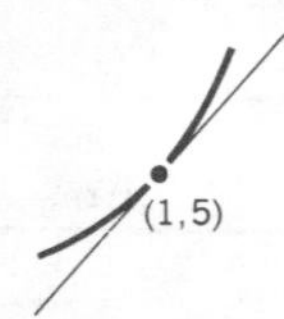

Figure 9

Having found that the concavity may or may not change as we move along a curve through a point where $f''(x) = 0$, we can begin to see why the second-derivative test applied at a critical point fails when $f''(a) = 0$. Figure 10 contains the graphs of two curves, $f(x) = x^3 + 2$ and $g(x) = x^4 + 2$, both of which have critical points at (0, 2). In addition, $f''(x) = g''(x) = 0$ at (0, 2). For the function $f(x) = x^3 + 2$, the critical point (0, 2) is also an inflection point. For the function $g(x) = x^4 + 2$, the critical point (0, 2) represents a minimum.

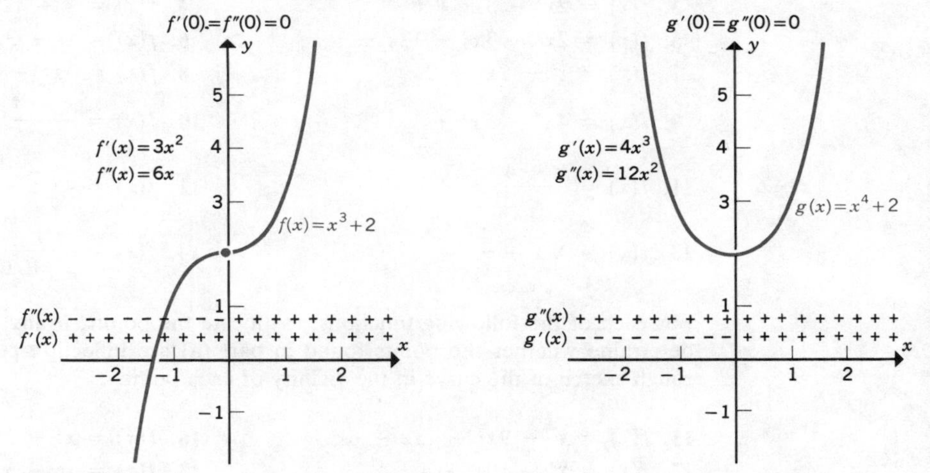

Figure 10

To see why inflection points play an important role in the fields of business and economics, let's look at the graph shown in Figure 11. The consumer price index (CPI) is plotted as a function of time during a runaway inflationary period (shown by the solid curve). The curve is rising and concave upward, which means that not only is the CPI increasing but the rate of increase, as measured by $\frac{d}{dt}$(CPI), is also increasing. Suppose that at $t = t_1$, the government takes action

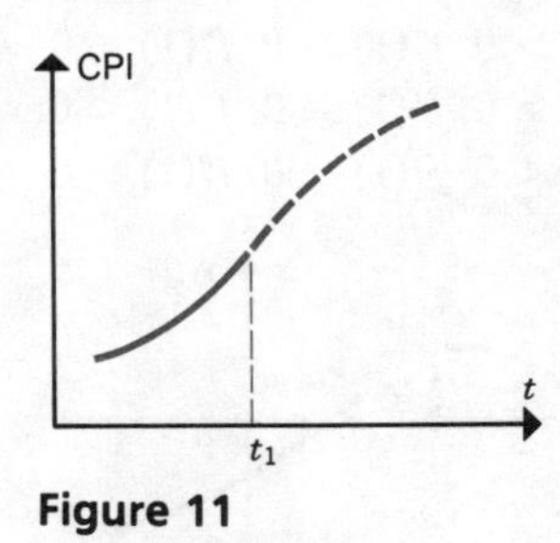

Figure 11

to brake the accelerating inflationary trend by instituting wage and price controls, sharply reducing the money supply, and raising taxes. The results of these actions, if effective, would cause a decrease in the rate at which the CPI increases, that is, the curve would become concave downward (shown by the dashed portion). The point at which the change from an accelerating to a decelerating rate of increase in the CPI occurs represents an inflection point.

4.3 EXERCISES

Find the critical points, if any, for each of the following functions. Use the second-derivative test to determine whether a critical point is a maximum or minimum. If the test fails, use the first-derivative test to determine the shape of the curve in the vicinity of each critical point.

1. $f(x) = x^2 + 2x + 3$

2. $f(x) = 4x - x^2 - 1$

3. $f(x) = 2x^3 - 3x^2 + 6$

4. $f(x) = 4 + 12x - x^3$

5. $f(x) = 2x^3 - 3x^2 - 12x + 1$

6. $f(x) = x^3 + 3x - 1$

7. $f(x) = x^4 - 2x^2 + 2$

8. $f(x) = 3x^4 - 4x^3$

9. $f(x) = 4x^5 - 5x^4 + 1$

10. $f(x) = \dfrac{x + 4}{x}$

11. $f(x) = \dfrac{x - 4}{x}$

12. $f(x) = 4\sqrt{x} - x$

13. $f(x) = \sqrt{x} + \dfrac{4}{x}$

14. $f(x) = \dfrac{1}{x^2 + 1}$

For each of the following functions, (a) locate the points, if any, where $f''(x) = 0$; (b) determine whether the points found in part (a) are inflection points; and (c) make a rough sketch of the curve in the vicinity of each point.

15. $f(x) = x^3 - 9x^2 + 15x + 6$

16. $f(x) = x^4 - 24x^2$

17. $f(x) = x^4 - 10x + 4$

18. $f(x) = x^3 + 3x - 2$

19. $f(x) = \dfrac{8}{x^2 + 3}$

20. $f(x) = \sqrt{x} + \dfrac{1}{x}$

21. $f(x) = 3x^5 - 10x^4 + 50$

22. $f(x) = 3x^5 - 5x^4 - 60x^3$

Match each of the functions described in Exercises 23–26 with one of the graphs in Figure 12.

23. $f(1) > 0, f'(1) < 0, f''(1) > 0$

24. $f(1) > 0, f'(1) > 0, f''(1) < 0$

25. $f(1) > 0, f'(1) < 0, f''(1) < 0$

26. $f(1) < 0, f'(1) < 0, f''(1) > 0$

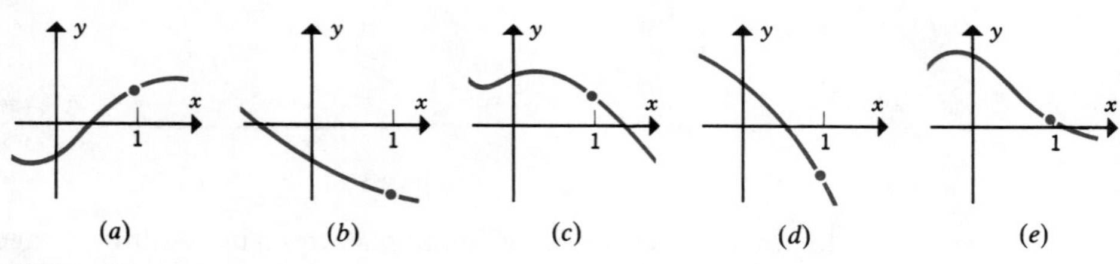

Figure 12

In Exercises 27 through 31, match each of the functions whose second derivatives are given with one of the graphs in Figure 13.

27. $f''(x) = x(x - 1)$ **28.** $f''(x) = x^2(x - 1)$ **29.** $f''(x) = x(x^2 - 1)$
30. $f''(x) = -x^{-5/3}$ **31.** $f''(x) = x^2(x^2 + 1)$

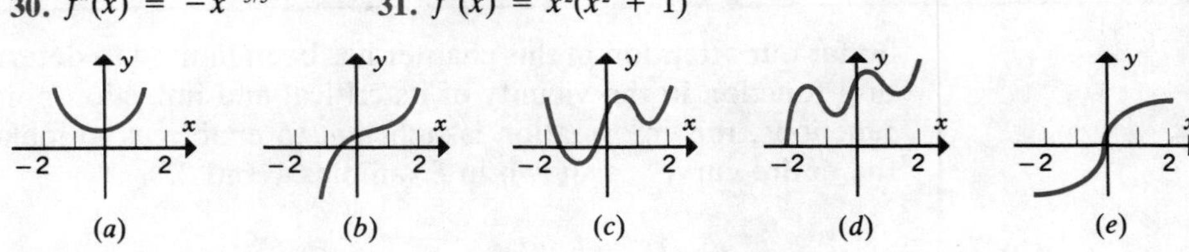

Figure 13

32. In recent years a well-known manufacturer of ballpoint pens has seen its sales decline because of heavy competition from a foreign manufacturer. The sales curve is shown in Figure 14. The company begins an intensive advertising campaign when $t = t_1$. What would you expect the shape of the curve to look like shortly after the advertising campaign begins should it be successful?

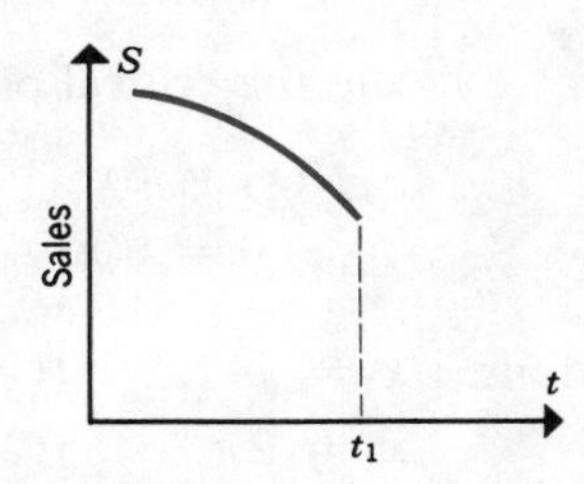

Figure 14

33. The following headline appeared in the real estate section of a local newspaper: "Price Increases of New Homes Are Slowing." If $P(t)$ represents the price as a function of t, the time, what does the headline tell us about $P''(t)$? Does it mean that $P'(t)$ is negative?

34. The graphs of the function $y = u(x)$ and the tangent line at the inflection point whose x-coordinate equals 2 are shown in Figure 15.

(a) If $f(x) = [u(x)]^2$, use the information shown on the graphs to find each of the following:
i. $f(2)$ ii. $f'(2)$ iii. $f''(2)$

(b) If $F(x) = xu(x)$, use the information shown on the graphs to find each of the following:
i. $F(2)$ ii. $F'(2)$ iii. $F''(2)$

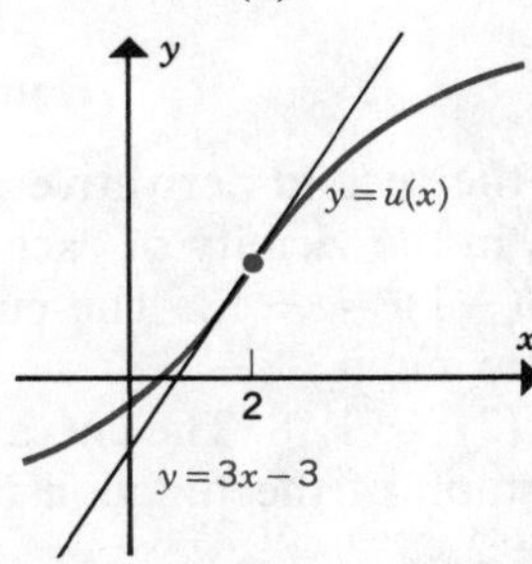

Figure 15

4.4 CURVE SKETCHING

So far our attention in this chapter has been limited to determining the behavior of a function in the vicinity of its critical and inflection points. For polynomial functions, this information is sufficient to enable us to make a rough sketch of the entire curve, as shown in Examples 1 and 2.

Polynomial Functions

Example 1 Make a rough sketch of the function

$$f(x) = 2x^3 - 3x^2 - 12x + 15$$

Solution 1. First, we locate the critical points

$$f'(x) = 6x^2 - 6x - 12$$
$$0 = 6(x^2 - x - 2) = 6(x - 2)(x + 1)$$

Solutions: $x_1 = -1, \quad f(-1) = 22$
$x_2 = 2, \quad f(2) = -5$

The critical points together with the horizontal tangent lines are shown in Figure 1.

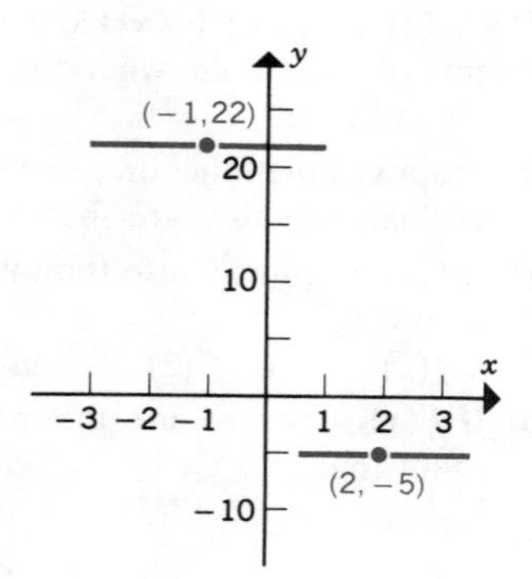

Figure 1

2. Next the second-derivative test will be used to determine the shape of the graph in the vicinity of each critical point. Since $f''(x) = 12x - 6$, we get:
(a) $f''(-1) = -18$. The curve is concave down; so $(-1, 22)$ is a relative maximum.
(b) $f''(2) = +18$. The curve is concave up; so $(2, -5)$ is a relative minimum.
The graph of the function in the vicinity of each critical point is shown in Figure 2.

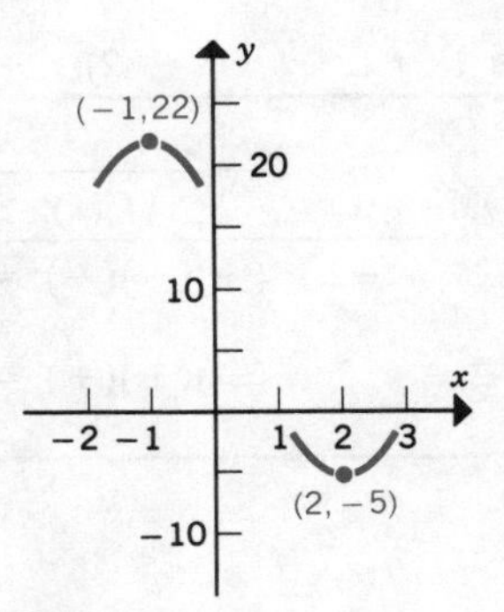

Figure 2

3. Inflection points are found by setting $f''(x)$ equal to zero and solving for x:

$$0 = 12x - 6$$

Solution: $x = 1/2, f(1/2) = 17/2$

It is not necessary to test this point to determine whether it is an inflection point. In step 2, we learned that the concavity to the left, that is, at $x = -1$, is different from that to the right, that is, at $x = 2$; therefore, we can conclude that $(\frac{1}{2}, \frac{17}{2})$ is an inflection point.

Incorporating the inflection point into our graph, we can draw the segment of the curve between the two critical points, as shown in Figure 3. The y intercept (0, 15) is also included because it is very easy to locate.

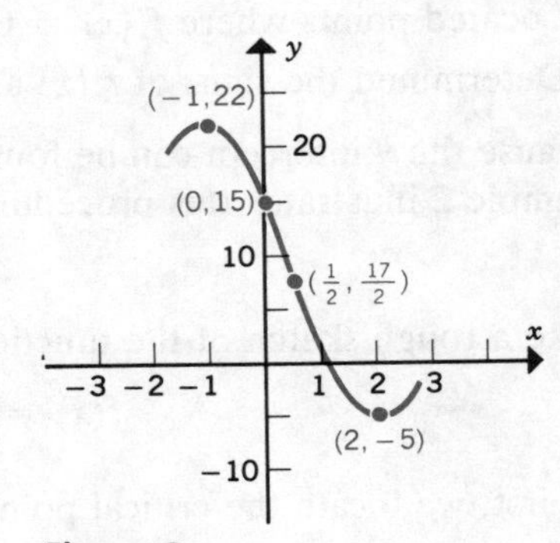

Figure 3

4. The graph can be completed once we determine the shape of the curve to the left of $(-1, 22)$ and to the right of $(2, -5)$. Because we have found the coordinates of all points at which the first and second derivative change sign, the behavior of the curve is uniform over each of the intervals, $x < -1$ and $x > 2$. The shape of the curve over each interval can be determined by finding the signs of $f'(x)$ and $f''(x)$ at arbitrary points on each as shown in Table 1. This analysis enables us to complete the graph, as shown in Figure 4.

Table 1 $f'(x) = 6(x - 2)(x + 1), f''(x) = 6(2x - 1)$

Interval	x	Signs of $f'(x)$	Signs of $f''(x)$	Curve	Graph
$x < -1$	-2	$(+)(-)(-) = +$	$(+)(-) = -$	Rising, concave down	
$x > 2$	3	$(+)(+)(+) = +$	$(+)(+) = +$	Rising, concave up	

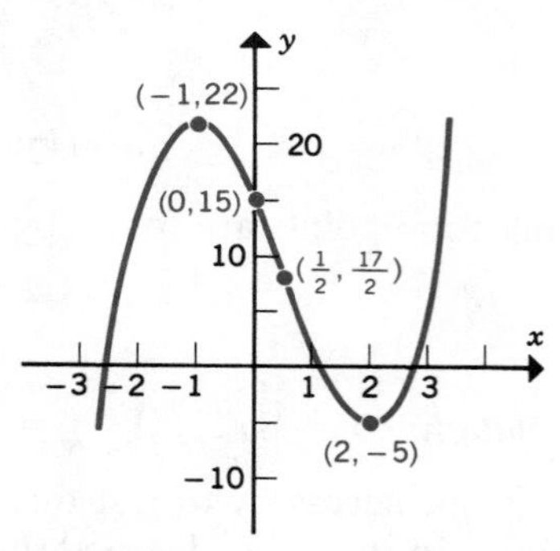

Figure 4

■

Ordinarily, we do not have to go through an analysis as detailed as that in the previous example. For a polynomial function, a rough sketch of the curves can be drawn once we have done the following.

1. Located points where $f'(x) = 0$ and where $f''(x) = 0$.
2. Determined the signs of $f'(x)$ and $f''(x)$ for all other values of x.

Because the y intercept can be found easily, it is advisable to include it as well. Example 2 illustrates this procedure.

Example 2 Make a rough sketch of the function

$$f(x) = 3x^4 - 4x^3 - 3$$

Solution 1. First, we locate the critical points and the points where $f''(x) = 0$:

$$f'(x) = 12x^3 - 12x^2 = 12x^2(x - 1)$$
$$f''(x) = 36x^2 - 24x = 12x(3x - 2)$$

Setting $f'(x)$ equal to zero, yields

$$0 = 12x^2(x - 1)$$

Solutions: $x_1 = 0, \quad f(0) = -3$

$x_2 = 1, \quad f(1) = -4$

Note: There are no points where $f'(x)$ does not exist.
Setting $f''(x)$ equal to zero and solving yields

$$0 = 12x(3x - 2)$$

Solutions: $x_1 = 0, \quad f(0) = -3$
$x_2 = 2/3, \quad f(2/3) = -97/27$

These three points are shown in Figure 5 together with the horizontal tangent lines at $x = 0$ and $x = 1$.

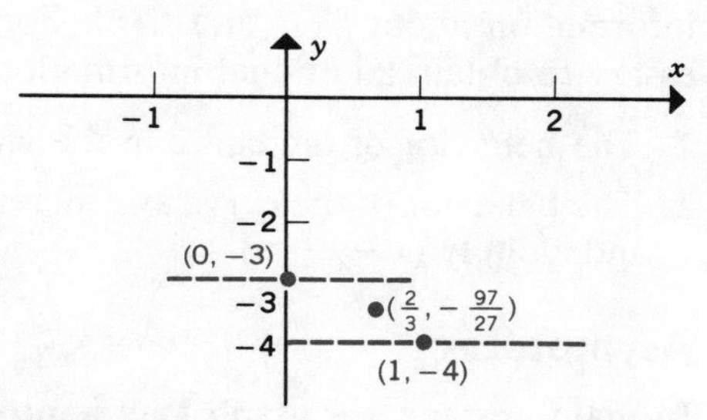

Figure 5

2. Since there are no other points where the signs of $f'(x)$ and $f''(x)$ can change, we can sketch the graph once the signs of $f'(x)$ and $f''(x)$ are found on the intervals shown in Figure 6. The analysis is shown in Table 2.

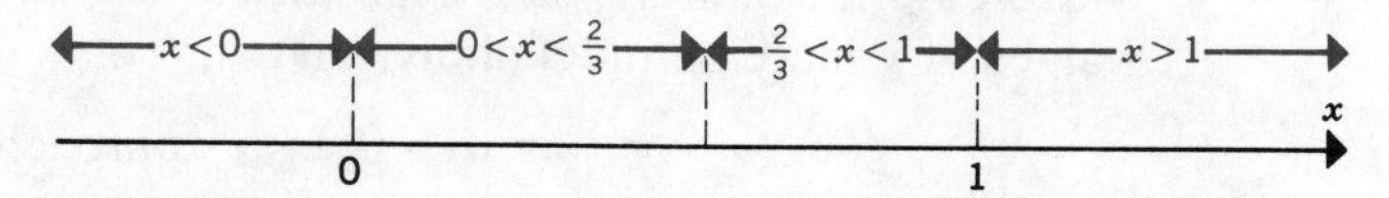

Figure 6

Table 2 $f'(x) = (12x^2)(x - 1)$, $f''(x) = (12x)(3x - 2)$

Interval	x	Signs of $f'(x)$	$f''(x)$	Curve	Graph
$x < 0$	-1	$(+)(-) = -$	$(-)(-) = +$	Falling, concave up	
$0 < x < 2/3$	$1/2$	$(+)(-) = -$	$(+)(-) = -$	Falling, concave down	
$2/3 < x < 1$	$3/4$	$(+)(-) = -$	$(+)(+) = +$	Falling, concave up	
$x > 1$	2	$(+)(+) = +$	$(+)(+) = +$	Rising, concave up	

Incorporating the information contained in Table 2, we can sketch the graph shown in Figure 7. Table 2 and the graph indicate that the function has a minimum at $(1, -4)$ and inflection points at $(0, -3)$ and $(\frac{2}{3}, -\frac{97}{27})$.

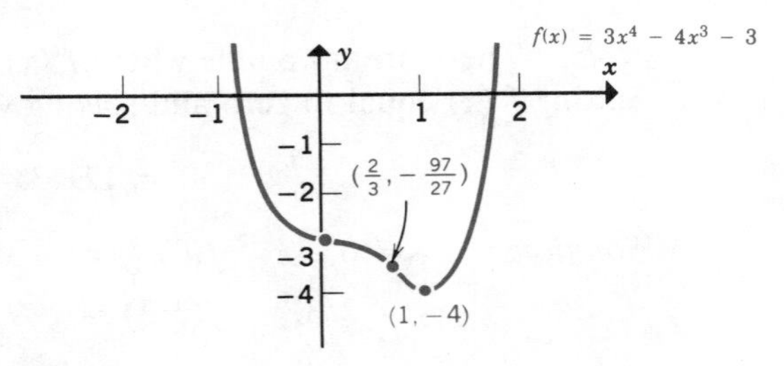

Figure 7

■

The graphs of the polynomials in Examples 1 and 2 were sketched using information about $f'(x)$ and $f''(x)$. For other types of functions it is often necessary to obtain additional information not related to the derivatives, such as

1. The behavior of the curve in the vicinity of a discontinuity.
2. The behavior of the curve as x increases indefinitely ($x \to \infty$) or as x decreases indefinitely ($x \to -\infty$).

Asymptotes

In many instances a graph may approach a straight line; this line is called an **asymptote** of the graph. For example, the graph of the function $f(x) = \dfrac{1}{x}$, shown in Figure 8, approaches the y axis as x approaches zero, at which the function is not defined. As x approaches zero from the left, the corresponding values of y decrease indefinitely; as x approaches zero from the right, the corresponding values of y increase indefinitely, that is,

$$f(x) \to -\infty \quad \text{as } x \to 0^- \quad (x \text{ approaches 0 from the left})$$
$$f(x) \to \infty \quad \text{as } x \to 0^+ \quad (x \text{ approaches 0 from the right})$$

The line $x = 0$ is called a **vertical asymptote.**

The graph also approaches the x axis as x increases indefinitely ($x \to \infty$) or as x decreases indefinitely ($x \to -\infty$), that is,

$$\lim_{x\to\infty} f(x) = \lim_{x\to\infty} \frac{1}{x} = 0, \qquad \lim_{x\to-\infty} f(x) = \lim_{x\to-\infty} \frac{1}{x} = 0$$

The line $y = 0$ is called a **horizontal asymptote.**

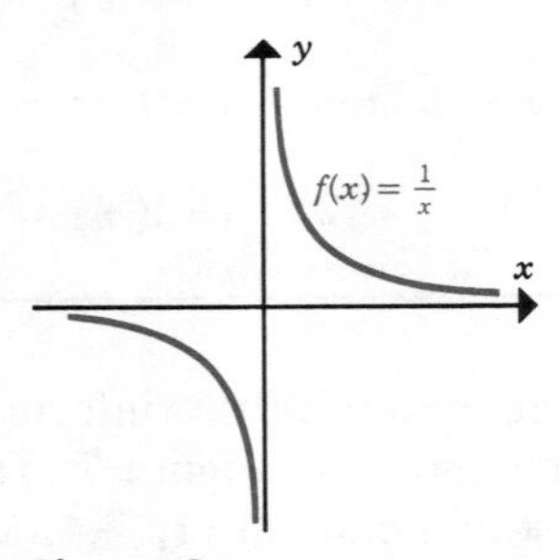

Figure 8

Asymptotes, if they exist, can be found for many functions as follows:

1. A function which can be written as a ratio has a vertical asymptote at $x = a$ when the denominator approaches zero, and the numerator approaches a nonzero value as x approaches a.
2. If $\lim_{x\to\infty} f(x) = L$ or $\lim_{x\to-\infty} f(x) = L$, then the line $y = L$ is a horizontal asymptote.

Example 3 illustrates how asymptotes are used to sketch the graph of a function that has neither critical nor inflection points.

Example 3 Make a rough sketch of the function

$$f(x) = \frac{2x}{x - 1}$$

Solution 1. Notice that the function is not defined at $x = 1$ and that the graph passes through (0, 0), as shown in Figure 9.

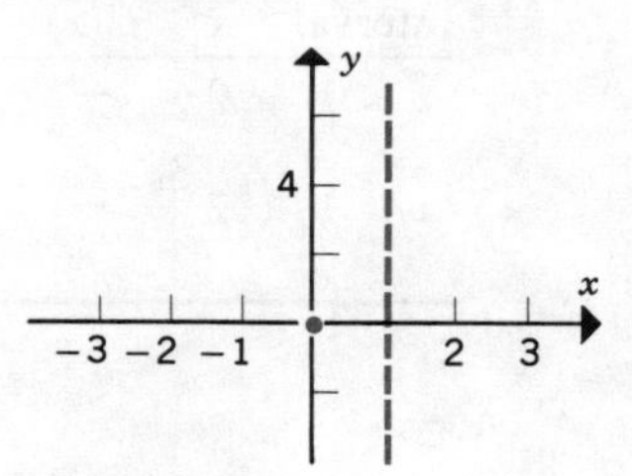

Figure 9

2. We look for points where $f'(x) = 0$ and those where $f''(x) = 0$. The quotient rule is used to find $f'(x)$:

$$f'(x) = \frac{2(x - 1) - (2x)(1)}{(x - 1)^2} = \frac{-2}{(x - 1)^2}$$

Writing $f'(x) = -2(x - 1)^{-2}$, we can use the general power rule to find $f''(x)$:

$$f''(x) = 4(x - 1)^{-3} = \frac{4}{(x - 1)^3}$$

First we set $f'(x)$ equal to zero to find any critical points:

$$0 = \frac{-2}{(x - 1)^2}$$

This equation has no solutions, so the function has no critical points. It should be noted that $f'(x) < 0$ for all x except $x = 1$, so the curve is falling at all points.

Next we look for any inflection points by setting $f''(x)$ equal to zero:

$$0 = \frac{4}{(x - 1)^3}$$

Again, this equation has no solutions, so there are no inflection points as well.

3. Since there are no critical or inflection points, the signs of either $f'(x)$ or $f''(x)$ can change only at $x = 1$, at which the function is not defined. Now we can determine the shape of the graph on the intervals $x < 1$ and $x > 1$ by finding the signs of $f'(x)$ and $f''(x)$ on each interval; the analysis is shown in Table 3. This information allows us to make a rough sketch of the curve (Figure 10) in the vicinity of (0, 0) and a second point (2, 4) that belongs to the interval $x > 1$.

Table 3 $f'(x = \frac{-2}{(x - 1)^2}$, $f''(x) = \frac{4}{(x - 1)^3}$

Interval	x	Signs of $f'(x)$	$f''(x)$	Curve	Graph
$x < 1$	0	−	−	Falling, concave down	╲
$x > 1$	2	−	+	Falling, concave up	╲

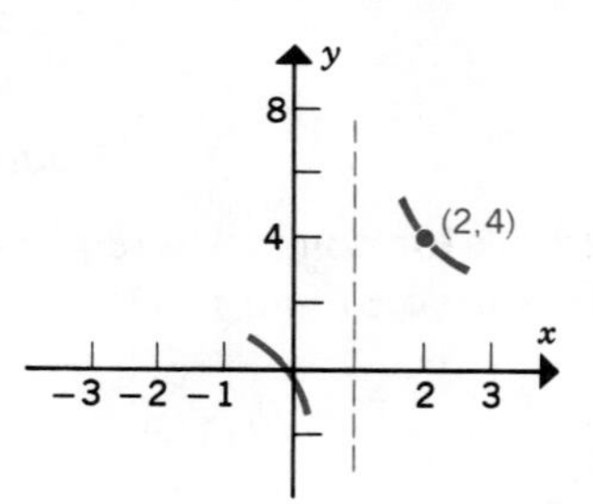

Figure 10

4. Vertical asymptotes. As x approaches 1, the denominator of $\frac{2x}{x - 1}$ approaches zero while the numerator approaches 4, suggesting that $x = 1$ is a vertical asymptote. Table 4 shows that

$$\frac{2x}{x - 1} \to \infty \quad \text{as } x \to 1^+$$

and that

$$\frac{2x}{x - 1} \to -\infty \quad \text{as } x \to 1^-$$

These features can be added to our graph, as shown in Figure 11.

Table 4

From the Left		From the Right	
x	$y = \dfrac{2x}{x-1}$	x	$y = \dfrac{2x}{x-1}$
0.50	−2.000	1.50	6.000
0.90	−18.000	1.10	22.000
0.99	−198.000	1.01	202.000
⋮	⋮	⋮	⋮

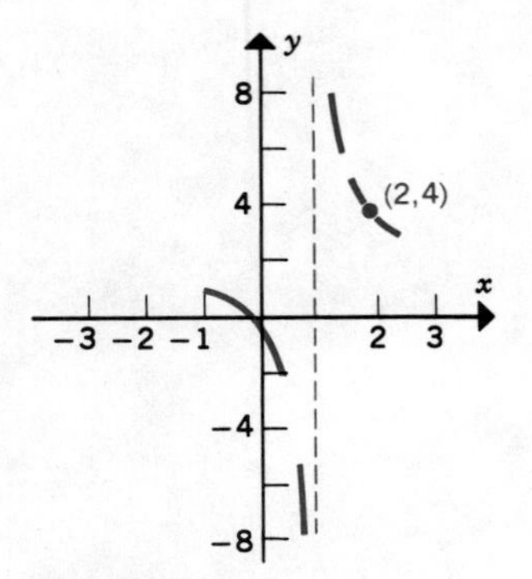

Figure 11

5. Horizontal asymptotes. Table 5 shows that as $x \to \pm\infty$, the corresponding values of y are approaching 2, or

$$\lim_{x\to-\infty} \frac{2x}{x-1} = 2 \quad \text{and} \quad \lim_{x\to\infty} \frac{2x}{x-1} = 2$$

The line $y = 2$ is a horizontal asymptote of the graph. These results are incorporated into the graph, as shown in Figure 12.

Table 5

$x \to -\infty$			$x \to \infty$		
x	$y = \dfrac{2x}{x-1}$		x	$y = \dfrac{2x}{x-1}$	
−10	$\frac{20}{11}$ =	1.818	10	$\frac{20}{9}$ =	2.222
−100	$\frac{200}{101}$ =	1.980	100	$\frac{200}{99}$ =	2.020
−1000	$\frac{2000}{1001}$ =	1.998	1000	$\frac{2000}{999}$ =	2.002

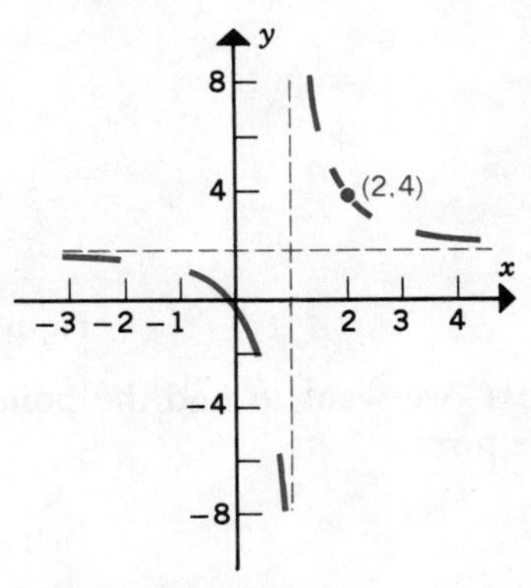

Figure 12

6. The last step is the easiest of all: connecting adjacent segments and smoothing out the graph. The result is shown in Figure 13.

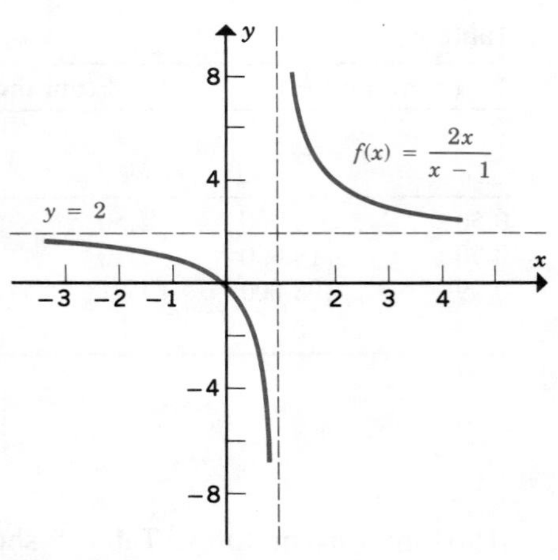

Figure 13

■

The next example is a continuation of a problem for which the critical points and the behavior of the function in the vicinity of each has been determined in an earlier example.

Example 4 Make a rough sketch of the function

$$f(x) = \frac{x^2}{x - 1}$$

Solution 1. The critical points and the behavior of the function in the vicinity of each were determined earlier, in Section 4.1, Example 6. The results of the analysis are shown in Figure 14.

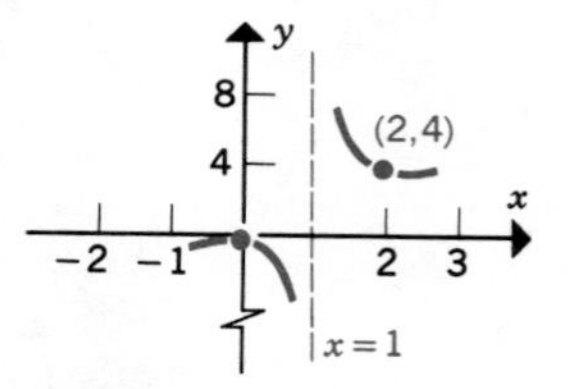

Figure 14

2. Next, we want to find the points where $f''(x) = 0$. Recalling that $f'(x)$ has the form

$$f'(x) = \frac{x^2 - 2x}{(x - 1)^2}$$

$f''(x)$ can be found using the quotient rule, giving

$$f''(x) = \frac{(2x - 2)(x - 1)^2 - (x^2 - 2x)2(x - 1)}{(x - 1)^4}$$

Working through the algebra gives, for the final version of $f''(x)$,

$$f''(x) = \frac{2}{(x - 1)^3}$$

Setting $f''(x)$ equal to zero yields

$$0 = \frac{2}{(x - 1)^3}$$

There are no solutions to this equation, indicating that there are no inflection points.

3. Asymptotes. As $x \to 1$, the denominator of the expression $\frac{x^2}{x - 1}$ approaches zero while the numerator approaches 1, indicating that $x = 1$ is a vertical asymptote. Columns 2 and 4 in Table 6 show that

$$f(x) \to -\infty \quad \text{as } x \to 1^- \quad (x \text{ approaches 1 from the left})$$
$$f(x) \to \infty \quad \text{as } x \to 1^+ \quad (x \text{ approaches 1 from the right})$$

Table 6

From the Left		From the Right	
x	$y = \frac{x^2}{x - 1}$	x	$y = \frac{x^2}{x - 1}$
0.50	−.500	1.50	4.500
0.90	−8.100	1.10	12.100
0.99	−98.010	1.01	102.010
⋮	⋮	⋮	⋮

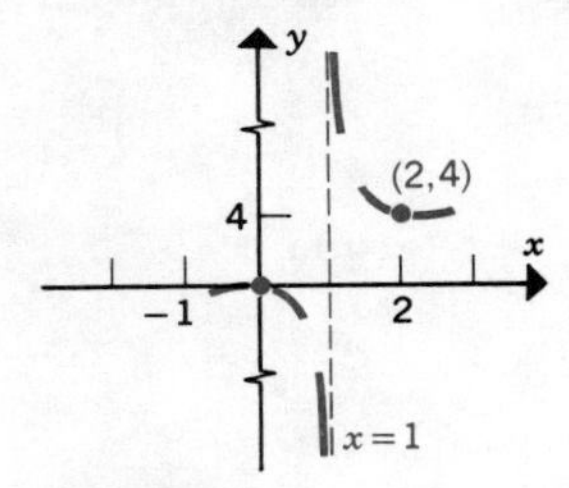

Figure 15

The graph in Figure 15 results when these features are included. The behavior of the function as $x \to \pm\infty$ can be determined by letting x increase indefinitely and decrease indefinitely, as shown in Table 7. The results given in the table indicate that

$$f(x) = \frac{x^2}{x - 1} \to -\infty \quad \text{as } x \to -\infty$$

$$f(x) = \frac{x^2}{x - 1} \to \infty \quad \text{as } x \to \infty$$

We can include these features in our graph, producing the segments shown in Figure 16.

Table 7

$x \to -\infty$		$x \to \infty$	
x	$y = \frac{x^2}{x - 1}$	x	$y = \frac{x^2}{x - 1}$
−10	−9.0900	10	11.1111
−100	−99.0099	100	101.010
⋮	⋮	⋮	⋮

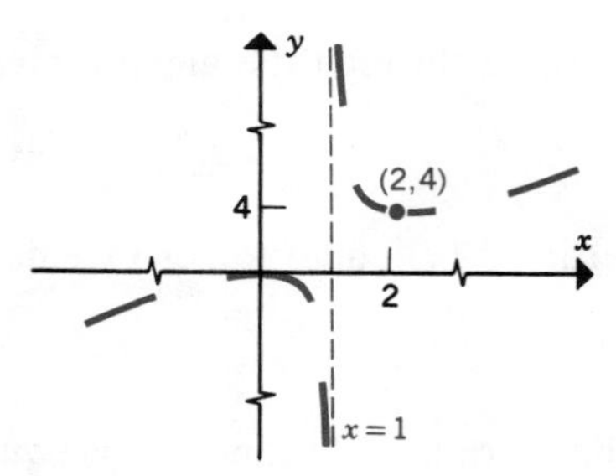

Figure 16

4. Because the function is differentiable everywhere on the intervals $x < 1$ and $x > 1$, we can connect adjacent segments on each branch. The graph shown in Figure 17 is the result. Note that the graph contains the coordinates of only two points, (0, 0) and (2, 4). ■

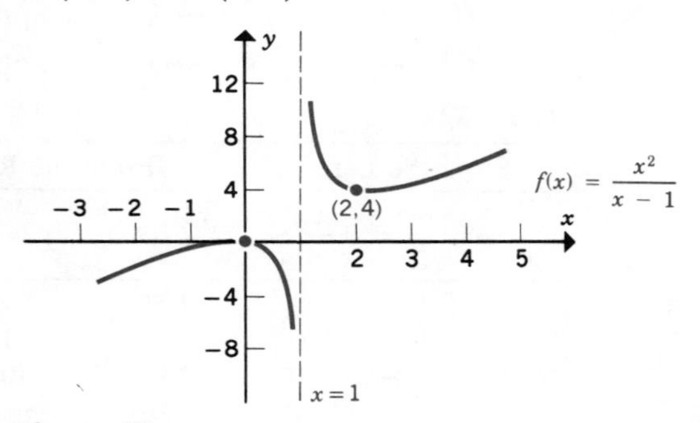

Figure 17

Procedure for Sketching a Graph

The procedures used to sketch the graph of a function are summarized in the following steps.

1. Find all the critical points.
2. Determine the behavior of the function in the vicinity of each critical point and sketch a graph of the curve in the vicinity of each.
3. Find all the inflection points.
4. Determine whether the graph has any vertical asymptotes by examining the behavior of the function in the vicinity of each value of x for which the function is not defined. Sketch the curve in the vicinity of each value.
5. Determine whether the graph has any horizontal asymptotes by studying the behavior of the function as $x \to \pm\infty$. Incorporate this information into the graph.

4.4 EXERCISES

Using the techniques described in this section, make a rough sketch of each of the following functions.

1. $f(x) = 2x^2 - 4x - 3$

2. $F(x) = x^2 + 6x + 3$

3. $f(x) = 8x - 2x^2 + 4$

4. $f(x) = x^2 + 2x + 1$

5. $g(x) = x^3 + 3x^2 - 1$

6. $f(x) = x^3 - 6x^2$

7. $f(x) = 2x^3 - 6x^2 - 18x + 24$

8. $F(x) = x^4 - 4x + 2$

9. $f(x) = x^3 + 6x^2 + 2$

10. $f(x) = x^3 + 2$

11. $g(x) = x^4 - 2x^2 + 1$

12. $f(x) = 3x^5 - 5x^3 + 1$

13. $f(x) = 2x^4 - 4x^2 + 3$

14. $f(x) = x^3 - 9x^2 + 15x$

15. $f(x) = 4\sqrt{x} - x$

16. $f(x) = 10x^3 - 6x^5$

17. $F(x) = 3x^4 - 4x^3 + 2$

18. $F(x) = x + \dfrac{1}{x}$

19. $f(x) = \dfrac{x^2}{x - 2}$

20. $g(x) = \dfrac{x}{x^2 + 1}$

21. $f(x) = \dfrac{2}{x^2 + 1}$

22. $f(x) = x^4 - 24x^2 + 10$

23. $f(x) = x - 2\sqrt{x}$

24. $F(x) = 4x + \dfrac{1}{x}$

25. $F(x) = (x - 1)^2(x - 4)$

26. $f(x) = x(x^2 + 1)^4$

27. $f(x) = 15x^{2/3} - 6x^{5/3}$

28. $f(x) = 2x^{1/3} + x^{4/3}$

29. A student was asked to sketch the graph of the function $f(x) = 4x^3 - 3x$. He constructed the following table of x and y coordinates to produce a few selected points and then drew a smooth curve through them. Does the curve shown in Figure 18 capture all the important features of the graph? If not, explain what is missing and correct the graph.

x	y
-2	-26
-1	-1
0	0
1	1
2	26

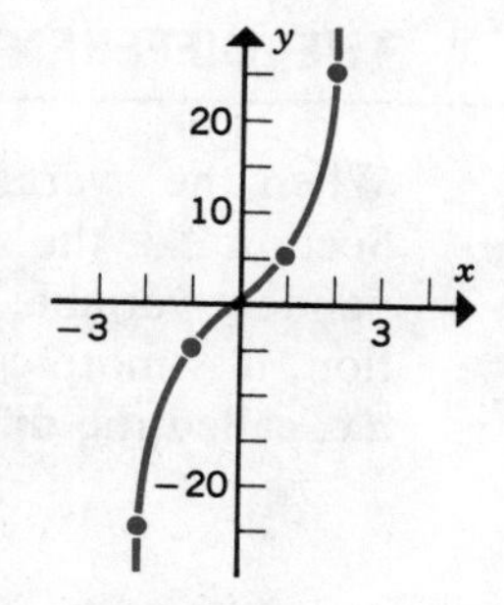

Figure 18

30. In a quality control department small samples are often selected from large batches of products shipped from a supplier; each item in the sample is then tested to determine whether or not an entire batch should be accepted. For example, suppose that three items are selected at random from a large batch of items, and that each is tested and then rated either as D (defective) or D* (nondefective). If the shipment is accepted when the number of defective items in the sample is fewer than two, the probability of accepting the entire batch is a function of x the fraction of defective items in the batch; the relation between the probability $P(x)$ and x is given by the equation

$$P(x) = (1 - x)^3 + 3x(1 - x)^2, \qquad 0 \le x \le 1$$

Find the critical points and the inflection points; then use this information to sketch the graph. The graph is called an **operating characteristic** (OC) curve for the sampling method.

31. (a) Find the critical points and inflection points, if any, for the function $f(x) = \dfrac{1}{\sqrt{1 - x^2}}$. Use this information to sketch the graph of the function. *Note:* It would be advisable to determine the domain of the function first.

(b) In relativity theory the relationship between $m(v)$, the mass of an object, and v, its velocity, is given by the equation

$$m(v) = \frac{m_0}{\sqrt{1 - \left(\dfrac{v}{c}\right)^2}}$$

where m_0 is the mass of the object when it is at rest ($v = 0$) and c is the velocity of light (3×10^{10} cm/sec). Use the results of part (a) to sketch the graph of this function.

32. The concentration $C(t)$ of a drug, in milligrams per liter, is given by the equation

$$C(t) = \frac{27t}{t^2 + 9}$$

where t is the time, in minutes, from the moment the drug is administered ($t = 0$). Find the critical points and inflection points, if any; then use this information to sketch the graph of the function.

4.5 THE DIFFERENTIAL

When the average rate of change of a function $y = f(x)$ was introduced in Section 2.3, the symbol h was used to denote $x_2 - x_1$, the change in the independent variable x. For our immediate purpose and for future work in integration, it is more convenient to designate an arbitrary change in x by the symbol dx, called the **differential of x** and defined as

$$\boxed{dx = x_2 - x_1}$$

An associated quantity dy, called the **differential of y,** is defined as

$$dy = f'(x)\,dx \tag{1}$$

What is the meaning and significance of dx and dy? Figure 1 contains a geometric representation of both quantities. The line segment PR coincides with the line tangent to the curve at an arbitrary point $(x, f(x))$. The base PQ of the triangle PQR represents the magnitude of dx, and QR, the height, represents the magnitude of dy. Basically, dy represents the change in the y coordinates on the tangent line when x is changed by the amount dx. The following examples illustrate how dy is found for a given function and how it is evaluated for specific values of x and dx.

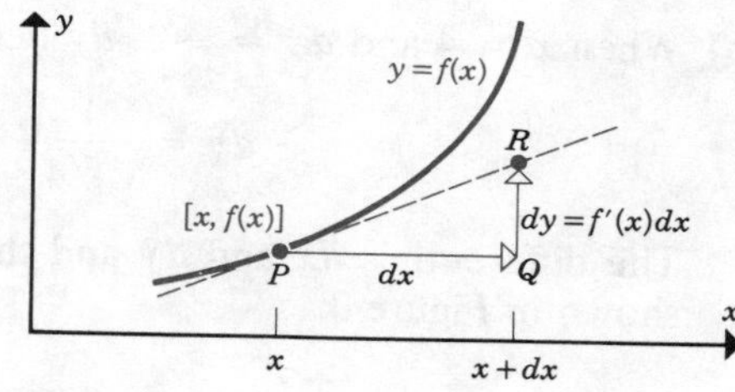

Figure 1

Example 1 (a) Find the differential dy for the function

$$f(x) = x^2$$

(b) Evaluate the differential when $x = 1$ and $dx = 0.75$.

Solution (a) The derivative $f'(x)$ is

$$f'(x) = 2x$$

so that the differential dy can be written as

$$dy = 2x\,dx$$

(b) When $x = 1$ and $dx = 0.75$, dy becomes

$$dy = 2(1)(0.75) = 1.50$$

The differentials dx and dy together with the function are displayed in Figure 2.

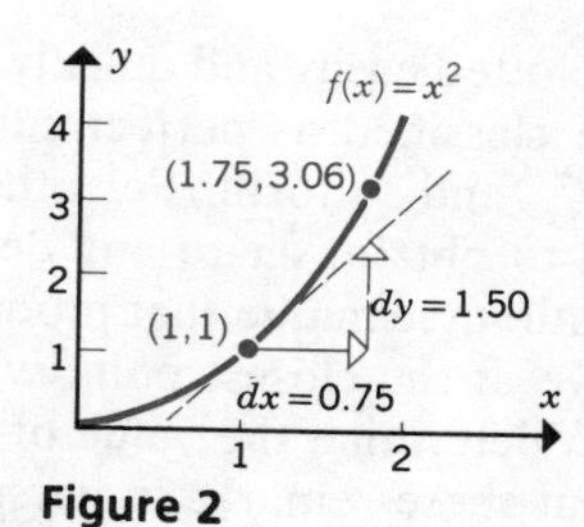

Figure 2

Example 2 (a) Find the differential dy for the function

$$f(x) = \sqrt{x} = x^{1/2}$$

(b) Evaluate the differential dy when $x = 4$ and $dx = -3$.

Solution (a) The derivative $f'(x)$ has the form

$$f'(x) = \frac{1}{2}x^{-1/2} = \frac{1}{2\sqrt{x}}$$

Applying the definition $dy = f'(x)\,dx$ gives

$$dy = \frac{1}{2\sqrt{x}}\,dx = \frac{dx}{2\sqrt{x}}$$

(b) When $x = 4$ and $dx = -3$, dy becomes

$$dy = \frac{1}{2\sqrt{4}}(-3) = -0.75$$

The differentials dx and dy and the graph of the function $f(x) = \sqrt{x}$ are shown in Figure 3.

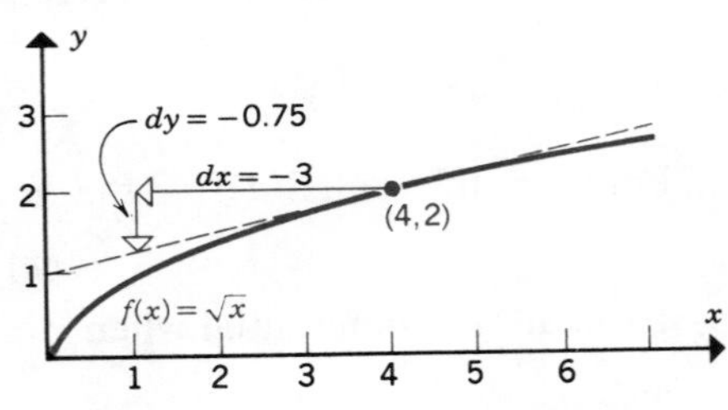

Figure 3

■

The graphs in each of the previous examples indicate that the tangent line and the curve almost coincide when dx is small. For this reason, the change in the y coordinate of a function often can be approximated by dy. This technique is useful in calculating approximate values of the y coordinates when a given function does not easily yield exact values of y except for selected values of x. For example, the function

$$f(x) = \sqrt[3]{x}$$

can be plotted easily and quickly if the independent variable x is assigned values that are classified as perfect cubes, for example, 0, ± 1, ± 8, ± 27, ± 64, $\pm \frac{1}{8}$, $\pm \frac{1}{27}$, $\pm \frac{27}{8}$, and so forth. For other values of x, approximation techniques must be used to obtain $\sqrt[3]{x}$ to any desired number of decimal places. A simple approximation technique that produces satisfactory results uses the tangent line to the curve at the closest point where x is a perfect cube, in place of the curve itself, to determine the value of y corresponding to the given value of x.

To put these remarks in perspective, suppose that $\sqrt[3]{8.5}$ is desired. The behavior of the function $f(x) = \sqrt[3]{x}$ together with the tangent line in the vicinity of (8, 2) is shown in Figure 4 (the curvature of the curve has been exaggerated purposely). Because the tangent line and the curve almost coincide when x is assigned values close to 8, the change in the function can be approximated by the differential dy for a given change dx in the x coordinate. In this case

$$dx = 8.5 - 8.0 = 0.5$$

$$f'(x) = \frac{1}{3(\sqrt[3]{x})^2}, \qquad f'(8) = \frac{1}{3(\sqrt[3]{8})^2} = \frac{1}{12}$$

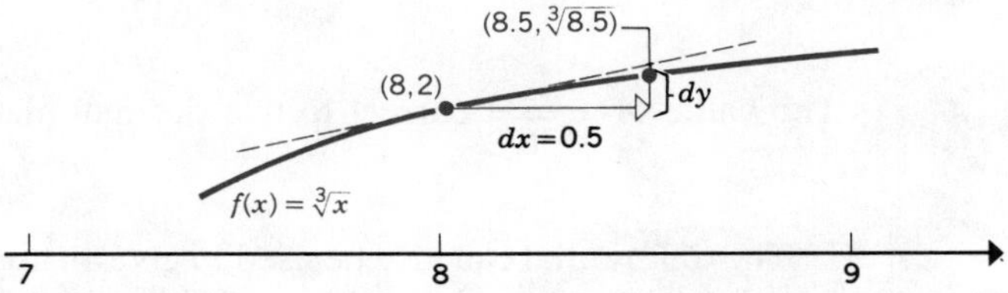

(8.5, $\sqrt[3]{8.5}$)
(8,2)
dy
$dx = 0.5$
$f(x) = \sqrt[3]{x}$
7 8 9

Figure 4

The differential dy then becomes

$$dy = f'(8)\,dx = \tfrac{1}{12}(0.5) \approx 0.042$$

so that we can now write

$$\sqrt[3]{8.5} \approx \sqrt[3]{8} + dy \approx 2 + 0.042 = 2.042$$

The value of $\sqrt[3]{8.5}$, correct to three decimal places, is 2.041. Thus the procedure provides a good estimate in this case. The generalized version of this procedure takes the form

$$\boxed{f(x + dx) \approx f(x) + dy = f(x) + f'(x)\,dx} \tag{2}$$

This approximation, called a **linear approximation,** is illustrated in Figure 5.

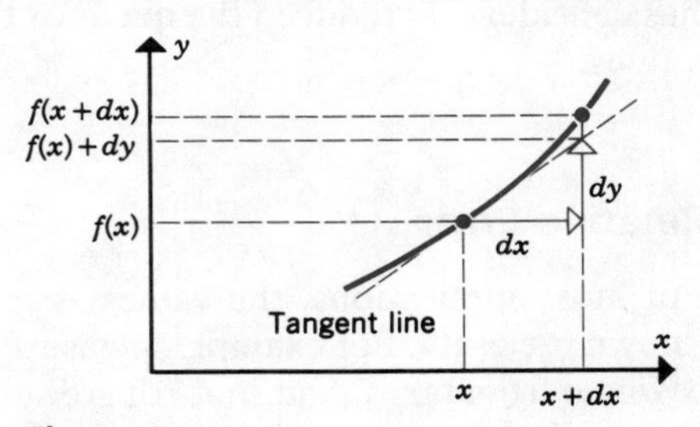

Figure 5

Example 3 Using differentials, find an approximate value of $\dfrac{1}{\sqrt{36.2}}$.

Solution Noting that $\dfrac{1}{\sqrt{36}} = \dfrac{1}{6}$, we can apply Equation 2 with

$$f(x) = \frac{1}{\sqrt{x}} = x^{-1/2}, \quad f'(x) = -\tfrac{1}{2}x^{-3/2}, \quad x = 36, \quad dx = 36.2 - 36.0 = 0.2$$

We then get the following result:

$$\begin{aligned}\frac{1}{\sqrt{36.2}} &\approx \frac{1}{6} + \frac{-1}{2}(36)^{-3/2}(0.2)\\ &\approx .16647\end{aligned}$$

The value of $\dfrac{1}{\sqrt{36.2}}$ correct to five decimal places, is 0.16621. ■

The differential can also be used to give an approximate value for a polynomial function when the independent variable x is assigned a value close to an integer, as shown in the next example.

Example 4 Given the function

$$f(x) = 2x^4 - 5x^2 + 9x$$

use differentials to approximate $f(3.04)$.

Solution Because 3.04 differs slightly from 3.00, Equation 2 is used with

$$x = 3.00, \qquad dx = 0.04, \qquad f'(x) = 8x^3 - 10x + 9$$

$$\begin{aligned}f(3.04) &\approx f(3.0) + f'(3.0)(0.04)\\ &\approx 144 + 195(0.04) = 151.80\end{aligned}$$

The value of $f(3.04)$, correct to two decimal places, is 151.97, so the error obtained by using the approximation is slight. The wide availability of calculators has considerably reduced the need for this approach when x assumes nonintegral values.

Relative Error

For most applications, the values assigned to the independent variable are not known precisely. For example, measurements of industrial output, income levels, revenue from taxes, and so forth are subject to error even when the observations are carried out carefully. When the error or uncertainty in the independent variable is given but is not large, the differential can be used to approximate the corresponding error in the dependent variable. The error in the independent variable is represented by dx, and the corresponding error in the dependent variable is represented by dy.

More important than the errors themselves are quantities called relative errors. The **relative error** in the **independent variable** x is defined as $\frac{dx}{x}$, while the **relative error** in the **dependent variable** y is defined as $\frac{dy}{y}$. For example, when a company forecasts annual sales of 6 million units, it is not expected that exactly 6 million units will be sold. If the uncertainty associated with the forecast is 300,000 or 0.3 million units, the number of units expected to be sold will be somewhere between 5.7 and 6.3 million. The relative error in x the number of units sold then becomes

$$\frac{dx}{x} = \pm\frac{0.30}{6.00} = \pm 0.05 \quad \text{or} \quad \pm 5\%$$

The next step is to determine what effect the relative error in x will have on, say, the revenue or profit for the firm. Suppose that the profit $P(x)$ is related to the number of units sold via the equation

$$P(x) = 14x - x^2 - 8$$

where $P(x)$ is expressed in millions of dollars. The company's profit, based on the forecast of 6 million units, is

$$P(6) = 84 - 36 - 8 = \$40 \text{ million}$$

The relative error in the profit function $\frac{dP}{P}$ can be found by first finding dP,

$$dP = P'(x)\,dx = (14 - 2x)\,dx$$

so that

$$\frac{dP}{P} = \frac{(14 - 2x)\,dx}{14x - x^2 - 8} = \frac{14x - 2x^2}{14x - x^2 - 8}\,\frac{dx}{x}$$

The last step was carried out because in most problems the relative error $\frac{dx}{x}$, or percentage relative error, $100\frac{dx}{x}$, is specified. The relative error $\frac{dP}{P}$ then becomes

$$\frac{dP}{P} = \frac{84 - 72}{40}(\pm 0.05) = \frac{12}{40}(\pm 0.05) = \pm 0.015 \quad \text{or} \quad \pm 1.5\%$$

This result indicates that profits will be somewhere within a range of approximately 1.5 percent above to 1.5 percent below the predicted \$40 million if the sales level turns out to be somewhere between 5 percent above to 5 percent below the sales forecast of 6 million units. The relative errors in x and y are defined as

$$\text{Relative error in } x = \frac{dx}{x}$$

$$\text{Relative error in } y = \frac{dy}{y} = \frac{f'(x)\,dx}{f(x)}$$

Example 5 The relationship between $p(x)$, the price per bushel of wheat, and x, the number of bushels harvested annually, is given by the equation

$$p(x) = \frac{2}{x+1}$$

where $p(x)$ is expressed in dollars per bushel and x in billions of bushels. Agricultural economists are forecasting a harvest of two billion bushels. The percentage error in their forecasts is usually 10 percent, that is,

$$\frac{dx}{x} = \pm 0.10$$

What is the relative error or uncertainty in the price per bushel of wheat?

Solution The differential dp is found first, using the definition $dp = p'(x)\,dx$:

$$dp = \frac{-2}{(x+1)^2}\,dx$$

Next the quantity $\frac{dp}{p}$ is formed, giving

$$\begin{aligned}\frac{dp}{p} &= \frac{-2}{(x+1)^2}\,\frac{dx}{p} \\ &= \frac{-2}{(x+1)^2}\,\frac{dx}{2/(x+1)} = \frac{-1}{x+1}\,dx \\ &= \frac{-x}{x+1}\,\frac{dx}{x}\end{aligned}$$

Substituting 2 for x and ± 0.10 for $\frac{dx}{x}$, we get the following result for $\frac{dp}{p}$:

$$\frac{dp}{p} = -\frac{2}{3}(\pm 0.10) \approx \mp 0.06667$$

From this we can conclude that if the wheat harvest is 10 percent above predicted levels, the price per bushel will drop about 6.7 percent and vice versa. ■

In Section 3.4 it was stated that the additional revenue obtained when production or sales increases from x units to $x + 1$ units can be approximated by $R'(x)$. The rationale for the statement stems from the approximation

$$R(x + dx) - R(x) \approx dR$$

when dx is small. Using the definition of the differential dR, we are able to write

$$R(x + dx) - R(x) \approx R'(x)\, dx$$

When $dx = 1$ and x is large, the approximation takes the form

$$R(x + 1) - R(x) \approx R'(x)$$

where the left-hand side represents the additional revenue when sales increase by one unit and the right-hand side is the marginal revenue.

4.5 EXERCISES

In Exercises 1 through 12, (a) find the differential dy and (b) evaluate dy for the given values of x and dx.

1. $f(x) = 6x^3$; $x = -1$, $dx = 0.10$

2. $f(x) = \dfrac{5}{x - 1}$; $x = 2$, $dx = -0.25$

3. $f(x) = x^2 - 5x - 2$; $x = 3$, $dx = 1$

4. $f(x) = \sqrt{x^2 + 1}$; $x = 0$, $dx = 0.01$

5. $f(x) = \dfrac{x}{3 - 2x}$; $x = 2$, $dx = -0.50$

6. $f(x) = \dfrac{2}{x^2 + 1}$; $x = -1$, $dx = 0.40$

7. $f(x) = (1 - x^2)^{1/3}$; $x = -3$, $dx = -0.30$

8. $f(x) = \dfrac{1}{\sqrt{x}} + x^2$; $x = 9$, $dx = -2$

9. $f(x) = \dfrac{x}{x + 1}$; $x = 2$, $dx = 1$

10. $f(x) = x^2(x + 2)$; $x = 4$, $dx = -2$

11. $f(x) = \dfrac{1}{\sqrt{2x}}$; $x = 2$, $dx = 0.25$

12. $f(x) = (x^3 + 1)^{1/2}$; $x = 2$, $dx = 0.50$

Use differentials to find approximate values for each of the quantities in Exercises 13 through 20.

13. $\sqrt{26}$

14. $\sqrt{8.9}$

15. $\sqrt[3]{28}$
16. $\sqrt[3]{63}$
17. $\sqrt[4]{15.7}$
18. $\sqrt[3]{124}$
19. $\dfrac{2}{\sqrt{50}}$
20. $\sqrt[4]{82}$

21. A 100-by-100-ft plot of land is being enclosed by a chain link fence. If the relative error in measuring the length of each side is ± 0.01, what is the relative error in the area enclosed by the fence?

22. The volume of a sphere V is given by the formula

$$V = \frac{4\pi R^3}{3}$$

where R is the radius of the sphere. Find the relative error in V when the relative error in R is ± 0.02.

23. Automobile production next year is expected to total nine million units. The Rain and Snow Tire Company, which produces original equipment tires for the major automobile companies, has found that $P(x)$, its annual profits, is related to x, total automobile production, by the equation

$$P(x) = 9.5x - \frac{x^2}{2} - 3.5$$

where x is expressed in millions of units and $P(x)$ in millions of dollars. If the forecast has a relative error of ± 0.15, what is the corresponding relative error in company profits?

24. If i, the annual rate of inflation, remains constant, homes that cost C dollars today will cost

$$C(1 + i)^5$$

dollars five years from now. Economists are forecasting an annual inflation rate of 6 percent for the next five years. If the relative error in this forecast is ± 0.30, find the relative error in the predicted price of a home five years from now if the home costs \$120,000 today.

25. Extend the analysis in Example 5 to determine the relative error in total farm revenue from wheat sales.

26. The time $T(L)$, in seconds, for a pendulum to make a complete swing and return to its original position is called the period; a pendulum's period is a function of its length L, in feet. When the displacement of the pendulum is small, the functional relation is described by the equation

$$T(L) = 2\pi\sqrt{\frac{L}{g}}$$

where g is the acceleration due to gravity ($g = 32$ ft/sec^2). Use the differential dT to approximate the change in the period when the length of a pendulum is increased from 4 to 4.1 ft.

27. The height $H(t)$, in inches, of a plant t weeks after it breaks ground is given by the equation

$$H(t) = 8\sqrt{t} - t, \qquad 0 \le t \le 20$$

Use the differential dH to approximate the change in the plant's height between $t = 9$ weeks and $t = 10$ weeks.

For the functions in Exercises 28 through 30, (a) find the differential dy, (b) evaluate dy for the given values of x and dx, and (c) sketch a triangle whose base equals dx, whose height equals dy, and whose hypotenuse coincides with the tangent line to the curve at the given value of x.

28. $f(x) = 4 - x^2;\quad x = -1,\ dx = 2$

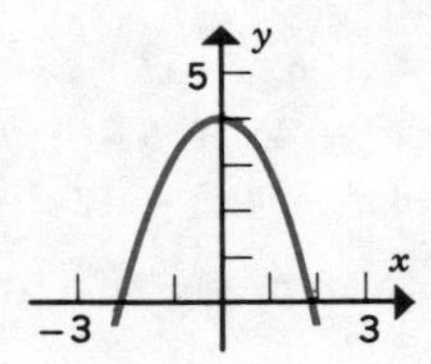

29. $f(x) = 5x^2 - 2x^3;\quad x = 2,\ dx = 1$

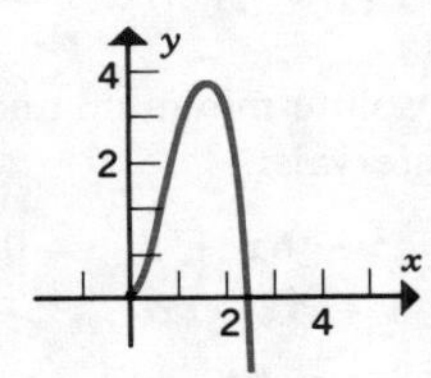

30. $f(x) = (2 - \sqrt{x})^3;\quad x = 1,\ dx = -1$

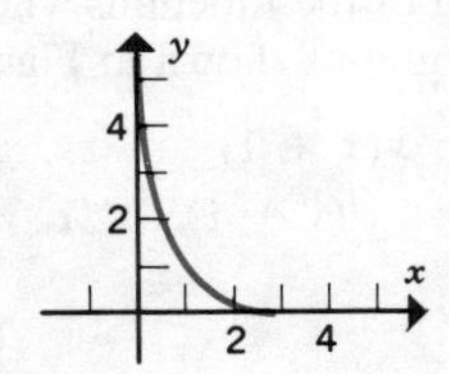

KEY TERMS

critical point
increasing function
decreasing function
first-derivative test
relative maximum
relative minimum
absolute maximum
absolute minimum
optimization

concave up
concave down
second-derivative test
inflection point
horizontal asymptote
vertical asymptote
differential
approximations using the differential
relative error

REVIEW PROBLEMS

In problems 1 through 12, find the critical points, if any, for each of the following functions. Use an appropriate test to determine the behavior of the function in the vicinity of each critical point. Finally, sketch the graph of the function in the vicinity of each critical point.

1. $f(x) = 4 + 6x - x^2$
2. $f(x) = x^3 - 3x^2 + 2$
3. $f(x) = x^2 - 4x + 5$
4. $F(x) = 5 + 12x + 3x^2 - 2x^3$
5. $f(x) = 3x^4 - 4x^3 - 12x^2 + 6$
6. $g(x) = (x - 2)^{2/3}$
7. $f(x) = x^2 + \dfrac{2}{x} + 3$
8. $f(x) = x^2 - 4\sqrt{x} + 2$
9. $f(x) = 12x^5 + 15x^4 - 40x^3 - 10$
10. $F(x) = \dfrac{x^2}{x^2 + 4}$
11. $f(x) = x^2(x - 7)^5$
12. $F(x) = (x - 2)^2(x - 3)^3$

Find the absolute minimum and maximum values of each of the following functions on the given intervals:

13. $f(x) = x^2 - 6x + 2, \quad 0 \le x \le 5$
14. $f(x) = 3 + 4x - x^2, \quad -1 \le x \le 3$
15. $f(x) = 2x^3 - 6x^2 + 4, \quad -1 \le x \le 4$
16. $F(x) = x^4 - x^3 + 5, \quad -2 \le x \le 2$

Match each of the functions whose derivatives are given in problems 17 through 22 with one of the graphs shown in Figure 1.

17. $f'(x) = x(x + 1)$
18. $g'(x) = x^2(x + 1)$
19. $F'(x) = x(x + 1)^2$
20. $G'(x) = x^2(x + 1)^2$
21. $h'(x) = -x(x + 1)$
22. $H'(x) = -x(x + 1)^2$

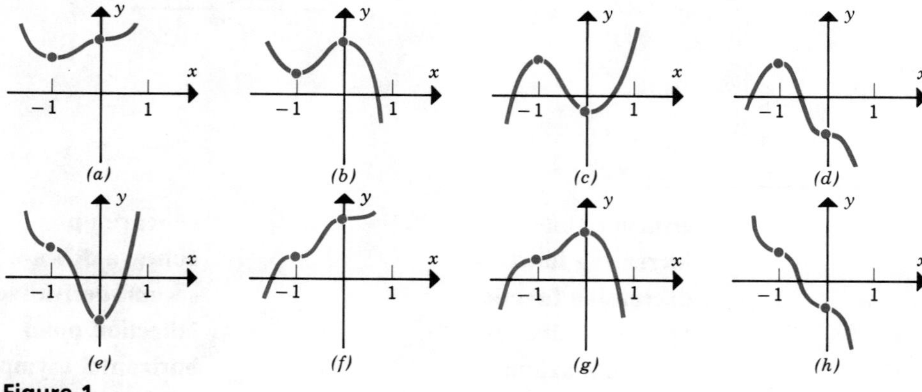

Figure 1

23. The graph of the function $f(x) = ax^2 + bx$ has a relative maximum at $(-1, 4)$. Find the values of a and b.

24. The graph of the function $F(x) = ax^3 + bx^2 + c$ contains critical points at $(-2, 5)$ and $(0, 1)$. Find the values of a, b, and c.

25. The graph of the function $f(x) = ax + \frac{b}{x}$ contains a critical point at $(2, 8)$. Find the values of a and b.

26. The point $(1, 2)$ is a critical point on the graph of the function $g(x) = \frac{ax}{x^2 + b}$. Find the values of a and b.

27. The graph of the function $f(x) = ax^3 + bx^2$ contains an inflection point at $(-1, 4)$. Find the values of a and b.

28. The graph of the function $f(x) = ax^3 + bx^2 + cx$ has a critical point at $(1, 6)$, which is also an inflection point. Find the values of a, b, and c.

Find the relative maxima and minima as well as the points of inflection, if any, for each of the following functions. Use the results of the analysis to sketch the graph of each function.

29. $f(x) = x^2 - 4x + 3$

30. $f(x) = 2 + 3x - x^3$

31. $g(x) = 3x^4 - 4x^3 + 5$

32. $F(x) = \frac{x - 1}{x + 1}$

33. $f(x) = x^2(x - 5)^3$

34. $g(x) = \frac{8x}{x^2 + 4}$

35. $f(x) = \frac{x^2}{x^2 + 4}$

36. $f(x) = 2\sqrt{x + 1} - x$

Match each of the functions, whose second derivatives are given in problems 37 through 40, with one of the graphs in Figure 2.

37. $f''(x) = x(x - 2)$

38. $F''(x) = x^2(x - 2)$

39. $g''(x) = x(x - 2)^2$

40. $G''(x) = x^2(x - 2)^2$

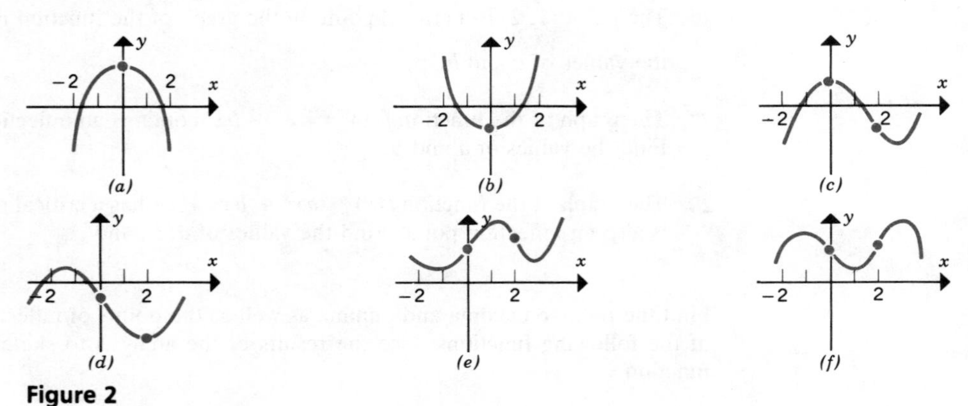

Figure 2

41. The owner of an apple orchard wants to determine the best time to pick and sell her apples. If she picks the apples this week, each tree will yield, on average, 170 pounds, and the apples can be sold for 50¢ per pound to a wholesaler. For each week she waits, the yield per tree will increase by 20 pounds, but the price that the wholesaler will be willing to pay will decrease by 4¢ per pound. Determine the number of weeks that she should wait before picking the apples if she wants to maximize her revenue.

42. A civic group is planning to build a rectangular play area, which is to be enclosed on three sides; an existing fence will form the fourth side. What is the maximum area that can be enclosed if the club has enough money to buy 180 feet of fencing?

43. The concentration of a drug in the bloodstream $C(t)$ at any time t, in minutes, is described by the equation

$$C(t) = \frac{90t}{t^2 + 9}$$

where $t = 0$ corresponds to the time at which the drug was swallowed. Determine how long it takes for the drug to reach its maximum concentration.

44. A landscape architect is designing a rectangular garden for a public park. The garden is to have an area of 720 square feet. His design must incorporate an 8-foot wide walkway on the north and south sides of the garden and a 10-foot walkway on the east and west sides, as shown in Figure 3. Find the dimensions of the garden that will minimize the total area allocated for the garden and the walkways.

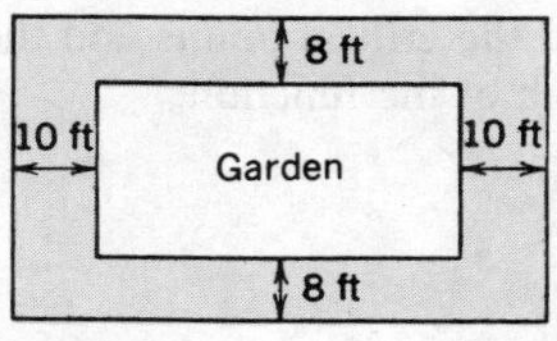

Figure 3

45. The demand for Fax machines is uniform throughout the year. Acme Office Supply Company sells 600 machines annually. Each machine costs the company \$500 to purchase from the manufacturer, the cost to reorder is \$25, and it costs the company \$3 to store a Fax machine for one year. Determine how often Acme should place an order for the Fax machines in order to minimize setup plus carrying costs.

46. A local cable television company has 5000 subscribers, each of whom pays \$18 per month. According to a survey conducted by a consultant, the company could gain an additional 500 subscribers for each decrease of \$1 in the monthly subscription rate. What rate should the company charge each month if it wants to maximize its revenues?

47. A state legislature is considering the imposition of a sales tax. State economists estimate that for each increase of 1¢ in the sales tax, there will be a decrease of \$50 million dollars in weekly purchases throughout the state. If the level of weekly purchases is currently \$600 million, determine how large the sales tax should be if the legislature wishes to maximize state income from the sales tax. If this tax is levied and the economists are correct, what would happen to the amount of weekly purchases in the state?

48. According to a model developed by a public health group, the number of people $N(t)$, in hundreds, who will be ill with the Manchurian flu at any time t, in months, next winter is described by the equation

$$N(t) = 100 + 30t - 10t^2, \qquad 0 \leq t \leq 4$$

where $t = 0$ corresponds to the beginning of December. Find the time when the flu will have reached its peak.

49. A large shipment of computer chips arrives at the receiving area of an electronics firm. A technician from the Quality Control Department selects four chips at random and tests each one. If the number of defective chips among those tested is fewer than two, the shipment is to be accepted. Under these conditions, the likelihood or probability that the shipment will bé accepted is a function of x, the fraction of the entire shipment that is defective. The relationship between $P(x)$, the probability, and x is given by the equation

$$P(x) = (1 - x)^4 + 4x(1 - x)^3, \qquad 0 \le x \le 1$$

Find the critical points and the inflection points; use this information to sketch the graph of the function.

5

EXPONENTIAL AND LOGARITHMIC FUNCTIONS

INTRODUCTION

In addition to the functions already studied, there are many other equally important functions that are needed in applications. Among them are the two types that will be presented in this chapter: **exponential** and **logarithmic** functions. Their properties are examined in Sections 5.1 and 5.2, their derivatives in Sections 5.3 and 5.4. Section 5.5 is devoted to the study of some applications requiring the use of these functions.

5.1 EXPONENTIAL FUNCTIONS

Definition An exponential function has the form

$$f(x) = b^x \qquad (b \neq 1) \tag{1}$$

where b is a positive constant.

The following are examples of exponential functions:

$$f(x) = 4^x, \qquad F(x) = (\tfrac{1}{2})^x, \qquad g(x) = (\sqrt{3})^x$$

Caution Do not confuse exponential functions with power functions. For an exponential function, the base is constant and the exponent can vary. The reverse is true for power functions, for example, $f(x) = x^2$ represents a power function and $g(x) = 2^x$ represents an exponential function.

To see how exponential functions come up in practice, let us consider how banks and other financial institutions compound interest on money deposited with them. Compound interest means that interest is paid not only on the money deposited, called the principal, but also on any interest that has accumulated. For example, suppose that \$500 is deposited in a bank that compounds interest once a year at 6 percent per year. Assuming that no withdrawals or additional deposits occur, the amount of money in the account grows year by year, as shown in Table 1, where the arithmetic has not been completed on purpose. Table 1 shows that the relationship between A, the amount of money in the account, and n, the number of years, is given by the equation

$$A = 500(1.06)^n \tag{2}$$

Table 1

Year	Balance at Beginning	Annual Interest	Balance at End
1	500	500(0.06)	500(1 + 0.06)
2	500(1 + 0.06)	500(1 + 0.06)(0.06)	$500(1 + 0.06)^2$
3	$500(1 + 0.06)^2$	$500(1 + 0.06)^2(0.06)$	$500(1 + 0.06)^3$
⋮	⋮	⋮	⋮
n	$500(1 + 0.06)^{n-1}$	$500(1 + 0.06)^{n-1}(0.06)$	$500(1 + 0.06)^n$

Note that unlike a power function, the base in Equation 2 is **constant** and the exponent can **vary.** Equation 2 can be used to find the amount in the account at the end of any given number of years; for example, the amount in the account after three years is found by setting $n = 3$, which yields

$$A = 500(1.06)^3 = 500(1.06)(1.06)(1.06)$$
$$= \$595.51$$

When n is large, $(1.06)^n$ can be found by a number of different approaches without formally carrying out the repeated multiplication.

1. Tables containing values of the compound interest factor $(1 + i)^n$ for various combinations of n and i have been tabulated. Table A at the end of the book is an example. Suppose that $n = 20$ and $i = 0.06$; the value of the quantity $(1.06)^{20}$ is located in the row corresponding to $n = 20$ and the column corresponding to $i = 0.06$:

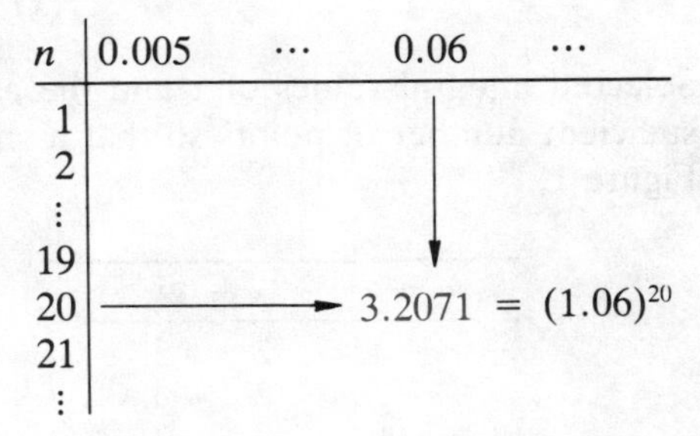

2. Logarithms, which will be introduced in the next section, can also be used to evaluate $(1.06)^{20}$.
3. A calculator with an exponential key y^x will give you the value of the factor $(1.06)^{20}$ quickly and easily. After 1.06 is entered as y, 20 is entered as x. Pressing the y^x key will cause the value of $(1.06)^{20}$ to appear in the display.

The relationship expressed in Equation 2 can be generalized to yield the **compound-interest equation.** Let P be the principal, i the interest rate per compounding period, and n the number of periods; the **future amount** A is given by the **compound-interest** formula

$$A = P(1 + i)^n \tag{3}$$

If the length of the compounding period changes, both i and n are affected, as shown in Example 1.

Example 1 Suppose that the bank compounds interest twice instead of once each year. How much will the deposit of $500 grow to in three years?

Solution Interest is compounded twice each year, so

$$n = 3(2) = 6 \text{ compounding periods}$$

The interest rate per period, i, is

$$i = \frac{0.06}{2} = 0.03$$

The compound-interest formula (Equation 3) yields

$$A = 500(1.03)^6 = \$597.03$$

■

Some of the properties of exponential functions can be illustrated by their graphs, as shown in Examples 2 and 3.

Example 2 Graph the exponential function

$$f(x) = 2^x$$

Solution Selected integral values of x and the associated values of y are used to plot a sufficient number of points so that a smooth curve can be drawn, as shown in Figure 1.

x	$y = 2^x$
-3	$2^{-3} = \frac{1}{8}$
-2	$2^{-2} = \frac{1}{4}$
-1	$2^{-1} = \frac{1}{2}$
0	$2^0 = 1$
1	$2^1 = 2$
2	$2^2 = 4$
3	$2^3 = 8$

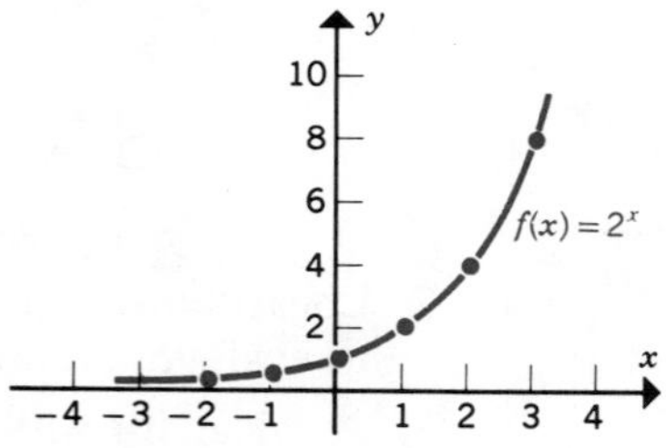

Figure 1

■

Example 3 Graph the exponential function

$$f(x) = (\tfrac{1}{2})^x = 2^{-x}$$

Solution The procedure used in Example 2 is also followed here; the graphical result is shown in Figure 2.

x	$y = (\frac{1}{2})^x$
-3	$(\frac{1}{2})^{-3} = 8$
-2	$(\frac{1}{2})^{-2} = 4$
-1	$(\frac{1}{2})^{-1} = 2$
0	$(\frac{1}{2})^0 = 1$
1	$(\frac{1}{2})^1 = \frac{1}{2}$
2	$(\frac{1}{2})^2 = \frac{1}{4}$
3	$(\frac{1}{2})^3 = \frac{1}{8}$

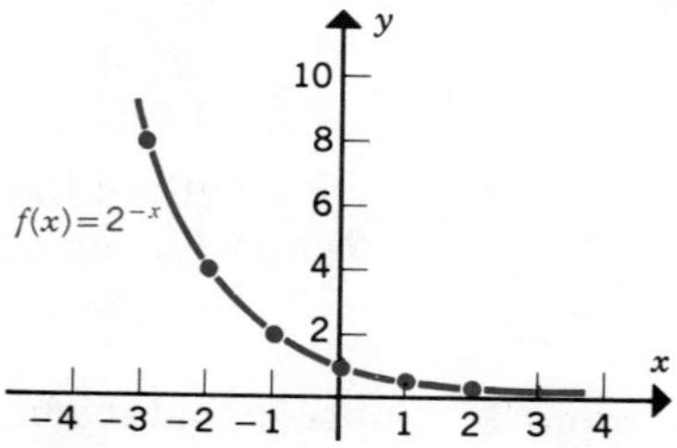

Figure 2

■

Figures 1 and 2 illustrate some of the more important features that exponential functions of the form $f(x) = b^x$ possess.

1. The point (0, 1) is common to all the functions.
2. The domain is the set of all real numbers; the corresponding values of y constitute the set of all positive numbers.
3. If $b > 1$, the function is increasing for all values of x, and

$$\lim_{x\to-\infty} b^x = 0 \quad \text{and} \quad \lim_{x\to\infty} b^x = \infty$$

4. If $0 < b < 1$, the function is decreasing for all values of x, and

$$\lim_{x\to-\infty} b^x = \infty \quad \text{and} \quad \lim_{x\to\infty} b^x = 0$$

These features are summarized graphically in Figure 3.

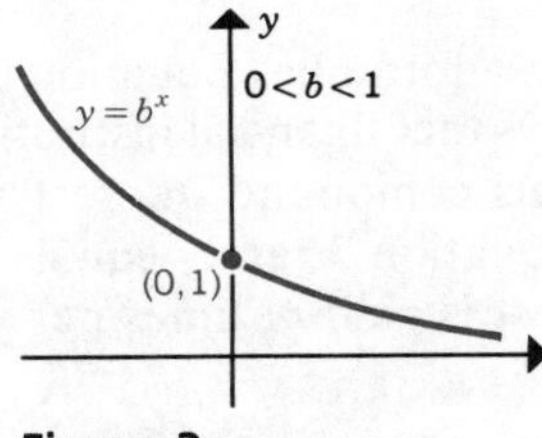

Figure 3

Since the basic properties of exponents will be needed in dealing not only with exponential functions but also with logarithmic functions in Section 5.2, we would be well served to recall them. For $a, b > 0$, we have

1. $b^m \cdot b^n = b^{m+n}$
2. $\dfrac{b^m}{b^n} = b^{m-n}$
3. $(b^m)^n = b^{mn}$
4. $(a \cdot b)^m = a^m \cdot b^m$
5. $\left(\dfrac{a}{b}\right)^m = \dfrac{a^m}{b^m}, \quad b \neq 0$
6. $b^0 = 1, \quad b \neq 0$
7. $b^{-n} = \dfrac{1}{b^n}, \quad b \neq 0$
8. $b^{1/n} = \sqrt[n]{b}$ provided b is not negative when n is an even integer.

These properties can be used to simplify or to rewrite exponential functions, as illustrated in Examples 4 and 5.

Example 4 Write the exponential function $f(x) = 3^{2x}$ in the form $f(x) = b^x$.

Solution Property 3 can be used to write

$$3^{2x} = (3^2)^x = 9^x$$

so the function can also be written as

$$f(x) = 3^{2x} = 9^x$$

■

Example 5 Write the function $f(x) = (4^x)(2^{-x})$ in the form

$$f(x) = b^x$$

Solution First, property 7 is used to write

$$2^{-x} = \frac{1}{2^x}$$

Noting that $(4^x)\left(\frac{1}{2^x}\right) = \frac{4^x}{2^x}$, we can next use property 5 to give

$$f(x) = \frac{4^x}{2^x} = \left(\frac{4}{2}\right)^x$$
$$= 2^x$$

■

The compound-interest formula (Equation 3) can be written in a way that reflects the variety of ways that financial institutions compound interest. Because most financial institutions compound interest more frequently that once each year, the variable i in Equation 3 rarely equals r, the *quoted* or *nominal* annual rate of interest. If interest is paid m times per year, then i and r are related by

$$i = \frac{r}{m}$$

For example, if a bank pays an 8 percent annual rate of interest ($r = 0.08$) compounded quarterly, then i, the interest rate per quarter, equals

$$i = \frac{0.08}{4} = 0.02$$

The exponent n in Equation 3 represents the total number of interest periods and, in general, does not equal the number of years that the money is left on deposit. If the length of time in years is designated t, the relation between t and n is given by

$$n = mt$$

For example, if money is left for $t = 5$ years in a bank that compounds interest quarterly ($m = 4$), the number of interest periods equals

$$n = (4)(5) = 20$$

Thus we see that the compound-interest formula can also be written in the form

$$\boxed{A = P\left(1 + \frac{r}{m}\right)^{mt}} \qquad (4)$$

Example 6 Two thousand dollars is deposited in an account paying 8 percent per year (the nominal rate). How much money is in the account at the end of three years if interest is compounded (a) annually? (b) semiannually? (c) quarterly?

Solution (a) In this case, $i = r = 0.08$, $m = 1$, and $n = t = 3$, so A can be written as

$$A = 2000(1.08)^3$$

Using Table A at the end of the book, we get

$$A = 2000(1.260) = \$2520$$

(b) Because interest is compounded twice each year,

$$m = 2, \quad i = \frac{0.08}{2} = 0.04, \quad n = mt = (2)(3) = 6$$

and we get

$$A = 2000(1.04)^6 = \$2530$$

(c) Interest is compounded four times each year,

$$m = 4, \quad i = \frac{0.08}{4} = 0.02, \quad n = mt = (4)(3) = 12$$

and we get

$$A = 2000(1.02)^{12} = \$2536$$ ■

Present Value

The compound-interest formula demonstrates that for positive values of i the dollar value of an investment increases with time. This means that money has a time value, that is, a dollar today is more valuable than a dollar at some future time. This concept is expressed as the **present value** of a given future amount. The present value can be obtained from Equation 3 by solving for P in terms of A, i, and n, yielding

$$P = \frac{A}{(1 + i)^n} = A(1 + i)^{-n} \tag{5}$$

When written in this form, P is called the **present value of a future amount** A; it gives the monetary value today of an amount A due n periods from now when the interest rate per period equals i. It allows us to place a monetary value on investments that are stated in terms of future dollars, as illustrated in Example 7.

Example 7 An eccentric aunt has set up a trust fund, the proceeds of which will pay you \$1500 three years from now. However, you find yourself in desperate need of cash immediately. A "friend" offers to purchase the rights to the proceeds of the trust fund; if he expects a rate of return of 10 percent annually on his money, how much money should he be willing to pay you today for the rights?

Solution Equation 5 is used to solve this problem; the present value of \$1500 due three years from now is to be found at a rate of return of 10 percent. Substituting 1500 for A, 0.10 for i, and 3 for n in Equation 5, we get

$$P = 1500(1.10)^{-3}$$
$$= \frac{1500}{(1.10)^3}$$

From Table A, we find that

$$(1.10)^3 = 1.331$$

so that

$$P = \$1126.97$$

If your friend offers you less, then he is going to earn an annual rate of return greater than 10 percent should you accept the offer. ■

Variations of Equation 3 can be used to describe the growth of systems other than monetary. Growth of populations such as people, animals, and bacteria can often be described by an equation of the form

$$P = P_0(1 + r)^t \tag{6}$$

where t is the time, P_o is the population when $t = 0$, and r is the rate at which the population is growing.

Example 8 The population of the world is approximately 5 billion people. If the annual rate of growth is 2 percent, what will the population be ten years from now?

Solution Using Equation 6, we find the population ten years from now equals

$$P = 5(1.02)^{10} = 5(1.2190) = 6.10 \text{ billion}$$ ■

Continuous Compounding and e^x

In calculus the most important exponential functions are those whose base equals e, an irrational number whose value to five decimal places is 2.71828. To see why the number e is important, let's look at the compound-interest equation written in the form

$$A = P\left(1 + \frac{r}{m}\right)^{mt}$$

and try to see what form it assumes as interest is compounded more and more frequently. To allow us to concentrate on the essential features of the process, we set the quantities P and t equal to \$1 and one year, respectively, giving

$$A = \left(1 + \frac{r}{m}\right)^m \tag{7}$$

This equation tells us the amount to which \$1 will grow in one year at the nominal rate r compounded m times per year. We want to see what happens as m increases indefinitely, that is, as $m \to \infty$. Before this is done, note that there are two processes competing with one another in Equation 7 as $m \to \infty$.

1. The base $1 + \dfrac{r}{m} \to 1$.
2. The exponent m increases without limit.

The right-hand side of Equation 7 is not in its optimum form to indicate clearly what happens as $m \to \infty$. If we multiply the exponent by r/r, Equation 7 can be written

$$A = \left(1 + \frac{r}{m}\right)^{m \cdot r/r} = \left[\left(1 + \frac{r}{m}\right)^{m/r}\right]^r$$

The quantity on which we now want to concentrate is

$$\left(1 + \frac{r}{m}\right)^{m/r} \tag{8}$$

Because r in expression 8 remains constant, the quantity m/r increases without limit as $m \to \infty$. In the interest of simplicity, a new variable, w, defined as

$$w = \frac{m}{r}$$

is introduced, so that expression 8 becomes

$$\left(1 + \frac{1}{w}\right)^w$$

To see what happens as $w \to \infty$, let us examine Table 2, in which the expression $(1 + 1/w)^w$ is evaluated for selected values of w.

Table 2

w	$\left(1 + \frac{1}{w}\right)^w$
1	2.00000
2	2.25000
3	2.37037
4	2.44141
5	2.48832
10	2.59374
100	2.70481
1,000	2.71692
10,000	2.71815
⋮	⋮

The values in the right-hand column appear to be approaching some limiting value. This limit is defined as e, that is,

$$e = \lim_{w \to \infty} \left(1 + \frac{1}{w}\right)^w = \lim_{m \to \infty} \left(1 + \frac{r}{m}\right)^{m/r} \tag{9}$$

Thus we find that in one year, with continuous compounding, the \$1 deposit has grown to become

$$A = e^r$$

For arbitrary values of P and t, the compound-interest equation, Equation 4, assumes the form

$$A = Pe^{rt} \tag{10}$$

when compounding occurs continuously.

Tables containing values of the quantity e^x have been developed to help in carrying out calculations with exponential expressions whose base is e. Table B at the back of the book represents a typical example where values of both e^x and $e^{-x} = 1/e^x$ are shown. Figure 4 shows the graphs of the two functions $f(x) = e^x$ and $g(x) = e^{-x}$.

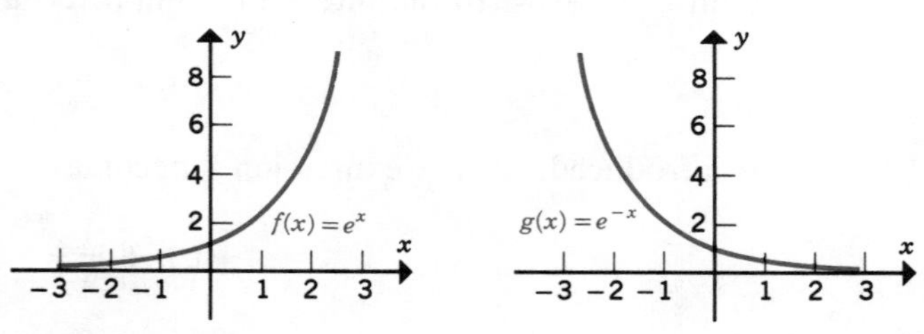

Figure 4

To see how Table B is used in making calculations, let us consider a few simple situations.

Example 9 What is the dollar value at the end of three years of \$500 deposited in a savings account for which interest is compounded continuously at a rate of 6 percent per year?

Solution Noting that $rt = (0.06)(3) = 0.18$, we use Equation 10:

$$A = 500e^{0.18}$$

From Table B or a calculator with an e^x key, we find that

$$e^{0.18} = 1.1972$$

so that

$$A = (500)(1.1972) = \$598.60$$ ■

Example 10 The price of suburban land is increasing continuously at a rate of 10 percent per year. How much money should you pay for an acre of land today if two and a half years from now the value of the land is expected to be $30,000?

Solution Equation 10 is used with $A = 30{,}000$, $i = 0.10$, $t = 2.5$, and P is unknown:

$$30{,}000 = Pe^{0.25}$$

or

$$P = 30{,}000e^{-0.25} = 30.000(0.7788) = \$23{,}364$$ ■

Equation 10 can also be used as a good approximation for cases in which interest is compounded daily.

Example 11 Three thousand dollars is deposited in an account paying 7 percent per year compounded daily. How much money is in the account at the end of two years?

Solution The approximate solution is obtained from Equation 10 with $r = 0.07$, $t = 2$, and $P = 3000$:

$$A = 3000e^{(0.07)(2)} = 3000e^{0.14}$$

From Table B, $e^{0.14} = 1.1503$; so we get

$$A = \$3450.90$$

The true value is obtained from Equation 4 with $P = 3000$, $i = 0.07/365$, and $n = 365(2)$, giving

$$A = 3000\left(1 + \frac{0.07}{365}\right)^{730} = \$3450.78$$ ■

The *present value* P of a future amount A due t years from now at a nominal rate r of interest compounded continuously can be obtained from Equation 10:

$$P = \frac{A}{e^{rt}} = Ae^{-rt} \tag{11}$$

Example 12 Find the present value of a work of art if its value five years from now is expected to be $40,000. Assume that the annual rate of return is 10 percent and that interest is compounded continuously.

Solution The present value can be found by setting $A = \$40{,}000$, $t = 5$, and $r = 0.10$, giving

$$P = 40{,}000e^{-(0.10)(5)} = 40{,}000e^{-0.50} = 40{,}000(0.6065) = \$24{,}260 \quad \blacksquare$$

Application: Double-Declining Balance Method of Depreciation

Accountants often use a method called **double-declining balance** to calculate the depreciation of equipment. Annual depreciation is calculated by multiplying the book value of an asset at the beginning of each year by the fraction $2/T$, where T is the useful life of the asset. To see how the book value V of a piece of equipment declines year by year, let us consider the following situation.

Darling Beer Company purchases a stainless steel vat that costs \$50,000 and has a useful life of 20 years with no salvage value. The annual depreciation and the book value at the end of each year are shown in Table 3; where again the arithmetic has purposely not been completed. Noting that

$$\frac{2}{T} = \frac{2}{20} = 0.1$$

we find that the value of the vat at the end of n years is

$$\begin{aligned} V &= 50{,}000(1 - 0.1)^n \\ &= 50{,}000(0.9)^n \end{aligned}$$

Table 3

Year	Value at Beginning	Annual Depreciation	Value at End
1	50,000	50,000(0.1)	50,000(1 − 0.1)
2	50,000(1 − 0.1)	50,000(1 − 0.1)(0.1)	$50{,}000(1 - 0.1)^2$
3	$50{,}000(1 - 0.1)^2$	$50{,}000(1 - 0.1)^2(0.1)$	$50{,}000(1 - 0.1)^3$
⋮	⋮	⋮	⋮
n	$50{,}000(1 - 0.1)^{n-1}$	$50{,}000(1 - 0.1)^{n-1}(0.1)$	$50{,}000(1 - 0.1)^n$
⋮	⋮	⋮	⋮

The total depreciation D over n years equals

$$\begin{aligned} D &= 50{,}000 - V = 50{,}000 - 50{,}000(0.9)^n \\ &= 50{,}000[1 - (0.9)^n] \end{aligned}$$

These results can be generalized for a piece of equipment whose initial cost is C, whose lifetime is T, and whose salvage value is zero; we get the following equations for V and D:

$$V = C\left(1 - \frac{2}{T}\right)^n \qquad (12)$$

$$D = C\left[1 - \left(1 - \frac{2}{T}\right)^n\right] \qquad (13)$$

5.1 EXERCISES

1. Draw the graphs of each of the following functions:
$f(x) = 3^x$ and $g(x) = (\frac{1}{3})^x = 3^{-x}$.
2. Draw the graphs of each of the following functions:
$f(x) = (\frac{2}{3})^x$ and $g(x) = (\frac{3}{2})^x$.
3. Draw the graphs of each of the following functions:
$f(x) = 2^x$ and $g(x) = x^2$.
4. Draw the graphs of each of the following functions:
$f(x) = (\frac{1}{2})^x = 2^{-x}$ and $g(x) = x^{1/2}$.
5. Draw the graph of the function $f(x) = 2^{|x|}$.

Using the properties of exponents, rewrite each of the following functions in the form $f(x) = b^x$ where $b > 0$.

6. $f(x) = 4^{x/2}$
7. $f(x) = 2^{3x}$
8. $f(x) = 3^{-x}$
9. $f(x) = 8^{x/3}$
10. $f(x) = (4^x)(2^x)$
11. $f(x) = (4^x)(2^{-x})$
12. $f(x) = \dfrac{8^x}{4^x}$
13. $f(x) = (2^x)(2^x)$
14. $f(x) = (3^{-x})^2$
15. $f(x) = (5^{2x})(25^x)$
16. $g(x) = \dfrac{3^{3x}}{27^{2x}}$
17. $F(x) = \dfrac{3^{4x}}{27^x}$
18. The graphs of the four functions
$F_1(x) = 2^x$, $F_2(x) = 3^{x/2}$, $F_3(x) = (\frac{2}{3})^x$, and $F_4(x) = 3^{-x}$
are shown in Figure 5. Match each function with its graph.

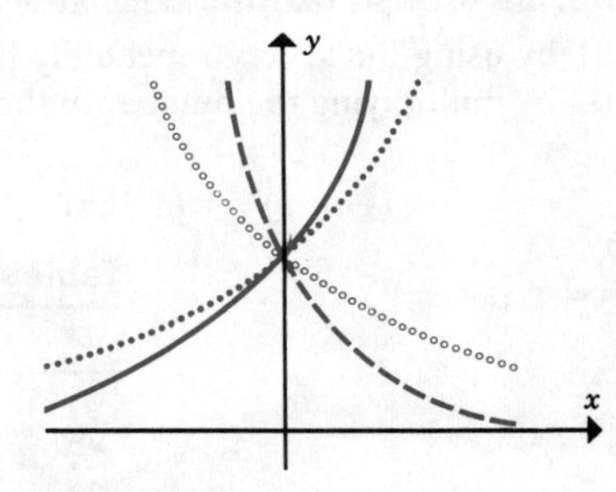

Figure 5

19. Three thousand dollars is deposited in a savings account paying 6 percent per year compounded annually. How much money is in the account at the end of three years?
20. Eight hundred dollars is deposited in a savings account paying 7 percent per year compounded annually. How much money is in the account at the end of ten years?
21. If $2000 is deposited in a money market account paying 8 percent interest per year, how much money is in the account at the end of six years when interest is compounded
(a) annually?
(b) semiannually?
(c) quarterly?

22. The number of bacteria in a culture increases at a rate of 10 percent every hour. If the culture now contains 10^6 bacteria, how many bacteria will be present at the end of
    (a) five hours? (b) ten hours?
23. Annual tuition at Ivy College is \$9000. If tuition increases at the rate of 6 percent per year, what will the annual tuition be ten years from now?
24. What is the present value of a \$1000 promissory note payable three years from now at an annual rate of interest of 12 percent compounded
    (a) annually? (b) semiannually? (c) monthly?
25. A stock analyst predicts that the price per share of IBM stock will equal \$300 five years from now. At a 10 percent rate of return, what is the present value of a share if the analyst is correct?
26. An investor has been told by her stock broker that the price of each share of Logical Computer will double in value over the next three years. Assuming that the annual rate of growth is constant, by how much will the value of each share increase over the next year?
27. A man has \$10,000 to invest and is considering two possibilities:
    (a) A five-year certificate of deposit for which the annual rate of interest is 10 percent compounded semiannually.
    (b) A mutual fund specializing in corporate bonds paying 12 percent interest compounded semiannually. The fund charges an initial, one-time investment fee equal to 4 percent of the initial investment.

    Assuming interest rates do not change, how much would each investment be worth at the end of five years?
28. Calculator exercise. Use a calculator to complete Table 4. Values of w that are powers of 2 are used so that the calculations can be carried out either
    (a) by using the x^2 key repeatedly if the calculator has one or
    (b) by multiplying the number in the display by itself repeatedly. For example, when $w = 8$,

$$(1 + \tfrac{1}{8})^8 = (1.125)^8 = [(1.125^2)^2]^2 = [1.26563^2]^2 \text{ etc.}$$

Table 4

w	$(1 + 1/w)^w$
1	
2	
4	
8	
16	
32	
64	
128	
⋮	

What is $\lim_{w\to\infty} (1 + 1/w)^w$?

Use Table B at the back of the book or a calculator with an e^x key to sketch the graph of each of the functions in Exercises 29 through 34.

29. $f(x) = e^{2x}$
30. $f(x) = e^{-2x}$
31. $f(x) = e^{x-1}$
32. $f(x) = e^{2-x}$
33. $f(x) = 1 - e^{x}$
34. $f(x) = 1 + e^{-x}$
35. If \$5000 is deposited in a savings account paying an annual interest of 8 percent compounded continuously, how much money will be in the account at the end of four years?
36. One hundred dollars is deposited in a savings account paying an annual interest of 7 percent compounded daily.
    (a) Use Equation 10 to approximate the amount in the account at the end of one year.
    (b) How much money would be in the account if interest were compounded once a year?
37. A man wants to deposit money in an account for his granddaughter's college education. If the bank pays 10 percent interest compounded continuously, how much should he deposit in order to have \$10,000 in five years?
38. Sales for a small cable television network have been growing continuously at an annual rate of 20 percent. Assuming that this rate of growth continues, find the dollar volume of sales five years from now if annual sales now are \$50 million.
39. The acquisition of a skill usually requires practice. The number of words per minute $N(t)$ that a trainee can type is given by the equation

$$N(t) = 120 - 90e^{-0.20t}$$

where t is the time, in weeks, from the beginning of the typing course.
    (a) How many words per minute could the trainee type at the beginning of the course?
    (b) How many words per minute can the trainee type at the end of four weeks? eight weeks?
40. Exponential functions are often used to describe the propagation of news through a community, state, or nation. The percentage of people $P(t)$ who have heard the news about a major disaster is given by the equation

$$P(t) = 100 - 100e^{-0.40t}$$

where t is the time in hours since the event occurred.
    (a) What percentage of the people had heard the news within one hour after the event?
    (b) What percentage had heard the news within ten hours after the event occurred?
41. A drug is injected into the bloodstream at time $t = 0$. The concentration $C(t)$, in milligrams per liter, t hours later is given by the equation $C(t) = 10e^{-0.20t}$.
    (a) What is the concentration of the drug one hour after the injection?
    (b) What is the concentration four hours after the injection?
42. An office copying machine with a useful life of eight years is purchased for \$160,000. Using the double-declining balance method,
    (a) Find the amount of depreciation during the third year.
    (b) Find the book value of the machine at the end of three years.
    (c) Plot the graph of V, the book value, as a function of n, the number of years from the date of purchase.
43. A manufacturer of gauges and dial indicators purchases five gear cutters at a cost of \$12,000 each. The company uses the double-declining balance method to calculate

annual depreciation. Assuming that the useful life of the equipment is ten years, find an equation that gives the book value of the five cutters as a function of t, the time in years from the date of purchase. What is the book value of the five cutters at the end of two years?

44. If depreciation occurs continuously, for the double-declining balance method, the equation that describes V as a function of t, the time from the date of purchase, assumes the form*

$$V = Ce^{-2t/T}$$

where C is the cost and T the useful life of the asset.

The owner of a restaurant purchases a dishwasher for $4000; if the expected life of the dishwasher is ten years, find the value of the dishwasher

(a) at the end of three years,
(b) four and a half years after purchase,
(c) nine months after purchase.

*Hegarty, J. C. "A Depreciation Model for Calculus Classes." The College Mathematics Journal, May 1987, Vol. 18, No. 3, pp. 219–221.

5.2 LOGARITHMIC FUNCTIONS

Another class of functions closely related to the exponential functions plays an important role in calculus; each function in this group is called a **logarithmic function.** To see how a logarithmic function originates, let's interchange the x and y coordinates for each point on the graph of the exponential equation $y = 2^x$; the resulting graph, shown as the solid curve in Figure 1, is described by the equation

$$x = 2^y \tag{1}$$

The graph of the equation $x = 2^y$ indicates that each value of x is paired with only one value of y, suggesting that y can be written as a function of x. This function is called the **logarithmic function to the base 2,** denoted by $\log_2 x$, and is written

$$y = f(x) = \log_2 x \tag{2}$$

x	y
$\frac{1}{8}$	-3
$\frac{1}{4}$	-2
$\frac{1}{2}$	-1
1	0
2	1
4	2
8	3

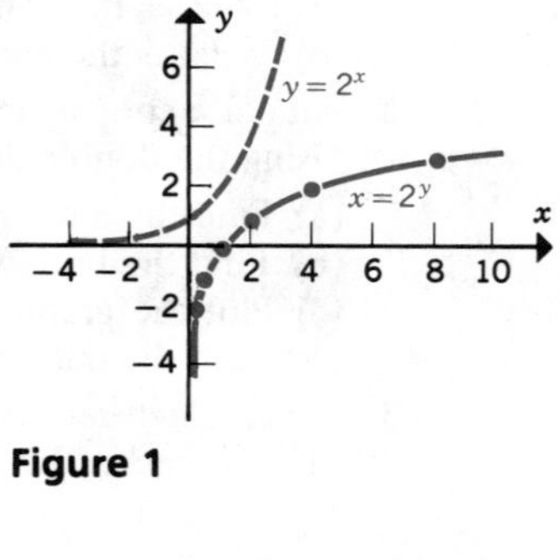

Figure 1

(Read $\log_{2x}$ as "log of x to the base 2.") Because each ordered pair (x, y) that satisfies Equation 1 also satisfies Equation 2, the two equations are equivalent, that is,

$$y = \log_2 x \quad \text{is equivalent to} \quad x = 2^y$$

The process of inverting the x and y coordinates for the exponential function $y = 2^x$ to produce the logarithmic function $y = \log_2 x$ can be extended to any exponential function $y = b^x$ and leads to the following definition.

Definition 1 $\quad y = \log_b x \quad$ is equivalent to $\quad x = b^y \qquad (b \neq 1) \qquad (3)$

At this point, it might be beneficial to show some simple equations in both their exponential and logarithmic forms (Table 1). Gaining proficiency in transforming from the logarithmic to the exponential form can be useful in solving some simple logarithmic equations.

Table 1

Exponential Form	Logarithmic Form
$4^3 = 64$	$\log_4 64 = 3$
$5^{-2} = \frac{1}{25}$	$\log_5 (\frac{1}{25}) = -2$
$(\frac{1}{3})^4 = \frac{1}{81}$	$\log_{1/3} (\frac{1}{81}) = 4$
$(\frac{1}{7})^{-2} = 49$	$\log_{1/7} 49 = -2$
$9^0 = 1$	$\log_9 1 = 0$
$6^1 = 6$	$\log_6 6 = 1$

Example 1 Find the solution to the equation

$$\log_6 x = 4$$

Solution Writing the equation as an equivalent exponential equation gives

$$x = 6^4 = 1296$$

■

Example 2 Solve the equation

$$\log_x 125 = 3$$

Solution Writing the equation in its exponential form,

$$x^3 = 125$$

and noting that $125 = 5^3$, we have

$$x^3 = 5^3$$

from which the solution $x = 5$ is obtained. ■

Example 3 Find the solution to the equation

$$\log_{1/2} 32 = x$$

Solution Rewriting the equation as

$$(\tfrac{1}{2})^x = 32$$

we can obtain the solution by noting that

$$32 = 2^5 = \frac{1}{2^{-5}} = \left(\frac{1}{2}\right)^{-5}$$

The equation can now be written as

$$(\tfrac{1}{2})^x = (\tfrac{1}{2})^{-5}$$

yielding the solution $x = -5$. ■

The graph of the function $y = \log_2 x$ was obtained from the graph of the function $y = 2^x$ by interchanging the x and y coordinates of each point on the curve $y = 2^x$. This technique can be applied to all exponential and logarithmic functions, enabling us to graph the function

$$y = f(x) = \log_b x \tag{4}$$

We use the graph of the equation $y = b^x$ (Figure 3 in Section 5.1) as a reference. The results are shown in Figure 2. The domain of a logarithmic function is the set of all positive real numbers; the corresponding values of y constitute the set of all real numbers.

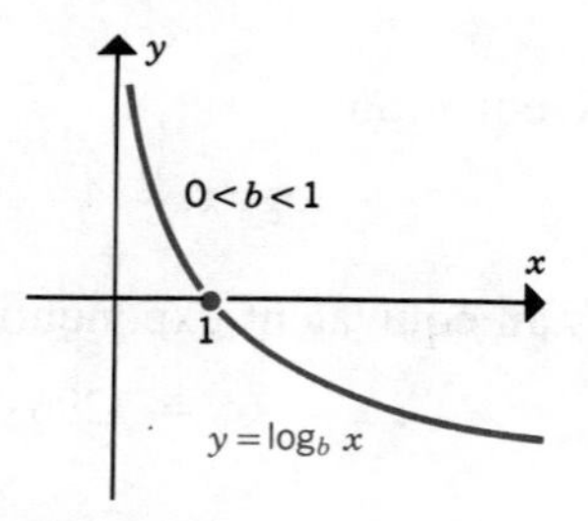

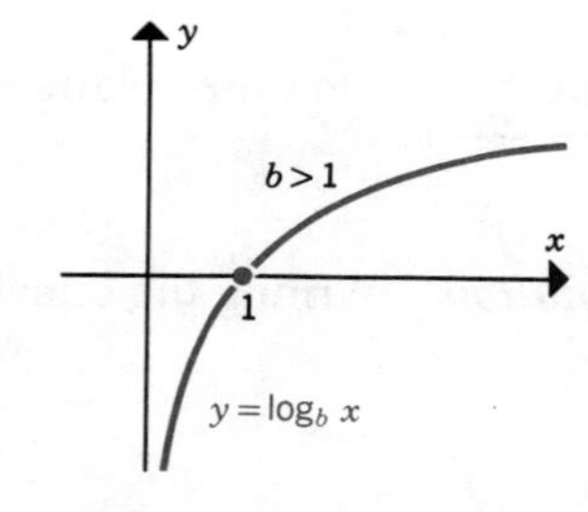

Figure 2

Definition 1 and the examples indicate that logarithms represent another way to write exponents. For example, the equivalence of the equations

$$4^3 = 64 \quad \text{and} \quad \log_4 64 = 3$$

shows that $\log_4 64$ represents the exponent to which 4 must be raised to obtain the number 64; this relationship can be stated formally as

$$4^{\log_4 64} = 64$$

From Definition 1, this property can be stated as an identity:

$$b \log_b x = x \tag{5}$$

Properties of Logarithms

Logarithmic functions possess a number of important properties that are very useful in carrying out calculations and in solving simple exponential and logarithmic equations. These properties together with their exponential counterparts are illustrated next.

Property I. $$\log_b(M \cdot N) = \log_b M + \log_b N, \qquad M, N > 0 \tag{6}$$

The **logarithm of a product equals the sum of the logarithms.** This property is based on the following law of exponents:

$$b^m b^n = b^{m+n}$$

and is illustrated in Example 4.

Example 4

$$\log_4(16 \cdot 64) = \log_4 16 + \log_4 64$$
$$= 2 + 3 = 5$$

The exponential version of the same relationship is

$$16 \cdot 64 = 4^2 4^3 = 4^{2+3} = 4^5$$ ■

Property II. $$\log_b \frac{M}{N} = \log_b M - \log_b N, \qquad M, N > 0 \tag{7}$$

The logarithm of a ratio equals the logarithm of the numerator minus the logarithm of the denominator. This property is based on the following law of exponents:

$$\frac{b^m}{b^n} = b^{m-n}$$

and is illustrated in Example 5.

Example 5

$$\log_2 \frac{128}{16} = \log_2 128 - \log_2 16$$
$$= 7 - 4 = 3$$

The exponential version is

$$\frac{128}{16} = \frac{2^7}{2^4} = 2^{7-4} = 2^3$$ ■

Property III.

$$\log_b(M^N) = N \cdot \log_b M, \qquad M > 0 \tag{8}$$

The logarithm of a power expression equals the exponent multiplied by the logarithm of the base. This property is based on the following law of exponents

$$(b^m)^n = b^{mn}$$

and is illustrated in Example 6.

Example 6

$$\log_2 16^3 = 3 \cdot \log_2 16$$
$$= 3 \cdot 4 = 12$$

The exponential version is

$$16^3 = (2^4)^3 = 2^{4 \cdot 3} = 2^{12}$$ ■

Two other important properties that follow directly from the definition of logarithms are

Property IV.

$$\log_b 1 = 0 \quad \text{since} \quad b^0 = 1 \tag{9}$$

Property V.

$$\log_b b = 1 \quad \text{since } b^1 = b \tag{10}$$

Example 7 If $\log_{10} 2 = 0.3010$ and $\log_{10} 3 = 0.4771$, use properties I through V to evaluate each of the following:

(a) $\log_{10} 6$, (b) $\log_{10} 1.5$, (c) $\log_{10} 81$.*

Solution

(a) $\log_{10} 6 = \log_{10} 2 \cdot 3 = \log_{10} 2 + \log_{10} 3$
$= 0.3010 + 0.4771 = 0.7781$

(b) $\log_{10} 1.5 = \log_{10}\left(\frac{3}{2}\right) = \log_{10} 3 - \log_{10} 2$
$= 0.4771 - 0.3010 = 0.1761$

(c) $\log_{10} 81 = \log_{10} 3^4 = 4 \log_{10} 3$
$= 4(0.4771) = 1.9084$ ■

*In computations, the base 10 is commonly used so that logarithms to the base 10 are called **common logarithms.**

The following two properties are useful in solving simple exponential and logarithmic equations.

Property VI.	If $M = N$, then $\log_b M = \log_b N$, $M, N > 0$	(11)
Property VII.	If $\log_b M = \log_b N$, then $M = N$	(12)

The use of these properties to solve exponential and logarithmic equations is illustrated in Examples 8 and 9.

Example 8 If $\log_{10} 2 = 0.3010$ and $\log_{10} 3 = 0.4771$, use properties I through VII to solve the equation $2^x = 3$.

Solution We can use property VI to write the equation $2^x = 3$ in terms of the logarithm to the base 10 of each side, yielding

$$\log_{10} 2^x = \log_{10} 3$$

Using property III, we can write the left-hand side as

$$\log_{10} 2^x = x \log_{10} 2$$

so that the equation now reads

$$x \log_{10} 2 = \log_{10} 3$$

Solution: $x = \dfrac{\log_{10} 3}{\log_{10} 2} = \dfrac{0.4771}{0.3010} = 1.585$ ■

Example 9 Use properties I through VII to solve the following equation for x:

$$\log_3 x + \log_3(x - 8) = 2$$

Solution Using property I, we can write the left-hand side as a single term as follows,

$$\log_3 x + \log_3(x - 8) = \log_3 x(x - 8) = \log_3(x^2 - 8x)$$

so the equation reads

$$\log_3(x^2 - 8x) = 2$$

The exponential version of this equation is

$$x^2 - 8x = 3^2 = 9$$

Writing this as a standard quadratic equation gives

$$x^2 - 8x - 9 = 0$$
$$(x - 9)(x + 1) = 0$$

Solution: $x = 9$

Why is the solution $x = -1$ not acceptable? ■

Natural Logarithms

In calculus, the most important logarithmic functions are those whose base equals e, namely those having the form

$$f(x) = \log_e x \tag{13}$$

Logarithms to the base e are called **natural logarithms.** Because this function appears frequently, its designation has been shortened to $\ln x$, that is,

$$\boxed{\ln x = \log_e x} \tag{14}$$

where $\ln x$ is read "natural log of x." Table C at the back of the book contains values of the quantity $\ln x$ for selected values of x. Using these values, we can sketch the graph of the natural logarithmic function $y = f(x) = \ln x$ shown in Figure 3.

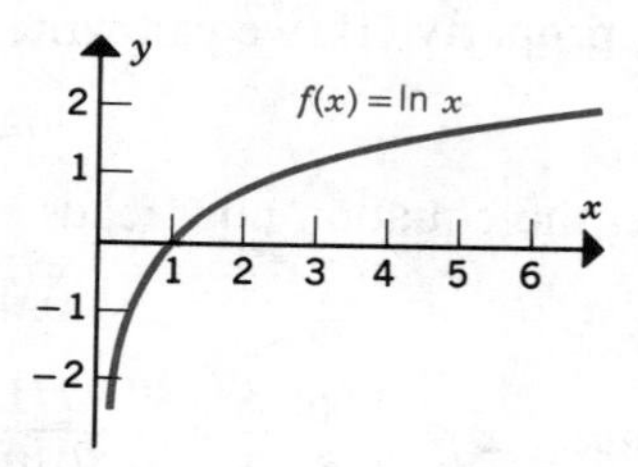

Figure 3

Properties I through VII of logarithms also apply to natural logarithms.

Properties of Natural Logarithms

I. $\ln (MN) = \ln M + \ln N;\ M, N > 0$

II. $\ln \left(\dfrac{M}{N}\right) = \ln M - \ln N;\ M, N > 0$

III. $\ln(M^N) = N(\ln M)$

IV. $\ln 1 = 0$

V. $\ln e = 1$

VI. If $M = N$, then $\ln M = \ln N,\ M, N > 0$

VII. If $\ln M = \ln N$, then $M = N$

These properties can be used to evaluate the natural logarithms of numbers that are not contained in Table C, as shown in the next example.

Example 10 Find ln(350).

Solution First the number is written in scientific notation:

$$350 = 3.5 \times 10^2$$

Next, using property I, we can write

$$\ln(3.5 \times 10^2) = \ln 3.5 + \ln 10^2$$

Property III can be used to rewrite the second term giving

$$\ln 350 = \ln 3.5 + 2 \ln 10$$

Finally, Table C is used to find ln(3.5) and ln(10),

$$\ln 350 = 1.2528 + 2(2.3026) = 5.8580$$

■

These properties are also useful in solving exponential and logarithmic equations, as shown in the next three examples.

Example 11 Solve the equation $3^x = 11$.

Solution Taking the natural logarithm of both sides gives

$$\ln 3^x = \ln 11$$

Using property III, we can rewrite the equation as

$$x(\ln 3) = \ln 11$$

Solving for x gives

$$x = \frac{\ln 11}{\ln 3} = \frac{2.3979}{1.0986} = 2.18$$

■

Example 12 Solve the equation $\ln(x + 4) - \ln x = 2$.

Solution The left-hand side of the equation can be rewritten with the aid of property II to give

$$\ln(x + 4) - \ln x = \ln \frac{x + 4}{x}$$

The equation now reads

$$\ln \frac{x + 4}{x} = 2$$

Transforming the equation to its equivalent exponential form gives

$$\frac{x + 4}{x} = e^2$$

This equation can now be solved for x:

$$x + 4 = e^2x$$
$$4 = x(e^2 - 1)$$

Solution: $x = \dfrac{4}{e^2 - 1}$ ■

Example 13 What is the length of time required for \$1000 to grow to \$2800 in an investment paying 20 percent per year compounded annually?

Solution Using the compound-interest equation $A = P(1 + i)^n$, we get

$$2800 = 1000(1.2)^n$$
$$2.8 = (1.2)^n$$

This equation can be solved by taking the natural logarithm of both sides:

$$\ln 2.8 = \ln(1.2)^n = n \ln 1.2$$

Solving for n yields

$$n = \frac{\ln 2.8}{\ln 1.2} = \frac{1.0296}{0.1823} = 5.65 \text{ years}$$

Since n is restricted to positive integral values, the solution is rounded up to $n = 6$ in order to ensure that the amount reaches our goal of \$2800. The amount at the end of six years equals

$$100(1.2)^6 = \$2986$$

If the money had been withdrawn after only five years, the amount would have been

$$1000(1.2)^5 = \$2488$$ ■

Application: Rule of Seventy

In finance, the *rule of seventy* enables a person to approximate the time required for an investment to double in value. According to the rule, T, the *doubling time* for an investment yielding p percent annually, can be approximated by

$$T \approx \frac{70}{p} \tag{15}$$

For example, money deposited in a savings account paying 10 percent interest per year will double in approximately $70/10 = 7$ years.

The rule can be derived by applying the properties of logarithms to the compound-interest equation $A = Pe^{rt}$. Setting $A = 2P$ and $t = T$, we get

$$2P = Pe^{rT}$$
$$2 = e^{rT} \tag{16}$$

The relation between r and T can be found writing Equation 16 in logarithmic form, yielding

$$\ln 2 = rT(\ln e) = rT$$

Solving for T gives

$$T = \frac{\ln 2}{r} = \frac{0.69}{r}$$

Since $p = 100r$, we can write

$$T = \frac{69}{p} \approx \frac{70}{p}$$

5.2 EXERCISES

Write each of the following exponential equations as an equivalent logarithmic equation.

1. $7^2 = 49$ **2.** $3^{-4} = \frac{1}{81}$ **3.** $(\frac{2}{3})^4 = \frac{16}{81}$
4. $(\frac{5}{4})^{-2} = \frac{16}{25}$ **5.** $4^x = 10$ **6.** $4^5 = 1024$
7. $a^3 = 12$ **8.** $b^3 = x$ **9.** $3^x = 2$

Solve each of the following equations for x.

10. $\log_2 16 = x$ **11.** $\log_{1/2} x = 3$ **12.** $\log_x 49 = 2$
13. $\log_x 625 = 4$ **14.** $\log_x \frac{16}{25} = 2$ **15.** $\log_x \frac{27}{8} = 3$
16. $\log_x \frac{1}{4} = -1$ **17.** $\log_4 2 = x$ **18.** $\log_4(\frac{1}{2}) = x$
19. $\log_{2/3}(\frac{4}{9}) = x$ **20.** $\log_{4/5} x = -3$ **21.** $\log_5 \sqrt{5} = x$
22. $\log_8 4 = x$ **23.** $\log_2(x + 1) = 3$ **24.** $\log_3(x^2 + 2) = 3$

If $\log_b 4 = 1.2042$ and $\log_b 3 = 0.9542$, use properties I, II, and III of logarithms to evaluate the quantities in Exercises 25 through 33.

25. $\log_b 12$ **26.** $\log_b 0.75$ **27.** $\log_b 16$
28. $\log_b 48$ **29.** $\log_b 9$ **30.** $\log_b 2$
31. $\log_b 36$ **32.** $\log_b(\frac{16}{9})$ **33.** $\log_b(\frac{1}{27})$

Use properties I through IV of logarithms to solve Exercises 34 through 38 for x.

34. $\log_3 x + \log_3 x = 4$ **35.** $\log_2(x + 1) + \log_2 6 = 1$
36. $\log_6 x = 2 \log_6 3 + \log_6 2$ **37.** $\log_7 x = 2 \log_7 5 - 3 \log_7 2$
38. $\log_5 x^2 = 2 \log_5 12 - \log_5 4$

Sketch the graph of each of the following functions.

39. $f(x) = \log_3 x$ **40.** $f(x) = \log_{1/3} x$

Use Table C at the end of the book to evaluate each of the following.

41. $\ln 470$ **42.** $\ln 0.89$ **43.** $\ln 1000$

44. ln 6500 **45.** ln 0.0046 **46.** ln 87,000
47. ln 0.056 **48.** ln 150

Solve each of the following equations for x.

49. $5^x = 12$ **50.** $4^{-x} = 3$ **51.** $3^{1-x} = 4$
52. $2^{x^2} = 9$ **53.** $8^x = 15$ **54.** $2^{-x} = 5$
55. $\ln x - \ln(x - 1) = 2$ **56.** $\ln x^2 = 6$
57. $\ln(x + e) + \ln(x - e) = 2$ **58.** $\ln(x + 2e) + \ln(x - e) = 2$

Use Table C to sketch the graph of each of the following functions.

59. $f(x) = 2 \ln x$ **60.** $f(x) = \ln(x - 1)$
61. $f(x) = \ln\sqrt{x}$ **62.** $f(x) = \ln(x + 2)$

63. How long will it take for an investment of \$1500 to increase in value to \$4000 if the annual rate of return is 30 percent compounded once a year?

64. If \$6000 is invested in a real estate limited partnership whose annual rate of return is 20 percent compounded once a year, how long will it take for the investment to increase to \$15,000?

65. A drug is injected into the bloodstream at time $t = 0$. The concentration $C(t)$, in milligrams per liter, t hours later is given by the equation $C(t) = 10e^{-0.40t}$. The drug is not effective when the concentration is less than 2 mg/L. When should the patient be given another injection to maintain the effectiveness of the drug?

66. The relationship between a normal young person's systolic blood pressure $P(w)$, in millimeters of mercury, and her or his body weight w, in pounds, is given by the equation

$$P(w) = 18 + 19.4 \ln w$$

(a) Find the systolic blood pressure for a child who weighs 80 lb.
(b) Find the systolic pressure for an older brother who weighs 100 lb.

67. The yield per tree $Y(x)$, in bushels, is related to x the number of trees in an apple orchard by the equation

$$Y(x) = 80 - 5 \ln x, \qquad x \geq 50$$

(a) What is the yield per tree when $x = 50$? $x = 100$?
(b) Find an equation that describes the relationship between $T(x)$ the total yield from the orchard and x.

68. The graphs of the functions

$$f_1(x) = \log_2 x, \qquad f_2(x) = \log_4 x, \qquad f_3(x) = \ln x$$

are shown in Figure 4. Match each function with its graph.

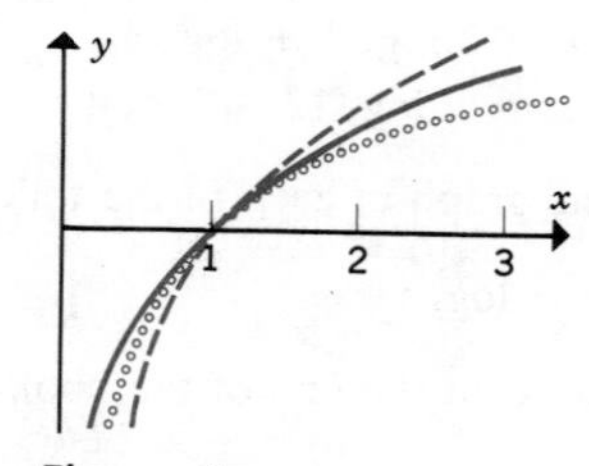

Figure 4

5.3 DERIVATIVES OF EXPONENTIAL FUNCTIONS

The derivative of the natural exponential function

$$f(x) = e^x$$

has a very simple form:

$$f'(x) = \frac{d(e^x)}{dx} = e^x \tag{1}$$

Equation 1 shows that the function $f(x) = e^x$ has the following property.

The function and its derivative are equal everywhere.

Graphically it means that the slope of the line tangent to the curve $y = e^x$ at the point (a, e^a) equals e^a, the y coordinate of the point of tangency, as shown in Figure 1.

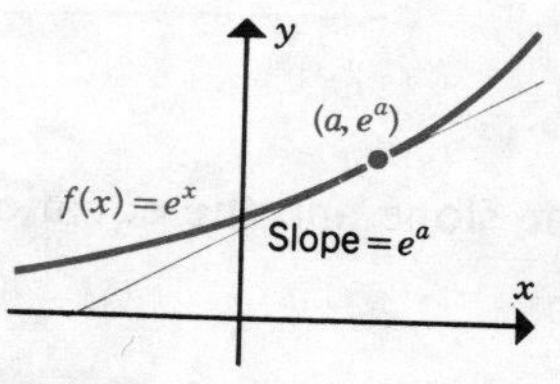

Figure 1

The derivative of an arbitrary exponential function

$$f(x) = b^x$$

has the form

$$f'(x) = \frac{d(b^x)}{dx} = b^x(\ln b) \tag{2}$$

The derivative of the function $f(x) = e^x$ can be found by using the definition of the derivative:

$$f'(x) = \lim_{h\to 0} \frac{f(x + h) - f(x)}{h}$$

For the function $f(x) = e^x$, we get

$$f'(x) = \lim_{h\to 0} \frac{e^{x+h} - e^x}{h} = \lim_{h\to 0} \frac{e^x e^h - e^x}{h}$$

$$= \lim_{h\to 0} \frac{e^x(e^h - 1)}{h} = e^x \lim_{h\to 0} \frac{e^h - 1}{h}$$

The limit can be found by setting up a table such as Table 1 and using Table B at the end of the book to generate the numbers in columns 2 and 4; these numbers appear to be approaching a unique value, namely 1, as $h \to 0$. This suggests that

$$\lim_{h \to 0} \frac{e^h - 1}{h} = 1$$

and that

$$\frac{de^x}{dx} = e^x$$

Table 1 $h \to 0$

From the Left		From the Right	
h	$(e^h - 1)/h$	h	$(e^h - 1)/h$
−0.05	0.97540	0.05	1.02540
−0.04	0.98025	0.04	1.02025
−0.03	0.98500	0.03	1.01500
−0.02	0.99000	0.02	1.01000
−0.01	0.99500	0.01	1.00500

Example 1 Find the slope and the equation of the line tangent to the curve

$$f(x) = e^x$$

at the point $(1, e)$.

Solution The slope of the tangent line at any point is given by $f'(x)$, which according to Equation 1 equals e at the point $(1, e)$. Using the point–slope formula $y - y_1 = m(x - x_1)$ with

$$m = e, \qquad x_1 = 1, \qquad y_1 = e$$

we get

$$y - e = e(x - 1)$$

or

$$y = ex$$

The graph of the curve and tangent line are shown in Figure 2.

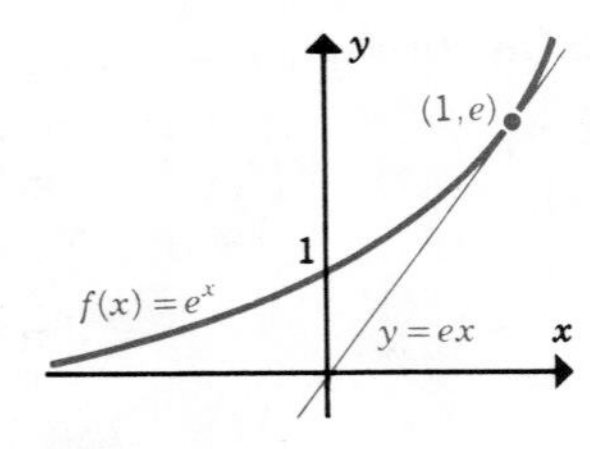

Figure 2

■

Many of the methods such as the product rule and quotient rule developed in Chapter 3 may be needed when differentiating functions containing exponential expressions, as shown in Examples 2 and 3.

Example 2 Find the derivative of the function

$$f(x) = x^2e^x$$

Solution The product rule is used, with

$$u(x) = x^2 \qquad v(x) = e^x$$
$$u'(x) = 2x \qquad v'(x) = e^x$$

Combining in the customary way gives

$$f'(x) = x^2e^x + 2xe^x$$

■

Example 3 Find the derivative of the function

$$f(x) = \frac{e^x - 1}{e^x + 1}$$

Solution The quotient rule is used, with

$$u(x) = e^x - 1 \qquad v(x) = e^x + 1$$
$$u'(x) = e^x \qquad v'(x) = e^x$$

Combining the expressions indicated by the arrows, we get

$$f'(x) = \frac{e^x(e^x + 1) - e^x(e^x - 1)}{(e^x + 1)^2}$$
$$= \frac{2e^x}{(e^x + 1)^2}$$

■

The curve-sketching methods developed in Section 4.4 can be used to sketch the graphs of functions that contain exponential expressions, as illustrated in Example 4.

Example 4 Sketch the graph of the function $f(x) = xe^x$.

Solution Because the factors x and e^x are continuous for all values of x, the function xe^x is continuous for all values of x. To locate the critical points, first we find $f'(x)$ using the product rule. The result is

$$f'(x) = xe^x + e^x = (1 + x)e^x$$

Next setting $f'(x)$ equal to zero gives

$$0 = (1 + x)e^x$$

Because $e^x > 0$ for all values of x, the only solution is

$$x = -1, \qquad f(-1) = -(e^{-1}) = -\frac{1}{e}$$

The second derivative $f''(x)$ is found next; the result is

$$f''(x) = (x + 2)e^x$$

Setting $f''(x)$ equal to zero gives

$$0 = (x + 2)e^x$$

Solution: $x = -2, \qquad f(-2) = \dfrac{-2}{e^2}$

These points together with the horizontal tangent line at $(-1, -1/e)$ are shown in Figure 3; they are used to divide the x axis into three distinct intervals. The signs of $f'(x)$ and $f''(x)$ on each interval are then found to determine the shape of the curve on each. The analysis, shown in Table 2, indicates that the point $(-1, -1/e)$ is a relative minimum and that $(-2, -2/e^2)$ is an inflection point.

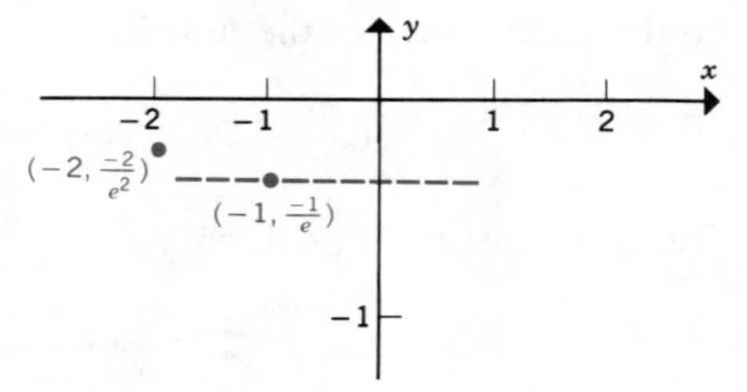

Figure 3

Table 2 $f'(x) = (1 + x)e^x$, $f''(x) = (x + 2)e^x$

Interval	x	Signs of $f'(x)$	Signs of $f''(x)$	Curve	Graph
$x < -2$	-3	$(-)(+) = -$	$(-)(+) = -$	Falling, concave down	
$-2 < x < -1$	$-\frac{3}{2}$	$(-)(+) = -$	$(+)(+) = +$	Falling, concave up	
$x > -1$	0	$(+)(+) = +$	$(+)(+) = +$	Rising, concave up	

If we add (0, 0) as a third point, we obtain the results shown on the graph in Figure 4. The graph can be completed once we have determined the behavior of xe^x as $x \to -\infty$, where two competing trends are at work,

$$\lim_{x \to -\infty} x = -\infty, \qquad \lim_{x \to -\infty} e^x = 0$$

The curve in Figure 4 indicates that xe^x appears to be approaching a finite value as $x \to -\infty$. A table such as Table 3 is helpful; the numbers in the second column indicate that

$$\lim_{x \to -\infty} (xe^x) = 0$$

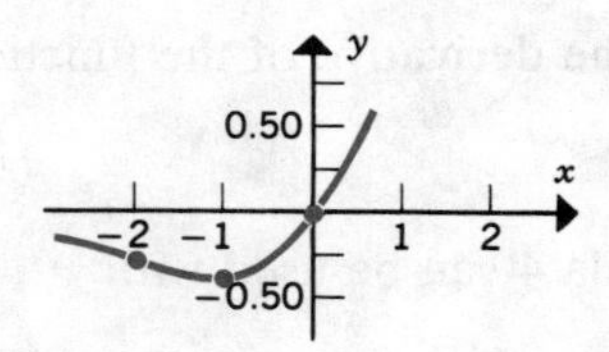

Figure 4

A complete graph of the curve can now be drawn and is shown in Figure 5.

Table 3

x	$y = xe^x$
-5	$-5e^{-5} = -\dfrac{5}{e^5} = -0.03369$
-10	$-10e^{-10} = -\dfrac{10}{e^{10}} = -0.00045$
-100	$-100e^{-100} = -\dfrac{100}{e^{100}} = 0.00000$[a]
$\vdots$	$\vdots$

[a]Calculated to five decimal places.

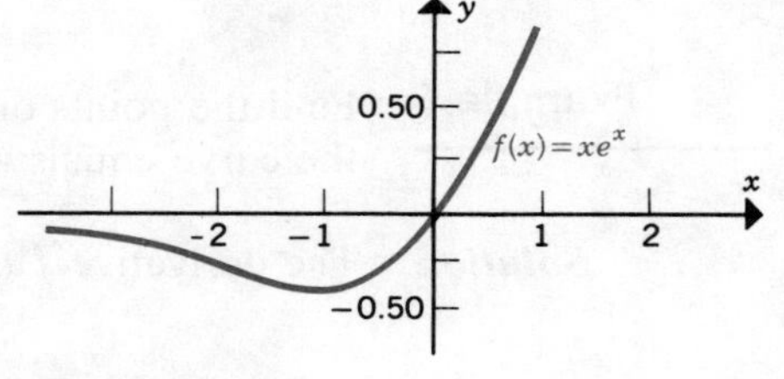

Figure 5

■

Often, it is necessary to differentiate functions of the type

$$f(x) = e^{u(x)} \tag{3}$$

For example,

$$f(x) = e^{x^2-3x+4}, \quad \text{where } u(x) = x^2 - 3x + 4$$

The derivative of $f(x) = e^{u(x)}$ is given by the formula

$$\boxed{\frac{d}{dx} e^{u(x)} = e^{u(x)} \cdot u'(x)} \tag{4}$$

Equation 4 can be obtained by noting first that $f(x) = e^{u(x)}$ can be written as a composite function (Section 3.3),

$$y = f(x) = g(u(x)), \qquad \text{where } g(u) = e^u$$

and then applying the chain rule,

$$\frac{dy}{dx} = \frac{dg}{du}\frac{du}{dx}$$

Noting that $\dfrac{dg}{du} = e^u$, we get

$$\frac{dy}{dx} = e^u \frac{du}{dx} = e^{u(x)}u'(x)$$

Example 5 Find the derivative of the function

$$f(x) = e^{x^2-3x+4}$$

Solution Formula 4 can be used with

$$u(x) = x^2 - 3x + 4$$
$$u'(x) = 2x - 3$$

so that

$$f'(x) = (2x - 3)e^{x^2-3x+4}$$ ■

Example 6 Find the points on the curve $f(x) = e^{2x}$ where the slope of the line tangent to the curve equals 4.

Solution The derivative $f'(x)$ is obtained from Equation 4 by noting that

$$u(x) = 2x$$
$$u'(x) = 2$$

So we get

$$f'(x) = 2e^{2x}$$

Setting $f'(x)$ equal to 4 gives

$$4 = 2e^{2x}$$
$$2 = e^{2x}$$

This equation can be solved by writing it in logarithmic form:

$$\ln 2 = \ln(e^{2x}) = 2x(\ln e)$$

Solving for x gives

$$x = \frac{\ln 2}{2} \approx 0.35, \qquad f(2) = e^{\ln 2} = 2$$ ■

Equation 4 can be used when implicit differentiation (Section 3.6) is needed to find dy/dx, as illustrated in Example 7.

Example 7 Find dy/dx for the curve defined by the equation

$$e^y + y^2 = 2x$$

Solution Since y is not expressed explicitly as a function of x, dy/dx is found by implicit differentiation. Equation 4 is used to differentiate e^y,

$$\frac{de^y}{dx} = e^y \frac{dy}{dx}$$

The other terms can be differentiated directly, producing

$$e^y \frac{dy}{dx} + 2y \frac{dy}{dx} = 2$$

Solving this equation for dy/dx gives

$$\frac{dy}{dx} = \frac{2}{2y + e^y}$$

■

Equation 4 can also be used to find the derivative of $f(x) = b^x$ (Equation 2) by writing b^x as an equivalent exponential expression in base e. Let

$$M = b^x$$

Taking the natural logarithm of both sides yields

$$\ln M = \ln(b^x) = x \ln b$$

Next rewriting this as an exponential equation in base e gives

$$e^{\ln M} = e^{x \ln b}$$

Noting that $e^{\ln M} = M$, the equation takes the form

$$M = b^x = e^{x \ln b}$$

This enables us to write the equation $f(x) = b^x$ as

$$f(x) = e^{x \ln b}$$

Now Equation 4 can be used with $u(x) = x \ln b$ to yield Equation 2:

$$f'(x) = (e^{x \ln b})(\ln b) = b^x \ln b$$

5.3 EXERCISES

Find the derivative of each of the following functions. **Warning:** Do not use the power rule (Section 3.1) when differentiating an exponential function, for example,

$$\frac{d(e^x)}{dx} \neq xe^{x-1}$$

1. $f(x) = e^{2x}$

2. $f(x) = e^{-2x}$

3. $f(x) = x^2 e^x$

4. $f(x) = e^x + e^{-x}$

5. $f(x) = 4^x$

6. $F(x) = 3^{-x}$

7. $f(x) = e^{x^2} + e^{-x^2}$

8. $f(x) = e^{-x^2+2x}$

9. $f(x) = \dfrac{e^x}{x}$

10. $F(x) = (e^x + x)^4$

11. $f(x) = x(2^x)$

12. $g(x) = \dfrac{e^x}{e^x + 1}$

13. $f(x) = xe^x - e^x - 1$

14. $f(x) = xe^{-x} + e^{-x}$

15. $f(x) = \sqrt{e^x}$

16. $f(x) = e^{\sqrt{x}}$

17. $f(x) = e^{e^x}$

18. $f(x) = x^2e^x - 2xe^x + 2e^x$

19. $F(x) = 2^{e^x}$

20. $f(x) = \dfrac{4}{e^x + e^{-x}}$

21. $f(x) = x^e$

22. $f(x) = e^2$

Find the slope and the equation of the line tangent to each of the following curves at the given points.

23. $f(x) = e^x$ at $(1, e)$

24. $f(x) = e^{-x^2}$ at $(1, 1/e)$

25. $f(x) = 2e^{x^2}$ at $(0, 2)$

26. $f(x) = e^x + e^{-x}$ at $(0, 2)$

27. $f(x) = x^2e^{-x}$ at $(-1, e)$

28. $f(x) = \dfrac{e^x}{x}$ at $(1, e)$

29. $f(x) = x^2 + e^{-x}$ at $(0, 1)$

30. $f(x) = xe^x - e^x - 1$ at $(1, -1)$

Find the points (if any) on each of the following curves where the slope of the tangent line has the given value.

31. $f(x) = e^x$, $m = 1$

32. $f(x) = e^{-x}$, $m = -2$

33. $f(x) = x^2e^x$, $m = 0$

34. $f(x) = e^{-x} + 3x$, $m = 2$

35. $f(x) = e^x - xe^x$, $m = 0$

36. $f(x) = e^{2x} + 4e^x$, $m = 6$
Hint: The expression $e^{2x} + 2e^x - 3$ is factorable:
$e^{2x} + 2e^x - 3 = (e^x + 3)(e^x - 1)$.

Find the critical and inflection points (if any) for each of the following functions. In addition, sketch the graph of each.

37. $f(x) = xe^{-x}$

38. $f(x) = e^x - x$

39. $f(x) = e^{-x^2/2}$

40. $f(x) = x^2e^x$

41. $f(x) = e^x - xe^x$

42. $f(x) = e^{2x} - 4e^x$

43. The graphs of the function $y = u(x)$ and the tangent line at $x = 1$ are shown in Figure 6. If $f(x) = e^{u(x)}$, find (a) $f(1)$ and (b) $f'(1)$.

44. The graphs of the function $y = g(x)$ and the tangent line at $x = 0$ are shown in Figure 7. If $F(x) = g(x)e^{-x}$, find (a) $F(0)$ and (b) $F'(0)$.

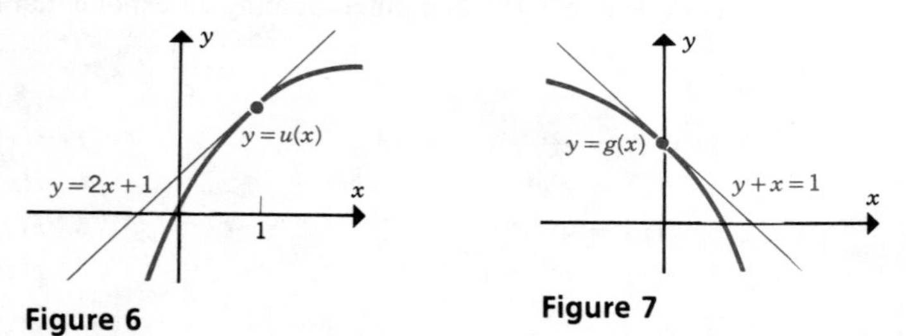

Figure 6

Figure 7

Use implicit differentiation to find dy/dx for each of the following equations.

45. $e^y + x^3 = 3x$

46. $xe^y = 2$

47. $ye^y = 4x$

48. $e^{y^2} = x^3 + 1$

49. $e^{y^2} = xy$

50. $xy^2 + e^{-y} = 2$

51. One thousand dollars is deposited in a savings account paying 6 percent annually compounded continuously.

(a) Find the equation that describes $A(t)$, the amount in the account as a function of t, the time, in years from the date of deposit.

(b) Use the derivative $A'(t)$ to determine how rapidly $A(t)$ is increasing when (i) $t = 1$ and (ii) $t = 5$.

52. The population of the world $P(t)$ (in billions) can be approximated by the equation $P(t) = 5e^{0.02t}$, where t is the time in years. Use the derivative $P'(t)$ to determine how rapidly the population is changing when (a) $t = 0$ and (b) $t = 10$.

53. A patient is being given glucose intravenously. The concentration of glucose $C(t)$, in milligrams per liter, at any time t, in hours, is given by the equation

$$C(t) = 10 - 8e^{-t}$$

where $t = 0$ corresponds to the time when the treatment began.

(a) Find the rate at which the concentration is changing with time when $t = 0$; $t = 5$.

(b) Show that the concentration is increasing for all values of $t > 0$.

54. An experiment is conducted in which the relationship between the number of bacteria $N(t)$ in a closed environment at any time t, in hours, is described by the equation

$$N(t) = t^2e^{-t}, \qquad t \geq 0$$

Find the time at which the number of bacteria reach its maximum value.

5.4 DERIVATIVES OF LOGARITHMIC FUNCTIONS

The derivative of the natural logarithmic function

$$f(x) = \ln x$$

has a very simple form:

$$f'(x) = \frac{d}{dx}(\ln x) = \frac{1}{x} \tag{1}$$

That is, the slope of the line tangent to the curve $f(x) = \ln x$ at the point $(a, \ln a)$ equals $1/a$, the **reciprocal of the x coordinate;** this result is shown graphically in Figure 1.

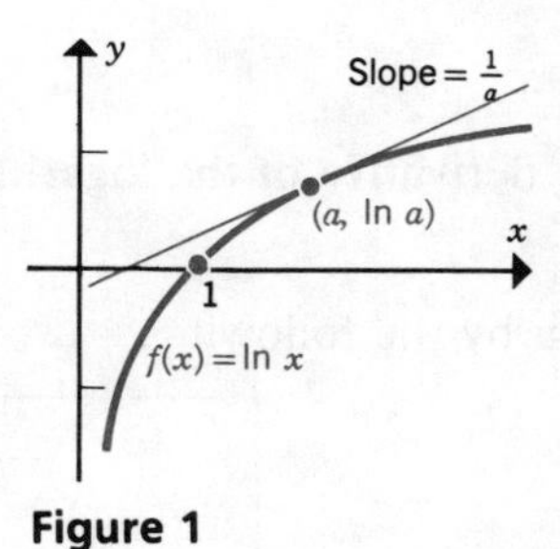

Figure 1

Example 1 Find the slope and equation of the line tangent to the curve $f(x) = \ln x$ at $(2, \ln 2)$.

Solution The slope m is found from Equation 1:

$$m = f'(2) = \tfrac{1}{2}$$

The point–slope equation $y - y_1 = m(x - x_1)$ can be used to find the equation of the tangent line,

$$y - \ln 2 = \tfrac{1}{2}(x - 2)$$

Simplifying gives

$$2y - x = 2 \ln 2 - 2$$

Since $\ln 2 = 0.6931$, the equation can be written as

$$2y = x - 0.6138$$

The curve and tangent line are shown in Figure 2.

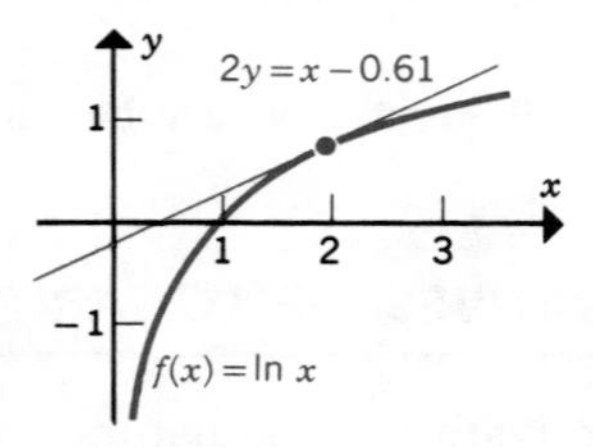

Figure 2 ■

Equation 1 can be derived in the following way. First write the equation $y = \ln x$ as an equivalent exponential equation

$$e^y = x$$

The derivative dy/dx can be found through the use of implicit differentiation (Section 3.6), yielding

$$e^y \frac{dy}{dx} = 1$$

Solving for dy/dx yields

$$\frac{dy}{dx} = \frac{1}{e^y} = \frac{1}{x}$$

The derivative of the logarithmic function

$$f(x) = \log_b x$$

is given by the following:

$$f'(x) = \frac{d}{dx}(\log_b x) = \frac{1}{x \ln b} \tag{2}$$

Example 2 Find the derivative of $f(x) = \log_2 x$.

Solution

$$f'(x) = \frac{1}{x \ln 2}$$

■

Equation 2 can be obtained by methods similar to those used to derive Equation 1, and its derivation is left as a problem in the exercises.

It should be noted that all the techniques, such as the sum, product, quotient, and power rules that have been developed earlier, are applicable in differentiating functions that contain logarithmic expressions.

Example 3 Find the second derivative of the function

$$f(x) = x^2 \ln x$$

Solution Using the product rule (Section 3.2) with

$$u(x) = x^2, \qquad v(x) = \ln x$$

$$u'(x) = 2x, \qquad v'(x) = \frac{1}{x}$$

we can combine the expressions in the usual way to give

$$f'(x) = x^2 \frac{1}{x} + 2x \ln x$$

$$= x + 2x \ln x$$

The second derivative is found by using the sum rule together with the product rule. The result is

$$f''(x) = 1 + 2x\frac{1}{x} + 2 \ln x = 2 \ln x + 3$$

■

Example 4 Find the derivative of the function

$$f(x) = (\ln x)^3$$

Solution Here we have a situation where the general power rule (Section 3.3) is called for, with

$$u(x) = \ln x, \qquad u'(x) = \frac{1}{x}, \qquad n = 3$$

Thus we get

$$f'(x) = 3(\ln x)^2 \frac{1}{x} = \frac{3(\ln x)^2}{x}$$

Note: Property III of logarithms (Section 5.2) cannot be used to find the derivative because

$$(\ln x)^3 \neq 3 \ln x$$

■

The techniques developed in Chapter 4 on curve sketching are also applicable when a function contains logarithmic expressions. However, the calculations are slightly more involved, and solving the equations

$$f'(x) = 0 \quad \text{and} \quad f''(x) = 0$$

may be more challenging.

Example 5 Find the critical points of the function

$$f(x) = x \ln x$$

In addition, sketch the graph of the function.

Solution Before we find the critical points, it is worthwhile to note that the domain of the function is limited to positive values of x. Awareness of this feature will make you less prone to accept spurious solutions of the equation

$$f'(x) = 0$$

The critical points are found by setting $f'(x)$ equal to zero and then solving the resulting equation for x. In this case, the product rule must be used to find $f'(x)$, yielding

$$f'(x) = \ln x + 1$$
$$0 = \ln x + 1$$

or

$$-1 = \ln x$$

The value of x that satisfies this equation can be found by using Table C or, preferably, by writing the equation in its equivalent exponential form to yield the solution

$$x = e^{-1} = \frac{1}{e}$$
$$y = \frac{1}{e} \ln \frac{1}{e} = \frac{-1}{e}$$

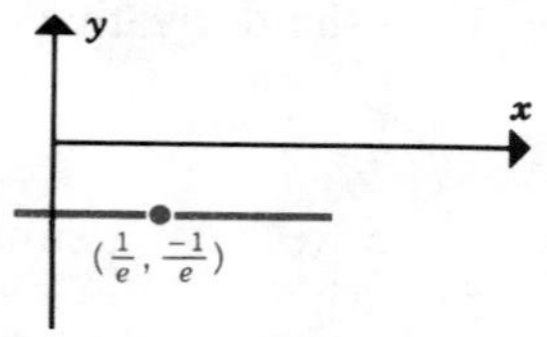

Figure 3

The point $(1/e, -1/e)$ together with the horizontal tangent line is shown in Figure 3. To sketch the graph, we first look for points where $f''(x) = 0$:

$$f''(x) = \frac{1}{x}$$
$$0 = \frac{1}{x}$$

This equation has no solutions, so the x-coordinate of the critical point is used to divide the positive x axis into two distinct intervals.

1. $0 < x < 1/e = 0.37$
2. $x > 1/e = 0.37$

The signs of $f'(x)$ and $f''(x)$ are then used to determine the shape of the curve on each interval. The analysis is shown in Table 1. The analysis shows that $(1/e, -1/e)$ is a relative minimum. If $(1, 0)$ is added as a second point, the result of the analysis is shown on the graph in Figure 4. The last step in our analysis is to determine what happens to $f(x) = x \ln x$ as $x \to 0$; There are two competing trends at work:

$$\lim_{x \to 0} x = 0, \qquad \lim_{x \to 0} \ln x = -\infty$$

Table 1 $f'(x) = \ln x + 1$. $f''(x) = 1/x$

Interval	x	$f'(x)$	$f''(x)$	Curve	Graph
$0 < x < 1/e$	0.2	−0.61	+5	Falling, concave up	
$x > 1/e$	1	+1	+1	Rising, concave up	

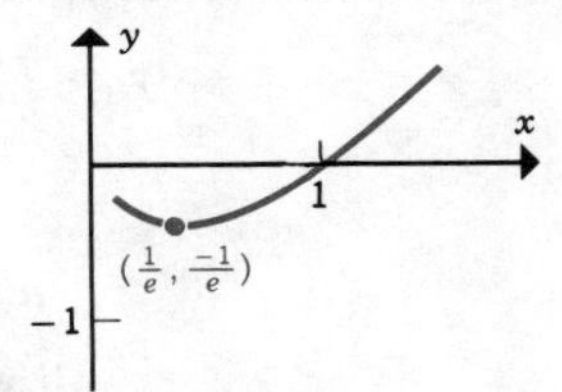

Figure 4

The curve in Figure 4 suggests that $f(x) \to 0$ as $x \to 0$. To see what happens as $x \to 0$, let us look at Table 2; the numbers in column 2 indicate that $x \to 0$ more rapidly than $\ln x \to -\infty$, so that

$$\lim_{x \to 0} (x \ln x) = 0$$

On the basis of this result, the rest of the curve can be sketched, as shown in Figure 5. The open circle at $(0, 0)$ indicates that the function is not defined at $x = 0$.

Table 2

x	$y = x(\ln x)$
0.30	$(0.30)(-1.20) = -0.36$
0.20	$(0.20)(-1.61) = -0.32$
0.10	$(0.10)(-2.30) = -0.23$
0.01	$(0.01)(-4.61) = -0.05$
⋮	⋮

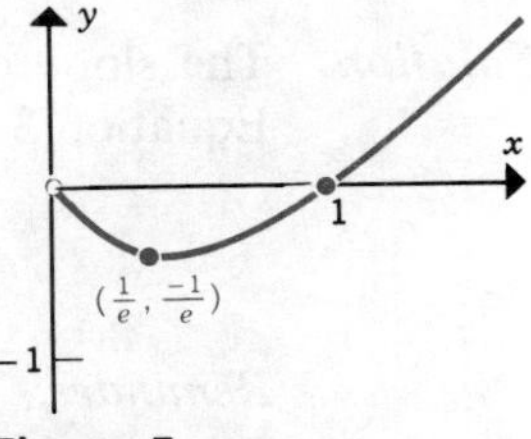

Figure 5

■

There are many occasions when it is necessary to differentiate functions of the type

$$f(x) = \ln u(x)$$

for example,

$$f(x) = \ln(2x^3 + x^2 + 1), \quad \text{where } u(x) = 2x^3 + x^2 + 1$$

The derivative of such functions can be found as follows:

$$f'(x) = \frac{d}{dx}[\ln u(x)] = \frac{u'(x)}{u(x)} \tag{3}$$

Equation 3 can be obtained by applying the chain rule (Section 3.3) to the function $y = \ln u(x)$; noting that $dy/du = 1/u$, we get

$$\frac{dy}{dx} = \frac{dy}{du}\frac{du}{dx} = \frac{1}{u(x)} \cdot u'(x) = \frac{u'(x)}{u(x)}$$

Example 6 Find the derivative of the function

$$y = f(x) = \ln(8x^2 + 5x - 3)$$

Solution Here

$$u(x) = 8x^2 + 5x - 3$$
$$u'(x) = 16x + 5$$

so

$$\frac{dy}{dx} = \frac{16x + 5}{8x^2 + 5x - 3}$$

■

Example 7 Find an equation of the line tangent to the curve

$$f(x) = \ln(2e - x)$$

at the point $(e, 1)$.

Solution The slope of the tangent line is found by evaluating $f'(x)$ at $x = e$. We use Equation 3 with

$$u(x) = 2e - x$$
$$u'(x) = -1$$

Reminder: The quantity e represents a number and is therefore treated as a constant when finding the derivative.

The derivative $f'(x)$ then becomes

$$f'(x) = \frac{-1}{2e - x}$$

At $x = e$, the slope $f'(e)$ of the tangent line is

$$f'(e) = \frac{-1}{e}$$

Next the point–slope formula $y - y_1 = m(x - x_1)$ with

$$m = \frac{-1}{e}, \qquad x_1 = e, \qquad y_1 = 1$$

is used to find an equation of the tangent line. The result is

$$y - 1 = \frac{-1}{e}(x - e)$$

or simplifying gives

$$ey + x = 2e$$

■

In many cases there is more than one way to find the derivative of a logarithmic function having the form $f(x) = \ln u(x)$.

1. Use Equation 3.
2. Use properties I through V of logarithms (Section 5.2) to rewrite the function and then differentiate.

The second approach is often faster, particularly when $u(x)$ is a product or ratio of two or more factors, as illustrated in the next example.

Example 8 Find the derivative of the function

$$f(x) = \ln \frac{x}{x^2 + 1}$$

Solution 1. Using Equation 3 with

$$u(x) = \frac{x}{x^2 + 1}, \qquad u'(x) = \frac{1 - x^2}{(x^2 + 1)^2}$$

we can write

$$f'(x) = \frac{1}{x/(x^2 + 1)} \frac{1 - x^2}{(x^2 + 1)^2} = \frac{1 - x^2}{x(x^2 + 1)}$$

2. First, rewrite the function using the property $\ln(M/N) = \ln M - \ln N$:

$$f(x) = \ln x - \ln(x^2 + 1)$$

Next, differentiating term by term gives

$$f'(x) = \frac{1}{x} - \frac{2x}{x^2 + 1} = \frac{1 - x^2}{x(x^2 + 1)}$$

which agrees with the result obtained in part 1. Notice that the differentiation can be carried out more quickly when $f(x)$ is written as the difference of two logarithmic expressions. ■

Equation 3 can also be used when finding dy/dx by implicit differentiation. This technique is illustrated in Example 9.

Example 9 Find dy/dx for the curve defined by the equation

$$\ln y + y^2 = 5x^2$$

Solution Since y is not written explicitly as a function of x, implicit differentiation can be used to find dy/dx. Equation 3 is used to differentiate $\ln y$ with respect to x:

$$\frac{d}{dx}(\ln y) = \frac{1}{y}\frac{dy}{dx}$$

After differentiating the other two terms with respect to x, we get

$$\frac{1}{y}\frac{dy}{dx} + 2y\frac{dy}{dx} = 10x$$

$$\frac{dy}{dx}\frac{1 + 2y^2}{y} = 10x$$

Solving this equation for dy/dx gives

$$\frac{dy}{dx} = \frac{10xy}{1 + 2y^2}$$

■

5.4 EXERCISES

Find the derivative of each of the following functions.

1. $f(x) = x + \ln x$
2. $f(x) = x^2 \ln x$
3. $f(x) = \ln x^2$
4. $f(x) = (\ln x)^2$
5. $f(x) = \dfrac{\ln x}{x}$
6. $f(x) = \dfrac{x}{\ln x}$
7. $f(x) = \ln x + \dfrac{1}{\ln x}$
8. $f(x) = \log_3 x$
9. $g(x) = \ln(x^2 + 1)$
10. $f(x) = \ln(-x)$
11. $f(x) = \log_4(x^2 - 1)$
12. $F(x) = \ln\sqrt{x^2 + 1}$
13. $f(x) = x^2 \ln x - x + 1$
14. $f(x) = x^n \ln x$
15. $f(x) = \sqrt{\ln x}$
16. $f(x) = \ln\sqrt{x}$

17. $f(x) = \ln \dfrac{x^2}{x^2 + 2}$

18. $f(x) = \ln[(x + 1)^4(x + 2)^6]$

19. $f(x) = e^{\ln x}$

20. $f(x) = \ln e^x$

21. $g(x) = x^{\ln 2}$

22. $f(x) = \ln 2^x$

Find the slope and the equation of the line tangent to each of the following curves at the given points.

23. $f(x) = \ln x$ at $(e, 1)$

24. $f(x) = \ln(x^2 - 3)$ at $(2, 0)$

25. $f(x) = x(\ln x)$ at $(1, 0)$

26. $f(x) = \dfrac{\ln x}{x}$ at $(e, 1/e)$

27. $f(x) = \ln(2x + 3)$ at $(0, \ln 3)$

28. $f(x) = \sqrt{\ln x}$ at $(e, 1)$

29. $f(x) = \log_2 x$ at $(4, 2)$

30. $f(x) = x^2\ln(x + 2)$ at $(0, 0)$

Find the points, if any, on each of the following curves where the slope of the tangent line has the given value.

31. $f(x) = \ln x, \quad m = \frac{1}{2}$

32. $f(x) = 2\ln(3x + 1), \quad m = \frac{3}{2}$

33. $f(x) = x \ln x, \quad m = 1$

34. $f(x) = \dfrac{\ln x}{x}, \quad m = 0$

35. $f(x) = x + \ln x, \quad m = 2$

36. $f(x) = x^2 + \ln x, \quad m = 3$

37. $f(x) = \dfrac{x}{\ln x}, \quad m = 0$

38. $f(x) = (\ln x)^2 + (\ln x)^3, \quad m = 0$

Find the critical and inflection points, if any, for each of the following functions. In addition, sketch the graph of each.

39. $f(x) = x^2 - 2\ln x$

40. $f(x) = x^2\ln x$

41. $f(x) = (\ln x)^2$

42. $f(x) = \ln x - x$

43. $f(x) = \dfrac{\ln x}{x}$

44. $f(x) = x(\ln x) - 2x$

45. The graphs of the function $y = u(x)$ and the tangent line at $x = 1$ are shown in Figure 6. If $f(x) = \ln u(x)$, find

(a) $f(1)$ (b) $f'(1)$

46. The graphs of the function $y = g(x)$ and the tangent line at $x = 0$ are shown in Figure 7. If $F(x) = g(x)\ln(x + 1)$, find

(a) $F(0)$ (b) $F'(0)$

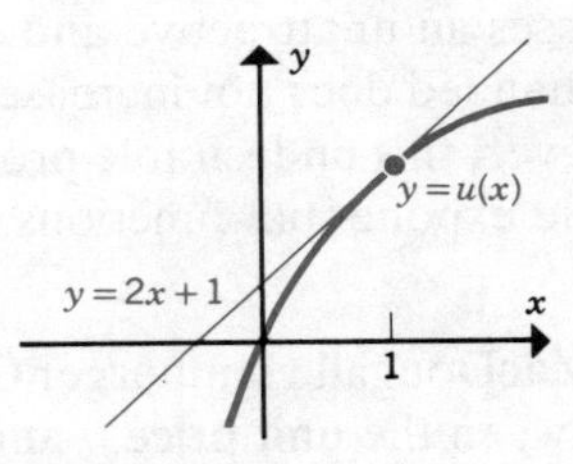

Figure 6

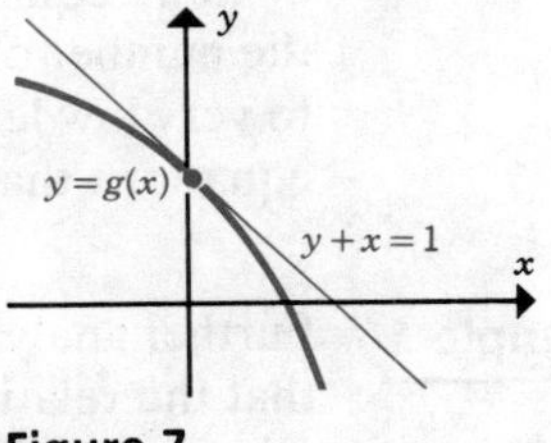

Figure 7

Use implicit differentiation to find dy/dx for each of the following equations.

47. $x \ln y + x = y^2$

48. $\ln(x^2 + y^2) = 3x$

49. $y \ln y = x^2$

50. $(\ln y)^2 + y = x^2 + 4$

51. $\ln(xy) = e^x$

52. $\ln(xy) = e^{xy}$

53. $e^y \ln y = x^2$

54. $y^2 \ln y = x + y$

55. The yield per tree $Y(x)$, in bushels, is related to x the number of trees in an apple orchard by the equation

$$Y(x) = 80 - 10 \ln x, \qquad x \geq 50$$

Determine the number of trees that should be planted if the owner wants to maximize $T(x)$, the total yield from the orchard.

56. The revenue $R(x)$ is related to x, the number of units sold, by the equation

$$R(x) = 10x - x \ln x$$

How many units should be sold to maximize revenue?

57. The relationship between the grade $G(t)$ that a student attains in a statistics course and the amount of time t, in hours, that the student studies each day can be approximated by the equation

$$G(t) = 40 + 30 \ln(1.5t + 1), \qquad t > 0$$

(a) If the student does not study, what grade should he or she expect to receive?
(b) Use the derivative $G'(t)$ to show that a student's grade increases as the amount of time spent in study increases.
(c) Show that the second derivative $G''(t) < 0$ for all $t > 0$. What is the significance of this result?

58. Show that the derivative of the function $f(x) = \log_b x$ equals $1/(x \ln b)$. *Hint:* $\log_b x = (\ln x)/(\ln b)$.

5.5 APPLICATIONS OF EXPONENTIAL AND LOGARITHMIC FUNCTIONS

A few simple illustrations will be presented in this section to demonstrate the usefulness and versatility of exponential and logarithmic functions.

Demand Equations

When the concept of a linear demand equation was introduced (Section 1.3), it was found necessary to restrict its domain to a narrow interval because a linear demand equation possesses an unattractive and unrealistic feature, namely that the number of items demanded does not increase dramatically as the price drops to very low levels. However, this undesirable property is not present for demand equations that are simple exponential functions, as illustrated in Example 1.

Example 1 Further analysis by the MacDougall Hamburger Company (Section 1.3) indicates that the relationship between the unit price p and the weekly sales q for its new Gluttonburger is given by the equation

$$p = 2.5e^{-0.02q} \tag{1}$$

whose graph is shown in Figure 1. This model indicates that demand for the product increases dramatically once the price drops close to the $1 level. Find the value of q for which revenue is maximized.

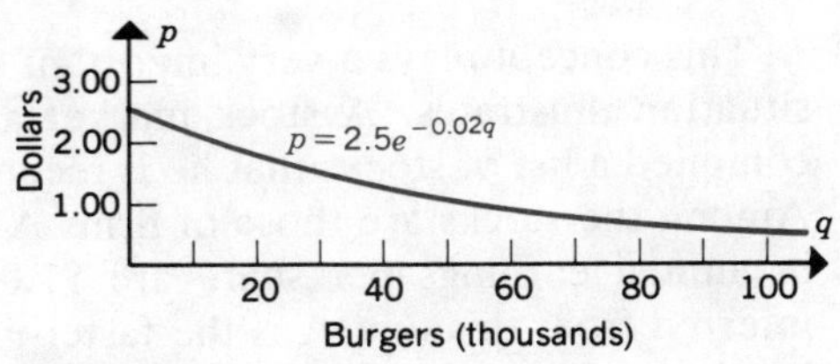

Figure 1

Solution The revenue function $R(q)$ is

$$R(q) = pq = 2.5qe^{-0.02q}$$

The value of q that maximizes $R(q)$ is found by setting the derivative $R'(q)$ equal to zero. Using the product rule, we get

$$R'(q) = 2.5e^{-0.02q} - 0.05qe^{-0.02q}$$

Next setting $R'(q)$ equal to zero gives

$$0 = (2.5 - 0.05q)e^{-0.02q}$$

Because $e^{-0.02q} > 0$ for all values of q, the only solution is obtained by solving

$$0 = 2.5 - 0.05q$$

Solution: $q = 50$ thousand burgers

$$R(50) = (2.5)(50)e^{-(0.02)(50)}$$
$$= \$45.98 \text{ thousand dollars}$$

Proving that the profit is maximized when $q = 50$ is left as an exercise. The corresponding value of p is found to be

$$p = 2.5e^{-0.02(50)} = 2.5e^{-1} = \$0.92$$

■

Relative Rate of Change

When the derivative of the function

$$f(x) = \ln u(x)$$

was found in Section 5.3, the result was

$$\frac{d}{dx}[\ln u(x)] = \frac{u'(x)}{u(x)}$$

The quantity $u'(x)/u(x)$ is called the **relative rate of change** of $u(x)$ with respect to x. The **percentage rate of change** is defined as

$$\frac{u'(x)}{u(x)} \times 100$$

This concept plays a very important role in decision making, as the following situation illustrates. A stock market analyst for a large brokerage house has compiled a list of stocks that he is recommending as "buys" for the near future. Among the stocks are those of firms A and B for which the expected increases in annual earnings per share are \$1.00 and \$0.25, respectively. It might be inferred from this that A is the faster-growing company and therefore its stock has better potential for price appreciation. However, suppose that further investigation reveals the data shown in Table 1. Using this data, we get the following relative rates of change in annual earnings:

$$\text{Company A:} \quad \frac{1.00}{5.00} = 0.20, \quad \text{or } 20\%$$

$$\text{Company B:} \quad \frac{0.25}{0.75} = 0.33, \quad \text{or } 33\%$$

From this perspective, the earnings of company B are growing at a faster annual rate, and its stock would appear to be the more attractive candidate for price appreciation.

Table 1

Company	Current Annual Earnings (dollars per share)	Estimated Increase in Annual Earnings
A	5.00	1.00
B	0.75	0.25

Example 2 It is estimated that the number of employees of the Malady Medical Company, a manufacturer of disposable syringes, thermometers, and so on, will grow linearly according to the following equation,

$$N(t) = 300 + 50t$$

where t represents the time in years. What is the relative rate of growth (a) at the end of one year and (b) at the end of five years?

Solution The annual rate of change in the number of employees is uniform, that is,

$$N'(t) = 50 \text{ employees/year}$$

However, the relative rate of change, $N'(t)/N(t)$, is far from constant:

$$\frac{N'(t)}{N(t)} = \frac{50}{300 + 50t}$$

(a) At the end of one year

$$\frac{N'(1)}{N(1)} = \frac{50}{350} = 0.14, \quad \text{or } 14\%$$

(b) At the end of five years

$$\frac{N'(5)}{N(5)} = \frac{50}{550} = 0.09, \quad \text{or } 9\%$$

Thus, although the annual increase is uniform, the relative rate of change is decreasing with time, as shown in Figure 2.

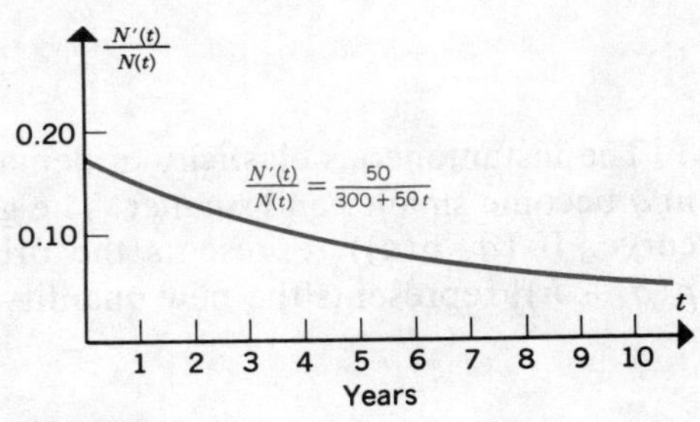

Figure 2 ■

Elasticity of Demand

The relative rate of change is used by economists in calculating a quantity called the **elasticity of demand;** for finite changes in price p and quantity q, it is defined as

$$\text{Elasticity of demand} = -\frac{\text{relative rate of change in quantity}}{\text{relative rate of change in price}}$$

$$= -\frac{\dfrac{q_2 - q_1}{q_1}}{\dfrac{p_2 - p_1}{p_1}} \tag{2}$$

Elasticity of demand is a ratio that describes the percentage change in sales in response to a given percentage change in price. The negative sign produces positive values of $E(q)$ since $q_2 - q_1$ is generally positive when $p_2 - p_1$ is negative and vice versa.

Example 3 A company can sell 300 lawn mowers a week when the price of each is \$200. When the price is reduced to \$180, the number sold rises to 360 lawn mowers per week. Find the elasticity of demand.

Solution When the price of each lawn mower is reduced from $p_1 = \$200$ to $p_2 = \$180$, the quantity sold increases from $q_1 = 300$ to $q_2 = 360$; this gives

$$\text{Elasticity of demand} = -\frac{\dfrac{360 - 300}{300}}{\dfrac{180 - 200}{200}} = \frac{0.20}{0.10} = 2.0$$

This result indicates that, at this price level, a 10 percent decrease in price, $(100)\left(\dfrac{180 - 200}{200}\right)$, causes a 20 percent increase in sales, $(100)\left(\dfrac{360 - 300}{300}\right)$. ■

The instantaneous elasticity of demand E can be found by letting the change in q become smaller and smaller. The graph in Figure 3 shows a typical demand curve. If $(q, p(q))$ represents the original quantity and price, and $[q + h, p(q + h)]$ represents the new quantity and price, then

$$\text{Elasticity of demand} = -\frac{\dfrac{q + h - q}{q}}{\dfrac{p(q + h) - p(q)}{p(q)}} = -\frac{\dfrac{h}{q}}{\dfrac{p(q + h) - p(q)}{p(q)}}$$

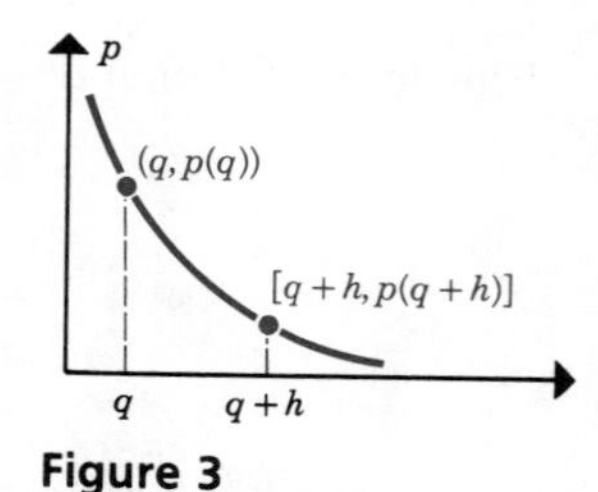

Figure 3

The right-hand side can be rewritten to give

$$\text{Elasticity of demand} = -\frac{p(q)}{q}\frac{1}{\dfrac{p(q + h) - p(q)}{h}}$$

As $h \to 0$, the ratio $\dfrac{p(q + h) - p(q)}{h}$ approaches $p'(q)$, resulting in the following definition of $E(q)$.

Definition The elasticity of demand $E(q)$ is defined as

$$E(q) = -\frac{p(q)}{q}\frac{1}{p'(q)} \tag{3}$$

Example 4 If the demand equation for lawn mowers is

$$p(q) = -\tfrac{1}{3}q + 300$$

find the elasticity of demand $E(q)$. Evaluate $E(q)$ when (a) $q = 300$ and (b) $q = 540$.

Solution Noting that $p'(q) = -\frac{1}{3}$, we find $E(q)$ from Equation 3,

$$E(q) = \frac{\frac{-q}{3} + 300}{q}\,\frac{1}{-\frac{1}{3}}$$

$$= \frac{900 - q}{q}$$

(a) When $q = 300$,

$$E(300) = \frac{900 - 300}{300} = 2$$

(b) When $q = 540$,

$$E(540) = \frac{900 - 540}{540} = \frac{2}{3}$$ ■

Economists classify the demand for a product according to the values of $E(q)$.

1. If $E(q) > 1$, demand is said to be **elastic.**
2. If $E(q) < 1$, demand is said to be **inelastic.**
3. If $E(q) = 1$, demand has unit elasticity.

When demand is elastic, the magnitude of the percentage change in q exceeds the magnitude of the percentage change in p. The reverse is true if demand is inelastic.

The value of $E(q)$ also provides information about the marginal revenue $R'(q)$.

Properties of $E(q)$

1. If demand is elastic, then $R'(q) > 0$.
2. If demand is inelastic, then $R'(q) < 0$.
3. If demand has unit elasticity, then $R'(q) = 0$.

Property 1 states that when demand is elastic, increasing q by lowering p will result in higher revenue. The reverse is true when prices are raised. This property can be demonstrated in the following way:

$$R(q) = p(q) \cdot q$$

Differentiating with respect to q yields

$$\begin{aligned} R'(q) &= p(q) + qp'(q) \\ &= p(q)\left(1 + \frac{q}{p(q)} p'(q)\right) \end{aligned}$$

The second term in the parentheses equals $-1/E(q)$, so we can write

$$R'(q) = p(q)\left(1 - \frac{1}{E(q)}\right)$$

When $E(q) > 1$, the factor $1 - 1/E(q) > 0$; since $p(q) > 0$, we can conclude that

$$R'(q) > 0$$

In the same way, when $0 < E(q) < 1$, the product $p(q)[1 - 1/E(q)] < 0$, so that $R'(q) < 0$.

Advertising

When a firm fails to promote its products, sales generally decline, owing in part to the advertising efforts of competing companies. Often the sales decline can be described by an exponential function of the form

$$S(t) = S_0 e^{-\lambda t}$$

where

$S(t)$ = level of sales at any time t

S_0 = level of sales when $t = 0$

λ = a positive constant called the *sales decay constant*

The value of λ will vary according to the nature of the product. For products in a very competitive market, λ will be large, but for products in a weakly competitive environment, λ will be small. The relative rate of decline in sales has a very simple form,

$$\frac{S'(t)}{S(t)} = \frac{-\lambda S_0 e^{-\lambda t}}{S_0 e^{-\lambda t}} = -\lambda$$

so that λ describes the constant relative rate of decline of sales under this model.

Example 5 The manufacturer of Sparkle, an all-purpose detergent, decides to cancel all promotional activities. A consultant predicts that annual sales will decline steadily according to the equation

$$S(t) = 25e^{-0.10t}$$

where $S(t)$ is given in millions of dollars.

(a) What will the sales level be at the end of five years?
(b) At what rate are sales declining at the end of the first year?
(c) What is the relative rate of change in sales?

Solution (a) The level of sales is found by evaluating $S(5)$, giving

$$S(5) = 25e^{-0.5} = \$15.2 \text{ million}$$

(b) The rate at which sales are declining at the end of year 1 is given by $S'(1)$.

$$S'(1) = -2.5e^{-0.10} = -\$2.3 \text{ million/year}$$

(c) The relative rate of change in sales equals $S'(t)/S(t)$:

$$\frac{S'(t)}{S(t)} = \frac{-2.5e^{-0.10t}}{25e^{-0.10t}} = -0.10 = -10\%/\text{year}$$

■

5.5 EXERCISES

1. Show that the solution $q = 50$, $R(50) = 45.98$ in Example 1 is a maximum by applying the second derivative test, that is, find $R''(q)$ and show that $R''(50) < 0$.
2. Solve the Gluttonburger problem (Example 1) by solving the demand equation
$$p = 2.5e^{-0.02q}$$
for q in terms of p and then expressing the revenue in terms of p. Find the value of p that maximizes the revenue. *Hint:* The demand equation can be rewritten by taking the natural logarithm of both sides:
$$\ln p = \ln 2.5e^{-0.02q} = \ln 2.5 + \ln e^{-0.02q}$$
$$\ln p = \ln 2.5 - 0.02q$$
3. The relationship between p and q for a new snowblower is given by the equation
$$p = 600e^{-0.05q}$$
Find the values of q and p for which the revenue is maximized.
4. If the relationship between p and q is given by the equation
$$p = ae^{-bq}, \quad \text{where } a, b > 0$$
show that the revenue is maximized when $q = 1/b$ and $p = a/e$.

Find the relative rate of change of each of the following functions at the indicated points.

5. $f(x) = x^4$ at (a) $x = 1$ and (b) $x = 3$

6. $f(t) = \sqrt{t}$ at (a) $t = 1$ and (b) $t = 9$

7. $F(t) = 1/t$ at (a) $t = 1$ and (b) $t = -1$

8. $f(x) = e^{0.20x}$ at (a) $x = 2$ and (b) $x = 6$

9. $g(s) = \ln(s^2 + 1)$ at (a) $s = 1$ and (b) $s = 3$

10. $f(t) = e^{-0.5x}$ at (a) $t = 2$ and (b) $t = 8$

11. The value $V(t)$, in dollars, of a 1940 comic book increases with time according to the equation

$$V(t) = 125 + 10t^{3/2}$$

where t is the time, in years, from today.

(a) What will be the value of the comic book four years from now?
(b) What will be the relative rate of change of $V(t)$ four years from now?
(c) What will be the relative rate of change of $V(t)$ nine years from now?

12. If $1000 is deposited in a savings account paying an annual rate of 10 percent interest compounded continuously, $A(t)$, the amount in the account at any time t, in years, is given by the equation

$$A(t) = 1000e^{0.10t}$$

Find the percentage rate of change of $A(t)$.

13. The number $N(t)$ of homes in a planned 3000-acre vacation community is expected to grow according to the equation

$$N(t) = 2000(1 - e^{-0.1t})$$

where t is the time, in years, from groundbreaking.

(a) Find the relative rate of change of $N(t)$.
(b) Find the relative rate of change of $N(t)$ when (i) $t = 1$ and (ii) $t = 10$.
(c) Find $\lim_{t\to\infty} N(t)$. What is the significance of this number?

14. Suppose that annual sales of posters of a popular television star are given by the equation

$$S(t) = 3t^2e^{-t}$$

where t is the time, in years, from the introduction to the public and $S(t)$ is the sales in millions of dollars.

(a) Find the relative rate of change when $t = 1$.
(b) Find the relative rate of change when $t = 3$.
(c) For what value of t is the relative rate of change equal to zero?

15. A sporting goods company has developed a new golf ball that carries farther than any competing model. An intense advertising campaign is begun to promote sales of the ball. As a result, sales increase and are described by the equation

$$S(t) = \frac{10}{1 + 4e^{-t}}$$

(a) Find the relative rate of change of $S(t)$ when $t = 1$.
(b) Find the relative rate of change of $S(t)$ when $t = 5$.

Find $E(q)$, the elasticity of demand, for each of the following demand equations. Determine whether demand is elastic, inelastic, or neither for the given value of q.

16. $p = 60 - 2q,\quad q = 20$

17. $p = \dfrac{10}{q + 1},\quad q = 4$

18. $p = \dfrac{12}{q^2 + 2}, \quad q = 2$

19. $p = e^{-0.04q}, \quad q = 3$

20. $p = \dfrac{16}{q^2}, \quad q = 4$

21. $p = 300 - 3q, \quad q = 40$

22. Sunshine Airlines charges \$99 for a flight from Chicago to Tucson. At this price the plane carries 120 passengers. Recently management dropped the price to \$79, and the number of passengers increased to 160. If the demand equation is linear,

(a) Find $E(q)$ and determine whether or not demand is elastic when $q = 120$.
(b) Find the value of q that maximizes the revenue.

23. Show that $E(q)$ is constant when a demand equation has the form

$$p = \frac{a}{q^n}$$

where a and n are positive numbers.

If a demand equation is written in the form $q = f(p)$, the elasticity of demand can be written as a function of p, and $E(p)$ assumes the form

$$E(p) = -\frac{p}{q}\frac{dq}{dp}$$

Find $E(p)$ for each of the functions in Exercises 24 through 29.

24. $q = -p + 10$

25. $q = \dfrac{5}{p}$

26. $q = -20 \ln p$

27. $q = \dfrac{8}{p - 1}$

28. $q = -30 \ln(p - 1)$

29. $q = 20 - 0.5p$

30. The Jiffy Button Company canceled all its promotional activities. The decision results in the following sales equation,

$$S(t) = 15e^{-0.01t}$$

where $S(t)$ represents the company's sales in millions.

(a) How fast are sales declining when $t = 2$?
(b) What is the relative rate of decline when $t = 2$?
(c) How long will it take for sales to reach a level equal to $S(0)/2$?

31. The advertising budget of the Round Tire Company was drastically reduced. Sales of its products are expected to be described by the equation

$$S(t) = 100e^{-0.05t}$$

(a) How fast are sales declining when $t = 1$? $t = 5$?
(b) What is the relative rate of change of $S(t)$ when $t = 1$? $t = 5$?
(c) How long will it take for sales to reach a level equal to $S(0)/2$?

5.6 NEWTON'S METHOD

This chapter concludes with a study of a widely used method for solving equations of the form

$$f(x) = 0 \tag{1}$$

Equations of this type are encountered in finding x intercepts, critical points, inflection points, points of intersection, and so on.* For example, the following equations are typical:

$$x^2 + 3x - 4 = 0$$
$$6x^5 + 7x^4 + 2 = 0$$
$$e^x + 2x = 0$$

When solutions to equations of the form of Equation 1 were sought in previous work, the equations were either

1. Linear or quadratic equations for which solutions could be obtained easily, or
2. Equations for which $f(x)$ was factorable; setting each factor equal to zero generated a set of more manageable equations.

When the equations are more complex, a technique utilizing the derivative, called **Newton's method,** can be used to find approximate values of the solutions.

The essentials of the method are illustrated in Figure 1, which shows the graph of a differentiable function $y = f(x)$ in the vicinity of the point $(t, 0)$ where the curve intersects the x axis. First, an initial estimate of t is made by selecting a value of x close to t; when possible, one of the two closest integral values is selected. This initial estimate is designated x_0. Next, the point $(x_0, f(x_0))$ is located on the graph, and the line tangent to the curve at this point is drawn. The next estimate, called x_1, is provided by the x-intercept of the tangent line. The equation used to find x_1 can be derived by noting that

1. The slope of the tangent line is given by $f'(x_0)$, and
2. The two points $(x_0, f(x_0))$ and $(x_1, 0)$ are on the tangent line.

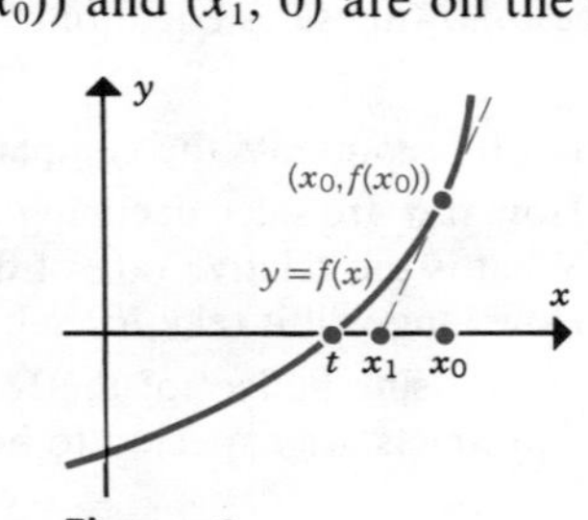

Figure 1

Incorporating this information into the equation $m = (y_2 - y_1)/(x_2 - x_1)$, which defines the slope of a line, we can write

$$f'(x_0) = \frac{f(x_0) - 0}{x_0 - x_1} = \frac{f(x_0)}{x_0 - x_1}$$

We now solve this equation for x_1 as follows,

$$(x_0 - x_1)f'(x_0) = f(x_0) \quad \text{or} \quad x_0 - x_1 = \frac{f(x_0)}{f'(x_0)}$$

*A calculator is recommended for the examples and exercises in this section.

from which we get

$$x_1 = x_0 - \frac{f(x_0)}{f'(x_0)}$$

This procedure is repeated, using x_1 as our new estimate of t, to generate another estimate x_2, as illustrated in Figure 2. The equation for x_2 is

$$x_2 = x_1 - \frac{f(x_1)}{f'(x_1)}$$

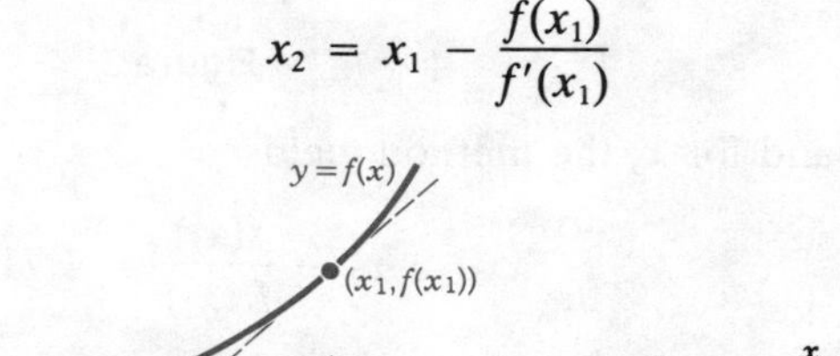

Figure 2

This procedure can be continued to generate still another estimate, x_3, defined in terms of x_2, $f(x_2)$, and $f'(x_2)$. The generalization of this technique enables us to express x_{n+1} in terms of x_n, $f(x_n)$, and $f'(x_n)$ for any integer $n \geq 0$. The result is

$$x_{n+1} = x_n - \frac{f(x_n)}{f'(x_n)}, \qquad n = 0, 1, 2, 3, \ldots \tag{2}$$

A number of examples illustrating Newton's method are presented next.

Example 1 Find an approximate solution of the equation

$$x^3 - 5 = 0$$

Solution For this case, $f(x) = x^3 - 5$ and $f'(x) = 3x^2$. A rough sketch of the function (Figure 3) shows that the solution is located between $x = 1$ and $x = 2$. Because the solution is closer to 2 than to 1, x_0 will be set equal to 2. The estimate x_1 is found using Equation (2),

$$x_1 = x_0 - \frac{f(x_0)}{f'(x_0)} = 2 - \frac{f(2)}{f'(2)} = 2 - \frac{3}{12} = 1.7500$$

where calculations are carried out to four decimal places. The estimate x_2 is found by repeating the process, yielding

$$x_2 = x_1 - \frac{f(x_1)}{f'(x_1)} = 1.75 - \frac{0.3594}{9.1875} = 1.7109$$

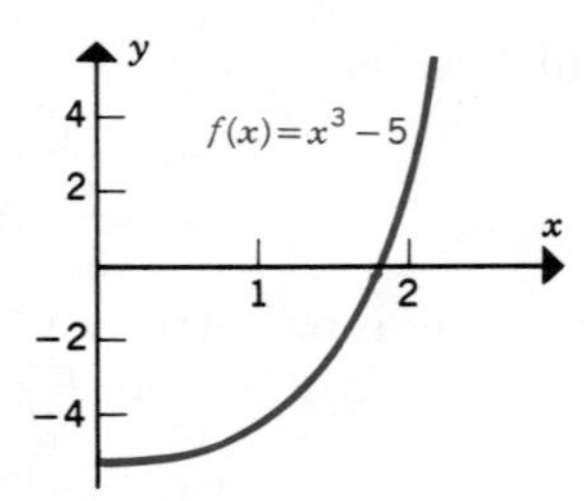

Figure 3

and for x_3 the method yields

$$x_3 = x_2 - \frac{f(x_2)}{f'(x_2)} = 1.7109 - \frac{0.0081}{8.7815} = 1.7100$$

Continuing, x_4 becomes

$$x_4 = x_3 - \frac{f(x_3)}{f'(x_3)} = 1.7100 - \frac{0.0002}{8.7723} = 1.7100$$

Because x_3 and x_4 are identical expressed to four decimal places, there is no need to continue the process any further. The value of t, to four decimal places, is

$$t = 1.7100$$

For the process to be carried out efficiently, it is advisable to set up a table such as Table 1 to obtain the successive approximations to t. It should be noted that the process just described is equivalent to finding the cube root of 5.

Table 1

n	x_n	$f(x_n) = x^3 - 5$	$f'(x) = 3x^2$	x_{n+1}
0	2.0000	3.0000	12.0000	1.7500
1	1.7500	0.3594	9.1875	1.7109
2	1.7109	0.0081	8.7815	1.7100
3	1.7100	0.0002	8.7723	1.7100

■

Example 2 Among the roots of the equation $x^4 + 5x - 3 = 0$, find the one that lies between $x = 0$ and $x = 1$.

Solution Noting that $f(x) = x^4 + 5x - 3$ and $f'(x) = 4x^3 + 5$, we can carry out the process as shown in Table 2, where $x_0 = 0$. Therefore, to four decimal places, the root is 0.5777.

Table 2

n	x_n	$f(x_n) = (x_n)^4 + 5(x_n) - 3$	$f'(x_n) = 4(x_n)^3 + 5$	x_{n+1}
0	0.0000	−3.0000	5.0000	0.6000
1	0.6000	0.1296	5.8640	0.5779
2	0.5779	0.0010	5.7720	0.5777
3	0.5777	−0.0001	5.7712	0.5777

■

The next example illustrates how Newton's method can be applied to the compound-interest formula

$$A = P(1 + i)^n$$

to find i, the interest rate per period, when P, A, and n are given.

Example 3 What annual rate of interest will cause a \$500 deposit to grow to \$750 in three years if interest is compounded semiannually?

Solution The equation to be solved has the form

$$750 = 500(1 + i)^6$$

or

$$(1 + i)^6 - 1.5 = 0$$

The value of i that satisfies this equation lies between $i = 0$ and $i = 1$. For this situation

$$f(i) = (1 + i)^6 - 1.5, \qquad f'(i) = 6(1 + i)^5$$

Using Equation 2 with $i_0 = 0$, we carry out the process as illustrated in Table 3. We then get an annual interest rate of

$$(0.0699)(2) = 0.1398 = 13.98\%$$

Table 3

n	i_n	$f(i_n) = (1 + i_n)^6 - 1.5$	$f'(i_n) = 6(1 + i_n)^5$	i_{n-1}
0	0.0000	−0.5000	6.0000	0.0833
1	0.0833	0.1162	8.9515	0.0703
2	0.0703	0.0033	·8.4271	0.0699
3	0.0699	−0.0001	8.4114	0.0699

■

Newton's method can also be used to locate the x coordinates of critical points when the expression for the first derivative cannot be factored, as the next example illustrates.

Example 4 Find the critical points for the function

$$g(x) = x^4 - 3x^2 + 5x - 2$$

Solution The critical points are found by solving the equation

$$g'(x) = 0$$

for x. Differentiating $g(x)$ term by terms gives

$$4x^3 - 6x + 5 = 0$$

as the equation to be solved. Newton's method, with $f(x) = 4x^3 - 6x + 5$, can be applied to find the solution or solutions, if any exist. A rough graph of

$y = f(x)$ (Figure 4) indicates that one solution exists in the interval between $x = -2$ and $x = -1$. *Note:* The graph does not represent $y = g(x)$ but $g'(x) = f(x)$. Table 4 summarizes the results obtained using Newton's method. There is one critical point, $(-1.523, -11.195)$, for the function $g(x) = x^4 - 3x^2 + 5x - 2$

Table 4

n	x_n	$f(x_n)$	$f'(x_n)$	x_{n+1}
0	−2.000	−15.000	42.000	−1.643
1	−1.643	−2.883	26.393	−1.534
2	−1.534	−0.235	22.238	−1.523
3	−1.523	0.007	21.834	−1.523

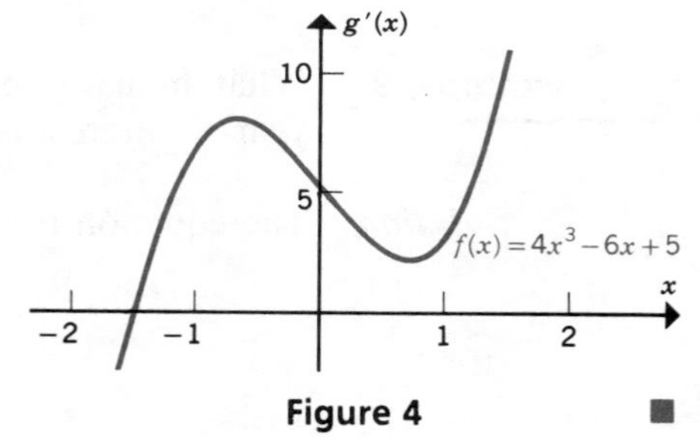

Figure 4 ■

Internal Rate of Return

Newton's method can be used to determine a quantity known as the internal rate of return on a capital or long-term investment, that is, one whose revenues and expenses are expected to continue for many years. Projects such as plant expansion, purchase of equipment, or development of a new product are examples of capital investments. If the initial cost of the project is designated C and the net cash flows for years 1 through m are designated $R_1, R_2, R_3, \ldots, R_m$, the **net present value** (NPV) of the investment is defined as

$$\text{NPV} = \frac{R_1}{1 + r} + \frac{R_2}{(1 + r)^2} + \frac{R_3}{(1 + r)^3} + \cdots + \frac{R_m}{(1 + r)^m} - C$$

where r is the rate of return the firm seeks to earn on the investment. The **internal rate of return** is defined as the rate of return r for which NPV equals zero, or the value of r that is a solution to the equation

$$0 = \frac{R_1}{1 + r} + \frac{R_2}{(1 + r)^2} + \frac{R_3}{(1 + r)^3} + \cdots + \frac{R_m}{(1 + r)^m} - \text{C} \tag{3}$$

If the value of r that satisfies Equation 3 is greater than the rate of return the firm requires on its investments, the project is considered a suitable investment.

To find the value of r that satisfies Equation 3, we can use Newton's method with $f(r)$ and $f'(r)$ defined as

$$f(r) = \frac{R_1}{1 + r} + \frac{R_2}{(1 + r)^2} + \cdots + \frac{R_m}{(1 + r)^m} - \text{C} \tag{4}$$

$$f'(r) = -\frac{R_1}{(1 + r)^2} - \frac{2R_2}{(1 + r)^3} - \cdots - \frac{mR_m}{(1 + r)^{m+1}} \tag{5}$$

For most projects r falls between zero and 1.

Example 5 A firm purchases a minicomputer whose initial cost is \$25,000. The net cash flows provided by this acquisition over the next three years are expected to be \$10,000 annually. Find the internal rate of return on the investment.

Solution The internal rate of return is determined by finding the value of r that satisfies the equation

$$0 = \frac{10,000}{1 + r} + \frac{10,000}{(1 + r)^2} + \frac{10,000}{(1 + r)^3} - 25,000$$

Newton's method can be used to solve this equation. Defining $f(r)$ and $f'(r)$ as

$$f(r) = \frac{10,000}{1 + r} + \frac{10,000}{(1 + r)^2} + \frac{10,000}{(1 + r)^3} - 25,000$$

$$f'(r) = -\frac{10,000}{(1 + r)^2} - \frac{20,000}{(1 + r)^3} - \frac{30,000}{(1 + r)^4}$$

We can use Equation 2 to find r. Beginning with $r_0 = 0$, the successive steps in the approximation are shown in Table 5. The internal rate of return is found to be 9.70 percent. The management of the firm must now decide whether this rate of return is sufficient to warrant approval of the project.

Table 5

n	r_n	$f(r_n)$	$f'(r_n)$	r_{n+1}
0	0.0000	5,000.00	−60,000	0.0833
1	0.0833	618.23	−46,036	0.0967
2	0.0967	13.70	−44,215	0.0970
3	0.0970	0.46	−44,175	0.0970

■

Although Newton's method generates values $x_0, x_1, x_2, x_3, \ldots, x_n$ that move successively closer to the desired solution $x = t$ for most of the functions encountered, there are occasions when each successive value may not be closer to the solution than its predecessor. Situations of this sort may arise if

1. The derivative $f'(x)$ equals zero at one or more points on the curve between $x = x_0$ and $x = t$. An illustration of this case is provided in the graph shown in Figure 5, where x_2 falls between x_0 and x_1.

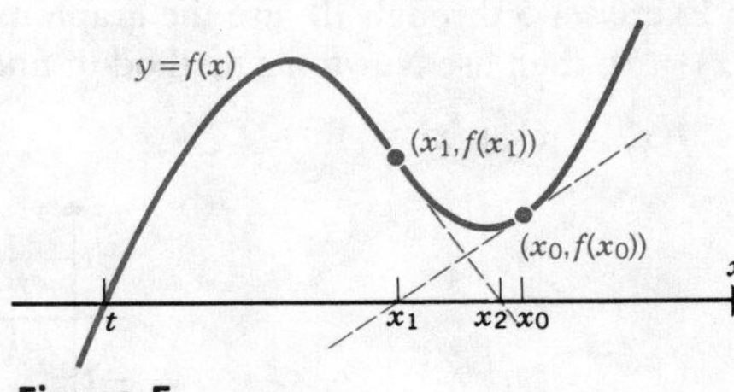

Figure 5

2. The second derivative $f''(x)$ changes sign at one or more points on the curve close to the root $x = t$. This case is illustrated graphically in Figure 6, where x_2 is located farther than x_0 or x_1 from the solution $x = t$.

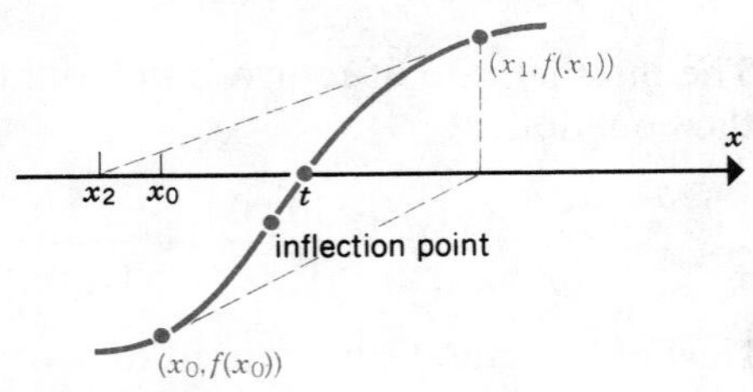

Figure 6

When either of these situations is encountered, it is advisable to examine the graph of the function in the vicinity of $x = t$ to determine the cause of the seemingly erratic behavior of the approximations. Sometimes selecting a new value of x_0 that is closer to the solution $x = t$ will remove the difficulty.

5.6 EXERCISES

1. Work out Example 1 using $x_0 = 1$.
2. Work out Example 2 using $x_0 = 1$.
3. Work out Example 3 using $i_0 = 1$.
4. (a) Work out Example 4 using $x_0 = -1$.
   (b) Determine the nature of the critical point found and make a rough sketch of the curve on the coordinate system of Figure 7.

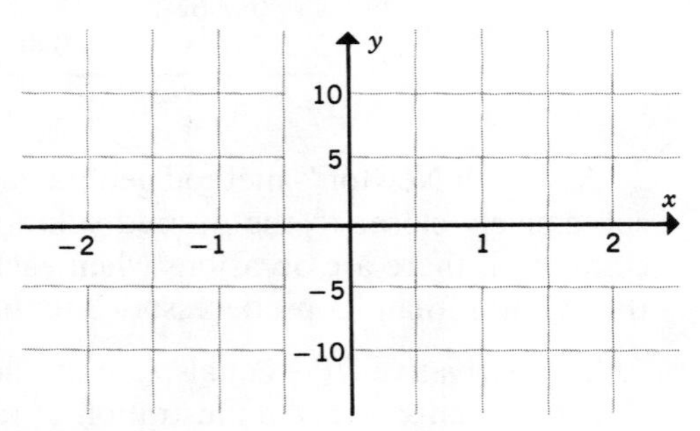

Figure 7

In Exercises 5 through 10, use the graph to find an approximate solution to the equation $f(x) = 0$; then use Newton's method to find the solution correct to three decimal places.

5. $f(x) = x^3 + x^2 - 1$

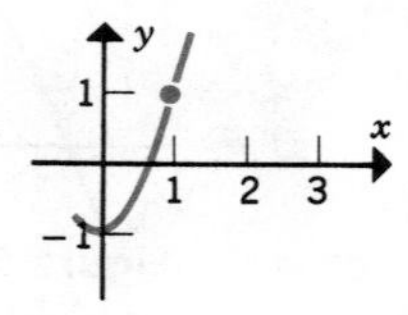

6. $f(x) = 1 + 2x - x^3$

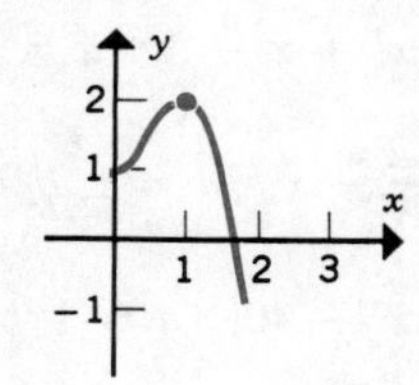

7. $f(x) = \sqrt{x + 1} - x$

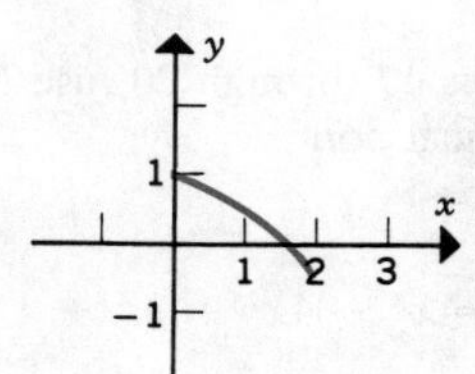

8. $f(x) = x^4 + 4x - 2$

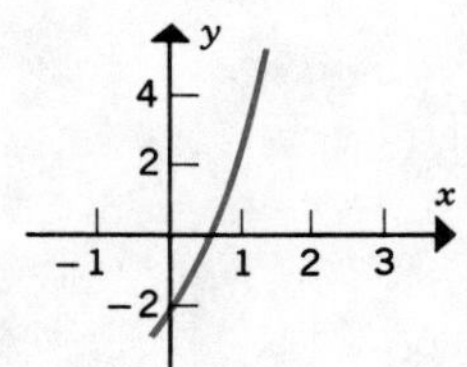

9. $f(x) = 1 + \sqrt{x} - x$

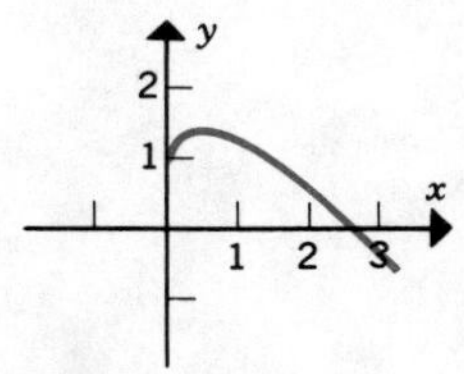

10. $f(x) = 2x + \dfrac{1}{x^2} - 4$

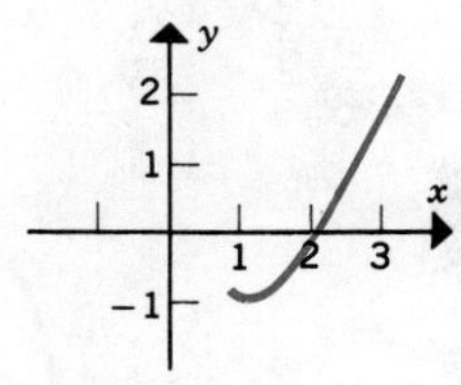

In Exercises 11 through 16, use Newton's method to find the root located between $x = a$ and $x = b$. Express your answer correct to three decimal places.

11. $f(x) = x^2 + 2x - 10$; $a = 2, b = 3$

12. $f(x) = x^3 + x^2 - 5x + 1; \quad a = -3, b = -2$

13. $f(x) = x^4 + x - 3; \quad a = 1, b = 2$

14. $f(x) = \dfrac{2}{x} - x; \quad a = 1, b = 2$

15. $f(x) = x^5 - 40; \quad a = 2, b = 3$

16. $f(x) = 3x^4 + 4x^3 - 6x^2 - 12x + 1; \quad a = 0, b = 1$

In exercises 17 through 20, use Newton's method together with the graphs to find each of the critical points.

17. $g(x) = x^3 - 4x^2 + 3x + 1$

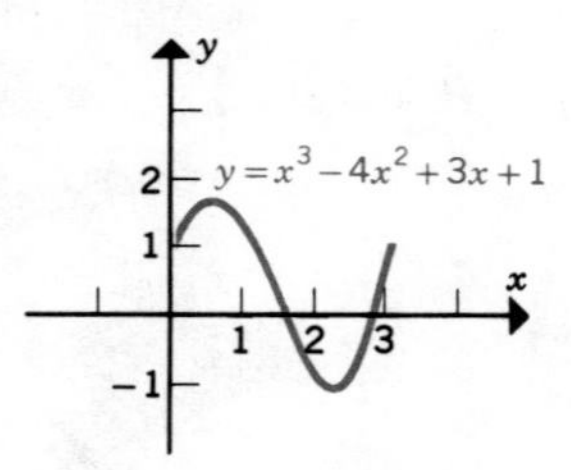

18. $g(x) = x^4 + 6x + 2$

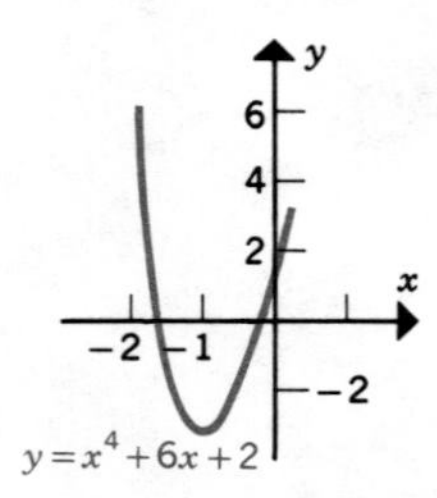

19. $g(x) = 9x^2 - x^4 - 2x + 5$

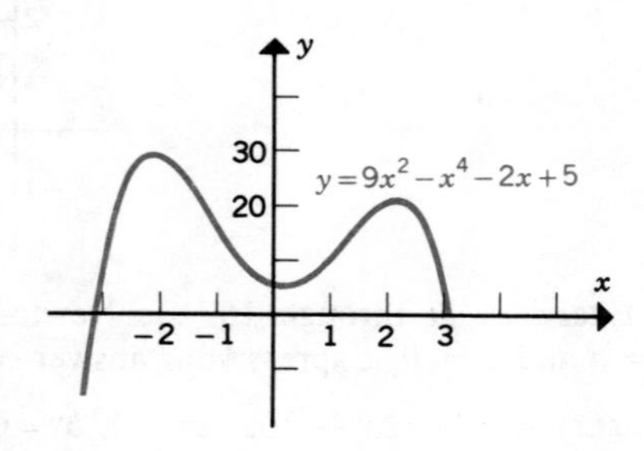

20. $g(x) = \dfrac{3}{x} + x$

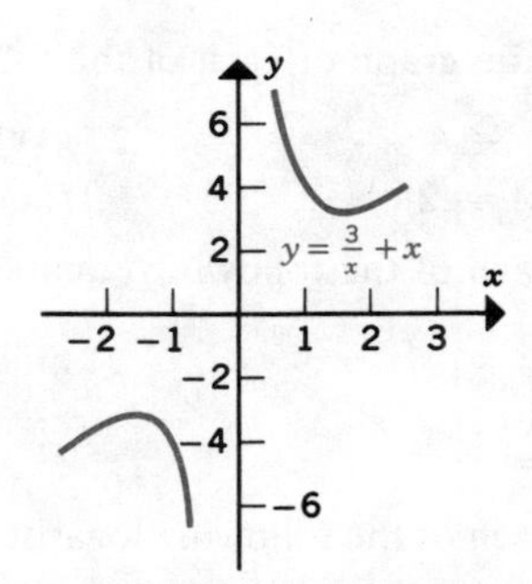

21. Five thousand dollars invested in the Doggie Bag Company grows to $9200 in five years. What is the average annual rate of return on this investment?
22. The neighborhood loan shark loans money at rather exorbitant interest rates. A $100 loan is to be repaid in seven days by paying him $110. Assuming that interest is compounded daily, determine the annual interest rate he is charging. (Calculate i to four decimal places.)
23. A two-year certificate of deposit (CD) purchased for $1000 is worth $1655 two years later. Find the annual rate of interest paid on the CD.
24. A retail store plagued by shoplifters and petty thieves installs a closed-circuit video security system at a cost of $25,000. The net annual savings are expected to be $8000 the first year and $15,000 per year thereafter. If the company is planning to move to new quarters three years from now, at which time a new system will have to be purchased, find the internal rate of return on the present purchase. Assume the salvage value is zero at the end of the three-year period.
25. A firm purchases an old, poorly insulated building that it plans to use for five years. Proper insulation will reduce heating costs by $500 per year. If installation of insulation costs $1000, what is the internal rate of return on the project?
26. A company is considering purchasing the patent rights to an electronic game from its inventor. If the inventor is willing to accept $50,000 for the patent rights and net income is expected to be $15,000 over each of the next five years, at which time the rights expire, find the internal rate of return on the investment.

KEY TERMS

exponential function
compound-interest equation
present value
continuous compounding
***e* and e^x**
properties of exponents
declining-balance method of depreciation
logarithmic function
properties of logarithms
natural logarithms
ln *x*
$\dfrac{d}{dx} e^{u(x)} = e^{u(x)}u'(x)$
$\dfrac{d}{dx} [\ln u(x)] = \dfrac{1}{u(x)}u'(x)$
relative rate of change
elasticity of demand
Newton's method
internal rate of return

REVIEW PROBLEMS

Sketch the graph of each of the following exponential functions.

1. $f(x) = 4^x$
2. $g(x) = 3^{-x}$
3. $f(x) = (\frac{2}{3})^x$
4. $F(x) = 2^{x-1}$
5. $f(x) = e^{-2x}$
6. $g(x) = \sqrt{e^x}$

Write each of the following equations as an equivalent logarithmic equation.

7. $3^4 = 81$
8. $9^{-3/2} = \dfrac{1}{27}$
9. $5^{1.25} = 7.477$
10. $7^0 = 1$
11. $e^2 = 7.389$
12. $e^{-1.4} = 0.2466$

Write each of the following logarithmic equations as an equivalent exponential equation.

13. $\log_2 64 = 6$
14. $\log_8 16 = \frac{4}{3}$
15. $\ln 10 = 2.303$
16. $\ln(0.001) = -6.908$

Evaluate each of the following.

17. $\log_3 81$
18. $\log_2 1$
19. $\log_{13} 13$
20. $\ln e^2$
21. $\log_2 \frac{1}{2}$
22. $\ln(1/e^2)$
23. $\log_7 49$
24. $\log_9 \frac{1}{27}$

Solve each of the following equations for x.

25. $\log_6 x + \log_6 3 = 2$
26. $\log_2 x + \log_2(x + 2) = 3$
27. $\ln x^2 = e$
28. $\ln(x + e) - \ln x = 1$
29. $2^x = 5$
30. $3^{-x} = 4$
31. $(1 + x)^{10} = 2$
32. $(1 + x)^{-5} = \dfrac{3}{4}$

Find the derivative of each of the following functions.

33. $f(x) = 2e^x$
34. $f(x) = x^2e^x$
35. $g(x) = 4e^{3x}$
36. $F(x) = x^2 \ln x$
37. $f(x) = \ln(x^2 + 1)$
38. $f(x) = x^e - e^x$
39. $f(x) = \dfrac{\ln x}{x}$
40. $f(x) = x \ln x - x + 2$
41. $g(x) = xe^{-x} + e^{-x} + 3$
42. $f(x) = e^{2x} + e^{-2x} + 5$

Match each of the functions whose derivatives are given in problems 41 through 44 with one of the graphs in Figure 1. (You should also find the second derivative of each function in order to exploit the information provided by the signs of both derivatives.)

43. $f'(x) = xe^x$
44. $g'(x) = xe^{-x}$
45. $F'(x) = x^2e^{2x}$
46. $G'(x) = x^2e^{-2x}$

Find the relative maxima, minima, and inflection points, if any, for each of the following functions. In addition, sketch the graph of each function.

47. $f(x) = x^2e^{2x}$
48. $f(x) = x \ln x - x + 2$
49. $g(x) = xe^x - e^x + 3$
50. $F(x) = x^2 - 8 \ln x + 1$

51. Two thousand dollars is invested in an account paying 12 percent interest annually. Find the amount in the account at the end of four years if interest is compounded
(a) annually, (b) semiannually, (c) monthly, (d) continuously

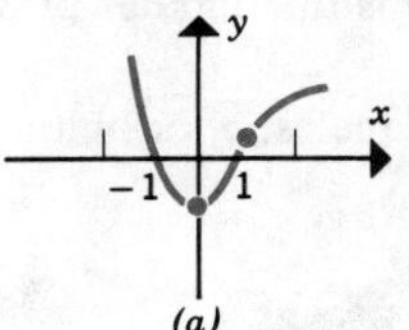

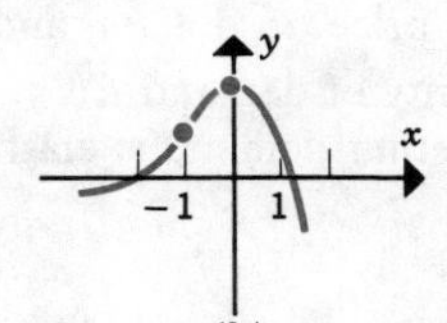

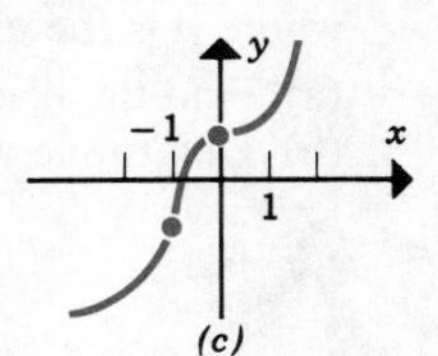

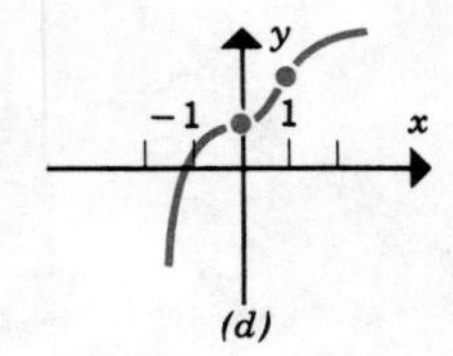

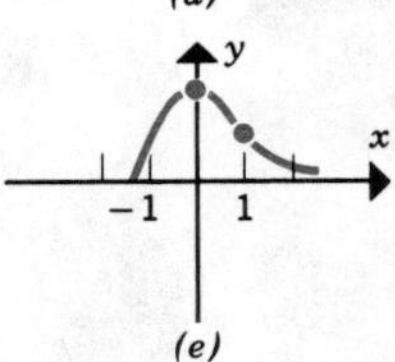

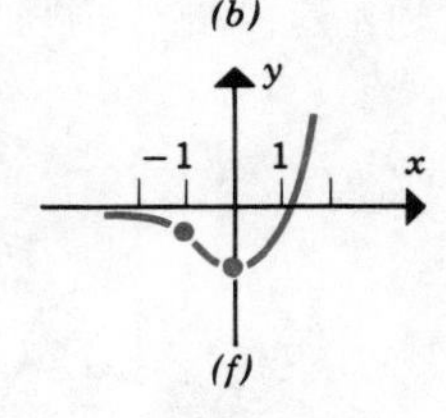

Figure 1

52. A woman owns a zero-coupon bond, which pays no interest. When the bond matures five years from now, its value will be \$6000. The woman needs money and is planning to sell the bond. How much can she expect to receive from a buyer who wants to receive a 10 percent annual return on his investment? (Assume interest is compounded annually.)

53. A biology student requires a large culture of *Escherichia coli* bacteria for a DNA experiment. The number of cells $N(t)$, in millions, increases exponentially with the time t, in days, according to the equation

$$N(t) = 2e^t$$

where $t = 0$ corresponds to the beginning of the experiment.
(a) How many cells are in the culture when $t = 2$?
(b) How fast is the culture growing when $t = 1$?
(c) What is the relative rate of change of $N(t)$?

54. A law firm whose business manager uses the declining balance method of depreciation has purchased a word processor for \$1500. The book value $V(t)$, in dollars, as a function of the time t, in years, can be described by the equation

$$V(t) = 1500e^{-2t/5}$$

where $t = 0$ corresponds to the date of purchase.
(a) What is the book value when $t = 2$?
(b) How fast is the book value changing when $t = 2$?
(c) What is the relative rate of change of $V(t)$ when $t = 2$?

55. The proportion of students in a statistics class who remembered a regression formula for t hours is given by the equation

$$p(t) = 0.3 - \ln t, \qquad \tfrac{1}{2} \le t \le \tfrac{4}{3} \text{ hours}$$

(a) Find the proportion of students who remembered the formula for one hour.
(b) How fast was $p(t)$ changing when $t = 1$?

56. The demand equation for a new Fax machine is

$$p = \frac{10{,}000}{4 + q^2}$$

where p is the unit price, in dollars, and q is the number of units sold each day.

(a) Find the elasticity of demand $E(q)$.

(b) Determine whether demand is elastic, inelastic, or neither when $q = 2$.

6

INTEGRATION

INTRODUCTION

Our attention so far has been concentrated on finding and then using the derivative of a given function in a variety of applications. The next stage, integral calculus, begins with a question: Suppose that we are given the derivative of an unknown function. Can we find the function itself? The process of finding a function when its derivative is given is called **antidifferentiation** or **integration.** Integration is used in many types of applications such as finding the area under a curve, determining how long a nonrenewable natural resource will last, calculating how many calories a runner has burned over a given length of time, and so on.

6.1 ANTIDERIVATIVES

Up to now, much of our attention has been focused on finding the derivative of a given function. Now, we begin to concentrate on the reverse, namely:

Given the function $y = f(x)$, find a second function $y = F(x)$ whose derivative equals $f(x)$, that is,

$$F'(x) = f(x)$$

The function $y = F(x)$ is called an **antiderivative** of $y = f(x)$.

The process of finding $F(x)$ is known as *antidifferentiation,* or, as it is more commonly called, *integration*.

Integration and antidifferentiation are synonyms.

For example, if $f(x) = x^2$, $F(x)$ can be found by working the power rule (Section 3.1) in reverse. By adding 1 to the exponent, we might expect $F(x)$ to have the form

$$F(x) = x^3$$

However, the derivative $F'(x) = 3x^2 \neq x^2$. The discrepancy is not difficult to correct; dividing our initial attempt, x^3, by 3 produces

$$F(x) = \frac{x^3}{3}$$

which satisfies the condition $F'(x) = x^2 = f(x)$. Because the derivative of a constant is zero, the following functions are also antiderivatives:

$$G(x) = \frac{x^3}{3} + 2, \qquad H(x) = \frac{x^3}{3} - 7$$

Therefore, whenever we write an antiderivative, an arbitrary constant term called the **constant of integration** will also be included. For example, it can be proved that all antiderivatives of the function $f(x) = x^2$ are represented by the function

$$F(x) = \frac{x^3}{3} + C$$

The procedure used to find an antiderivative of $f(x) = x^2$ can be used to find an antiderivative of a function having the form $f(x) = x^n$ $(n \neq -1)$. This method is called the power rule for antiderivatives.

Power Rule for Antiderivatives

If $f(x) = x^n$, $n \neq -1$, all antiderivatives are represented by the function

$$F(x) = \frac{x^{n+1}}{n+1} + C \tag{1}$$

where C is an arbitrary constant called the constant of integration.

Equation 1 can be checked directly by differentiating $F(x)$:

$$F'(x) = \frac{n+1}{n+1} x^{n+1-1} + 0 = x^n = f(x)$$

Example 1 Find all antiderivatives of

(a) $f(x) = \dfrac{1}{x^5}$ (b) $g(x) = \sqrt[3]{x}$

Solution (a) First, $f(x)$ is written as a power function

$$f(x) = \frac{1}{x^5} = x^{-5}$$

Applying the power rule (Equation 1) gives

$$F(x) = \frac{x^{-5+1}}{-5+1} + C = -\frac{x^{-4}}{4} + C$$
$$= -\frac{1}{4x^4} + C$$

This result can be checked by differentiating to show that $F'(x) = f(x)$:

$$F'(x) = -\frac{(-4)x^{-5}}{4} + 0 = \frac{1}{x^5} = f(x)$$

(b) The function $g(x) = \sqrt[3]{x} = x^{1/3}$. Applying the power rule gives

$$G(x) = \frac{x^{4/3}}{\frac{4}{3}} + C = \frac{3}{4}x^{4/3} + C$$

Again, we can differentiate to show that $G'(x) = g(x)$:

$$G'(x) = (\tfrac{3}{4})(\tfrac{4}{3})x^{1/3} + 0 = x^{1/3} = g(x)$$ ■

The family of all antiderivatives of a function $y = f(x)$ is also represented by the symbol

$$\int f(x)\,dx$$

which is called the **indefinite integral** of $f(x)$.

Definition If $F'(x) = f(x)$, then

$$\int f(x)\,dx = F(x) + C \tag{2}$$

In this notation $\int$ is the **integral sign** and $f(x)$ the **integrand.** The power rule for antiderivatives can be written in integral form as:

$$\int x^n\,dx = \frac{x^{n+1}}{n+1} + C, \qquad n \neq -1 \tag{3}$$

Example 2 Find each of the following:

(a) $\int x^5\,dx$ (b) $\int \frac{1}{x^3}\,dx$

Solution (a) Using Equation 3, we get

$$\int x^5\,dx = \frac{x^6}{6} + C$$

(b) Writing the integrand $1/x^3$ as x^{-3}, we get

$$\int x^{-3}\,dx = \frac{x^{-2}}{-2} + C = -\frac{1}{2x^2} + C$$ ■

If both sides of Equation 2 are differentiated with respect to x, the resulting equation emphasizes the relationship between differentiation and integration, namely that differentiation is the inverse of integration and vice versa,

$$\frac{d}{dx}\int f(x)\,dx = F'(x) + 0$$

Since $F'(x) = f(x)$, we get

$$\boxed{\frac{d}{dx}\int f(x)\,dx = f(x)}$$

In words, this equation states that the derivative of an indefinite integral equals the integrand.

Equation 3 cannot be used to find an antiderivative of the function $f(x) = x^{-1}$ because the right-hand side is not defined when $n = -1$. However, in Chapter 5 we found that

$$\frac{d}{dx}\ln x = \frac{1}{x} \quad \text{and} \quad \frac{d}{dx}\ln(-x) = \frac{1}{x}$$

which leads us to write

$$\int \frac{1}{x}\,dx = \begin{cases} \ln x + C, & x > 0 \\ \ln(-x) + C, & x < 0 \end{cases}$$

or

$$\boxed{\int \frac{1}{x}\,dx = \ln|x| + C} \tag{4}$$

An antiderivative of the function $f(x) = e^{ax}$ has a simple form:

$$\boxed{\int e^{ax}\,dx = \frac{e^{ax}}{a} + C} \tag{5}$$

Equation 5 can be verified by differentiation:

$$\frac{d}{dx}\left(\frac{e^{ax}}{a} + C\right) = \frac{1}{a}e^{ax}(a) + 0 = e^{ax}$$

Example 3 Find $\int e^{4x}\,dx$.

Solution Using Equation 5, we get

$$\int e^{4x}\,dx = \frac{e^{4x}}{4} + C$$

■

Since integration is the inverse of differentiation, some of the differentiation rules, such as the constant factor and sum rules, have integral counterparts.

Constant Factor Rule

$$\int kf(x)\,dx = k\int f(x)\,dx, \quad \text{where } k \text{ is a constant.} \tag{6}$$

Sum Rule

$$\int [f(x) + g(x)]\,dx = \int f(x)\,dx + \int g(x)\,dx \tag{7}$$

Example 4 Find $\int 8x^3\,dx$.

Solution According to the constant factor rule,

$$\int 8x^3\,dx = 8\int x^3\,dx$$

Next, applying the power rule gives

$$8\int x^3\,dx = 8\,\frac{x^4}{4} + C = 2x^4 + C$$

■

Example 5 Find $\int (9x^2 + 8x)\,dx$.

Solution Using the sum rule, we can write

$$\begin{aligned}\int (9x^2 + 8x)\,dx &= \int 9x^2\,dx + \int 8x\,dx \\ &= 9\int x^2\,dx + 8\int x\,dx \quad \text{(constant factor rule)} \\ &= 9\,\frac{x^3}{3} + C_1 + \frac{8x^2}{2} + C_2 \quad \text{(power rule)} \\ &= 3x^3 + 4x^2 + C\end{aligned}$$

where $C = C_1 + C_2$. Finally, it is good practice to check all your answers by differentiating:

$$\frac{d}{dx}(3x^3 + 4x^2 + C) = 9x^2 + 8x$$

■

Example 6 Find $\int \left(e^x - \frac{3}{x^2}\right) dx.$

Solution

$$\begin{aligned}\int \left(e^x - \frac{3}{x^2}\right) dx &= \int e^x\, dx + \int \frac{-3}{x^2}\, dx \quad \text{(sum rule)}\\ &= \int e^x\, dx - 3\int \frac{1}{x^2}\, dx \quad \text{(constant factor rule)}\\ &= \int e^x\, dx - 3\int x^{-2}\, dx\\ &= e^x - 3\,\frac{x^{-1}}{-1} + C \quad \text{(Equation 5 and power rule)}\\ &= e^x + \frac{3}{x} + C\end{aligned}$$

Check the answer:

$$\frac{d}{dx}(e^x + 3x^{-1} + C) = e^x - 3x^{-2} + 0 = e^x - \frac{3}{x}$$

■

If the integrand is a ratio, an antiderivative can sometimes be found by rewriting the integrand as a sum of two or more terms, as illustrated in the next example.

Example 7 Find $\int \frac{x^2 - 4}{x^3}\, dx.$

Solution An antiderivative can be found by writing the integrand as a sum and then applying the sum rule:

$$\begin{aligned}\int \frac{x^2-4}{x^3}\, dx &= \int \left(\frac{x^2}{x^3} - \frac{4}{x^3}\right) dx = \int \left(\frac{1}{x} - \frac{4}{x^3}\right) dx\\ &= \int \frac{1}{x}\, dx - 4\int x^{-3}\, dx = \ln|x| - \frac{4x^{-2}}{-2} + C\\ &= \ln|x| + \frac{2}{x^2} + C\end{aligned}$$

Again, we check our answer:

$$\frac{d}{dx}\left(\ln|x| + \frac{2}{x^2} + C\right) = \frac{1}{x} - \frac{4}{x^3} = \frac{x^2-4}{x^3}$$

■

The constant of integration C appears in all antiderivatives because the derivative, which represents the slope of a tangent line, does not uniquely define a function or curve. Inspection of Figure 1 shows that the slopes of the tangent lines are identical at $x = a$ for each of the three curves shown; each represents an equation of the form

$$F(x) = x^2 + C$$
$$F'(x) = 2x$$

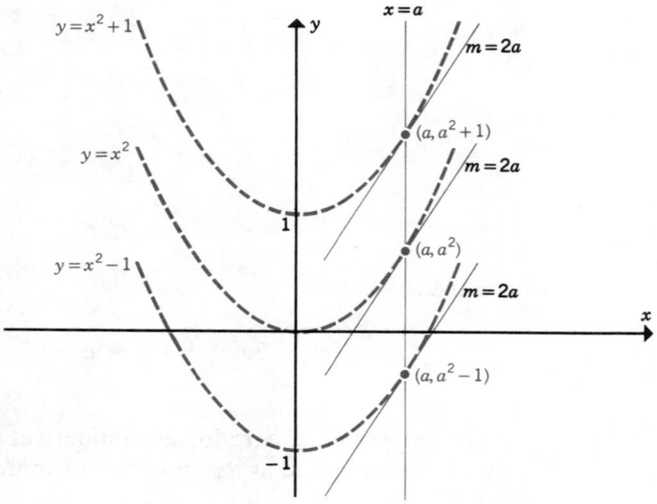

Figure 1

In order to define an antiderivative uniquely, we need additional information in the form of the coordinates of a point on the curve, as illustrated in the next example.

Example 8 The slope of the line tangent to the curve $y = F(x)$ at any point is given by the equation

$$F'(x) = 3x^2$$

If the curve passes through the point $(-1, 6)$, find the equation of the function $y = F(x)$.

Solution An equation for $F(x)$ is found by integrating:

$$F(x) = \int 3x^2\,dx = \frac{3x^3}{3} + C = x^3 + C$$

Since $F(-1) = 6$, the value of C can be found from the equation

$$6 = (-1)^3 + C$$

or

$$C = 7$$

The equation defining $F(x)$ becomes

$$F(x) = x^3 + 7$$

■

Applications

Integration or, equivalently, antidifferentiation is needed when a derivative representing a rate of change such as the velocity or marginal cost is given and we want to find an equation for the distance or the cost, respectively. These and other examples are shown in Table 1.

Table 1

Derivative		Antiderivative	
Name	Symbol	Name	Symbol
Velocity	$v(t) = s'(t)$	Distance	$s(t)$
Acceleration	$a(t) = v'(t)$	Velocity	$v(t)$
Marginal profit	$P'(x)$	Profit	$P(x)$
Marginal revenue	$R'(x)$	Revenue	$R(x)$
Marginal cost	$C'(x)$	Cost	$C(x)$

Applications that require integration are illustrated in Examples 9 through 12.

Example 9 (a) The marginal cost for a company that produces outboard motors has the form

$$C'(x) = 300 + 0.02x$$

If fixed costs are \$10,000, find $C(x)$, the cost function.

(b) How much does it cost to produce 200 motors?

Solution (a) The cost $C(x)$ is found by means of integration:

$$C(x) = \int (300 + 0.02x)\,dx$$
$$= 300x + 0.01x^2 + C$$

The constant of integration is found by noting that $C(0)$ represents the fixed costs, so $C(0) = 10{,}000$; substituting this information gives

$$10{,}000 = 0 + 0 + C$$

so

$$C = 10{,}000$$

The cost equation becomes

$$C(x) = 300x + 0.1x^2 + 10{,}000$$

(b) The cost to produce 200 motors is

$$C(200) = 300(200) + 0.1(200)^2 + 10{,}000$$
$$= \$74{,}000$$

■

Example 10 (a) An object is thrown up from the top of a 400-ft building. The velocity as a function of t, the time, in seconds, is given by the equation

$$v(t) = 160 - 32t$$

where $t = 0$ corresponds to the moment of release. Find an equation for $s(t)$, the distance of the object above the ground.

(b) How high is the object above the ground when $t = 3$ sec?

Solution (a) Because $v(t) = s'(t)$, $s(t)$ can be found by integration:

$$s(t) = \int (160 - 32t)\, dt$$
$$= 160t - 16t^2 + C$$

Since $s(0) = 400$ ft, this information can be used to find the constant of integration,

$$400 = 160(0) - 16(0) + C$$

giving

$$C = 400$$

so

$$s(t) = 160t - 16t^2 + 400$$

(b) When $t = 3$,

$$s(3) = 160(3) - 16(3)^2 + 400$$
$$= 736 \text{ ft}$$

■

In many situations the net change in an antiderivative is desired; in such situations it is not necessary to find the constant of integration, as illustrated in the next example.

Example 11 A furniture company that manufactures recliners finds that $P'(x)$, the marginal profit, is given by the equation

$$P'(x) = 200 - 0.4x$$

Current sales are 100 units. How much do profits change if the number of units sold increases to 120 units?

Solution We want to determine the quantity $P(120) - P(100)$. The profit $P(x)$ is found by integrating $P'(x)$ with respect to x,

$$\begin{aligned} P(x) &= \int (200 - 0.4x)\,dx \\ &= 200x - 0.2x^2 + C \end{aligned}$$

The change in profit $P(120) - P(100)$ is found next:

$$\begin{aligned} P(120) - P(100) &= [200(120) - 0.2(120)^2 + C] \\ &\quad - [200(100) - 0.2(100)^2 + C] \\ &= 200(120 - 100) - 0.2(120^2 - 100^2) + (C - C) \\ &= 200(20) - 0.2(120 + 100)(120 - 100) \\ &= 200(20) - 0.2(220)(20) = \$3120 \end{aligned}$$

Note that the terms containing C cancel, so that it is not necessary to find C to determine the difference $P(120) - P(100)$. ■

Sum-of-the-Years-Digits Depreciation*

We have examined two methods of depreciation used frequently in accounting:

1. The straight-line method (Section 1.3), and
2. The double-declining balance method (Section 5.1).

A third method called the sum-of-the-years-digits (SYD) method is frequently used. Like the double-declining balance method, it is an accelerated method of depreciation, that is, the decline in value is greater in the early years than in the later years of the life of an asset.

Let T be the useful life, in years, and $V(t)$ the value, in dollars, of an asset. According to the SYD method, $V'(t)$, the time rate of change of $V(t)$, is proportional to $T - t$, the remaining life of an asset, that is,

$$\boxed{V'(t) = k(T - t), \qquad 0 \le t \le T} \tag{8}$$

where k is the constant of proportionality. Equation 8 indicates that the magnitude of $V'(t)$ is largest when $t = 0$ and approaches zero as t approaches T. An equation for $V(t)$ can be found by integrating, yielding

$$\begin{aligned} V(t) &= \int k(T - t)\,dt = \int kT\,dt - \int kt\,dt \\ &= kT\,t - \frac{kt^2}{2} + C \end{aligned} \tag{9}$$

where C is the constant of integration. The determination of k and C for a given asset is illustrated in the following example.

*Hegarty, J. C., Calculus Model of Sum-of-the-Years-Digits Depreciation, Mathematics and Computer Education, Fall 1987, Vol. 21, No. 3, pp. 159–61.

Example 12 A marketing research firm purchases a minicomputer for \$100,000. Its useful life is five years, and its salvage value is zero.

(a) Find an equation for $V(t)$.

(b) By how much does the value of the computer change from $t = 0$ to $t = 3$ years?

Solution (a) In this case $T = 5$, so Equation 9 can be written as

$$V(t) = 5kt - \frac{kt^2}{2} + C \tag{10}$$

The constants k and C are determined as follows:

1. When $t = 0$, $V(0) = 100{,}000$. Substituting these values into Equation 10 yields for C

$$C = 100{,}000$$

so the equation for $V(t)$ can be written

$$V(t) = 5kt - \frac{kt^2}{2} + 100{,}000 \tag{11}$$

2. When $t = 5$, $V(5) = 0$ (asset has no salvage value). Substituting these values into equation 11 gives

$$V(5) = 0 = 25k - \frac{25k}{2} + 100{,}000$$

Solving for k yields $k = -8{,}000$. The equation for $V(t)$ can be written

$$\begin{aligned} V(t) &= 100{,}000 - 40{,}000t + 4000t^2 \\ &= 4000(25 - 10t + t^2) \\ &= 4000(5 - t)^2 \end{aligned}$$

(b) The change in the value of the minicomputer from $t = 0$ to $t = 3$ equals $V(3) - V(0)$, yielding

$$\begin{aligned} V(3) - V(0) &= 4000(5 - 3)^2 - 4000(5)^2 \\ &= 4000(4 - 25) \\ &= -\$84{,}000 \end{aligned}$$

This result tells us that the value of the minicomputer declines by \$84,000 during the first three years; the remaining \$16,000 is depreciated during the last two years. ■

6.1 EXERCISES

Find all antiderivatives for each of the following.

1. $f(x) = 5$

2. $f(x) = -2x$

3. $f(t) = 6\sqrt[3]{t}$
4. $F(x) = 4e^{2x}$
5. $f(x) = \frac{2}{x} + 8x^3$
6. $g(s) = \sqrt{2}$
7. $f(x) = 10e^{-5x}$
8. $f(x) = \frac{x^2 + 1}{x}$

Find each of the following indefinite integrals. Check your answer by differentiating.

9. $\int 4\, dx$
10. $\int (12x^5 - 4x^3)\, dx$
11. $\int \frac{10}{x^6}\, dx$
12. $\int (15x^2 + 6x - 3)\, dx$
13. $\int \left(4x^3 - \frac{1}{x}\right) dx$
14. $\int (e^x - e^{-x})\, dx$
15. $\int 10x^{3/2}\, dx$
16. $\int \left(4e^{2x} - \frac{2}{x} + 3\right) dx$
17. $\int 9\sqrt{x}\, dx$
18. $\int \left(\frac{6}{\sqrt{x}} + 2x\right) dx$
19. $\int (x^3 + 1)^2\, dx$
20. $\int (x + 1)(x - 1)\, dx$
21. $\int \frac{3x^3 + 2x^2 + x}{x}\, dx$
22. $\int \frac{x^4 - x^3 + x^2}{x^3}\, dx$
23. $\int (\sqrt{x} + 2)(\sqrt{x} - 2)\, dx$
24. $\int \frac{x^2 + 4x + 3}{x + 1}\, dx$
25. $\int \frac{e^{2x} + e^x + 1}{e^x}\, dx$
26. $\int \frac{e^{2x} - 2e^x + 1}{e^x - 1}\, dx$
27. $\int nx^{n-1}\, dx$
28. $\int e^{nx}\, dx$

Determine whether each of the following equations is true or false.

29. $\int xe^x\, dx = xe^x - e^x + C$
30. $\int xe^{-2x}\, dx = -\frac{xe^{-2x}}{2} + \frac{e^{-2x}}{4} + C$
31. $\int \ln x\, dx = x \ln x - x + C$
32. $\int x \ln x\, dx = \frac{x^2 \ln x}{2} - \frac{x^2}{2} + C$
33. $\int \frac{1}{1 + e^x}\, dx = x + \ln(1 + e^x) + C$
34. $\int (\ln x)^2\, dx = 2x - 2x \ln x + x(\ln x)^2 + C$
35. $\int \frac{x^2}{\sqrt{x^2 + 1}}\, dx = \frac{1}{2}x\sqrt{x^2 + 1} - \frac{1}{2}\ln|x + \sqrt{x^2 + 1}| + C$
36. $\int \frac{x}{(x + 1)^3}\, dx = -\frac{1}{x + 1} + \frac{1}{2(x + 1)^2} + C$

Find each of the functions described in Exercises 37 through 42.

37. $F'(x) = 2x - 3$, $F(-1) = 3$

38. $g'(x) = 4\sqrt{x}$, $g(1) = 0$

39. $f'(t) = 6e^{3t}$, $f(0) = 4$

40. $f'(x) = \dfrac{2}{x} - 6x$, $f'(1) = 2$

41. $F'(s) = 3s^2 - \dfrac{1}{s^2}$, $F(-1) = 2$

42. $g'(x) = \dfrac{1}{x} + 4e^{-2x}$, $g(1) = 0$

43. Find an equation of the curve that passes through (1, 2) if the slope of the tangent line at any point equals $3x^2 - 4x$.

44. Find the equation of the curve that passes through (−1, 3) if the slope of the tangent line at any point equals $4x^3 - 1/x$.

45. An object is dropped from the top of a 256-ft building. Its velocity as a function t, the time (in seconds), is given by the equation

$$v(t) = -32t \quad \text{ft/sec}$$

(a) Find an equation for $s(t)$, the distance of the object above the ground.
(b) How long does it take for the object to hit the ground?

46. A toy rocket is launched from the top of a 200-m hill. Its velocity $v(t)$, (in m/sec) is described by the equation

$$v(t) = 98 - 9.8t$$

(a) Find an equation for $s(t)$, the distance of the rocket above the ground as a function of t.
(b) How high above the ground does the rocket go?

47. The marginal revenue is given by the equation

$$R'(x) = 120 - 2x \quad \text{dollars/unit}$$

(a) Find an equation for $R(x)$ the total revenue.
(b) By how much does the revenue change if the number of units sold increases from 15 to 20 units?

48. The marginal profit is given by the equation

$$P'(x) = 200 - 0.2x \quad \text{dollars/unit}$$

By how much does the profit change if the number of units sold changes from 1000 to 1200 units?

49. The number of employees $N(t)$ at the E Coli Engineering Company is growing at a rate described by the equation

$$N'(t) = 10e^{0.10t} \quad \text{people/year}$$

The number of employees today is 200.
(a) Find an equation for $N(t)$.
(b) How long will it take for the work force to reach 600 employees?

50. Acme Realty purchases a new copier for \$10,000. The life of the copier is five years and the machine has no salvage value.
(a) Find $V(t)$, the book value at any time t, if the company uses SYD depreciation.
(b) What is the book value at the end of four years?

51. Dumbo's Pizza purchases an oven for \$18,000. The estimated life of the oven is ten years, and it has no salvage value.
(a) Find $V(t)$, the book value of the oven at any time t, if the company uses SYD depreciation.
(b) By how much does the value of the oven decrease from $t = 2$ to $t = 6$?

52. Annual output from the HiHo Silver Mines is given by the equation

$$f(t) = 200e^{-0.05t} \text{ tons/year}$$

Find the amount of silver mined from $t = 0$ to $t = 4$.

53. (a) The rate at which the number of people $N(t)$ in a small community are becoming ill with a new type of influenza is given by the equation $N'(t) = 25 + 6t$, where t is the time, in days, from today ($t = 0$). If 500 people have already come down with the illness, find an equation that predicts the number of people who become ill as a function of the time t for $t \geq 0$.
(b) If the number of residents in the community is 1050, how long will it be before each resident has become ill with the flu. (Assume that no one catches the flu more than once.)

54. A chemical pollutant enters a stream at a rate, in gallons per hour, given by the equation $f(t) = 5e^{-0.20t}$ where t is the time in hours from the moment the leak began ($t = 0$). Assuming that the stream contains none of the chemical before the leaking begins, find an equation for the number of gallons in the stream at any time t. How many gallons are in the stream at the end of two days?

Find the derivative of each of the following functions.

55. $F(x) = \int e^{-x^2}\,dx$

56. $F(x) = \int \ln(x^2 + 1)\,dx$

6.2 AREA UNDER A CURVE

Antiderivatives can be used to solve an important problem in calculus, namely determining the area of a plane region R, such as that shaded in Figure 1; the region is bounded by the vertical lines $x = a$ and $x = b$, the x axis, and the graph of the function $y = f(x)$, which is nonnegative and continuous over the interval $a \leq x \leq b$. The area of the shaded region in Figure 1 is called the area under the curve, whose equation is $y = f(x)$, from $x = a$ to $x = b$.*

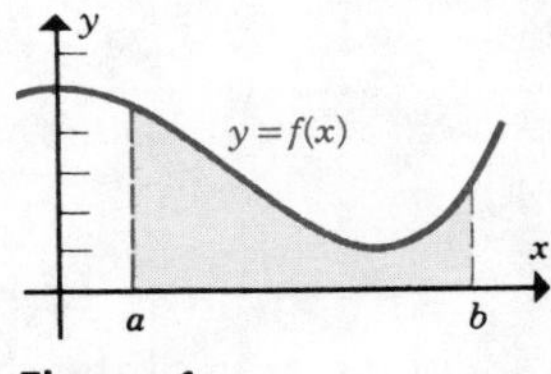

Figure 1

The importance of this problem goes beyond the field of geometry. For example, suppose that $C'(x)$, the marginal cost, for a medical supply company

*Hereafter the area will be referred to simply as the area under the curve $y = f(x)$ from $x = a$ to $x = b$.

that produces digital thermometers is \$2 per thermometer. If the daily output is 100 thermometers and management is considering an increase in output to 125 thermometers, the additional cost of production will equal $(125 - 100)(2) = \$50$. The additional cost is represented by the area of the shaded rectangle in Figure 2, which is also the area under the curve $C'(x) = 2$ from $x = 100$ to $x = 125$.

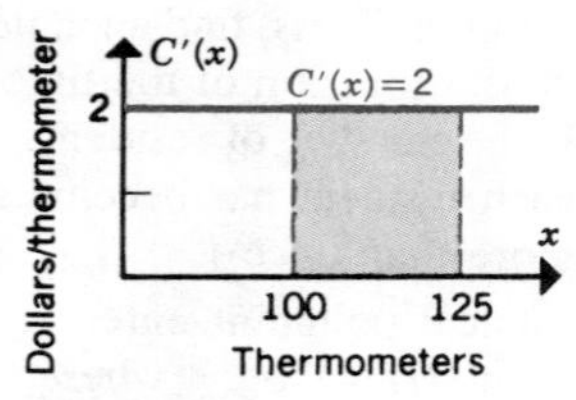

Figure 2

Now, suppose that the marginal cost is not constant for all x but varies as shown in Figure 3. The area under the marginal cost curve $C'(x) = f(x)$ between $x = 100$ and $x = 125$ still represents the additional cost, now unknown, of increasing production from $x = 100$ to $x = 125$ thermometers.

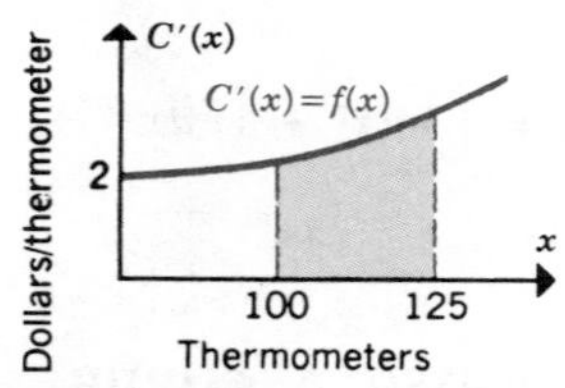

Figure 3

The relationship between the area under a curve and the equation of the curve will be explored for two simple functions. First, let's look at the rectangular region shown in Figure 4. The area $A(x)$ of the region under the curve $f(x) = k (k > 0)$ between $x = 0$ and an arbitrary value of $x > 0$ is given by the equation

$$A(x) = kx \tag{1}$$

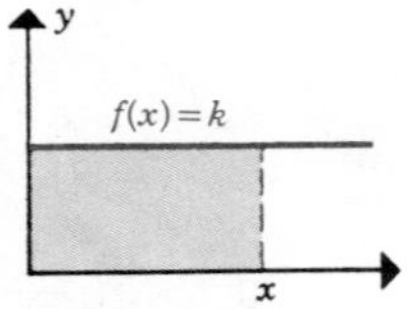

Figure 4

Next, let's find the area of the shaded triangle shown in Figure 5. Since the area of a triangle is given by the formula

$$\text{Area} = \tfrac{1}{2}(\text{base})(\text{height})$$

$A(x)$, the area under the curve $f(x) = kx$ between $x = 0$ and an arbitrary value of $x > 0$, is given by

$$\begin{aligned} A(x) &= \tfrac{1}{2}(x)(kx) \\ &= \tfrac{1}{2}kx^2 \end{aligned} \tag{2}$$

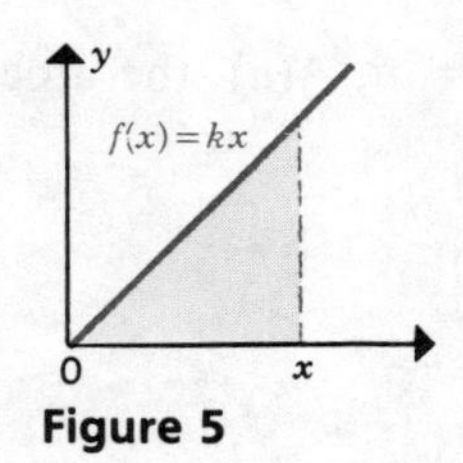

Figure 5

The relationship between $A(x)$ and $f(x)$ for each of the curves, can be found by differentiating $A(x)$ in Equations 1 and 2:

$$\text{Equation 1:}\quad A'(x) = k \;\; = f(x)$$
$$\text{Equation 2:}\quad A'(x) = kx = f(x)$$

Although we looked at only two cases, the relationship $A'(x) = f(x)$ is true in general.

Theorem 1 Let $y = f(x)$ be a nonnegative, continuous function on the interval $a \leq x \leq b$. Let $A(x)$ be the area of the region under the curve $y = f(x)$ between $x = a$ and an arbitrary value of x, shown in Figure 6. Then

$$\boxed{A'(x) = f(x)} \tag{3}$$

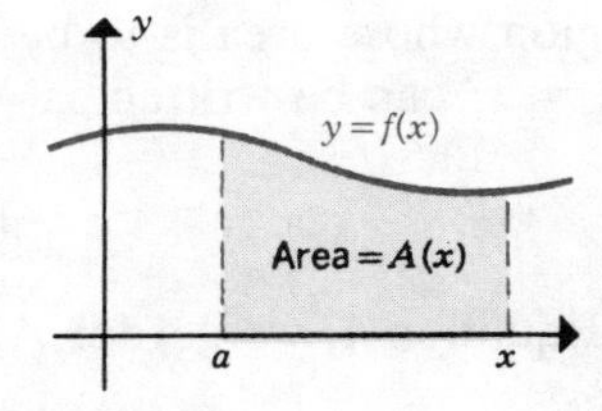

Figure 6

In words, Equation 3 says that the rate of change of the area equals the y coordinate of the right-hand edge of the region. The validity of this theorem is demonstrated at the end of this section. Equation 3 now can be used to find a formula for the area under a curve $y = f(x)$ between $x = a$ and $x = b$ (Figure 7). If $F(x)$ denotes an antiderivative of $f(x)$, we can integrate both sides of Equation (3) and get the following result:

$$A(x) = \int f(x)\, dx = F(x) + C$$

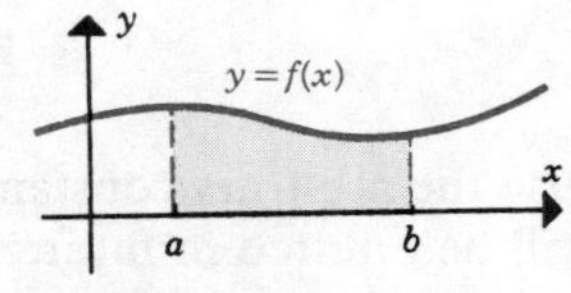

Figure 7

When $x = a$, $A(a)$, the area under the curve, equals zero, so

$$A(a) = 0 = F(a) + C$$

Solving for C gives

$$C = -F(a)$$

So

$$A(x) = F(x) - F(a)$$

When $x = b$, the quantity $A(b)$ represents the area under the curve, and we have the following important result:

Area Under a Curve

If $y = f(x)$ is a nonnegative continuous function on the closed interval $a \leq x \leq b$, the area A of the region under the curve between $x = a$ and $x = b$ is given by the equation

$$\boxed{A = F(b) - F(a)} \tag{4}$$

where $F'(x) = f(x)$.

Example 1 Find the area under the curve $f(x) = x^2$ between $x = 1$ and $x = 2$.

Solution The region whose area is to be found is shown in Figure 8. An antiderivative of $f(x) = x^2$ can be written as

$$F(x) = \frac{x^3}{3} + C$$

Using Equation 4, we get for A, the area,

$$\begin{aligned} A &= F(2) - F(1) \\ &= (\tfrac{8}{3} + C) - (\tfrac{1}{3} + C) \\ &= \tfrac{7}{3} \end{aligned}$$

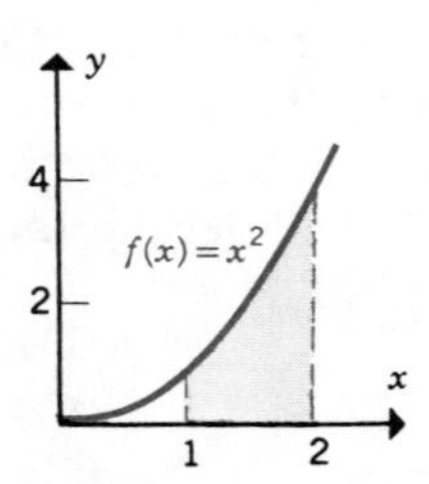

Figure 8

■

Because the arbitrary constant C is a term of both $F(b)$ and $F(a)$ and cancels out, it will be omitted in future calculations.

Example 2 Find the area under the curve

$$f(x) = \frac{3\sqrt{x}}{2} + 1$$

between $x = 4$ and $x = 9$.

Solution The region whose area is to be determined is shown in Figure 9. Noting that an antiderivative of $f(x)$ is

$$F(x) = x^{3/2} + x$$

and using Equation 4, we get for the area,

$$\begin{aligned} A &= F(9) - F(4) \\ &= (9^{3/2} + 9) - (4^{3/2} + 4) = 24 \end{aligned}$$

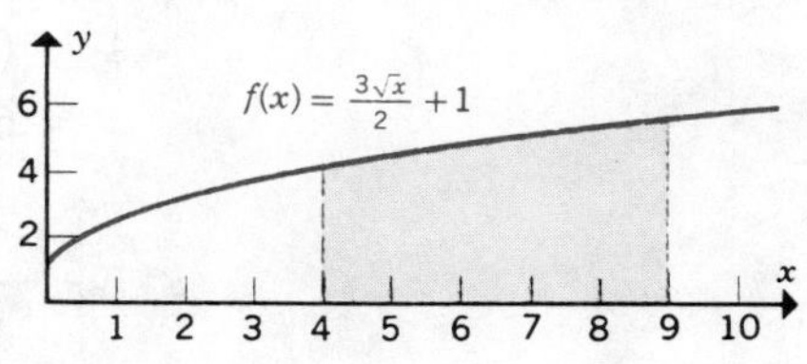

Figure 9 ■

In many situations, the area of a region will be given, and we have to find the value of either a, the left-hand boundary, or b, the right-hand boundary. The next example is an illustration.

Example 3 Find the value of b for which the area of the region beneath the curve $f(x) = 1/x^2$ between $x = \frac{1}{2}$ and $x = b$ equals $\frac{7}{4}$.

Solution The curve and the region are shown in Figure 10. An antiderivative of $1/x^2$ is found:

$$F(x) = -1/x$$

Using Equation 4, we can write

$$\begin{aligned} A = \tfrac{7}{4} &= F(b) - F(\tfrac{1}{2}) \\ &= -\frac{1}{b} + \frac{1}{\frac{1}{2}} \\ &= -\frac{1}{b} + 2 \end{aligned}$$

Solving for b gives

$$b = 4$$

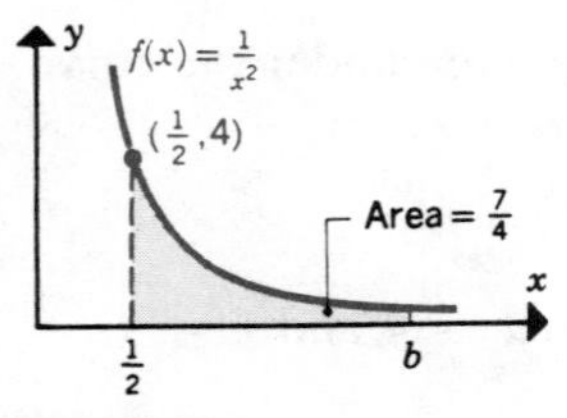

Figure 10

Units

When working through any problem, the units associated with each variable should be noted. Units are particularly important in area problems because the value of the area will depend upon the units of both the independent and dependent variables. For example, the area of the rectangular region in Figure 11, showing the daily output of an oil well as a function of t, equals

$$\begin{aligned} A &= (10)(8 - 2) \\ &= 60 \end{aligned}$$

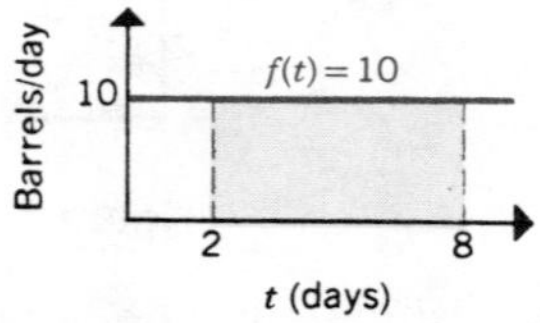

Figure 11

The number 60 represents the total number of barrels of oil pumped between $t = 2$ and $t = 8$ days. The units associated with the area are found by multiplying the units of the independent variable t (days) by the units of the dependent variable $f(t)$ (barrels per day), so we get

$$\begin{aligned} \text{Units of } A &= (\text{units of } t)[\text{units of } f(t)] \\ &= (\text{days})\left(\frac{\text{barrels}}{\text{day}}\right) \\ &= \text{barrels} \end{aligned}$$

Therefore, for the shaded area in Figure 11 we write

$$A = 60 \text{ barrels}$$

In general, for the units of the area A we have

$$\text{Units of } A = (\text{units of independent variable})(\text{units of dependent variable})$$

Applications

The area under a curve is an important quantity when the dependent variable, $f(x)$, represents a rate of change or derivative of a function. If $f(x) = F'(x)$

and is nonnegative for $a \le x \le b$, the area under the curve $y = f(x)$ between $x = a$ and $x = b$ represents the **net change in $F(x)$** between $x = a$ and $x = b$. This concept is illustrated in Examples 4 and 5.

Example 4 The marginal cost for a company making stratoloungers is given by the equation

$$C'(x) = 25 + 0.02x \quad \text{dollars/item}$$

where x represents the number of items produced. The present level of production is 100 units. What is the additional cost if production is increased to 150 items?

Solution The graph of the marginal cost function is shown in Figure 12. The shaded area represents the additional cost of increasing production. An antiderivative of $C'(x) = 25 + 0.2x$ can be found:

$$C(x) = 25x + 0.1x^2$$

We now get

$$\begin{aligned} \text{Additional cost} &= C(150) - C(100) \\ &= (3750 + 2250) - (2500 + 1000) \\ &= \$2500 \end{aligned}$$

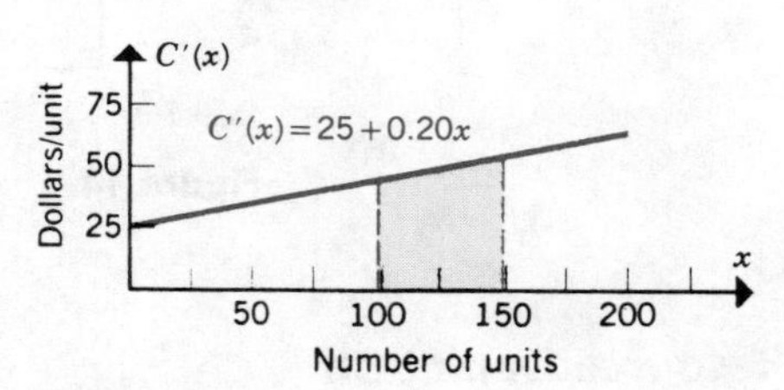

Figure 12

This answer could also have been obtained by breaking the region up into a rectangle and triangle, as shown in Figure 13, and then calculating the sum of the areas.

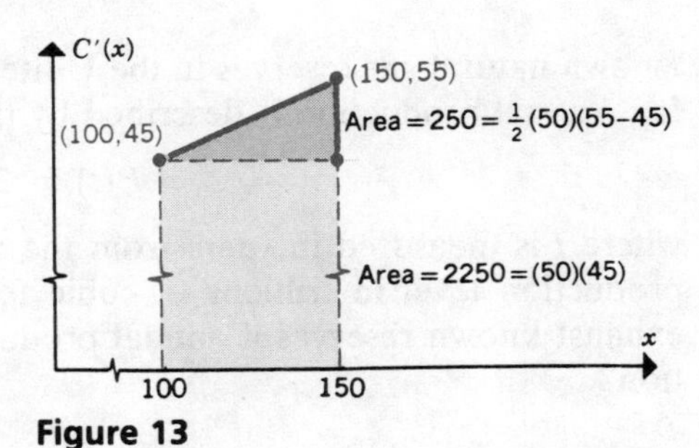

Figure 13

■

Example 5 The expected number of acres consumed annually by forest fires is given by the equation

$$f(t) = 10{,}000e^{0.05t} \quad \text{acres/year}$$

where t is the time, in years, from today ($t = 0$). How many acres will be consumed during the next three years?

Solution Since $f(t)$ represents a rate of change, the area under the curve in Figure 14 from $t = 0$ to $t = 3$ equals the total number of acres burned over the next three years. The area A is given by

$$A = F(3) - F(0)$$

where $F'(t) = f(t)$. By integrating, we find

$$F(t) = \frac{10{,}000}{0.05} e^{0.05t} = 200{,}000e^{0.05t}$$

The difference $F(3) - F(0)$ gives the area A:

$$A = F(3) - F(0) = 200{,}000(e^{0.15} - e^{0})$$
$$= 200{,}000(0.1618) = 32{,}360 \text{ acres}$$

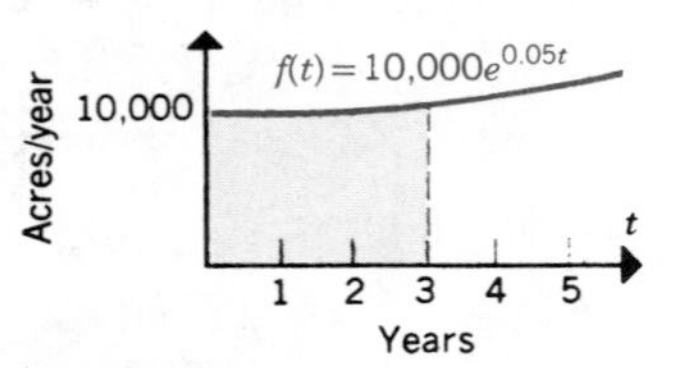

Figure 14

■

Resource Depletion

We encounter many applications in which we are given the total known amount A of a natural resource, such as oil, zinc, or copper, and the rate $f(t)$ at which it is being consumed. We want to determine how many more years the resource will last. The next example is an illustration of this kind of problem.

Example 6 Known natural gas reserves in the United States are estimated to be 225 trillion ft^3. Annual production is described by the equation

$$P(t) = 20e^{-0.08t}$$

where t is measured in years from the present ($t = 0$) and $P(t)$ is the annual production level in trillions of cubic feet per year. How long will it take to exhaust known reserves if annual production continues according to the equation?

Solution The curve showing the annual production $P(t)$ as a function of time is shown in Figure 15. We want to find the value of t, called T, for which the area under the curve equals the known reserves, namely 225 trillion ft^3. Noting that an antiderivative of $20e^{-0.08t}$ is given by

$$F(t) = \frac{20e^{-0.08t}}{-0.08} = -250e^{-0.08t}$$

we use Equation 4, with $A = 225$, $a = 0$, and $b = T$, and get

$$\begin{aligned} 225 &= F(T) - F(0) \\ &= -250e^{-0.08T} + 250 \\ -25 &= -250e^{-0.08T} \\ 0.10 &= e^{-0.08T} \end{aligned}$$

This equation can be solved for T by using the methods described in Section 5.2. Taking the natural logarithm of both sides gives

$$\ln(0.10) = \ln(e^{-0.08T}) = (-0.08T)(\ln e)$$

Noting that $\ln e = 1$ and that $\ln(0.10) = -2.3$ (Table C), we get

$$-0.08T = -2.3$$

The solution is

$$T = 28.75 \text{ years}$$

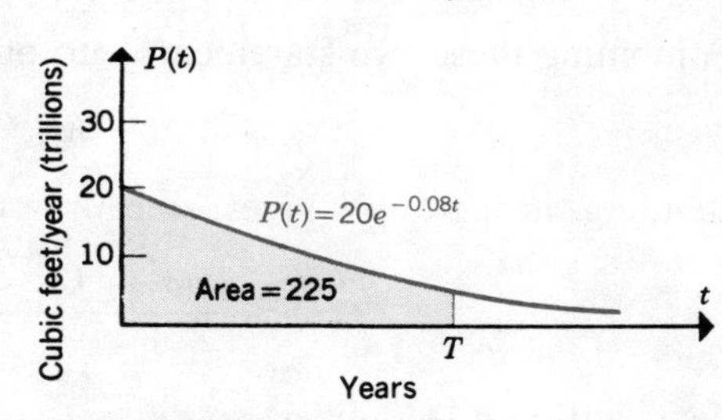

Figure 15

■

A Proof of Theorem 1

Let $A(x)$ be the area under the curve $y = f(x)$ between $x = a$ and an arbitrary value of x, as shown in Figure 16. Next the right-hand boundary line is shifted

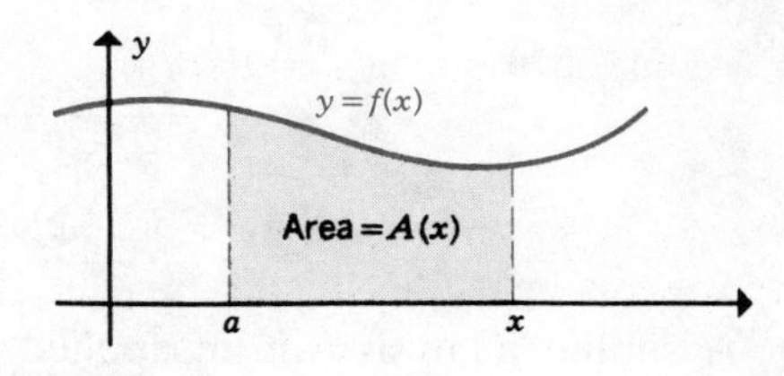

Figure 16

slightly to the right to $x + h$; $A(x + h)$ is the new area under the curve between a and $x + h$, as shown in Figure 17.

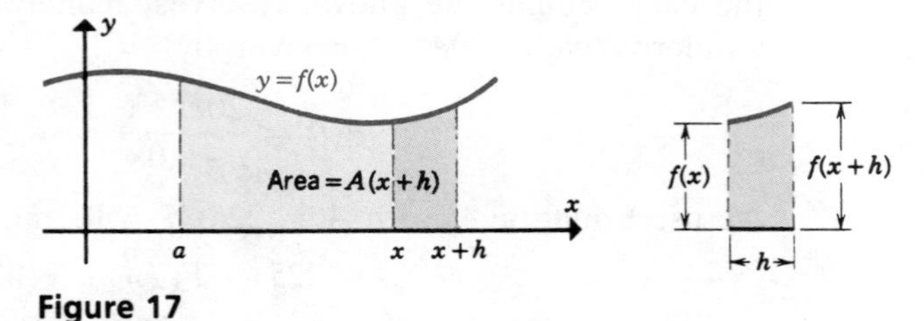

Figure 17

Our analysis will be limited to the situation in which $f(x)$ is increasing on the interval from x to $x + h$, such as the function whose graph is shown in Figure 17; the area of the small shaded strip, $A(x + h) - A(x)$, is greater than or equal to the area of the rectangle whose base is h and whose height equals $f(x)$, that is,

$$hf(x) \leq A(x + h) - A(x)$$

Similarly, the area of the strip is less than or equal to the area of the rectangle whose base is h and whose height equals $f(x + h)$, that is,

$$A(x + h) - A(x) \leq hf(x + h)$$

Combining these two statements into one gives

$$hf(x) \leq A(x + h) - A(x) \leq hf(x + h)$$

Dividing through by h gives, since $h > 0$,

$$f(x) \leq \frac{A(x + h) - A(x)}{h} \leq f(x + h)$$

Next, letting $h \to 0$ gives

$$\lim_{h\to 0} f(x) \leq \lim_{h\to 0} \frac{A(x + h) - A(x)}{h} \leq \lim_{h\to 0} f(x + h)$$

From this, we get

$$f(x) \leq A'(x) \leq f(x)$$

so

$$\boxed{A'(x) = f(x)}$$

A similar analysis can be applied to the situation in which x is decreasing everywhere on the interval from x to $x + h$.

6.2 EXERCISES

Find the area of each of the shaded regions.

1. 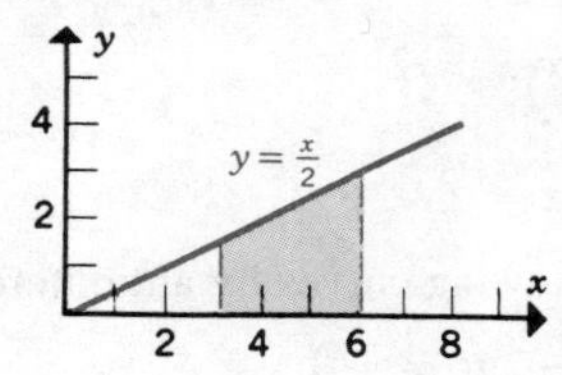

2. 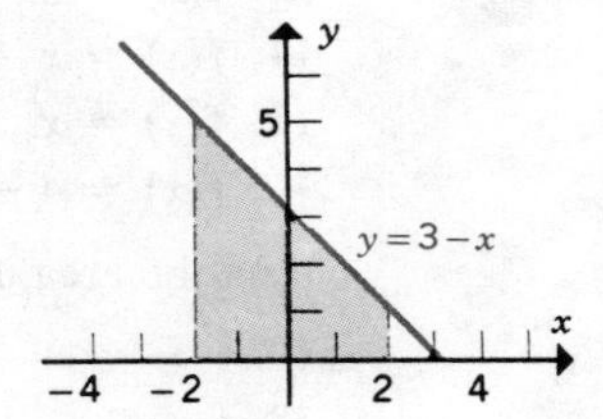

3. 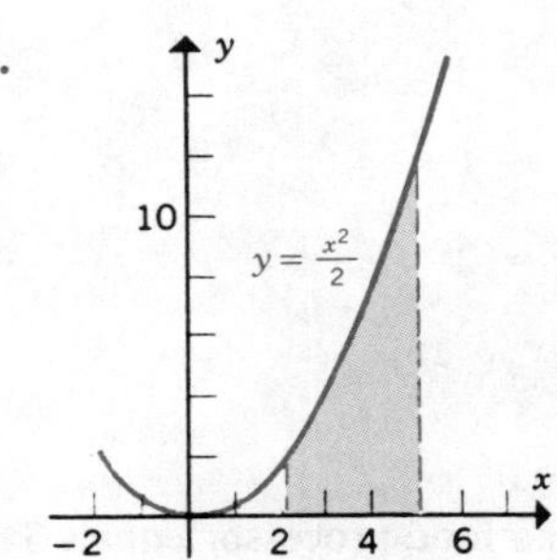

4. 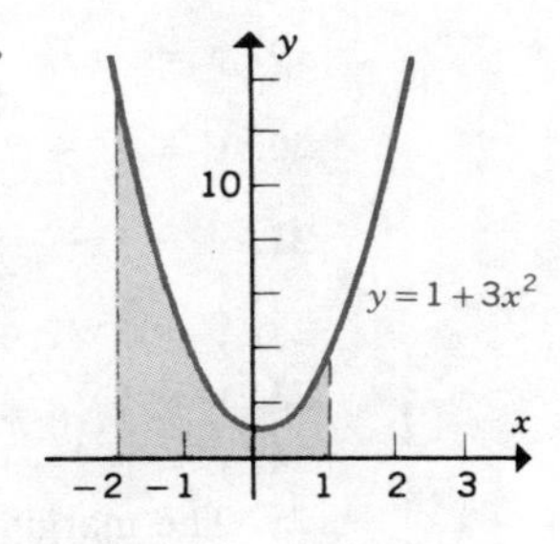

5. 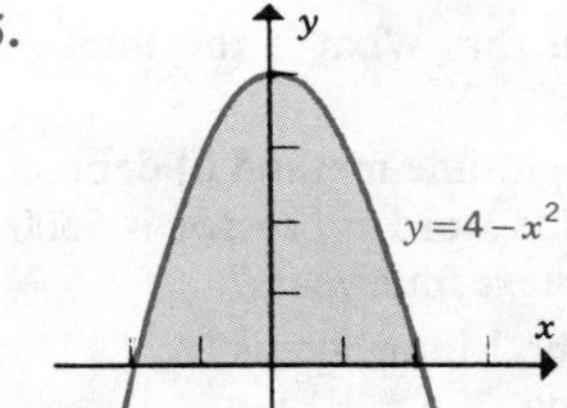

6. 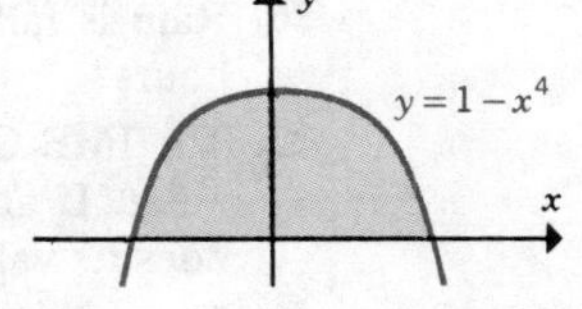

7. 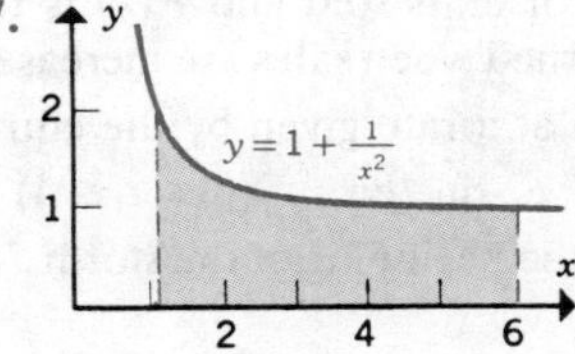

8. 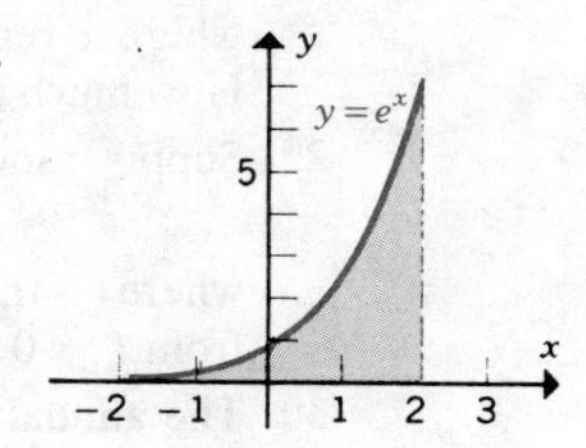

9. 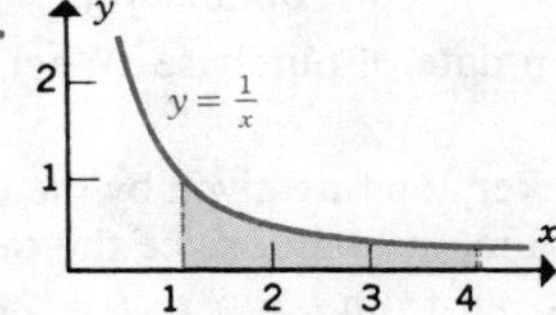

10. 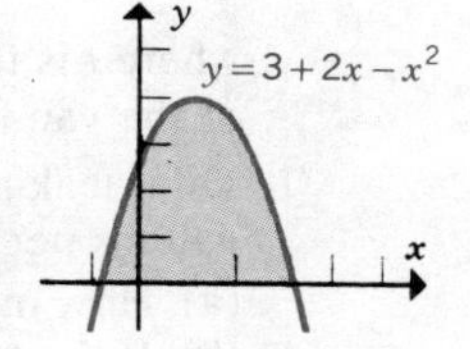

Find the area under each of the following curves by antidifferentiation; in addition, sketch each curve and use elementary geometry to find the area.

11. $f(x) = 3$ from $x = 0$ to $x = 3$
12. $f(x) = \frac{5}{2}$ from $x = -1$ to $x = 5$
13. $f(x) = 2x$ from $x = 0$ to $x = 4$
14. $f(x) = x + 1$ from $x = -1$ to $x = 2$
15. $f(x) = x$ from $x = 1$ to $x = 4$
16. $f(x) = 4 - x$ from $x = 0$ to $x = 6$

Find the area under each of the following curves by antidifferentiation.

17. $y = x^2 - 2x + 2$ from $x = -2$ to $x = 1$
18. $y = 2 - 4x^3$ from $x = -2$ to $x = 0$
19. $y = 3x - \dfrac{1}{x^4}$ from $x = 1$ to $x = 2$
20. $y = e^x - 2x$ from $x = 0$ to $x = 1$
21. $y = 3x^2 - \dfrac{1}{x}$ from $x = 1$ to $x = 2$
22. $y = 3\sqrt{x} + 2$ from $x = 4$ to $x = 16$
23. $y = 5\sqrt[3]{x^2}$ from $x = 1$ to $x = 8$
24. $y = e^{-2x}$ from $x = -1$ to $x = 0$
25. The marginal cost for the Eureka food processor equals \$40 per unit. If production is increased from 1000 units to 1200 units, by how much does the total cost increase?
26. Rain is falling at a rate of 0.1 in./hr. What is the total rainfall over the next five hours?
27. The IMB Company uses the straight-line method of depreciation for its office equipment. If annual depreciation on its word processor is \$500 per year, by how much does its value decrease over the next four years?
28. The marginal profit for the Rambo Motorbike is
$$P'(x) = 300 - 0.2x \quad \text{dollars/unit}$$
where x represents the number of units sold and $P(x)$ is the corresponding profit. How much additional profit is earned when sales are increased from 100 to 120 units?
29. Suppose snow falls on a ski area at a rate given by the equation
$$f(t) = 4t - t^2 \quad \text{in./hr} \qquad (0 \leq t \leq 4)$$
where t is the time in hours from the beginning of the storm. What is the total snowfall from $t = 0$ to $t = 4$?
30. The annual depreciation of a delivery van is described by the equation
$$f(t) = 2000 - 400t \quad \text{dollars/year}$$
where t is the time, in years, from date of purchase. What is the total depreciation of the van from $t = 0$ to $t = 3$?
31. Oil is leaking from a damaged tanker at a rate given by the equation $f(t) = 500e^{-0.02t}$ gallons per hour, where t is the time, in hours, since the tanker began leaking.
    (a) How many gallons of oil leak out from $t = 0$ to $t = 24$?
    (b) If the tanker held 20,000 gallons when the leaking began, how long does it take before the tanker is empty?
32. The annual depreciation of a milling machine is given by the equation
$$f(t) = 10{,}000 - 2000t \quad \text{dollars/year}$$
where t is time, in years, from the date of purchase.

(a) By how much does the value decline from $t = 0$ to $t = 2$?
(b) Determine the decline in value from $t = 0$ to $t = 5$.
(c) If the milling machine has a five-year lifetime, under what conditions would the answer to part (b) represent the purchase cost of the asset?

33. Under the current budget, Regency Electronics can produce ten VCRs per day. If the daily production budget is increased by $244 and if the marginal cost is given by the equation

$$C'(x) = 100 + 2x$$

determine how many additional units can be produced.

34. In a suburban community the number of children $N(t)$ who attend public schools is growing at a rate given by the equation $N'(t) = 50 + 30t$, where t is the time, in years, from today ($t = 0$).
(a) How many additional children will be attending public schools four years from now?
(b) The superintendent estimates that the present school system can accommodate 840 additional students. When will the present system reach capacity?

6.3 DEFINITE INTEGRAL

In the last section, we found that the area of the region under the curve $y = f(x)$ between $x = a$ and $x = b$ equals $F(b) - F(a)$, where $F'(x) = f(x)$. In this section another method for determining the area will be presented. This method is illustrated in Figure 1, where the area under the curve is subdivided into a large number of rectangles, the sum of whose areas can be found directly. The area under the curve is then determined by finding the limit of this sum as the number of rectangles becomes infinite. More important, this process leads to a definition of the **definite integral,**

$$\int_a^b f(x)\,dx$$

following which the **fundamental theorem of calculus** will be presented.

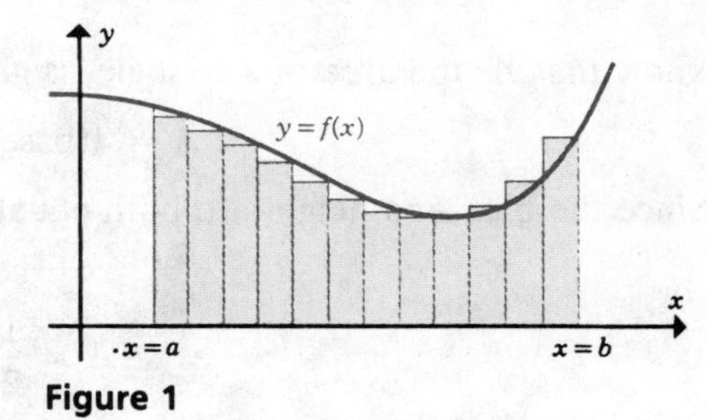

Figure 1

The process of finding the area will require the addition of a large number of terms. Special notation, called sigma notation, is used to describe the addition; the symbol Σ (Greek capital sigma) is used to designate summation. Given the

n numbers $x_1, x_2, x_3, \ldots, x_n$, their sum is represented by the expression

$$\sum_{k=1}^{n} x_k$$

That is,

$$\boxed{\sum_{k=1}^{n} x_k = x_1 + x_2 + x_3 + \cdots + x_n} \tag{1}$$

The letter k is called an *index*. The following illustrate how sigma notation is used:

$$\sum_{k=1}^{6} k = 1 + 2 + 3 + 4 + 5 + 6 = 21$$

$$\sum_{k=1}^{10} 3 = 3 + 3 + 3 + \cdots + 3 = 30$$

$$\sum_{k=1}^{5} k^2 = 1 + 2^2 + 3^2 + 4^2 + 5^2 = 55$$

$$\sum_{k=1}^{4} f(x_k) = f(x_1) + f(x_2) + f(x_3) + f(x_4)$$

We begin by finding the area of the triangular region under the curve $f(x) = x$ between $x = 0$ and $x = 1$, shown in Figure 2. From geometry we

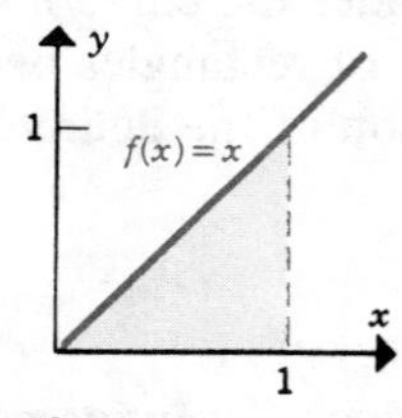

Figure 2

know that A, the area of a triangle, is given by the formula

$$A = \tfrac{1}{2}(\text{base})(\text{height})$$

Since the base and height are both equal to 1, we find that

$$\boxed{A = \frac{1}{2}} \tag{2}$$

The process we are about to describe to find A is fairly lengthy compared to the simple geometric analysis just employed. Our objective is to describe the process represented by the definite integral; this can be accomplished more effectively if the geometry of the region is not complex.

Now suppose we divide the interval [0, 1]* into two equal subintervals $[0, \frac{1}{2}]$ and $[\frac{1}{2}, 1]$ and construct two rectangles whose heights equal the y coordinates at the right-hand endpoints, that is, $f(\frac{1}{2})$ and $f(1)$, and whose bases equal $\frac{1}{2}$. The rectangles are shown in Figure 3. The sum of the areas can be calculated

$$A_1 + A_1 = (\tfrac{1}{2})(\tfrac{1}{2}) + (\tfrac{1}{2})(1) = \frac{3}{4}$$

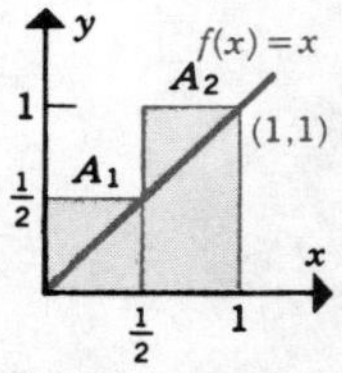

Figure 3

Obviously, this approximation exceeds $\frac{1}{2}$, the true area under the curve. The approximation can be improved by subdividing the interval [0, 1] into more subintervals, say four, and repeating the process, as shown in Figure 4. The sum of the areas of the four rectangles is

$$\sum_{k=1}^{4} A_k = A_1 + A_2 + A_3 + A_4$$

$$= \left(\frac{1}{4}\right)\left(\frac{1}{4}\right) + \left(\frac{1}{4}\right)\left(\frac{1}{2}\right) + \left(\frac{1}{4}\right)\left(\frac{3}{4}\right) + \left(\frac{1}{4}\right)(1)$$

$$= \left(\frac{1}{4}\right)\frac{1 + 2 + 3 + 4}{4} = \frac{10}{16} = \frac{5}{8}$$

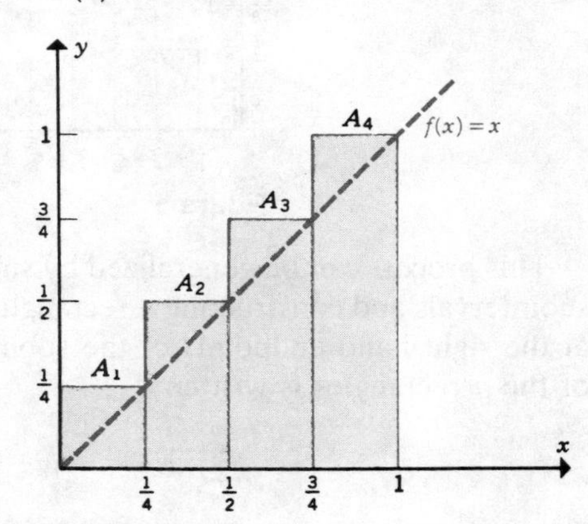

Figure 4

Again, the approximation yields a result that is too large, but it is an improvement over the preceding approximation. A better approximation can be obtained by

*Recall that [0, 1] represents the closed interval $0 \leq x \leq 1$. In general, $[a, b]$ stands for the closed interval $a \leq x \leq b$.

subdividing the interval [0, 1] into an even larger number of subintervals, say eight, as shown in Figure 5, and constructing rectangles on each subinterval, as described previously.

The sum of the areas of the eight rectangles then becomes

$$\begin{aligned}\sum_{k=1}^{8} A_k &= A_1 + A_2 + \cdots + A_8 \\ &= \left(\frac{1}{8}\right)\left(\frac{1}{8}\right) + \left(\frac{1}{8}\right)\left(\frac{1}{4}\right) + \cdots + \left(\frac{1}{8}\right)\left(\frac{7}{8}\right) + \left(\frac{1}{8}\right)(1) \\ &= \left(\frac{1}{8}\right)\left(\frac{1 + 2 + 3 + 4 + 5 + 6 + 7 + 8}{8}\right) \\ &= \left(\frac{1}{8}\right)\left(\frac{1}{8}\right)(1 + 2 + 3 + \cdots + 8) = \frac{9}{16}\end{aligned}$$

Note that the sum can also be written in summation notation:

$$1 + 2 + 3 + \cdots + 8 = \sum_{k=1}^{8} k$$

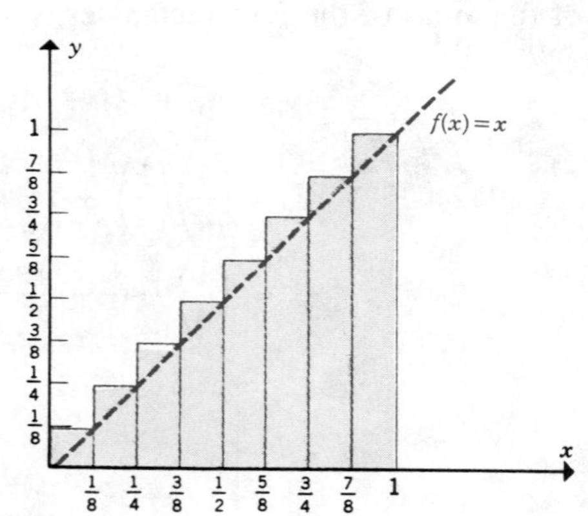

Figure 5

This process can be generalized by subdividing the interval [0, 1] into n equal subintervals and constructing n rectangles whose heights equal the y coordinates at the right-hand endpoints of the subinterval, as shown in Figure 6. The sum of the n rectangles is written

$$\sum_{k=1}^{n} A_k = A_1 + A_2 + \cdots + A_n$$

This sum can also be written as

$$\begin{aligned}\sum_{k=1}^{n} A_k &= \left(\frac{1}{n}\right)\left(\frac{1}{n}\right) + \left(\frac{1}{n}\right)\left(\frac{2}{n}\right) + \left(\frac{1}{n}\right)\left(\frac{3}{n}\right) + \cdots \\ &\quad + \left(\frac{1}{n}\right)\left(\frac{n + 1}{n}\right) + \left(\frac{1}{n}\right)\left(\frac{n}{n}\right) \\ &= \frac{1}{n^2}[1 + 2 + 3 + \cdots + (n - 1) + n]\end{aligned}$$

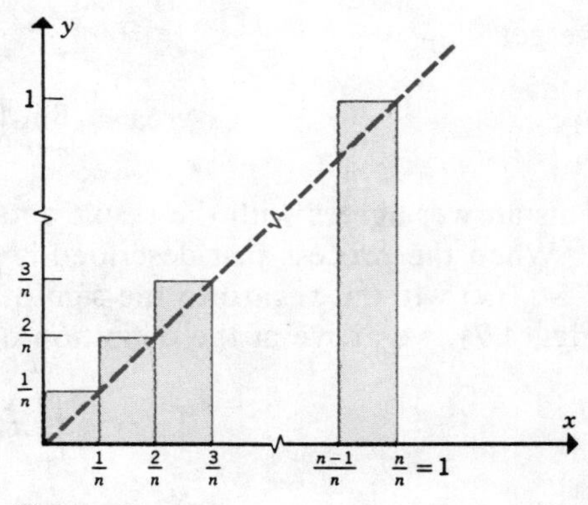

Figure 6

The sum on the right-hand side can be written as a single term; let

$$S = 1 + 2 + 3 + \cdots + (n - 1) + n \tag{3}$$

By rearranging terms, we can write

$$S = n + (n - 1) + (n - 2) + \cdots + 2 + 1 \tag{4}$$

Note that the sum of the first terms on the right-hand sides of Equations 3 and 4 equals $n + 1$; the same also holds true for the second through the nth items. Therefore adding the left- and right-hand sides of Equations 3 and 4 gives

$$\begin{aligned} 2S &= (n + 1) + (n + 1) + (n + 1) + \cdots + (n + 1) + (n + 1) \\ &= n(n + 1) \end{aligned}$$

so

$$S = \frac{n(n + 1)}{2}$$

This result allows us to write the sum of the areas of the n rectangles in the following way:

$$\sum_{k=1}^{n} A_k = \frac{1}{n^2}\frac{n(n + 1)}{2} = \frac{n + 1}{2n}$$

Finally, the area under the curve $f(x) = x$ can be found by letting n, the number of rectangles, become infinite ($n \to \infty$). So we get

$$\begin{aligned} \text{Area} &= \lim_{n\to\infty}\left(\sum_{k=1}^{n} A_k\right) \\ &= \lim_{n\to\infty} \frac{n + 1}{2n} \\ &= \lim_{n\to\infty}\left(\frac{1}{2} + \frac{1}{2n}\right) \end{aligned}$$

As n becomes infinite, the second term in the parentheses approaches zero, and we get

$$\text{Area} = \lim_{n\to\infty}\left(\frac{1}{2} + \frac{1}{2n}\right) = \frac{1}{2}$$

This answer agrees with the result obtained in Equation 2.

When the process just described is generalized and applied to any function $y = f(x)$ without regard to the sign of $f(x)$ over the closed interval $[a, b]$ (see Figure 7), we arrive at the definition of the *definite integral*, symbolized by

$$\int_a^b f(x)\,dx$$

where a is the **lower limit** of integration and b the **upper limit** of integration.

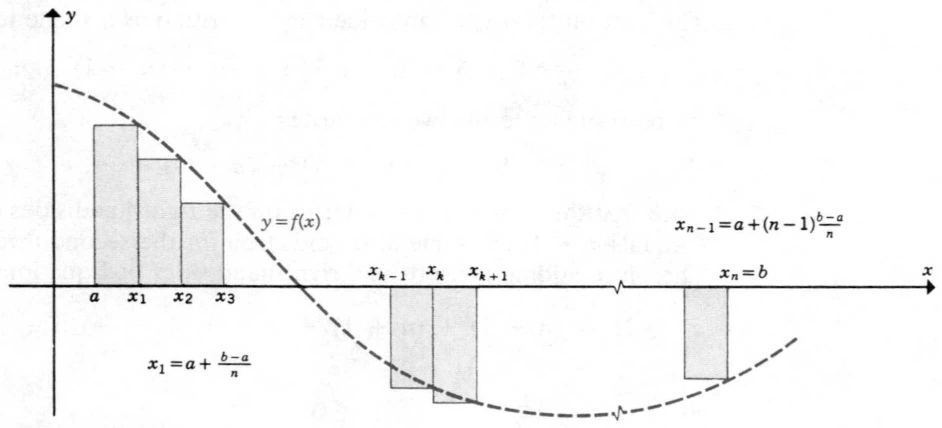

Figure 7

Definition Suppose that $y = f(x)$ is a continuous function on the closed interval $[a, b]$. Let the interval be subdivided into n equal subintervals, each of whose widths is $(b - a)/n$. Designating the right-hand endpoint in each subinterval as x_1, $x_2, \ldots, x_n = b$, we can define the **definite integral** $\int_a^b f(x)\,dx$ as

$$\int_a^b f(x)\,dx = \lim_{n\to\infty} \frac{b-a}{n}[f(x_1) + f(x_2) + f(x_3) + \cdots + f(x_n)] \quad (5)$$

As was true of the derivative, the application of Equation 5 is often difficult and the effort required to obtain a result is generally time-consuming. Fortunately, the effort can be reduced considerably for situations in which an antiderivative of $f(x)$ is known by employing the fundamental theorem of calculus.

Fundamental Theorem of Calculus

If the function $y = f(x)$ is continuous on the closed interval $[a, b]$, then

$$\int_a^b f(x)\,dx = F(b) - F(a) \tag{6}$$

where $F'(x) = f(x)$.

This result is not surprising in view of a similar result in Section 6.2 (Equation 4) where the area under the curve $y = f(x)$ between $x = a$ and $x = b$ was found to equal $F(b) - F(a)$. The equation, Area $= F(b) - F(a)$, is a special case of the fundamental theorem of calculus; it is valid when $f(x) \geq 0$ everywhere on $[a, b]$.

The difference $F(b) - F(a)$ can also be written in the form

$$F(b) - F(a) = F(x)\Big|_a^b$$

so that the fundamental theorem of calculus takes the form

$$\int_a^b f(x)dx = F(x)\Big|_a^b = F(b) - F(a) \tag{7}$$

Example 1 Evaluate $\int_1^2 x^2\,dx$.

Solution Since $F(x) = x^3/3$ is an antiderivative of x^2, we get

$$\int_1^2 x^2\,dx = \frac{x^3}{3}\Big|_1^2 = \frac{8}{3} - \frac{1}{3} = \frac{7}{3}$$

■

It should be noted that the definite integral $\int_a^b f(x)\,dx$ represents a number whereas the indefinite integral $\int f(x)\,dx$ stands for a family of functions. The difference between the two types of integrals can be demonstrated by differentiating each type:

$$\frac{d}{dx}\left[\int_a^b f(x)\,dx\right] = \frac{d}{dx}[F(b) - F(a)] = 0$$

$$\frac{d}{dx}\left[\int f(x)\,dx\right] = \frac{d}{dx}[F(x) + C] = F'(x) = f(x)$$

Example 2 illustrates a situation in which the definite integral is used explicitly instead of its numerical value.

Example 2 Rob, a tax resister, uses definite integrals in place of numbers to complete his federal income tax return. On line 7 of his 1040 form, the following entry appears

7. Wages, salaries, tips, etc., $\int_2^4 200x^3\,dx$.

How much money did Rob earn last year?

Solution Noting that $50x^4$ is an antiderivative of $200x^3$, we find

$$\int_2^4 200x^3\,dx = 50x^4\Big|_2^4 = 50(216 - 16) = \$10{,}000$$

Although it is legitimate mathematically to represent real numbers as definite integrals, we do not recommend that you use them on your tax return, particularly if you do not want to draw attention to yourself. ■

The definite integral possesses some useful properties

Properties of Definite Integrals

1. $\int_a^b kf(x)\,dx = k\int_a^b f(x)\,dx$, where k is a constant.
2. $\int_a^b [f(x) + g(x)]\,dx = \int_a^b f(x)\,dx + \int_a^b g(x)\,dx$.
3. $\int_a^b f(x)\,dx = \int_a^c f(x)\,dx + \int_c^b f(x)\,dx$, where $a \le c \le b$.

Example 3 Evaluate $\int_0^1 (5x^4 + e^x)\,dx$.

Solution Property 2 allows us to integrate term by term:

$$\int_0^1 (5x^4 + e^x)\,dx = \int_0^1 5x^4\,dx + \int_0^1 e^x\,dx$$

Property 1 tells that constant factors can be moved outside the integral, so we have

$$\begin{aligned}\int_0^1 (5x^4 + e^x)\,dx &= 5\int_0^1 x^4\,dx + \int_0^1 e^x\,dx\\ &= 5\,\frac{x^5}{5}\Big|_0^1 + e^x\Big|_0^1\\ &= (1 - 0) + (e - e^0) = 1 + e - 1 = e\end{aligned}$$

■

Property 3 is very useful when $f(x)$, the integrand, is defined by two or more equations on the interval $a \leq x \leq b$.

Example 4 Evaluate $\int_1^4 f(x)\, dx$, where

$$f(x) = \begin{cases} 3x^2, & 1 \leq x \leq 2 \\ 48/x^2, & 2 < x \leq 4 \end{cases}$$

Solution According to property 3, we can write the integral as a sum:

$$\begin{aligned} \int_1^4 f(x)\, dx &= \int_1^2 f(x)\, dx + \int_2^4 f(x)\, dx \\ &= \int_1^2 3x^2\, dx + \int_2^4 \frac{48}{x^2}\, dx \\ &= x^3 \Big|_1^2 - \frac{48}{x}\Big|_2^4 \\ &= (8 - 1) - (12 - 24) = 19 \end{aligned}$$

■

If $f(x) \geq 0$ on $[a, b]$, the definite integral $\int_a^b f(x)\, dx$ can be interpreted as the area between the curve $y = f(x)$ and the x axis from $x = a$ to $x = b$. If $f(x) < 0$ for some x on $[a, b]$, the definite integral represents the area above the x axis minus the area beneath the x axis, as illustrated in Example 5.

Example 5 Evaluate $\int_{-1}^1 x\, dx$ and compare with the area between the curve $f(x)$ and the x axis.

Solution The graph of the function $f(x) = x$ is shown in Figure 8. The area of the shaded region between the curve and the x axis equals the sum of the areas of the two triangles:

$$\text{Area} = \tfrac{1}{2} + \tfrac{1}{2} = 1$$

Evaluating the definite integral yields

$$\begin{aligned} \int_{-1}^1 x\, dx &= \tfrac{1}{2}x^2 \Big|_{-1}^1 \\ &= \tfrac{1}{2} - \tfrac{1}{2} = 0 \end{aligned}$$

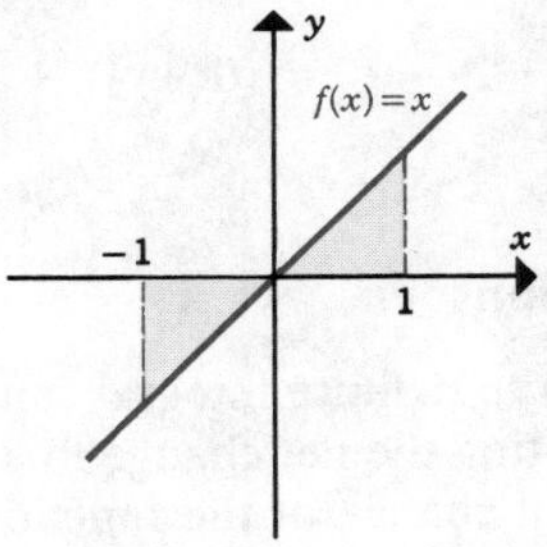

Figure 8

In this instance the negative contribution to the definite integral $\int_{-1}^{0} x\,dx = -\frac{1}{2}$ cancels the positive contribution $\int_{0}^{1} x\,dx = \frac{1}{2}$. ■

If $f(x)$ is negative for some x on $[a, b]$, the area of the region between the curve and the x axis can be found from the formula

$$\text{Area} = \int_a^b |f(x)|\,dx \tag{8}$$

The absolute value guarantees that all contributions to the integral will be nonnegative.

Example 6 Use Equation 8 to find the area between the curve $f(x) = x$ and the x axis between $x = -1$ and $x = 1$.

Solution The graph of the function $g(x) = |f(x)| = |x|$ is shown in Figure 9. Since $g(x) \geq 0$ for all x in the interval $-1 \leq x \leq 1$, the definite integral

$$\int_{-1}^{1} g(x)\,dx = \int_{-1}^{1} |x|\,dx$$

will yield the area. Since

$$|x| = \begin{cases} -x, & x < 0 \\ x, & x \geq 0 \end{cases}$$

property 3 of definite integrals is used:

$$\begin{aligned} \int_{-1}^{1} |x|\,dx &= \int_{-1}^{0} |x|\,dx + \int_{0}^{1} |x|\,dx \\ &= \int_{-1}^{0} -x\,dx + \int_{0}^{1} x\,dx \\ &= -\tfrac{1}{2}x^2 \Big|_{-1}^{0} + \tfrac{1}{2}x^2 \Big|_{0}^{1} \\ &= (0 + \tfrac{1}{2}) + (\tfrac{1}{2} - 0) = 1 \end{aligned}$$ ■

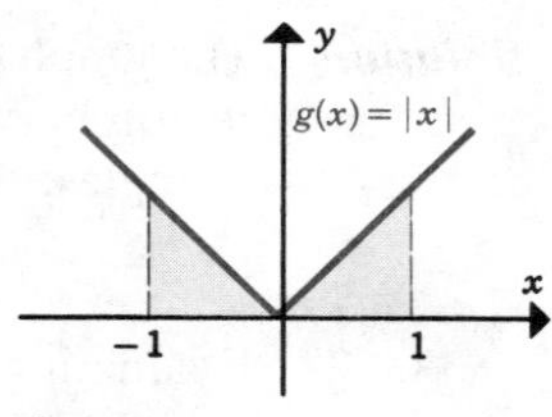

Figure 9

Applications

Because the definite integral can be either positive or negative, it is very useful in calculating the net change in a function, particularly in applications for which the rate of change of the function is given. Example 7 illustrates the use of the definite integral to calculate the net change in a company's revenue as production is changed.

Example 7 The marginal revenue $R'(x)$ for a firm producing skateboards is given by the equation

$$R'(x) = 120 - 2x \quad \text{dollars/unit}$$

where x is the number of units produced (in thousands). Find the change in revenue as production is increased from 50 to 65 thousand units.

Solution The change in revenue is given by

$$\int_{50}^{65} R'(x)\,dx = \int_{50}^{65} (120 - 2x)\,dx$$

which becomes

$$(120x - x^2)\Big|_{50}^{65} = [(120)(65) - 65^2] - [(120)(50) - 50^2]$$

$$= \$75 \text{ thousand}$$

The change in revenue can be viewed graphically as the net of (1) a $100 thousand *increase* as production is raised from 50 to 60 thousand units and (2) a $25 thousand *decrease* as production is raised further from 60 to 65 thousand units (Figure 10).

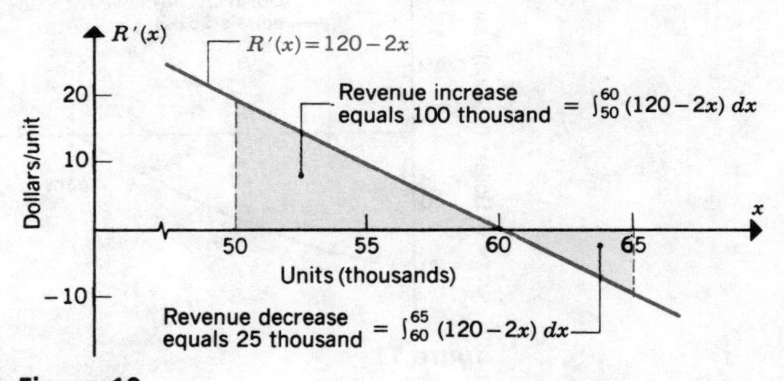

Figure 10

The definite integral can be used to calculate the change in a nation's balance of payments over a given period of time. If the definite integral is positive, the dollar value of exported goods and services exceeds that of imported goods and services; the reverse is true if the definite integral is negative.

Example 8 Suppose that the annual balance of payments over the next ten years is expected to follow the equation

$$f(t) = -20.4 + 1.6t + 0.3t^2 \quad \text{billion dollars/year}$$

(a) What will be the change in the nation's balance of payments (BP) over the next three years?

(b) What will be the change in the balance of payments from $t = 5$ to $t = 10$ years?

Solution (a) The definite integral $\int_0^3 f(t)\, dt$ is used to calculate the change:

$$\text{Change} = \int_0^3 (-20.4 + 1.6t + 0.3t^2)\, dt$$

Working through the details gives

$$\text{Change} = (-20.4t + 0.8t^2 + 0.1t^3)\Big|_0^3$$

$$= (-61.2 + 7.2 + 2.7) = -\$51.3 \text{ billion}$$

The negative sign indicates that \$51.3 billion more will leave the country than will enter over the next three years. The area of the shaded region in Figure 11 equals \$51.3 billion.

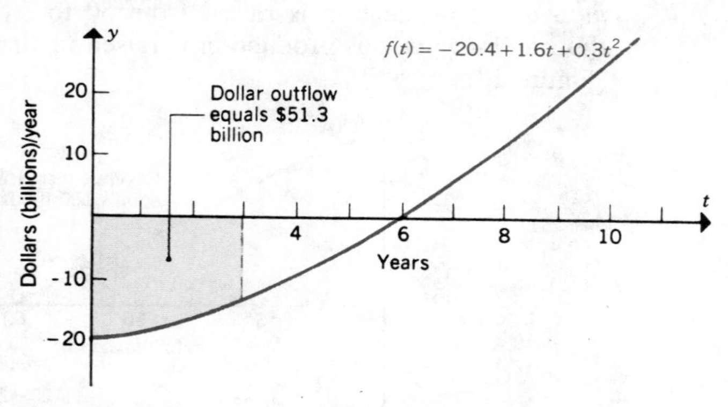

Figure 11

(b) From $t = 5$ to $t = 10$, the change is

$$\text{Change} = \int_5^{10} (-20.4 + 1.6t + 0.3t^2)\, dt$$

$$= (-20.4t + 0.8t^2 + 0.1t^3)\Big|_5^{10}$$

$$= (-204 + 80 + 100) - (-102 + 20 + 12.5)$$

$$= \$45.5 \text{ billion}$$

Thus, over this five-year period the dollar value of exported goods and services is expected to exceed that of imported goods and services by \$45.5 billion. Figure 12 shows that the \$45.5 billion is a net, resulting from an outflow of \$2.5 billion from year 5 to year 6 and a \$48 billion inflow during years 6 through 10.

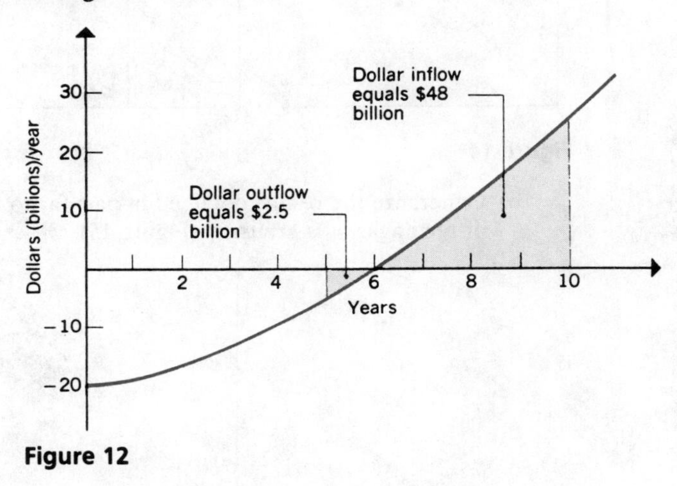

Figure 12

■

6.3 EXERCISES

1. When we demonstrated the process for approximating the area under the curve $f(x) = x$ by subdividing the interval $[0, 1]$ into equal subintervals, the case $n = 3$ was not treated. On the coordinate system of Figure 13, sketch the three rectangles and calculate the sum of the areas.

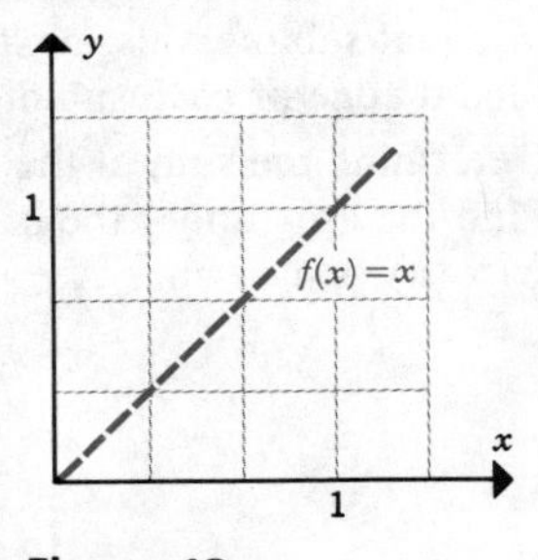

Figure 13

2. When we find the area under a curve, it is not necessary to restrict the analysis to situations in which the heights of the rectangles correspond to the right-hand endpoints. Other methods for accomplishing this can also be developed. Suppose that for the curve $f(x) = x$, the heights of the rectangles are determined at the left-hand endpoints.

   (a) Calculate approximations of the area for $n = 2, 4, 8$ as shown in Figure 14.

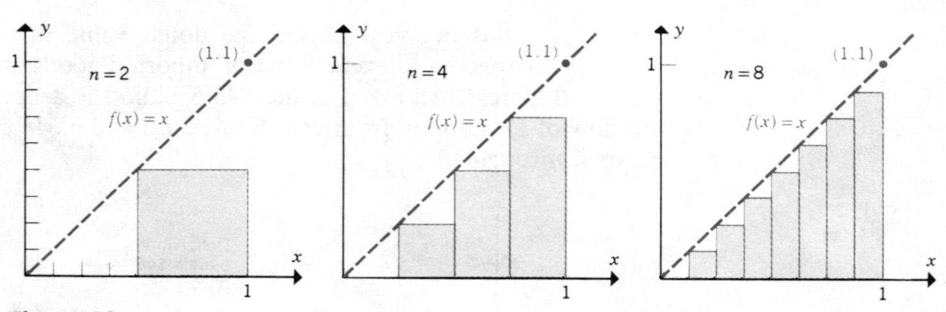

Figure 14

(b) Generalize the results obtained in part (a) to the situation in which the number of rectangles n is arbitrary (Figure 15).

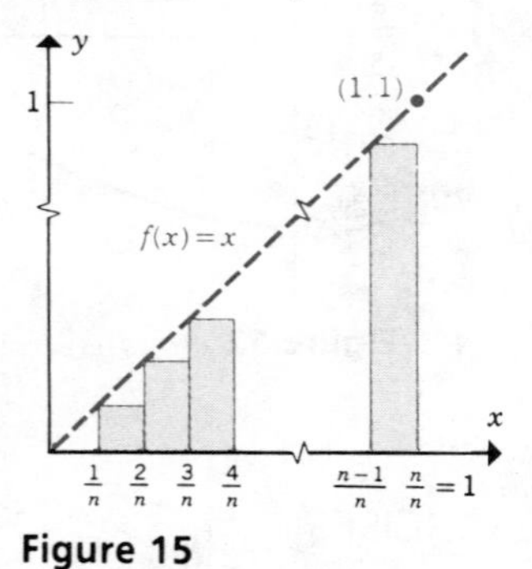

Figure 15

(c) Find the limit of the result obtained in part (b) as $n \to \infty$ to obtain the area beneath the curve.

3. Extend the method used in the text to find the area beneath the curve $f(x) = x$ between $x = 0$ and $x = 2$. As shown in Figure 16, the interval [0, 2] is subdivided into n equal subintervals; next, n rectangles are constructed using the value of the right-hand edge of each subinterval to find the height.
   (a) Determine the sum of the areas $A_1 + A_2 + A_3 + \cdots + A_n$.
   (b) Find the area under the curve by determining the limit of the result obtained in part (a) as $n \to \infty$.

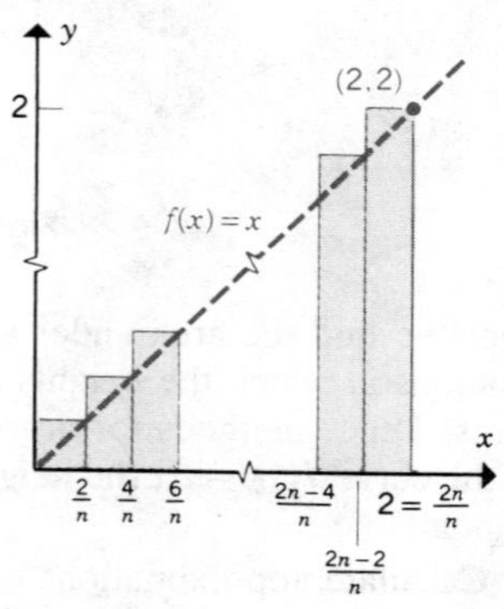

Figure 16

Evaluate each of the following definite integrals.

4. $\int_1^4 (5 + 3x^2)\, dx$

5. $\int_{-1}^0 (2 - x)\, dx$

6. $\int_{-2}^1 4x^2\, dx$

7. $\int_1^3 \frac{6}{x}\, dx$

8. $\int_{-3}^3 (x^2 - 7)\, dx$

9. $\int_1^9 \left(\frac{1}{2\sqrt{x}} + x^2 - 3\right) dx$

10. $\int_0^1 (e^{2x} + x^3 + 1)\, dx$

11. $\int_{-2}^{-1} \frac{1}{x^2}\, dx$

12. $\int_{-1}^2 (3t^2 - t + 2)\, dt$

13. $\int_1^8 2\sqrt[3]{x}\, dx$

14. $\int_1^2 \left(6e^{3x} - \frac{1}{x}\right) dx$

15. $\int_1^{16} 5\sqrt[4]{t}\, dt$

16. $\int_0^1 (x^3 + 1)^2\, dx$

17. $\int_{-1}^1 (x + 2)(x - 2)\, dx$

18. $\int_1^2 \frac{3x^3 + 2x^2 + x}{x}\, dx$

19. $\int_0^1 \frac{x^2 + 4x + 3}{x + 1}\, dx$

20. $\int_0^1 x^e\, dx$

21. $\int_{-2}^{-1} \frac{1}{x}\, dx$

22. $\int_0^1 nx^{n-1}\, dx$

23. $\int_0^1 e^{nx}\, dx$

24. $\int_1^3 f(x)\, dx$, where $f(x) = \begin{cases} x^2, & 0 \le x \le 2 \\ 6 - x, & x > 2 \end{cases}$

25. $\int_{-1}^1 f(x)\, dx$, where $f(x) = \begin{cases} 1 - x^3, & x \le 0 \\ e^x, & x > 0 \end{cases}$

26. $\int_0^4 f(t)\, dt$, where $f(t) = \begin{cases} 7 - 2t, & t \le 3 \\ t^2 - 2t - 2, & t > 3 \end{cases}$

In Exercises 27 through 32, find the values of d that satisfy the given equation.

27. $\int_0^d 3x\, dx = \frac{27}{2}$

28. $\int_0^d (6x + 3)\, dx = 6$

29. $\int_1^d 3x^2\, dx = 7$

30. $\int_d^1 \sqrt{x}\, dx = \frac{7}{12}$

31. $\int_0^d (2x + 1)\, dx = 12$

32. $\int_d^2 \frac{1}{x}\, dx = \ln 2$

In Exercises 33 through 36 find the areas of the shaded regions.

33.

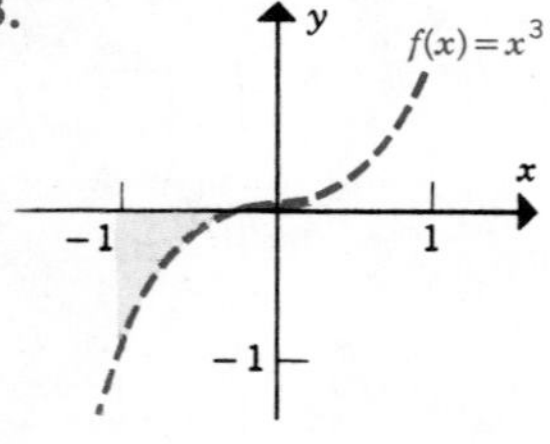

34.

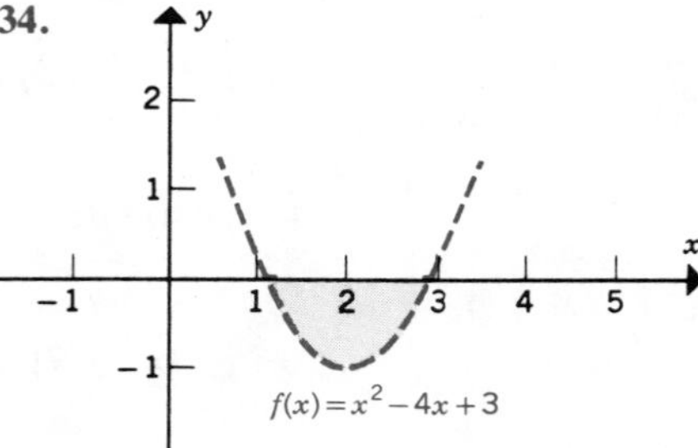

35.

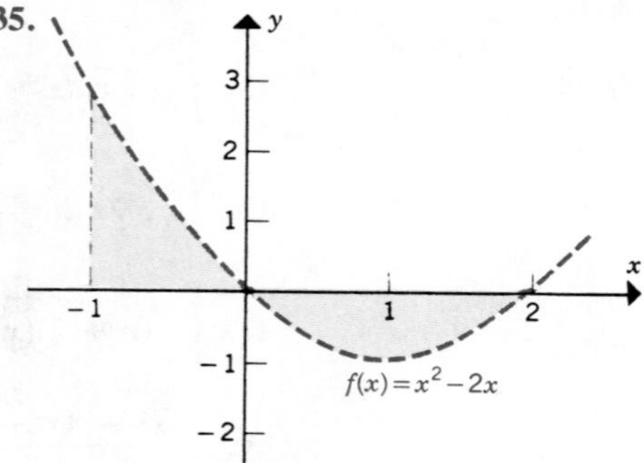

36.

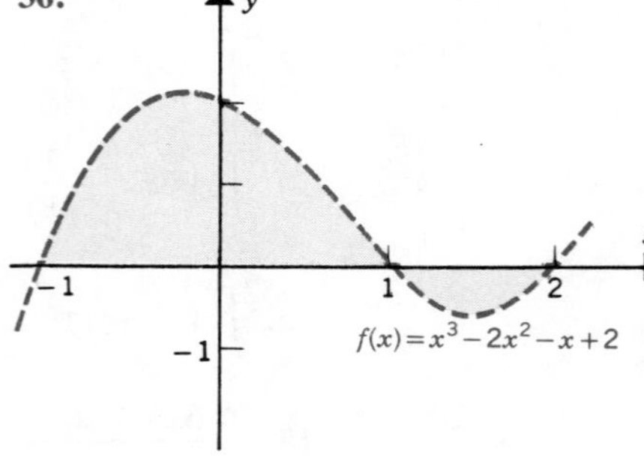

37. A mathematics professor returned a set of exams to the students in her calculus class. The grades on the papers were written as definite integrals. Rodney's grade was

$$\int_5^6 3x^2\,dx$$

What grade did he receive on the examination?

38. The marginal profit for the Eclipse Watch Company is given by the equation $P'(x) = 30 - 0.04x$, where x represents the number of watches sold and $P(x)$ is the corresponding profit.
   (a) Current sales are 750 watches. What is the change in profits if sales increase to 800 watches?
   (b) Does the negative answer in part (a) imply that selling 800 watches is not profitable for Eclipse? Explain.

39. The XYZ Company uses the continuous declining balance method (Section 5.1) of depreciation for its printing press. If annual depreciation is given by the equation

$$f(t) = 10{,}000e^{-0.20t} \quad \text{dollars/year}$$

where t is the time, in years, from date of purchase, determine the total depreciation from $t = 0$ to $t = 5$.

40. (a) Ten million acres are destroyed each year by forest fires. How many acres will be destroyed by fire during the next three years?
   (b) Suppose the number of acres destroyed each year by forest fires is given by the

equation

$$f(t) = 10e^{0.05t} \text{ million acres/year}$$

where t is the time, in years, from today. How many acres will be destroyed during the next three years?

41. A television station has purchased a mobile unit that gets 16 miles per gallon of gasoline. The unit travels 2000 miles per month. If the price of a gallon of gasoline is described by the equation

$$p(t) = 0.95 + 0.05\sqrt{t}$$

where t is the time from the present in months and $p(t)$ is the price in dollars, find the total amount of money that will be spent on gasoline during the next three years.

42. During the recessionary period of the mid-1970s, the annual inventory accumulation $A(t)$ of U.S. manufacturers in billions of dollars per year was given by the equation

$$A(t) = t^3 + 7t^2 - 28t + 20 \text{ billion dollars/year} \qquad (0 \leq t \leq 3)$$

where $t = 0$ corresponds to the beginning of 1974. The graph of the equation is shown in Figure 17. What was the change in inventories between the beginning of 1974 and the end of 1975?

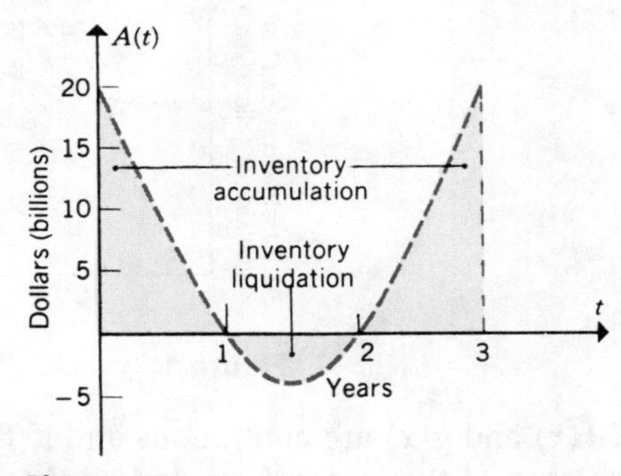

Figure 17

43. The amount of water $A(t)$, in millions of gallons, in a small reservoir is changing at a rate given by the equation $A'(t) = -18 + 6\sqrt{t}$, where t is the time, in years, from today.

(a) By how much will the amount of water in the reservoir change over the next four years?

(b) When will the amount of water in the reservoir reach a minimum?

(c) How much water should the reservoir contain now if it is not to run dry while the amount of water is decreasing?

44. Medication has been prescribed for a man who has high blood pressure. The patient's blood pressure is changing at a rate given by the equation $P'(t) = -5e^{-0.10t}$, where $P(t)$ is the patient's blood pressure, in millimeters of mercury (Hg), and t is the time, in months, from the beginning of the treatment.

(a) By how much does the patient's blood pressure change over the first two months he is on medication?

(b) The doctor plans to discontinue the medication when the patient's blood pressure is 30 mm lower than it was when the treatment began. When will the patient stop taking medication?

6.4 AREA BETWEEN TWO CURVES

In many applications it is necessary to determine the area between two given curves. For example, suppose the graph $y = f(t)$ in Figure 1 represents the annual consumption of gasoline, with t the time in years from today ($t = 0$). A tax increase on gasoline is proposed in order to reduce consumption. An analysis of the effect of the tax on gasoline consumption yields the lower curve $y = g(t)$. The graphs indicate that the tax is very effective initially in reducing consumption, but within ten years consumption returns to the same level it would have reached without the tax. The area of the shaded region in Figure 1 represents the total number of gallons of gasoline saved over the ten-year period owing to the imposition of the tax. The area can be written as the definite integral

$$\int_0^{10} [f(t) - g(t)]\, dt$$

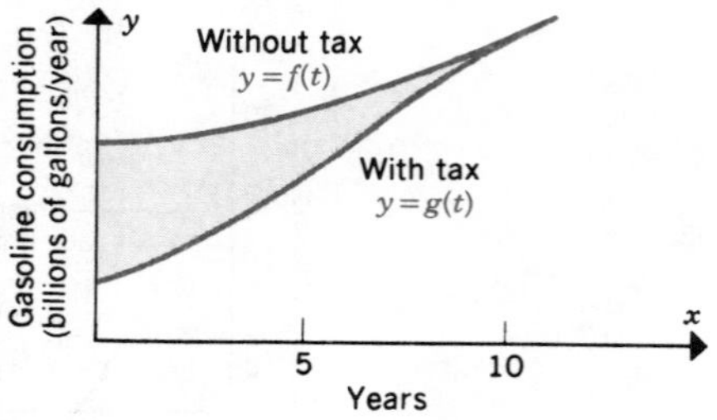

Figure 1

If $f(x)$ and $g(x)$ are continuous and if $f(x) \geq g(x)$ over the interval $a \leq x \leq b$, the **area of the region bounded by the curves $y = f(x)$ and $y = g(x)$,** and the vertical lines $x = a$ and $x = b$, shown in Figure 2, is given by the equation

$$\text{Area} = \int_a^b [f(x) - g(x)]\, dx \tag{1}$$

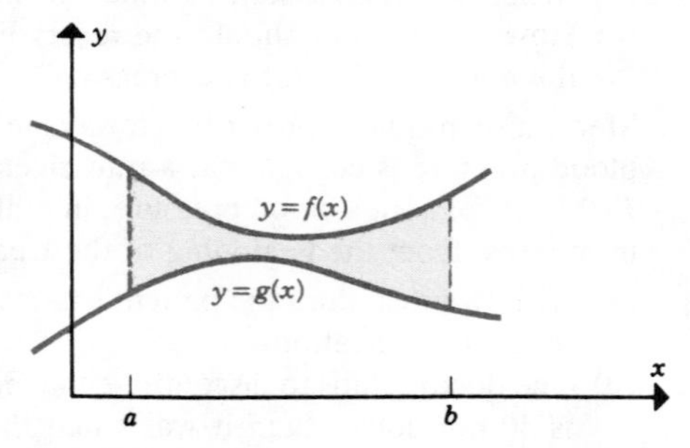

Figure 2

Equation 1 can be derived by examining Figure 3, in which the area beneath each curve and the x axis is shown. The area between the two curves then becomes

$$A_I - A_{II} = \int_a^b f(x)\,dx - \int_a^b g(x)\,dx$$

$$= \int_a^b [f(x) - g(x)]\,dx$$

When the reverse is true, that is, $g(x) \geq f(x)$ over the interval $a \leq x \leq b$,

$$\text{Area} = \int_a^b [g(x) - f(x)]\,dx \tag{2}$$

Equations 1 and 2 can be summarized as

$$\text{Area between curves} = \int_a^b (\text{upper} - \text{lower})\,dx$$

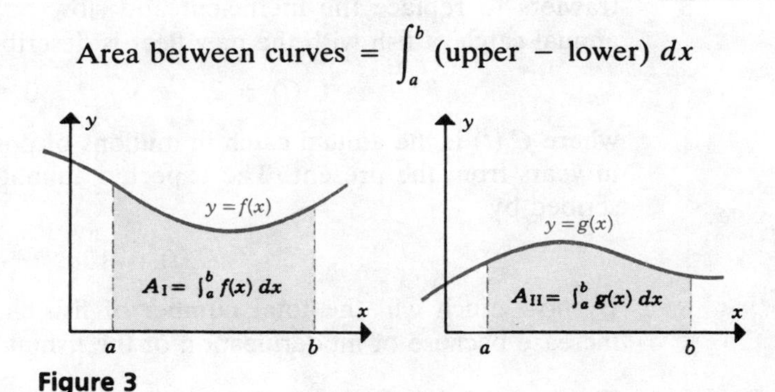

Figure 3

Example 1 Find the area of the region bounded by the two curves

$$f(x) = 8 - x^2 \quad \text{and} \quad g(x) = x + 1$$

and by the vertical lines $x = -1$ and $x = 2$.

Solution It is advisable to make a rough sketch of the two curves to determine the relative positions of the curves $y = f(x)$ and $y = g(x)$. Figure 4 shows the two curves where it can be seen that $f(x) > g(x)$ over the interval $-1 \leq x \leq 2$. The area of the shaded region can be determined as follows:

$$\text{Area} = \int_{-1}^{2} [(8 - x^2) - (x + 1)]\,dx = \int_{-1}^{2} (7 - x^2 - x)\,dx$$

$$= \left(7x - \frac{x^3}{3} - \frac{x^2}{2}\right)\Bigg|_{-1}^{2} = \left(14 - \frac{8}{3} - 2\right) - \left(-7 + \frac{1}{3} - \frac{1}{2}\right) = \frac{33}{2}$$

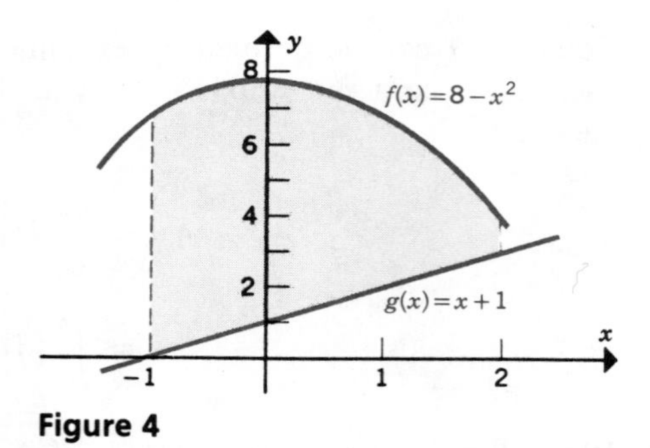

Figure 4

When the dependent variable represents a rate of change, the area between the curves can be used to describe the difference between two alternative courses of action, as illustrated in the next example.

Example 2 The Acme Fish Company is about to take delivery on a fleet of modern fishing trawlers to replace the inefficient and slow boats now in use. The expected annual catch of fish with the new fleet is described by the equation

$$C_1(t) = 25 - \sqrt{t}, \qquad 0 \leq t \leq 16$$

where $C_1(t)$ is the annual catch in millions of pounds per year and t is the time in years from the present. The expected annual catch with the old fleet is described by

$$C_2(t) = 10e^{-0.12t}$$

By how much will the total number of fish caught over the next nine years increase because of modernization of the fishing fleet?

Solution The area of the shaded region in Figure 5 represents the expected additional catch over the next nine years. The additional catch can be written as the definite integral

$$\int_0^9 (25 - \sqrt{t} - 10e^{-0.12t})\, dt$$

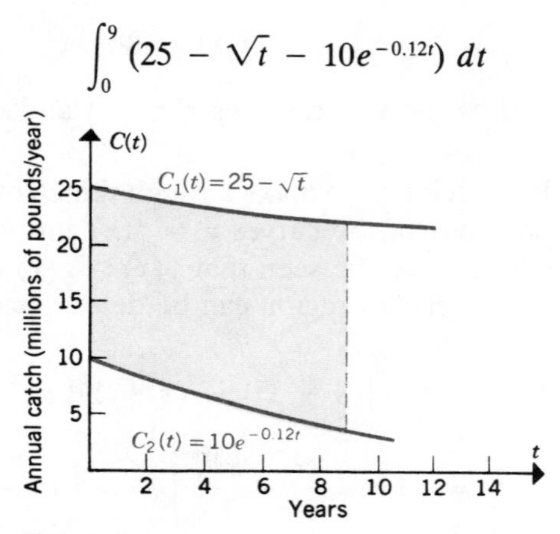

Figure 5

Working through the details gives

$$\text{Additional catch} = \left(25t - \frac{2\sqrt{t^3}}{3} + 83.3e^{-0.12t}\right)\Bigg|_0^9$$

$$= 152 \text{ million pounds}$$ ■

When we find the area of the region between two intersecting curves, the values of a and b represent the x coordinates of the points of intersection and must often be determined before the area can be calculated, as the next example illustrates.

Example 3 Find the area between the two curves

$$f(x) = 4 - x^2 \quad \text{and} \quad g(x) = -x + 2$$

Solution The region whose area we wish to find is shown in Figure 6. The area in question is given by the definite integral

$$\int_a^b [(4 - x^2) - (2 - x)]\, dx$$

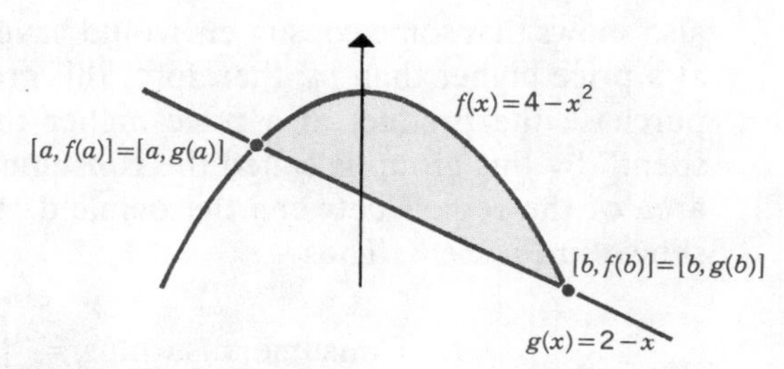

Figure 6

where a and b represent the x coordinates of the points where the two curves intersect. Since the two functions are equal at the points of intersection, we can set $f(x) = g(x)$ and solve the resulting equation for x to yield the x coordinates of the points of intersection

$$4 - x^2 = -x + 2$$

$$x^2 - x - 2 = 0$$

$$(x - 2)(x + 1) = 0$$

from which we get the solutions

$$x_1 = -1, \qquad x_2 = 2$$

so that the area can be written as

$$\int_{-1}^{2} (2 + x - x^2)\, dx$$

which gives

$$\left(2x + \frac{x^2}{2} - \frac{x^3}{3}\right)\Bigg|_{-1}^{2} = \left(4 + 2 - \frac{8}{3}\right) - \left(-2 + \frac{1}{2} + \frac{1}{3}\right) = \frac{9}{2}$$ ■

Applications

Consumers' Surplus

A demand equation describes the relationship between p, the unit price of a product, and q, the quantity or number of items that can be sold (Section 1.3). A typical demand curve is shown in Figure 7. Suppose the unit price of a product is p_0 and the corresponding quantity is q_0. The area of the shaded rectangle in Figure 7 equals p_0q_0, the revenue generated from the sale of q_0 units. The curve

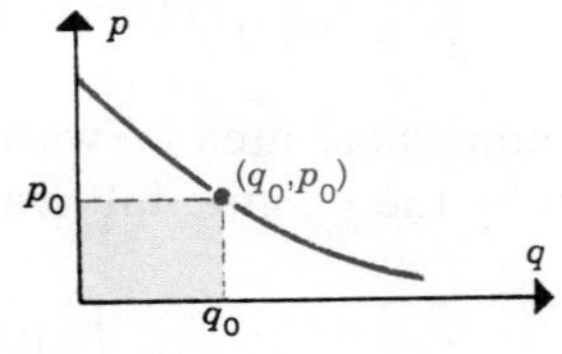

Figure 7

also shows that some consumers would have been willing to purchase the product at a price higher than p_0; therefore, this group "saves" money by not having to purchase the product at a price higher than p_0. The amount of money "not spent" by this group is called the **consumers' surplus** and is represented by the area of the region between the demand curve and the horizontal line $p = p_0$ shown in Figure 8. Thus

$$\text{Consumers' surplus} = \int_0^{q_0} [p(q) - p_0]\, dq$$

$$\text{Consumers' surplus} = \int_0^{q_0} p(q)\, dq - p_0q_0 \qquad (3)$$

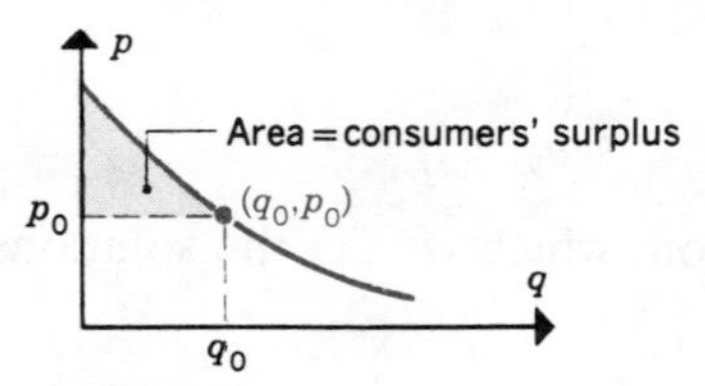

Figure 8

Example 4 The Gremlin Company has developed a new watch that not only gives the time and date but also forecasts the weather and measures the wearer's pulse rate.

The relationship between $p(q)$, the unit price in hundreds of dollars, and q, the quantity, in thousands of units, is given by the equation $p(q) = 0.5q + 10$. What is the consumers' surplus if the company sells the watch for four hundred dollars?

Solution The demand curve is shown in Figure 9; when $p(q) = 4$, the corresponding value of q is 12. The area of the shaded region, which represents the consumers' surplus, is

$$\int_0^{12} (10 - 0.5q)\,dq - (12)(4)$$

Evaluating this yields

$$\begin{aligned} \text{Consumers' surplus} &= \left(10q - \frac{0.5q^2}{2}\right)\Bigg|_0^{12} - 48 \\ &= 120 - 36 - 48 = \$36 \text{ hundred thousand} \\ &= \$3.6 \text{ million} \end{aligned}$$

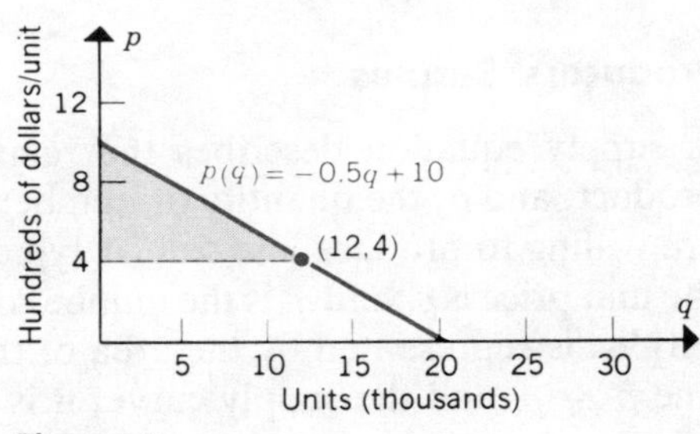

Figure 9

■

Finding the consumers' surplus may require the use of differential calculus, as shown in the next example.

Example 5 Suppose the Gremlin Company (Example 4) prices the watch to yield maximum revenue. What is the consumers' surplus under those conditions?

Solution The first matter that must be dealt with is determining the quantity and hence the price at which the company's revenue will be maximized. The company's revenue $R(q)$ is given by the equation

$$\begin{aligned} R(q) &= p(q)q \\ &= (10 - 0.5q)q = 10q - 0.5q^2 \end{aligned}$$

Using the optimization techniques described in Chapter 4, we first differentiate, getting

$$R'(q) = 10 - q$$

Next, setting $R'(q)$ equal to zero and solving for q gives

$$10 - q = 0$$

The solution is

$$q = 10, \qquad p(10) = 5$$

Because $R''(q) = -1$ for all values of q, the revenue is maximized when $q = 10$. The consumers' surplus can now be found by evaluating

$$\int_0^{10} (10 - 0.5q)\, dq - (10)(5)$$

Working through the details gives us

$$\begin{aligned} \text{Consumers' surplus} &= \left(10q - \frac{0.5q^2}{2}\right)\Bigg|_0^{10} - 50 \\ &= \$25 \text{ hundred thousand} \\ &= \$2.5 \text{ million} \end{aligned}$$

■

Producers' Surplus

A supply equation describes the relationship between p, the unit price of a product, and q, the quantity or number of items that manufacturers or suppliers are willing to produce and sell. A typical supply curve is shown in Figure 10. If the unit price is p_0 and q_0 is the number of items produced and sold, the **producers' surplus** is represented by the area of the shaded region between the horizontal line $p = p_0$ and the supply curve; it is defined as

$$\text{Producers' surplus} = \int_0^{q_0} [p_0 - p(q)]\, dq$$

$$\boxed{\text{Producers' surplus} = p_0 q_0 - \int_0^{q_0} p(q)\, dq}$$ (

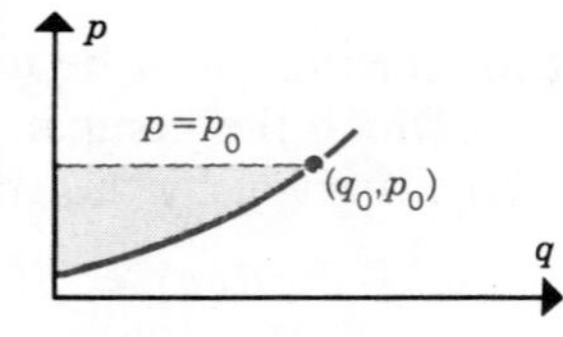

Figure 10

The supply curve in Figure 10 indicates that some sellers or producers would be willing to provide the item at prices lower than p_0. The additional revenue that this group realizes when the price equals p_0 is the producers' surplus and is represented by the area of the shaded region between the horizontal line $p = p_0$ and the supply curve.

Example 6 The supply equation for instant coffee is

$$p(q) = e^{0.04q}$$

where $p(q)$ is the price (dollars per pound) and q is the amount sold (millions of pounds). What is the producers' surplus when 20 million pounds are sold?

Solution When $q_0 = 20$, $p_0 = e^{0.04(20)} = \$2.23/\text{lb}$. Equation 4 is used to find the producers' surplus, giving

$$\begin{aligned}\text{Producers' surplus} &= (20)(2.23) - \int_0^{20} e^{0.04q}\,dq \\ &= 44.60 - \left.\frac{e^{0.04q}}{0.04}\right|_0^{20} \\ &= 44.60 - \frac{e^{0.8} - 1}{0.04} \\ &= 44.60 - 30.64 = \$13.96 \text{ million}\end{aligned}$$

■

6.4 EXERCISES

Find the area of each of the following shaded regions.

1.

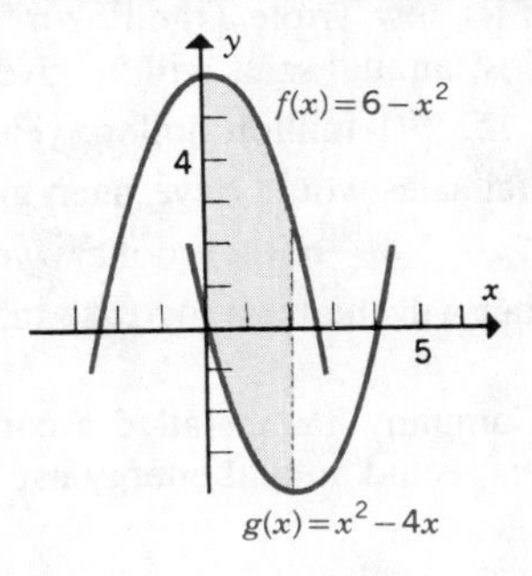

2.

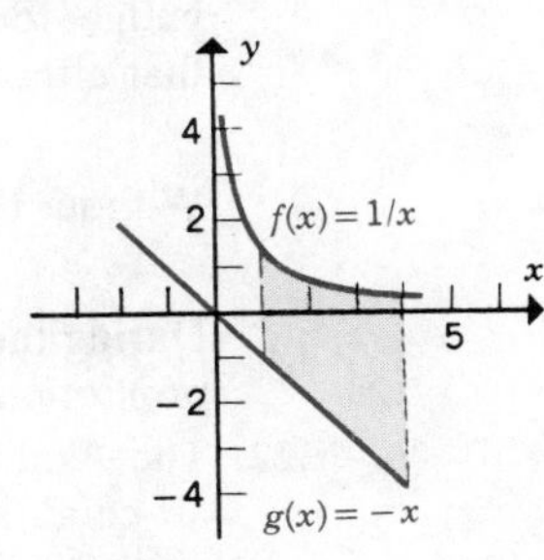

3.

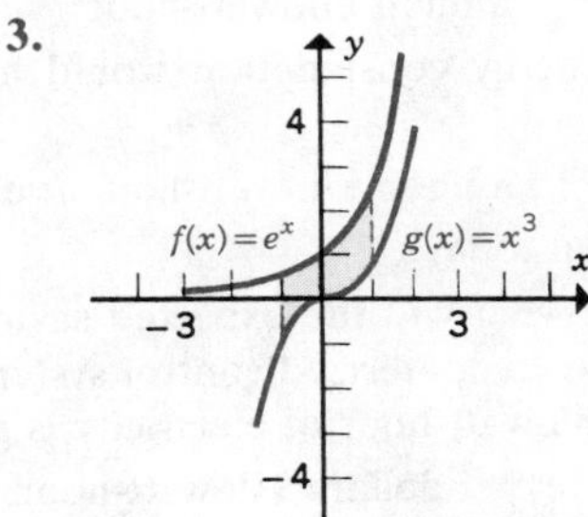

4.

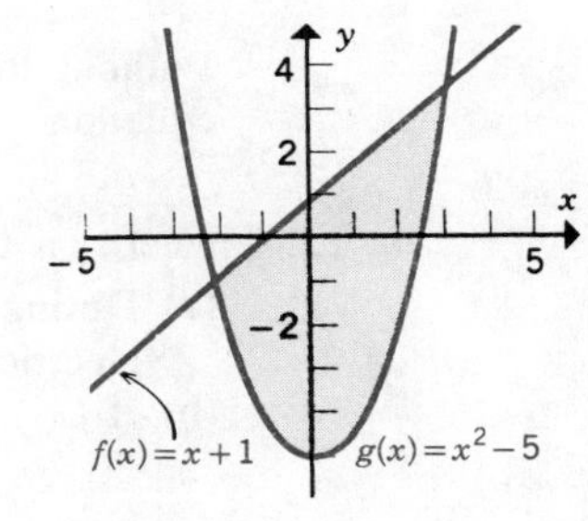

In Exercises 5 through 12, find the area between the curves.

5. $f(x) = 5 - x$, $g(x) = x$; $a = -1, b = 2$
6. $f(x) = \dfrac{8}{x^2}$, $g(x) = \dfrac{x^2}{4}$; $a = 1, b = 2$
7. $f(x) = e^x$, $g(x) = x^4 - 1$; $a = 0, b = 1$
8. $f(x) = 6x - 3x^2$, $g(x) = -x$; $a = \frac{1}{2}, b = 2$
9. $f(x) = e^x$, $g(x) = \dfrac{1}{x}$; $a = 1, b = 2$
10. $f(x) = -1$, $g(x) = x^2 - 8$; $a = -1, b = 2$
11. $f(x) = x^2$, $g(x) = -x^2$; $a = -2, b = 2$
12. $f(x) = e^{-x}$, $g(x) = \dfrac{1}{x}$; $a = -2, b = -1$

In Exercises 13 through 20, find the points of intersection and determine the area of the region between the intersecting curves.

13. $f(x) = 2x + 3$, $g(x) = x^2$
14. $f(x) = x^2$, $g(x) = x^3$
15. $f(x) = 1 - x^2$, $g(x) = x^2 - 1$
16. $f(x) = \sqrt{x}$, $g(x) = x^2$
17. $f(x) = 5 + 6x - x^2$, $g(x) = x^2 + 5$
18. $f(x) = -x + 4$, $g(x) = 3/x$
19. $f(x) = 17$, $g(x) = x^4 + 1$
20. $f(x) = 2$, $g(x) = e^{|x|}$
21. The Clean Cut Razor Blade Company plans to allocate a large part of its advertising budget to promote the sale of its new Triple Trac Razor. A market analysis predicts that after the promotion begins, annual sales will be given by the equation

$$S_1(t) = 15e^{0.05t} \quad \text{million dollars/year}$$

Without the promotion, annual sales would have been given by the equation

$$S_2(t) = 12 + 0.6t \quad \text{million dollars/year}$$

During the first two years, what will the total increase in sales be in response to the promotion?

22. The Arnold Life Insurance Company has installed a computerized heating and an air-conditioning system. The expected annual energy use during the next 15 years is given by the equation

$$E_1(t) = 30e^{0.02t} \quad \text{million kilowatt-hours/year}$$

Without the system, annual energy consumption would have been given by the equation

$$E_2(t) = 40e^{0.03t} \quad \text{million kilowatt-hours/year}$$

where t is the time in years from today.

(a) During the next ten years, what will the expected savings in kilowatt-hours of electricity be because of the computerized control system?

(b) If $C(t)$, the unit cost of a kilowatt-hour of electricity is given by the equation

$$C(t) = 0.05e^{0.06t} \quad \text{dollars/kilowatt-hour}$$

find the total savings (in dollars) over the next ten years.

23. The minimum hourly wage, $W_1(t)$, can be approximated by the equation

$$W_1(t) = 0.25 + 0.06t \quad \text{dollars/hour}$$

where t is the time in years from 1938 when the minimum-wage law went into effect. The average manufacturing hourly wage, $W_2(t)$, can be approximated by the equation

$$W_2(t) = 0.50 + 0.125t \quad \text{dollars/hour}$$

During the period 1938 through 1978, how much more money has a worker in manufacturing earned than a worker who has received only the minimum hourly rate? Assume that each worked 40 hours per week and 52 weeks each year.

24. Acme Press has two printing presses, one a small, older unit for which the marginal cost is

$$C_1'(x) = 0.50 + 0.01x \quad \text{dollars/unit}$$

and a new, large, high-speed unit for which the marginal cost is

$$C_2'(x) = 0.30 + 0.002x \quad \text{dollars/unit}$$

Setup costs for the larger press are $200 higher than those for the smaller press. Assuming that cost is the only factor in our considerations, for what size jobs should the small press be used? The large press?

Find the consumers' surplus for each of the following problems.

25. Demand equation is $p(q) = 6 - 0.02q$; $q_0 = 10$

26. Demand equation is $p(q) = 10e^{-0.5q}$; $q_0 = 4$

27. Demand equation is $p(q) = 20 - \sqrt{q}$; $q_0 = 100$

28. Demand equation is $p(q) = 12 - q$; $q_0 = 8$

Find the producers' surplus for each of the following problems.

29. Supply equation is $p(q) = 10 + 0.4q$; $q_0 = 20$

30. Supply equation is $p(q) = 30 + 2q$; $q_0 = 40$

31. Supply equation is $p(q) = 6e^{0.05q}$; $q_0 = 4$

32. Supply equation is $p(q) = 8 + \sqrt{q}$; $q_0 = 25$

Find the consumers' surplus and the producers' surplus when supply equals demand for each of the following situations.

33. Demand equation is $p(q) = 10 - 0.5q$; supply equation is $p(q) = 4 + q$

34. Demand equation is $p(q) = 24 - 0.6q$; supply equation is $p(q) = 8 + 0.4q$

35. Demand equation is $p(q) = 20e^{-q}$; supply equation is $p(q) = 5e^{q}$

36. Demand equation is $p(q) = 30 - 3\sqrt{q}$; supply equation is $p(q) = 12 + q$

37. A new jogging shoe developed by a sporting goods company has received favorable comments from reviewers in national running magazines. The demand equation is

$$p(q) = 75 - 10q \quad \text{dollars/pair}$$

where q represents the number sold (in millions).

(a) Find the consumers' surplus if the selling price is $40 a pair.

(b) Find the consumers' surplus if the company sets the selling price to maximize its revenue.

38. If the demand equation for a product has the form $p(q) = a + bq$, and the company sets the selling price to maximize its revenue, show that the consumers' surplus equals $-a^2/(8b)$.

39. The rate at which a wound heals with one type of treatment is given by the equation

$$A_1'(t) = -\frac{3\sqrt{t}}{2}$$

where $A_1(t)$ is the area, in square millimeters, of the wound and t is the time, in days, from the injury ($t = 0$). With a different treatment, an identical wound heals at a rate given by the equation $A_2'(t) = -2t$. Find the difference in the amount of healing, as measured by the difference in the areas, between the two treatments from $t = 1$ to $t = 9$ days.

40. The rate at which the population of an underdeveloped country is growing is given by the equation $P_1'(t) = 2e^{0.06t}$, where $P_1(t)$ is the population, in millions, and t is the time, in years, from today. If the survival rate at birth for this country equaled that of the developed countries, the population would grow at a rate given by the equation $P_2'(t) = 2e^{0.10t}$. Find the decrease in the country's population over the next five years because the survival rate at birth does not equal that of the developed nations.

6.5 ADDITIONAL APPLICATIONS OF THE DEFINITE INTEGRAL

There are many other applications of the definite integral in addition to those described in Sections 6.1 through 6.4. Some of the applications described in this section will require the use of the following formula:

$$\int te^{at}\,dt = \frac{e^{at}}{a^2}(at - 1) \tag{1}$$

The techniques for deriving this formula will be presented in Chapter 7. However, you can verify its correctness by differentiating the right-hand side to obtain the expression te^{at}.

The definite integral is very useful in approximating expressions containing many terms such as those encountered when calculating the amount or present value of an annuity.

Approximating the Amount of an Annuity

When interest is compounded continuously, a deposit or principal of P dollars grows to an amount $A(t)$ according to Equation 9 in Section 5.1,

$$A(t) = Pe^{rt}$$

where r represents the quoted annual interest rate and t is the time in years. It was assumed that in obtaining this result no withdrawals or additional deposits were made. Because this assumption is highly restrictive, more general situations will now be considered.

An **annuity** is defined as a set of **equal payments** made at **equal time intervals.** For example, if money is borrowed to finance a home, an auto, or an education, the repayment process generally takes the form of equal monthly payments for a specified length of time. This set of equal monthly payments forms an annuity. Social Security payments and pensions are other examples of annuities.

To keep matters simple, suppose that \$100 is deposited in a savings account at the beginning of each year for six years; the annual rate of interest is r. The initial deposit ($t = 0$) is in the account for six years and grows to $100e^{6r}$. The second deposit ($t = 1$) is in the account for five years $(6 - 1)$ and grows to $100e^{(6-1)r} = 100e^{5r}$. Proceeding in this way, we obtain for S the total amount in the account at the end of six years

$$S = \underset{\substack{\uparrow \\ t=0}}{100e^{(6-0)r}} + \underset{\substack{\uparrow \\ t=1}}{100e^{(6-1)r}} + \cdots + \underset{\substack{\uparrow \\ t=k}}{100e^{(6-k)r}} + \cdots + \underset{\substack{\uparrow \\ \text{last deposit}}}{100e^{(6-5)r}}$$

$$= 100e^{6r} + 100e^{5r} + \cdots + 100e^{(6-k)r} + \cdots + 100e^{r} \tag{2}$$

Each term is shown in the graph of Figure 1.

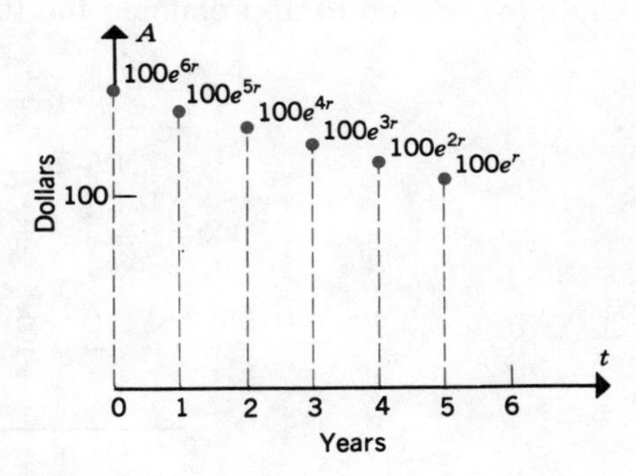

Figure 1

The total amount in the account can also be regarded as the sum of the areas of the six rectangles shown in Figure 2 because the base of each rectangle equals 1. Superimposed on the rectangles is the graph of the equation

$$A(t) = 100e^{(6-t)r}$$

The area under the curve between $t = 0$ and $t = 6$ is almost equal to the area of the six rectangles. Therefore, the area under the curve can be used to approximate $S(6)$, the total amount or sum in the account at the end of six years, that is,

$$S(6) \approx \int_0^6 100e^{(6-t)r}\,dt = \frac{-100}{r}\,e^{(6-t)r}\Big|_0^6$$

$$= \frac{100}{r}\,(e^{6r} - 1) \tag{3}$$

For example, suppose the annual rate of interest is 5 percent; then

$$\int_0^6 100e^{0.05(6-t)}\,dt = \frac{100}{0.05}\,(e^{0.30} - 1) = 2000(1.3499 - 1) = \$699.80$$

whereas the value given by Equation 2 is

$$100e^{0.30} + 100e^{0.25} + \cdots + 100e^{0.05} = \$717.36$$

This procedure can be generalized to yield an approximation for S, the **future value of an annuity,**

$$S \approx \int_0^T Re^{r(T-t)}\,dt = \frac{R}{r}(e^{rT} - 1) \tag{4}$$

where R is the size of each payment, r is the interest rate, and T is the duration of the annuity.

It is important to keep the units of time associated with R, r, and T consistent when (4) is used to approximate the future value of an annuity.

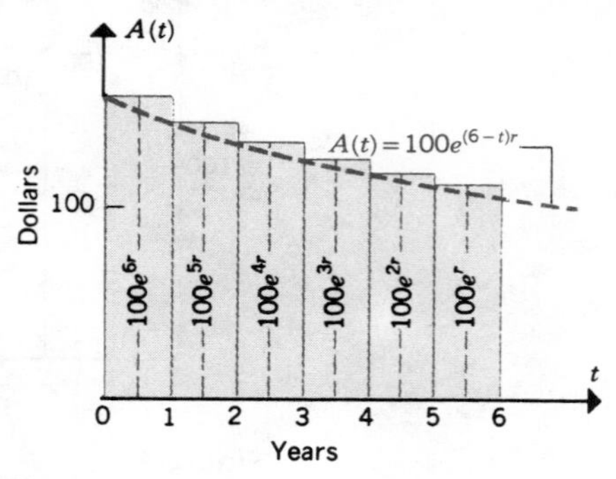

Figure 2

Example 1 A person deposits \$50 per month in a savings account paying 6 percent per year compounded continuously. How much is in the account at the end of ten years?

Solution If we set R in (4) equal to \$50 per month, then r and T must be expressed in terms of months as well, that is

$$r = \frac{.06}{12} = .005 \text{ per month}, \qquad T = 10(12) = 120 \text{ months}$$

The approximate value of S becomes

$$S \approx \int_0^{120} 50e^{0.005(120-t)}\,dt$$

$$= \frac{50}{0.005}(e^{0.60-1})$$

$$= 10{,}000(1.8221 - 1) = \$8221$$

■

Present Value of an Annuity

The present value of a single amount A due t years in the future is given by Equation 10 in Section 5.1,

$$P = Ae^{-rt}$$

when interest is compounded continuously. Suppose we want to determine the present value of six equal annual payments of \$100, the first to begin one year from now; again r is the annual interest rate. The sum of the present values, that is, the present value of the annuity, equals

$$\underset{\text{1st payment}}{\underset{\uparrow}{100e^{-r}}} + \underset{\text{2nd payment}}{\underset{\uparrow}{100e^{-2r}}} + \cdots + \underset{\text{6th and last payment}}{\underset{\uparrow}{100e^{-6r}}}$$

This sum can be represented geometrically as the sum of the areas of the six rectangles shown in Figure 3; also shown is the graph of the equation

$$P = 100e^{-rt}$$

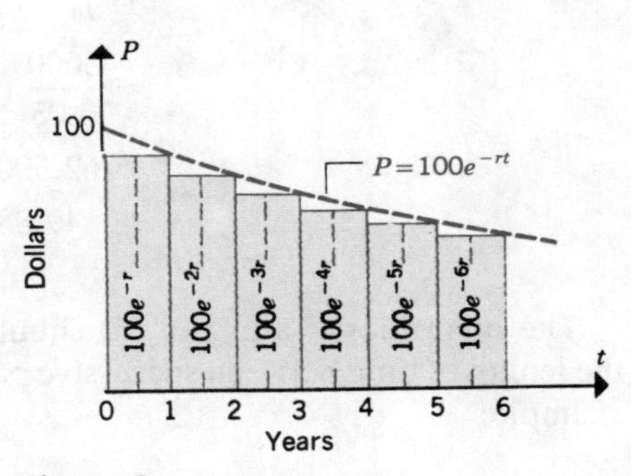

Figure 3

The area under the curve between $t = 0$ and $t = 6$ is approximately equal to the sum of the areas of the six rectangles; therefore, it can be used to approximate the present value of the six equal payments:

$$\text{Present value} \approx \int_0^6 100e^{-rt}\,dt = \frac{100}{r}(1 - e^{-6r})$$

If $r = 0.05$, we get

$$\text{Present value} \approx \frac{100}{0.05}(1 - e^{0.30}) = \$518.36$$

whereas the sum of the present values of the six equal payments is

$$100e^{-0.05} + 100e^{-0.10} + \cdots + 100e^{-0.30} = \$505.51$$

This procedure can be generalized to yield an approximation for P, the **present value of the annuity,**

$$P \approx \int_0^T Re^{-rt}\,dt = \frac{R}{r}(1 - e^{-rT}) \qquad (5)$$

where R represents the amount of each payment, T is the duration of the annuity, and r is the interest rate.

Example 2 An executive wants to establish a fund from which she will withdraw \$5000 annually for the next 20 years. If the annual interest rate is 8 percent, how much money should she place in the fund?

Solution We use (5) to find an approximate value for the present value of a series of 20 equal payments, each equal to \$5000. We get

$$\begin{aligned} P &\approx \int_0^{20} 5000e^{-0.08t}\,dt \\ &\approx \frac{5000}{0.08}(1 - e^{-1.6}) \\ &\approx 62{,}500(1 - 0.2019) \\ &\approx \$49{,}881.25 \end{aligned}$$

■

The quantities R and r in (5) should be expressed in units compatible with the length of time between successive payments, as demonstrated in the following example.

Example 3 A state lottery commission pays \$1000 per week for 20 years to each of its grand prize winners. How much money should the state deposit with an insurance company that pays $7\frac{1}{2}$ percent annually to meet the weekly payments for 20 years?

Solution The equally spaced payments constitute an annuity for which $R = 1000$, $T = 52(20) = 1040$, and $r = 0.075/52$ (we leave it in this form). Using (5), we get

$$\begin{aligned} P &\approx \int_0^{1040} 1000e^{-(0.075t/52)}\,dt \\ &\approx \frac{(1000)(52)}{0.075}(1 - e^{-[(0.075)(1040)]/52}) \\ &\approx 693{,}333(1 - e^{-1.5}) = \$538{,}630 \end{aligned}$$

We get the same result if we let $R = 1000(52)$, $r = .075$ and $T = 20$. ■

Discounted Cash Flows

In most situations requiring financial analysis, the cash flows are not uniform as they are for annuities. If, in addition, the cash flows are continuous, the definite integral can be employed to obtain the present value of the set of uneven cash flows by replacing the constant factor R in (5) with $R(t)$, which expresses the cash flows as a function of time. The equation for the present value P becomes

$$P = \int_0^T R(t)e^{-rt}\,dt \tag{6}$$

Example 4 Monthly production at a Yukon oil well is expected to follow the equation

$$B(t) = 3000, \qquad 0 \le t \le 360$$

where t is the time from the present expressed in months and $B(t)$ is the monthly output in barrels per month. The price, in dollars, of a barrel of crude oil is expected to increase according to the equation

$$p(t) = 13 + 0.10t, \qquad 0 \le t \le 360$$

Find the present value of the revenue generated during the next 30 years from the sale of oil pumped from the well if the owner expects a 15 percent annual return on his investment.

Solution The revenue generated each month from the sale of the crude oil is

$$R(t) = B(t)p(t) = 39{,}000 + 300t$$

Letting $r = 0.15/12 = 0.0125$, we find the present value of the monthly revenues from Equation 6:

$$\text{Present value} = \int_0^{360} (39{,}000 + 300t)e^{-0.0125t}\,dt$$

$$= \int_0^{360} 39{,}000e^{-0.0125}\,dt + \int_0^{360} 300te^{-0.0125t}\,dt$$

The evaluation can be carried out term by term, noting that Equation 1 is needed to evaluate the second term. We then get

$$\int_0^{360} 39{,}000e^{-0.0125t}\,dt = \frac{39{,}000}{0.0125}(1 - e^{-(360)(0.0125)})$$

$$= 3{,}120{,}000(1 - e^{-4.5}) = 3{,}085{,}368$$

$$\int_0^{360} 300te^{-0.0125t}\,dt = 300\left[\frac{e^{-0.0125t}}{(0.0125)^2}(-0.0125t - 1)\right]\Bigg|_0^{360}$$

$$= 300(-391.04 + 6400)$$

$$= 1{,}802{,}689$$

The present value then becomes

$$\text{Present value} = 3{,}085{,}368 + 1{,}802{,}689 = \$4{,}888{,}057$$

■

Example 5 The XYZ Company uses the continuous declining balance method of depreciation (Section 5.1) for its printing press. If the press depreciates at a rate given by the equation

$$R(t) = 10{,}000e^{-0.20t} \quad \text{dollars/year}$$

find the present value of the depreciation from $t = 0$ to $t = 5$. Assume that XYZ expects a 10 percent annual return on its investments.

Solution Equation 6 is used to find the present value of the total depreciation from $t = 0$ to $t = 5$:

$$\begin{aligned}\text{Present value} &= \int_0^5 (10{,}000e^{-0.20t})e^{-0.10t}\,dt \\ &= 10{,}000 \int_0^5 e^{-0.30t}\,dt \\ &= 10{,}000 \left(\frac{e^{-0.30t}}{-0.30}\right)\Bigg|_0^5 \\ &= 10{,}000 \left(\frac{1 - e^{-1.50}}{0.30}\right) = \$25{,}896\end{aligned}$$

■

Average Value of a Function

The definite integral can also be used to calculate the average value of a continuous function $y = f(x)$ over a given interval $a \leq x \leq b$. First, let's look at the six temperature readings spaced one hour apart as shown in Table 1. The average temperature $\overline{T}$ is found by adding the temperature values and dividing the sum by the number of observations, yielding

$$\overline{T} = \frac{10 + 12 + 14 + 17 + 14 + 11}{6} = 13°\text{C}$$

For a finite set of values $y_1, y_2, y_3, \ldots, y_n$, the average value $\overline{y}$, is defined as

$$\overline{y} = \frac{\sum_{k=1}^{n} y_k}{n} = \frac{y_1 + y_2 + y_3 + \cdots + y_n}{n}$$

The average temperature T can be viewed graphically in a way that leads to the definition of the average value of a continuous function. The temperature values in Table 1 are shown in Figure 4 together with $\overline{T}$. Figure 5 shows that the sum of the six temperatures equals the sum of the areas of the six rectangles, each of whose bases equals 1.

Table 1

Time	Temperature (°C)
1 P.M.	10°
2 P.M.	12°
3 P.M.	14°
4 P.M.	17°
5 P.M.	14°
6 P.M.	11°

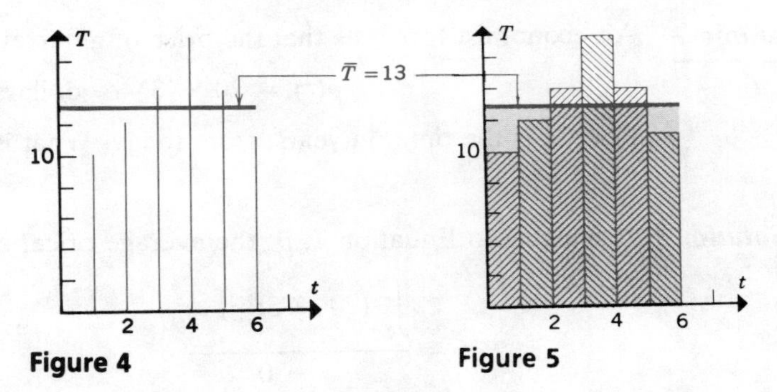

Figure 4 **Figure 5**

In addition, the area of the shaded rectangle under the average temperature curve $\overline{T} = 13$ equals the sum of the areas of the six rectangles. This concept is used in defining the average value of a continuous function.

Definition If the function $y = f(x)$ is continuous on the interval $a \le x \le b$, then $\overline{y}$, the average value, is defined as

$$\overline{y} = \frac{\int_a^b f(x)\,dx}{b - a} \tag{7}$$

Example 6 Find the average value of the function $f(x) = 3x^2$ on the interval $1 \le x \le 3$.

Solution The average value is found from Equation 7 to be

$$\overline{y} = \frac{\int_1^3 3x^2\,dx}{3 - 1} = \frac{x^3\Big|_1^3}{2} = \frac{27 - 1}{2} = 13$$

The area under the curve $f(x) = 3x^2$ in Figure 6 equals the area of the rectangle, whose height equals $\overline{y}$ and whose base equals 2. ■

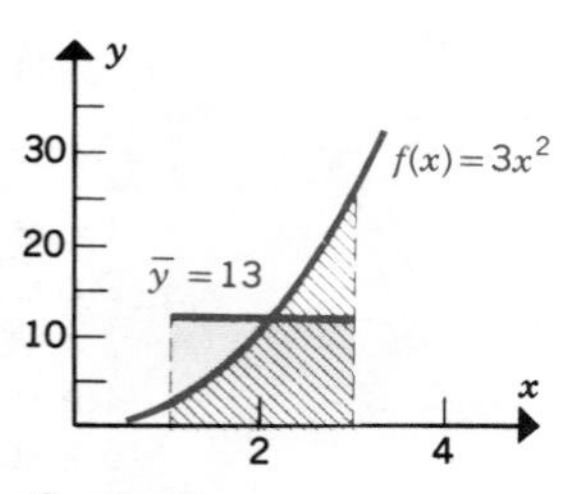

Figure 6

Example 7 An economist forecasts that the price of a barrel of oil will follow the equation

$$p(t) = 20 + 3\sqrt{t} \quad \text{dollars/barrel}$$

where t is the time, in years, from today. What is the average price from $t = 0$ to $t = 4$?

Solution According to Equation 7, $\overline{p}$, the average price, equals

$$\overline{p} = \frac{\int_0^4 (20 + 3t^{1/2})\,dt}{4 - 0}$$

$$= \frac{1}{4}(20t + 2t^{3/2})\Big|_0^4 = \frac{1}{4}[80 + 2(8)] = \$24/\text{barrel}. \quad \blacksquare$$

6.5 EXERCISES

1. A family deposits \$100 per month in a savings account paying 6 percent annually compounded continuously. Use (4) to approximate the total amount in the account at the end of 15 years.
2. If the family in Exercise 1 had invested its money in an account paying 8 percent per year, how much additional money would have been in the account at the end of 15 years?
3. If the family in Exercise 1 deposits \$100 per month for 20 years, approximately how much money will be in the account at the end of 20 years?
4. A worker and his employer each contribute \$10 per week to a pension plan for which the annual rate of return is 8 percent compounded continuously. If the worker retires after 25 years, approximately how much money will be in his account?
5. Another worker in the pension plan described in Exercise 4 wants to have \$30,000 at the end of 20 years. Determine approximately how much money he should contribute each week to the pension fund.
6. The Family Finance Company charges 12 percent annual interest compounded continuously on home improvement loans. A woman borrows \$3000. Determine her monthly payments approximately if the term of the loan is four years.
7. A retired college professor is notified that her pension account contains \$100,000. If the insurance company that supervises the pension fund pays 8 percent annually, what monthly payment should the professor receive over the next 20 years, assuming that there is no money left in her account at the end of 20 years.

8. Bionics will use the continuous double-declining balance method (Section 5.1) to depreciate a new electron microscope. The microscope depreciates at a rate given by the equation

$$R(t) = 40{,}000e^{-0.40t} \quad \text{dollars/year}$$

If Bionics expects a 12 percent annual return on its investments, find the present value of the depreciation occurring from $t = 0$ to $t = 5$.

9. Annual output at the Acme Copper Mine as a function of time is expected to follow the equation

$$A(t) = 10 + 0.5t \quad \text{thousand tons/year}$$

where t is the time in years. If the price of a ton of copper remains constant at $1500 and if the company seeks a 12 percent annual return on its investments, find the present value of the copper mined during the next four years.

10. Amalgamated Industries uses the sum-of-the-years-digits (SYD) method (Section 6.1) to depreciate its corporate jet. If annual depreciation is given by the equation

$$R(t) = 100(5 - t), \quad \text{thousand dollars/year}$$

find the present value of the depreciation from $t = 0$ to $t = 5$, assuming the company earns a 15 percent return on its investments.

Find the average value of each of the following functions over the given interval.

11. $f(x) = 5;\ 0 \le x \le 2$
12. $f(x) = 6 - 2x;\ -1 \le x \le 1$
13. $f(x) = 9 - 3x^2;\ 0 \le x \le 1$
14. $f(x) = \sqrt{x} + x;\ 0 \le x \le 9$
15. $f(x) = 4e^{-2x};\ -1 \le x \le 0$
16. $f(x) = \dfrac{x^3 + x^2}{x};\ 1 \le x \le 2$
17. $f(x) = \dfrac{1}{x};\ 1 \le x \le 3$
18. $f(x) = x^e;\ 0 \le x \le e$
19. $f(x) = nx^{n-1};\ 0 \le x \le 2$
20. $f(x) = 2x - e^x;\ 0 \le x \le 1$
21. An annual output at the Hi-Ho Silver Mines is given by the equation

$$A(t) = 20 - 0.2t \quad \text{thousand tons/year}$$

Find the average output from $t = 1$ to $t = 5$.

22. Annual profits at the Dyspeptic Cola Company are described by the equation

$$P(t) = 10e^{0.05t} \quad \text{million dollars/year}$$

Find the average profit from $t = 0$ to $t = 2$.

KEY TERMS

antiderivative
indefinite integral
integrand
sum-of-the-years-digits depreciation
area under a curve
definite integral
limits of integration
fundamental theorem of calculus
area between two curves
consumers' surplus
producers' surplus
amount of an annuity
present value of an annuity
average value of a function

REVIEW PROBLEMS

Find each of the following indefinite integrals.

1. $\int 8\,dx$

2. $\int 2x\,dx$

3. $\int \sqrt{5}\,dx$

4. $\int (4x + 3)\,dx$

5. $\int 6x^2 - 2x + 3)\,dx$

6. $\int 3\sqrt{x}\,dx$

7. $\int \left(8x^3 + \frac{1}{x^2}\right)dx$

8. $\int (e^x + e^{-x})\,dx$

9. $\int \left(4e^{2x} + \frac{2}{x}\right)dx$

10. $\int \left(6e^{-3x} + \frac{1}{2\sqrt{x}}\right)dx$

11. $\int (x^{-e} + e^{-x})\,dx$

12. $\int \frac{e^{2x} + e^x + 1}{e^x}\,dx$

Determine whether each of the following equations is true (T) or false (F).

13. $\int x(x - 1)^4\,dx = \frac{(x - 1)^6}{6} + \frac{(x - 1)^5}{5} + C$

14. $\int x\sqrt{x - 2}\,dx = \frac{2(x - 2)^{5/2}}{5} + \frac{2(x - 2)^{3/2}}{3} + C$

15. $\int 4xe^{2x}\,dx = 2xe^{2x} - e^{2x} + C$

16. $\int x\ln(x + 2)\,dx = \frac{(x + 2)^2\ln(x + 2)}{2} - \frac{(x + 2)^2}{4} - 2(x + 2)\ln(x + 2) + 2x + C$

In problems 17 through 20, find the area of the region bounded by the curve $y = f(x)$ the x axis, and the vertical lines $x = a$ and $x = b$.

17. $f(x) = 1 + e^{-x};\quad a = -1, b = 1$

18. $f(x) = 9 - x^2;\quad a = -2, b = 3$

19. $f(x) = x^2 - 4;\quad a = -1, b = 1$

20. $f(x) = 1 + \frac{2}{x};\quad a = 1, b = e$

Evaluate each of the following definite integrals.

21. $\int_1^5 3\,dx$

22. $\int_0^4 2x^3\,dx$

23. $\int_{-1}^1 (5 - 3x^2)\,dx$

24. $\int_0^1 6e^{-2x}\,dx$

25. $\int_1^e \left(2 - \frac{1}{x}\right) dx$

26. $\int_0^{\ln 2} \frac{e^{3x} + 2e^{2x} - 2}{e^x} dx$

27. $\int_0^4 (\sqrt{x} - 2)\, dx$

28. $\int_0^1 (x^{2e} + e^{2x})\, dx$

29. $\int_{-2}^{-1} \left(\frac{1}{x} - 2x\right) dx$

30. $\int_{-2}^{-1} \left(\frac{x^4 + x^3 - x^2}{x^2}\right) dx$

Find the derivative of each of the following functions.

31. $g(x) = \int (x^2 e^{-x} + \sqrt{xe^{-x}})\, dx$

32. $f(x) = \int_2^1 x \ln(x^2 + x - 1)\, dx$

33. $F(t) = \int_1^t (3x^2 - 2x + 4)\, dx$

34. $g(t) = \int_t^3 (4x^3 - 4x + 1)\, dx$

Find the area of the region between each pair of intersecting curves.

35. $f(x) = 7 - x^2,\ g(x) = 3x + 3$

36. $f(x) = \sqrt{x},\ g(x) = \frac{x}{2}$

37. $f(x) = 3 - 2x^2,\ g(x) = x^4$

38. $f(x) = 3 - x,\ g(x) = \frac{2}{x}$

39. Use the graphs shown in Figure 1 to rank the four integrals

$$\int_a^b f(x)\, dx, \qquad \int_a^b g(x)\, dx, \qquad \int_a^b h(x)\, dx, \qquad \int_a^b k(x)\, dx$$

from largest (1) to smallest (4). *Note:* The scale on the y axis is the same for all four graphs.

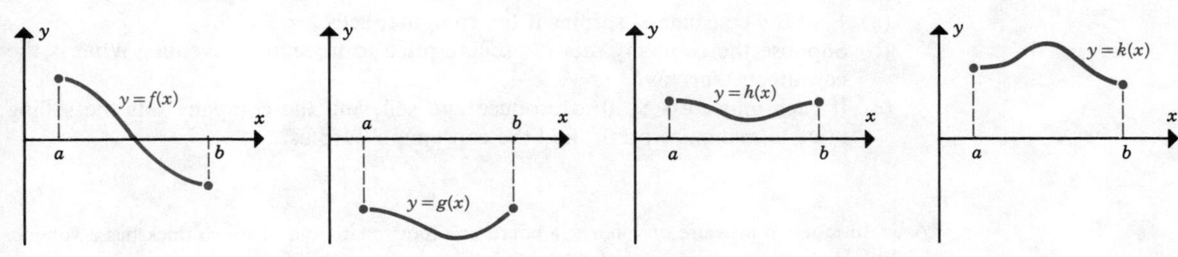

Figure 1

40. The population of a small town in the Sunbelt is growing at a rate described by the equation

$$R(t) = 200e^{0.04t} \text{ people per year}$$

where t is the time, in years, from today ($t = 0$).

(a) How much larger will the population of the town be three years from now?

(b) The town planner feels that the population can increase by 2000 people without adversely affecting the quality of life for the town's residents. How many more years before the population will have increased by this number?

41. At the beginning of 1980, surgeons at a major hospital were performing 5000 heart bypass operations per year. Since then, the number has been decreasing at a rate of 200 operations per year.

(a) Express the total number of heart bypass operations performed during the 1980s as a definite integral; evaluate this integral.

(b) If each operation produces $10,000 in revenue for the hospital, calculate the total revenue generated from heart bypass operations during the 1980s.

(c) Suppose the revenue from each operation is not constant, but is described by the equation $R(t) = 10{,}000 - 500t$, where t is the time, in years, measured from the beginning of 1980 ($t = 0$). Determine how much revenue was produced during the 1980s from heart bypass operations.

42. Acme Lumber Company is harvesting lumber from a tract of land in the Northwest at a rate given by the equation

$$R(t) = 2e^{-0.2t}$$

where $R(t)$ is the number of board-feet* (in millions) harvested per year and t is the time, in years, from today.

(a) How much lumber will be harvested during the next five years?

(b) An aerial survey indicates that the tract contains 8 million board-feet of lumber. How many years will it take to harvest this lumber?

43. The demand equation for a laptop computer has the form

$$p(q) = 3000 - \frac{q}{2}$$

where $p(q)$ is the unit price, in dollars, and q is the number of units sold each week.

(a) Find the consumers' surplus if the computer sells for $1600.

(b) Suppose the company sets the selling price to maximize revenue. What is the consumers' surplus?

(c) If each unit costs $800 to produce and sell, and the company sets the selling price to maximize profit, find the consumers' surplus.

*A board-foot is a measure of volume; a board one foot square and one inch thick has a volume equal to one board-foot, or, equivalently, 144 in^3.

44. The demand equation for a new Fax machine is $p(q) = 2200 - q$ and, the supply equation is $p(q) = 1000 + q/2$, where $p(q)$ is the unit price, in dollars, and q is the quantity (items) sold. Find the consumers' and producers' surpluses when supply and demand are equal.

45. The demand equation for an exercise machine can be written

$$p(q) = 2000e^{-0.04q}$$

where $p(q)$ is the unit price, in dollars, and q is the number of machines that can be sold.

(a) Find the consumers' surplus when 20 units are sold.
(b) Find the consumers' surplus if the company sets the selling price to produce maximum revenue.

46. During the next 25 days the blood supply at a local hospital is expected to change at a rate given by the equation $f(t) = 3\sqrt{t} - 12$, where t is the time, in days, from today ($t = 0$) and $f(t)$ is the rate in pints per day.

(a) What is the change in the blood supply over the next four days?
(b) When will the hospital's supply of blood reach its minimum level?
(c) By how much does the hospital's blood supply drop between today and the day when it reaches its minimum level?

47. The population $P(t)$ of a retirement community at any time t, in years, is given by the equation

$$P(t) = 800 + 600e^{-0.20t}$$

where $t = 0$ corresponds to today. Find the average population of the community over the next two years.

48. The value $V(t)$, in thousands of dollars, of an asset at any time t, in years, is described by the equation

$$V(t) = t^2 - 12t + 36$$

Find the average value of the asset from $t = 0$ to $t = 4$.

49. According to a model developed by a public health group, the number of people $N(t)$, in hundreds, who will be ill with the Manchurian flu at any time t, in months, next winter is described by the equation

$$N(t) = 50 + 30t - 10t^2, \qquad 0 \le t \le 4$$

where $t = 0$ corresponds to the beginning of December. Find the average number of people who will be ill with the flu during this four-month period.

50. The annual profit $P(t)$, in millions of dollars, of a bioengineering company at any time t, in years, is given by the equation

$$P(t) = 10e^{0.25t} - 30$$

where $t = 0$ corresponds to the start-up time of the company. Find the average annual profit over the first three years of operation.

7

TECHNIQUES OF INTEGRATION

INTRODUCTION

In Chapter 6, we learned that integration is the inverse of differentiation. However, as an operation, integration is generally more difficult to carry out than is differentiation. As a result, the techniques needed for integration are more numerous and complex than those for differentiation. We restrict our attention to two of the most widely used methods, substitution and integration by parts, in Sections 7.1 and 7.2, respectively. When none of the available techniques is effective, often tables of integrals can be used to find an antiderivative; their use is demonstrated in Section 7.3.

In many situations a definite integral cannot be evaluated because it may be very difficult or even impossible to express an antiderivative in terms of simple functions. Approximation methods such as Simpson's rule or the trapezoidal rule can be employed to evaluate a definite integral to any desired degree of accuracy; these approximation methods are studied in Section 7.4. Section 7.5 is devoted to examining definite integrals for which one or both limits of integration are infinite; such integrals belong to a class known as improper integrals. Evaluating improper integrals requires limits to be determined; Section 7.6 presents a method called L'Hôpital's rule for finding limits when the standard methods described in Chapter 2 are not effective.

7.1 INTEGRATION BY SUBSTITUTION

The integrals we encountered in Chapter 6 fall into one of the following forms:

$$\int x^n\,dx = \frac{x^{n+1}}{n+1} + C, \qquad (n \neq -1)$$

$$\int e^{ax}\,dx = \frac{e^{ax}}{a} + C \tag{1}$$

$$\int \frac{1}{x}\,dx = \ln|x| + C$$

If a given integral is not one of these simple types, other methods must be used to find an antiderivative.

Suppose that we have been given an integral

$$\int f(x)\,dx$$

but are unable to find an antiderivative of $f(x)$. In many instances the **method of substitution** can be applied to find an antiderivative. The method consists of defining a new variable u as a function of x, that is, $u = u(x)$, and then writing the given integral in terms of u and du, yielding

$$\int f(x)\,dx = \int g(u)\,du$$

If we make a wise choice in defining u, we may be able to find an antiderivative of $g(u)$.

To see how this method works, suppose we want to find

$$\int 2x(x^2 + 1)^6 \, dx$$

In its present form, the integral does not fall into one of the simple forms shown in Equation 1. Noting that the factor $2x$ is the derivative of the base $x^2 + 1$, let us define u as

$$u = x^2 + 1$$

Since the differential $du = 2x \, dx$, the original integral can be written in terms of u and du as

$$\int 2x(x^2 + 1)^6 \, dx = \int (x^2 + 1)^6(2x \, dx) = \int u^6 \, du$$

The integration with respect to u can be carried out to give

$$\int u^6 \, du = \frac{u^7}{7} + C$$

Rewriting the result in terms of x gives

$$\int 2x(x^2 + 1)^6 \, dx = \tfrac{1}{7}(x^2 + 1)^7 + C$$

We can check our answer by differentiating:

$$\frac{d}{dx}\left[\frac{(x^2 + 1)^7}{7} + C\right] = 2x(x^2 + 1)^6$$

Many of the integrals for which integration by substitution is an effective technique will assume one of the following forms when u is defined appropriately:

$$\boxed{\int u^n \, du = \frac{u^{n+1}}{n + 1} + C \qquad (n \neq -1)} \tag{2}$$

$$\boxed{\int e^u \, du = e^u + C} \tag{3}$$

$$\boxed{\int \frac{1}{u} \, dx = \ln|u| + C} \tag{4}$$

Integration by substitution can be an effective technique when the integrand contains two factors, one of which is a multiple of the derivative of another expression in the integrand. Examples 1 through 4 illustrate situations of this kind.

Example 1 Find $\int (2x + 3)(x^2 + 3x - 5)^4\, dx$.

Solution Noting that $2x + 3$ equals the derivative of the base $x^2 + 3x - 5$, we define u as

$$u = x^2 + 3x - 5$$

with du becoming

$$du = (2x + 3)\, dx$$

The integral in terms of u now becomes

$$\int u^4\, du$$

which can be integrated according to Equation 2 as

$$\int u^4\, du = \frac{u^5}{5} + C$$

Now expressing the relationship in terms of x, we get

$$\int (2x + 3)(x^2 + 3x - 5)\, dx = \frac{(x^2 + 3x - 5)^5}{5} + C$$

Again, the expression on the right-hand side can be differentiated with respect to x to show that it is an antiderivative of the integrand

$$(2x + 3)(x^2 + 3x - 5)^4.$$ ■

Example 2 Find $\int x^2(x^3 + 1)^5\, dx$.

Solution Noting that the factor x^2 equals $\frac{1}{3}$ times the derivative of $x^3 + 1$, we define u as

$$u = x^3 + 1$$

The differential du is

$$du = 3x^2\, dx$$

or

$$x^2\, dx = \frac{du}{3}$$

The integral can now be written in terms of u and du:

$$\int (x^3 + 1)^5(x^2\,dx) = \int u^5\left(\frac{du}{3}\right) = \frac{1}{3}\int u^5\,du = \frac{1}{18}u^6 + C$$

Therefore, we conclude that

$$\int x^2(x^3 + 1)^5\,dx = \frac{1}{18}(x^3 + 1)^6 + C$$

■

Example 3 Find $\int xe^{x^2}\,dx$.

Solution Observing that $x = \frac{1}{2}\frac{d}{dx}(x^2)$, we anticipate that the integral might assume a form like that given in Equation 3. We define u as

$$u = x^2$$
$$du = 2x\,dx$$

or

$$x\,dx = \frac{1}{2}\,du$$

The integral can now be written as

$$\int e^u\left(\frac{du}{2}\right) = \frac{1}{2}\int e^u\,du = \tfrac{1}{2}e^u + C$$

Substituting x^2 for u gives

$$\int xe^{x^2}\,dx = \tfrac{1}{2}e^{x^2} + C$$

■

Example 4 Find $\int \frac{6x^2 + 14}{x^3 + 7x - 1}\,dx$.

Solution Noting that the numerator, $6x^2 + 14$, is two times the derivative of the denominator, $x^3 + 7x - 1$, we define u as follows:

$$u = x^3 + 7x - 1$$
$$du = (3x^2 + 7)\,dx$$

or

$$(6x^2 + 14)\,dx = 2\,du$$

The integral can now be written as

$$\int \frac{6x^2 + 14}{x^3 + 7x - 1}\,dx = \int 2\,\frac{du}{u} = 2\int \frac{du}{u}$$
$$= 2\ln|u| + C$$
$$= 2\ln|x^3 + 7x - 1| + C$$ ■

The method of substitution, like many techniques of integration, is a trial-and-error process. Success in using this method to find an integral sometimes requires a little ingenuity, as illustrated in the next example.

Example 5 Find $\int x\sqrt{x + 1}\,dx$.

Solution If we let $u = x + 1$, we quickly find that $du = dx$, which tells us that du is not a multiple of $x\,dx$. At this point the integral can be written as

$$\int x\sqrt{u}\,du$$

Since our ultimate goal is to express the integral in terms of u and du, we want to express the factor x in terms of u, if possible. This can be done by noting that

$$x = u - 1$$

so that the integral can be written in terms of u and du as

$$\int (u - 1)\sqrt{u}\,du = \int (u^{3/2} - u^{1/2})\,du$$
$$= \int u^{3/2}\,du - \int u^{1/2}\,du$$
$$= \frac{2u^{5/2}}{5} - \frac{2u^{3/2}}{3} + C$$
$$= \frac{2(x + 1)^{5/2}}{5} - \frac{2(x + 1)^{3/2}}{3} + C$$

This result can be checked by differentiating:

$$\frac{d}{dx}\left[\frac{2(x + 1)^{5/2}}{5} - \frac{2(x + 1)^{3/2}}{3} + C\right] = x\sqrt{x + 1}$$

This problem also illustrates one of the features that causes integration to be more difficult than differentiation, namely the necessity to reverse, or "unscramble," the algebraic steps that were carried out in simplifying the expression obtained from the differentiation process. ■

When we use the method of substitution to evaluate a definite integral, the process can be carried out by employing either of the following procedures.

1. After defining u, carry out the integration with respect to u and then rewrite your answer in terms of x. Finally, evaluate the definite integral using the given limits of integration.
2. After defining u, rewrite both the integrand and the limits of integration in terms of the variable u. Next, carry out the integration with respect to u and leave your answer expressed in terms of the variable u. Finally, evaluate the definite integral, using the new limits of integration.

The two alternative methods are illustrated in the next example.

Example 6 Evaluate $\int_0^1 \frac{2x}{(x^2 + 1)^2}\,dx$.

Solution Letting $u = x^2 + 1$, $du = 2x\,dx$, we can proceed in either of the following ways.

1. $$\int \frac{2x}{(x^2 + 1)^2}\,dx = \int \frac{du}{u^2} = \int u^{-2}\,du = \frac{u^{-1}}{-1} = \frac{-1}{x^2 + 1}$$

   Now, making use of the fundamental theorem of calculus, we have

   $$\int_0^1 \frac{2x}{(x^2 + 1)^2}\,dx = \left.\frac{-1}{x^2 + 1}\right|_0^1 = -\frac{1}{2} + 1 = \frac{1}{2}$$

2. Noting that $u = 1$ when $x = 0$ and $u = 2$ when $x = 1$, we can write the definite integral as

   $$\int_0^1 \frac{2x}{(x^2 + 1)^2}\,dx = \int_1^2 \frac{du}{u^2}$$

   Note the change in the limits of integration in this equation. The integral on the right-hand side can be found directly:

   $$\int_1^2 \frac{du}{u^2} = \left.\frac{-1}{u}\right|_1^2 = -\frac{1}{2} + 1 = \frac{1}{2}$$ ■

7.1 EXERCISES

Find each of the following integrals, using the method of substitution where necessary.

1. $\int (x - 5)^3\,dx$

2. $\int \sqrt{x + 2}\,dx$

3. $\int e^{x+2}\,dx$

4. $\int \frac{1}{x + 6}\,dx$

5. $\int 2(2x+1)^5\,dx$

6. $\int 5e^{(5x-1)}\,dx$

7. $\int \frac{4}{4x+3}\,dx$

8. $\int \sqrt{4x-5}\,dx$

9. $\int x(x^2-1)^4\,dx$

10. $\int x^2e^{x^3}\,dx$

11. $\int xe^{-x^2}\,dx$

12. $\int (x+2)(x^2+4x-3)^3\,dx$

13. $\int \frac{5x}{(3x^2+7)^2}\,dx$

14. $\int \frac{x}{x^2+1}\,dx$

15. $\int x^5(x^6+2)\,dx$

16. $\int \frac{1}{x^2}\sqrt{1+x^{-1}}\,dx$

17. $\int xe^{1-x^2}\,dx$

18. $\int (3x^2-4x+1)(x^3-2x^2+x-7)^2\,dx$

19. $\int \frac{e^x}{2+e^x}\,dx$

20. $\int (x+\sqrt{x-1})\,dx$

21. $\int \left(xe^{x^2}-\frac{2x}{x^2+1}\right)dx$

22. $\int \left[x(x^2+3)-\frac{3}{(x+5)^4}\right]dx$

23. $\int \frac{e^{1/x}}{x^2}\,dx$

24. $\int \frac{2x}{1+x}\,dx$

25. $\int (x+1)\sqrt{x+3}\,dx$

26. $\int \frac{\ln x}{x}\,dx$

27. $\int \frac{(\ln x)^2}{x}\,dx$

28. $\int \frac{1}{x\ln x}\,dx$

29. $\int x^2\sqrt{x+1}\,dx$

30. $\int x^3\sqrt{x^2+1}\,dx$

Evaluate each of the following definite integrals.

31. $\int_0^2 x(x^2-2)^3\,dx$

32. $\int_1^2 xe^{x^2}\,dx$

33. $\int_0^2 (2x+3)(x^2+3x-5)^2\,dx$

34. $\int_1^3 \frac{2}{x+1}\,dx$

35. $\int_2^{\sqrt{12}} x\sqrt{x^2-3}\,dx$

36. $\int_0^1 \frac{x}{1+3x^2}\,dx$

37. $\int_2^4 \frac{x}{1+x}\,dx$

38. $\int_0^1 \frac{e^x}{2+e^x}\,dx$

39. Find the area of the shaded region shown in Figure 1.

40. Find the average value of the function $f(x) = xe^{x^2}$ between $x = 0$ and $x = 2$.

41. The demand equation for tickets to a college football game is

$$p(q) = \frac{2000}{q+1}$$

where $p(q)$ is the unit price, in dollars, and q is the number of tickets sold, in thousands. Find the consumers' surplus when the ticket price is $4.00.

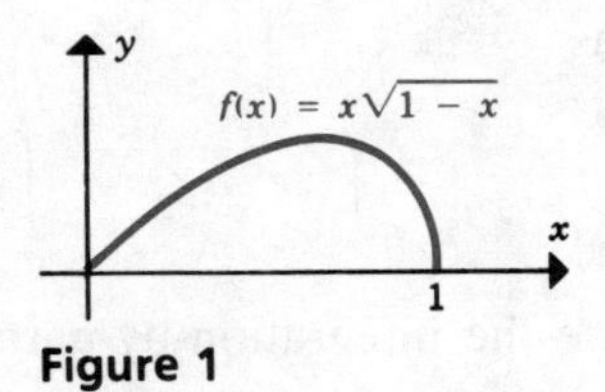

Figure 1

42. The pulse rate of a person startled by a loud noise is described by the equation

$$B(t) = 60 + 200te^{-0.50t^2}, \qquad t \geq 0$$

where $B(t)$ is the pulse rate, in beats per minute, and t is the time, in minutes, from the moment the noise occurred. Find the average pulse rate from $t = 0$ to $t = 1$.

7.2 INTEGRATION BY PARTS

The method of substitution, presented in Section 7.1, is not always effective when trying to find an integral because we may not be able to find an antiderivative of the integrand $g(u)$. For example, suppose that we want to use substitution to find

$$\int xe^x\, dx \tag{1}$$

How should $u(x)$ be defined? If we let

$$u = x, \qquad du = dx$$

we get the integral

$$\int ue^u\, du$$

which, except for the variable of integration, is identical to (1). We can make another attempt and let

$$u = e^x, \qquad du = e^x\, dx$$

If we note that the equation $u = e^x$ is equivalent to $x = \ln u$, the integral can be written in terms of u and du as

$$\int xe^x\, dx = \int \ln u\, du$$

At this stage the new integral is no easier to find than the original, so the method of substitution has not been particularly useful.

Another method, **integration by parts,** is often effective when the integrand is a product of two functions. For example, it works quite well with integrals such as

$$\int xe^x\,dx, \qquad \int x^2 \ln x\,dx, \qquad \int x(3x+5)^6\,dx$$

because the integration-by-parts method is based on the product rule (Section 3.3):

$$\frac{d}{dx}(uv) = u\frac{dv}{dx} + v\frac{du}{dx}$$

Writing this equation as

$$u\frac{dv}{dx} = \frac{d}{dx}(uv) - v\frac{du}{dx}$$

and integrating both sides with respect to x gives

$$\int u\frac{dv}{dx}\,dx = \int \frac{d}{dx}(uv)\,dx - \int v\frac{du}{dx}\,dx$$

The first term on the right-hand side can be written in a simpler form because an antiderivative of the derivative of a function equals the function, that is,

$$\int \frac{d}{dx}f(x)\,dx = f(x)$$

Or in this case

$$\int \frac{d}{dx}(uv)\,dx = uv$$

Now it is possible to write

$$\int u\frac{dv}{dx}\,dx = uv - \int v\frac{du}{dx}\,dx \tag{2}$$

Although Equation 2 contains the essential features of the integration-by-parts technique, a simpler and more widely used version can be derived if Equation 2 is written in terms of the differentials du and dv

$$du = \frac{du}{dx}\,dx \qquad dv = \frac{dv}{dx}\,dx$$

giving

$$\boxed{\int u\,dv = uv - \int v\,du} \tag{3}$$

In Equation 3, $\int u\, dv$ represents the given integral. The first thing that must be done is to select the expressions that define u and dv. If we are wise or lucky in our selection, $\int v\, du$, the integral on the right-hand side of Equation 3, should be less complex and easier to integrate than $\int u\, dv$. The method is illustrated in Example 1 for various choices of u and dv.

Example 1 Find $\int xe^x\, dx$.

Solution First we have to define u and dv so that we can write the integral in the form

$$\int u\, dv$$

We will look at two choices for u and dv:

1. $u = x,\ dv = e^x\, dx.$
2. $u = e^x,\ dv = x\, dx.$

Let us begin by selecting choice 1 and carrying it through according to Equation 3:

1. Let $u = x$ and $dv = e^x\, dx$. The next step is to find the differential $du = u'(x)\, dx$ and an antiderivative of $v'(x) = e^x$; we get

$$du = (1)\, dx, \qquad v = e^x$$

Following the procedure described in Equation 3, we have

$$\int \underbrace{x}_{u}\ \underbrace{e^x\, dx}_{dv} = \underbrace{x}_{u}\ \underbrace{e^x}_{v} - \int \underbrace{e^x}_{v}\ \underbrace{dx}_{du}$$

The integral on the right-hand side is known so that we can now write

$$\int xe^x\, dx = xe^x - e^x + C$$

As usual, this result can be checked by differentiating the function on the right-hand side.

2. Suppose we had defined u and dv as

$$u = e^x \quad \text{and} \quad dv = x\, dx$$

The quantities du and v become

$$du = e^x\, dx \qquad v = \frac{1}{2}x^2$$

Again proceeding as described by Equation 3, we get

$$\int \underbrace{e^x}_{u} \underbrace{x\,dx}_{dv} = \underbrace{e^x}_{u} \underbrace{\frac{1}{2}x^2}_{v} - \int \underbrace{\frac{1}{2}x^2}_{v} \underbrace{(e^x\,dx)}_{du}$$

Although formally correct, this choice of u and dv generates an integral on the right-hand side that is more complex than that on the left-hand side. When this happens, it is usually an indication that u and dv should be redefined. ■

Example 2 Find $\int \ln x\,dx$.

Solution The choice of u and dv in this case is straightforward:

$$u = \ln x, \quad dv = dx$$
$$du = \frac{1}{x}\,dx, \quad v = x$$

Integrating according to Equation 3 gives

$$\int \ln x\,dx = x \ln x - \int x \frac{1}{x}\,dx$$
$$= x \ln x - \int dx$$

yielding the result

$$\int \ln x\,dx = x \ln x - x + C$$

Question: Why would we not select the following for u and dv?

$$u = 1, \qquad dv = \ln x\,dx$$ ■

Often, integration by parts must be carried out more than once in order to find a given integral, as illustrated in the following example.

Example 3 Find $\int x^2 e^{-x}\,dx$.

Solution We let

$$u = x^2, \qquad dv = e^{-x}\,dx$$
$$du = 2x\,dx, \qquad v = -e^{-x}$$

Then using Equation 3, we get

$$\int x^2 e^{-x}\,dx = x^2(-e^{-x}) - \int (-e^{-x})2x\,dx$$

$$= -x^2 e^{-x} + 2\int xe^{-x}\,dx \qquad (4)$$

The integral on the right-hand side can be found by applying integration by parts once more; we let

$$u = x, \qquad dv = e^{-x}\,dx$$
$$du = dx, \qquad v = -e^{-x}$$

We get

$$2\int xe^{-x}\,dx = 2\left[-xe^{-x} + \int e^{-x}\,dx\right] = -2xe^{-x} - 2e^{-x} + C$$

Substituting this result in Equation 4 gives

$$\int x^2 e^{-x}\,dx = -x^2 e^{-x} - 2xe^{-x} - 2e^{-x} + C$$

■

Sometimes it is necessary to employ two or more methods in order to find an antiderivative, as the next example illustrates.

Example 4 Find $\int x \ln(x^2 + 1)\,dx$.

Solution Although this integral looks like an ideal candidate for integration by parts, you should note that one of the factors, namely x, equals one-half times the derivative of $x^2 + 1$. This suggests that substitution may be a more appropriate method to try first. If we let

$$u = x^2 + 1, \qquad du = 2x\,dx$$

the integral becomes

$$\frac{1}{2}\int \ln u\,du$$

Integration by parts was applied to an integral of this type in Example 2. We can use the result to write

$$\frac{1}{2}\int \ln u\,du = \frac{1}{2}u \ln u - \frac{1}{2}u + C$$

Substituting $x^2 + 1$ for u gives

$$\int x \ln(x^2 + 1)\, dx = \frac{1}{2}(x^2 + 1)\ln(x^2 + 1) - \frac{1}{2}(x^2 + 1) + C$$ ■

For some situations, integration by parts and substitution are equally effective.

Example 5 Find $\int x\sqrt{x + 1}\, dx$.

Solution The method of substitution was used to find this integral in Section 7.1 (Example 5). Integration by parts can also be used. Letting

$$u = x, \qquad dv = (x + 1)^{1/2}\, dx$$
$$du = dx, \qquad v = \frac{2(x + 1)^{3/2}}{3}$$

we get

$$\int x\sqrt{x + 1}\, dx = \frac{2x(x + 1)^{3/2}}{3} - \frac{2}{3}\int (x + 1)^{3/2}\, dx$$
$$= \frac{2x(x + 1)^{3/2}}{3} - \frac{4}{15}(x + 1)^{5/2} + C$$

If you compare this answer with that obtained using substitution, they do not appear to be equivalent. The difference is only superficial because the derivative of each equals $x\sqrt{x + 1}$. You can also show algebraically that the two answers are equivalent. ■

Application

Total sales of a product over a given period of time can often be described by a definite integral. Integration by parts is sometimes needed to evaluate the integral.

Example 6 Monthly sales of a best-selling novel are given by the equation

$$S(t) = 100te^{-t} \quad \text{thousand copies/month}$$

where t is the time, in months, from the date the book was published. How many books are sold during the first two months?

Solution Since $S(t)$ represents the monthly rate at which the books are sold, the total number sold over the first two months equals the definite integral

$$\int_0^2 100te^{-t}\, dt = 100 \int_0^2 te^{-t}\, dt$$

also shown as the shaded area under the curve in Figure 1. Integration by parts can be used to find $\int te^{-t}\,dt$. Let

$$u = t, \qquad dv = e^{-t}\,dt$$
$$du = dt, \qquad v = -e^{-t}$$

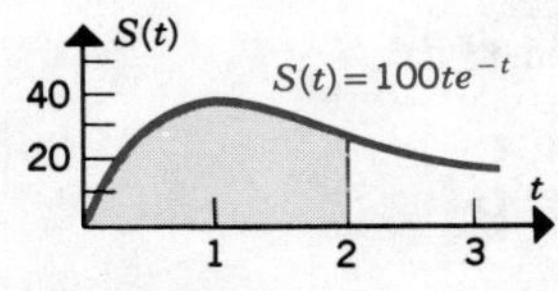

Figure 1

Now we can use Equation 3 to write

$$\int te^{-t}\,dt = -te^{-t} + \int e^{-t}\,dt$$
$$= -te^{-t} - e^{-t} = -e^{-t}(t + 1)$$

The definite integral can now be evaluated, yielding

$$100\int_0^2 te^{-t}\,dt = -100e^{-t}(t+1)\Big|_0^2$$
$$= -100e^{-2}(3) + 100$$
$$= 100 - \frac{300}{e^2} = 59.4 \text{ thousand copies}$$ ■

7.2 EXERCISES

Use integration by parts to find each of the following integrals.

1. $\int xe^{2x}\,dx$

2. $\int xe^{-3x}\,dx$

3. $\int x^2 e^{4x}\,dx$

4. $\int x^2 e^x\,dx$

5. $\int \ln 3x\,dx$

6. $\int x \ln x\,dx$

7. $\int \sqrt{x} \ln x\,dx$

8. $\int x^2 \ln x\,dx$

9. $\int \frac{\ln x}{\sqrt{x}}\,dx$

10. $\int x(x-1)^8\,dx$

11. $\int x\sqrt[3]{x+2}\,dx$

12. $\int x^2(x-2)^5\,dx$

13. $\int \ln(x+1)\,dx$

14. $\int x\ln(x+1)\,dx$

Hint: Let $w = x + 1$. Write each integral in terms of w and dw.

15. $\int \frac{\ln x}{x}\, dx$

16. $\int \sqrt{x}\, e^{\sqrt{x}}\, dx$ *Hint:* Let $w = \sqrt{x}$.

Evaluate each of the following definite integrals.

17. $\int_{-1}^{1} xe^{-x}\, dx$

18. $\int_{1}^{2} \ln x\, dx$

19. $\int_{2}^{3} x \ln x^2\, dx$

20. $\int_{0}^{2} \ln (x + 1)\, dx$

21. $\int_{0}^{1} xe^{x}\, dx$

22. $\int_{0}^{2} xe^{-x}\, dx$

23. $\int_{1}^{2} \frac{\ln x}{x}\, dx$

24. $\int_{0}^{2} x(x + 1)^4\, dx$

25. Find the area of the region under the curve $f(x) = xe^x$ between $x = 0$ and $x = 1$.

26. Find the area of the region under the curve $f(x) = \ln x$ between $x = 1$ and $x = e$. Sketch the graph of the region.

27. Annual output from the Hi-Ho Silver Mines is given by the equation

$$f(t) = 200e^{-0.05t} \quad \text{tons/year}$$

If $p(t)$, the price of a ton of silver, is given by the equation

$$p(t) = 400 + 10t \quad \text{thousand dollars/ton}$$

determine the total revenue earned from $t = 0$ to $t = 4$.

28. An experiment is conducted in which the relationship between $N(t)$, the number of bacteria (in millions) in a closed environment, and the time t, in hours, is described by the equation

$$N(t) = t^2e^{-t}, \qquad t \geq 0$$

Find the average number of bacteria between $t = 0$ and $t = 1$.

29. The marginal revenue function $R'(q)$ for a company selling large-screen television sets is

$$R'(q) = 2000e^{-0.1q} - 200qe^{-0.1q} \quad \text{dollars/unit}$$

Find the change in revenue as the number of sets sold increases from $q = 6$ to $q = 8$.

30. Find the average value of the function $f(x) = x \ln x$ over the interval $1 \leq x \leq 5$.

7.3 TABLES OF INTEGRALS

Because integration requires more ingenuity, time, and effort than does differentiation, we often find ourselves facing an integral that we cannot "crack." When this happens, a list of integration formulas or table of integrals can be helpful. Table D in the Appendix represents a short version of such a table; more extensive tables can be found in more advanced textbooks or in mathematical handbooks.

When we use a table of integrals, the objective is to match the integrand in the problem with the appropriate integrand in the table. The matching has been made easier by grouping the formulas into categories according to the structure of the integrand. For example, there are integrands categorized as "forms involving $a + bx$, forms involving $x^2 - a^2$, and so on.

We should point out in advance that you may have to modify a given integral or use the method of substitution in order to match one of the forms in the table. The following examples are intended to illustrate the use of the table and to point out some of the modifications that you may have to employ to match one of the forms.

Example 1 Use the table of integrals to find $\int \frac{1}{x(2x + 3)}\,dx$.

Solution The integrand contains the linear factor $2x + 3$, which is of the form $ax + b$. Among the set of formulas under the heading "forms involving $ax + b$" (Table D) is formula 5, whose integrand has the same form as ours:

$$\int \frac{1}{x(ax + b)}\,dx = \frac{1}{b}\ln\left|\frac{x}{ax + b}\right| + C$$

Setting $a = 2$ and $b = 3$, we use this formula to give

$$\int \frac{1}{x(2x + 3)}\,dx = \frac{1}{3}\ln\left|\frac{x}{2x + 3}\right| + C$$

■

In many situations, the method of substitution may be required to modify a given integral so that the integrand can be matched with one in the table, as illustrated in the next example.

Example 2 Use the table of integrals to find $\int \frac{x}{x^4 - 16}\,dx$.

Solution None of the formulas in the table falls into the category $x^4 - a^4$. However, the expression $x^4 - 16$ also represents a difference of two squares:

$$x^4 - 16 = (x^2)^2 - 4^2$$

Noting also that the derivative of x^2 equals $2x$, which is twice the numerator, we try the substitution

$$u = x^2, \qquad du = 2x\,dx$$

When written in terms of u, the integral becomes

$$\frac{1}{2}\int \frac{1}{u^2 - 16}\,du$$

In this new form the integrand matches that in formula 9 (Table D) with $a = 4$:

$$\int \frac{1}{u^2 - a^2}\,du = \frac{1}{2a}\ln\left|\frac{u - a}{u + a}\right| + C$$

So we get

$$\frac{1}{2}\int \frac{1}{u^2 - 4^2}\,du = \frac{1}{16}\ln\left|\frac{u-4}{u+4}\right| + C$$

Finally, substituting x^2 for u, we get

$$\int \frac{x}{x^4 - 16}\,dx = \frac{1}{16}\ln\left|\frac{x^2-4}{x^2+4}\right| + C$$

■

Formulas 17 and 20 in Table D are called **recursion formulas.** For any permissible value of n, the given integral is expressed in terms of a second integral whose integrand is identical to that in the original integral except for the replacement of n by $n - 1$. The process of integration is repeated as often as necessary until the last integral is found. The next example illustrates this technique.

Example 3 Use the table of integrals to find $\int (\ln x)^3\,dx$.

Solution Using formula 20 from Table D with $n = 3$, we have

$$\int (\ln x)^3\,dx = x(\ln x)^3 - 3\int (\ln x)^2\,dx$$

The integral on the right-hand side is found by the same method, that is, applying formula 20 with $n = 2$ this time:

$$\int (\ln x)^2\,dx = x(\ln x)^2 - 2\int \ln x\,dx$$

Putting together the results obtained so far, we have

$$\int (\ln x)^3\,dx = x(\ln x)^3 - 3x(\ln x)^2 + 6\int \ln x\,dx$$

The integral on the right-hand side can be found by either applying formula 20 again, this time with $n = 1$, or applying formula 10. In either case, the final result becomes

$$\int (\ln x)^3\,dx = x(\ln x)^3 - 3x(\ln x)^2 + 6x\ln x - 6x + C$$

■

Although Table D does not contain forms involving the quadratic expression $x^2 + bx + c$, the algebraic method of completing the square can be used to rewrite the expression as a sum or difference of two squares; this step may bring a given integral into agreement with one of the forms in the table. We write the quadratic expression $x^2 + bx + c$ as $(x^2 + bx) + c$. The expression within the parentheses can be written as a perfect square by adding $(b/2)^2$ inside the parentheses and at the same time subtracting it from c to give

$$\left(x^2 + bx + \left(\frac{b}{2}\right)^2\right) + c - \left(\frac{b}{2}\right)^2 = \left(x + \frac{b}{2}\right)^2 + c - \frac{b^2}{4}$$

If we let $u = x + \dfrac{b}{2}$, the expression $x^2 + bx + c$ can be written in the form $u^2 \pm a^2$, and we may be able to match the integral with one of those in the table. This approach is illustrated in the next example.

Example 4 Use the table of integrals to find $\displaystyle\int \frac{1}{\sqrt{x^2 + 4x - 5}}\,dx$.

Solution The integrand, as written, does not match any of the forms in the table. The method of completing the square can be used to rewrite $x^2 + 4x - 5$.

$$\begin{aligned} x^2 + 4x - 5 &= (x^2 + 4x + 4) - 5 - 4 \\ &= (x + 2)^2 - 3^2 \end{aligned}$$

so the integral assumes the form $\displaystyle\int \frac{1}{\sqrt{(x + 2)^2 - 9}}\,dx$. Letting $u = x + 2$ and $a = 3$, we can write the integral as

$$\int \frac{1}{\sqrt{u^2 - a^2}}\,du$$

This integral matches one of those shown in formula 12 (Table D), namely

$$\int \frac{1}{\sqrt{u^2 - a^2}}\,du = \ln|u + \sqrt{u^2 - a^2}| + C$$

Setting $u = x + 2$ and $a = 3$, we get

$$\int \frac{1}{\sqrt{x^2 + 4x - 5}}\,dx = \ln|x + 2 + \sqrt{x^2 + 4x - 5}| + C$$ ■

7.3 EXERCISES

Use the integration formulas in Table D to find the following integrals.

1. $\displaystyle\int x^3 \ln x\,dx$
2. $\displaystyle\int \frac{5}{2x(x + 1)}\,dx$
3. $\displaystyle\int \frac{1}{\sqrt{9x^2 + 9}}\,dx$
4. $\displaystyle\int \frac{7}{x^2 - 3}\,dx$
5. $\displaystyle\int \frac{x}{(4x + 1)^2}\,dx$
6. $\displaystyle\int \sqrt{25x^2 + 9}\,dx$

7. $\int \frac{3x}{5x + 2}\,dx$

8. $\int \frac{6}{2x^2 + x}\,dx$

9. $\frac{1}{x\sqrt{4 - 9x^2}}\,dx$

10. $\int x^3 e^{2x}\,dx$

11. $\int \frac{1}{16 - x^2}\,dx$

12. $\int \frac{4}{3x(2x + 7)^2}\,dx$

13. $\int \frac{x}{x^4 - 1}\,dx$

14. $\int \frac{x^2}{\sqrt{x^6 - 9}}\,dx$

15. $\int \frac{4x}{\sqrt{2x + 3}}\,dx$

16. $\int e^x \sqrt{e^{2x} + 1}\,dx$ *Hint:* Use substitution with $u = e^x$.

17. $\int \frac{2}{\sqrt{x^2 - 9}}\,dx$

18. $\int (\ln x)^4\,dx$

19. $\int \frac{1}{x^2 + 2x - 3}\,dx$

20. $\int \frac{1}{x^2 + 6x - 7}\,dx$

Evaluate each of the following definite integrals.

21. $\int_0^1 \frac{1}{4 - x^2}\,dx$

22. $\int_1^6 \frac{x}{\sqrt{x + 3}}\,dx$

23. $\int_0^1 \frac{3x}{2x + 1}\,dx$

24. $\int_1^2 \frac{1}{x^2 + x}\,dx$

25. $\int_0^3 \sqrt{x^2 + 7}\,dx$

26. $\int_1^2 (\ln x)^2\,dx$

27. $\int_1^3 \frac{x}{(x + 1)^2}\,dx$

28. $\int_3^8 \frac{x}{\sqrt{x + 1}}\,dx$

29. $\int_1^e (\ln x)^3\,dx$

30. $\int_{-1}^1 \frac{1}{x\sqrt{9 - x^2}}\,dx$

31. Show that formula 3 in Table D can be obtained by means of substitution, with $u = ax + b$ and $du = a\,dx$.

32. Show that formula 7 in Table D can be obtained by means of substitution, with $u = ax + b$ and $du = a\,dx$.

33. Show that formula 17 in Table D can be obtained by integration by parts, with $u = x^n$, $dv = e^{ax}\,dx$.

34. Show that formula 19 in Table D can be obtained by integration by parts, with $u = \ln x$, $dv = x^n\,dx$.

35. Show that formula 20 in Table D can be obtained by integration by parts, with $u = (\ln x)^n$, $dv = dx$.

7.4 NUMERICAL INTEGRATION

There are many occasions when a definite integral $\int_a^b f(x)\,dx$ cannot be evaluated because an antiderivative of $f(x)$ cannot be found, even in a table; a well-known example is the function $f(x) = e^{-x^2}$, which plays an important role in mathematical statistics. However, a definite integral can be approximated to any desired degree of accuracy by numerical methods, two of which, the **trapezoidal rule** and **Simpson's rule,** will be studied in this section.

Trapezoidal Rule

To explain the rationale behind the trapezoidal rule more effectively, we look at a function $y = f(x)$, which is positive everywhere over the closed interval $a \leq x \leq b$, as shown in Figure 1. When we employ the trapezoidal rule, the

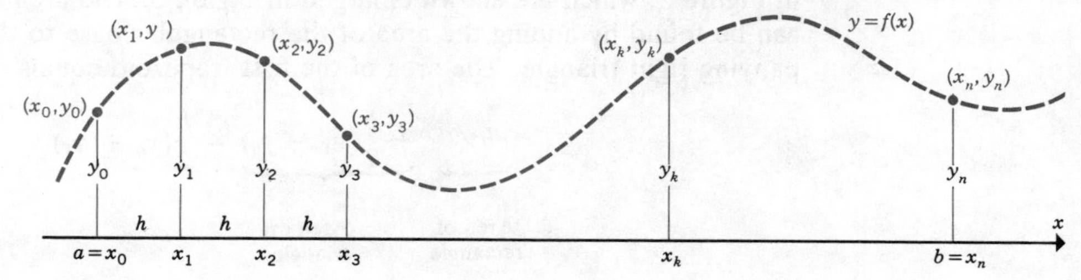

Figure 1

first step is to subdivide the closed interval $a \leq x \leq b$ into n equal subintervals each of whose widths h equals

$$h = \frac{b - a}{n}$$

The x coordinates of the endpoints of the subintervals are

$$x_0 = a, \quad x_1 = a + h, \quad x_2 = a + 2h, \ldots,$$
$$x_k = a + kh, \ldots, x_n = a + nh = b$$

Next, the y coordinate associated with each x_k is determined. The resulting set of points is shown in Figure 1. Finally, the curve $y = f(x)$ is replaced by a set of straight-line segments connecting adjacent pairs of points, as shown in Figure 2. The area under the curve $y = f(x)$ from $x = a$ to $x = b$ is then approximated by the sum of the areas of the n trapezoids, shown in Figure 2. The analytical result is stated as the *trapezoidal rule.*

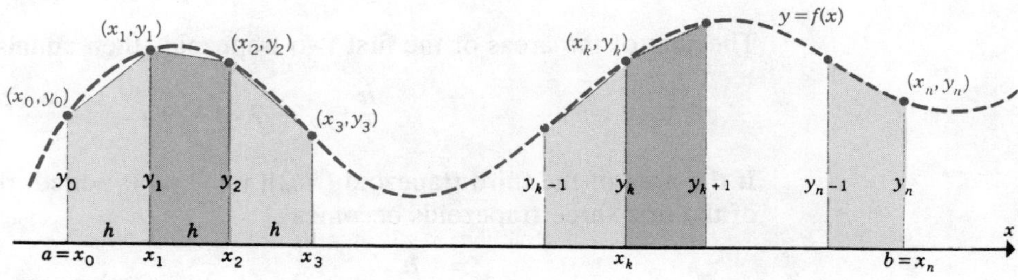

Figure 2

Trapezoidal Rule

If $y = f(x)$ represents a continuous function over the closed interval $a \leq x \leq b$, then

$$\int_a^b f(x)\, dx \approx \frac{h}{2}(y_0 + 2y_1 + 2y_2 + \cdots + 2y_k + \cdots + y_n) \qquad (1)$$

Equation 1 can be obtained by examining the first two trapezoids on the left in Figure 2, which are shown enlarged in Figure 3. The area of each trapezoid can be found by adding the area of the rectangular base to that of the accompanying right triangle. The area of the first trapezoid equals

$$\underbrace{hy_0}_{\text{Area of rectangle}} + \underbrace{\frac{h}{2}(y_1 - y_0)}_{\text{Area of triangle}} = \frac{h}{2}(y_0 + y_1)$$

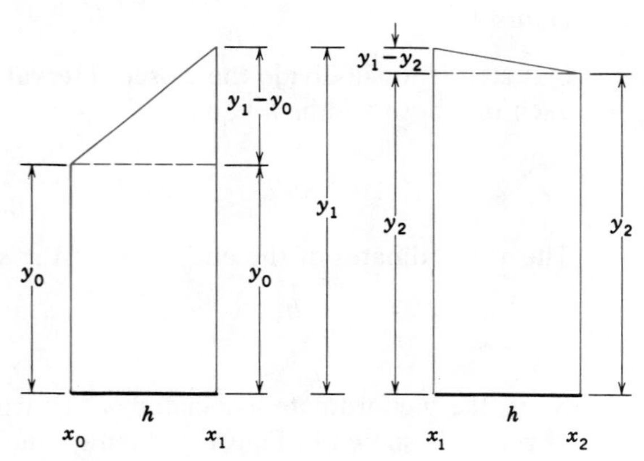

Figure 3

Similarly, the area of the second trapezoid equals

$$hy_2 + \frac{h}{2}(y_1 - y_2) = \frac{h}{2}(y_1 + y_2)$$

The sum of the areas of the first two trapezoids then equals

$$\frac{h}{2}(y_0 + 2y_1 + y_2)$$

If the area of the third trapezoid $(h/2)(y_2 + y_3)$ is added, the sum of the areas of the first three trapezoids becomes

$$\frac{h}{2}(y_0 + 2y_1 + 2y_2 + y_3)$$

If we continue in this manner until the area of the last, or nth, trapezoid is included, we get Equation 1 for the area of the n trapezoids.

In the first example, the trapezoidal rule will be demonstrated for a situation in which the fundamental theorem of calculus can also be used so that the difference between the two methods can be determined.

Example 1 Use the trapezoidal rule to find an approximate value for the definite integral:

$$\int_1^3 x^2\,dx \quad \text{with } n = 4$$

Solution The length of each subinterval, h, is found first:

$$h = \frac{b - a}{n} = \frac{3 - 1}{4} = 0.50$$

The sum within the brackets in Equation 1 can be found with the aid of a table such as Table 1, where the last column contains each term in the sum. Calculations are carried out to four decimal places. Using Equation 1, we get

$$\int_1^3 x^2\,dx \approx \frac{0.50}{2}(35) = 8.7500$$

Using the fundamental theorem of calculus, we get

$$\int_1^3 x^2\,dx = \frac{x^3}{3}\bigg|_1^3 = \frac{27}{3} - \frac{1}{3} = 8.6667$$

Table 1

k	x_k	$y_k = (x_k)^2$	Coefficient	Term
0	1.0	1.0000	1	1.0000
1	1.5	2.2500	2	4.5000
2	2.0	4.0000	2	8.0000
3	2.5	6.2500	2	12.5000
4	3.0	9.0000	1	9.0000
				35.000 = sum

■

Example 2 Use the trapezoidal rule to find an approximate value of

$$\int_0^1 \sqrt{2 + x^2}\,dx \quad \text{with } n = 10$$

Solution The length of each subinterval is given by

$$h = \frac{b - a}{n} = \frac{1 - 0}{10} = 0.10$$

Again, we set up a table (Table 2) to handle the sum in the brackets of Equation 1.

Table 2

k	x_k	$y_k = \sqrt{2 + (x_k)^2}$	Coefficient	Term
0	0.0000	1.4142	1	1.4142
1	0.1000	1.4177	2	2.8354
2	0.2000	1.4283	2	2.8566
3	0.3000	1.4457	2	2.8914
4	0.4000	1.4697	2	2.9394
5	0.5000	1.5000	2	3.0000
6	0.6000	1.5362	2	3.0724
7	0.7000	1.5780	2	3.1560
8	0.8000	1.6248	2	3.2496
9	0.9000	1.6763	2	3.3526
10	1.0000	1.7321	1	1.7321
				30.4997 = sum

Then, using Equation 1, we obtain the following:

$$\int_0^1 \sqrt{2 + x^2}\, dx \approx \frac{0.10}{2}(30.4997) = 1.5250$$

The definite integral $\int_0^1 \sqrt{2 + x^2}\, dx$ can be found using formula 11 in Table D:

$$\int_0^1 \sqrt{2 + x^2}\, dx = \left[\frac{x}{2}\sqrt{x^2 + 2} + \ln\left|x + \sqrt{x^2 + 2}\right|\right]_0^1 = 1.5245 \quad ■$$

Simpson's Rule

Simpson's rule is a technique in which the curve $y = f(x)$ is replaced by a set of adjacent parabolic segments over the closed interval $a \le x \le b$. The sum of the areas beneath these parabolic segments serves as the approximate value of $\int_a^b f(x)\, dx$.

First, the closed interval $a \le x \le b$ is subdivided into an **even** number n of equal subintervals, each of whose widths h equals

$$h = \frac{b - a}{n}$$

As shown in Figure 4, a parabola is drawn through the three successive points (x_0, y_0), (x_1, y_1) and (x_2, y_2). A second parabola is constructed through the three points (x_2, y_2), (x_3, y_3) and (x_4, y_4). This procedure is continued until the curve is covered with parabolic segments. When the sum of the areas under the para-

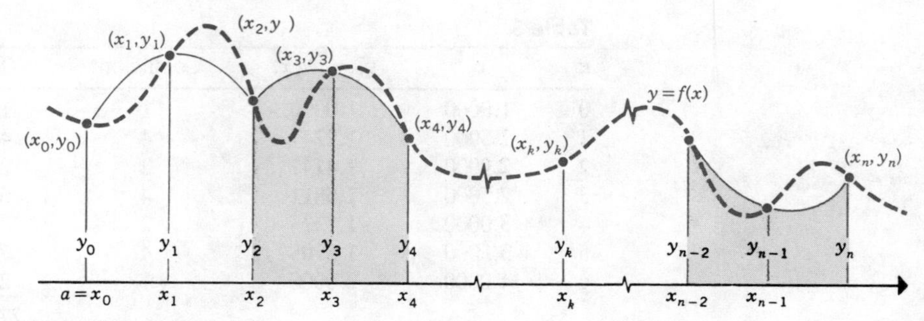

Figure 4

bolic segments is calculated, the approximation known as *Simpson's rule* emerges.*

Simpson's Rule (n is even)

$$\int_a^b f(x)\,dx \approx \frac{h}{3}(y_0 + 4y_1 + 2y_2 + 4y_3 + 2y_4 + \cdots + 4y_{n-1} + y_n) \quad (2)$$

Simpson's rule is illustrated in Example 3 for an integral for which the fundamental theorem of calculus can also be used.

Example 3 Use Simpson's rule to find an approximate value of $\int_1^4 \sqrt{x}\,dx$, with $n = 6$.

Solution The length h of each subinterval equals

$$h = \frac{4 - 1}{6} = 0.50$$

Table 3 has been set up to evaluate each term within the brackets in Equation 2. Then, according to Simpson's rule, we get

$$\int_1^4 \sqrt{x}\,dx \approx \frac{0.50}{3}(27.990) = 4.6665$$

Using the fundamental theorem of calculus, we obtain

$$\int_1^4 \sqrt{x}\,dx = \tfrac{2}{3}(x^{3/2})\Big|_1^4 = \tfrac{2}{3}(8 - 1) = 4.6667$$

Even for this crude approximation, the error (0.0002) is extremely small.

*A derivation of Simpson's rule can be found in *Calculus and Analytic Geometry*, Fourth edition, by Al Shenk, Scott, Foresman and Company, Glenview, Illinois, 1988 pp. 276–278.

Table 3

k	x_k	$y_k = \sqrt{x_k}$	Coefficient	Term
0	1.0000	1.0000	1	1.0000
1	1.5000	1.2247	4	4.8988
2	2.0000	1.4142	2	2.8284
3	2.5000	1.5811	4	6.3244
4	3.0000	1.7321	2	3.4642
5	3.5000	1.8708	4	7.4832
6	4.0000	2.0000	1	2.0000
				27.9990 = sum

■

Example 4 Use Simpson's rule to approximate $\int_1^2 \frac{1}{1 + x^2}\,dx$, with $n = 10$.

Solution First, h, the length of each subinterval, is found:

$$h = \frac{2 - 1}{10} = 0.10$$

We proceed as before by completing Table 4. Then, using Simpson's rule, we get

$$\int_1^2 \frac{1}{1 + x^2}\,dx \approx \frac{0.10}{3}(9.6522) = 0.3217$$

Table 4

k	x_k	$y_k = \frac{1}{1 + (x_k)^2}$	Coefficient	Term
0	1.00	0.5000	1	0.5000
1	1.10	0.4525	4	1.8100
2	1.20	0.4098	2	0.8196
3	1.30	0.3717	4	1.4868
4	1.40	0.3378	2	0.6756
5	1.50	0.3077	4	1.2308
6	1.60	0.2809	2	0.5618
7	1.70	0.2571	4	1.0284
8	1.80	0.2358	2	0.4716
9	1.90	0.2169	4	0.8676
10	2.00	0.2000	1	0.2000
				9.6522 = sum

■

Error Estimates

When an approximation method is used, it is always useful to have an estimate of the error associated with the method. If the error E is defined as

$$E = |(\text{actual}) - (\text{approximation})|$$

the maximum value of E can be found from formulas developed for both the trapezoidal rule and Simpson's rule.

Trapezoidal Rule

If M equals the maximum value of $|f''(x)|$ on the interval $a \leq x \leq b$, then

$$E \leq \frac{M(b - a)^3}{12n^2} \tag{3}$$

Simpson's Rule

When Simpson's rule is used to approximate a definite integral, the formula for the maximum error E depends on $|f^{(4)}(x)|$, the absolute value of the fourth derivative of $f(x)$. If M equals the maximum value of $|f^{(4)}(x)|$ on the interval $a \leq x \leq b$, then

$$E \leq \frac{M(b - a)^5}{180n^4} \tag{4}$$

Example 5 Find the maximum error in the estimate of $\int_1^3 \ln x \, dx$

(a) Using the trapezoidal rule with $n = 10$.

(b) Using Simpson's rule with $n = 10$.

Solution (a) First, we want to determine the maximum value of $|f''(x)|$ over the interval $1 \leq x \leq 3$. We differentiate the function $f(x) = \ln x$ twice to get

$$f''(x) = -\frac{1}{x^2} \quad \text{or} \quad |f''(x)| = \frac{1}{x^2}$$

Over the interval $1 \leq x \leq 3$, the quantity $1/x^2$ decreases as x increases; therefore, the maximum value of $1/x^2$ equals 1 when $x = 1$, so

$$E \leq \frac{(1)(3 - 1)^3}{12(10)^2} = \frac{8}{12(100)} = 0.0067$$

(b) First, we have to find $|f^{(4)}(x)|$,

$$f^{(3)}(x) = \frac{2}{x^3}, \qquad f^{(4)}(x) = -\frac{6}{x^4}$$

so $|f^{(4)}(x)| = 6/x^4$. Over the interval $1 \le x \le 3$, the maximum value of $6/x^4$ equals 6, so

$$E \le \frac{6(3-1)^5}{180(10^4)} = 0.00011$$ ■

In Example 5, the maximum error associated with Simpson's rule is much less than that for the trapezoidal rule. This is so because the maximum error in Simpson's rule is proportional to $1/n^4$ versus $1/n^2$ for the trapezoidal rule.

Formulas 3 and 4 are very useful in determining the minimum value of n required to keep the error E at or below some predetermined level, as the next example illustrates.

Example 6 If the trapezoidal rule is used to approximate $\int_1^3 \ln x\, dx$, find n, the number of subintervals needed to keep E at or below 0.01.

Solution In Example 5a, we found that $M = 1$. We want to find the values of n that satisfy the inequality

$$E_{\max} = \frac{(1)(2)^3}{12n^2} \le 0.01$$

Writing the inequality as

$$\frac{2}{3n^2} \le \frac{1}{100}$$

or, equivalently, as $n^2 \ge 66.7$, we can solve for n, getting

$$n \ge 8.17$$

The smallest integral value of n that satisfies the inequality is

$$\boxed{n = 9}$$ ■

7.4 EXERCISES

Use the trapezoidal rule to find an approximate value for each of the following definite integrals. Find the error associated with this method by evaluating each integral by means of the fundamental theorem of calculus.

1. $\int_0^2 x^2\, dx, \quad n = 4$

2. $\int_{-1}^3 x^3\, dx, \quad n = 8$

3. $\int_0^1 e^x\, dx, \quad n = 4$

4. $\int_1^4 \frac{1}{x}\, dx, \quad n = 6$

5. $\int_3^4 \sqrt{x}\, dx, \quad n = 4$

6. $\int_1^3 \frac{1}{x^2}\, dx, \quad n = 8$

7. $\int_0^1 \frac{x}{1 + x^2}\, dx, \quad n = 4$

8. $\int_0^1 x^2\sqrt{1 + x^3}\, dx, \quad n = 4$

Use Simpson's rule to find an approximate value for each of the following definite integrals. In addition, find the error associated with this method by evaluating each integral by means of the fundamental theorem of calculus.

9. $\int_0^3 2x\, dx, \quad n = 6$

10. $\int_1^9 \sqrt{x}\, dx, \quad n = 8$

11. $\int_1^2 \ln x\, dx, \quad n = 4$

12. $\int_2^5 \frac{1}{x^2}\, dx, \quad n = 6$

13. $\int_0^2 xe^{x^2}\, dx, \quad n = 8$

14. $\int_0^3 \sqrt{1 + x}\, dx, \quad n = 6$

Find approximate values for each of the following definite integrals using both the trapezoidal rule and Simpson's rule.

15. $\int_0^1 \frac{1}{1 + x^2}\, dx, \quad n = 4$

16. $\int_0^1 e^{x^2}\, dx, \quad n = 10$

17. $\int_1^4 \frac{2}{1 + \sqrt{x}}\, dx, \quad n = 6$

18. $\int_1^3 \frac{e^x}{x}\, dx, \quad n = 4$

19. $\int_1^2 \frac{\sqrt{1 + x}}{x}\, dx, \quad n = 8$

20. $\int_1^3 \frac{5}{1 + x^3}\, dx, \quad n = 8$

21. $\int_0^2 \ln(1 + x^2)\, dx, \quad n = 8$

22. $\int_0^5 \sqrt{x}\, e^{-x}\, dx, \quad n = 10$

In exercises 23 through 28, determine the largest possible error when each integral is approximated by (a) the trapezoidal rule and (b) Simpson's rule.

23. $\int_0^1 x^4\, dx, \quad n = 4$

24. $\int_1^4 \sqrt{x}\, dx, \quad n = 6$

25. $\int_0^1 e^{-x}\, dx, \quad n = 4$

26. $\int_0^1 \sqrt{2 - x}\, dx, \quad n = 4$

27. $\int_0^1 \frac{x}{1 + x}\, dx, \quad n = 4$

28. $\int_0^1 xe^x\, dx, \quad n = 4$

29. If the trapezoidal rule is used to approximate $\int_0^3 \ln(1 + x)\, dx$, find n, the number of subintervals needed to keep the maximum possible error below 0.0001.

30. If the trapezoidal rule is used to approximate $\int_1^2 e^{-x^2}\, dx$, find n, the number of subintervals needed to keep the maximum error below 0.005.

7.5 IMPROPER INTEGRALS

In studying and working with the definite integral $\int_a^b f(x)\,dx$, we have assumed, until now, that

1. The function $f(x)$ is continuous over the closed interval $a \le x \le b$, and
2. Both a and b are finite.

On the other hand, an integral $\int_a^b f(x)$ is called **improper** if

1. The integrand $f(x)$ is unbounded at one or more values of x in the interval $a \le x \le b$, or
2. Either a or b is infinite.

In this section we will study integrals of the second type, which assume one of the following forms:

$$\int_a^\infty f(x)\,dx, \qquad \int_{-\infty}^b f(x)\,dx, \qquad \int_{-\infty}^\infty f(x)\,dx$$

Improper integrals are encountered when we attempt to calculate the area under a curve, such as $f(x) = 1/x^2$, to the right of the vertical line $x = 1$, as shown in Figure 1. For a situation such as this, we want (1) to determine whether the area in question is finite and (2) to evaluate the area if it is finite. To deal with this problem, we use the following procedure.

A. The right boundary is set equal to some finite value, say $x = b$, and the area between $x = 1$ and $x = b$ is found by evaluating the integral

$$\int_1^b \frac{1}{x^2}\,dx.$$

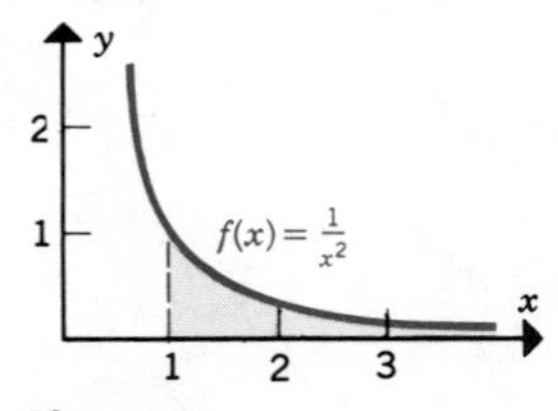

Figure 1

B. Next, the limit of the definite integral as $b \to \infty$ is determined, that is, we look for

$$\lim_{b\to\infty} \int_1^b \frac{1}{x^2}\,dx$$

This procedure is shown graphically in Figure 2.

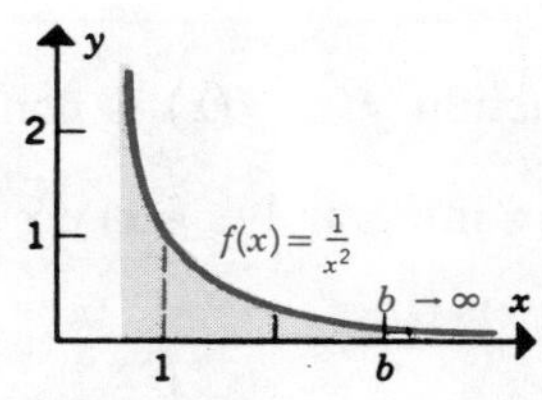

Figure 2

The process just described enables us to find the *improper integral* $\int_1^{\infty} \frac{1}{x^2}\,dx$, defined as

$$\int_1^{\infty} \frac{1}{x^2}\,dx = \lim_{b\to\infty} \int_1^{b} \frac{1}{x^2}\,dx$$

Following the steps described in A and B yields

A. $\int_1^{b} \frac{1}{x^2}\,dx = \frac{-1}{x}\Big|_1^b = 1 - \frac{1}{b}$

B. $\lim_{b\to\infty} \int_1^{b} \frac{1}{x^2}\,dx = \lim_{b\to\infty}\left(1 - \frac{1}{b}\right)$

As b is assigned values that increase without limit, the quantity $1/b$ approaches zero, that is, $1/b \to 0$ as $b \to \infty$. Thus we get

$$\lim_{b\to\infty}\left(1 - \frac{1}{b}\right) = 1$$

so now we can write

$$\int_1^{\infty} \frac{1}{x^2}\,dx = 1$$

The generalization of this process leads to the following definitions.

Definition 1 If a function $y = f(x)$ is continuous on the interval $a \leq x < \infty$, the **improper integral** $\int_a^{\infty} f(x)\,dx$ is defined as

$$\int_a^{\infty} f(x)\,dx = \lim_{b\to\infty} \int_a^{b} f(x)\,dx \tag{1}$$

If the limit exists, the integral is said to be **convergent;** if it does not, the integral is **divergent.**

In the same way, it is possible to define the improper integrals $\int_{-\infty}^{b} f(x)\,dx$ and $\int_{-\infty}^{\infty} f(x)\,dx$.

Definition 2 If a function $y = f(x)$ is continuous over the interval $-\infty < x \leq b$, the **improper** integral $\int_{-\infty}^{b} f(x)\,dx$ is defined as

$$\int_{-\infty}^{b} f(x)\,dx = \lim_{a \to -\infty} \int_{a}^{b} f(x)\,dx \tag{2}$$

Definition 3 If a function $y = f(x)$ is continuous over the interval $-\infty < x < \infty$, the **improper** integral $\int_{-\infty}^{\infty} f(x)$ is defined as

$$\int_{-\infty}^{\infty} f(x)\,dx = \lim_{a \to -\infty} \int_{a}^{0} f(x)\,dx + \lim_{b \to \infty} \int_{0}^{b} f(x)\,dx \tag{3}$$

Note: In order for $\int_{-\infty}^{\infty} f(x)\,dx$ to be convergent, both limits in Equation 3 must exist.

Example 1 Find $\int_{1}^{\infty} \frac{1}{\sqrt{x^3}}\,dx$, if it exists.

Solution Geometrically, we want to determine whether the area beneath the curve $f(x) = 1/\sqrt{x^3}$ to the right of the vertical line $x = 1$ is finite and, if it is, to evaluate it. The region in question is shown in Figure 3. According to Equation 1, we first find $\int_{1}^{b} \frac{1}{\sqrt{x^3}}\,dx$. Working this through gives

$$\int_{1}^{b} \frac{1}{\sqrt{x^3}}\,dx = \int_{1}^{b} x^{-3/2}\,dx = -2x^{-1/2}\Big|_{1}^{b} = 2 - \frac{2}{\sqrt{b}}$$

Next, we find the limit as $b \to \infty$, obtaining

$$\int_{1}^{\infty} \frac{1}{\sqrt{x^3}}\,dx = \lim_{b \to \infty} \left(2 - \frac{2}{\sqrt{b}}\right)$$

As b is assigned values that increase without limit, the quantity $2/\sqrt{b}$ approaches zero, so we get

$$\lim_{b \to \infty} \left(2 - \frac{2}{\sqrt{b}}\right) = 2$$

Using this result, we can write

$$\int_1^\infty \frac{1}{\sqrt{x^3}}\,dx = 2$$

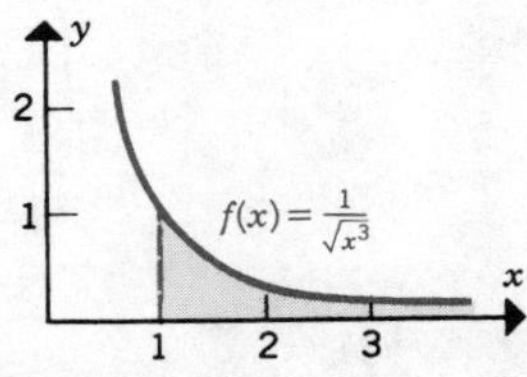

Figure 3

■

Example 2 Find $\int_1^\infty \frac{1}{\sqrt{x}}\,dx$, if it exists.

Solution Again, we seek to determine whether the area under the curve $f(x) = 1/\sqrt{x}$ to the right of the line $x = 1$ is finite and then to evaluate it if it is finite. The area of the shaded region in Figure 4 represents the integral $\int_1^\infty \frac{1}{\sqrt{x}}\,dx$. Proceeding as we did in Example 1, we first evaluate $\int_1^b \frac{1}{\sqrt{x}}\,dx$:

$$\int_1^b \frac{1}{\sqrt{x}}\,dx = \int_1^b x^{-1/2}\,dx = 2x^{1/2}\Big|_1^b = 2\sqrt{b} - 2$$

Next, taking the limit as $b \to \infty$, we have

$$\int_1^\infty \frac{1}{\sqrt{x}}\,dx = \lim_{b\to\infty} (2\sqrt{b} - 2)$$

As b is assigned values that increase indefinitely, the quantity $(2\sqrt{b} - 2)$ also increases without limit; that is, $(2\sqrt{b} - 2) \to \infty$ as $b \to \infty$. Thus the limit does not exist, so we conclude that $\int_1^\infty \frac{1}{\sqrt{x}}\,dx$ is divergent.

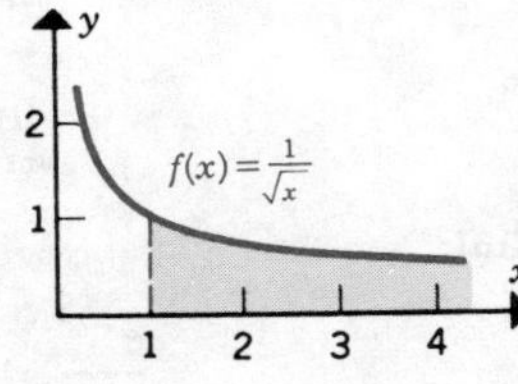

Figure 4

■

Example 3 Evaluate $\int_{-\infty}^{0} e^x\,dx$.

Solution The graph of the function $f(x) = e^x$ together with the shaded region representing the integral is shown in Figure 5. Using Equation 2, we write

$$\int_{-\infty}^{0} e^x\,dx = \lim_{a\to-\infty} \int_{a}^{0} e^x\,dx$$

$$= \lim_{a\to-\infty} e^x \Big|_a^0$$

$$= \lim_{a\to-\infty} (1 - e^a)$$

Noting that $e^a \to 0$ as $a \to -\infty$, we obtain

$$\int_{-\infty}^{0} e^x\,dx = 1$$

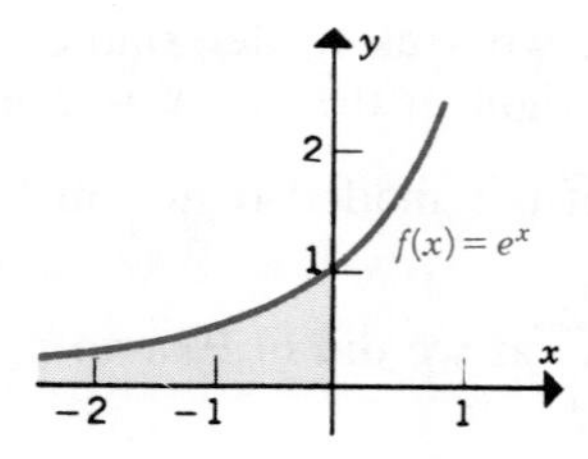

■

Example 4 Evaluate $\int_{-\infty}^{\infty} xe^{-x^2}\,dx$.

Solution The graph of the function $f(x) = xe^{-x^2}$ is shown in Figure 6. Proceeding according to Equation 3, we have

$$\int_{-\infty}^{\infty} xe^{-x^2}\,dx = \lim_{a\to-\infty} \int_{a}^{0} xe^{-x^2}\,dx + \lim_{b\to\infty} \int_{0}^{b} xe^{-x^2}\,dx$$

$$= \lim_{a\to-\infty} \frac{-e^{-x^2}}{2}\Big|_a^0 + \lim_{b\to\infty} \frac{-e^{-x^2}}{2}\Big|_0^b$$

$$= \lim_{a\to-\infty} \left(-\frac{1}{2} + \frac{e^{-a^2}}{2}\right) + \lim_{b\to\infty} \left(\frac{1}{2} - \frac{e^{-b^2}}{2}\right)$$

Noting that

$$\frac{e^{-a^2}}{2} = \frac{1}{2e^{a^2}} \to 0 \quad \text{as } a \to -\infty$$

and

$$\frac{e^{-b^2}}{2} = \frac{1}{2e^{b^2}} \to 0 \quad \text{as } b \to \infty$$

we have

$$\int_{-\infty}^{\infty} xe^{-x^2}\,dx = -\frac{1}{2} + \frac{1}{2} = 0$$

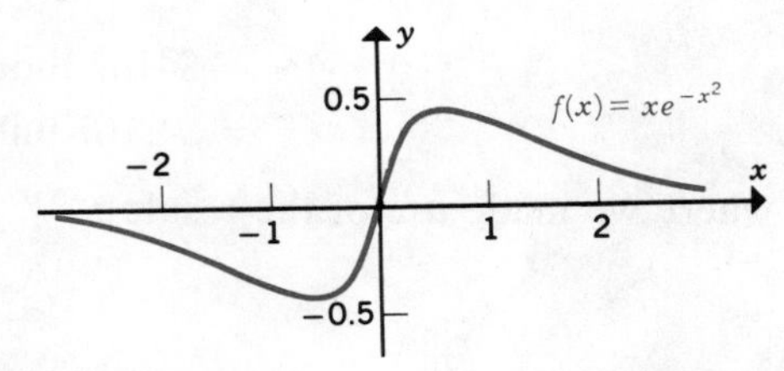

Figure 6

Application

The present value of a stream of cash flows that continues indefinitely can be evaluated by means of an improper integral. The present value is obtained by letting t, the upper limit in Equation 6, Section 6.5, become infinite, yielding

$$\text{Present value} = \int_{0}^{\infty} R(t)e^{-rt}\,dt \tag{4}$$

where $R(t)$ is the cash flow and r is the rate of return per unit time.

Example 5 Annual production at the Wildcat Coal Mine is expected to follow the equation

$$M(t) = 50e^{-0.03t}$$

where t is the time from the present expressed in years and $M(t)$ is the annual output in thousands of tons per year. If the per-ton price of coal remains constant at \$15 per ton, what is the present value of the revenue generated from the sale of all the coal to be mined, assuming that the company expects a 15 percent annual return on its investments?

Solution The annual revenue $R(t)$ from mining and selling the coal is given by the equation

$$R(t) = M(t)p(t) = (50e^{-0.03t})(15) = 750e^{-0.03t} \quad \text{thousand dollars/year}$$

The present value of the revenue obtained from all future sales of coal is

$$\begin{aligned}
\text{Present value} &= \int_0^\infty R(t)e^{-rt}\,dt = \int_0^\infty 750e^{-0.03t}e^{-0.15t}\,dt \\
&= \int_0^\infty 750e^{-0.18t}\,dt = \lim_{b\to\infty}\int_0^b 750e^{-0.18t}\,dt \\
&= \lim_{b\to\infty}\left[\frac{750}{-0.18}(e^{-0.18b} - 1)\right] \\
&= \$4167 \text{ thousand} \\
&= \$4.167 \text{ million}
\end{aligned}$$

where we made use of the result $e^{-0.18b} = 1/e^{0.18b} \to 0$ as $b \to \infty$. ■

7.5 EXERCISES

In Exercises 1 through 18, evaluate the improper integrals that converge.

1. $\int_3^\infty \frac{2}{x^2}\,dx$
2. $\int_1^\infty \frac{1}{\sqrt{x^3}}\,dx$
3. $\int_1^\infty \frac{4}{x^3}\,dx$
4. $\int_2^\infty x\,dx$
5. $\int_2^\infty \frac{1}{(x-1)^2}\,dx$
6. $\int_{-\infty}^\infty e^{2x}\,dx$
7. $\int_{-\infty}^{-2} \frac{1}{x^2}\,dx$
8. $\int_1^\infty \frac{dx}{x}$
9. $\int_{-\infty}^{-1} \frac{2}{\sqrt[3]{x^4}}\,dx$
10. $\int_{-\infty}^{-1} \frac{2}{\sqrt[3]{x^2}}\,dx$
11. $\int_0^\infty e^{-2x}\,dx$
12. $\int_{-\infty}^\infty x^2\,dx$
13. $\int_{-\infty}^0 \frac{x}{(x^2+2)^2}\,dx$
14. $\int_0^\infty \frac{x}{x^2+1}\,dx$
15. $\int_{-\infty}^0 xe^{-3x^2}\,dx$
16. $\int_{-\infty}^\infty xe^{-3x^2}\,dx$
17. $\int_0^\infty \frac{1}{\sqrt{x+1}}\,dx$
18. $\int_1^\infty \frac{e^{1/x}}{x^2}\,dx$

In Exercises 19 through 24, find the value of k that satisfies each equation.

19. $\int_0^\infty e^{-kx}\,dx = 1$
20. $\int_1^\infty \frac{k}{x^2}\,dx = 4$
21. $\int_{-\infty}^0 xe^{kx^2}\,dx = 1$
22. $\int_1^\infty \frac{1}{x^k}\,dx = \frac{1}{4}$

23. $\int_{-\infty}^{-1} \frac{1}{x^k}\, dx = \frac{1}{3}$ **24.** $\int_0^{\infty} \frac{x}{(x^2+1)^k}\, dx = \frac{1}{6}$

25. Refer to Example 5. What is the minimum number of tons of coal that the Wildcat Coal Mine must contain if the company is to continue production indefinitely at a rate given by the equation $M(t) = 50e^{-0.03t}$?

26. Geologists estimate that the oil reserves at the Swamp Oil Field are 20,000 barrels. The company is considering two drilling plans to recover the oil. The recovery rates for each plan are given by the following equations:

Plan 1. $R_1(t) = 1000e^{-0.10t}$ barrels/year.

Plan 2. $R_2(t) = 500e^{-0.01t}$ barrels/year.

Under which plan would the company be able to operate indefinitely? Explain.

27. Oil is leaking from a damaged oil tanker at a rate given by the equation

$$f(t) = 500e^{-0.10t} \quad \text{gallons/hour}$$

where $t = 0$ corresponds to the time when the leak began. If the leak continues unchecked, how many gallons of oil will eventually spill from the tanker?

28. (a) An annuity that continues indefinitely is called a **perpetuity.** Use Equation 5, Section 6.5, to show that the present value of a perpetuity equals R/r.

(b) Mr. B. G. Bucks, a rich alumnus of Android University, wants to establish a student scholarship in his name. If the annual proceeds of the scholarship are \$6000, and the annual rate of return is 12 percent, find the amount needed to fund the scholarship.

7.6 L'HÔPITAL'S RULE

In Section 2.1, we encountered indeterminate forms such as 0/0. The word *indeterminate* indicated that, based on this result, we were unable to determine whether or not a limit existed. When we evaluate improper integrals, other indeterminate forms such as ∞/∞ or $\infty - \infty$ can result. For example, suppose that we want to evaluate the improper integral

$$\int_0^{\infty} xe^{-x}\, dx = \lim_{b\to\infty} \int_0^b xe^{-x}\, dx$$

Integration by parts can be used to carry out the integration, yielding

$$\int_0^b xe^{-x}\, dx = (-xe^{-x} - e^{-x})\Big|_0^b$$

$$= 1 - e^{-b} - be^{-b} = 1 - \frac{1}{e^b} - \frac{b}{e^b}$$

If we let $b \to \infty$, we get

$$\int_0^{\infty} xe^{-x}\, dx = \lim_{b\to\infty}\left(1 - \frac{1}{e^b} - \frac{b}{e^b}\right)$$

As $b \to \infty$, the second term $1/e^b$ approaches zero. On the other hand, both the numerator and denominator of the term b/e^b increase without limit as $b \to \infty$. The result ∞/∞ is also called an indeterminate form because it is not possible, at this point, to determine whether or not the limit exists; we have to determine which of the two, the numerator or denominator, increases more rapidly as $b \to \infty$. As we have done before, we can construct a table and evaluate b/e^b as b increases indefinitely. The numbers in row 2 of Table 1 suggest that $b/e^b \to 0$ as $b \to \infty$. Therefore, we can say that

$$\lim_{b\to\infty} \frac{b}{e^b} = 0$$

and that

$$\int_0^\infty xe^{-x}\,dx = 1$$

Table 1

b	1	10	100	⋯
b/e^b	0.3679	0.0005	0.0000[a]	

[a]To four decimal places.

Although constructing a table of values is helpful, the method has limitations. It is time-consuming and inefficient and may lead to incorrect conclusions based on the limited set of values selected. A method called **L'Hôpital's rule** enables us to determine the limit, if it exists, when we encounter one of the indeterminate forms 0/0 or ∞/∞.

L'Hôpital's Rule

Let a be a real number, ∞, or $-\infty$. If $\lim_{x\to a} [f(x)/g(x)]$ leads to one of the ratios 0/0 or ∞/∞, then

$$\lim_{x\to a} \frac{f(x)}{g(x)} = \lim_{x\to a} \frac{f'(x)}{g'(x)} \tag{1}$$

In words, the rule states that $\lim_{x\to a} [f(x)/g(x)]$ depends on the relative rate, as measured by the ratio $f'(x)/g'(x)$ at which $f(x)$ and $g(x)$ approach zero or infinity as $x \to a$.

Example 1 Use L'Hôpital's rule to find $\lim_{x\to\infty} \dfrac{x}{e^x}$.

Solution 1. First, we check to make sure that L'Hôpital's rule applies:

$$\lim_{x\to\infty} x = \infty, \qquad \lim_{x\to\infty} e^x = \infty$$

L'Hôpital's rule can be used.

2. Apply L'Hôpital's rule:

$$f(x) = x, \qquad f'(x) = 1$$
$$g(x) = e^x, \qquad g'(x) = e^x$$
$$\lim_{x \to \infty} \frac{f'(x)}{g'(x)} = \lim_{x \to \infty} \frac{1}{e^x} = 0$$

So

$$\lim_{x \to \infty} \frac{x}{e^x} = 0$$

■

Example 2 Use L'Hôpital's rule to evaluate $\lim_{x \to 1} \dfrac{x^2 - 1}{\sqrt{x} - 1}$.

Solution Since $\lim_{x \to 1}(x^2 - 1) = 0$ and $\lim_{x \to 1}(\sqrt{x} - 1) = 0$, we get the indeterminate form 0/0, so we can apply L'Hôpital's rule:

$$f(x) = x^2 - 1, \qquad f'(x) = 2x$$
$$g(x) = \sqrt{x} - 1, \qquad g'(x) = 1/(2\sqrt{x})$$

Applying L'Hôpital's rule gives

$$\lim_{x \to 1} \frac{x^2 - 1}{\sqrt{x} - 1} = \lim_{x \to 1} \frac{2x}{(1/2\sqrt{x})} = \lim_{x \to 1} (4x^{3/2}) = 4$$

This result can also be obtained algebraically by multiplying numerator and denominator by $\sqrt{x} + 1$, thus rationalizing the denominator:

$$\frac{x^2 - 1}{\sqrt{x} - 1} \frac{\sqrt{x} + 1}{\sqrt{x} + 1} = \frac{(x + 1)(x + 1)(\sqrt{x} + 1)}{(x + 1)} = (x + 1)(\sqrt{x} + 1)$$

so we get

$$\lim_{x \to 1} (x + 1)(\sqrt{x} + 1) = 2 \times 2 = 4$$

■

If the indeterminate form does not take the form 0/0 or ∞/∞, then the expression should be rewritten so that L'Hôpital's rule can be applied, as illustrated in the next example.

Example 3 Find $\lim_{x \to 0} (x \ln x)$.

Solution As $x \to 0$, the factors x and $\ln x$ are in competition:

$$\lim_{x \to 0} x = 0, \qquad \lim_{x \to 0} \ln x = -\infty$$

yielding the indeterminate form $(0)(-\infty)$. In order to apply L'Hôpital's rule, the indeterminate form should be 0/0 or ∞/∞. This can be accomplished by writing

$$\lim_{x\to 0} x \ln x = \lim_{x\to 0} \frac{\ln x}{1/x}$$

which yields the indeterminate form $-\infty/\infty$. L'Hôpital's rule can be applied; writing

$$f(x) = \ln x, \qquad f'(x) = 1/x$$
$$g(x) = 1/x, \qquad g'(x) = -1/x^2$$

we get

$$\lim_{x\to 0} \frac{\ln x}{1/x} = \lim_{x\to 0} \frac{1/x}{-1/x^2} = \lim_{x\to 0} (-x) = 0$$

Therefore, we can conclude that

$$\lim_{x\to 0} x \ln x = 0$$

This result tells us that x approaches zero more rapidly than $\ln x$ approaches $-\infty$ as $x \to 0$. ■

If $\lim_{x\to a} [f'(x)/g'(x;]$ is also indeterminate, L'Hôpital's rule is applied again. It is applied as often as needed until we can determine whether or not the limit exists.

Example 4 Find $\lim_{x\to 1} \frac{(\ln x)^2}{(x-1)^2}$

Solution As $x \to 1$, $\ln x \to 0$ and $(x - 1)^2 \to 0$, leading to the indeterminate form 0/0. Letting

$$f(x) = (\ln x)^2, \qquad f'(x) = \frac{2 \ln x}{x}$$
$$g(x) = (x-1)^2, \qquad g'(x) = 2(x-1)$$

we can apply L'Hôpital's rule to give

$$\lim_{x\to 1} \frac{(\ln x)^2}{(x-1)^2} = \lim_{x\to 1} \frac{\ln x}{x(x-1)}$$

As $x \to 1$, both $\ln x$ and $x(x - 1)$ approach zero, producing 0/0 again. L'Hôpital's Rule is applied a second time, with

$$f(x) = \ln x, \qquad f'(x) = \frac{1}{x}$$
$$g(x) = x(x-1), \qquad g'(x) = 2x - 1$$

so we get

$$\lim_{x\to 1} \frac{\ln x}{x(x-1)} = \lim_{x\to 1} \frac{1/x}{2x-1} = 1$$

Therefore, we can conclude that

$$\lim_{x \to 1} \frac{(\ln x)^2}{(x - 1)^2} = 1$$

■

It is important to remember that L'Hôpital's rule is applied only to indeterminate forms. If it is used in situations in which an indeterminate form does not occur, it can lead to erroneous answers, as the next example illustrates.

Example 5 Find $\lim_{x \to 1} \frac{e^x}{x - 1}$.

Solution As $x \to 1$,

$$e^x \to e \quad \text{and} \quad (x - 1) \to 0$$

leading to the ratio $e/0$, which is not indeterminate. The ratio $e/0$ tells that the *limit does not exist.* If we were to apply L'Hôpital's rule to this situation, with

$$f(x) = e^x, \qquad f'(x) = e^x$$
$$g(x) = x - 1, \qquad g'(x) = 1$$

we would find that

$$\lim_{x \to 1} \frac{f'(x)}{g'(x)} = \lim_{x \to 1} \frac{e^x}{1} = e \neq \lim_{x \to 1} \frac{f(x)}{g(x)}$$

■

Application

L'Hôpital's rule is often needed in the evaluation of present values of cash flows that continue indefinitely (Equation 4, Section 7.4).

Example 6 Annual revenue from the sale of silver at the Hi-Ho Silver Mine is given by the equation

$$R(t) = 50 + t \quad \text{thousand dollars/year}$$

Find the present value of the revenue generated from the sale of all the silver to be mined if the company expects a 10 percent annual return on its investments?

Solution The present value (in thousands of dollars) of all future revenues is given by the equation

$$\begin{aligned}
\text{Present value} &= \int_0^\infty (50 + t)e^{-0.10t}\,dt \\
&= \int_0^\infty 50e^{-0.10t}\,dt + \int_0^\infty te^{-0.10t}\,dt \\
&= \lim_{b \to \infty} \left(\int_0^b 50e^{-0.10t}\,dt + \int_0^b te^{-0.10t}\,dt \right)
\end{aligned}$$

Noting that the second integral requires integration by parts, we get

$$\text{Present value} = \lim_{b\to\infty} [-500e^{-0.10t} - 10te^{-0.10t} - 10e^{-0.10t}]\Big|_0^b$$
$$= \lim_{b\to\infty} [-510e^{-0.10t} - 10te^{-0.10t}]\Big|_0^b$$
$$= \lim_{b\to\infty} (-510e^{-0.10b} + 510 - 10be^{-0.10b} + 0)$$

As $b \to \infty$, the term $-510e^{-0.10b} \to 0$, but the term $-10be^{-0.10b}$ leads to the indeterminate form $(-\infty)(0)$. The limit can be found by writing $-10be^{-0.10b}$ as a ratio $-10b/e^{0.10b}$ and applying L'Hôpital's rule:

$$\lim_{b\to\infty} \frac{-10b}{e^{0.10b}} = \lim_{b\to\infty} \frac{-100}{e^{-0.10b}} = 0$$

So we get

$$\text{Present value} = \$510 \text{ thousand}$$

■

7.6 EXERCISES

Use L'Hopital's rule, when appropriate, to determine each of the following limits.

1. $\lim_{x\to\infty} \frac{x^2}{e^x}$
2. $\lim_{x\to 1} \frac{x^2 - 1}{x^2 - 2x - 3}$
3. $\lim_{x\to 0^+} \frac{\ln x}{x}$
4. $\lim_{x\to 0} \frac{\sqrt[3]{x} - 2}{x^2 - 9x + 8}$
5. $\lim_{x\to\infty} xe^{-x}$
6. $\lim_{x\to 4} \frac{\sqrt{x} - 2}{x^3 - 3x^2 - 5x - 4}$
7. $\lim_{x\to -\infty} \frac{2x + 3}{x - 1}$
8. $\lim_{x\to 3} \frac{x^2 - 2x - 3}{x + 1}$
9. $\lim_{x\to -\infty} x^3e^x$
10. $\lim_{x\to -1} \frac{x^2 + 2x + 1}{x^3 + 1}$
11. $\lim_{x\to\infty} e^x \ln x$
12. $\lim_{x\to\infty} e^{-x} \ln x$
13. $\lim_{x\to 0} \frac{e^x - 1}{x}$
14. $\lim_{x\to\infty} \frac{2x^3 + 6x}{x^3 + 5}$
15. $\lim_{x\to e} \frac{\ln x - 1}{x - e}$
16. $\lim_{x\to 0} \frac{e^{2x} - 2e^x + 1}{e^x - 1}$
17. $\lim_{x\to\infty} \frac{e^{2x} - 2e^x + 1}{e^x - 1}$
18. $\lim_{x\to\infty} x(\ln x)^2$
19. $\lim_{x\to\infty} (x^3e^{-2x})$
20. $\lim_{x\to\infty} \frac{e^{4x}}{x^2}$
21. $\lim_{x\to 1} \frac{\ln x}{x - 1}$
22. $\lim_{x\to 1} \frac{(\ln x)^2}{x - 1}$
23. $\lim_{x\to 1} \frac{\sqrt[4]{x} - 1}{x^2 - 4x + 3}$
24. $\lim_{x\to 0} \frac{e^x - x + 1}{x^2}$

Evaluate each of the following improper integrals when they exist.

25. $\int_0^\infty xe^{-x}\,dx$

26. $\int_1^\infty x^2e^{-x}\,dx$

27. $\int_{-\infty}^0 xe^{2x}\,dx$

28. $\int_0^2 x\ln x\,dx$

29. $\int_0^\infty xe^x\,dx$

30. $\int_0^1 \ln x\,dx$

31. Annual production at the Wildcat Coal Mine is given by the equation

$$M(t) = 50e^{-0.03t} \quad \text{tons/year}$$

If $p(t)$, the price of coal, is given by the equation

$$p(t) = 40 + 2t \quad \text{dollars/ton}$$

find the total revenue generated from $t = 0$, assuming that the mine operates indefinitely?

32. Find the present value of the revenue generated in Exercise 31 if the company's annual rate of return on investment is 10 percent.

33. Oil is leaking from a damaged oil tanker at a rate given by the equation

$$R(t) = 200te^{-0.1t} \quad \text{barrels/hour} \qquad t \geq 0$$

Determine how many barrels of oil leak into the harbor if the leak goes unchecked.

KEY TERMS

integration by substitution
integration by parts
table of integrals
numerical approximation
trapezoidal rule
Simpson's rule
error estimates
improper integrals
convergent integral
divergent integral
L'Hôpital's rule

REVIEW PROBLEMS

Use the method of substitution to find each of the following indefinite integrals.

1. $\int 10x(x^2 + 1)^4\,dx$

2. $\int \frac{3x^2}{x^3 + 1}\,dx$

3. $\int 2xe^{-x^2}\,dx$

4. $\int 6x\sqrt{x^2 - 2}\,dx$

5. $\int \frac{6x + 6}{x^2 + 2x + 2}\,dx$

6. $\int x(3x + 6)(x^3 + 3x^2 + 4)^2\,dx$

7. $\int x\sqrt{2x + 3}\,dx$

8. $\int x^2e^{x^3}\,dx$

9. $\int (6x^2 + 2)(x^3 + x + 5)^4\,dx$

10. $\int \frac{x^2}{\sqrt{x + 2}}\,dx$

Use integration by parts to find each of the following indefinite integrals.

11. $\int xe^{-2x}\,dx$

12. $\int 3x \ln x\,dx$

13. $\int \ln 4x\,dx$

14. $\int x^2\sqrt{x+2}\,dx$

15. $\int 4x^2e^{-2x}\,dx$

16. $\int \frac{\ln x}{x^2}\,dx$

17. $\int \frac{\ln x}{x^n}\,dx$

18. $\int 2xe^{x+1}\,dx$

19. $\int x^2\sqrt{2x+3}\,dx$

20. $\int x^{n+1}\ln x\,dx$

Use the formulas in Table D to find each of the following indefinite integrals.

21. $\int \frac{5}{2x^2+3x}\,dx$

22. $\int \frac{3}{x^2-16}\,dx$

23. $\int \frac{2}{\sqrt{4x^2+25}}\,dx$

24. $\int \frac{4}{x(3x+2)^2}\,dx$

In problems 25 through 30, find an approximate value for each definite integral using (a) the trapezoidal rule and (b) Simpson's rule, with $n = 4$.

25. $\int_0^4 \sqrt{x^2+1}\,dx$

26. $\int_{-2}^{2} \sqrt{4-x^2}\,dx$

27. $\int_0^2 \frac{1}{1+x^2}\,dx$

28. $\int_0^4 \frac{x}{x^4+2}\,dx$

29. $\int_0^2 e^{\sqrt{x}}\,dx$

30. $\int_1^9 (\ln x)^2\,dx$

31. (a) Use Simpson's rule, with $n = 8$, to find an approximate value for the area of the region bounded by the semicircle $y = \sqrt{1-x^2}$ and the x axis.
(b) Compare your answer to part (a) with that given by the formula for the area of a semicircle, $A = \pi r^2/2$.

32. In statistics the area under the standard normal curve $f(x) = \frac{1}{\sqrt{2\pi}}e^{-x^2/2}$ between two given values of x is often used to calculate the probability of an event. Use Simpson's rule, with $n = 8$, to find an approximate value of

$$\frac{1}{\sqrt{2\pi}}\int_{-1}^{1} e^{-x^2/2}\,dx$$

In problems 33 through 36, evaluate each convergent improper integral.

33. $\int_0^\infty 6e^{-2x}\,dx$

34. $\int_0^\infty \frac{4x}{(x^2+1)^{3/2}}\,dx$

35. $\int_1^\infty 6xe^{-2x}\,dx$

36. $\int_{-\infty}^{0} 6xe^{x^2-1}\,dx$

In problems 37 through 40, use L'Hôpital's rule, when needed, to find each limit.

37. $\lim_{x\to 0} \frac{e^{-x}-1}{x}$

38. $\lim_{x\to\infty} \frac{\ln x^2}{e^x}$

39. $\lim_{x\to\infty} \frac{e^{2x}}{x}$

40. $\lim_{x\to 0} \frac{2e^x - x^2 - 2}{x^2}$

41. The concentration $C(t)$ of an antibiotic drug at any time t, in hours, is given by the equation

$$C(t) = \frac{t}{t^2 + 1}, \qquad t \geq 0$$

What is the average concentration from $t = 0$ to $t = 4$?

42. An analyst for an auto parts manufacturer predicts that annual sales $S(t)$, in thousands, of a new carburetor at any time t, in years, is given by the equation $S(t) = 100te^{-0.25t}$, where $t = 0$ corresponds to today.
   (a) How many carburetors are sold between $t = 0$ and $t = 4$?
   (b) How many carburetors should the company produce now to take care of all future demand?

43. A company that uses the declining-balance method of depreciation finds that the book value of an asset is changing at a rate given by the equation

$$f(t) = -6000e^{-0.2t}$$

where $f(t)$ is the rate in dollars per year and t is the time, in years, from the date of purchase of the asset.
   (a) By how much does the book value change from $t = 0$ to $t = 5$?
   (b) If the purchase cost C of the asset is defined by the equation

$$C = \int_0^\infty |f(t)|\, dt$$

   find C.

44. The concentration of cholesterol in a patient has been declining at a rate given by the equation

$$C'(t) = \frac{-50}{\sqrt{t}(1 + \sqrt{t})^2}$$

where $C(t)$ is the concentration, in milligrams per deciliter, and t is the time, in months, from the beginning of the treatment ($t = 0$).
   (a) Determine the change in her cholesterol level over the first four weeks of treatment.
   (b) What is the long-term ($t \to \infty$) change in her cholesterol level?

45. Acme Lumber Company is harvesting lumber from a tract of land in the Northwest at a rate given by the equation $f(t) = 2e^{-0.2t}$ million board-feet per year, where t is the time, in years, from today ($t = 0$).
   (a) At this rate, how much lumber can eventually be harvested from the tract?
   (b) If the price p, in dollars, of 100 board-feet is expected to increase with time according to the equation $p = 1 + 0.04t$, determine the total revenue that the harvested lumber will produce.
   (c) If the management of Acme expects an annual rate of return of 15 percent compounded continuously on their investments, find the present value of all the revenue that will be generated from harvesting lumber on the tract.

46. The demand equation for a new compact-disk player is

$$p(q) = 300e^{1-0.02q}$$

where $p(q)$ is the unit selling price, in dollars, and q is the number of items sold each week.

(a) Find the consumers' surplus when $q = 75$.
(b) Find the consumer's surplus if the unit selling price is set to maximize revenue.

47. A pollutant from a chemical plant is leaking into a river at a rate given by the equation

$$f(t) = \frac{2000t}{(t^2 + 1)^2}, \qquad t \geq 0$$

where $f(t)$ is the rate in gallons per month and t is the time, in months, from today ($t = 0$).

(a) How many gallons leak into the river during the next ten months?
(b) Over the long term ($t \to \infty$), how many gallons leak into the river?

48. A large drug company uses the sum-of-the-years-digits method to determine the depreciation of an electron microscope. The book value $V(t)$, in dollars, at any time t, in years, is described by the equation

$$V(t) = 6000(5 - t)^2, \qquad 0 \leq t \leq 5$$

Find the average value of the electron microscope from $t = 0$ to $t = 5$.

8

MULTIVARIABLE CALCULUS

INTRODUCTION

Up to this point we have focused our attention upon functions of a single independent variable. However, most applications involve quantities that are functions of two or more independent variables. This chapter serves as an introduction to functions of this type and to applications in which the derivatives and integrals of multivariable functions are needed.

8.1 FUNCTIONS OF TWO VARIABLES

Suppose, for simplicity, that a retailer sells only two different types of cameras, one an inexpensive instamatic that sells for \$20 and the second a 35-mm camera that retails for \$125. If x represents the number of instamatics sold and y the number of 35-mm cameras sold, the revenue R can be expressed mathematically as a function of x and y by the equation

$$R = 20x + 125y$$

For each pair of values assigned to the *independent* variables x and y, the *dependent* variable R assumes a single value. In this situation R is said to be a *function* of the two independent variables x and y.

Definition An equation of the form $z = f(x, y)$ represents a **function of two variables** if the equation determines a unique real value of z for each ordered pair of real numbers (x, y) for which the equation is defined.

The variables x and y are called the **independent variables** and z is the **dependent variable.** The **domain** is the set of all ordered pairs (x, y) for which the equation is defined.

Example 1 If the function $z = f(x, y)$ is defined by the equation

$$z = f(x, y) = 3x^2 - 2y^2 + xy$$

find each of the following:

(a) $f(1, 4)$
(b) $f(4, 1)$
(c) $f(a, b)$
(d) $f(x + h, y)$
(e) $f(x, y + h)$

Solution The value assigned to the dependent variable z is found by substituting the given values of x and y into the equation that defines the function, so we get

(a) $z = f(1, 4) = 3(1)^2 - 2(4)^2 + 1(4) = -25$

(b) $z = f(4, 1) = 3(4)^2 - 2(1)^2 + 4 = 50$

(c) $z = f(a, b) = 3a^2 - 2b^2 + ab$

(d) $z = f(x + h, y)$
$= 3(x + h)^2 - 2y^2 + (x + h)y$
$= 3x^2 + 6xh + 3h^2 - 2y^2 + xy + yh$

(e) $z = f(x, y + h)$
$= 3x^2 - 2(y + h)^2 + x(y + h)$
$= 3x^2 - 2y^2 - 4yh - 2h^2 + xy + xh$ ■

Example 2 Suppose that $C(W, L)$, the cost, in dollars, of manufacturing a refrigerator is given by the equation

$$C(W, L) = 20 + W + 6L$$

where W is the weight, in pounds, and L is the amount of labor, in hours, used to produce one refrigerator. What is the cost of a refrigerator that weighs 125 pounds and requires 10 hours of labor?

Solution The unit cost of each refrigerator is found by setting $W = 125$ and $L = 10$ and evaluating. We get

$$C(125, 10) = 20 + 125 + 6(10) = \$205$$ ■

Three-Dimensional Coordinate Systems

To plot the graph of a function of two variables $z = f(x, y)$, we need a coordinate system whose points are represented by ordered triples (x, y, z) of real numbers. A three-dimensional coordinate system can be formed by setting three number lines, usually denoted as the x, y, and z axes, at right angles to one another and aligning them so that the common point of intersection, called the *origin*, corresponds to the number 0 on all three axes. The customary alignment of the three axes is shown in Figure 1. The three planes formed by each pair of coordinate axes are called the xy, xz, and yz-planes, portions of which are shown in Figure 2.

Each ordered triple (x, y, z) of real numbers can be represented as a point P in a three-dimensional coordinate system and vice versa; that is, each point can be represented as an ordered triple. The first element x of an ordered triple (x, y, z) is called the *x coordinate* and represents the directed distance of the point P from the yz-plane. In the same way, the second and third elements of the ordered triple are called the y and z *coordinates* and represent the directed

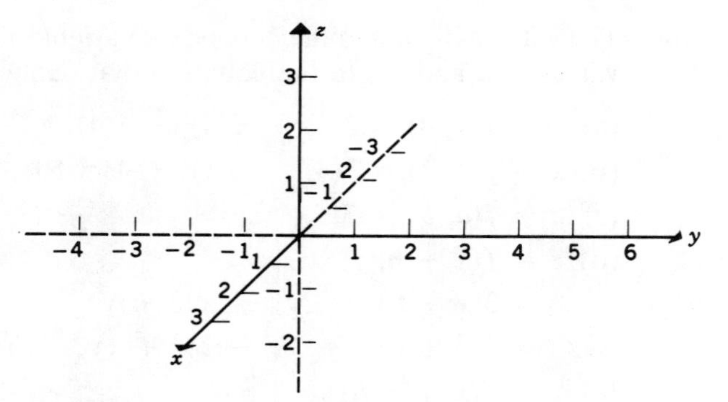

Figure 1

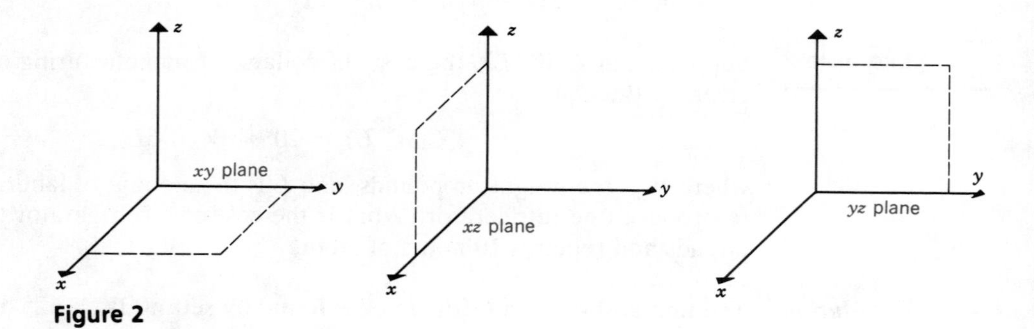

Figure 2

distances of P from the xz and xy-planes, respectively. Figure 3 shows the point P corresponding to the ordered triple (2, 4, 5). The xz, yz, and xy-planes divide space into eight regions known as *octants*. The x, y, and z coordinates of all points in the first octant are positive. Points on the x, y, and z axes are char-

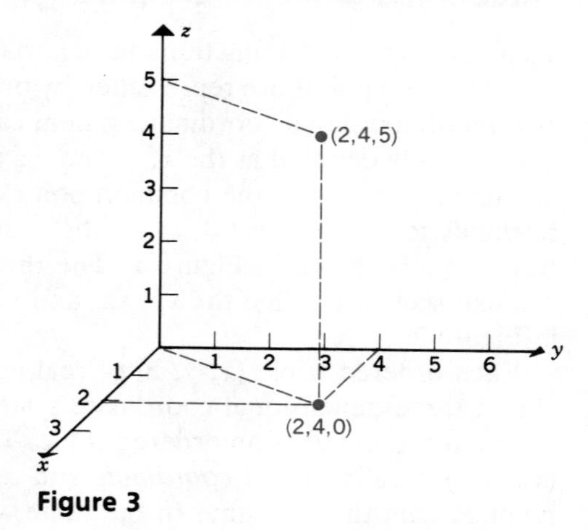

Figure 3

acterized by the fact that the coordinates of the other two variables equal zero; a group of points located on the axes is shown in Figure 4.

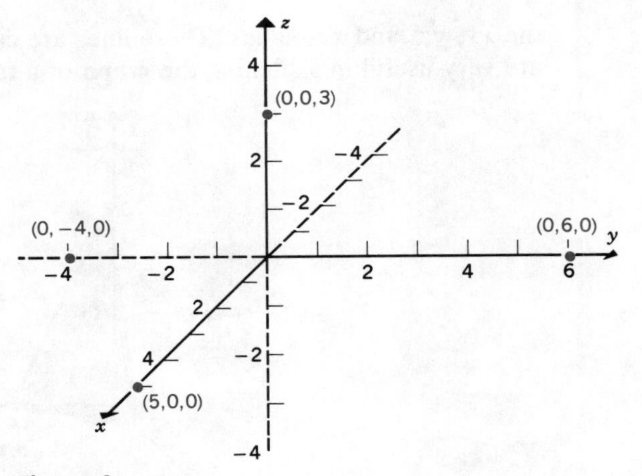

Figure 4

As in a two-dimensional system, the graph of a function of two variables is the set of all points (x, y, z) satisfying the equation $z = f(x, y)$, which defines the function. Sketching the graph of a function of two variables is generally difficult and time-consuming because the graph usually is a three-dimensional surface, whereas the graph of a function of one variable is generally a two-dimensional curve. However, there is one type of function whose graph can be sketched easily, the linear function

$$z = f(x, y) = ax + by + c \quad (a, b, \text{ and } c \text{ are constant}) \tag{1}$$

The **graph** of a **linear function** of two variables is a **plane.** Noting that a plane is determined uniquely by three noncollinear points, we can construct a partial representation of its graph by locating, whenever possible, the points where the plane intersects the three coordinate axes. The point where the plane intersects the x axis is called the $\boldsymbol{x}$ **intercept** of the plane; the points where the plane intersects the y and z axes are called the **y** and **z intercepts,** respectively. The process of sketching a partial graph is illustrated in the next example.

Example 3 Sketch the graph of the function

$$z = f(x, y) = -4x - 2y + 8$$

Solution The intercepts are found by setting two of the three variables equal to zero and then solving the resulting equation for the third variable. This procedure yields the following set of ordered triples: (0, 0, 8), (0, 4, 0), and (2, 0, 0). These points are shown in Figure 5 together with the shaded surface which represents the portion of the plane located in the first octant. The three lines shown with their equations on the graph result from the intersection of the given plane with the xy, yz, and xz-planes. These lines are called **traces** of the given plane and are very useful in sketching the graph of a function.

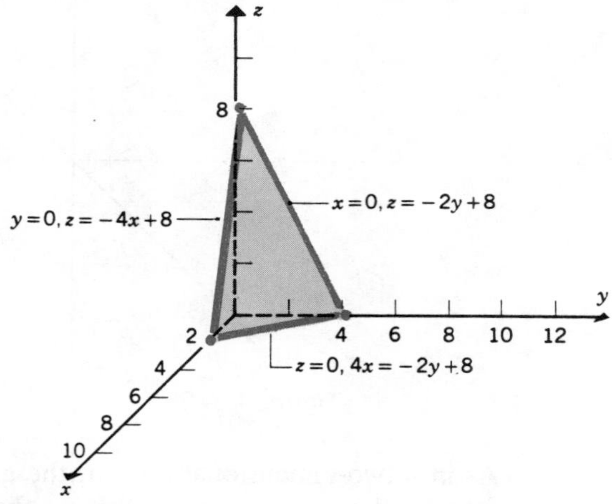

Figure 5

If the coefficient of one of the variables in Equation 1 equals zero, the graph is a plane parallel to the axis represented by this variable, as illustrated in the next example.

Example 4 Sketch the graph of the function

$$z = f(x, y) = \frac{-2y}{3} + 2 \quad \text{or} \quad 2y + 3z = 6$$

Solution The graph is found by plotting the equation $2y + 3z = 6$ in the yz-plane $(x = 0)$ and then extending the line indefinitely parallel to the x axis, as shown in Figure 6, to generate the graph of the equation.

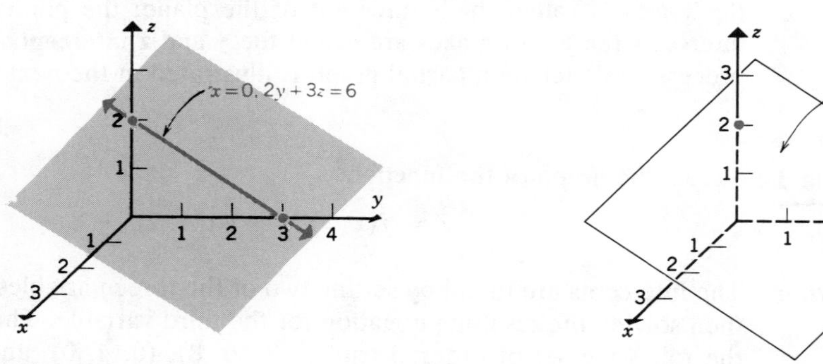

Figure 6 ■

If the coefficients of two of the variables in Equation 1 equal zero, the graph is a plane parallel to the coordinate plane represented by those variables, as illustrated in the next example.

Example 5 Sketch the graph of the equation $y = 4$.

Solution The equation $y = 4$ is satisfied by all ordered triples having the form $(x, 4, z)$, that is, the set of all points whose y coordinate equals 4. This set of points lies in a plane which is parallel to the xz-plane and whose y intercept equals 4, as shown in Figure 7.

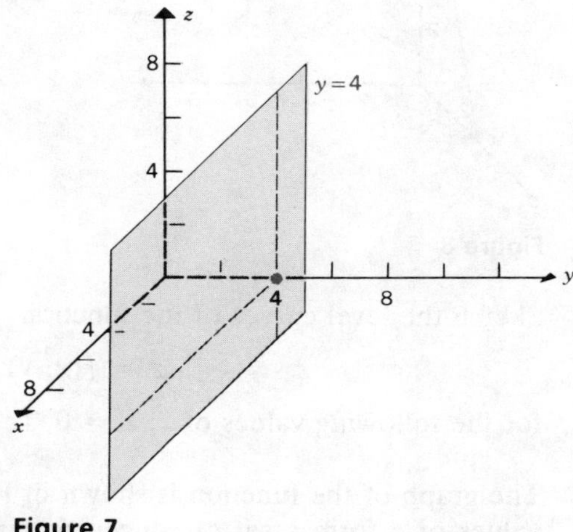

Figure 7

Level Curves

Functions of two variables can also be represented graphically in two dimensions by level curves. A **level curve** is the curve that results from vertically projecting onto the xy-plane all points on the surface $z = f(x, y)$ that have the same z-coordinate. Level curves are generally easier to sketch than the surface itself and often can provide a great deal of information about the shape of the surface.

Example 6 Sketch the level curves of the function

$$z = f(x, y) = -2x - 4y + 8$$

for the following values of z: $z = 0, 2,$ and 4.

Solution A partial graph of the function is shown in Figure 8*a*. When the value of z is fixed, the graph of the resulting equation is a straight line parallel to the xy-plane. For example, when $z = 4$, the equation becomes $4 = -2x - 4y + 8$; its graph is a straight line located four units above and parallel to the xy-plane. A portion of this graph is also shown in Figure 8*a* together with the graphs of the equations

$$z = 0 \quad \text{and} \quad 0 = -2x - 4y + 8$$
$$z = 2 \quad \text{and} \quad 2 = -2x - 4y + 8$$

The corresponding level curves are found by vertically projecting the points on these lines onto the xy-plane and are shown in Figure 8*b*.

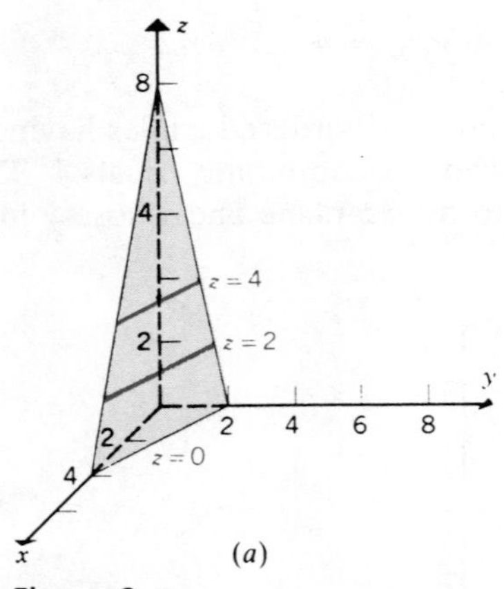

(b)

Figure 8

Example 7 Sketch the level curves of the function

$$z = f(x, y) = x^2 + y^2$$

for the following values of z: $z = 0, 1, 4, 9$, and 16.

Solution The graph of the function is shown in Figure 9. The level curves for the given values of z form a set of concentric circles about the origin (0, 0) in the xy-plane, as shown in Figure 10. ■

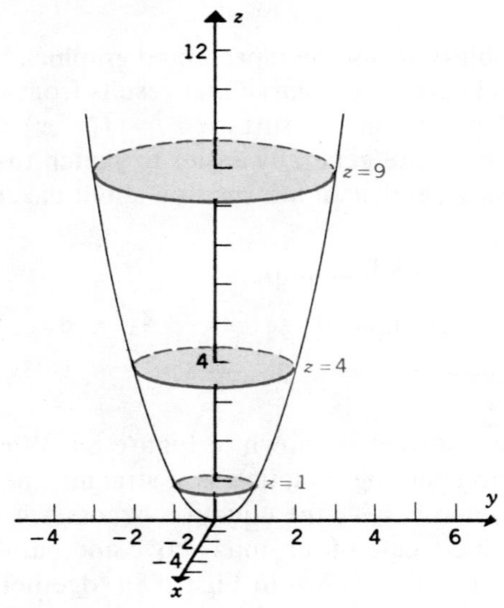

Figure 9

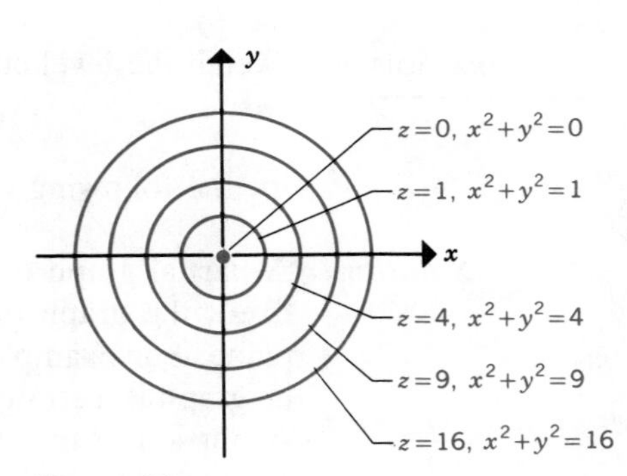

Figure 10

Example 8 Sketch the level curves of the function

$$z = f(x, y) = \frac{1}{xy} \quad \text{for } z = \frac{1}{4}, \frac{1}{2}, 1, 2$$

Solution For each value of z, we find the equation that describes the relationship between x and y. For example, when $z = \frac{1}{4}$, we get the equation

$$\frac{1}{4} = \frac{1}{xy}$$

or

$$xy = 4$$

The graph of this equation is shown in Figure 11 along with those for $z = \frac{1}{2}$, 1, 2. ■

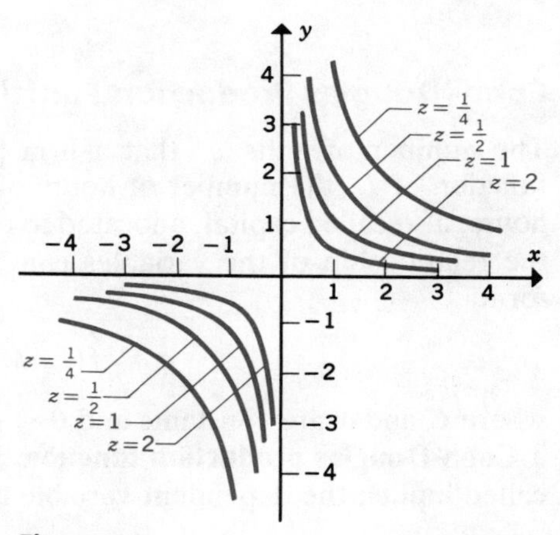

Figure 11

Applications

Level curves are useful in evaluating the allocation of available resources among competing demands, as illustrated in the next example.

Example 9 (a) The developers of a new shampoo, Fair Hair, have decided to promote the product via television and magazine advertising. If a 30-second television spot costs \$1000 and a full-page magazine spread \$4000, what is the monthly cost in terms of x, the number of television spots, and y, the number of magazine advertisements?

(b) If monthly advertising costs have been set at \$60,000, what does the corresponding level curve look like?

Solution (a) The cost $C(x, y)$ is described by the equation

$$C(x, y) = 1000x + 4000y$$

(b) When $C(x, y) = 60{,}000$, the equation becomes

$$60{,}000 = 1000x + 4000y$$

and its graph is shown in Figure 12. The points on the line represent the various combinations of television and magazine promotions that satisfy the $60,000 total cost requirement.

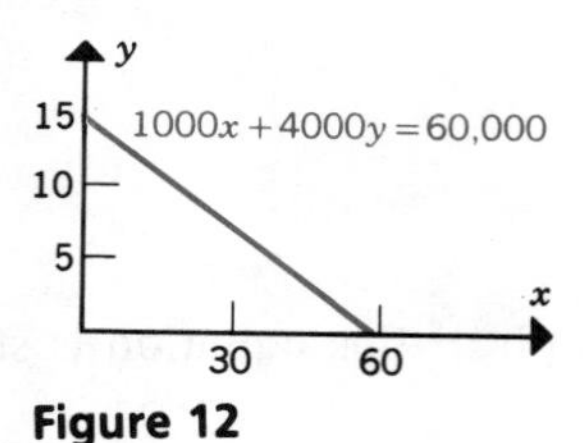

Figure 12

Cobb–Douglas Production Function

The number of units q that a firm produces in a given length of time is a function of L, the number of hours of labor, and M, the number of machine-hours, also called capital, allocated to producing the product. In many situations the relationship of the variables can be described by an equation having the form

$$q = f(L, M) = CL^{\alpha}M^{1-\alpha} \tag{2}$$

where C and α are constants and $0 < \alpha < 1$. An equation of this type is called a **Cobb-Douglas production function.** The independent variables L and M are called **inputs;** the dependent variable q is the **output** for the process.

Example 10 The production function for a steel mill is given by the equation

$$q = 100L^{0.75}M^{0.25}$$

where q is the output in tons.

(a) What is the output when $L = 625$ and $M = 10{,}000$?

(b) What is the output when $L = 1250$ and $M = 20{,}000$?

Solution (a) The output can be calculated directly:

$$\begin{aligned} q &= 100(625)^{3/4}(10{,}000)^{1/4} \\ &= 100(125)(10) = 125{,}000 \text{ tons} \end{aligned}$$

(b)
$$\begin{aligned} q &= 100(1250)^{3/4}(20{,}000)^{1/4} \\ &= 100(625)^{3/4}(2)^{3/4}(10{,}000)^{1/4}(2)^{1/4} \\ &= 100(125)(10)(2)^{3/4}(2)^{1/4} = 250{,}000 \text{ tons} \end{aligned}$$

In part (b), where the inputs were doubled, the resulting output was also doubled. If each input in a Cobb–Douglas production function is multiplied by a constant k, the output is also multiplied by k. In economics, a production function with this property is said to have *constant returns to scale*.

8.1 EXERCISES

1. If $z = f(x, y) = 2x + 3y$, find each of the following:
(a) $f(2, 1)$ (b) $f(-1, 4)$
(c) $f(0, -1)$ (d) $f(\sqrt{2}, \sqrt{2})$
(e) $f(1 + h, 2)$ (f) $f(1 + h, 2) - f(1, 2)$

2. If $z = f(x, y) = x^2 - y^2$, find each of the following.
(a) $f(1, 1)$ (b) $f(2, 3)$
(c) $f(1 + h, 1)$ (d) $f(1 + h, 1) - f(1, 1)$
(e) $\dfrac{f(1 + h, 1) - f(1, 1)}{h}$ (f) $\lim\limits_{h \to 0} \dfrac{f(1 + h, 1) - f(1, 1)}{h}$

3. If $z = f(x, y) = \dfrac{y}{x}$, find each of the following.
(a) $f(1, 2)$ (b) $f(2, 1)$
(c) $f(1, 2 + k)$ (d) $f(1, 2 + k) - f(1, 2)$
(e) $\dfrac{f(1, 2 + k) - f(1, 2)}{k}$ (f) $\lim\limits_{k \to 0} \dfrac{f(1, 2 + k) - f(1, 2)}{k}$

4. If $z = f(x, y) = 2x\sqrt{y} - 3y\sqrt{x} + 1$, find each of the following.
(a) $f(1, 4)$ (b) $f(4, 9)$
(c) $f(0, 16)$ (d) $f(\frac{1}{4}, 1)$
(e) $f(a, a)$ (f) $f(4a, a)$

5. If $z = f(x, y) = x \ln y - y \ln x$, find each of the following.
(a) $f(e, e)$ (b) $f(1, 1)$
(c) $f(1, e)$ (d) $f(e, 1)$
(e) $f(1, e^2)$ (f) $f(e^2, 1)$

6. If $z = f(x, y) = \sqrt{x + y}$, find each of the following.
(a) $f(1, 0)$ (b) $f(1, 3)$
(c) $f(5, 4)$ (d) $f(-4, 5)$
(e) $f(2, 2)$ (f) $f(3a^2, 6a^2)$

7. If $z = f(x, y) = ye^x$, find each of the following.
(a) $(0, 1)$ (b) $(1, 0)$
(c) $(1, e)$ (d) $\left(1, \dfrac{1}{e}\right)$
(e) $(\frac{1}{2}, \sqrt{e})$ (f) $(\ln 2, 2)$

8. If $f(x, y) = e^y \ln x$, find each of the following.
(a) $f(1, 1)$ (b) $f(e, 0)$
(c) $f(1, 0)$ (d) $f(1, a)$
(e) $f(2, \ln 2)$ (f) $f(e, \ln 3)$

9. The volume of a right-circular cylinder (Figure 13) can be found from the equation

$$V(r, h) = \pi r^2 h$$

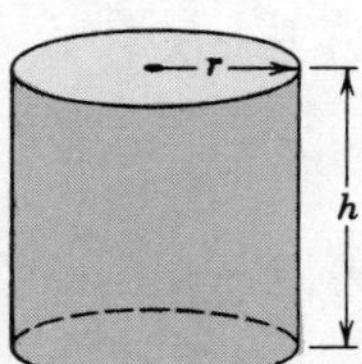

Figure 13

where r is the radius and h is the height. Find each of the following:

(a) $V(2, 9)$ (b) $V(3, 4)$
(c) $V(5, 5)$ (d) $V(10, 10)$

10. One thousand dollars is deposited in a savings account whose annual rate of interest is r, compounded continuously. The amount in the account after t years is given by the equation

$$A(r, t) = 1000e^{rt}$$

Find each of the following:

(a) $A(0.10, 5)$ (b) $A(0.20, 10)$
(c) $A(0.06, 15)$ (d) $A(0.15, 4)$

Sketch the graph of each of the following functions.

11. $z = 9 - 3x - 6y$

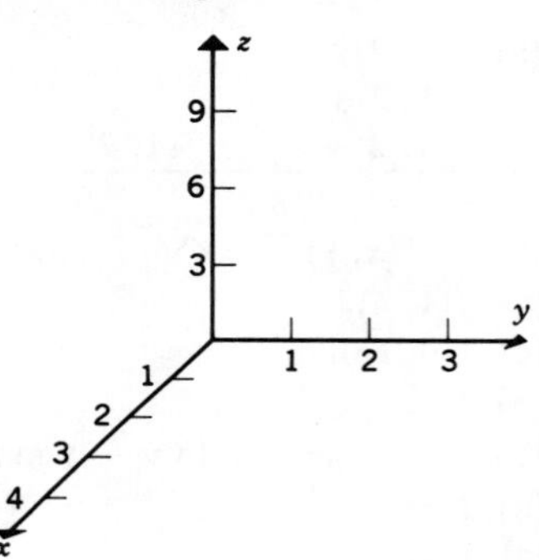

12. $z = 8 - 4x - 2y$

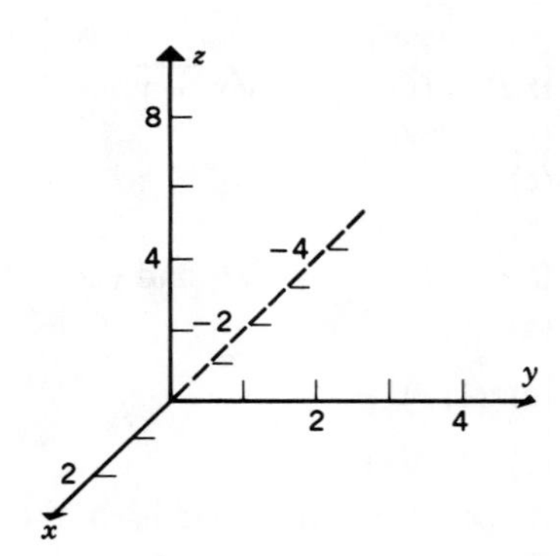

Sketch the level curves of each of the following functions for the given values of z.

13. $z = f(x, y) = x + y$, $z = -1, 0, 1, 2$

14. $z = f(x, y) = -x^2 + y$, $z = -1, 0, 4$

15. $z = f(x, y) = xy$, $z = -1, 0, 1$

16. $z = f(x, y) = \ln(x - y)$, $z = -1, 0, 1, 2$

17. $z = f(x, y) = \dfrac{x}{y}$, $z = -2, -1, 0, 1, 2$

18. $z = f(x, y) = \dfrac{1}{x^2y}$, $z = \dfrac{1}{2}, 1, 2$

19. $z = f(x, y) = \sqrt{x + y}$, $z = 0, 1, 2$

20. $z = f(x, y) = x^2 - y^2, \qquad z = -1, 0, 1$

21. The Tasty Hamburger Company sells hamburgers for 75¢ each and cheeseburgers for 85¢ each. Find an equation for the company's revenue R in terms of x, the number of hamburgers sold, and y, the number of cheeseburgers sold.

22. The Office Remodeling Company charges \$2 per square foot to carpet a floor, \$1 per square foot to panel a wall, and \$1.50 per square foot to install an acoustical ceiling. Find an equation describing the cost C of completely remodeling a rectangular office whose length is L, width is W, and height is H. What is the cost of remodeling an office whose dimensions are 20 by 15 by 8 ft?

23. An electronics company produces two types of compact-disk players, brands X and Y. Both brands are sold to wholesalers, brand X for \$200 and brand Y for \$300. Unit production costs (labor and material) are \$100 for brand X and \$175 for brand Y; overhead costs are \$10,000 per week. If x and y are the numbers of brand X and Y disk players produced each week, find an equation for

(a) The weekly revenue $R(x, y)$.
(b) The weekly cost $C(x, y)$.
(c) The weekly profit $P(x, y)$.

24. The present value of a perpetuity (fixed payment that continues indefinitely) is given by the equation

$$PV = \frac{R}{r}$$

where R is the fixed payment and r is the rate of return. Sketch the level curves for a perpetuity that has the following present values: (a) \$1000, (b) \$2000.

25. The cans for Burpy Cola are right-circular cylinders (see Figure 14). Material for the top and bottom surfaces costs 10¢ per square foot, and material for the cylinder costs 4¢ per square foot. Find an equation for the cost C in terms of r, the radius, and h, the height. (The surface area of a closed circular cylinder is $2\pi rh + 2\pi r^2$.)

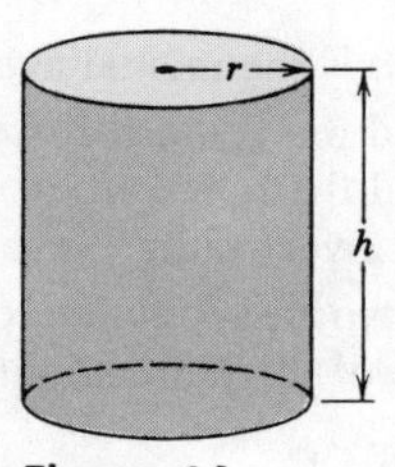

Figure 14

26. The production function for a furniture manufacturer is given by the equation

$$q = 500L^{0.50}M^{0.50}$$

where q is the output in units, L is the labor in hours, and M is the number of machine hours.

(a) What is the output when $L = 16$ and $M = 100$?
(b) What is the output when $L = 48$ and $M = 75$?

27. The production function for a firm producing machine tools is

$$q = f(L, M) = 50L^{1/5}M^{4/5}$$

(a) Find the output when $L = 243$ and $M = 32$.

(b) Find the output when $L = 486$ and $M = 64$.
(c) Show that the output q is doubled when both L and M are doubled.

28. Given the production function

$$q = CL^{\alpha}M^{1-\alpha}$$

show that multiplying both L and M by k means that the output q is multiplied by k.

29. Both the Schmaltz and Bash breweries have recently introduced their own brands of low-calorie beer. Let p_1 and q_1 denote the unit price and quantity of beer sold by Schmaltz, and let p_2 and q_2 represent the unit price and quantity sold by Bash. Suppose the quantity q_1 depends on both p_1 and p_2 as described by the equation

$$q_1 = f(p_1, p_2) = 10 - p_1 + 0.5p_2$$

and the dependence of q_2 upon p_1 and p_2 is given by the equation

$$q_2 = g(p_1, p_2) = 8 + 0.4p_1 - p_2$$

Find the revenues R_1 and R_2 for each brewery in terms of the unit prices p_1 and p_2.

30. The Fine Furniture Company manufactures two types of student desks: one is completely assembled and finished; the other is unassembled, unfinished, and sold in the form of a kit. The number of each kind that can be sold monthly is a function not only of its own unit price but also of the unit price of the other. Let q_1 represent the number of finished desks and q_2 the number of kits; let p_1 and p_2 represent the unit prices of each. The dependence of q_1 and q_2 on the unit prices p_1 and p_2 is described by the equations

$$q_1 = 500 - 5p_1 + 2p_2$$
$$q_2 = 400 + 2p_1 - 4p_2$$

Find the monthly revenue $R(p_1, p_2)$.

31. The following equation is used to measure the intelligence quotient (IQ) of a child,

$$\text{IQ} = f(m, c) = \frac{100m}{c}$$

where m is the mental age and c is the chronological age of the child.
(a) Find the IQ of a 10-year-old child whose mental age is 12.
(b) Find the IQ of a boy whose mental age equals his chronological age.
(c) A 12-year-old girl has an IQ of 130. What is her mental age?

32. When two resistors are connected in parallel, as shown in Figure 15, the total resistance of the pair, in ohms, is given by the equation

$$R = f(R_1, R_2) = \frac{R_1R_2}{R_1 + R_2}$$

(a) Find the resistance when a 50-ohm resistor and a 100-ohm resistor are connected in parallel.
(b) Find the resistance when R_1 equals R_2.

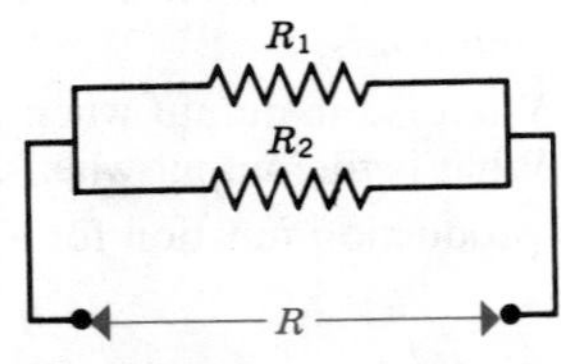

Figure 15

8.2 PARTIAL DERIVATIVES

The derivative of a function of a single variable $y = f(x)$ can be used to determine the slope of the line tangent to the corresponding curve at any point. When the graph of a function in two variables $z = f(x, y)$ is a smooth surface, an infinite number of lines tangent to the surface can be drawn at any point. This property suggests that a single derivative is not sufficient to describe the rate of change of a function of two variables. However, the rate of change of a function of two variables can be determined if only one variable is permitted to change while the other is kept constant. When the independent variable y is held constant, the function $z = f(x, y)$ can be treated as a function of x alone. The derivative with respect to the variable x then can be found by using the methods developed in Chapter 3. This derivative, called the **partial derivative of *f* with respect to *x*** is written

$$f_x(x, y) \quad \text{or} \quad \frac{\partial z}{\partial x}$$

For example, if $z = f(x, y) = x^3 + 2xy + y^4$, $f_x(x, y)$ is found by treating y as a constant and differentiating with respect to x, giving

$$f_x(x, y) = 3x^2 + 2y$$

In the same way, the partial derivative with respect to y is found by keeping the variable x constant while y is permitted to change. This derivative is written

$$f_y(x, y) \quad \text{or} \quad \frac{\partial z}{\partial y}$$

For the function $z = f(x, y) = x^3 + 2xy + y^4$, we have

$$f_y(x, y) = 2x + 4y^3$$

Definition Let $z = f(x, y)$ be a function of the independent variables x and y. The partial derivatives with respect to x and y, respectively, are defined as

$$f_x(x, y) = \frac{\partial z}{\partial x} = \lim_{h \to 0} \frac{f(x + h, y) - f(x, y)}{h} \tag{1}$$

$$f_y(x, y) = \frac{\partial z}{\partial y} = \lim_{k \to 0} \frac{f(x, y + k) - f(x, y)}{k} \tag{2}$$

The definitions will not be needed to find the partial derivatives; the methods developed in Chapter 3 are sufficient to enable us to find $f_x(x, y)$ and $f_y(x, y)$.

However, it is essential that you remember to treat y as a constant when finding $f_x(x, y)$ and that x be treated as a constant when finding $f_y(x, y)$.

Example 1 Find $f_x(x, y)$ and $f_y(x, y)$ for the function

$$z = f(x, y) = x^3 - 6xy + 2y^2$$

Solution The partial derivative with respect to x is found by differentiating each term with respect to x while keeping y constant, yielding

$$f_x(x, y) = 3x^2 - 6y$$

When we find $f_y(x, y)$, the roles of x and y are reversed, so we get

$$f_y(x, y) = -6x + 4y$$

■

Example 2 Find $f_x(x, y)$ and $f_y(x, y)$ for the function

$$f(x, y) = y \ln x - xe^{2y}$$

Solution We find $f_x(x, y)$ by differentiating each term with respect to x while keeping y constant

$$f_x(x, y) = y\frac{1}{x} - (1)e^{2y} = \frac{y}{x} - e^{2y}$$

Similarly, $f_y(x, y)$ can be found by differentiating with respect to y while keeping x constant; we get

$$f_y(x, y) = (1) \ln x - xe^{2y}(2) = \ln x - 2xe^{2y}$$

■

Let us now turn our attention to the question of the geometric interpretation of the partial derivatives $f_x(x, y)$ and $f_y(x, y)$. In particular, let us examine $f_x(x, y)$ and $f_y(x, y)$ for the function

$$z = f(x, y) = 9 - x^2 - y^2$$

at the point (1, 2, 4). To see what the partial derivative $f_x(1, 2)$, also written

$$\left.\frac{\partial z}{\partial x}\right|_{(1,2)}$$

represents geometrically, let us consider Figure 1, which shows the graph of the function together with a portion of the plane $y = 2$ [remember that y is kept constant when finding $f_x(x, y)$]. The intersection of the surface $z = 9 - x^2 - y^2$ with the plane $y = 2$ generates a curve called the *trace* of the surface in the plane. The equation of the trace is

$$z = f(x, 2) = 9 - x^2 - 2^2 = 5 - x^2$$

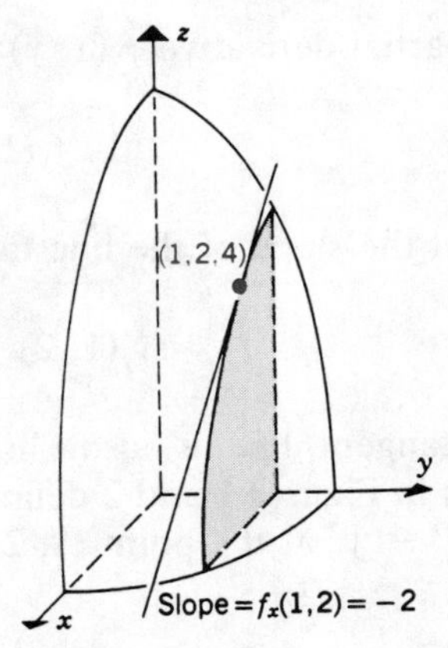

Figure 1

The partial derivative of z with respect to x in the plane $y = 2$ is

$$f_x(x, 2) = \left.\frac{\partial z}{\partial x}\right|_{y=2} = -2x$$

and represents the **slope** of a line tangent to the trace. The slope of the line tangent to the trace at the point (1, 2, 4), also shown in Figure 1, equals

$$f_x(1, 2) = \left.\frac{\partial z}{\partial x}\right|_{(1,2)} = -2(1) = -2$$

In the same way, $f_y(1, 2)$ represents the slope of the line tangent to the trace that results from the intersection of the surface $z = f(x, y) = 9 - x^2 - y^2$ with the plane $x = 1$, shown in Figure 2. The relationship between the y and z coordinates along the trace is described by the equation

$$z = f(1, y) = 9 - 1^2 - y^2 = 8 - y^2$$

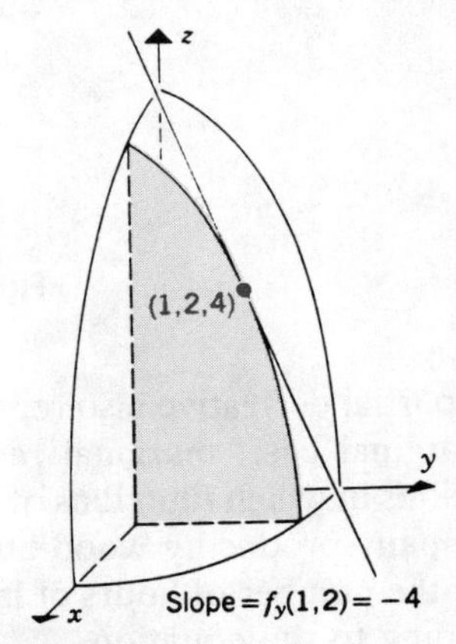

Figure 2

The partial derivative $f_y(1, y)$ takes the form

$$f_y(1, y) = \left.\frac{\partial z}{\partial y}\right|_{x=1} = -2y$$

so that the slope of the line tangent to the trace at (1, 2, 4) becomes

$$f_y(1, 2) = \left.\frac{\partial z}{\partial y}\right|_{(1,2)} = -2(2) = -4$$

This tangent line is also shown in Figure 2. The two perpendicular lines shown in Figures 1 and 2 define the plane tangent to the surface $z = f(x, y) = 9 - x^2 - y^2$ at the point (1, 2, 4) which is shown in Figure 3.

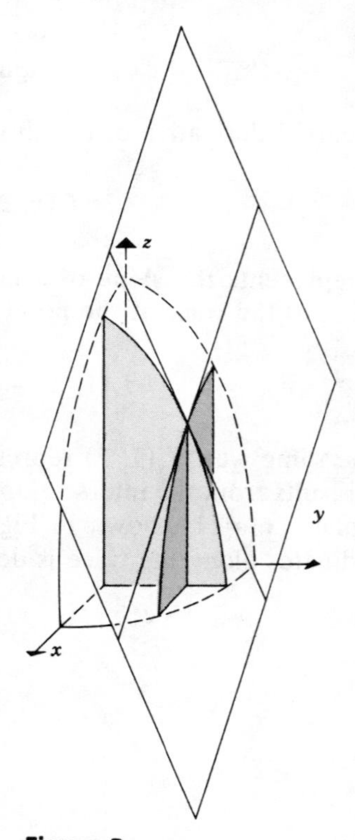

Figure 3

A partial derivative also represents a rates of change. Therefore, the concepts of marginal cost, marginal revenue, and marginal profit can be extended to situations in which functions of two or more variables are required. For example, a company producing wood-burning stoves finds that its monthly costs depend on L, the number of hours of labor, and W, the number of pounds of steel used, according to the equation

$$C(L, W) = 1500 + 7L + 2W$$

The partial derivative $\frac{\partial C}{\partial L} = C_L(L, W)$ represents the **marginal cost of labor,** and $\frac{\partial C}{\partial W} = C_W(L, W)$ represents the **marginal cost of material.** For this situation

$$\frac{\partial C}{\partial L} = \$7 \text{ per hour}$$

$$\frac{\partial C}{\partial W} = \$2 \text{ per pound of steel}$$

These results indicate that an additional hour of labor costs \$7 and each additional pound of steel costs \$2.

Second-Order Partial Derivatives

Each of the partial derivatives $f_x(x, y)$ and $f_y(x, y)$ can be differentiated with respect to the variables x and y, giving rise to four second partial derivatives defined as follows.

Definition If $z = f(x, y)$ is a function of the independent variables x and y, then

$$\frac{\partial}{\partial x}\left(\frac{\partial z}{\partial x}\right) = \frac{\partial^2 z}{\partial x^2} = f_{xx}(x, y) = f_{xx} \tag{3}$$

$$\frac{\partial}{\partial y}\left(\frac{\partial z}{\partial y}\right) = \frac{\partial^2 z}{\partial y^2} = f_{yy}(x, y) = f_{yy} \tag{4}$$

$$\frac{\partial}{\partial x}\left(\frac{\partial z}{\partial y}\right) = \frac{\partial^2 z}{\partial x\, \partial y} = f_{yx}(x, y) = f_{yx} \tag{5}$$

$$\frac{\partial}{\partial y}\left(\frac{\partial z}{\partial x}\right) = \frac{\partial^2 z}{\partial y\, \partial x} = f_{xy}(x, y) = f_{xy} \tag{6}$$

Although the order of differentiation is the same for both $\frac{\partial^2 z}{\partial x\, \partial y}$ and f_{yx}, that is, differentiating first with respect to y and second with respect to x, the order of the subscripts x and y is not the same for the two expressions. The order is different because both notations are abbreviations for expressions in which the correct order is displayed explicitly, that is,

$$\frac{\partial^2 z}{\partial x\, \partial y} = \frac{\partial}{\partial x}\left(\frac{\partial z}{\partial y}\right), \qquad f_{yx} = (f_y)_x$$

Fortunately, it is not necessary to concern yourself with the order in which the differentiation is carried out because $f_{yx} = f_{xy}$ for all the functions we shall encounter as well as for those used in most applications; however, it should be noted that $f_{xy} = f_{yx}$ is not true for all functions.

Example 3 Find the second partial derivatives of the function

$$z = f(x, y) = x^3 - 6xy + 2y^2$$

given in Example 1.

Solution Using Equations 3 through 6 and the results from Example 1, we get

$$f_{xx} = \frac{\partial^2 z}{\partial x^2} = \frac{\partial}{\partial x}(3x^2 - 6y) = 6x$$

$$f_{yy} = \frac{\partial^2 z}{\partial y^2} = \frac{\partial}{\partial y}(-6x + 4y) = 4$$

$$f_{yx} = \frac{\partial^2 z}{\partial x\, \partial y} = \frac{\partial}{\partial x}(-6x + 4y) = -6$$

$$f_{xy} = \frac{\partial^2 z}{\partial y\, \partial x} = \frac{\partial}{\partial y}(3x^2 - 6y) = -6$$

■

Example 4 Find the second partial derivatives of the function

$$z = f(x, y) = x^2y - xy^3$$

Solution The partial derivatives $f_x(x, y)$ and $f_y(x, y)$ are found by differentiating with respect to x and y, respectively, giving

$$\frac{\partial z}{\partial x} = f_x(x, y) = 2xy - y^3, \qquad \frac{\partial z}{\partial y} = f_y(x, y) = x^2 - 3xy^2$$

Using Equations 3 through 6, we obtain the following for the second partial derivatives:

$$f_{xx} = \frac{\partial}{\partial x}(2xy - y^3) = 2y$$

$$f_{yy} = \frac{\partial}{\partial y}(x^2 - 3xy^2) = -6xy$$

$$f_{yx} = \frac{\partial}{\partial x}(x^2 - 3xy^2) = 2x - 3y^2$$

$$f_{xy} = \frac{\partial}{\partial y}(2xy - y^3) = 2x - 3y^2$$

■

Application

For a Cobb–Douglas production function,

$$q = f(L, M) = CL^{\alpha}M^{1-\alpha}, \qquad 0 < \alpha < 1$$

the partial derivatives $\dfrac{\partial q}{\partial L}$ and $\dfrac{\partial q}{\partial M}$ are called the **marginal productivities of labor**

and capital, respectively. They measure the approximate change in output per unit increases in labor and capital.

Example 5 The weekly output q of a firm producing skis is described by the equation

$$q = 8L^{0.5}M^{0.5}$$

where L represents the number of hours of labor per week and M the number of machine-hours per week; q is output in hundreds of units per week. Find the marginal productivities of labor and capital, respectively, when $L = 900$ and $M = 400$.

Solution The marginal productivities of labor and capital are found by taking the partial derivatives with respect to L and M, respectively, giving

$$\frac{\partial q}{\partial L} = 4\left(\frac{M}{L}\right)^{0.5}, \quad \frac{\partial q}{\partial M} = 4\left(\frac{L}{M}\right)^{0.5}$$

When $L = 900$ and $M = 400$, the marginal productivities $\frac{\partial q}{\partial L}$ and $\frac{\partial q}{\partial M}$ become

$$\left.\frac{\partial q}{\partial L}\right|_{(900,400)} = 4\left(\frac{900}{400}\right)^{0.5} = 6 \text{ units/hour of labor}$$

$$\left.\frac{\partial q}{\partial M}\right|_{(900,400)} = 4\left(\frac{400}{900}\right)^{0.5} = \frac{8}{3} \text{ units/machine-hour}$$

These results indicate that an additional hour of labor will produce an increase of approximately six units in the output if the number of machine-hours remains constant at 400. Similarly each additional hour of machine time increases the output by approximately $2\frac{2}{3}$ units if the number of hours of labor remains constant at 900 hours. ■

8.2 EXERCISES

Find $f_x(x, y)$ and $f_y(x, y)$ for each of the functions in Exercises 1 through 18.

1. $f(x, y) = x^2 - y^2$
2. $f(x, y) = 3x^4 - 6y^3$
3. $f(x, y) = x^2y^3$
4. $f(x, y) = x^2y^2 - 6x^3 + 7y$
5. $f(x, y) = \frac{x}{y}$
6. $f(x, y) = \frac{y}{x - y}$
7. $f(x, y) = \sqrt{xy}$
8. $f(x, y) = \ln(x - y)$
9. $f(x, y) = xe^y$
10. $f(x, y) = (x^2 - y^2)^3$
11. $f(x, y) = e^{xy}$
12. $f(x, y) = e^x \ln y$
13. $f(x, y) = \ln \frac{y - x}{y + x}$
14. $f(x, y) = \ln(x + e^y)$
15. $f(x, y) = \ln\sqrt{x + y}$
16. $f(x + y) = \frac{xy}{y - x}$

17. $f(x, y) = \sqrt{x^2 + y^2}$

18. $f(x, y) = \ln(xy^2)$

19. Find the slope of the line tangent to the curve formed by the intersection of the surface

$$z = 2x^2 + y^2$$

and the plane $x = 1$ at the point (1, 1, 3).

20. Repeat exercise 19 with the surface

$$z = e^{xy}$$

the plane $y = 2$, and the point (0, 2, 1).

21. Repeat exercise 19 with the surface $z = \sqrt{xy}$, the plane $y = 4$, and the point (1, 4, 2).

22. (a) Repeat exercise 19 with the surface $z = \frac{x}{y}$, the plane $x = 6$, and the point (6, −2, −3).
(b) Repeat part (a) with the point (6, 3, 2) replacing (6, −2, −3).

23. Repeat exercise 19 with the surface $z = \ln(xy)$, the plane $y = e$, and the point $(e, e, 2)$.

24. Repeat exercise 19 with the surface $z = x^2y^2$, the plane $x = 2$, and the point (2, −2, 16).

Find f_{xx}, f_{yy}, f_{xy}, and f_{yx} for each of the functions in Exercises 25 through 34.

25. $f(x, y) = 3x^2 + 2xy - 7y^2$

26. $f(x, y) = (x^2 - 3y)^2$

27. $f(x, y) = e^x + \ln(xy)$

28. $f(x, y) = xe^{y^2}$

29. $f(x, y) = y\sqrt{x}$

30. $f(x, y) = \frac{x}{y}$

31. $f(x, y) = \ln(x + y)$

32. $f(x, y) = e^x \ln y$

33. $f(x, y) = \frac{1}{x - y}$

34. $f(x, y) = \frac{x}{x - y}$

35. The volume of a right-circular cylinder is given by the equation

$$V(r, h) = \pi r^2 h$$

where r is the radius and h is the height (Figure 4).
(a) Find $V_r(r, h)$ and $V_h(r, h)$.
(b) Find the values of $V_r(r, h)$ and $V_h(r, h)$ when $r = 8$ and $h = 10$.

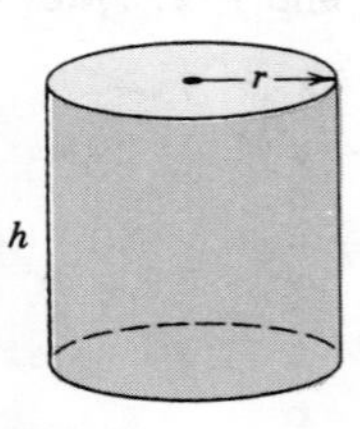

Figure 4

36. If P dollars are deposited in a savings account whose annual interest rate is r, compounded continuously, the amount in the account at the end of three years, is given by the equation

$$A(P, r) = Pe^{3r}$$

(a) Find $A_p(P, r)$ and $A_r(P, r)$.
(b) Find the values of $A_p(P, r)$ and $A_r(P, r)$ when $P = 1000$ and $r = 0.15$.

37. The present value of a perpetuity (fixed payment that continues indefinitely) is given by the equation

$$PV(R, r) = \frac{R}{r}$$

where R is the fixed payment and r the rate of return.
(a) Find $PV_R(R, r)$ and $PV_r(R, r)$.
(b) Find the values of $PV_R(R, r)$ and $PV_r(R, r)$ when $R = 500$ and $r = 0.10$.

38. If automobiles drive up to a bank teller's window at a rate of x automobiles per hour on average, and if the teller can service, on average, y cars per hour, the average number of cars in the line is given by the formula

$$N(x, y) = \frac{x}{y - x}$$

(a) Find $N_x(x, y)$ and $N_y(x, y)$.
(b) Find the values of $N_x(x, y)$ and $N_y(x, y)$ when $x = 20$ and $y = 30$.

39. The production function for a manufacturer of exercise equipment is given by the equation

$$q = 4L^{1/3}M^{2/3}$$

where q is the output in hundreds of units, L is the number of hours of labor, and M is the number of machine-hours. Find the marginal productivity of labor and capital, respectively,
(a) when $L = 125$ and $M = 216$,
(b) when $L = 216$ and $M = 125$,
(c) when $L = M$.

40. The production function for a firm manufacturing machine tools is

$$q = 50L^{1/5}M^{4/5}$$

where q is the daily output and L is the number of hours of labor, and M is the number of machine-hours. Find the marginal productivity of labor and capital, respectively,
(a) when $L = 243$ and $M = 32$,
(b) when $L = 64$ and $M = 486$,
(c) when $L = M$.

41. The surface area of a person in square meters, can be determined from the empirical equation

$$A(w, h) = 2.024w^{0.425}h^{0.725}$$

where w is the person's weight, in kilograms, and h is the height, in meters. Find A_w and A_h for a child who weighs 30 kilograms and whose height is 1 meter. The following information will speed your analysis if you do not have a y^x key on your calculator.

$$30^{0.425} = 4.244, \qquad 30^{-0.575} = 0.1415$$

42. The resistance of a blood vessel, in dynes, can be calculated from the equation

$$R(L, r) = \frac{kL}{r^4}$$

where L and r are the length and the radius, respectively, in centimeters, and k is a constant, the viscosity of blood.

(a) Find R_L and R_r.
(b) What is the significance of the negative sign associated with R_r?

8.3 MAXIMA AND MINIMA

Partial derivatives are very useful in locating and analyzing the shape of a surface at points that are **relative maxima** and **minima.** Figure 1*a* illustrates a case in which the point $(a, b, f(a, b))$ represents a relative maximum for the surface whose equation is $z = f(x, y)$. A case in which $(a, b, f(a, b))$ represents a relative minimum is shown in Figure 1*b*. At points where the surface "peaks" or "bottoms out," the plane tangent to the surface is parallel to the xy-plane; at such points the partial derivatives $f_x(x, y)$ and $f_y(x, y)$, if they exist, equal zero, that is,

$$f_x(x, y) = 0 \quad \text{and} \quad f_y(x, y) = 0 \tag{1}$$

Points satisfying Equation 1 are called **critical points.** The first step in locating relative maxima and minima is to find the coordinates of all critical points.

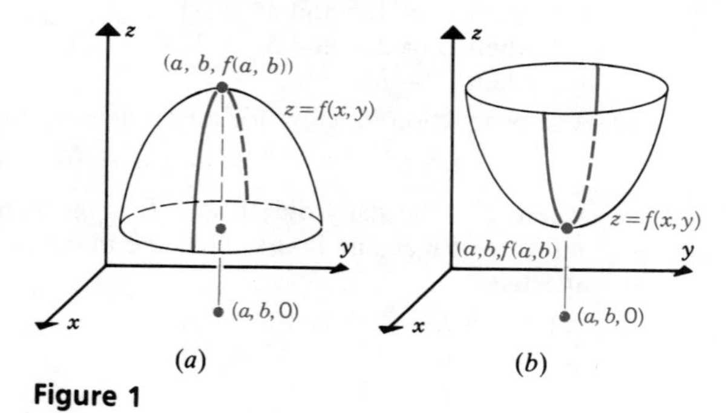

Figure 1

It should be noted that a critical point is not necessarily a relative maximum or minimum. For example a surface may assume a shape similar to that shown in Figure 2 where the critical point $(a, b, f(a, b))$ is neither a relative maximum nor a relative minimum; a critical point such as that shown in Figure 2 is called a **saddle point.** A saddle point is the three-dimensional counterpart of an inflection point, namely, it is a relative minimum when you move toward it in one direction and a relative maximum when you move toward it from another direction.

After all the critical points have been found, the next step is to determine the shape of the surface in the vicinity of each one. A test, called the **second-**

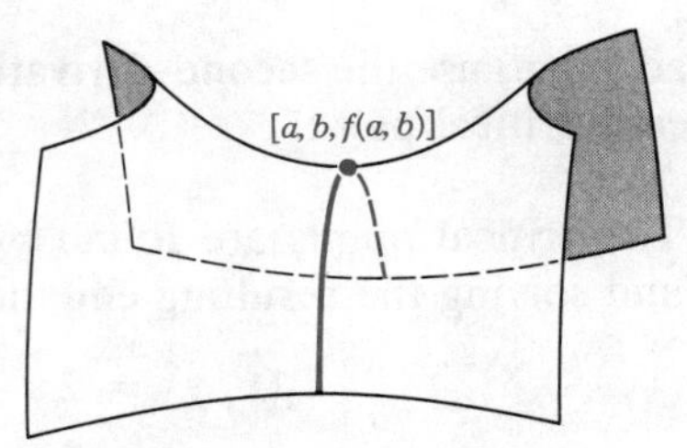

Figure 2

derivative test, enables you to determine the shape of most surfaces in the vicinity of their critical points.

Second-Derivative Test

Let $(a, b, f(a, b))$ be a point on the surface $z = f(x, y)$ for which

$$f_x(a, b) = 0 \quad \text{and} \quad f_y(a, b) = 0$$

Define the numbers A, B and C as

$$A = f_{xx}(a, b), \qquad B = f_{xy}(a, b), \qquad C = f_{yy}(a, b)$$

The second-derivative test takes the following form.

1. If $AC - B^2 > 0$ and $A > 0$, then $(a, b, f(a, b))$ is a relative **minimum.**
2. If $AC - B^2 > 0$ and $A < 0$, then $(a, b, f(a, b))$ is a relative **maximum.**
3. If $AC - B^2 < 0$, then $(a, b, f(a, b))$ is a **saddle point.**
4. If $AC - B^2 = 0$, the test **fails,** that is, no conclusions about the nature of the surface in the vicinity of the critical point can be drawn.

The second-derivative test is summarized in Table 1.

Table 1

$AC - B^2$	A	$(a, b, f(a, b))$
+	+	Relative minimum
+	−	Relative maximum
−		Saddle point
0		Test fails

The following examples illustrate how the critical points and the behavior of the function in the neighborhood of each are determined.

Example 1 Find the critical points for the function

$$z = f(x, y) = x^2 + y^2 - 8x + 2y + 7$$

In addition, use the second-derivative test to determine the shape of the surface at each critical point.

Solution 1. The critical points are found by setting $f_x(x, y)$ and $f_y(x, y)$ equal to zero and solving the resulting equations for x and y:

$$f_x(x, y) = 2x - 8, \qquad f_y(x, y) = 2y + 2$$
$$0 = 2x - 8, \qquad 0 = 2y + 2$$

Solutions: $x = 4, \quad y = -1$

The z coordinate is calculated next:

$$z = (4)^2 + (-1)^2 - 8(4) + 2(-1) + 7 = -10$$

We find that $(4, -1, -10)$ is the only critical point.

2. Next, the second-derivative test is applied. It is necessary first to find $f_{xx}(x, y)$, $f_{xy}(x, y)$, and $f_{yy}(x, y)$ in order to calculate A, B, and C:

$$f_{xx} = 2, \qquad f_{xy} = 0, \qquad f_{yy} = 2$$

We have then $A = 2$, $B = 0$, and $C = 2$. Thus

$$AC - B^2 = (2)(2) - 0 = 4 > 0 \quad \text{and} \quad A = 2 > 0$$

so we conclude that the point $(4, -1, -10)$ is a relative minimum. ■

Example 2 Find the critical points and determine the shape of the surface in the vicinity of each for the function

$$z = f(x, y) = y^2 - x^2 + 1$$

Solution 1. The critical points are found first:

$$f_x(x, y) = -2x, \qquad f_y(x, y) = 2y$$

Setting $f_x(x, y)$ and $f_y(x, y)$ equal to zero gives

$$0 = -2x, \qquad 0 = 2y$$

Solving yields the critical point $(0, 0, 1)$.

2. The shape of the surface in the vicinity of the critical point is found by applying the second-derivative test:

$$f_{xx} = -2, \qquad f_{xy} = 0, \qquad f_{yy} = 2$$

We get $AC - B^2 = -4$, so the point is a saddle point. ■

Example 3 Find the critical points and determine the shape of the surface in the vicinity of each for the function

$$z = f(x, y) = x^2 + y^2 - xy + 3y$$

Solution 1. Again, the critical points are found by setting both $f_x(x, y)$ and $f_y(x, y)$ equal to zero and solving the resulting set of equations.

$$f_x(x, y) = 2x - y, \qquad f_y(x, y) = 2y - x + 3$$

Setting $f_x(x, y)$ and $f_y(x, y)$ equal to zero gives

$$0 = 2x - y \tag{2}$$
$$0 = 2y - x + 3 \tag{3}$$

We are looking for the values of x and y that satisfy both Equations 2 and 3. One method of finding the solutions is to solve Equation 2 for y in terms of x, yielding,

$$y = 2x \tag{4}$$

and then to substitute $2x$ for y in Equation 3. Carrying this out generates a single equation containing the variable x.

$$0 = 2(2x) - x + 3 \quad \text{or} \quad 0 = 3x + 3$$

Solving this equation yields the solution

$$x = -1$$

Substituting this result into Equation 4 to find y gives

$$y = -2$$

Finally, the value of z is found to be

$$\begin{aligned} z &= f(-1, -2) \\ &= (-1)^2 + (-2)^2 - (-1)(-2) + 3(-2) = -3 \end{aligned}$$

2. Thus, the point $(-1, -2, -3)$ is the only critical point. The second-derivative test is applied to determine the shape of the surface in the vicinity of the critical point. The second partial derivatives f_{xx}, f_{xy}, and f_{yy} have the following forms,

$$f_{xx} = 2, \qquad f_{xy} = -1, \qquad f_{yy} = 2$$

so

$$A = 2, \qquad B = -1, \qquad C = 2$$

Applying the second-derivative test, we get

$$AC - B^2 = (2)(2) - 1 = 3 > 0 \quad \text{and} \quad A = +2 > 0$$

which leads us to conclude that $(-1, -2, -3)$ is a relative minimum. ■

Applications

These methods can also be used to maximize revenues or profits or, on the other hand, to minimize costs, as illustrated in Examples 4 and 5.

Example 4 The Television Hut Company sells two types of portable color television sets. The demand equations for each are

$$q_1 = 20 - 8p_1 + 2p_2$$
$$q_2 = 15 + 2p_1 - 3p_2$$

where the subscript 1 refers to the less expensive model and the subscript 2 to the more expensive model; p is expressed in hundreds of dollars and q is in thousands of sets. Find the revenue R and determine at what price levels the revenue is maximized.

Solution The revenue R can be expressed as

$$R = q_1p_1 + q_2p_2$$

Using the demand equations allows us to express the revenue as a function of p_1 and p_2:

$$\begin{aligned} R &= (20 - 8p_1 + 2p_2)p_1 + (15 + 2p_1 - 3p_2)p_2 \\ &= 20p_1 - 8p_1^2 + 4p_2p_1 + 15p_2 - 3p_2^2 \end{aligned}$$

The values of p_1 and p_2 that maximize R can be found by setting both $\dfrac{\partial R}{\partial p_1}$ and $\dfrac{\partial R}{\partial p_2}$ equal to zero and solving the resulting equations. First, we find the partial derivatives, getting

$$\frac{\partial R}{\partial p_1} = 20 - 16p_1 + 4p_2$$

$$\frac{\partial R}{\partial p_2} = 15 + 4p_1 - 6p_2$$

Setting each equal to zero generates the set of equations

$$0 = 20 - 16p_1 + 4p_2 \tag{4}$$
$$0 = 15 + 4p_1 - 6p_2 \tag{5}$$

Solving Equation 4 for p_2 gives

$$p_2 = -5 + 4p_1$$

Substituting the expression $-5 + 4p_1$ for p_2 in Equation 5 gives

$$\begin{aligned} 0 &= 15 + 4p_1 - 6(-5 + 4p_1) \\ &= 45 - 20p_1 \end{aligned}$$

From this equation we then get the solution

$$p_1 = \$2.25 \text{ hundred}$$

Substituting this value of p_1 into either Equation 4 or 5 gives

$$p_2 = \$4 \text{ hundred}$$

from which we then obtain the revenue R:

$$R = 20(2.25) - 8(2.25)^2 + 4(4)(2.25) + 15(4) - 3(4)^2$$
$$= \$52.5 \text{ hundred thousand} = \$5.25 \text{ million}$$

To show that this result represents a maximum, we can use the second-derivative test:

$$\frac{\partial^2 R}{\partial p_1^2} = -16, \qquad \frac{\partial^2 R}{\partial p_1\, \partial p_2} = 4, \qquad \frac{\partial^2 R}{\partial p_2^2} = -6$$

Thus $A = -16$, $B = 4$, and $C = -6$; so we get

$$AC - B^2 = +80 > 0 \quad \text{and} \quad A = -16 < 0$$

We conclude that the revenue is maximized. ■

Example 5 A home improvement firm has decided to market in kit form an inexpensive rectangular shed that has no floor and one open side. If material for the three sides and the roof costs 25¢ per square foot, find the dimensions of the shed that minimize the cost of material if the volume of the shed equals 686 ft³.

Solution A sketch of the shed when assembled is shown in Figure 1. Denoting the dimensions of the shed as x, y, and w, the cost of materials can be written as

$$C = 25xy + 50yw + 25xw \tag{6}$$

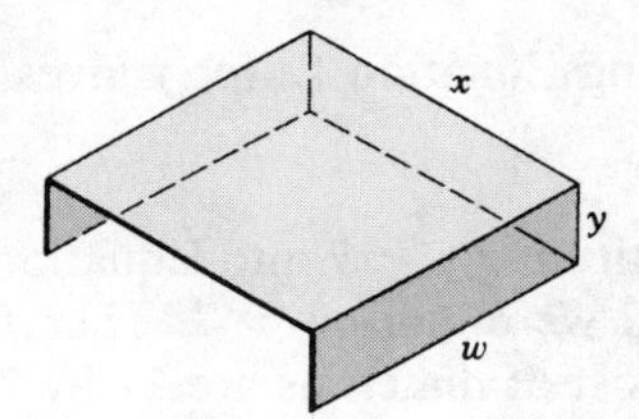

Figure 1

The fact that the volume is fixed at 686 ft³ can be represented by the equation

$$xyw = 686 \tag{7}$$

The restriction expressed by Equation 7 indicates that x, y, and w are not independent variables. We can solve Equation 7 for one of the variables in terms of the other two and substitute this result into Equation 6. Solving for w gives

$$w = \frac{686}{xy}$$

and substituting $686/(xy)$ for w in Equation 6 gives the cost as a function of x and y:

$$C(x, y) = 25xy + 50y\,\frac{686}{xy} + 25x\,\frac{686}{xy}$$

$$C(x, y) = 25xy + \frac{(50)(686)}{x} + \frac{(25)(686)}{y}$$

The values of x and y that minimize $C(x, y)$ can be found by solving the equations $C_x(x, y) = 0$ and $C_y(x, y) = 0$:

$$C_x(x, y) = 25y - \frac{(50)(686)}{x^2} = 0 \tag{8}$$

$$C_y(x, y) = 25x - \frac{(25)(686)}{y^2} = 0 \tag{9}$$

This system of equations can be solved by first writing Equation 8 as

$$yx^2 - 2(686) = 0 \tag{10}$$

Next we solve Equation 9 for x in terms of y gives

$$x = \frac{686}{y^2} \tag{11}$$

Substituting $\frac{686}{y^2}$ for x in Equation 10 gives

$$y\left(\frac{686}{y^2}\right)^2 - 2(686) = 0$$

or

$$686 - 2y^3 = 0 \tag{12}$$

Solving Equation 12 for y gives

$$y = \sqrt[3]{343} = 7$$

Substituting $y = 7$ into Equation 11 yields the solution $x = 14$, and from Equation 7 we obtain $w = 7$. Therefore, the dimensions of the shed that minimize the cost of materials are 14 by 7 by 7 ft. The minimum cost C equals

$$\begin{aligned} C &= 25(14)(7) + 50(7)(7) + 25(14)(7) \\ &= 7350¢ = \$73.50 \end{aligned}$$

Proving that the cost is minimized when the dimensions are 14 by 7 by 7 ft is left to an exercise. ■

8.3 EXERCISES

For the functions in Exercises 1 through 20, find the critical points and use the second-derivative test to determine the shape of the curve in the vicinity of each.

1. $f(x, y) = x^2 + 2y^2 + 1$
2. $f(x, y) = 1 - x^2 - 2y^2$
3. $f(x, y) = x^2 - y^2 + 3$
4. $f(x, y) = x^2 + y^2 - 2x - 4y - 3$
5. $f(x, y) = x^2 + y^2 - 8x + 2y + 5$

6. $f(x, y) = 2 - 4x + 4y - x^2 - y^2$
7. $f(x, y) = x^2 - y^2 + xy + 2x - 9y + 3$
8. $f(x, y) = x^2 + y^2 - xy - 3x - y + 2$
9. $f(x, y) = x^3 + 3x^2 - y^2 + 2y + 4$
10. $f(x, y) = y^3 - 6y^2 + x^2 - 2x + 1$
11. $f(x, y) = x^3 + y^3 - 3xy - 5$
12. $f(x, y) = x^3 - y^3 - 3xy + 2$
13. $f(x, y) = 2x^3 - 3x^2 - 12x + y^2 - 2y + 3$
14. $f(x, y) = y^3 - 3y^2 - 9y + x^2 - 2x + 6$
15. $f(x, y) = 4 + 6xy - x^3 - y^2$
16. $f(x, y) = 2xy + \frac{4}{x} - \frac{1}{y}$
17. $f(x, y) = xy + \frac{8}{x} + \frac{27}{y}$
18. $f(x, y) = e^{x^2} + y^2$
19. $f(x, y) = xy + e^x$
20. $f(x, y) = 2y + e^{xy}$
21. Find three positive numbers x, y, and w whose sum is 45 and whose product xyw is a maximum.
22. Find three positive numbers x, y, and w whose product is 27 and whose sum $x + y + w$ is a minimum.
23. A builder has a contract to build a one-story rectangular warehouse that will have a volume of 375,000 ft³. Material for the floor costs \$2 per square foot. Material for three of the walls and the roof costs \$1 per square foot. The fourth wall, consisting of sliding doors, costs \$3 per square foot. What are the dimensions of the building that minimize the cost of material? Although building costs are minimized, does this building have some features that might make it difficult to operate as a warehouse?
24. Complete the analysis in Example 5 by showing that the cost of material is minimized for a shed that measures 14 by 7 by 7 ft.
25. A furniture designer wants to construct a rectangular box that will have a volume of 32 ft³ and will be open at the top. Find the dimensions of the box that will minimize the surface area and thereby require the least amount of material.
26. A closed rectangular box is to be built of two different kinds of material. Material for the top and bottom costs \$2 per square foot whereas material for the four sides costs \$1 per square foot. If the budgeted cost is \$12, find the dimensions of the box that maximize the volume.
27. The owner of a sporting goods store carries two kinds of jogging shoes. He pays \$10 per pair for the less expensive shoe and \$15 per pair for the more expensive shoe. The demand equations for the two types of shoes are

$$q_1 = 150 - 10p_1 + 4p_2$$
$$q_2 = 93 + 4p_1 - 5p_2$$

where the subscript 1 refers to the less expensive shoes and 2 to the more expensive shoes. Set up the profit function $P(p_1, p_2)$ and determine the unit prices p_1 and p_2 that maximize his profit.

28. The length plus the girth of a package mailed through the U.S. postal service cannot exceed 84 in. What are the dimensions of the rectangular package that maximize the volume?

Note: In Figure 2 length plus girth equals $l + (2w + 2h) = 84$.

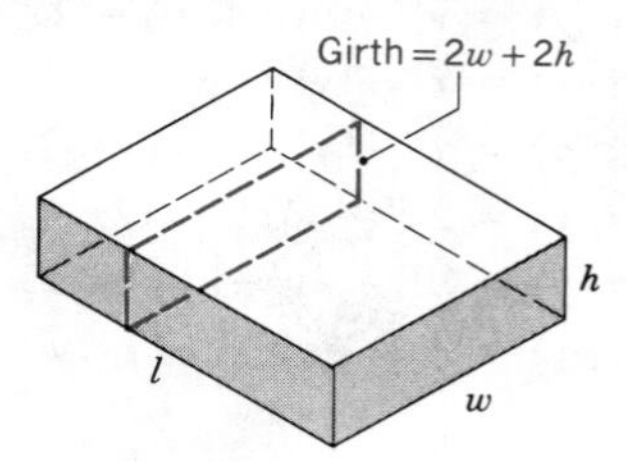

Figure 2

8.4 LAGRANGE MULTIPLIERS

Most maximization or minimization problems require that the optimization be carried out subject to constraints on the independent variables. For example, a utility that burns coal to generate electricity cannot operate its furnaces to generate the maximum energy possible from the coal because too many pollutants would be emitted from its stacks in the process. The company must try to generate the maximum amount of energy consistent with Environmental Protection Agency (EPA) guidelines on pollution control. In the economic realm, all companies have finite resources that serve as constraints on the operations of the firm. In many situations, it is the function of the firm's management to allocate the available resources so that the company's profits are maximized.

Geometrically, the introduction of constraints is demonstrated in Figure 1. Figure 1*a* shows a surface $z = f(x, y)$ whose minimum point is found by the method described in Section 8.3, that is, solving the equations $f_x(x, y) = 0$ and $f_y(x, y) = 0$. The minimum in such a situation is said to be **unconstrained.** If, in addition, the variables x and y must satisfy an equation of the form $g(x, y) = 0$, called a **constraint,** the ordered triple (x, y, z) that satisfies both $z = f(x, y)$ and $g(x, y) = 0$ will lie on a curve such as that shown in Figure 1*b*, which forms the outer boundary of the shaded region. The minimum value of z on the curve (the **constrained minimum**) will generally differ from the minimum point on the surface $z = f(x, y)$.

When constraints were encountered previously in Section 4.2 and Example 5 of Section 8.3, the constraint equation was solved for one of the variables in terms of the remaining variables. Substituting this result into the function to be optimized reduced the total number of variables by one. Another approach, known as the method of *Lagrange multipliers,* can be used when the constraint equation is unwieldy or difficult to solve for one variable in terms of the remaining variables. The method works as follows.

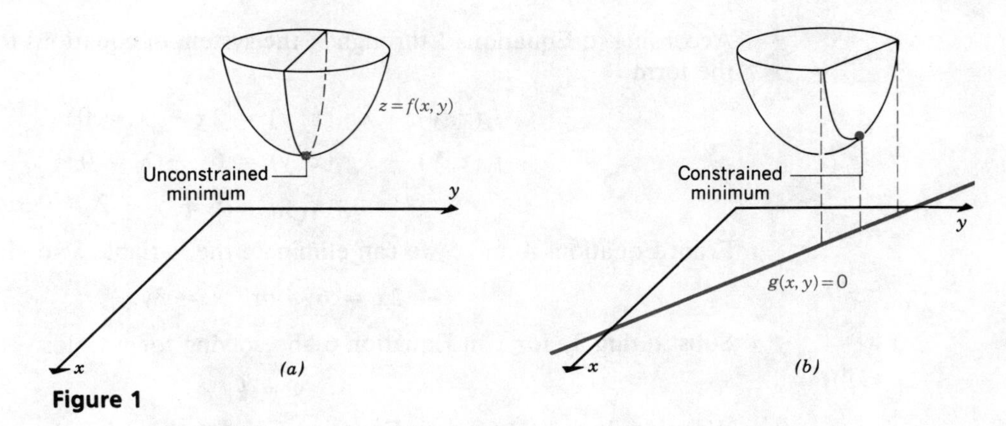

Figure 1

A new function, defined as follows, is introduced

$$F(x, y, \lambda) = f(x, y) - \lambda g(x, y)$$

where the quantity λ is known as a **Lagrange multiplier.** The values of x and y that optimize the function $z = f(x, y)$ subject to the constraint $g(x, y) = 0$ are found by solving the system of equations

$$F_x(x, y, \lambda) = 0 \qquad F_y(x, y, \lambda) = 0 \qquad F_\lambda(x, y, \lambda) = 0$$

or equivalently,

$$\boxed{f_x(x, y) - \lambda g_x(x, y) = 0} \tag{1}$$

$$\boxed{f_y(x, y) - \lambda g_y(x, y) = 0} \tag{2}$$

$$\boxed{g(x, y) = 0} \tag{3}$$

The method is illustrated in the following examples.

Example 1 Minimize the function $f(x, y) = x^2 + 3y^2$ subject to the constraint

$$x + y = 2$$

Solution The constraint equation $g(x, y) = 0$ is written as

$$g(x, y) = x + y - 2 = 0$$

According to Equations 1 through 3, the system of equations to be solved takes the form

$$f_x(x, y) - \lambda g_x(x, y) = 2x - \lambda = 0 \quad (4)$$
$$f_y(x, y) - \lambda g_y(x, y) = 6y - \lambda = 0 \quad (5)$$
$$g(x, y) = x + y - 2 = 0 \quad (6)$$

From Equations 4 and 5 we can eliminate the variable λ to yield

$$2x = 6y \quad \text{or} \quad x = 3y$$

Substituting $3y$ for x in Equation 6 and solving for y yields

$$y = \tfrac{1}{2}$$

We can substitute $\frac{1}{2}$ for y in Equation 6 and get

$$x = \tfrac{3}{2}$$

The corresponding value of z is

$$z = f(\tfrac{3}{2}, \tfrac{1}{2}) = (\tfrac{3}{2})^2 + 3(\tfrac{1}{2})^2 = 3$$ ■

Although the proof of the Lagrange multiplier method is beyond the scope of this course, the method can be demonstrated graphically for the function $z = f(x, y) = x^2 + 3y^2$ subject to the constraint $x + y = 2$ (Example 1). A number of level curves are shown in Figure 2 together with the graph of the constraint equation. The Lagrange multiplier selects from among all the points common to both types of curves those for which the constraint curve is tangent to the level curves. The optimum values of the function $z = f(x, y)$ are located at these points.

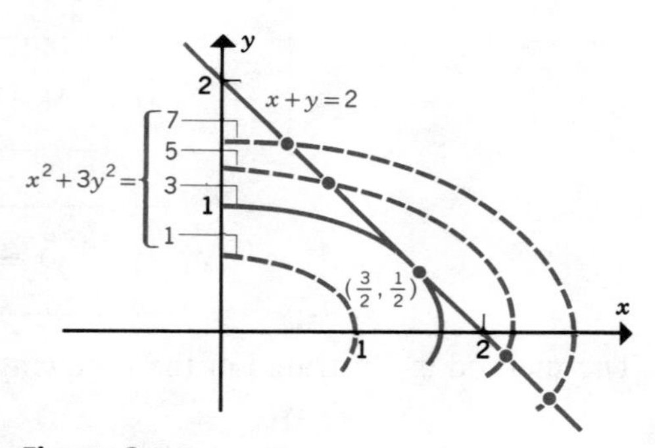

Figure 2

Example 2 Maximize the function $f(x, y) = xy$ subject to the constraint

$$x^2 + y^2 = 8$$

Solution Again, the constraint equation is written as

$$g(x, y) = x^2 + y^2 - 8 = 0$$

Equations 1 through 3 yield the following:

$$f_x(x, y) - \lambda g_x(x, y) = y - 2\lambda x = 0 \tag{7}$$

$$f_y(x, y) - \lambda g_y(x, y) = x - 2\lambda y = 0 \tag{8}$$

$$g(x, y) = x^2 + y^2 - 8 = 0 \tag{9}$$

If Equation 7 is solved for λ, we get

$$\lambda = \frac{y}{2x}$$

Substituting $y/(2x)$ for λ in Equation 8 gives

$$x - 2\frac{y}{2x}y = 0$$

or

$$x^2 = y^2 \tag{10}$$

Substituting this result into Equation 9 yields the equation

$$2x^2 = 8$$

whose solutions are

$$x_1 = 2 \quad \text{and} \quad x_2 = -2$$

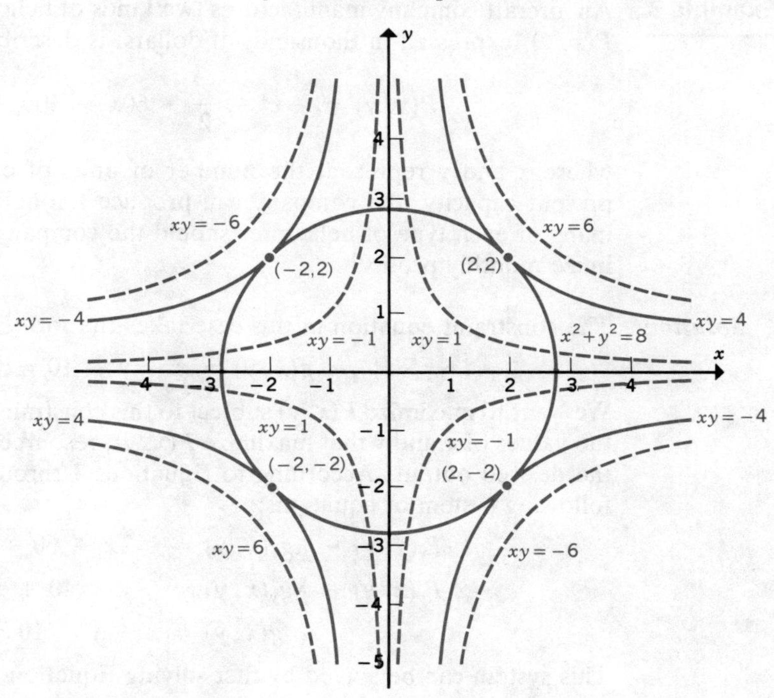

Figure 3

Equation 10 indicates that for each of these solutions, there are two values of y, that is, $y = \pm 2$, so there are four ordered pairs that satisfy Equations 7 through 9; (2, 2), (2, −2), (−2, 2), and (−2, −2). The corresponding values of z are

$$z_1 = f(2, 2) = 4$$
$$z_2 = f(2, -2) = -4$$
$$z_3 = f(-2, 2) = -4$$
$$z_4 = f(-2, -2) = 4$$

The maximum value of $f(x, y)$, that is, 4, occurs at (2, 2, 4) and (−2, −2, 4). The remaining two solutions represent minimum values. These results are shown graphically in Figure 3, where various level curves for the function $z = f(x, y)$ are shown together with the constraint curve $g(x, y) = x^2 + y^2 - 8 = 0$. As before, the optimum values of $f(x, y)$ occur where the level curves are tangent to the constraint curve. ■

Application

The method of Lagrange multipliers can also be used to determine how resources should be allocated to maximize profits.

Example 3 An aircraft company manufactures two kinds of helicopters. The monthly profit $P(x, y)$, expressed in thousands of dollars, is described by the equation

$$P(x, y) = -x^2 - \frac{y^2}{2} + 60x + 40y + xy - 100$$

where x and y represent the number of units of each type produced. At its present capacity, the company can produce ten helicopters each month. How many of each type of helicopter should the company produce in order to maximize monthly profits?

Solution The constraint equation in this case takes the form

$$g(x, y) = x + y - 10 = 0$$

We want to maximize $P(x, y)$ subject to this constraint. When the company finds the values of x and y that maximize $P(x, y)$, resources can be allocated to attain the desired output. According to Equations 1 through 3 we want to solve the following system of equations:

$$P_x(x, y) - \lambda g_x(x, y) = -2x + 60 + y - \lambda = 0 \qquad (11)$$
$$P_y(x, y) - \lambda g_y(x, y) = -y + 40 + x - \lambda = 0 \qquad (12)$$
$$g(x, y) = x + y - 10 = 0 \qquad (13)$$

This system can be solved by first solving Equation 11 for λ, obtaining

$$\lambda = -2x + 60 + y$$

Substituting this result into Equation 12 yields

$$-y + 40 + x - (-2x + 60 + y) = 0 \quad \text{or} \quad 3x - 2$$

Solving this equation for x in terms of y gives

$$x = \frac{20}{3} + \frac{2y}{3}$$

Substituting this result into Equation 13 gives

$$\left(\frac{20}{3} + \frac{2y}{3}\right) + y - 10 = 0$$

$$20 + 2y + 3y - 30 = 0$$

or

$$5y = 10$$

The solution is

$$y = 2$$

The corresponding value of x can be found from the constraint $x + y - 10 = 0$ to give

$$x = 8$$

We can now calculate $P(8, 2)$

$$P(8, 2) = -64 - 2 + 480 + 80 + 16 - 100 = \$410 \text{ thousand}$$ ■

8.4 EXERCISES

Use the method of Lagrange multipliers to solve Exercises 1 through 24.

1. Find the maximum value of the function $f(x, y) = xy$ subject to the constraint $x + y = 2$. In addition, draw the level curve corresponding to the maximum value of $f(x, y)$ together with a sketch of the constraint equation on the coordinate system of Figure 4.

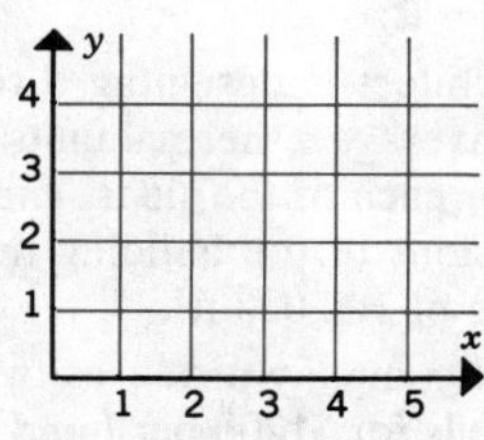

Figure 4

2. Find the maximum and minimum values of the function $f(x, y) = x + y$ subject to the constraint $x^2 + y^2 = 2$. Again, draw the level curves for both the minimum and maximum values of $f(x, y)$ together with a sketch of the constraint equation on the coordinate system of Figure 5.

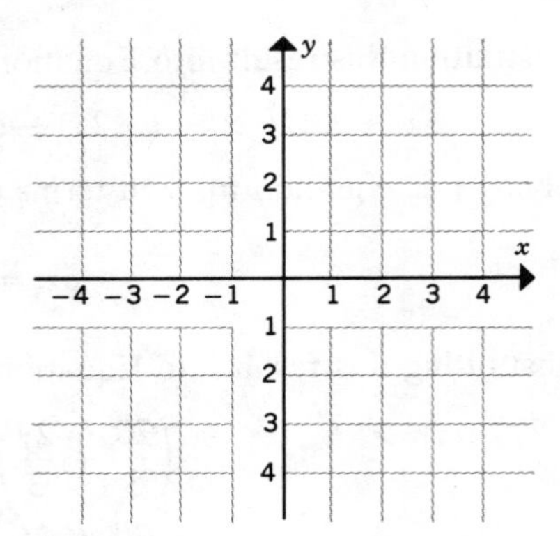

Figure 5

3. Find the maximum value of the function $f(x, y) = 3xy + x$ subject to the k constraint $x + y = 1$.
4. Find the minimum value of the function $f(x, y) = x^2 + y^2$ subject to the constraint $x + y = 8$.
5. Find the minimum value of the function $f(x, y) = x^2 + 2y^2$ subject to the constraint $2x + 3y = 9$.
6. Find the maximum value of the function $f(x, y) = x^2 - y^2$ subject to the constraint $3x + 3y = 4$.
7. Find the minimum value of the function $f(x, y) = 2x^2 + y^2 - xy$ subject to the constraint $2x + 3y = 14$.
8. Find the maximum value of the function $f(x, y) = 2x^2 - y^2 + xy$ subject to the constraint $y + 2x = 4$.
9. Find the maximum value of the function $f(x, y) = 2x + 3y - x^2 - y^2$ subject to the constraint $2x + y = 6$.
10. Find the maximum value of the function $f(x, y) = e^{xy}$ subject to the constraint $x^2 + y^2 = 8$.
11. Find the minimum value of the function $f(x, y) = \sqrt{x^2 + y^2}$ subject to the constraint $x + y = 6$
12. Find the minimum value of the function $f(x, y) = e^{x^2+y^2+xy}$ subject to the constraint $x + y = 2$.
13. An architect is designing a rectangular warehouse with a square base. Daily heat losses average 9 thermal units per square foot for the roof, 5 thermal units per square foot for each of the sides, and 1 thermal unit per square foot for the floor. Find the dimensions of the building that minimize total daily heat loss if the building has a volume of 108,000 ft³.
14. A steel company produces two grades of steel at one of its plants. The better-grade steel sells for \$1200 per hundred tons, whereas the standard grade sells for \$900 per hundred tons. Daily plant output is currently 148 hundred tons. The daily cost function described by the equation

$$C(q_1, q_2) = 2q_1^2 + q_2^2 - q_1q_2$$

where q_1 and q_2 represent the daily outputs in hundreds of tons of the better and standard grades, respectively. Find the amount of each grade of steel that should be produced to maximize daily profits.

15. The Cobb–Douglas production function (Section 8.1) for a company producing steel I-beams is given by the equation

$$q = 40L^{0.25}M^{0.75}$$

where L is the number of units of labor, M is the number of units of capital, and q equals the output (number of I-beams). Labor costs \$20 per unit and capital costs \$60 per unit. If management sets a production goal of 25,000 I-beams, find the number of units of labor and capital that minimizes production costs while meeting the production goal.

16. Suppose that the management of the company in the previous problem has a production budget of \$60,000. How many units of labor and capital are needed to maximize output?

17. Find two numbers whose sum is 80 and whose product is a maximum.

18. Find two numbers whose sum is 55 and whose product is a maximum.

19. From the set of all pairs of numbers whose difference is 14, find that pair whose product is a minimum.

20. One hundred feet of fencing are available for enclosing a rectangular plot to be used for a small vegetable garden. What are the dimensions of the plot that will maximize the area of the garden?

21. The Supreme Boat Rental Company wants to install a fence around a rectangular plot of land where its canoes and rowboats can be stored. If 1800 ft^2 of land are to be enclosed and if fencing is needed on only three sides, what are the dimensions of the plot that will require the minimum length of fence?

22. A closed rectangular box is to be made with a square base and a volume of 5000 in ^{3}. If the material for the sides and the top costs 2¢ per square inch and material for the base costs 3¢ per square inch, find the dimensions of the box that minimize the cost of material.

23. An open rectangular box is to be made with a square base and a volume of 1440 in.^{3}. If material for the sides costs 3¢ per square inch and that for the base cost 5¢ per square inch, what are the dimensions of the box that minimize the cost of material?

24. Identical square pieces are cut out from the four corners of a thin piece of cardboard measuring 25 by 25 in. The flaps are then turned up to make an open rectangular box, as shown in Figure 6, where x represents the height and y the length of the square base. Find the dimensions x and y that maximize the volume of the box.

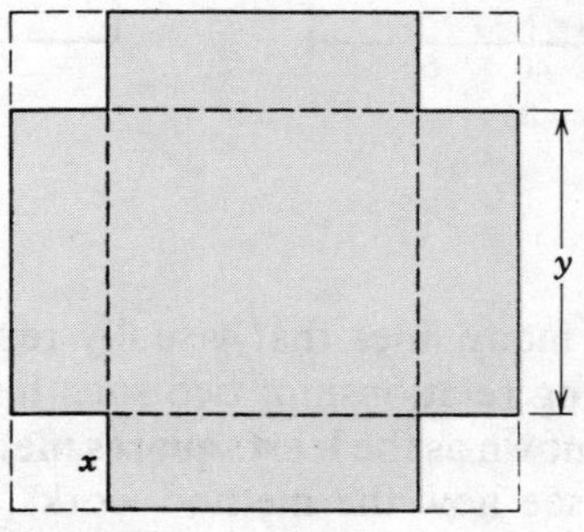

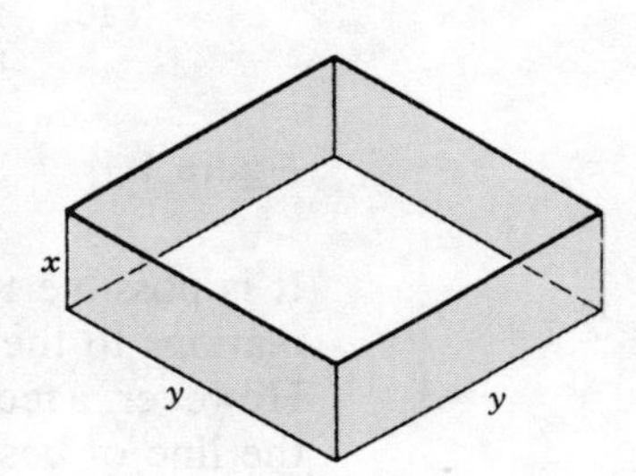

Figure 6

8.5 LEAST SQUARES METHOD

Until now, we have purposely avoided discussing how the functions describing the relationships between two variables are determined. Very often they are determined by finding what is called a "curve of best fit" to experimental data. To see what this means, suppose the cost accountant of a firm has gathered the data shown in Table 1, which gives the company's total cost C for each of five different orders for the same item where x represents the number of items sold.

Table 1

Order Size, x (thousands)	Total Cost, C \$ (thousand)
30	40
40	45
25	33
45	54
35	47

For planning purposes, the company's management would like to develop a method for determining the total cost of each order as a function of the order size x. The data in Table 1 are shown graphically in Figure 1a. Inspection of the graph indicates that the underlying relationship between C and x could be approximated reasonably well by a linear relationship of the form

$$C = mx + b$$

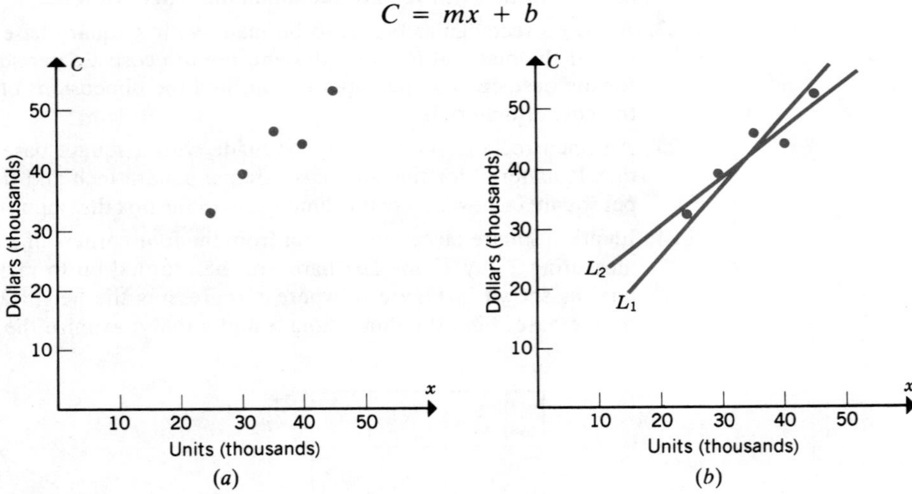

Figure 1

It is possible to sketch many lines that visually represent reasonable approximations to the underlying relationship; two such lines are shown in Figure 1b. However, a technique known as the **least squares method** enables us to determine the line of best fit. To see how the method works, let us suppose that we are given a set of n points such as those shown in Figure 2 for which a linear relationship seems to exist.

The method of least squares locates the line for which the sum of the squares of the vertical distances between the data points and the line is minimized. Using the slope–intercept formula $y = mx + b$, we now show how the method of least squares enables us to find m and b. The square of the vertical distance from (x_1, y_1) to the line equals $(mx_1 + b - y_1)^2$. The sum of the squares S of the distances between all n points and the line is given by the equation

$$S = (mx_1 + b - y_1)^2 + (mx_2 + b - y_2)^2 + \cdots + (mx_n + b - y_n)^2 \quad (1)$$

Equation 1 can be written in a more compact form using the sigma notation introduced in Section 6.3:

$$S = \sum_{k=1}^{n} (mx_k + b - y_k)^2 \quad (2)$$

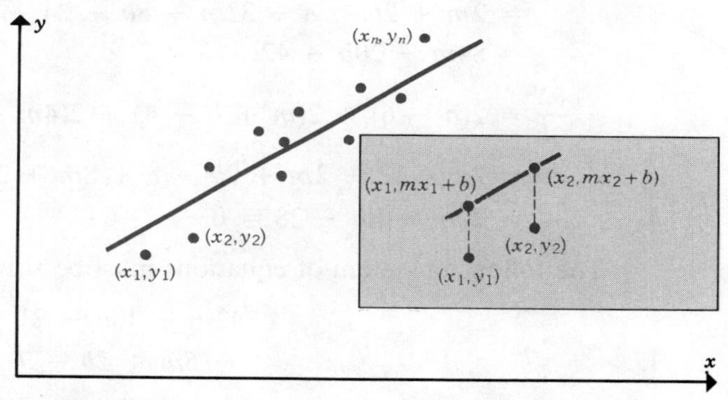

Figure 2

The next step is to find the values of m and b that minimize S. This is accomplished by setting both $\frac{\partial S}{\partial m}$ and $\frac{\partial S}{\partial b}$ equal to zero and then solving the resulting system of equations for m and b. This procedure is illustrated in Example 1, following which formulas for finding m and b in terms of $x_1, x_2, \ldots, x_n$ and $y_1, y_2, \ldots, y_n$ will be developed.

Example 1 Find the line of best fit for the points (0, 6), (1, 4), (4, 3), and (5, 1).

Solution We want to minimize the sum of the squares of the vertical distances from the four points to the unknown line L as shown in Figure 3. According to Equation 1, the sum S is given by the equation

$$\begin{aligned} S &= (0m + b - 6)^2 + (1m + b - 4)^2 + (4m + b - 3)^2 + (5m + b - 1)^2 \\ &= (b - 6)^2 + (m + b - 4)^2 + (4m + b - 3)^2 + (5m + b - 1)^2 \end{aligned}$$

The values of m and b that minimize S are determined by finding $\frac{\partial S}{\partial m}$ and $\frac{\partial S}{\partial b}$ and setting each equal to zero:

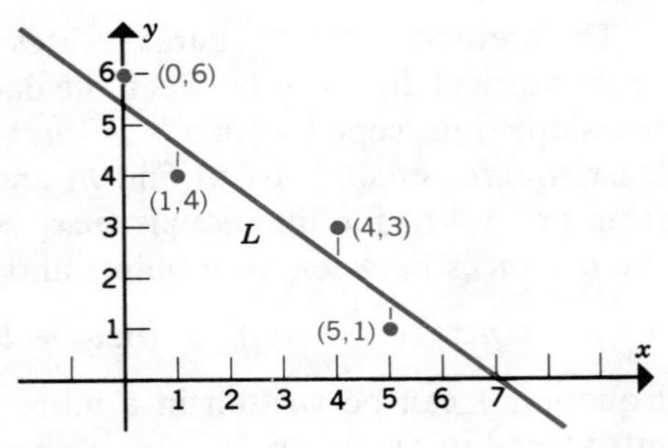

Figure 3

$$\begin{aligned}\frac{\partial S}{\partial m} &= 2(m + b - 4) + 2(4m + b - 3)(4) + 2(5m + b - 1)(5)\\ &= 2m + 2b - 8 + 32m + 8b - 24 + 50m + 10b - 10\\ &= 84m + 20b - 42 = 0\end{aligned}$$

$$\begin{aligned}\frac{\partial S}{\partial b} &= 2(b - 6) + 2(m + b - 4) + 2(4m + b - 3) + 2(5m + b - 1)\\ &= 2b - 12 + 2m + 2b - 8 + 8m + 2b - 6 + 10m + 2b - 2\\ &= 20m + 8b - 28 = 0\end{aligned}$$

The following system of equations must be solved to find m and b:

$$42m + 10b - 21 = 0 \qquad (3)$$

$$5m + 2b - 7 = 0 \qquad (4)$$

The solution for m can be obtained if we multiply each term in Equation 4 by 5, thereby making the coefficients of b identical for both equations. We now have the equivalent system

$$42m + 10b - 21 = 0 \qquad (5)$$

$$25m + 10b - 35 = 0 \qquad (6)$$

Subtracting Equation 6 from Equation 5 gives us a single equation in m:

$$17m + 14 = 0 \quad \text{or} \quad m = -\tfrac{14}{17} = -0.82$$

Substituting this result into Equation 5 or 6 and solving for b yields

$$b = 5.56$$

The equation of the least squares line has the form

$$\boxed{y = -0.82x + 5.56}$$

■

It is not necessary to follow the procedure described in Example 1 each time the least squares line is desired. Formulas are available to enable you to find m

and b in terms of $x_1, x_2, \ldots, x_n$ and $y_1, y_2, \ldots, y_n$ and n. The formulas are generalizations of the procedure described in Example 1 and are obtained by differentiating S in Equation 1 with respect to both m and b and setting each equal to zero:

$$\frac{\partial S}{\partial m} = 2x_1(mx_1 + b - y_1) + 2x_2(mx_2 + b - y_2) + \cdots + 2x_n(mx_n + b - y_n) = 0$$

Noting that m and b are the only variables on the right-hand side, we can group like terms together:

$$\frac{\partial S}{\partial m} = 2m(x_1^2 + x_2^2 + \cdots + x_n^2) + 2b(x_1 + x_2 + \cdots + x_n) - 2(x_1y_1 + x_2y_2 + \cdots + x_ny_n) = 0$$

The second equation required in the analysis is found by setting $\partial S/\partial b$ equal to zero:

$$\frac{\partial S}{\partial b} = 2(mx_1 + b - y_1) + 2(mx_2 + b - y_2) + \cdots + 2(mx_n + b - y_n) = 0$$

Grouping all terms that contain m as a factor and doing the same for those that contain b as a factor gives

$$(2mx_1 + 2mx_2 + \cdots + 2mx_n) + (2b + 2b + \cdots + 2b) - (2y_1 + 2y_2 + \cdots + 2y_n) = 0$$

$$2m(x_1 + x_2 + \cdots + x_n) + 2nb - 2(y_1 + y_2 + \cdots + y_n) = 0$$

Now we have the system

$$(x_1 + x_2 + \cdots + x_n)m + nb = y_1 + y_2 + \cdots + y_n \tag{7}$$

$$(x_1^2 + x_2^2 + \cdots + x_2^2)m + (x_1 + x_2 + \cdots + x_n)b = x_1y_1 + x_2y_2 + \cdots + x_ny_n \tag{8}$$

To avoid becoming immersed in too much detail when we solve the system for m and b, we write Equations 7 and 8 in terms of four quantities A, B, C, and D defined as follows:

$$A = x_1 + x_2 + \cdots + x_n = \sum_{k=1}^{n} x_k$$

$$B = y_1 + y_2 + \cdots + y_n = \sum_{k=1}^{n} y_k$$

$$C = x_1^2 + x_2^2 + \cdots + x_n^2 = \sum_{k=1}^{n} x_k^2$$

$$D = x_1y_1 + x_2y_2 + \cdots + x_ny_n = \sum_{k=1}^{n} x_ky_k$$

Equations 7 and 8 can now be written as

$$Am + nb = B \tag{9}$$
$$Cm + Ab = D \tag{10}$$

We can eliminate the variable b in Equations 9 and 10 by multiplying each term in Equation 9 by A and each term in Equation 10 by n to produce the equivalent system

$$A^2m + nAb = AB \tag{11}$$
$$nCm + nAb = nD \tag{12}$$

Subtracting Equation 11 from 12 gives

$$(nC - A^2)m = nD - AB$$

Solving for m gives

$$m = \frac{nD - AB}{nC - A^2}$$

or

$$m = \frac{n \sum_{k=1}^{n} x_k y_k - \left(\sum_{k=1}^{n} x_k\right)\left(\sum_{k=1}^{n} y_k\right)}{n\left(\sum_{k=1}^{n} x_k^2\right) - \left(\sum x_k\right)^2} \tag{13}$$

Similarly, we find that b, the y intercept, is given by the formula

$$b = \frac{BC - AD}{nC - A^2}$$

$$b = \frac{\left(\sum_{k=1}^{n} y_k\right)\left(\sum_{k=1}^{n} x_k^2\right) - \left(\sum_{k=1}^{n} x_k\right)\left(\sum_{k=1}^{n} x_k y_k\right)}{n\left(\sum_{k=1}^{n} x_k^2\right) - \left(\sum_{k=1}^{n} x_k\right)^2} \tag{14}$$

Example 2 Find the equation of the least squares line for the points (0, 3), (1, 5), (2, 4), (3, 3), and (4, 7) using Equations 13 and 14.

Solution It is advisable to set up a table such as Table 2 so that the calculation can be carried out more quickly. The quantities m and b can now be found using Equations 13 and 14:

$$m = \frac{(5)(50) - (10)(22)}{5(30) - (10)^2} = 0.60, \qquad b = \frac{(22)(30) - (10)(50)}{5(30) - (10)^2} = 3.20$$

The line $y = 0.60x + 3.20$ and the data points are shown in Figure 4.

Table 2

k	x_k	y_k	x_k^2	$x_k y_k$
1	0	3	0	0
2	1	5	1	5
3	2	4	4	8
4	3	3	9	9
5	4	7	16	28
Totals	10	22	30	50

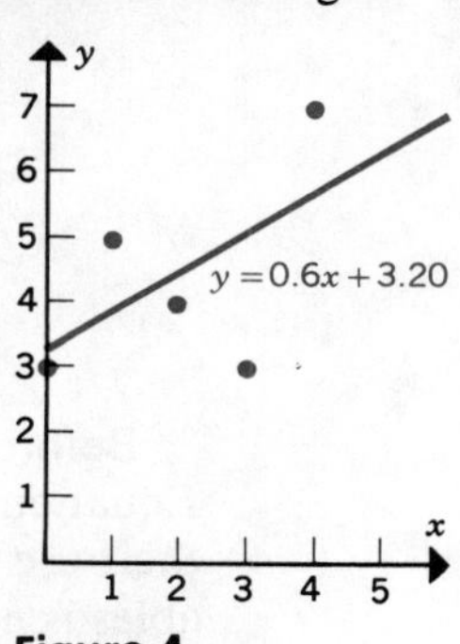

Figure 4 ■

Example 3 Use Equations 13 and 14 to find the least squares line for the cost accounting data shown in Table 1.

Solution The total cost data represent the y coordinates for each ordered pair in Table 3. Constructing the table enables us to carry out the calculations. Using Equations 13 and 14, we get

$$m = \frac{5(7900) - (175)(219)}{5(6375) - (175)^2} = \frac{1175}{1250} = 0.94$$

$$b = \frac{(219)(6375) - (175)(7900)}{5(6375) - (175)^2} = \frac{13{,}625}{1250} = 10.90$$

Table 3

k	x_k	C_k	x_k^2	$x_k C_k$
1	30	40	900	1200
2	40	45	1600	1800
3	25	33	625	825
4	45	54	2025	2430
5	35	47	1225	1645
Totals	175	219	6375	7900

So the cost equation takes the form

$$C = 0.94x + 10.90$$

The line and the data points are shown in Figure 5.

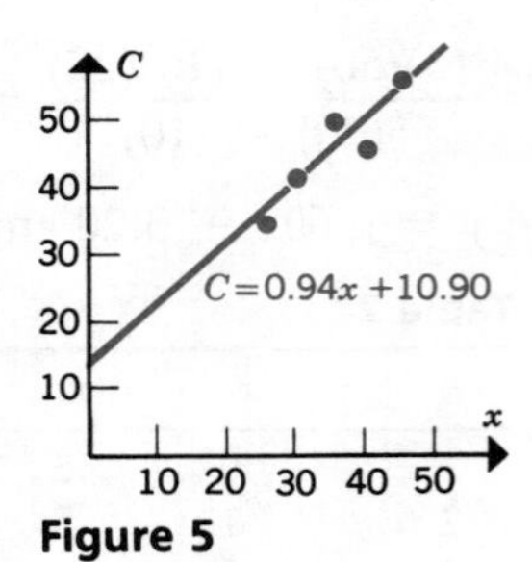

Figure 5

■

Before we conclude this section, we want to inject a word of caution. Although Equations 13 and 14 enable you to fit a least squares line to any set of data, their use is not appropriate if the underlying relationship between the two variables is not linear. For example, the points shown in Figures 6*a* and 6*b* indicate that the relationship between the variables seems to be nonlinear, and, therefore, the equations developed in this section are not applicable. Once a particular functional relationship has been discovered or assumed, methods similar to those just studied will enable you to obtain a curve of best fit.

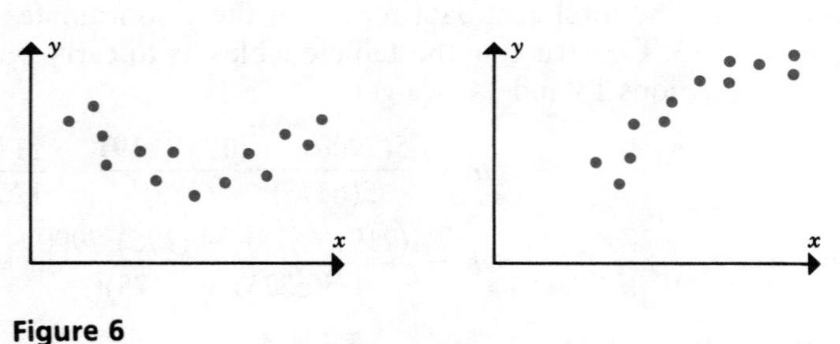

Figure 6

8.5 EXERCISES

Use the techniques described in this section to find an equation of the line of best fit for the sets of points in Exercises 1 through 3. In addition, plot the points together with the line on the accompanying coordinate system.

1. $\{(2, 1), (3, 3), (1, 2), (5, 4)\}$

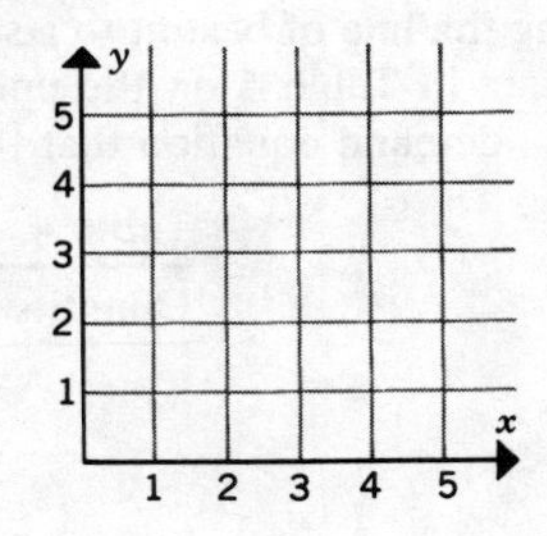

2. $\{(0, 5), (3, 4), (5, 2), (2, 3)\}$

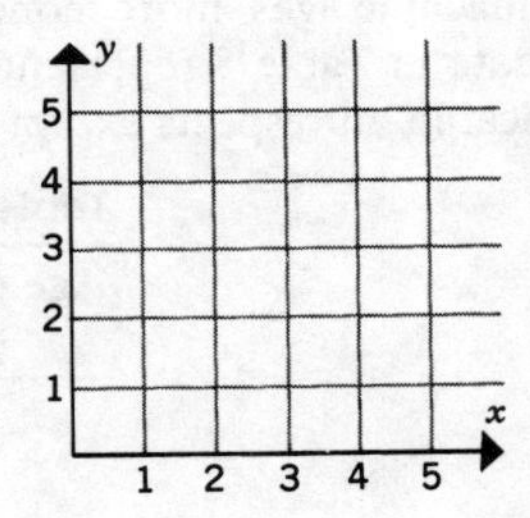

3. $\{(0, 1), (2, 2), (5, 4), (3, 3)\}$

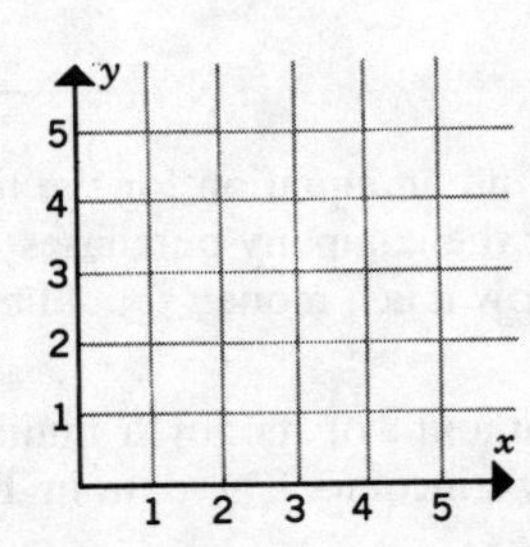

Find the equation of the line of best fit for the sets of points in exercises 4 through 10.

4. $\{(1, 3), (3, 4), (5, 5), (6, 4)\}$

5. $\{(2, 4), (3, 3), (5, 2), (7, 1)\}$

6. $\{(1, -2), (3, 0), (4, 1), (5, 4)\}$

7. $\{(1, 5), (2, 3), (3, 1), (5, 0)\}$

8. $\{(2, 4), (3, 2), (4, 0), (6, -1)\}$

9. $\{(1, 5), (2, 6), (3, 3), (4, 1), (5, 0)\}$

10. $\{(0, -2), (2, 0), (3, 1), (4, 3), (5, 4)\}$

11. Demand-and-supply equations can be approximated in many situations by determining the line of best fit to a set of data. Suppose a market research study provides the data in Table 4 on the unit price and quantity for a new smooth-writing pen. Find a demand equation that represents the line of best fit for this data.

Table 4

Quantity (millions)	Price per Unit
4	$2.00
3	2.40
2	2.75
1	3.00

12. As a machine ages, more money must be spent annually on maintenance and repair. The data in Table 5 represent annual maintenance and repair costs for six pumps, identical in all respects except for age, in a sewage treatment plant.

Table 5

Age (years), x	Annual Costs, y
5	$150
4	160
6	175
2	140
9	180
1	125

(a) Find an equation for the line of best fit.
(b) If the company purchases (secondhand) a seventh pump that is eight years old, how much money should be budgeted for repair and maintenance of the pump?

13. The amount of money a family of four spends on entertainment is related to the family's income. The data in Table 6 represent the result of a study of six different families.

Table 6

Annual Income (1000s), x	Entertainment Expenditures (1000s), y
$13	$1.0
20	1.5
22	1.9
30	3.0
35	2.7
15	0.7

(a) Find an equation for the line of best fit.
(b) How much money does a typical family of four with an annual income of \$25,000 spend on entertainment?

14. Monthly sales of a popular soft drink are affected by the outside temperature. Table 7 shows monthly sales data for six consecutive months together with the average daytime temperature for the corresponding month. Find an equation of the line of best fit.

Table 7

Temperature (°F)	Monthly Sales (1000s)
55	10
62	12
69	16
78	20
75	17
70	16
66	14

15. The heights (inches) and weights (pounds) of eight randomly selected Canadian adult males are given in Table 8.

Table 8

Height, x	65	73	70	67	72	66	74	65
Weight, y	145	190	175	180	210	155	200	170

(a) Find an equation of the least squares line for this data.
(b) Use the equation to estimate the weight of a man who is 5 ft 11 in. tall.

16. Table 9 shows the grades of six randomly selected students on the midterm and final examinations in a psychology course.

Table 9

Midterm examination, x	82	91	60	75	84	65
Final examination, y	85	84	68	69	93	60

(a) Find an equation of the least squares line for this data.
(b) Use this equation to predict the grade on the final examination for a student who received a 78 on the midterm examination.

8.6 TOTAL DIFFERENTIAL

In Section 4.6 the differential of a function of a single independent variable was defined; if $y = f(x)$, the differential dy is defined as

$$dy = f'(x)\,dx$$

where dx stands for an arbitrary change in the independent variable x. Because x and dx are independent variables, dy is in reality a function of two variables.

The concept of the differential can be extended to a function of two independent variables. If $z = f(x, y)$ and dx and dy are arbitrary changes in x and y, respectively, then the **differential** dz is defined as

$$dz = f_x(x, y)\,dx + f_y(x, y)\,dy \tag{1}$$

Example 1 Find the differential dz for each of the following functions:

(a) $f(x, y) = x^3y^2$

(b) $f(x, y) = \dfrac{x}{y}$

Solution (a) Since $f_x(x, y) = 3x^2y^2$ and $f_y = 2x^3y$, we get

$$dz = 3x^2y^2\,dx + 2x^3y\,dy$$

(b) In this case $f_x(x, y) = \dfrac{1}{y}$ and $f_y(x, y) = -\dfrac{x}{y^2}$; so

$$dz = \frac{1}{y}\,dx - \frac{x}{y^2}\,dy$$

■

Any value that dz assumes depends not only on the values assigned to x and y but also on the values assigned to dx and dy. Thus dz is in reality a function of four independent variables namely x, y, dx, and dy, as the next example illustrates.

Example 2 Evaluate the differential dz for the function $z = f(x, y) = x^3y^2$ when $x = 1$, $y = 2$, $dx = -1$, and $dy = \frac{1}{2}$.

Solution From Example 1(a) we know that

$$dz = 3x^2y^2\,dx + 2x^3y\,dy$$

Substituting the given values of x, y, dx, and dy yields the following value of dz:

$$\begin{aligned} dz &= 3(1)^2(2)^2(-1) + 2(1)^3(2)(\tfrac{1}{2}) \\ &= -12 + 2 = -10 \end{aligned}$$

■

In Section 4.5 we learned that for given values of x and dx, the corresponding value of dy represents the change in the variable y along the line tangent to the curve $y = f(x)$ at the point (x, y). In the same way, for a given function $z = f(x, y)$ and given values of x, y, dx, and dy, the differential dz represents the change in the variable z along the plane tangent to the surface $z = f(x, y)$ at the point (x, y, z). The geometric meaning of dz is illustrated in Figure 1, where the plane tangent to a surface $z = f(x, y)$ at an arbitrary point $(a, b, f(a, b))$ together with the plane parallel to the x, y plane formed by the differentials dx and dy is shown.

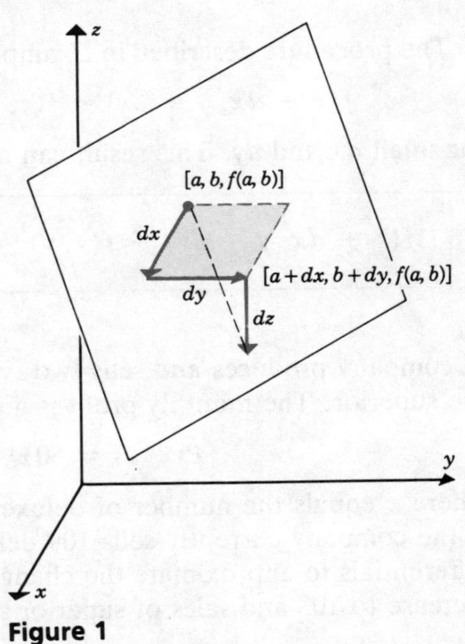

Figure 1

Note: The surface is not shown. The differential dz represents the vertical displacement between the point $(a + dx, b + dy, f(a, b))$ and the tangent plane.

Approximations Using the Differential

A surface and the plane tangent to it at a given point $(a, b, f(a, b))$ almost coincide for small values of dx and dy. As a result, the change in the dependent variable z can be approximated by the differential dz for given changes in x and y, denoted by dx and dy. This method is illustrated in the next example.

Find $f(1.02, 3.01)$ for the function $f(x, y) = \sqrt{x^2 + y}$.

Solution Noting that $f(1, 3) = 2$, we can use the point $(1, 3, 2)$ as a reference point and use the differential dz to approximate the difference between $f(1.02, 3.01)$ and $f(1, 3)$; for this calculation we have $dx = 0.2$ and $dy = 0.1$. The differential dz is given by the equation

$$dz = \frac{x}{\sqrt{x^2 + y}}\,dx + \frac{1}{2\sqrt{x^2 + y}}\,dy$$

Substituting $x = 1$, $y = 3$, $dx = 0.2$, and $dy = 0.1$, we get

$$dz = \tfrac{1}{2}(0.02) + \tfrac{1}{4}(0.01) = 0.0125$$

So $\sqrt{(1.02)^2 + 3.01} \approx f(1, 3) + dz = 2.0125$. Using a calculator, we find that $\sqrt{(1.02)^2 + 3.01} = 2.0126$ to four decimal places. ■

The procedure described in Example 3 can be generalized and takes the form

$$f(x + dx, y + dy) \approx f(x, y) + f_x(x, y)\,dx + f_y(x, y)\,dy \qquad (2)$$

for small dx and dy. This result can also be written as

$$\boxed{f(x + dx, y + dy) - f(x, y) \approx f_x(x, y)\,dx + f_y(x, y)\,dy = dz} \qquad (3)$$

Example 4 A company produces and sells two types of cross-country skis, the deluxe and the superior. The monthly profit is given by the equation

$$P(x, y) = 50x^{1/2} + 90y^{2/3} - 1{,}000$$

where x equals the number of deluxe and y the number of superior skis sold. If the company currently sells 100 deluxe and 64 superior skis per month, use differentials to approximate the change in monthly profit if sales of deluxe skis increase to 103 and sales of superior skis decrease to 62 skis.

Solution The current values of x and y are 100 and 64, respectively. The changes in x and y are given by the differentials $dx = 3$ and $dy = -2$. The differential dP approximates the change in the profit and can be found directly:

$$\begin{aligned} dP &= P_x(x, y)\,dx + P_y(x, y)\,dy \\ &= 25x^{-1/2}\,dx + 60y^{-1/3}\,dy \end{aligned}$$

Substituting $x = 100$, $y = 64$, $dx = 3$, and $dy = -2$, we get

$$\begin{aligned} dP &= 25(100)^{-1/2}(3) + 60(64)^{-1/3}(-2) \\ &= -\$22.50 \end{aligned}$$

This result indicates that the company's profit will decrease by roughly \$22.50 when these changes in sales occur. ■

Relative Error

The concept of relative error, first introduced in Section 4.5 for functions of a single variable, can be extended to functions of two or more variables. As before, the relative errors in the independent variable x and y are defined as follows:

$$\text{Relative error in } x = \frac{dx}{x}$$

$$\text{Relative error in } y = \frac{dy}{y}$$

These errors reflect the accuracy of the measurement process or the forecasting method used to provide values for x and y. The **relative error** in the **dependent variable** z is defined in the same way:

$$\boxed{\text{Relative error in } z = \frac{dz}{z}} \tag{4}$$

The relative error in z is determined from the function that defines the relationsip between z and the variables x and y together with their relative errors. The next example illustrates how the relative error in a dependent variable is found.

Example 5 A furniture manufacturer finds that $P(x, y)$, her annual profit (in millions of dollars) depends on x, the number of new homes sold, and y, the number of existing homes sold, according to the equation

$$P(x, y) = 50x^2 + 25y^2 - 10xy$$

where x and y are in millions of units. Economic forecasters are predicting that sales of new homes will equal 1 million units and sales of existing homes 3 million units in the coming year. If the relative error in their forecasts of x and y is ± 0.15, find the relative error in the predicted annual profit.

Solution The predicted annual profit can be calculated, yielding

$$P(1, 3) = 50(1)^2 + 25(3)^2 - 10(1)(3) = \$245 \text{ million}$$

The differential dP has the form

$$\begin{aligned} dP &= (100x - 10y)\, dx + (50y - 10x)\, dy \\ &= (100x - 10y)x\frac{dx}{x} + (50y - 10x)y\frac{dy}{y} \end{aligned}$$

where the second equation has been written to take advantage of our knowledge of the relative errors in x and y. Enough information is available to find the relative error in $P(x, y)$, and we get

$$\frac{dP}{P} = \frac{100(1) - 10(3)}{245}(1)(\pm 0.15) + \frac{50(3) - 10(1)}{245}(3)(\pm 0.15)$$
$$= \pm(0.043 + 0.257) = \pm .300$$

Thus the percentage error in the predicted profit is roughly $\pm 30\%$. ■

When the dependent variable is determined by a product or ratio of powers of the independent variables, the relative error in the dependent variable can be determined from a knowledge of the relative errors in the independent variables alone. In addition, the analysis indicates clearly how the relative error in each independent variable is magnified or reduced when determining the relative error in the dependent variable. The next example illustrates this procedure.

Example 6 The resistance R of a cylindrical pipe to the flow of fluid is given by the equation

$$R = \frac{kL}{r^4}$$

where L is the length and r the radius of the pipe and k is a constant of proportionality. Find the relative error in R if the relative error in L is ± 0.02 and the relative error in r is ± 0.01.

Solution The differential dR is found first, yielding

$$dR = \frac{k}{r^4}\,dL - \frac{4kL}{r^5}\,dr$$

The relative error dR/R is found next:

$$\frac{dR}{R} = \frac{k}{r^4 R}\,dL - \frac{4kL}{r^5 R}\,dr$$

Substituting kL/r^4 for R on the right-hand side and simplifying gives

$$\frac{dR}{R} = \frac{dL}{L} - 4\,\frac{dr}{r}$$

Note that the relative error in R is written in terms of the relative errors in L and r alone, yielding

$$\frac{dR}{R} = \pm 0.02 - 4(\pm 0.01)$$

The upper and lower limits on dR/R occur when all the terms on the right-hand side reinforce each other, that is,

$$\frac{dR}{R} = \pm 0.02 \pm 0.04 = \pm 0.06$$

This result also illustrates that the relative error in r is magnified by a factor of 4, the exponent associated with r. ■

8.6 EXERCISES

Find the differential for each of the functions in exercises 1 through 20.

1. $f(x, y) = x^2y^4$

2. $f(x, y) = \dfrac{1}{x + 2y}$

3. $f(x, y) = e^{y-x}$

4. $f(x, y) = \ln(x - 3y)$

5. $f(x, y) = \sqrt{x^3 + y^2}$

6. $f(s, t) = se^t$

7. $f(x, y) = y \ln x$

8. $f(x, y) = xy + \sqrt{xy}$

9. $f(x, y) = \dfrac{2}{x^2 + y}$

10. $g(u, v) = \ln(u^2v^3)$

11. $f(x, y) = (\ln x)e^y$

12. $f(x, y) = e^{x/y}$

13. $f(x, y) = \sqrt[3]{2x + 5y}$

14. $f(x, y) = \dfrac{x + y}{x - y}$

15. $f(x, y) = \dfrac{1}{x + y} + \ln(x + y)$

16. $f(p, q) = pe^{p+q}$

17. $f(x, y) = 4x^{1/3}y^{2/3}$

18. $f(x, y) = \sqrt{\ln(2x + y)}$

19. $f(x, y) = \dfrac{1 + \ln x}{1 + \ln y}$

20. $f(x, y) = x(\ln y) + y(\ln x)$

Evaluate the differential dz for each of the functions in Exercises 21 through 30.

21. $z = f(x, y) = x^2 + y^3$; $x = 1, y = -2, dx = 0.25, dy = 0.50$

22. $z = f(x, y) = \dfrac{y}{x}$; $x = 2, y = 1, dx = 0.1, dy = 0.20$

23. $z = f(x, y) = \ln(x + y)$; $x = 0, y = 1, dx = -1, dy = 2$

24. $z = f(x, y) = e^{x+y}$; $x = y = 0, dx = 2, dy = -1$

25. $z = f(x, y) = \sqrt{x - y}$; $x = -1, y = -2, dx = 2, dy = 1$

26. $z = f(x, y) = \dfrac{2}{x + y}$; $x = 1, y = 1, dx = 0.1, dy = -0.2$

27. $z = f(x, y) = x \ln y$; $x = 2, y = e, dx = 0.25, dy = -0.50$

28. $z = f(x, y) = e^{-(y/x)}$; $y = 3, x = -3, dx = 0.30, dy = 0.10$

29. $z = f(x, y) = \dfrac{x - y}{x + y}$; $x = y = 2, dx = 0.30, dy = -0.10$

30. $z = f(x, y) = \ln(x^2 + y^2)$; $x = e, y = -e, dx = 1, dy = -1$

31. The XYZ heating company has received a contract to install heating and air-conditioning equipment for a rectangular building whose dimensions are 200 by 50 by 20 ft. The president of XYZ has been notified by the architect that the length of the proposed building has been increased from 200 to 205 ft and the height from 20 to 21 ft. Use the differential to approximate the change in the volume of the building caused by the changes in the length and height.

32. The resistance R of an electric circuit, in ohms, can be determined from the equation

$$R = \frac{V}{i}$$

where V is the voltage and i is the current (amperes). Use the differential to approximate the change in R if V is increased from 120 to 125 volts while i decreases from 10 to 9.5 amperes.

33. The surface area of a person can be approximated from the empirical equation

$$A = 2.024w^{0.425}h^{0.725}$$

where A is the area, in square meters, w is the weight, in kilograms, and h is the height, in meters. Use the differential to approximate the change in the surface area of a child whose height changes from 1 to 1.2 meters and whose weight changes from 30 to 34 kg. The following information will speed your analysis:

$$30^{0.425} = 4.224, \qquad 30^{-0.575} = 0.1415$$

34. (See Example 6, Section 8.2.) The weekly output q of a firm manufacturing skis is given by the equation

$$q = 8L^{0.5}M^{0.5}$$

where L represents the number of hours of labor per week and M the number of machine-hours per week used in the production process; q is the weekly output in hundreds. Use the differential to approximate the change in output if L decreases from 900 to 880 and M increases from 400 to 410.

35. A contractor building a cylindrical rocket for NASA has been informed that the dimensions of the rocket have changed slightly; the length has been increased from 100 to 105 ft and the radius decreased from 20 to 19 ft. Use the differential to approximate the change in the volume of the rocket. The volume of a right-circular cylinder is given by the equation

$$V(l, r) = \pi r^2 l$$

where l is the length and r the radius of the rocket.

36. A manufacturer is building triangular trusses for a residential housing development; the dimensions of each are shown in Figure 2. Use the differential to approximate the additional length of each truss if the base of the triangle increased from 24 to 25 ft and the height from 9 to 10 ft.

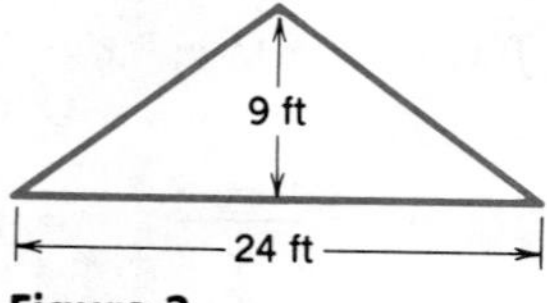

Figure 2

37. In Exercise 32, find the relative error in the calculation of the resistance R if the relative error in the measurement of the voltage is ± 0.01 and the relative error in the current measurement is ± 0.005.

38. Find the relative error in the output predicted in Exercise 34 if the relative error in L (caused, e.g., by absenteeism) is ± 0.02 and the relative error in M (caused by machine breakdowns) is ± 0.03.

39. Find the relative error in determining the volume of a rectangular box if the relative error in measuring the length, width, and height equals ± 0.01 for each dimension.

40. In Exercise 33, find the relative error in the calculation of A if the relative errors in measuring w and h are ± 0.01 and ± 0.02, respectively.

8.7 DOUBLE INTEGRALS

Differentiating a function of two variables is carried out by differentiating with respect to one variable while the other variable is kept constant. The same approach is used in integrating a function of two variables. For example, if $f_x(x, y) = 6x^2y$, we can antidifferentiate or integrate with respect to x while keeping y constant to yield

$$f(x, y) = \int 6x^2y\,dx = y\int 6x^2\,dx = 2x^3y + C(y)$$

where $C(y)$ is an arbitrary function of y, the variable held constant during this process. This process is also known as **partial integration** with respect to x. Partial integration with respect to y can be carried out in the same way, as illustrated in the next example.

Example 1 Find each of the following:

(a) $\int (6x^2y + 12xy^2)\,dx$

(b) $\int (6x^2y + 12xy^2)\,dy$

Solution (a) The variable y is kept constant while integration with respect to x is carried out in the usual manner:

$$\int (6x^2y + 12xy^2)\,dx = \frac{6x^3y}{3} + \frac{12x^2y^2}{2} + C(y)$$

$$= 2x^3y + 6x^2y^2 + C(y)$$

(b) In this case, the opposite is true; x is held constant and integration is carried out with respect to y:

$$\int (6x^2y + 12xy^2)\,dy = \frac{6x^2y^2}{2} + \frac{12xy^3}{3} + C(x)$$
$$= 3x^2y^2 + 4xy^3 + C(x)$$ ■

The process can be extended to include definite integrals of functions of two variables.

Example 2 Find each of the following:

(a) $\int_1^2 (6x^2y + 12xy^2)\,dx$ (b) $\int_0^1 (6x^2y + 12xy^2)\,dy$

Solution (a) In Example 1(a) we found the partial integral of $(6x^2y + 12xy^2)$ with respect to x; so we can write

$$\int_1^2 (6x^2y + 12xy^2)\,dx = [2x^3y + 6x^2y^2 + C(y)]\Big|_{x=1}^{x=2}$$
$$= [2(2)^3y + 6(2)^2y^2 + C(y)] - [2(1)^3y + 6(1)^2y^2 + C(y)]$$
$$= 14y + 18y^2$$

Notice that $C(y)$, the arbitrary function of y, cancels out; since this is true in general, it will not be included in future definite integrals.

(b) From Example 1(b), we can write

$$\int_0^1 (6x^2y + 12xy^2)\,dy = [3x^2y^2 + 4xy^3]\Big|_{y=0}^{y=1}$$
$$= 3x^2(1)^2 + 4x(1)^3 - [3x^2(0) + 4x(0)]$$
$$= 3x^2 + 4x$$ ■

The integrations carried out in Example 2 resulted in a function of x alone or y alone. Each of these functions can serve as the integrand in a second definite integral, as illustrated in the next example.

Example 3 Evaluate each of the following:

(a) $\int_0^1 \left[\int_1^2 (6x^2y + 12xy^2)\,dx\right] dy$ (b) $\int_1^2 \left[\int_0^1 (6x^2y + 12xy^2)\,dy\right] dx$

Solution (a) From Example 2(a) we have

$$\int_1^2 (6x^2y + 12xy^2)\,dx = 14y + 18y^2$$

So we can write

$$\int_0^1 \left[\int_1^2 (6x^2y + 12xy^2)\,dx\right] dy = \int_0^1 (14y + 18y^2)\,dy$$

$$= (7y^2 + 6y^3)\Big|_{y=0}^{y=1}$$

$$= 7 + 6 = 13$$

(b) From Example 2(b), we have

$$\int_0^1 (6xy^2 + 12xy^2)\,dy = 3x^2 + 4x$$

Following the same procedure as in part (a), we get

$$\int_1^2 \left[\int_0^1 (6x^2y + 12xy^2)\,dy\right] dx = \int_1^2 (3x^2 + 4x)\,dx$$

$$= (x^3 + 2x^2)\Big|_{x=1}^{x=2}$$

$$= (8 + 8) - (1 + 2) = 13 \quad ■$$

Integrals of the form

$$\int_a^b \left[\int_c^d f(x, y)\,dy\right] dx \quad \text{and} \quad \int_c^d \left[\int_a^b f(x, y)\,dx\right] dy$$

are examples of **double integrals.** They are also called **iterated integrals** because the operation of integration is carried out first with respect to one variable and then repeated with respect to the second variable. The integrals are usually written without the brackets:

$$\int_a^b \int_c^d f(x, y)\,dy\,dx = \int_a^b \left[\int_c^d f(x, y)\,dy\right] dx$$

$$\int_c^d \int_a^b f(x, y)\,dx\,dy = \int_c^d \left[\int_a^b f(x, y)\,dx\right] dy$$

where the order of integration is indicated by the order of dx and dy.

Example 4 Evaluate each of the following double integrals:

$$\text{(a)} \int_0^2 \int_{-1}^0 10x^4y \, dy \, dx \qquad \text{(b)} \int_{-1}^0 \int_0^2 10x^4y \, dx \, dy$$

Solution (a) Integration with respect to y is carried out first:

$$\int_0^2 \int_{-1}^0 10x^4y \, dy \, dx = \int_0^2 \left(5x^4y^2 \Big|_{y=-1}^{y=0} \right) dx = \int_0^2 -5x^4 \, dx$$

$$= -x^5 \Big|_0^2 = -32$$

(b) In this case integration with respect to x is carried out first:

$$\int_{-1}^0 \int_0^2 10x^4y \, dx \, dy = \int_{-1}^0 \left(2x^5y \Big|_{x=0}^{x=2} \right) dy$$

$$= \int_{-1}^0 64y \, dy = 32y^2 \Big|_{-1}^0 = -32$$

■

Examples 3 and 4 illustrate an important feature of double integrals of the form

$$\int_a^b \int_c^d f(x, y) \, dy \, dx \quad \text{and} \quad \int_c^d \int_a^b f(x, y) \, dx \, dy$$

namely the order of integration does not affect the value of the integral.

Theorem 1 If $z = f(x, y)$ is a continuous function over the region defined by the inequalities $a \leq x \leq b$ and $c \leq y \leq d$, the following two double integrals are equal:

$$\boxed{\int_a^b \int_c^d f(x, y) \, dy \, dx = \int_c^d \int_a^b f(x, y) \, dx \, dy}$$

Volumes on Rectangular Regions

The next question that we want to address is: what geometric significance can be attached to a double integral such as

$$\int_a^b \int_c^d f(x, y) \, dy \, dx$$

First, notice that the region R in the xy-plane defined by the inequalities $a \leq x \leq b$ and $c \leq y \leq d$ represents a rectangle, as shown in Figure 1. If

$f(x, y)$ is nonnegative everywhere over the region, the double integral

$$\int_a^b \int_c^d f(x, y)\, dy\, dx$$

represents the **volume of the solid region** that lies between the surface $z = f(x, y)$ and the xy-plane bounded by the four vertical walls $x = a$, $x = b$, $y = c$, and $y = d$, as shown in Figure 2.

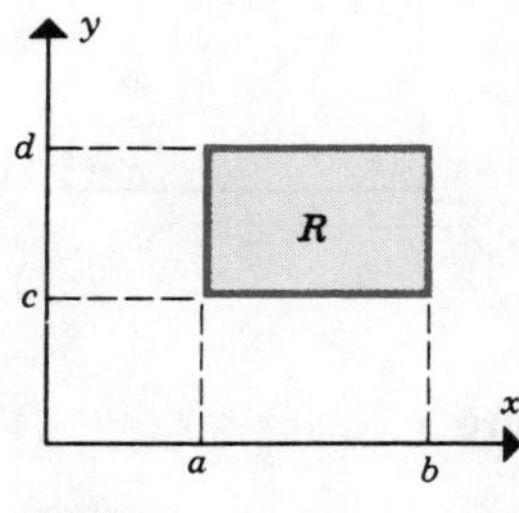

Figure 1

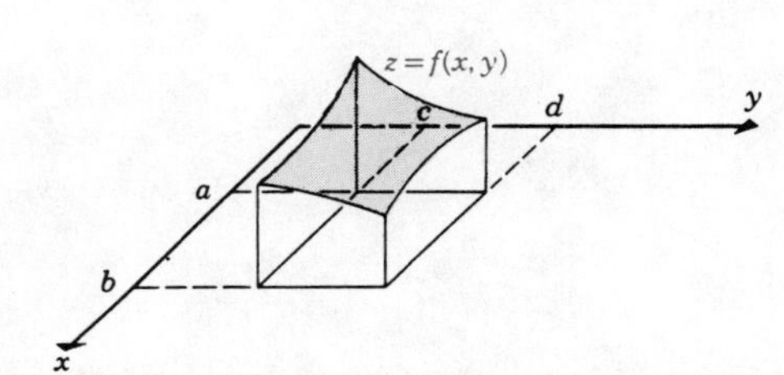

Figure 2

Theorem 2 If $f(x, y) \geq 0$ everywhere over the rectangular region R in the xy-plane defined by the inequalities $a \leq x \leq b$ and $c \leq y \leq d$, then V, the volume of the solid region lying between $z = f(x, y)$ and R, is

$$V = \int_a^b \int_c^d f(x, y)\, dy\, dx$$

Example 5 Find the volume of the solid region beneath the surface

$$f(x, y) = 9 - x^2 - y^2$$

and bounded by the four sides $x = 0$, $x = 1$, $y = 0$, and $y = 2$.

Solution The solid bounded above by the surface $f(x, y) = 9 - x^2 - y^2$, below by the rectangular region defined by the inequalities $0 \leq x \leq 1$, $0 \leq y \leq 2$, and by the four vertical walls $x = 0$, $x = 1$, $y = 0$, and $y = 2$ is shown in Figure 3. The volume of the solid is found by evaluating the following double integral:

$$\begin{aligned}
V &= \int_0^1 \int_0^2 (9 - x^2 - y^2)\, dy\, dx \\
&= \int_0^1 \left(9y - x^2y - \frac{y^3}{3}\right)\bigg|_{y=0}^{y=2} dx = \int_0^1 \left(18 - 2x^2 - \frac{8}{3}\right) dx \\
&= \int_0^1 \left(\frac{46}{3} - 2x^2\right) dx = \left(\frac{46}{3}x - \frac{2x^3}{3}\right)\bigg|_0^1 \\
&= \frac{44}{3}
\end{aligned}$$

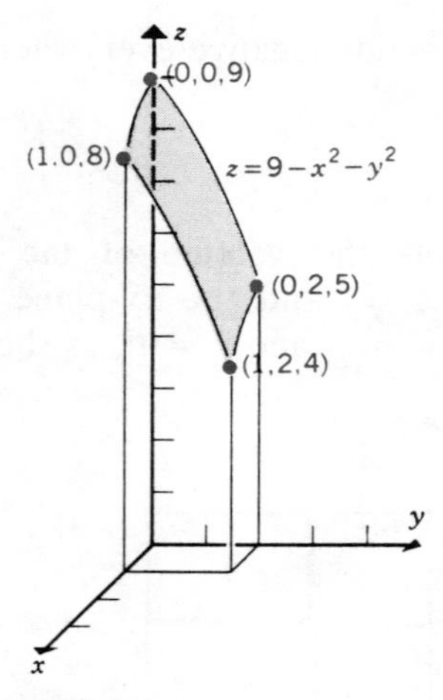

Figure 3

The same result is obtained by evaluating the double integral

$$\int_0^2 \int_0^1 (9 - x^2 - y^2)\, dx\, dy$$

■

Example 6 Find the volume of the solid lying between the surface

$$f(x, y) = 15\sqrt{1 + x + y}$$

and the rectangular region in the xy-plane defined by the inequalities $2 \leq x \leq 7$ and $0 \leq y \leq 1$.

Solution Since $f(x, y) \geq 0$ everywhere over the region, the volume can be found by evaluating the following integral:

$$\begin{aligned}
V &= \int_2^7 \int_0^1 15\sqrt{1 + x + y}\, dy\, dx \\
&= \int_2^7 10(1 + x + y)^{3/2} \Big|_{y=0}^{y=1} dx \\
&= \int_2^7 [10(2 + x)^{3/2} - 10(1 + x)^{3/2}]\, dx \\
&= [4(2 + x)^{5/2} - 4(1 + x)^{5/2}] \Big|_{x=2}^{x=7} \\
&= 4(9)^{5/2} - 4(8)^{5/2} - 4(4)^{5/2} + 4(3)^{5/2} \\
&\approx 182.3
\end{aligned}$$

■

8.7 EXERCISES

Find each of the following integrals.

1. $\int 10x^4y^2\,dx$
2. $\int (8xy^3 + 2)\,dy$
3. $\int \left(\frac{2x+1}{y}\right) dx$
4. $\int \frac{2x+1}{y}\,dy$
5. $\int \left(\frac{6x^3 + 3y^2}{x}\right) dx$
6. $\int \left(\frac{6x^3 + 3y^2}{x}\right) dy$
7. $\int ye^{xy}\,dx$
8. $\int ye^{xy}\,dy$
9. $\int \frac{x}{x^2 - y^2}\,dx$
10. $\int \frac{x}{x^2 - y^2}\,dy$
11. $\int \frac{y\sqrt{x}}{y+1}\,dx$
12. $\int \frac{y\sqrt{x}}{y+1}\,dy$

Find each of the following definite integrals.

13. $\int_0^2 12x^3y^2\,dx$
14. $\int_1^3 12x^3y^2\,dy$
15. $\int_1^9 6y\sqrt{x}\,dx$
16. $\int_2^4 6y\sqrt{x}\,dy$
17. $\int_1^2 \frac{y + x^2}{x}\,dx$
18. $\int_0^2 \frac{y + x^2}{x}\,dy$
19. $\int_0^1 ye^{-xy}\,dx$
20. $\int_0^1 ye^{-xy}\,dy$
21. $\int_1^3 \frac{6y}{\sqrt{x - y^2}}\,dx$
22. $\int_{-1}^1 \frac{6y}{\sqrt{x - y^2}}\,dy$
23. $\int_2^3 ye^{x-y}\,dx$
24. $\int_0^1 ye^{x-y}\,dy$

Evaluate each of the following double integrals.

25. $\int_0^2 \int_{-1}^1 (6x^2 - y)\,dx\,dy$
26. $\int_{-2}^0 \int_0^1 8xy^3\,dy\,dx$
27. $\int_{-2}^2 \int_{-1}^1 (4xy + 3y^2)\,dx\,dy$
28. $\int_0^1 \int_0^1 xe^{x+y}\,dy\,dx$
29. $\int_1^2 \int_0^3 \frac{x}{y + x^2}\,dx\,dy$
30. $\int_0^3 \int_1^2 \frac{x}{y + x^2}\,dy\,dx$
31. $\int_1^3 \int_0^2 \frac{3y}{\sqrt{x + y^2}}\,dx\,dy$
32. $\int_0^2 \int_1^3 \frac{3y}{\sqrt{x + y^2}}\,dy\,dx$
33. $\int_{-1}^1 \int_0^1 yxe^{-x^2y}\,dx\,dy$
34. $\int_1^2 \int_0^1 x^3e^{-x^2y}\,dy\,dx$

Find the volume of the solid between each of the following surfaces and the given rectangular region in the xy-plane.

35. $f(x, y) = 2x + 6y$; rectangle R: $-1 \le x \le 1, 1 \le y \le 2$
36. $f(x, y) = 8xy$; rectangle R: $0 \le x \le 2, 0 \le y \le 1$
37. $f(x, y) = e^{x+y}$; rectangle R: $-1 \le x \le 1, 0 \le y \le 2$
38. $f(x, y) = 12 - 3x^2 - 3y^2$; rectangle R: $-2 \le x \le 2, 1 \le y \le 1$
39. $f(x, y) = 27 - 3x^2$; rectangle R: $-3 \le x \le 3, -1 \le y \le 1$
40. $f(x, y) = \dfrac{y + x}{x}$; rectangle R: $1 \le x \le 2, 1 \le y \le 2$

8.8 MORE DOUBLE INTEGRALS

The double integrals considered to this point have been restricted to those for which the limits of integration are constant, namely those having the form

$$\int_c^d \int_a^b f(x, y)\, dx\, dy \tag{1}$$

Geometrically, this restriction means that R, the region of integration in the x, y plane, is a rectangle defined by the inequalities

$$a \le x \le b \quad \text{and} \quad c \le y \le d$$

as shown in Figure 1.

Integration over more general regions such as those shown in Figures 2 and 3 can be defined and carried out by employing the same methods used in eval-

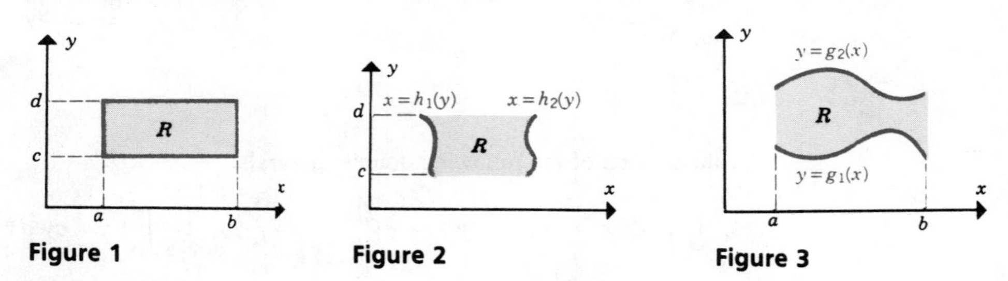

Figure 1 Figure 2 Figure 3

uating double integrals such as those in Equation 1. The double integrals take the form

$$\int_c^d \int_{h_1(y)}^{h_2(y)} f(x, y)\, dx\, dy = \int_c^d \left[\int_{h_1(y)}^{h_2(y)} f(x, y)\, dx \right] dy \tag{2}$$

or

$$\int_a^b \int_{g_1(x)}^{g_2(x)} f(x, y)\, dy\, dx = \int_a^b \left[\int_{g_1(x)}^{g_2(x)} f(x, y)\, dy \right] dx \tag{3}$$

The closed region R in Figure 2 is defined by the inequalities

$$h_1(y) \leq x \leq h_2(y) \quad \text{and} \quad c \leq y \leq d$$

and the closed region R in Figure 3 is defined by the inequalities

$$a \leq x \leq b, \qquad g_1(x) \leq y \leq g_2(x).$$

When we extend the concept of the double integral to more general regions, it is important to note that the limits of integration on the inner integral may be variable; the limits on the outer integral are constant. Examples 1 and 2 illustrate how the concept of a double integral is extended to include more general regions and to show how a region of integration can be sketched.

Example 1 Evaluate $\int_1^2 \int_0^y 15x^2y \, dx \, dy$. In addition, sketch R, the region of integration in the xy-plane.

Solution Noting that

$$\int_1^2 \int_0^y 15x^2y \, dx \, dy = \int_1^2 \left[\int_0^y 15x^2y \, dx \right] dy$$

we carry out the evaluation in stepwise fashion. First, partial integration with respect to x is carried out:

$$\int_1^2 \left[\int_0^y 15x^2y \, dx \right] = \int_1^2 5x^3y \Big|_{x=0}^{x=y} dy$$

$$= \int_1^2 5y^4 \, dy \tag{4}$$

The corresponding region of the xy-plane represented by the limits of integration with respect to x is shown in Figure 4, where the shaded region represents the set of points that lie between the lines $x = 0$ and $x = y$. Next, partial integration

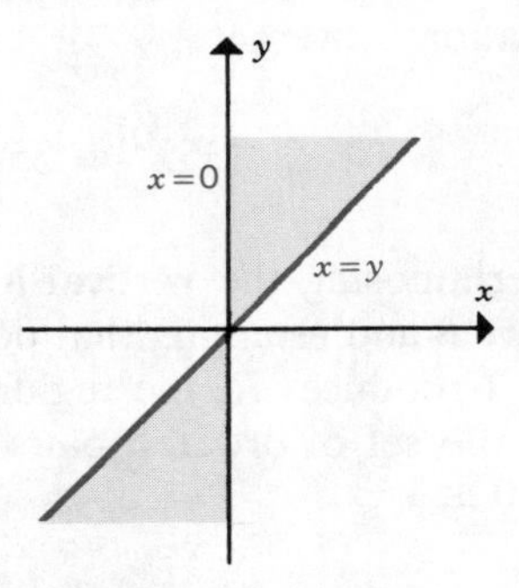

Figure 4

with respect to y is carried out on the integral shown in Equation 4:

$$\int_1^2 5y^4\,dy = y^5 \Big|_{y=1}^{y=2} = 32 - 1 = 31$$

The region of integration R can be shown by superimposing the horizontal lines $y = 1$ and $y = 2$ on the graph in Figure 4. The result is the closed shaded region in Figure 5 whose points satisfy the inequalities

$$0 \le x \le y \quad \text{and} \quad 1 \le y \le 2$$

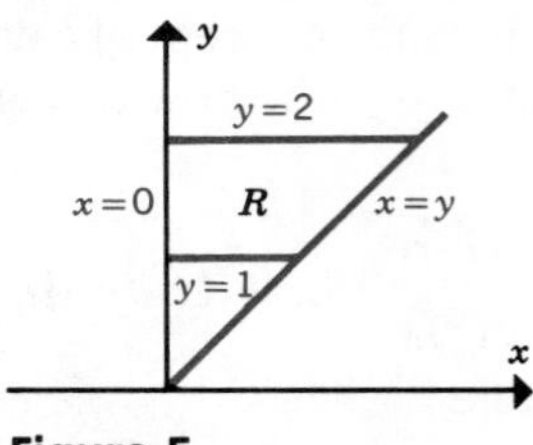

Figure 5

■

Example 2 Evaluate $\int_0^1 \int_{x^2}^x 70x^2y\,dy\,dx$. In addition, sketch the region of integration.

Solution As before, the integration is carried out in stepwise fashion.

1. Partial integration with respect to y:

$$\int_0^1 \left[\int_{x^2}^x 70x^2y\,dy\right] dx = \int_0^1 35x^2y^2 \Big|_{y=x^2}^{y=x} dx$$

$$= \int_0^1 (35x^4 - 35x^6)\,dx \tag{5}$$

   The region between the curves $y = x$ and $y = x^2$ which represents the limits of integration with respect to y, is shown in Figure 6.

2. Partial integration with respect to x: Carrying out the integration shown in Equation 5, we get

$$\int_0^1 (35x^4 - 35x^6)\,dx = (7x^5 - 5x^7) \Big|_{x=0}^{x=1} = 2$$

   Superimposing the vertical lines $x = 0$ and $x = 1$ on the graph shown in Figure 6 and retaining that portion of the shaded region between $x = 0$ and $x = 1$ produces R, the region of integration shown in Figure 7. The region R is the set of ordered pairs (x, y) that satisfy the inequalities $x^2 \le y \le x$ and $0 \le x \le 1$.

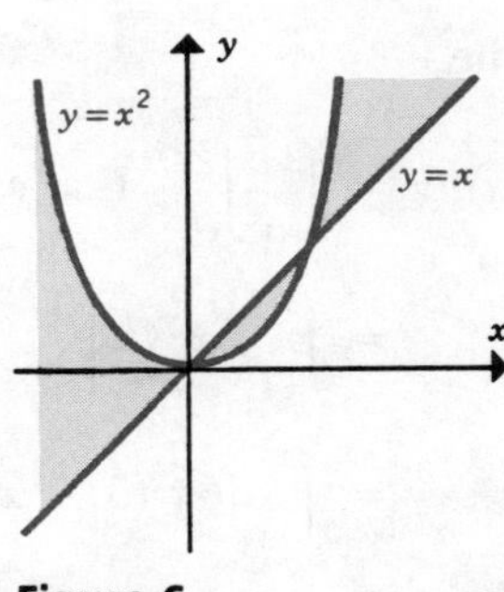

Figure 6

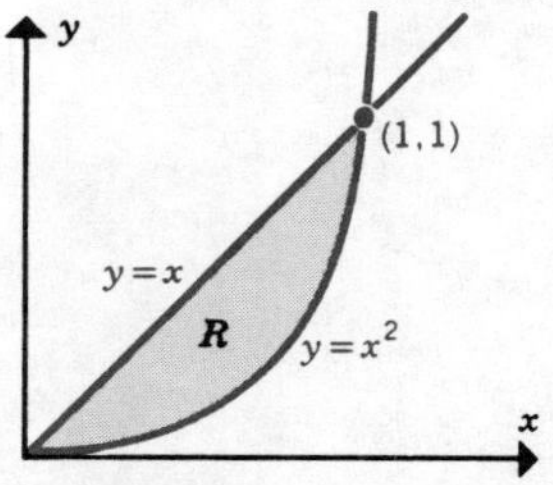

Figure 7

The double integrals $\int_c^d \int_{h_1(y)}^{h_2(y)} f(x, y)\, dx\, dy$ and $\int_a^b \int_{g_1(x)}^{g_2(x)} f(x, y)\, dy\, dx$ are often written in a simpler form as $\iint_R f(x, y)\, dx\, dy$; that is,

$$\iint_R f(x, y)\, dx\, dy = \begin{cases} \displaystyle\int_c^d \int_{h_1(y)}^{h_2(y)} f(x, y)\, dx\, dy \\ \displaystyle\int_a^b \int_{g_1(x)}^{g_2(x)} f(x, y)\, dy\, dx \end{cases}$$

where R represents the region of integration. Note that the constant limits are associated with the outer integral, the variable limits with the inner integral.

Example 3 Find $\iint_R 12x^2y\, dx\, dy$, where R is the region of the xy-plane bounded by the curves $y = x^2$, $y = 3 - x$, $x = -1$ and $x = 1$.

Solution The region of integration is shown in Figure 8. The points (x, y) that belong to R satisfy the inequalities

$$-1 \leq x \leq 1 \quad \text{and} \quad x^2 \leq y \leq 3 - x$$

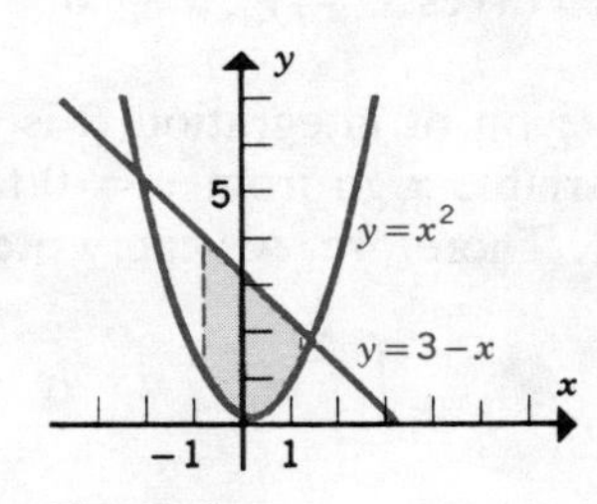

Figure 8

Since the limits on x are constant and those on y are variable in the inequalities that define R, we can write

$$\iint_R 12x^2y\,dx\,dy = \int_{-1}^{1}\int_{x^2}^{3-x} 12x^2y\,dy\,dx$$

$$= \int_{-1}^{1} 6x^2y^2\Big|_{y=x^2}^{y=3-x} dx$$

$$= \int_{-1}^{1} [6x^2(3-x)^2 - 6x^4]\,dx$$

$$= \int_{-1}^{1} [6x^2(9-6x+x^2) - 6x^4]\,dx$$

$$= \int_{-1}^{1} (54x^2 - 36x^3)\,dx$$

$$= (18x^3 - 9x^4)\Big|_{x=-1}^{x=1} = 36$$

■

Two important steps must be taken before you can evaluate a definite integral of the form $\iint_R f(x, y)\,dy\,dx$:

(1) The order in which the integration is to be carried out must be determined, and

(2) The limits of integration for each variable must be defined

Drawing a graph of R, the region of integration in the xy-plane, can be very helpful in carrying out these steps, as the next two examples illustrate.

Example 4 Evaluate $\iint_R 60xy\,dx\,dy$ where R is the closed region in the x, y plane bounded by the curves $x = y^2$, $x = 0$, and $y = 1$.

Solution The region of integration R is shown in Figure 9. The limits of integration on the variable x go from $x = 0$ to $x = y^2$, and the limits on y go from $y = 0$ to $y = 1$. Therefore, we can write

$$\iint_R 60xy\,dx\,dy = \int_0^1\int_0^{y^2} 60xy\,dx\,dy$$

$$= \int_0^1 30x^2y\Big|_{x=0}^{x=y^2} dy = \int_0^1 30y^5\,dy$$

$$= 5y^6\Big|_0^1 = 5$$

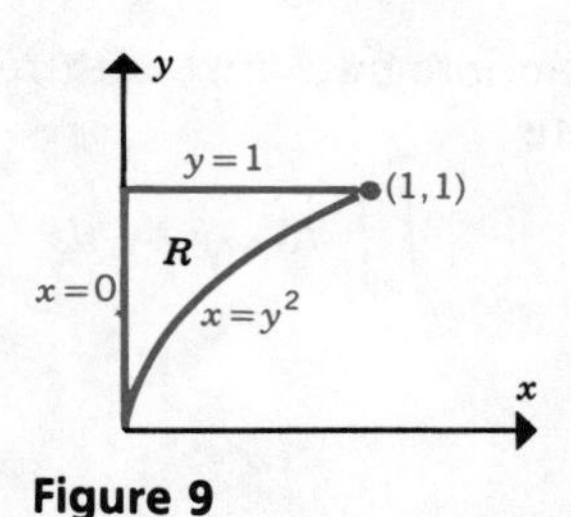

Figure 9

■

Reversing the Order of Integration

Example 4 represents a situation in which the order of integration can be reversed. Noting that the portion of the curve $x = y^2$ shown in Figure 9 can also be written as

$$y = \sqrt{x}$$

we can find $\iint_R 60xy\,dx\,dy$ by integrating first with respect to y and then with respect to x as follows:

$$\begin{aligned}\iint_R 60xy\,dx\,dy &= \int_0^1 \int_{\sqrt{x}}^1 60xy\,dy\,dx \\ &= \int_0^1 30xy^2 \Big|_{y=\sqrt{x}}^{y=1} dx \\ &= \int_0^1 (30x - 30x^2)\,dx \\ &= (15x^2 - 10x^3)\Big|_0^1 = 5\end{aligned}$$

which is the result we obtained in Example 4. The next example also illustrates a problem in which the order of integration can be reversed easily.

Example 5 Find $\iint_R 70x^2y\,dx\,dy$ where R is the region shown in Figure 10.

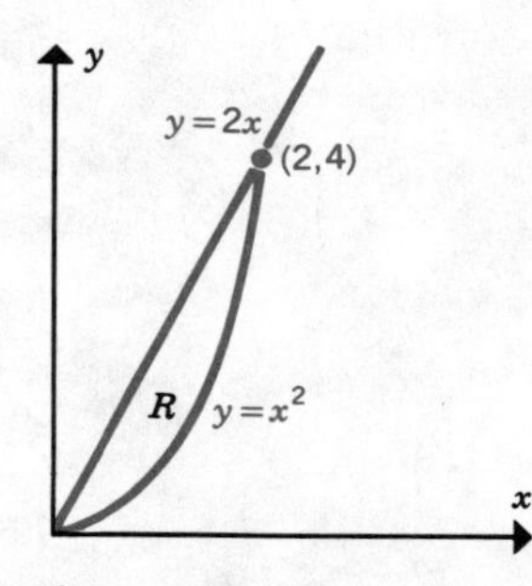

Figure 10

Solution 1. If we integrate first with respect to y and second with respect to x, we can write

$$\iint_R 70x^2y\,dy\,dx = \int_0^2 \int_{x^2}^{2x} 70x^2y\,dy\,dx$$
$$= \int_0^2 35x^2y^2 \Big|_{y=x^2}^{y=2x} dx$$
$$= \int_0^2 [140x^4 - 35x^6]\,dx$$
$$= (28x^5 - 5x^7)\Big|_0^2 = 896 - 640 = 256$$

2. The equations of the curves which form the boundary of R can be written as $x = y/2$ and $x = \sqrt{y}$. This enables us to integrate first with respect to x and we can write the integral as

$$\iint_R 70x^2y\,dx\,dy = \int_0^4 \int_{y/2}^{\sqrt{y}} 70x^2y\,dx\,dy$$
$$= \int_0^4 \frac{70x^3y}{3} \Big|_{x=y/2}^{x=\sqrt{y}} dy$$
$$= \int_0^4 \left(\frac{70y^{5/2}}{3} - \frac{70y^4}{24}\right) dy$$
$$= \left(\frac{20y^{7/2}}{3} - \frac{7y^5}{12}\right)\Big|_0^4$$
$$= \frac{(20)(4)^{7/2}}{3} - \frac{(7)(4)^5}{12} = 256$$ ■

When $f(x, y)$ is nonnegative everywhere over the region R, a double integral such as $\int_R\int f(x, y)\,dx\,dy$ can be used to find the volume of a solid bounded above by the surface $z = f(x, y)$, below by the xy-plane, and on the sides by the vertical walls defined by the boundary of the region R as illustrated in Figure 11.

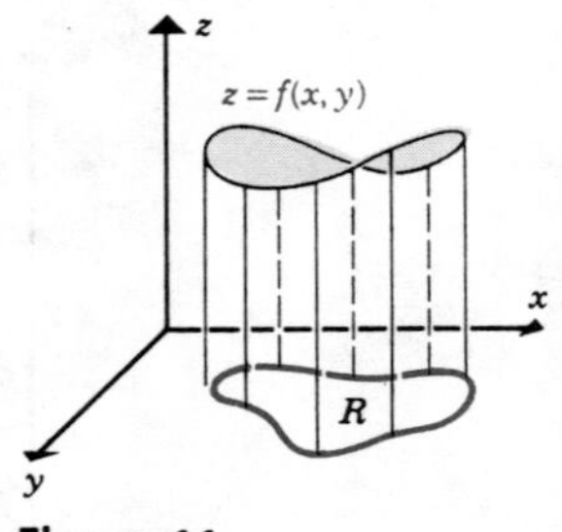

Figure 11

Theorem If $f(x, y) \geq 0$ everywhere over a closed region R in the xy-plane, then V, the volume of the solid contained between the surface $z = f(x, y)$ and the region R, is

$$V = \iint_R f(x, y)\, dy\, dx \tag{6}$$

Example 6 Find the volume of the solid lying between the surface $f(x, y) = 1 + x^2 + y^2$ and the region R bounded by the curves $y = \sqrt{x}$ and $y = x^2$.

Solution The region R is shown in Figure 12. The volume of the solid can be found by evaluating the double integral $\int_R\int (1 + x^2 + y^2)\, dy\, dx$. If we decide to integrate

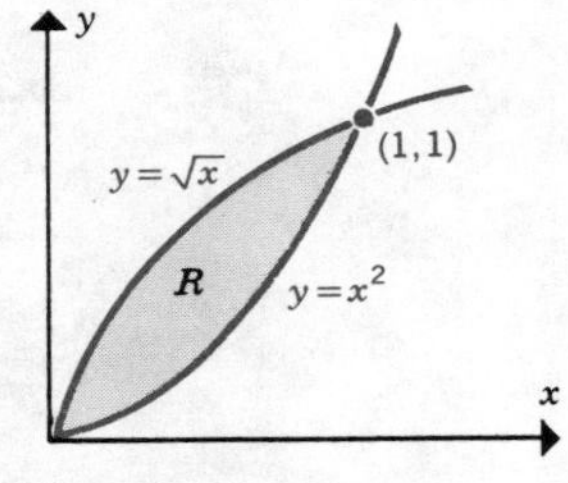

Figure 12

partially first with respect to y, we can write the double integral in the following form:

$$\begin{aligned}
\iint_R (1 - x^2 + y^2)\, dy\, dx &= \int_0^1 \int_{x^2}^{\sqrt{x}} (1 + x^2 + y^2)\, dy\, dx \\
&= \int_0^1 \left(y + x^2 y + \frac{y^3}{3} \right) \Bigg|_{y=x^2}^{y=\sqrt{x}} dx \\
&= \int_0^1 \left[\left(\sqrt{x} + x^{5/2} + \frac{x^{3/2}}{3} \right) - \left(x^2 + x^4 + \frac{x^6}{3} \right) \right] dx \\
&= \int_0^1 \left(x^{1/2} + x^{5/2} + \frac{x^{3/2}}{3} - x^2 - x^4 - \frac{x^6}{3} \right) dx \\
&= \left[\frac{2x^{3/2}}{3} + \frac{2x^{7/2}}{7} + \frac{2x^{5/2}}{15} - \frac{x^3}{3} - \frac{x^5}{5} - \frac{x^7}{21} \right]_{x=0}^{x=1} \\
&= \frac{53}{105}
\end{aligned}$$

■

Application: Crop Yields

When crops are planted, the yield per acre is generally not uniform but is usually largest near the center of a field and lowest near the boundary. If $f(x, y)$ represents the yield per acre, then the total yield from a region R is found by evaluating the double integral $\iint_R f(x, y)\,dx\,dy$, that is,

$$\text{Total yield} = \iint_R f(x, y)\,dx\,dy$$

Example 7 The yield per acre from corn planted in the region R, shown in Figure 13, is given by the equation

$$f(x, y) = 200 - 3x^2 - 3y^2 \text{ bushels/acre}$$

Find and evaluate the double integral that gives the total yield from the region.

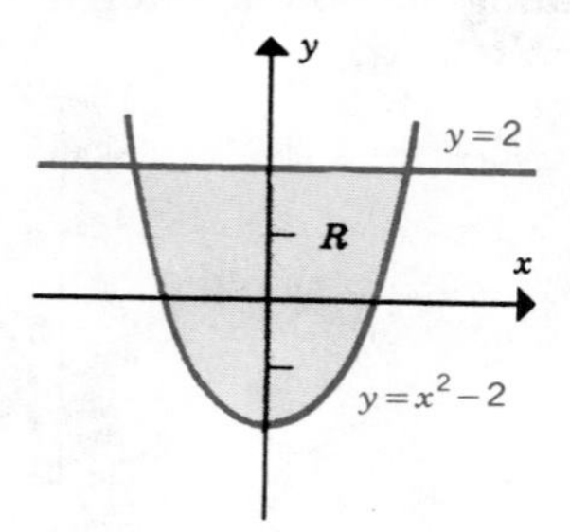

Figure 13

Solution The total yield is given by the double integral

$$\iint_R f(x, y)\,dy\,dx = \int_{-2}^{2}\int_{x^2-2}^{2} (200 - 3x^2 - 3y^2)\,dy\,dx$$

$$= \int_{-2}^{2} (200y - 3x^2y - y^3)\Big|_{y=x^2-2}^{y=2}\,dx$$

$$= \int_{-2}^{2} (784 - 200x^2 - 3x^4 + x^6)\,dx$$

When we evaluate the integral, we get

$$\text{Total yield} = 2067.6 \text{ bushels}$$ ■

8.8 EXERCISES

Evaluate each of the following integrals. In addition, sketch the graph of R, the region in the xy-plane over which the integration is carried out.

1. $\displaystyle\int_0^1\int_0^y 48xy\,dx\,dy$

2. $\int_{-1}^{0} \int_{0}^{x^2} (2x - 2y)\, dy\, dx$

3. $\int_{-1}^{0} \int_{x^3}^{0} 4\, dy\, dx$

4. $\int_{0}^{2} \int_{x}^{\sqrt{x}} 60x^3y\, dy\, dx$

5. $\int_{1}^{2} \int_{0}^{y^2} y\sqrt{x}\, dx\, dy$

6. $\int_{0}^{-1} \int_{0}^{y^2} (x + y)\, dx\, dy$

7. $\int_{1}^{2} \int_{x}^{x^2} (2x + 3y^2)\, dy\, dx$

8. $\int_{0}^{2} \int_{-y}^{0} e^{x-y}\, dx\, dy$

Evaluate each of the following integrals over the given region R.

9. $\iint_R 30x^2y\, dx\, dy$; R is the region bounded by the curves $y = x$, $x = 0$, and $y = 2$

10. $\iint_R 24(x - y)^2\, dx\, dy$; R is the region bounded by the curves $y = 1 - x$, $x = -1$, and $y = 0$

11. $\iint_R \frac{1}{x}\, dx\, dy$; R is the region bounded by the curves $x = y$, $y = 2$, and $x = 1$

12. $\iint_R xy\, dx\, dy$; R is the region bounded by the curves $y = \sqrt{x + 1}$, $y = 0$, and $x = 0$

13. $\iint_R x\, dy\, dx$; R is the region bounded by the curve $y = \sqrt{1 - x^2}$ and the x axis

14. $\iint_R xy\sqrt{x^2 - y^2}\, dy\, dx$; R is the region bounded by the curves $y = x$, $y = 0$, and $x = 1$

15. $\iint_R xe^{-y}\, dx\, dy$; R is the region bounded by the curve $y = 1 - x^2$ and the x and y axes

16. $\iint_R yx^2\, dx\, dy$; R is the region between the curve $y = x - x^2$ and the x axis

Find each of the following integrals over the regions shown.

17. $\iint_R (4x + 12y)\, dx\, dy$

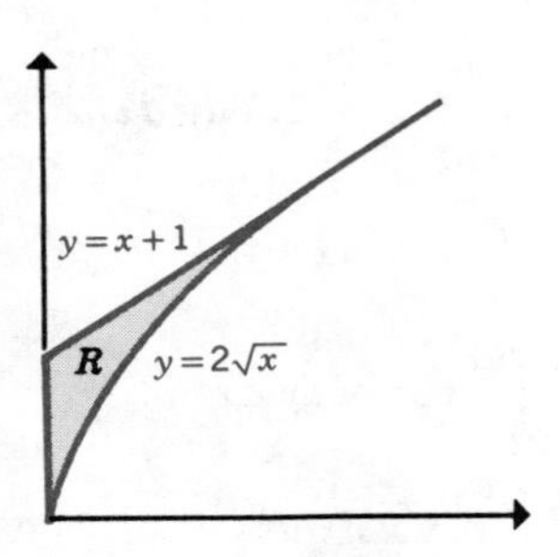

18. $\displaystyle\iint_R 30x^2y\,dx\,dy$

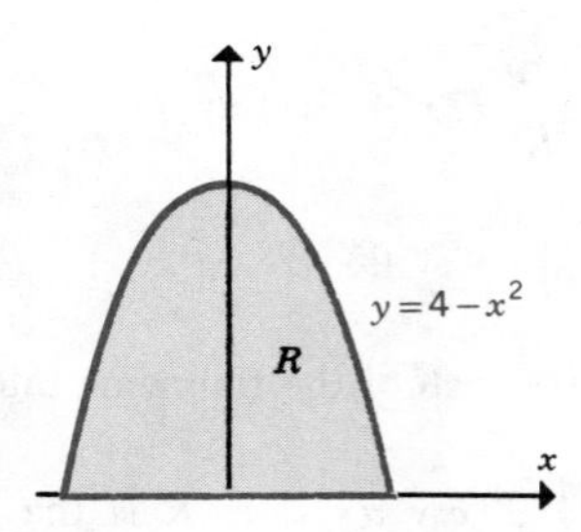

19. $\displaystyle\iint_R 24xy\,dx\,dy$

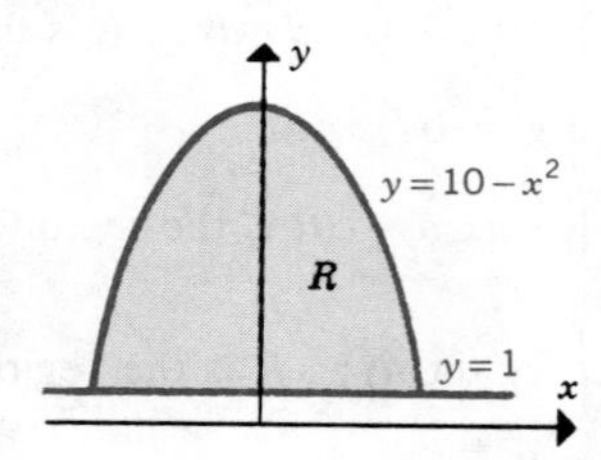

20. $\displaystyle\iint_R (12y^3 - 12xy)\,dy\,dx$

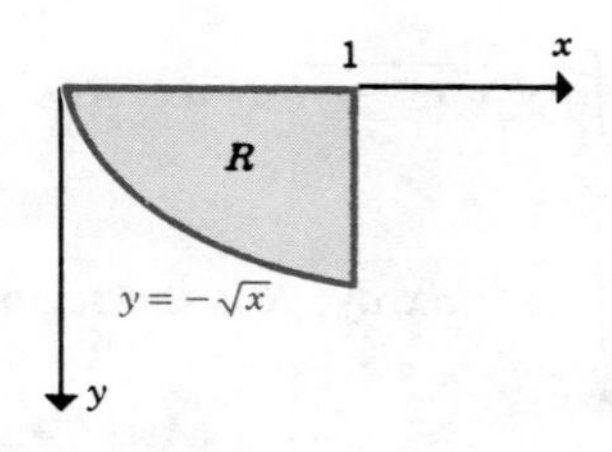

KEY TERMS

function of two variables
ordered triple
critical point
relative maximum

octant
surface
level curve
trace
plane
production function
inputs
outputs
partial derivative
marginal cost of labor
marginal cost of material
second-order partial derivative
marginal productivity
relative minimum
saddle point
second-derivative test
constraint
Lagrange multiplier
method of least squares
total differential
relative error
double integral
iterated integral
region of integration
volume of a solid under a surface

REVIEW PROBLEMS

1. If $f(x, y) = 3x^2 - 4xy + y^2$, find

(a) $f(1, 2)$
(b) $f(2, 1)$
(c) $f(1 + h, 2)$
(d) $\dfrac{f(1 + h, 2) - f(1, 2)}{h}$
(e) $\lim_{h \to 0} \dfrac{f(1 + h) - f(1, 2)}{h}$
(f) $f_x(1, 2)$

2. If $g(x, y) = \dfrac{x - y}{x^2 + y^2}$, find

(a) $g(1, 1)$
(b) $g(2, 1)$
(c) $g(1, 2)$
(d) $g(\frac{1}{2}, \frac{1}{4})$

3. If $f(x, y) = xe^{-y} + ye^{-x}$, find

(a) $f(1, 1)$
(b) $f(0, 1)$
(c) $f(1, 0)$
(d) $f(-1, 1)$

4. If $F(x, y) = \ln(x + y)$, find

(a) $F(-1, 2)$
(b) $F(3, -2)$
(c) $F(e, e)$
(d) $F(2e^2, -e^2)$

Find $f_x(x, y)$ and $f_y(x, y)$ for each of the following functions.

5. $f(x, y) = x^2 - y^2$

6. $f(x, y) = \dfrac{3x}{x + y}$

7. $f(x, y) = 3x^2e^y$

8. $f(x, y) = \ln(x^2 + 2y)$

9. $f(x, y) = \sqrt{x^2 - y^3}$

10. $f(x, y) = e^{2x/y}$

11. $f(x, y) = (e^x + e^{-y})^2$

12. $f(x, y) = (\ln x - \ln y)^2$

Find f_{xx}, f_{yy}, and f_{xy} for each of the following functions

13. $f(x, y) = x^2 + 4xy - y^3$

14. $f(x, y) = yx^3 + 2x^2y^2$

15. $f(x, y) = e^{xy}$
16. $f(x, y) = \ln(2x + 3y)$
17. $f(x, y) = xe^{-y}$
18. $f(x, y) = y^2\sqrt{x}$

In problems 19 through 26, find all the critical points on each surface defined by the given equation. Use the second-derivative test, when possible, to determine the shape of the surface in the neighborhood of each critical point.

19. $f(x, y) = x^2 + 2y^2 - 4x - 12y + 3$
20. $f(x, y) = x^3 + 6x^2 + 2y^2 - 8y + 5$
21. $f(x, y) = e^{-x^2-y^2}$
22. $f(x, y) = y^3 - 6x^2 + 12xy + 2$
23. $f(x, y) = -x^2 - 4xy + 8x + 3y^2 - 12y - 7$
24. $f(x, y) = 2x^2 + 2xy + 3y^2 - 8x + 6y + 5$
25. $f(x, y) = \ln(x^2 + y^2 + 1)$
26. $f(x, y) = x^3 + y^3 - 6x^2 - 9y^2 + 10$

Use the method of Lagrange multipliers in problems 27 through 30.

27. Minimize $f(x, y) = x^2 + y^2$ subject to the constraint $x + y - 6 = 0$.
28. Maximize $f(x, y) = y^2 - x^2$ subject to the constraint $x - 2y + 9 = 0$.
29. Maximize $f(x, y) = x + 2xy + 2y$ subject to the constraint $x + 2y = 50$.
30. Minimize $f(x, y) = x + 2y$ subject to the constraint $xy = 98$ for $x, y > 0$.

In problems 31 through 34 use the information provided to evaluate the differential dz for each function.

31. $z = f(x, y) = 3x^2 - 4y^2 + 2xy; \quad x = -1, y = 2, dx = 0.20, dy = -0.10$
32. $z = f(x, y) = e^{xy}; x = -1, y = -1, dx = 0.05, dy = 0.02$
33. $z = f(x, y) = \dfrac{y}{x^2}; x = 1, y = 2, dx = -0.01, dy = 0.04$
34. $z = f(x, y) = \ln(x - y); \quad x = -e, y = -2e, dx = 1, dy = 2.$

Find each of the following integrals.

35. $\int 20x^3y^3\, dy$
36. $\int xe^{-xy}\, dx$
37. $\int y \ln x\, dy$
38. $\int y \ln x\, dx$

Find each of the following.

39. $\int_1^4 (6x^2y^2 + 3y)\, dx$
40. $\int_0^2 (x + 2y - 9y^2)\, dy$
41. $\int_1^2 ye^x\, dx$
42. $\int_1^e x \ln y\, dy$

Evaluate each of the following iterated integrals.

43. $\int_1^2 \left[\int_0^1 (6xy^2 - 2y)\, dx\right] dy$
44. $\int_0^1 \left[\int_0^4 (6\sqrt{x} - 2\sqrt{y})\, dy\right] dx$
45. $\int_1^2 \left[\int_0^1 4ye^{2x}\, dx\right] dy$
46. $\int_1^e \left[\int_0^1 \ln y\, dx\right] dy$

In problems 47 through 50, find the volume of the solid region between each of the surfaces and the given rectangular region R.

47. $f(x, y) = 16 - 3x^2 - 3y^2$; $R: 0 \le x \le 1, -1 \le y \le 1$
48. $f(x, y) = 12 - 6x - 2y + xy$; $R: 0 \le x \le 1, 0 \le y \le 2$
49. $f(x, y) = e^{x-y}$; $R: 0 \le x \le 2, 1 \le y \le 2$
50. $f(x, y) = 2x + 6y$; $R: 1 \le x \le 3, 0 \le y \le 4$

Evaluate each of the following double integrals.

51. $\int_1^3 \int_0^{2y} (3x + 4y)\, dx\, dy$

52. $\int_0^1 \int_0^{y^2} 6x\sqrt{y}\, dx\, dy$

53. $\int_0^1 \int_y^{y^2} x^2y^2\, dx\, dy$

54. $\int_1^4 \int_0^{\sqrt{y}} 12xy^2\, dx\, dy$

Evaluate each of the following double integrals over the given region R.

55. $\iint_R 36xy^2\, dx\, dy$; R is the region bounded by the curves $y = x$, $y = 0$ and $x = 1$.

56. $\iint_R (8 - 3x^2 - 3y^2)\, dx\, dy$; R is the region bounded by the curves $y = -x$, $x = 1$, and $y = 2$.

57. If a bank pays interest at an annual rate of 10 percent compounded continuously, $A(P, t)$, the amount in the account, is given by the equation
$$A(P, t) = Pe^{0.10t}$$
where P is the principal, in dollars, and t is the time, in years, from the date of deposit.
(a) Find $A(1000, 2)$.
(b) Find $A_P(P, t)$ and $A_t(P, t)$.
(c) Evaluate $A_P(1000, 2)$ and $A_t(1000, 2)$.

58. Poiseuille's law states that the resistance, in dynes, of a blood vessel is a function of the length L, in centimeters, and the radius r, in centimeters. The relationship is described by the equation
$$R(L, r) = \frac{kL}{r^4}$$
where k is a constant, the viscosity of blood.
(a) In terms of k, what is the resistance when $L = 10$ cm and $r = 0.20$ cm?
(b) In terms of k find $R_L(L, r)$ and $R_r(L, r)$.
(c) Evaluate $R_L(10, 0.20)$ and $R_r(10, 0.20)$.
(d) Use the differential dR to find the change in R when the radius of a blood vessel decreases from 0.20 to 0.18 cm and the length increases from 10 to 10.1 cm.

59. A company produces two products, A and B. The monthly profit as a function of x, the number of units of A, and y, the number of units of B, is given by the equation
$$P(x, y) = -4x^2 + 12xy - 10y^2 + 12x + 16y - 4$$
Find the number of units of A and B that should be produced to maximize monthly profit.

60. A boatbuilding company estimates its annual production q from the equation

$$q = 8L^{3/4}M^{1/4}$$

where L is the number of units of labor and M is the number of units of capital. Labor costs are \$600 per unit and capital costs are \$1200 per unit. If the company has allocated a total of \$768,000 for both costs, find the values of L and M for which production is maximized. Find the maximum production.

61. Five patients were given a new drug which reduces high blood pressure. The systolic blood pressure, in millimeters of mercury, was measured before the drug treatment began and after it was concluded. The results are shown in Table 1.

Table 1

Before	x	190	215	180	230	210
After	y	125	150	130	140	135

(a) Use the method of least squares to find a linear relationship between x and y.

(b) A patient whose present blood pressure is 200 mm is going to take the drug. Use the least squares equation to predict his systolic pressure at the conclusion of the treatment.

62. An analyst for a soft-drink company wants to determine the relation between annual sales y, in millions of dollars, and the amount of money budgeted for advertising x, in millions of dollars. Table 2 shows the data for six major soft-drink companies.

Table 2

Advertising	x	2	10	30	15	40	25
Sales	y	30	75	200	75	250	160

(a) Find the equation of the least squares line.

(b) Use the least squares equation to predict the annual sales of a company that budgets \$25 million for advertising.

9

DIFFERENTIAL EQUATIONS

INTRODUCTION

Since any system, be it economic, biological, chemical, or social, is continually changing, any attempt to describe it mathematically must contain information on the rates at which relevant quantities are changing. This means that our equations must contain derivatives if we hope to come close to representing the system. Equations of this type are called *differential equations*. In this chapter we look at some elementary types of differential equations, methods of solving them, and some applications.

9.1 SOLUTIONS OF DIFFERENTIAL EQUATIONS

A **differential equation** of a single independent variable x is an equation that contains one or more of the derivatives y', y'', y''', . . . , of an unknown function $y = f(x)$. For example, each of the following is a differential equation:

$$y' = 2x \tag{1}$$

$$xy' + y^2 = 1 \tag{2}$$

$$y'' + y' = xy \tag{3}$$

Differential equations can be classified according to the highest-order derivative that appears in an equation. For example, Equations 1 and 2 are called *first-order differential equations* because the highest derivative is the first. Equation 3 is a *second-order differential equation* because y'' is the highest derivative. In short, the **order of a differential equation** equals the **order of the highest derivative** in the equation.

A function $y = f(x)$ is called a **solution** of a given differential equation if substituting $f(x)$ and its derivatives in the equation produces a true statement. For example, the function

$$y = \frac{1}{x^2} \tag{4}$$

is a solution to the differential equation

$$xy' + 2y = 0 \tag{5}$$

This can be checked easily. Differentiating both sides of Equation 4 gives

$$y' = -\frac{2}{x^3}$$

Substituting this result together with the original function $y = 1/x^2$ into Equation 5 gives

$$x\frac{-2}{x^3} + 2\frac{1}{x^2} = 0$$

which is a true statement.

The function $y = 1/x^2$ is not the only solution to Equation 5. You can verify that each of the following is also a solution:

$$y = \frac{5}{x^2}, \qquad y = \frac{-3}{x^2}, \qquad y = \frac{100}{x^2}$$

In fact, any function of the form

$$y = \frac{C}{x^2} \tag{6}$$

where C is any real number, satisfies Equation 5. The solution $y = C/x^2$ is called the **general solution** of Equation 5; when individual values are assigned to C, the resulting solution is called a **particular solution.**

A general solution such as that given in Equation 6 can be represented geometrically by a family of curves such as those shown in Figure 1. Particular solutions are represented by curves having the same value of C, some of which are also shown in Figure 1.

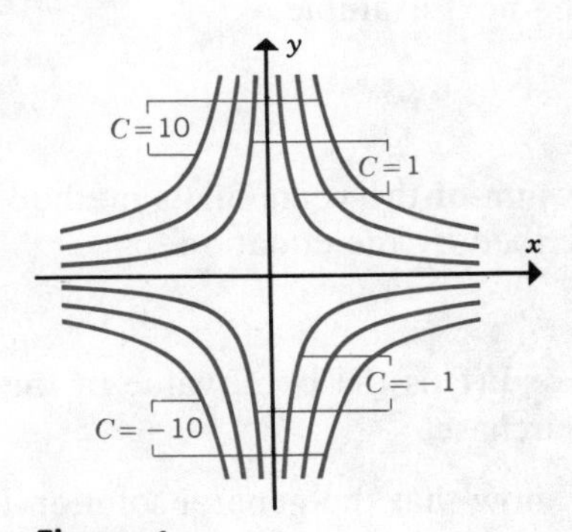

Figure 1

A particular solution is obtained from the general solution by providing additional information, generally in the form of **initial conditions.** Initial conditions provide values of the unknown function and some of its derivatives for a given value of x, as shown in Example 1.

Example 1 Find the particular solution to the differential equation

$$xy' + 2y = 0$$

that satisfies the condition $y = f(3) = 1$.

Solution The general solution to the differential equation was found to be

$$y = \frac{C}{x^2} \tag{6}$$

From this family of functions, we want to select one whose graph passes through (1, 3). Substituting this information into Equation 6 gives

$$1 = \frac{C}{(3)^2}$$

or $C = 9$, yielding the particular solution

$$y = f(x) = \frac{9}{x^2}$$ ■

The amount of information needed to obtain a particular solution from a general solution depends on the order of the differential equation. If the given, differential equation is of the nth order, the general solution will contain n arbitrary constants; so a general solution to a first-order differential equation contains one arbitrary constant, to a second-order differential equation two arbitrary constants, and so on. This means that enough information must be given so that we can assign a unique value to each arbitrary constant, as illustrated in the next example.

Example 2 The sum-of-the-years-digits method of depreciation for a minicomputer can be described by the equation

$$V''(t) = \$8000 \text{ per year}^2$$

where $V(t)$ is the book value of the minicomputer and t is the time from date of purchase

(a) Show that the general solution to this equation has the form

$$V(t) = 4000t^2 + C_1 t + C_2 \qquad (7)$$

(b) Find the particular solution corresponding to the conditions

$$V(0) = 64{,}000, \qquad V(4) = 0$$

Solution (a) Differentiating both sides of Equation (7) gives

$$V'(t) = 8000t + C_1$$

Differentiating a second time yields

$$V''(t) = 8000$$

Since the function given in Equation 7 satisfies the differential equation and contains two arbitrary constants, it is the general solution.

(b) A particular solution can be found by first using the condition $V(0) = 64{,}000$, yielding

$$C_2 = 64{,}000$$

Next the condition $V(4) = 0$ yields the equation

$$0 = 4000(16) + 4C_1 + 64{,}000$$

from which we get

$$C_1 = -32{,}000$$

So we get for the particular solution

$$V(t) = 4000t^2 - 32{,}000t + 64{,}000$$

■

Example 3 Determine which of the following are solutions to the differential equation

$$y'' - y = 2e^x \tag{8}$$

(a) $y = xe^x$
(b) $y = xe^{-x}$
(c) $y = xe^x + C_1e^x + C_2e^{-x}$

Solution (a) Differentiating $y = xe^x$ twice gives

$$y'' = 2e^x + xe^x$$

Substituting the expressions for y'' and y into Equation 8 gives

$$\underbrace{2e^x + xe^x}_{y''} - \underbrace{xe^x}_{y} = 2e^x$$

which is a true statement. This tells us that $y = xe^x$ is a solution. However, it is not the general solution because it does not contain any arbitrary constants.

(b) Differentiating $y = xe^{-x}$ twice gives $y'' = -2e^{-x} + xe^{-x}$. Substituting for y'' and y gives

$$\underbrace{-2e^{-x} + xe^{-x}}_{y''} - \underbrace{xe^{-x}}_{y} = -2e^{-x} \neq 2e^x$$

so $y = xe^{-x}$ is not a solution to Equation 8.

(c) If we differentiate $y = xe^x + C_1e^x + C_2e^{-x}$ twice, we get

$$y'' = 2e^x + xe^x + C_1e^x + C_2e^{-x}.$$

We can now write the differential equation as

$$\underbrace{2e^x + xe^x + C_1e^x + C_2e^{-x}}_{y''} - \underbrace{xe^x + C_1e^x + C_2e^{-x}}_{y} = 2e^x$$

which is a true statement. The function

$$y = xe^x + C_1e^x + C_2e^{-x}$$

is a solution of Equation 8. In addition, the presence of the arbitrary constants C_1 and C_2 indicates that the function represents a general solution.

■

Example 4 Find the particular solution of the differential equation

$$y'' - y = 2e^x$$

(Example 3) that satisfies the initial conditions

$$y(0) = 4, \qquad y'(0) = -1$$

Solution From Example 3, we have the general solution

$$y = xe^x + C_1e^x + C_2e^{-x}$$

whose derivative is $y' = e^x + xe^x + C_1e^x - C_2e^{-x}$
We can use the initial conditions to write

$$y(0) = 4 = C_1 + C_2$$

$$y'(0) = -1 = 1 + C_1 - C_2$$

Solving this system for C_1 and C_2 gives

$$C_1 = 1, \qquad C_2 = 3$$

so the particular solution assumes the form

$$y = xe^x + e^x + 3e^{-x}$$

■

Although it is not possible in this chapter to describe all the methods used to find solutions of differential equations, there are some simple differential equations for which the method of solution is straightforward. For example, a differential equation of the form

$$y' = f(x)$$

can be solved directly by integrating both sides to yield

$$\boxed{y = \int f(x)\,dx}$$

The challenge in this case is to find an antiderivative of $f(x)$. Similarly, for the differential equation

$$y'' = g(x)$$

a solution is obtained by integrating twice. The first integration yields

$$y' = \int g(x)\,dx$$

Integrating a second time yields

$$\boxed{y = \int \left[\int g(x)\,dx \right] dx}$$

If we have a differential equation of the form $y^{(n)} = g(x)$, a solution can be obtained by carrying out n integrations, yielding

$$y = \int \left\{ \int \cdots \left[\int g(x)\, dx \right] dx \right\} dx$$

Example 5 (a) Find the general solution to the differential equation

$$y'' = 4e^{2x}$$

(b) Find the particular solution that satisfies the initial conditions

$$y(0) = 2, \qquad y'(0) = 3$$

Solution (a) Integrating once gives

$$y' = 2e^{2x} + C_1$$

Integrating a second time gives the general solution

$$y = e^{2x} + C_1 x + C_2$$

(b) A particular solution is found by determining the constants C_1 and C_2 from the initial conditions $y(0) = 2$ and $y'(0) = 3$. These conditions yield the equations

$$3 = 2e^0 + C_1 \tag{9}$$

$$2 = e^0 + 0 + C_2 \tag{10}$$

From Equation 9 we get $C_1 = 1$. Equation 10 gives

$$2 = 1 + C_2$$

from which we can conclude that $C_2 = 1$. Thus the particular solution becomes

$$\boxed{y = e^{2x} + x + 1}$$

■

9.1 EXERCISES

In Exercises 1 through 10, show that each equation satisfies the given differential equation.

1. $y'' = 12$; $y = 6x^2 + 3x + 2$

2. $y' + y = 2e^x$; $y = e^x + e^{-x}$

3. $xy' + y = 1$; $y = \dfrac{2 + x}{x}$

4. $xy'' - y' = 2x$; $y = x^2 \ln x$

5. $(x + 1)y'' + y' = 2$; $y = \ln(x + 1) + 2x$

6. $y'' - 2xy' - 2y = 0$; $y = e^{x^2}$

7. $xy' - y = x^2e^x; \quad y = xe^x + x$
8. $y'' - y' - 2y = 0; \quad y = e^{2x} + e^{-x}$
9. $xy' - ny = x^n; \quad y = x^n \ln x$
10. $xy'' - y' - 4x^3y = 0; \quad y = e^{x^2} + e^{-x^2}$

In Exercises 11 through 20, verify that each general solution satisfies the given differential equation. In addition, find the particular solution that satisfies the given initial conditions.

11. $y'' = 4; \quad y = 2x^2 + C_1x + C_2; \quad y'(0) = 3, y(0) = 1$
12. $y' + 3y = 0; \quad y = Ce^{-3x}; \quad y(0) = 2$
13. $xy'' = 2; \quad y = 2x \ln x + C_1x + C_2; \quad y'(1) = 2, y(1) = -1$
14. $y'' = (x + 1)^{1/2}; \quad y = \dfrac{4(x + 1)^{5/2}}{15} + C_1x + C_2; \quad y'(0) = \frac{2}{3}, y(0) = \frac{1}{15}$
15. $y'' - y' - 2y = 0; \quad y = C_1e^{-x} + C_2e^{2x}; \quad y'(0) = 2, y(0) = 1$
16. $y'' - y = 0; \quad y = C_1e^x + C_2e^{-x}; \quad y'(0) = 2, y(0) = -2$
17. $y' + y^2 = 0; \quad y = \dfrac{1}{x + C}; \quad y(0) = \frac{1}{4}$
18. $3y^2y' = 2x + 1; \quad y = (x^2 + x + C)^{1/3}; \quad y(0) = -1$
19. $y' - 2xy = 0; \quad y = Ce^{x^2}; \quad y(0) = 3$
20. $y'' + 2y' + y = 0; \quad y = C_1e^{-x} + C_2xe^{-x}; \quad y'(0) = 1, y(0) = -1$

In Exercises 21 through 30, solve each of the differential equations subject to the given initial conditions

21. $y' = 2x, y(0) = 3$
22. $y'' = \dfrac{1}{x^2}, y'(1) = 2, y(1) = -1$
23. $y'' = e^x + e^{-x}, y(0) = 2, y'(0) = 1$
24. $y''' = 4, y''(0) = -1, y'(0) = 3, y(0) = 2$
25. $y'' = \dfrac{1}{6(x)^{1/2}}, y(1) = 2, y'(1) = 3$
26. $y'' = -2, y(0) = 3, y'(0) = 4$
27. $y'' = 8e^{-2x}, y'(0) = 12, y(0) = 4$
28. $y'' = xe^x, y'(0) = 1, y(0) = 2$
29. $y'' = 12(x + 1)^2, y'(0) = 3, y(0) = -1$
30. $y'' = 0, y'(0) = 4, y(0) = -2$
31. The acceleration a of an object moving under the influence of gravity as a function of time is given by the equation
$$a = s''(t) = -32 \text{ ft/sec}$$
where $s(t)$ is the distance of the object above the surface of the earth and t is the time in seconds. Find the particular solution that satisfies the initial conditions $s'(0) = 128$ ft/sec and $s(0) = 600$ ft.
32. The differential equation $V''(t) = \$80$ per year^2 describes the depreciation of an office copier under the sum-of-the-years-digits (SYD) method of depreciation; $V(t)$ is the book value (in dollars) and t is the time (in years) from the date of acquisition. Find the particular solution that satisfies the initial conditions $V'(0) = -\$800$ per year and $V(0) = \$4000$.

33. A six-pack of cola immersed in a stream cools off at a rate given by the differential equation $T'(t) = -\dfrac{T - 50}{100}$, where T is the temperature (in °F) of the cola and t is the time (in minutes) from the moment of immersion. Show that the function

$$T(t) = 50 + Ce^{-t/100}$$

where C is an arbitrary constant, is the general solution of the differential equation.

9.2 SEPARATION OF VARIABLES

Although it is difficult to solve most differential equations, there are some important types that can be solved directly. Let's look at the first-order differential equation

$$y' = \frac{dy}{dx} = \frac{3x^2}{2y} \tag{1}$$

If we multiply both sides of this equation by the product $2y\ dx$, we can write

$$2y\left(\frac{dy}{dx}\,dx\right) = 3x^2\,dx \tag{2}$$

Since $\dfrac{dy}{dx}\,dx = dy$, Equation 2 can be written as

$$2y\,dy = 3x^2\,dx \tag{3}$$

Notice that the variables have been **"separated"**; the left-hand side contains $2y$ and the differential dy, the right-hand side $3x^2$ and the differential dx. It is now possible to integrate each side, yielding

$$\int 2y\,dy = \int 3x^2\,dx$$

$$y^2 + C_1 = x^3 + C_2$$

$$y^2 = x^3 + C \tag{4}$$

where $C = C_2 - C_1$. Through implicit differentiation it is possible to show that Equation 4 is the general solution of Equation 1. Equation 4 also indicates that the general solution to a differential equation may not be defined explicitly.

The **separation of variables** method is applicable whenever a first-order differential equation can be written in the form $y' = f(x)/g(x)$, or

$$\boxed{\frac{dy}{dx} = \frac{f(x)}{g(y)}} \tag{5}$$

Multiplying both sides by $g(y)\,dx$ gives

$$g(y)\left(\frac{dy}{dx}\,dx\right) = f(x)\,dx$$
$$g(y)\,dy = f(x)\,dx$$

Integrating both sides yields

$$\boxed{\int g(y)\,dy = \int f(x)\,dx} \tag{6}$$

The challenge at this stage is to find antiderivatives for $g(y)$ and $f(x)$.

Example 1 (a) Find the general solution of the differential equation

$$y' = \frac{y^2}{x+1}$$

(b) Find the particular solution corresponding to the initial condition $y(0) = 1$.

Solution (a) Writing y' as $\frac{dy}{dx}$, the equation becomes $\frac{dy}{dx} = \frac{y^2}{x+1}$. Next, multiplying both sides by dx/y^2, we get the equation

$$\frac{dy}{y^2} = \frac{dx}{x+1}$$

in which the variables are separated. Integration can be carried out on each side to give

$$\int \frac{dy}{y^2} = \int \frac{dx}{x+1}$$
$$-\frac{1}{y} = \ln|x+1| + C$$

Again, a constant of integration is not needed on both sides of the equation. We can also write the general solution as

$$y = \frac{-1}{\ln|x+1| + C}$$

(b) The particular solution can be found by substituting $x = 0$, $y = 1$ into the general solution, yielding

$$1 = \frac{-1}{\ln 1 + C}$$

from which we get $C = -1$. The particular solution then becomes

$$y = \frac{-1}{\ln|x + 1| - 1}$$

■

Example 2 According to the declining balance method (DBM) of depreciation, $\frac{dV}{dt}$, the time rate of change in V, the book value of an asset, is proportional to the current value of V, that is,

$$\frac{dV}{dt} = kV$$

where k is the constant of proportionality. If the acquisition cost of a minicomputer is \$30,000 and if the book value at the end of two years is \$10,000, find an equation for V as a function of t, the time from the date of purchase.

Solution Multiplying both sides of the differential equation by the quantity $\frac{dt}{V}$, we get

$$\frac{dV}{V} = k\,dt$$

$$\int \frac{dV}{V} = \int k\,dt$$

We can integrate both sides to yield

$$\ln|V| = kt + C_1$$

Because $V > 0$, $\ln|V| = \ln V$. If we let $C_1 = \ln C$, where C is a constant, we can write

$$\ln V = kt + \ln C$$

or

$$\ln \frac{V}{C} = kt$$

This equation can be written as the following exponential equation

$$V = Ce^{kt}$$

This equation contains two constants C and k, whose values can be determined from the conditions

$$V(0) = 30{,}000 \quad \text{and} \quad V(2) = 10{,}000$$

From the first, we can write

$$30{,}000 = Ce^0$$

so $C = 30{,}000$ and our equation takes the form

$$V(t) = 30{,}000e^{kt}$$

Using the information $V(2) = 10{,}000$, we get

$$10{,}000 = 30{,}000e^{2k}$$
$$\tfrac{1}{3} = e^{2k}$$

If we take the natural logarithm of both sides, we get

$$2k = \ln \tfrac{1}{3} = \ln 1 - \ln 3$$
$$= -\ln 3$$

yielding

$$k = -\frac{\ln 3}{2} = -0.5493$$

so the equation for V becomes

$$\boxed{V = 30{,}000e^{-0.5493t}}$$

■

Example 3 (a) Find an equation of the family of curves for which the slope of the tangent line at any point is given by the equation

$$\frac{dy}{dx} = 2xy^2$$

(b) From this family find the curve that passes through $(1, -3)$.

Solution (a) The separation of variables method can be used; multiplying both sides by dx/y^2 gives

$$\frac{dy}{y^2} = 2x\,dx$$

Integrating both sides gives

$$\int \frac{dy}{y^2} = \int 2x\,dx$$
$$-\frac{1}{y} = x^2 + C$$

Solving for y gives

$$y = \frac{-1}{x^2 + C}$$

(b) An equation of the curve that passes through $(1, -3)$ can be found by substituting $x = 1$, $y = -3$ into the general solution, yielding

$$-3 = \frac{-1}{1 + C}$$

Solving for C yields $C = -2/3$, so the equation becomes

$$y = \frac{-1}{x^2 - 2/3} = \frac{3}{2 - 3x^2}$$

■

Example 4 In a community of 20,000 people, the time rate at which a rumor spreads is proportional to the number of people who have not heard it and can be described by the differential equation

$$\frac{dN}{dt} = k(20{,}000 - N)$$

where N equals the number of people who have heard the rumor and k is the constant of proportionality. If $N(0) = 0$ and $N(2) = 4000$, find an equation that describes N as a function of t.

Solution The differential equation can be solved using the separation-of-variables method. Multiplying both sides by $dt/(20{,}000 - N)$ gives

$$\frac{dN}{20{,}000 - N} = k\,dt$$

$$\int \frac{dN}{20{,}000 - N} = \int k\,dt$$

$$\ln(20{,}000 - N) = -kt + \ln C$$

The absolute value is not needed on the left-hand side because $(20{,}000 - N) > 0$ for all t. This equation can be written in exponential form as

$$20{,}000 - N = Ce^{-kt}$$

or as

$$N = 20{,}000 - Ce^{-kt}$$

The constants C and k can be evaluated from the conditions

$$N(0) = 0, \qquad N(2) = 4000$$

From the condition $N(0) = 0$, we get $C = 20{,}000$. The condition $N(2) = 4000$ yields the following:

$$4000 = 20{,}000 - 20{,}000e^{-2k} \quad \text{or}$$
$$e^{-2k} = 0.80$$

We can solve this equation for k by taking the natural logarithm of both sides, yielding

$$-2k = \ln(0.80)$$

From this, we get

$$k = -\frac{\ln(0.80)}{2} = 0.1116$$

so the equation for N becomes

$$\boxed{N = 20{,}000 - 20{,}000e^{-0.1116t}}$$

■

9.2 EXERCISES

In Exercises 1 through 20 use the separation-of-variables method to find the general solution of each differential equation.

1. $y' = 4x$
2. $y' = 3x^2$
3. $y' = y$
4. $xy' = y$
5. $3y^2y' = 2x$
6. $x^2y' - y^2 = 0$
7. $y' = 2yx$
8. $e^yy' = 3x^2$
9. $2yy' = 3x^2$
10. $y' = \dfrac{y}{x + 1}$
11. $y' = 2x(y - 1)$
12. $3y\sqrt{y} = 2x$
13. $y' = \dfrac{xy}{x^2 + 1}$
14. $\dfrac{yy'}{y^2 + 1} = 2x$
15. $2yy' = \ln(x)$
16. $2e^xyy' = 1$
17. $y' = 2e^yx$
18. $(6y^2 + 2)y' = 8x + 3$
19. $y' = \dfrac{e^x}{e^y}$
20. $y' \ln(y) = \ln(x)$

In Exercises 21 through 30, use the separation-of-variables method to find the particular solution of each differential equation.

21. $3y^2y' = 4x;\quad y(0) = 1$
22. $y' - (y + 1) = 0;\quad y(0) = 2$
23. $2yy' - 3x^2 = 0;\quad y(0) = 4$
24. $yy' = xe^{x^2};\quad y(0) = 1$
25. $2y\sqrt{x^2 + 1}\, y' = 3x;\quad y(0) = 2$
26. $x^2y' - y = 0;\quad y(1) = -2$
27. $3y'\sqrt{y} = 4x;\quad y(0) = 4$
28. $\dfrac{dy}{dt} = 10 - 2y;\quad y(0) = 20$
29. $(1 + x)\dfrac{dy}{dx} = y^2;\quad y(0) = 1$

30. $(x^2 + 1)\dfrac{dy}{dx} = xy;\ \ y(0) = 2$
31. Find an equation of the curve that passes through (1, 3) and whose slope $\dfrac{dy}{dx}$ equals $\dfrac{25x}{9y}$.
32. Find an equation of the curve that passes through (2, −1) and whose slope $\dfrac{dy}{dx}$ equals $\dfrac{3x^2}{2y}$.
33. Find an equation of the curve that passes through (−1, 0) and whose slope $\dfrac{dy}{dx}$ equals $\dfrac{3x^2}{2y + 2}$.
34. Find an equation of the curve that passes through (0, 1) and whose slope $\dfrac{dy}{dx}$ equals $2x(y + 1)$.
35. According to Newton's law of cooling, $\dfrac{dT}{dt}$, the time rate of change in T, the temperature of an object, is proportional to the difference between its temperature and the temperature of the surrounding environment. When placed in a room whose temperature is 50°C, a cup of coffee cools from 85° to 70°C in 20 minutes. Find an equation that gives the temperature of the coffee as a function of time; how long does it take for the coffee to cool to 60°F?
36. When interest is compounded continuously, $A'(t)$, the time rate of change of the amount in the account, is proportional to $A(t)$, the amount at any time t. If an initial deposit of \$5000 grows to \$5700 in two years, find an equation that describes A as a function of t. How much money is in the account at the end of five years?

9.3 EXACT EQUATIONS

The separation-of-variables method is not effective in solving all first-order differential equations. For example, the differential equation

$$\frac{dy}{dx} = \frac{2xy}{2y - x^2} \tag{1}$$

cannot be solved by the separation-of-variables method. If we multiply both sides of Equation 1 by $(2y - x^2)\,dx$, we get

$$(2y - x^2)\,dy = 2xy\,dx \tag{2}$$

Notice that the variables are not "separated"; they are mixed together. As a result, it is not possible to integrate each side separately. However, suppose we rewrite Equation 2 as

$$2xy\,dx + (x^2 - 2y)\,dy = 0 \tag{3}$$

The left-hand side of Equation 3 looks like the differential of some unknown function $z = F(x, y)$. Recall from Section 8.6 that dz, the differential of a function $z = F(x, y)$, is defined as

$$dz = \frac{\partial F}{\partial x}\,dx + \frac{\partial F}{\partial y}\,dy \tag{4}$$

At this point we do the following.

1. Determine whether or not the left-hand side of Equation 3 represents the differential of some function $z = F(x, y)$; that is, is
$$2xy = \frac{\partial F}{\partial x} \quad \text{and is} \quad x^2 - 2y = \frac{\partial F}{\partial y}\,?$$
2. Find the function $z = F(x, y)$ if $xy\,dx + (x^2 - 2y)\,dy = dz$.

Step 1 is carried out by taking advantage of the following relationship which the mixed second partial derivatives satisfy (Section 8.2):

$$\frac{\partial^2 F}{\partial x\,\partial y} = \frac{\partial^2 F}{\partial y\,\partial x} \tag{5}$$

Equation 5 tells us the following. If there is a function $z = F(x, y)$ for which $\frac{\partial F}{\partial x} = 2xy$ and $\frac{\partial F}{\partial y} = x^2 - 2y$, the mixed partial derivatives $\frac{\partial^2 F}{\partial y\,\partial x}$ and $\frac{\partial^2 F}{\partial x\,\partial y}$ must be equal. Since

$$\frac{\partial}{\partial y}(2xy) = 2x \quad \text{and} \quad \frac{\partial}{\partial x}(x^2 - 2y) = 2x$$

we can move on to step 2 to see whether we can find a function whose differential dz can be written as

$$dz = 2xy\,dx + (x^2 - 2y)\,dy \tag{6}$$

Equations 3 and 6 tell us that $dz = 0$. It can be shown that the constant function is the only function whose differential equals zero for all values of x and y. This means that the general solution of Equation 3 can be written

$$F(x, y) = C \tag{7}$$

where C is an arbitrary constant. The mathematical form of $F(x, y)$ can be found by means of partial integration. Since we know that

$$\frac{\partial F}{\partial x} = 2xy \tag{8}$$

and

$$\frac{\partial F}{\partial y} = x^2 - 2y \tag{9}$$

we can integrate both sides of Equation 8 with respect to x to give

$$F(x, y) = x^2 y + g(y) \tag{10}$$

where $g(y)$, a function of y alone, replaces the constant of integration as the added term. To find $g(y)$, we differentiate both sides of Equation 10 with respect to y and set the results equal to the right-hand side of Equation 9, producing

$$\frac{\partial F}{\partial y} = x^2 + g'(y) = x^2 - 2y \quad \text{or} \quad g'(y) = -2y$$

Integrating both sides with respect to y gives

$$g(y) = -y^2 + C_1$$

where C_1 is a constant of integration. Substituting this result into Equation 10 gives

$$F(x, y) = x^2y - y^2 + C_1 \tag{11}$$

Thus the solution of Equation 1 can be written as

$$F(x, y) = x^2y - y^2 + C_1 = C$$

or, equivalently, as

$$x^2y - y^2 = K \tag{12}$$

where $K = C - C_1$.

The critical step in carrying out the procedure just described was showing that

$$2xy\,dx + (x^2 - 2y)\,dy$$

is *exact,* that is, it is the differential of a function.

Definition A given differential form $P(x, y)\,dx + Q(x, y)\,dy$ is said to be **exact** if there exists a function $z = F(x, y)$ that has the property

$$dz = P(x, y)\,dx + Q(x, y)\,dy$$

namely

$$P(x, y) = \frac{\partial F}{\partial x} \quad \text{and} \quad Q(x, y) = \frac{\partial F}{\partial y}$$

The following test is used to determine whether or not a given differential form is exact.

Test: The differential form $P(x, y)\,dx + Q(x, y)\,dy$ is exact if the following is true:

$$\boxed{\frac{\partial P}{\partial y} = \frac{\partial Q}{\partial x}} \tag{13}$$

Note: If the equality in Equation 13 does not hold, then $P(x, y)\,dx + Q(x, y)\,dy$ does not represent the differential of a function. This means that some other method must be used to solve the differential equation

$$P(x, y) + Q(x, y)\frac{dy}{dx} = 0$$

If a differential form $P(x, y)\,dx + Q(x, y)\,dy$ is exact, it can be shown that the general solution to the differential equation

$$P(x, y) + Q(x, y)\frac{dy}{dx} = 0$$

can be written as

$$\boxed{F(x, y) = C} \tag{14}$$

where C is a constant. The next step is to find $F(x, y)$, as illustrated in the next example.

Example 1 Determine whether or not the following differential equation is exact; if it is exact, find the general solution.

$$\frac{dy}{dx} = \frac{x^2 - y}{x} \tag{15}$$

Solution The equation is first written in the form

$$(y - x^2)\,dx + x\,dy = 0$$

Letting $P(x, y) = y - x^2$ and $Q(x, y) = x$, we find that

$$\frac{\partial P}{\partial y} = 1 = \frac{\partial Q}{\partial x}$$

This result tells us that the given differential equation is exact, which in turn means that there exists a function $z = F(x, y)$ whose differential dz satisfies the condition

$$dz = (y - x^2)\,dx + x\,dy = 0$$

Since

$$\frac{\partial F}{\partial x} = y - x^2 \tag{16}$$

and

$$\frac{\partial F}{\partial y} = x \tag{17}$$

we can integrate both sides of Equation 16 with respect to x to get

$$F(x, y) = \int (y - x^2)\,dx = xy - \frac{x^3}{3} + g(y) \tag{18}$$

where $g(y)$ is a function of y alone. Next, we differentiate $F(x, y)$ with respect to y, getting

$$\frac{\partial F}{\partial y} = x + g'(y) \tag{19}$$

Equating the right-hand sides of (17) and (19) and simplifying yields

$$g'(y) = 0$$

from which we can conclude that

$$g(y) = C_1$$

where C_1 is an arbitrary constant. Therefore, the general solution to Equation 15 becomes

$$F(x, y) = xy - \frac{x^3}{3} + C_1 = C$$

or

$$xy - \frac{x^3}{3} = K$$

where $K = C - C_1$. ■

Example 2 Determine whether or not the following differential equation is exact. If it is, find its general solution.

$$\frac{dy}{dx} = \frac{2 - 2xe^y}{x^2e^y} \tag{20}$$

Solution To see whether the equation is exact, we write it in differential form as

$$(2xe^y - 2)\,dx + x^2e^y\,dy = 0$$

Letting $P(x, y) = 2xe^y - 2$ and $Q(x, y) = x^2e^y$, we find that

$$\frac{\partial P}{\partial y} = 2xe^y = \frac{\partial Q}{\partial x}$$

Therefore, the equation is exact, and its general solution has the form $F(x, y) = C$, where $F_x = 2xe^y - 2$ and $F_y = x^2e^y$. Integrating F_y with respect to y yields

$$F(x, y) = \int x^2e^y\,dy = x^2e^y + G(x) \tag{21}$$

where $G(x)$ is a function of x alone. If we differentiate $F(x, y)$ with respect to x, we get

$$F_x = 2xe^y + G'(x)$$

In testing for exactness, we found that $F_x = 2xe^y - 2$; setting the two expressions for F_x equal to one another yields

$$2xe^y + G'(x) = 2xe^y - 2$$

or

$$G'(x) = -2$$

Integrating with respect to x yields

$$G(x) = -2x + C_1$$

Substituting this result in Equation 21 yields

$$F(x, y) = x^2e^y - 2x + C_1 = C$$

The general solution to Equation 20 becomes

$$x^2e^y - 2x = K$$

where $K = C - C_1$. ■

A particular solution can be obtained from the general solution by providing additional information such as the coordinates of a point through which the curve passes, as illustrated in the next example.

Example 3 Find an equation of the curve whose slope is given by the equation

$$\frac{dy}{dx} = \frac{2 - 2xe^y}{x^2e^y}$$

and which passes through the point (1, 0).

Solution From Example 2, we know that the general solution has the form

$$x^2e^y - 2x = K$$

Since the curve passes through the point (1, 0), we can find K by substituting (1, 0) into the general solution, yielding

$$K = 1e^0 - 2 = -1$$

The equation of the curve is

$$x^2e^y - 2x + 1 = 0$$

■

Application: Finding a Demand Equation from Elasticity of Demand

In Section 5.5, E, the elasticity of demand, was defined as

$$E = -\frac{p}{q}\frac{dq}{dp}$$

Generally, E is a function of both p and q; from this relationship it is possible to find dq/dp in terms of p and q. If the resulting differential equation is exact, the procedure developed in this section can be used to find a demand equation, as shown in the next example.

Example 4 Find the relationship between p and q when E, the elasticity of demand, has the form

$$E = 1 + \frac{p}{2q}$$

Solution Using the definition

$$E = -\frac{p}{q}\frac{dq}{dp}$$

we get the differential equation

$$\frac{dq}{dp} = -\frac{p + 2q}{2p}$$

To see whether this equation is exact, we write it as

$$2p\,dq + (p + 2q)\,dp = 0$$

Letting $P(p, q) = 2p$ and $Q(p, q) = p + 2q$, we find that

$$\frac{\partial P}{\partial p} = 2 = \frac{\partial Q}{\partial q}$$

so the equation is exact and its general solution has the form $F(x, y) = C$. Noting that $\partial F/\partial q = 2p$, we can integrate with respect to q to yield

$$F(q, p) = 2pq + G(p)$$

Differentiating F with respect to p and setting the result equal to $p + 2q$ gives

$$2q + G'(p) = p + 2q$$

From this we get $G'(p) = p$; integrating with respect to p gives

$$G(p) = \frac{p^2}{2} + C_1$$

so our solution becomes

$$2pq + \frac{p^2}{2} + C_1 = C$$

or

$$4pq + p^2 = K$$

where $K = 2(C - C_1)$. ■

9.3 EXERCISES

Determine whether or not each of the following differential equations is exact; if it is exact, find the general solution.

1. $\frac{dy}{dx} = -\frac{y}{x}$

2. $\frac{dy}{dx} = \frac{e^x}{e^y}$

3. $\frac{dy}{dx} = \frac{x - y}{1 + x}$

4. $\frac{dy}{dx} = \frac{1 + y^2}{1 - 2xy}$

5. $\frac{dy}{dx} = \frac{1 + 2xy}{3y^2 - x}$

6. $\frac{dy}{dx} = \frac{x + y}{y - x}$

7. $\frac{dy}{dx} = \frac{4x^3 + y}{3y^2 - x}$

8. $\frac{dy}{dx} = \frac{\ln(x + y)}{x - y}$

9. $\frac{dy}{dx} = -\frac{2xy + 1}{x^2}$

10. $\frac{dy}{dx} = -\frac{\sqrt{x} + y}{x}$

11. $\frac{dy}{dx} = \frac{2x - ye^x}{e^x + 2y}$

12. $\frac{dy}{dx} = \frac{x^2 + y^2x}{4y^3 - x^2y}$

Find an equation of each of the curves whose slopes are given in Exercises 13 through 20 and which pass through the given points.

13. $\frac{dy}{dx} = \frac{x^2}{y^2}$, (1, 1)

14. $\frac{dy}{dx} = \frac{e^x}{e^y}$, (2, 1)

15. $\frac{dy}{dx} = \frac{1 - y}{1 + x}$, (0, 2)

16. $\frac{dy}{dx} = \frac{2 - ye^x}{1 + e^x}$, (0, 1)

17. $\frac{dy}{dx} = \frac{y + x}{y - x}$, (1, −1)

18. $\frac{dy}{dx} = \frac{2 - 3y\sqrt{x}}{1 + 2x^{3/2}}$, (1, 1)

19. $\frac{dy}{dx} = \frac{3 - 2xy^3}{1 + 3x^2y^2}$, (2, 2)

20. $\frac{dy}{dx} = \frac{x - y^2}{1 + 2xy}$, (1, −2)

21. The rate of change of q, the quantity, with respect to p, the price, is given by the differential equation

$$\frac{dq}{dp} = -\frac{q^2 + 1}{2pq}$$

Find an equation that describes the relationship between p and q if $q = 5$ when $p = 2$.

9.4 APPLICATIONS

One or more differential equations are often required to describe mathematically the dynamics underlying many phenomena in business and the sciences. This section is devoted to illustrating some frequently encountered situations, ones that can be described or modeled by simple first-order differential equations.

Continuous Compound Interest

When P dollars are deposited in an account paying a quoted annual interest rate r, compounded continuously, the rate at which A, the amount in the account, grows is described by the differential equation

$$\frac{dA}{dt} = rA \tag{1}$$

that is, the rate of growth is proportional to the amount in the account. The separation-of-variables method can be used to solve this equation. Writing it as

$$\frac{dA}{A} = r\,dt$$

each side can be integrated to yield

$$\ln A = rt + K \tag{2}$$

where K is an arbitrary constant; notice that $|A|$ is not required because $A > 0$ for all t. When $t = 0$, $A = P$ and Equation 2 yields

$$K = \ln P$$

So we have

$$\ln A = rt + \ln P$$

Combining the logarithmic terms and converting to an exponential equation gives

$$A = Pe^{rt} \tag{3}$$

which is the continuous version of the discrete compound-interest formula (Equation 4, Section 5.1). (See Figure 1.)

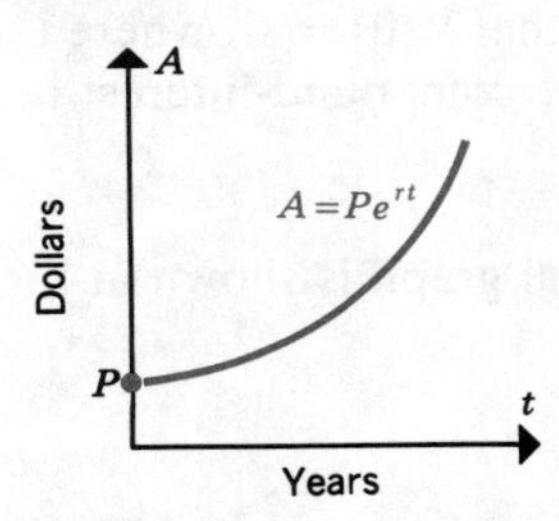

Figure 1

Example 1 Five thousand dollars is deposited in an account paying 8 percent annual interest, compounded continuously.

(a) How much is in the account at the end of two years?

(b) How long does it take for the amount in the account to double?

Solution (a) The amount A can be found by using Equation 3 with $P = \$5000$, $r = 0.08$, and $t = 2$ years to give

$$A = 5000e^{(0.08)(2)} = 5000e^{0.16} = \$5867.55$$

(b) The length of time it takes for A to double is independent of P. Setting $A = 2P$ and $r = 0.08$, we can use Equation 3 to yield

$$2 = e^{0.08t}$$

This equation can be solved by taking the natural logarithm of both sides,

$$\ln 2 = 0.08t$$

from which we get $t = 8.66$ years. ■

Depreciation: Continuous Declining Balance Method

The declining balance method is an accelerated method of depreciation for which the time rate of change of V, the book value, is proportional to the current value of V,

$$\frac{dV}{dt} = -kV \tag{4}$$

where $k > 0$ is the constant of proportionality. According to current tax practice, k is given by the ratio

$$k = \frac{r}{T}$$

where T is the useful life of an asset, in years, and r is a constant satisfying the condition $1 < r \leq 2$. The differential equation becomes

$$\frac{dV}{dt} = -\frac{r}{T}V \tag{5}$$

Noting that $V(0) = C$, where C is the acquisition cost of an asset, we can proceed as in the compound-interest model to get

$$V(t) = Ce^{-rt/T} \tag{6}$$

A typical graph is shown in Figure 2.

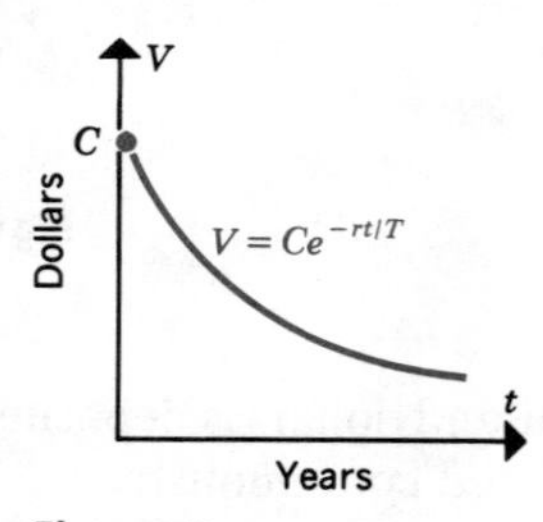

Figure 2

Example 2 (a) A regional medical center purchases a dialysis machine for \$80,000. If the useful life of the machine is five years, find its book value at the end of three years when $r = 2$ (double-declining balance).

(b) How long does it take for its book value to decline to one-half its acquisition cost?

Solution (a) We can use Equation 6 with $r = 2$ and $T = 5$ to get

$$V(3) = 80000e^{-2(3)/5} = \$24{,}095$$

(b) Setting $V = C/2$, we get

$$\tfrac{1}{2} = e^{-2t/5}$$

We can use natural logarithms to solve this equation, obtaining

$$t = 1.73 \text{ years}$$

■

Radioactive Decay

A radioactive substance decays at a rate that is proportional to the amount of the substance present at any time. If N equals the number of radioactive atoms at any time t, the time rate of change of N is described by the equation

$$\frac{dN}{dt} = -kN$$

where k is the constant of proportionality. The procedure used in the two previous cases can be applied again; letting N_0 be the number of radioactive atoms present when $t = 0$, we get

$$N(t) = N_0 e^{-kt} \tag{7}$$

Each radioactive substance is characterized by its half-life t^*, the length of time required for the number of radioactive atoms to decrease by a factor of 2. Equation 7 can be written in terms of t^* by noting that $N(t^*) = N_0/2$; using this information, we can write Equation 7 as

$$\frac{N_0}{2} = N_0 e^{-kt^*}$$

Again, using natural logarithms, we get

$$k = \frac{\ln 2}{t^*} = \frac{0.69}{t^*}$$

Equation 7 can be written as

$$N(t) = N_0 e^{-0.69t/t^*} \tag{8}$$

Radiocarbon dating is a technique used by archaeologists to determine the ages of fossils uncovered during excavations. The age is determined by measuring the amount of radioactive carbon-14 present in a substance. Because carbon-14 has a long half-life (5568 years), it decays slowly; even trace amounts are enough to permit determination of the age of the remains of a plant or animal.

Example 3 At an excavation site in the Middle East, the amount of carbon-14 found in a fossil was 30 percent of the amount originally present. Estimate the age of the fossil.

Solution Equation 8 can be used to determine the age. If $t = 0$ corresponds to the time when the plant or animal died and $t = T$ represents today,

$$\frac{N(T)}{N_0} = 0.30$$

and Equation 8 becomes

$$0.30 = e^{-0.69T/5568}$$

Again, natural logarithms can be used to solve this equation, yielding

$$T = -\frac{5568}{0.69} \ln 0.30 = 9716 \text{ years}$$

■

Falling Objects and Air Resistance

As an object falls from great heights, its velocity increases rapidly, but as its velocity increases, so does the resistance of the air through which it passes. Experimentally, it has been found that the force of this resistance is proportional to the velocity of the object. As a result, the dynamics of a falling object can be described by the differential equation

$$\frac{dv}{dt} = g - kv \tag{9}$$

where v is the velocity of the object, t is the time, in seconds, g is the acceleration caused by gravity (-32 ft/sec^2), and $k > 0$ is a constant of proportionality.

The functional relationship between v and t can be found by applying the separation-of-variables technique to Equation 9. Writing the equation in the form

$$\int \frac{dv}{g - kv} = \int dt$$

we get

$$\ln(g - kv) = -kt + \ln K \tag{10}$$

where $\ln K$ is the constant of integration. If the initial ($t = 0$) velocity of the object is v_0, we find that

$$K = g - kv_0$$

enabling us to write Equation 10 as

$$\ln \frac{g - kv}{g - kv_0} = -kt \tag{11}$$

Converting Equation 11 to exponential form allows us to write v as a function of t, yielding

$$v(t) = \frac{g}{k} - \left(\frac{g}{k} - v_0\right)e^{-kt} \tag{12}$$

Figure 3 contains a typical graph of Equation 12 together with a horizontal asymptote, which represents the terminal or limiting velocity v_L of the object. As $t \to \infty$, $e^{-kt} \to 0$; therefore, the limiting velocity becomes

$$v_L = \lim_{t\to\infty} v(t) = \frac{g}{k} \tag{13}$$

This feature of the equation enables one to determine k from a measurement of the limiting velocity.

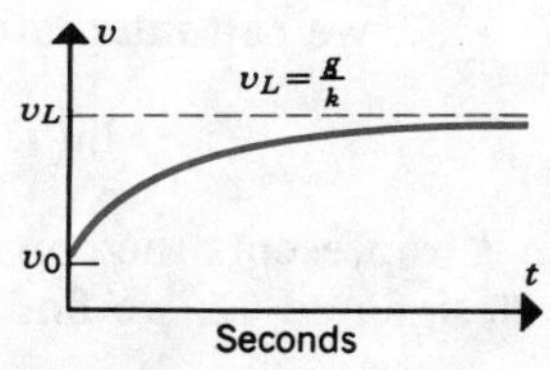

Figure 3

Example 4 (a) A piece of luggage falls out of the storage bay of a jetliner with zero initial velocity. If the limiting velocity is -320 ft/sec*, find an equation that describes the velocity as a function of t.

(b) What is the velocity at the end of 10 sec?

Solution (a) Since the terminal velocity $v_L = -320$ ft/sec, we can find k from Equation 13:

$$k = \frac{-32}{-320} = \frac{1}{10} = 0.1$$

Because the object falls from rest, $v_0 = 0$, and we get

$$v(t) = -320\,(1 - e^{-0.1t})$$

(b) At the end of 10 sec, the luggage is moving with velocity

$$v(10) = -320(1 - e^{-1}) = -202 \text{ ft/sec}$$

■

Newton's Law of Cooling

Newton's law of cooling states that the time rate of change of T, the temperature of an object, is proportional to the difference between its temperature and T_s,

*A negative value for the velocity indicates that an object is falling; a positive value indicates that it is rising.

the temperature of its surroundings. The differential equation describing this type of situation is

$$\frac{dT}{dt} = -k(T - T_s) \tag{14}$$

where $k > 0$ is the constant of proportionality. Notice that Equation 14 has the same structure as Equation 9. Therefore, their solutions should have the same form. The separation-of-variables technique can be applied to Equation 14, yielding

$$\int \frac{dT}{T - T_s} = -\int kt$$

When $T > T_s$, we can integrate to get

$$\ln(T - T_s) = -kt + \ln K \tag{15}$$

where $\ln K$ represents the constant of integration. If the initial ($t = 0$) temperature is designated T_0, we find

$$\ln K = \ln(T_0 - T_s)$$

Combining the logarithmic terms in (15) gives

$$\ln \frac{T - T_s}{T_0 - T_s} = -kt$$

This can be written as an exponential equation, yielding

$$T = T_s + (T_0 - T_s)e^{-kt} \tag{16}$$

The relationship described in Equation 16 also holds when $T_0 < T_s$. Typical graphs are shown in Figure 4 for $T_0 > T_s$ and $T_0 < T_s$. Again, we encounter a limiting situation; as $t \to \infty$, $T \to T_s$, that is, the temperature of the object approaches that of its surroundings. In this case, k, the constant of proportionality, can be determined by specifying the temperature at some time $t > 0$, as illustrated in the next example.

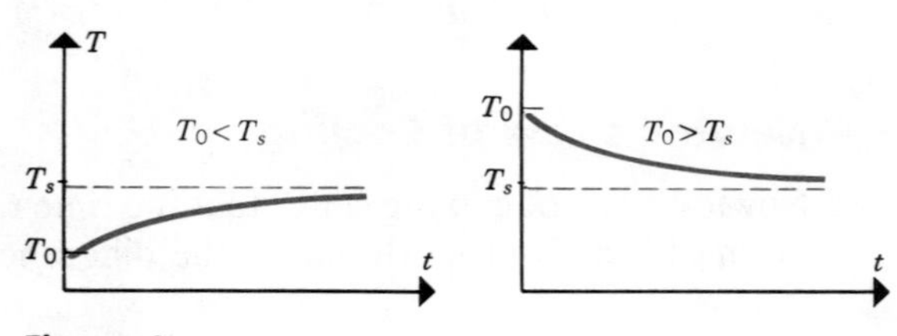

Figure 4

Example 5 In a local restaurant, a bowl of hot soup ($T = 90°C$) is placed on the table in front of a customer. The customer is engrossed in reading his newspaper and does not touch his soup for 15 minutes by which time it has cooled to 40°C and he complains that it is not hot enough. If the air temperature is 25°C, find the equation that describes the relationship between T, the temperature of the soup, and t, the length of time, in minutes, that the soup sits in front of the customer.

Solution For this situation $T_s = 25°C$, $T_0 = 90°C$, and $T = 40°C$ when $t = 15$. This information can be used to solve Equation 16 for k,

$$40 = 25 + (90 - 25)e^{-15k}$$

or

$$e^{-15k} = \tfrac{15}{65}$$

We can use natural logarithms to find k, giving

$$k = -\tfrac{1}{15}(\ln 15 - \ln 65) = 0.0978$$

The equation that describes T as a function of t becomes

$$T(t) = 25 + 65e^{-0.0978t}$$

■

Continuous Annuity

The compound-interest formula assumes that no additional deposits or withdrawals are made after the initial deposit is made. Suppose that money is added to the account at a rate of R dollars per year. How does A, the amount in the account, grow? In this situation the differential equation describing the time rate of growth of A is

$$\frac{dA}{dt} = rA + R \tag{17}$$

(growth due to interest: rA; growth due to new money: R)

Again the separation-of-variables technique can be used to solve this equation by first writing it as

$$\frac{dA}{rA + R} = dt$$

Integrating yields

$$\ln(rA + R) = rt + \ln K \tag{18}$$

where $\ln K$ is the constant of integration. If A_0 represents the amount when $t = 0$, we find that

$$K = rA_0 + R$$

so that Equation 18 can be written as

$$\ln \frac{rA + R}{rA_0 + R} = rt$$

Writing this equation in exponential form yields

$$A(t) = A_0e^{rt} + \frac{R}{r}(e^{rt} - 1) \qquad (19)$$

Because the two terms on the right-hand side in Equation 17 reinforce one another, the amount in the account grows without limit. A typical graph is shown in Figure 5.

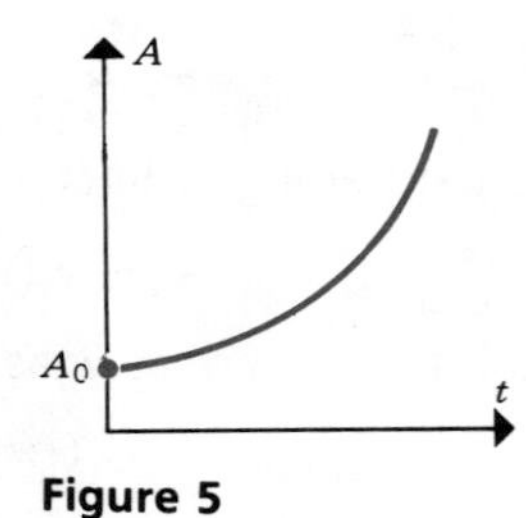

Figure 5

Example 6 A woman is advised by her financial advisor to invest \$1200 per year in tax-free bonds yielding 6 percent per year. If she has \$5000 already invested in tax-free bonds, how much will her investment be worth 20 years from now?

Solution Equation 19 with $A_0 = \$5000$, $R = \$1200$, and $r = 0.06$, can be used. The value of her investment in tax-free bonds at any time t can be written

$$A(t) = 5000e^{0.06t} + 20{,}000(e^{0.06t} - 1)$$

At the end of 20 years, the total value of her investments is given by

$$\begin{aligned} A(20) &= 5000e^{1.20} + 20{,}000(e^{1.20} - 1) \\ &= \$63{,}002 \end{aligned}$$

■

Mixture Problem

In a meat-packing plant, a tank has one pound of salt dissolved in 200 gal of water. In order to increase the saline content, water containing 0.1 lb of salt per gallon is pumped into the tank at the rate of 25 gal/hr, is stirred with the existing mixture, and the new mixture is drained off at the same rate. Find a differential equation that describes how A, the amount of salt in the tank, changes with time.

The time rate of change of A is the difference between the rate at which salt enters and leaves the tank:

$$\text{Rate at which salt enters} = (25 \text{ gal/hr})(0.1 \text{ lb/gal}) = 2.5 \text{ lb/hr}$$

$$\text{Rate at which salt leaves} = (25 \text{ gal/hr})\left(\frac{A}{200} \text{ lb/gal}\right) = \frac{25A}{200} \text{ lb/hr}$$

The differential equation becomes

$$\frac{dA}{dt} = 2.5 - \frac{25A}{200} = \frac{20 - A}{8}$$

This equation has a structure similar to that for a continuous annuity. The separation-of-variables method gives

$$\frac{dA}{20 - A} = \frac{1}{8}\, dt$$

Since $20 > A$, we can integrate to get

$$\ln(20 - A) = -\frac{t}{8} + \ln C$$

where $\ln C$ is a constant of integration. We can write this result as an equivalent exponential equation

$$(20 - A) = Ce^{-t/8}$$

Solving for A gives

$$A = 20 - Ce^{-t/8}$$

If $A = 1$ when $t = 0$, we can find C from the equation

$$1 = 20 - C$$
$$C = 19$$

Finally we get

$$A(t) = 20 - 19e^{-t/8} \tag{20}$$

The graph of Equation 20 is shown in Figure 6.

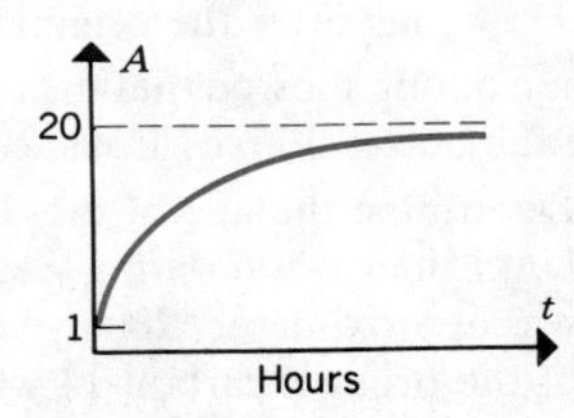

Figure 6

9.4 EXERCISES

1. Ten thousand dollars is deposited in an account that pays interest at an annual rate of 6 percent compounded continuously.
   (a) Find the equation that describes A, the amount, as a function of t.
   (b) How long does it take for the amount in the account to double? to quadruple?
2. A retired truck driver deposits P dollars in an account that pays interest at an annual rate of r, compounded continuously. Find the value of r that will double the amount in the account in six years.
3. A woman deposits P dollars in a trust fund for her grandson. If the annual interest rate is 7.5 percent compounded continuously, how long will it take for the initial deposit to triple in value?
4. (a) A professional baseball team purchases a whirlpool at a cost of \$4000. If the team's accountant uses the continuous declining balance method with $r = 2$ and $T = 5$ to depreciate the asset, find the equation that describes V, the book value of the asset, as a function of t, the time in years from the date of purchase.
   (b) How long does it take for the book value of the asset to decline to three-fourths of its original cost?
5. A printing company purchases a collator for \$10,000. If the continuous declining balance method, with $r = 1.75$, is used to depreciate the asset, determine the useful life T if $V(3) = \$4725$.
6. A laundry has purchased a heavy-duty dryer for \$2000. If the lifetime of the dryer is five years and the company uses the continuous declining balance method of depreciation, with $r = 2$, find the following:
   (a) An equation that describes V, the book value, as a function of t, the time in years from date of purchase.
   (b) The time t when the book value is 25 percent of its original cost.
7. (a) The level of radioactivity at a nuclear testing site is too high for human occupancy. If the level of radioactivity decreases to 80 percent of its original intensity in 20 days, determine the half-life of the radioactive substance.
   (b) If human occupancy of the site is deemed safe when the level of radioactivity drops to 20 percent of its original intensity, how long must the site remain unoccupied?
8. Vandalism was frequent at the sites of the ancient pyramids because of the large amounts of gold and jewelry that were buried with the dead. At an archaeological site in Egypt, scientists discovered one set of bones in which the amount of carbon-14 was 75 percent of the amount originally present and a second set in which the amount of carbon-14 was 80 percent of the amount originally present. The scientists hypothesized that the second set belonged to a thief who was killed during a "break-in." How long after the original burial did the robbery occur?
9. Carbon dating showed that the amount of carbon-14 in the Shroud of Turin, purported to be the burial cloth of Jesus, contained 93 percent of the amount originally present.
   (a) Determine the age of the Shroud.
   (b) Until the carbon dating was carried out, many people believed that the Shroud was approximately 2000 years old. If the Shroud were 2000 years old, how much of the original carbon-14 would have been present?
10. A skydiver falls from a plane with zero initial velocity. After ten seconds of free fall, she opens her chute, at which time her velocity is 80 percent of her limiting velocity.

Find an equation that describes her velocity v as a function of t, the time in seconds, during free fall.

11. Integrate both sides of Equation 12 with respect to t to find an equation for s, the distance above the ground, as a function of t. *Note:* Let $s_0 = s(t = 0)$. If the skydiver in exercise 10 exits the plane at a height of 10,000 ft, find an equation that gives s as a function of t.

12. A skydiver falls from a plane with zero initial velocity. After ten seconds of free fall, his velocity is 90 percent of the limiting velocity. Find the limiting velocity.

13. (a) An automobile engine whose temperature is 150°F is turned off and allowed to cool. The air temperature is 80°F and the engine temperature is 120°F ten minutes after the engine is turned off. Find an equation that describes T, the temperature, as a function of t, the time.

(b) How long would it take for the engine to cool to 120°F if the air temperature were 50°F?

14. Newton's law of cooling can be used to estimate the time at which death occurred. The body of a reputed mobster is pulled from Lake Michigan at 6 A.M. The body temperature of the victim is 85°F and the water temperature of the lake is 65°F. The victim's body is removed immediately to a morgue, which is kept at a temperature of 40°F; two hours later the temperature of the body is 75°F. Assuming that the victim's body temperature was 98°F when he was dumped into the lake, use Newton's law of cooling to determine how long his body was in the water before it was recovered.

15. A turkey is taken from a refrigerator whose temperature is 35°F and placed in an oven whose temperature is 350°F. A cooking thermometer inserted into the turkey reads 80°F 30 minutes after the turkey has been placed in the oven. If the turkey is to be removed from the oven when the thermometer reads 160°F, how long will the turkey be left in the oven?

16. Suppose the financial advisor in Example 6 can put the client's money in tax-free bonds yielding 7 percent interest per year. How much will the client's investment be worth 20 years from now?

17. In developing Equation 19, we assumed that $R > 0$. However, the equation is also valid when $R < 0$, that is, when money is being withdrawn at a constant rate. Suppose a 60-year-old man has \$100,000 invested in an IRA paying 8 percent per year. Find an equation that describes A, the amount, as a function of t if he withdraws \$10,000 annually from the account. How long will it take for A to become zero?

18. Suppose the woman in Example 6 wishes to have \$80,000 at the end of 20 years. Assuming that all other conditions remain unchanged, by how much should she increase her annual contribution to attain her goal?

19. A tunnel whose volume is 20,000 ft^3 is free of carbon monoxide. At 6 A.M. heavy traffic begins flowing through the tunnel and fumes containing 0.0001 percent carbon monoxide enter the tunnel at a rate of 100 ft^3/min. Fans throughout the tunnel pump out the well-circulated mixture at the rate of 100 ft^3/min. Find an equation that describes V, the volume of carbon monoxide, as a function of t, the time in minutes.

20. A 10,000-gal water tank contains 0.5 g of copper sulfate. Water whose concentration of copper sulfate is twice as large as that of the water in the tank is pumped in at the rate of 100 gal/min, and the thoroughly mixed water is pumped out of the tank at the same rate. Find an equation that gives the concentration of copper sulfate as a function of t, the time in minutes from the introduction of the water with the higher concentration of copper sulfate.

KEY TERMS

Differential equation
General solution
Particular solution
Initial conditions
First-order linear differential equation
Separation of variables method
Exact differential equation

REVIEW PROBLEMS

In problems 1 through 4, show that each equation is a solution of the accompanying differential equation.

1. $y'' = 8; \quad y = 4x^2 - 5x + 1$

2. $xy' - 3y - x^3 = 0; \quad y = x^3 \ln x$

3. $y'' - 4y = 0; \quad y = e^{2x} + e^{-2x}$

4. $y' + 2xy^2 = 0; \quad y = \dfrac{1}{x^2 + 3}$

In problems 5 through 8, verify that each general solution satisfies the given differential equation. In addition, find the particular solution that satisfies the given initial conditions.

5. $y' + 2xy^2 = 0; \quad y = \dfrac{1}{x^2 + C}; \quad y(0) = \frac{1}{2}$

6. $xy'' - 1 = 0; \quad y = x \ln|x| + C_1x + C_2; \quad y(1) = 2, y'(1) = -1$

7. $y'' - 3y' + 2y = 0; \quad y = C_1e^{2x} + C_2e^x; \quad y(0) = 1, y'(0) = 2$

8. $yy' = x; \quad y = \sqrt{x^2 + C}, y(0) = 2$

In problems 9 through 12, solve each differential equation subject to the given initial conditions.

9. $y'' = 6; \quad y'(0) = -2, y(0) = 4$

10. $y'' = 4e^{2x} - 9e^{-3x};$
$y'(0) = -2, y(0) = 4$

11. $y' = x \ln x; \quad y(1) = 3$

12. $y'' = xe^x; \quad y'(0) = 1, y(0) = 2$

In problems 13 through 18, use the separation-of-variables method to find the particular solution of each differential equation that satisfies the given initial conditions.

13. $8y^3y' = 3x^2; \quad y(1) = 2$

14. $e^{2x}y' = 6; \quad y(0) = -3$

15. $e^yy' = \ln x; \quad y(1) = 0$

16. $xy' + y = 2; \quad y(1) = 4$

17. $9y'\sqrt{y} = 2x; \quad y(0) = 4$

18. $x^2y' - y^2 = 0; \quad y(1) = 2$

19. Find an equation for the curve that passes through (2, 1) and whose slope $\dfrac{dy}{dx}$ equals $\dfrac{3x^2}{4y^3}$.

20. Find an equation for the curve that passes through (1, 3) and whose slope $\dfrac{dy}{dx}$ equals $4xe^y$.

In problems 21 through 24, determine whether each differential equation is exact. If it is, find the general solution.

21. $\dfrac{dy}{dx} = \dfrac{3x^2}{4y}$

22. $\dfrac{dy}{dx} = -\dfrac{2xy + y^2}{x^2 + 2xy}$

23. $\dfrac{dy}{dx} = \dfrac{ye^x - e^y}{xe^y - e^x}$

24. $\dfrac{dy}{dx} = \dfrac{xy \ln y - y^2}{x^2 - xy \ln x}$

25. A drug is administered intravenously at a rate of 4 mg/hr. Simultaneously, the drug is removed through diffusion at a rate that is proportional to the amount $A(t)$ of the drug in the patient's blood at any time.
   (a) Write a differential equation for the amount of the drug $A(t)$ in the bloodstream at any time t.
   (b) Find the general solution to the differential equation in part (a).
   (c) Find the particular solution that satisfies the conditions $A(0) = 0$, and $\lim_{t\to\infty} A(t) = 20$.

26. A grandfather adds money continuously to a trust fund for his granddaughter's education at a rate of $2000 per year. The money is invested in a mutual fund paying interest at a rate of 10 percent per year compounded continuously. No money is to be withdrawn until the granddaughter begins college 18 years from now.
   (a) Write a differential equation for $A(t)$, the amount in the account at any time t.
   (b) Solve the differential equation in part (a), assuming that $A(0) = 0$.
   (c) Determine the amount of money in the trust 18 years from now ($t = 0$) when the granddaughter begins her college education.

27. An air purifier has been installed in a cocktail lounge. The rate at which cigarette smoke can be removed is proportional to the amount $A(t)$ of cigarette smoke in the lounge.
   (a) Set up a differential equation for $A(t)$ when the lounge is empty and no additional smoke is being added to the air. Find the solution to the differential equation if the purifier can reduce the amount of cigarette smoke from 30 ppm (parts per million) to 10 ppm within two hours after closing.
   (b) When the lounge is open to the public, cigarette smoke is being added to air at a rate of 20 ppm per hour. Set up a differential equation for $A(t)$ when the lounge is open to the public and the purifier is operating. Solve this differential equation for $A(t)$ if the amount of smoke is 5 ppm when the lounge opens.

28. Electric power lines are downed during a severe snowstorm in Minnesota. The outside temperature is 10°F. The temperature inside a home whose only sources of heat are electric was 70°F when the power went out; two hours later the temperature was 65°F.
   (a) The temperature T inside the house changes at a rate that is proportional to the difference between the inside and outside temperatures. Write a differential equation for $T(t)$.
   (b) Solve the differential equation to find T as a function of the time t, in hours, from the moment when the power goes out.

29. A company that manufactures circuit breakers uses the continuous declining balance method with $r = 2$ to depreciate its assets. It purchases a circuit tester for $6000. Determine the useful life of the asset T, in years, if its book value two years after it is purchased equals $3640.

30. Epidemiologists have developed a model for the spread of a disease. In a community in which there are L susceptible people, the rate at which the disease spreads, dN/dt, is proportional to the product of N, the number of people who have contracted the disease, and $L - N$, the number who remain susceptible to the disease.

(a) Set up a differential equation for $N(t)$.

(b) Find a particular solution to the differential equation when $L = 5000$, $N(0) = 5$, and $N(2) = 100$. *Hint:* To carry out the integration, you will need to use the equation

$$\frac{1}{N(L - N)} = \left(\frac{1}{N} + \frac{1}{L - N}\right)\frac{1}{L}$$

(c) How long will it take for 1000 people to contract the disease?

10

PROBABILITY AND CALCULUS

INTRODUCTION

Probability is the branch of mathematics that deals with uncertainty or risk. Typical questions that probability theory attempts to answer are "What is the probability or likelihood of getting three heads in three tosses of a coin?" "What is the probability that a new tire will go 40,000 miles or more without a blowout?" "What is the average or expected rate of return on money invested in a new bioengineering company?"

10.1 DISCRETE RANDOM VARIABLES

Probability is needed when dealing with experiments or situations for which there are two or more outcomes, none of which occurs with certainty. The set S of all possible outcomes is called a **sample space** for the experiment. For example, if a woman is about to give birth, a possible sample space is

$$S_1 = \{B, G\}$$

where B stands for "boy" and G for "girl." Your favorite football team is playing next Saturday afternoon; a sample space for the experiment is

$$S_2 = \{W, L, T\}$$

where W stands for "wins," L for "losses," and T for "ties." For more complex experiments, the sample space may require many more elements. For example, if a couple plans to have three children, a sample space is

$$S_3 = \{BBB, BBG, BGB, BGG, GBB, GBG, GGB, GGG\}$$

where each element in S_3 indicates not only the sex of a child but also the order in which each child is born. In the football team example, if we consider the possible outcomes over the next two weeks, a sample space is

$$S_4 = \{WW, WL, WT, LW, LL, LT, TW, TL, TT\}$$

After a sample space has been defined, the next step is to assign to each outcome of S a **probability value,** which is a number representing the likelihood that the outcome will occur. For example, if we assume that a boy is as likely as a girl to be born, then for S_1, which has two outcomes, each of which is equally likely, we could say that

$$P(B) = P(G) = \frac{1}{2}$$

In the same way, for S_3, we could assign a probability or likelihood value of one-eighth to each outcome:

$$P(BBB) = P(BBG) = P(BGB) = \cdots P(GGG) = \frac{1}{8}$$

For sample spaces S_2 and S_4, the assignment of probabilities is not so straightforward. Although S_2 contains three outcomes, they are not, in general, equally likely; that is,

$$P(W) \neq P(L) \neq P(T)$$

How then should probabilities be assigned? Probability theory does not tell you how the assignment should be made; it merely states that if S contains n outcomes, that is,

$$S = \{e_1, e_2, \ldots, e_n\}$$

then the probabilities assigned to $e_1, e_2, \ldots, e_n$ must have the following properties.

Rules of Probability

1. $0 \leq P(e_k) \leq 1, \qquad k = 1, 2, \ldots, n$
2. $\sum_{k=1}^{n} P(e_k) = 1$

For example, if your team was playing a much weaker team, you might assign the following probabilities for the outcomes in S_2:

$$P(W) = 0.80, \qquad P(L) = 0.15, \qquad P(T) = 0.05$$

Of course, many other assignments are possible.

Probabilities can be assigned to outcomes based on data obtained from a probability experiment. This means that probabilities are assigned to the possible outcomes, based on the relative frequency of each, as illustrated in Example 1.

Example 1 An outdoor barbeque is scheduled for a day in September. Weather records indicate that measurable rain falls on 12 of the 30 days of September. What probability should be assigned to the outcome "rain on the day of the barbeque"?

Solution Let S, the sample space, be defined as

$$S = \{R, \overline{R}\}$$

where R means "rain" and $\overline{R}$ means "no rain." The weather data is summarized in Table 1. The numbers in the relative-frequency column show the fraction or proportion associated with each outcome; these numbers represent reasonable values to assign for probabilities, that is,

$$P(R) = \frac{12}{30} = \frac{2}{5}, \qquad P(\overline{R}) = \frac{18}{30} = \frac{3}{5}$$

■

Table 1

Outcome	Frequency (days)	Relative Frequency
R	12	$\frac{12}{30}$
$\overline{R}$	18	$\frac{18}{30}$

Probabilities can also be assigned to outcomes on the basis of subjective assessments of the relative likelihoods of each outcome of a sample space. This approach is illustrated in Example 2.

Example 2 Four players A, B, C, and D reach the same semifinal round of a major tennis tournament. Tennis expert Bud Love provides the following assessment of the players: "Players C and D are about equal in ability; player B is superior to both and is twice as likely as either to win the tournament. Player A is the best of the four and is three times as likely to win as either C or D." Use this assessment to assign a value to the probability of winning the tournament for each player.

Solution If we write the sample space for this experiment as

$$S = \{A, B, C, D\}$$

we can use Bud's assessment to express $P(A)$, $P(B)$, and $P(C)$ in terms of $P(D)$, the probability that D wins the tournament; Bud's assessment gives

$$P(C) = P(D)$$
$$P(B) = 2P(D)$$
$$P(A) = 3P(D)$$

Since $P(A) + P(B) + P(C) + P(D) = 1$, we can write

$$3P(D) + 2P(D) + P(D) + P(D) = 1$$

yielding $P(D) = \frac{1}{7}$. The other probabilities follow directly:

$$P(C) = \frac{1}{7}, \quad P(B) = \frac{2}{7}, \quad P(A) = \frac{3}{7}$$

■

Random Variables

In many experiments the number of outcomes is extremely large. As a result, a complete analysis of the sample space and assignment of probabilities can become cumbersome and unwieldy. In most experiments the important information can be represented as a number. For example, in a family with three children, we may be interested in knowing how many of the children are girls without knowing the order in which the children are born. A blackjack player is usually interested in the total point count of the cards rather than the face value or suit of each card.

The important numerical information is provided by a function called a random variable.

Definition A **random variable** is a function that assigns a number to each outcome of a sample space.

For example, in a family with three children where

$$S = \{BBB, BBG, BGB, BGG, GBB, GBG, GGB, GGG\}$$

let's define the random variable X as the number of girls in the family. Therefore, X can take on the values 0, 1, 2, 3, as shown in Table 2.

Table 2

Outcome	X	Probability
BBB	0	$\frac{1}{8}$
BBG	1	$\frac{1}{8}$
BGB	1	$\frac{1}{8}$
BGG	2	$\frac{1}{8}$
GBB	1	$\frac{1}{8}$
GBG	2	$\frac{1}{8}$
GGB	2	$\frac{1}{8}$
GGG	3	$\frac{1}{8}$

Probability Distribution of a Random Variable

The data in Table 2 can be used to assign probabilities to each value of the random variable X, that is, $P(X = 0)$, $P(X = 1)$, $P(X = 2)$, and $P(X = 3)$. It is easy to see that

$$P(X = 0) = \frac{1}{8}, \qquad P(X = 3) = \frac{1}{8}$$

For the value $X = 1$ there are three outcomes, each having a probability equal to $\frac{1}{8}$; the probability $P(X = 1)$ is found by adding the probabilities of the three outcomes:

$$P(X = 1) = P(BBG) + P(BGB) + P(GBB) = \frac{3}{8}$$

In the same way

$$P(X = 2) = P(BGG) + P(GBG) + P(GGB) = \frac{3}{8}$$

The set of all values of the random variable X and the probability function $P(X)$ define a **probability distribution** of X, as shown in Table 3 and Figure 1.

Table 3

X	$P(X)$
0	$\frac{1}{8}$
1	$\frac{3}{8}$
2	$\frac{3}{8}$
3	$\frac{1}{8}$

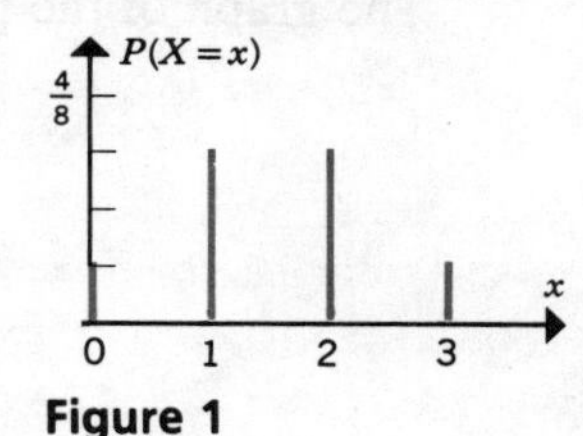

Figure 1

Table 3 and Figure 1 illustrate the properties of a probability distribution. If the values that a random variable X can assume are denoted as $x_1, x_2, \ldots x_n$, the following conditions must be satisfied by $P(X)$.

Properties of a Probability Distribution

1. $0 \le P(X = x_k) \le 1, \qquad k = 1, 2, \ldots, n$

2. $\sum_{k=1}^{n} P(X = x_k) = 1$

Example 3 An unfair coin is loaded so that heads (H) are twice as likely to come up as tails (T). The coin is tossed twice and the outcome on each toss recorded. A sample space for the experiment is

$$S = \{HH, HT, TH, TT\}$$

and the probabilities for each outcome are shown in Table 4. Let the random variable X represent the number of heads that appear in the two tosses.

(a) What values can X assume?

(b) What is the probability distribution for X?

Table 4

Outcome	Probability
HH	$\frac{4}{9}$
HT	$\frac{2}{9}$
TH	$\frac{2}{9}$
TT	$\frac{1}{9}$

Solution (a) Since $X(HH) = 2$, $X(HT) = 1$, $X(TH) = 1$, and $X(TT) = 0$, X can assume the values 0, 1, 2.

(b) The probability distribution for X becomes

$$P(X = 0) = P(TT) = \frac{1}{9}$$

$$P(X = 1) = P(HT) + P(TH) = \frac{4}{9}$$

$$P(X = 2) = P(HH) = \frac{4}{9}$$

The graph of the probability distribution is shown in Figure 2.

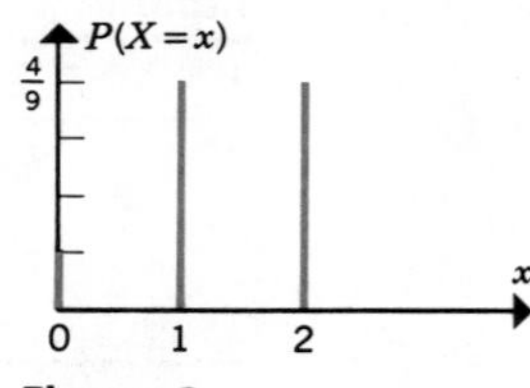

Figure 2

■

Although a probability distribution provides valuable information regarding a probability experiment, many people prefer numerical information in a more compact and manageable form. The *expected value* and *standard deviation* of a random variable serve this purpose; the expected value is a value around which the distribution is balanced, whereas the standard deviation is a measure of the spread or variability of the distribution.

Expected Value of a Random Variable

Suppose that we were to survey hundreds of families with three children and count the number of girls in each family. What would we expect the average number of girls per family to be? Using the probability distribution from Table 3, we would expect one out of every eight families to have no girls, three out of eight to have one girl, and so on. Thus,

$$\begin{aligned}\text{Average number of girls per family} &= 0(\tfrac{1}{8}) + 1(\tfrac{3}{8}) + 2(\tfrac{3}{8}) + 3(\tfrac{1}{8})\\ &= 1.5 \text{ girls per family}\end{aligned}$$

The average value found in this way also represents $E(X)$, the expected value of the random variable X, which is defined as follows.

Definition Let X be a random variable whose values are $x_1, x_2, \ldots, x_n$, and let $P(X = x_k)$ be the corresponding probability values. The **expected value** of X is defined as

$$E(X) = \sum_{k=1}^{n} x_k P(X = x_k) = x_1 P(X = x_1) + \cdots + x_n P(X = x_n) \quad (1)$$

Example 4 A company produces five jet planes each day for the Department of Defense. The probability distribution for X, the number of planes with major defects, is shown in Table 5. Find $E(X)$, the expected number of planes that have major defects.

Table 5

X	0	1	2	3	4	5
$P(X)$	0.990	0.007	0.003	0	0	0

Solution The expected number is found by using Equation 1:

$$\begin{aligned}E(X) &= 0(0.99) + 1(0.007) + 2(0.003)\\ &= 0.013 \text{ planes per day}\end{aligned}$$

■

The concept of expected value plays an important role in decision making. As illustrated in the next example, $E(X)$ is one of the first criteria used in deciding whether or not to undertake a risky project.

Example 5 A venture capitalist has been asked to invest $1 million in a wildcat oil-drilling operation. If oil is found, the investor will receive $3 million from the drillers; if oil is not found, he receives nothing. If the probability of striking oil is 0.6, find $E(X)$, where the random variable X represents the net gain to the investor.

Solution The random variable X can assume two values, $x_1 = -\$1$ million, and $x_2 = \$2$ million; the probabilities are

$$P(X = -1) = 0.4, \qquad P(X = 2) = 0.6$$

Then, we have

$$E(X) = -(0.4) + 2(0.6) = \$0.8 \text{ million}$$

Generally, an investment in a risky project is not made unless $E(X) > 0$. ■

Variance and Standard Deviation

In addition to $E(X)$, other measures are needed to describe a probability distribution. The inadequacy of $E(X)$ alone can be demonstrated by the distributions shown in Figure 3; $E(X)$ is the same for the two distributions, but it is obvious that the spread or variability of the distribution in Figure 3*b* is greater than that in Figure 3*a*. A quantity called the *variance* is often used as a measure of the variability.

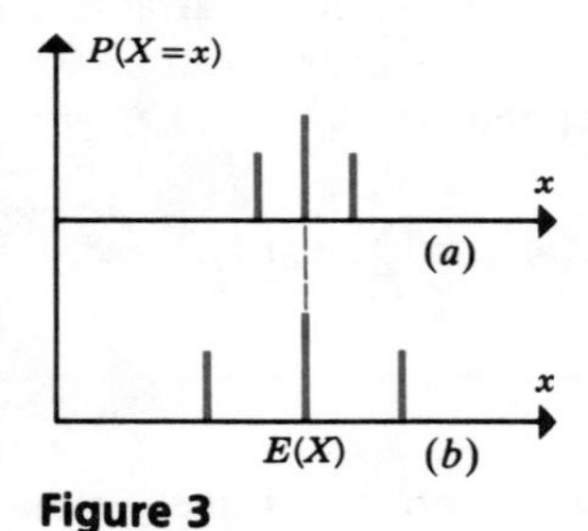

Figure 3

Definition Let X be a random variable that assumes values $x_1, x_2, \ldots, x_n$ and let $P(X = x_1), P(X = x_2), \ldots, P(X = x_n)$ be the corresponding probabilities. The **variance** is defined as

$$\begin{aligned} V(X) &= \sum_{k=1}^{n} [x_k - E(X)]^2 P(X = x_k) \\ &= [x_1 - E(X)]^2 + [x_2 - E(X)]^2 + \cdots + [x_n - E(X)]^2 \end{aligned} \tag{2}$$

A more widely used measure of the variability of a probability distribution is the **standard deviation,** $\sigma(X)$, defined as

$$\sigma(X) = \sqrt{V(X)} \tag{3}$$

One of the advantages in using $\sigma(X)$ as a measure of variability is the fact that it has the same units as $E(X)$.

Example 6 Find $V(X)$ and $\sigma(X)$ for the random variable X whose probability distribution is given in Table 6.

Table 6

X	0	1	2	3	4
$P(X)$	0.20	0.10	0.40	0.10	0.20

Solution First we find $E(X)$

$$E(X) = 0(0.2) + 1(0.1) + 2(0.4) + 3(0.1) + 4(0.2) = 2.0$$

We use the definition to find $V(X)$

$$\begin{aligned} V(X) &= (0 - 2)^2(0.2) + (1 - 2)^2(0.1) + (2 - 2)^2(0.4) \\ &\quad + (3 - 2)^2(0.1) + (4 - 2)^2(0.2) \\ &= 4(0.2) + (1)(0.1) + 0 + 1(0.1) + 4(0.2) \\ &= 1.8 \end{aligned}$$

The standard deviation can be found directly:

$$\sigma(X) = \sqrt{1.8} \approx 1.34$$

■

Example 7 An investor is considering two investments whose annual rates of return are denoted by the random variables X and Y. The probability distributions of X and Y are given in Tables 7 and 8. Find the expected value and standard deviation for X and Y.

Table 7

	Investment 1		
X	11%	12%	13%
$P(X)$	0.20	0.60	0.20

Table 8

	Investment 2				
Y	10%	11%	12%	13%	14%
$P(Y)$	0.10	0.20	0.40	0.20	0.10

Solution For investment 1 we find

$$\begin{aligned} E(X) &= 11(0.2) + 12(0.6) + 13(0.2) = 12\% \\ V(X) &= (11 - 12)^2(0.2) + (12 - 12)^2(0.6) + (13 - 12)^2(0.2) = 0.4 \\ \sigma(X) &= \sqrt{0.4} \approx 0.63\% \end{aligned}$$

For investment 2 we find

$$\begin{aligned} E(Y) &= (10)(0.1) + 11(0.2) + 12(0.4) + 13(0.2) + 14(0.1) \\ &= 12\% \\ V(Y) &= (10 - 12)^2(0.1) + (11 - 12)^2(0.2) \\ &\quad + (12 - 12)^2(0.4) + (13 - 12)^2(0.2) + (14 - 12)^2(0.1) \\ &= 0.4 + 2 + 0.2 + 4 = 1.2 \\ \sigma(Y) &= \sqrt{1.2} \approx 1.10\% \end{aligned}$$

Notice that both investments provide the same return, namely 12 percent. Which of the two investments has the greater uncertainty or risk? ■

10.1 EXERCISES

1. The following experiment is performed. A fair coin is tossed, following which one card is drawn from a deck of 52. Find the sample space whose elements show the outcome of the coin toss and the suit of the card drawn. What probability would you assign to each element of the sample space?
2. A fair die is rolled, and the number of dots on the upper face is recorded.
   (a) Find a sample space for this experiment and assign probabilities to each of the outcomes.
   (b) Let E be the event "Number is greater than 4." List the elements of the set E and then find $P(E)$.
   (c) Let F be the event "Number is divisible by two." List the elements of F and find $P(F)$.
3. Three companies A, B, and C have submitted bids to the Pentagon for the contract to design and build a new ground-to-air missile. A defense expert assesses the relative likelihoods of success for each company: "Company A is most likely to receive the contract; its chances are twice as good as those of company B and three times as good as those of company C." List a sample space for this experiment and assign probabilities to each of the outcomes.

Two unfair coins are tossed, and the result of each toss (heads or tails) is recorded. The sample space S is

$$S = \{HH, HT, TH, TT\}$$

Which of the probability assignments in Exercises 4 through 9 are in accord with the rules of probability? If an assignment is not in accord, indicate why it is not.

4. $P(HH) = \frac{3}{16}, P(HT) = \frac{1}{16}, P(TH) = \frac{9}{16}, P(TT) = \frac{3}{16}$
5. $P(HH) = 1, P(HT) = P(TH) = P(TT) = 0$
6. $P(HH) = \frac{3}{8}, P(HT) = \frac{3}{16}, P(TH) = \frac{1}{8}, P(TT) = \frac{1}{16}$
7. $P(HH) = \frac{3}{8}, P(HT) = \frac{5}{16}, P(TH) = \frac{1}{4}, P(TT) = \frac{1}{8}$
8. $P(HH) = \frac{3}{16}, P(HT) = \frac{7}{16}, P(TH) = \frac{1}{2}, P(TT) = -\frac{1}{8}$
9. $P(HH) = \dfrac{a}{a+b}, P(HT) = 0, P(TH) = 0, P(TT) = \dfrac{b}{a+b} \qquad a, b > 0$
10. Five cards are dealt without replacement from a standard deck of 52 cards. The probability distribution for the random variable X, where X is the number of aces dealt, is shown in Table 9. Find $E(X)$, $V(X)$, and $\sigma(X)$.

Table 9

X	0	1	2	3	4
$P(X)$	0.659	0.299	0.040	0.002	0.000[a]

[a]To three decimal places.

11. A car salesman has kept a record of the number of cars he has sold each day during the past 200 days. His record is summarized in Table 10. If the random variable X represents the number of cars sold each day, find the probability distribution of X. In addition, find $E(X)$, $V(X)$, and $\sigma(X)$.

Table 10

Number of Cars Sold	Frequency (days)
0	40
1	70
2	50
3	30
4	10
	Total = 200

If a random variable X can assume the values 1, 2, 3, 4, which of the distributions shown in Exercises 12 through 18 qualify as probability distributions? If a distribution is not a probability distribution, explain why it is not.

12.

X	1	2	3	4
$P(X)$	0.5	0.3	0.15	0.05

13.

X	1	2	3	4
$P(X)$	0	1	0	0

14.

X	1	2	3	4
$P(X)$	0	0	0.99	0.01

15.

X	1	2	3	4
$P(X)$	0.6	0.3	−0.1	0.2

16.

X	1	2	3	4
$P(X)$	0.23	0.10	0.37	0.09

17.

X	1	2	3	4
$P(X)$	0.05	0.71	0.20	0.05

18.

X	1	2	3	4
$P(X)$	0.05	0.86	−0.04	0.15

19. A boosters' club sponsors a lottery that has only one prize, \$1000 in cash; 2000 tickets are sold at \$1 each. If the random variable X is the net gain for a person who purchases one ticket, find $E(X)$ and $V(X)$.

20. The probability distributions for the random variables X, Y, and Z are shown in Figure 4. If $E(X) = E(Y) = E(Z)$ and the scales on the x, y, and z axes are identical, use the graphs to rank the variances $V(X)$, $V(Y)$, and $V(Z)$ from largest to smallest.

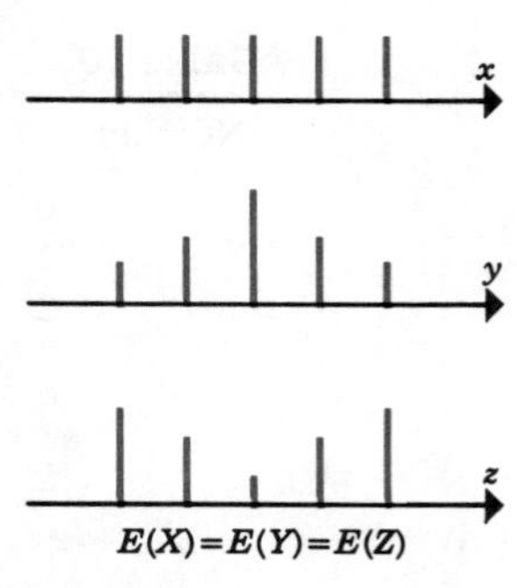

Figure 4

10.2 CONTINUOUS RANDOM VARIABLES

There are many situations in which the values assumed by a random variable are not discrete. For example, if a friend plans to pick you up tomorrow at 8 A.M. and drive you to the airport for a 9 A.M. flight, you would be interested in estimating $P(0 < T \leq 1)$, the probability that T, the time to drive to the airport, will be less than or equal to one hour. In this instance the random variable T is called a **continuous random variable** because there is an infinite number of values that T can assume between any two values $T = t_1$ and $T = t_2$.

In a situation in which T is a continuous random variable, probabilities can be represented as areas under a curve such as that shown in Figure 1, where the shaded area represents the probability that the driving time to the airport will be less than one hour.

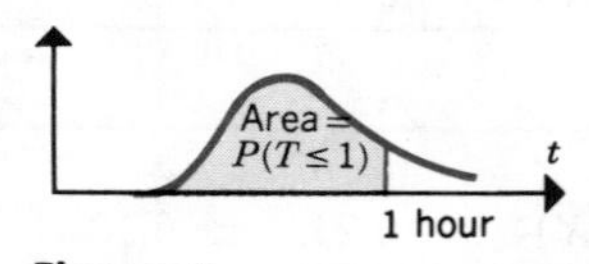

Figure 1

Probability Density Function

We have seen that the area of the region beneath a nonnegative function $y = f(x)$ from $x = a$ to $x = b$ is given by the definite integral

$$\int_a^b f(x)\,dx$$

Now suppose that we have an experiment in which the random variable X can assume any value in the interval $L \leq X \leq M$. A major effort is made to find a nonnegative function $y = f(x)$ for which the probability that X lies between a and b can be written as

$$P(a \leq X \leq b) = \int_a^b f(x)\,dx$$

and can be represented geometrically as the area under the curve $y = f(x)$ from $x = a$ to $x = b$, as shown in Figure 2.

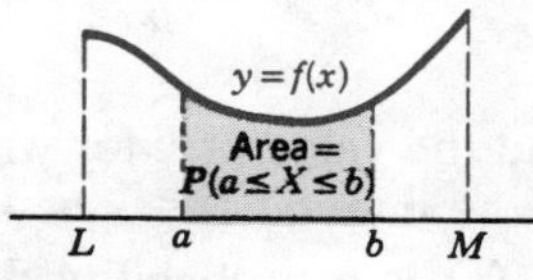

Figure 2

Here a and b are values of X in the interval $L \le X \le M$. The function $y = f(x)$ is called a **probability density function** of the random variable X and possesses the following properties.

> **Properties of a Probability Density Function**
>
> 1. $f(x) \ge 0, \qquad L \le x \le M$
> 2. $\displaystyle\int_L^M f(x)\,dx = 1$

These properties are shown graphically in Figure 3. Property 1 says that the probability associated with a given event will never be negative, and property 2 guarantees that observed values of X will fall within the range of permissible values of X.

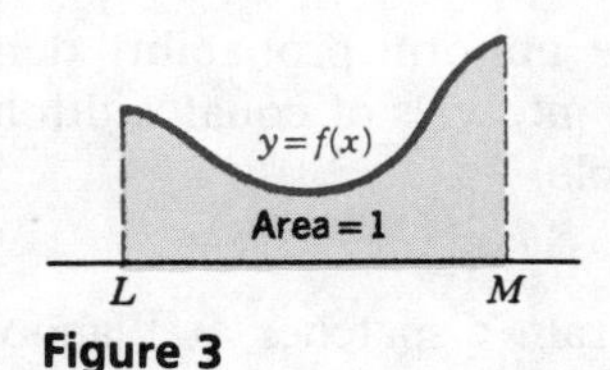

Figure 3

Example 1 1. Let

$$f(x) = \begin{cases} \dfrac{x}{2}, & 0 \le x \le 2 \\ 0, & \text{elsewhere} \end{cases}$$

(a) Show that $f(x)$ is a probability density function.

(b) If $f(x)$ is a probability density function for the random variable X, find $P(0 \le X \le 1)$.

Solution (a) Since $x/2 \ge 0$ for all x in $[0, 2]$, $f(x)$ satisfies property 1, the nonnegativity condition. Next, we evaluate the integral:

$$\int_0^2 \frac{x}{2}\,dx = \left.\frac{x^2}{4}\right|_0^2 = \frac{4}{4} = 1$$

Therefore, $f(x)$ represents a probability density function.

(b) $P(0 \le X \le 1) = \int_0^1 \frac{x}{2}\,dx = \left.\frac{x^2}{4}\right|_0^1 = \frac{1}{4} = 0.25$ ■

Example 2 (a) Find the value of k for which $f(x) = k\sqrt{x}$ is a probability density function on the interval $1 \le x \le 4$.

(b) If $f(x)$ is a probability density function for the random variable X, find $P(1 \le X \le 2)$.

Solution (a) Since $\sqrt{x} > 0$ on the interval $1 \le x \le 4$, then $f(x) > 0$, provided $k > 0$. The value of k is found by evaluating the integral:

$$\int_1^4 k\sqrt{x}\,dx = \left.\frac{2kx^{3/2}}{3}\right|_1^4 = \frac{14k}{3}$$

Setting this result equal to 1 yields

$$k = \frac{3}{14}$$

(b)
$$P(1 \le X \le 2) = \int_1^2 \frac{3x^{1/2}}{14}\,dx = \left.\frac{x^{3/2}}{7}\right|_1^2$$
$$= \frac{1}{7}[2^{3/2} - 1] \approx 0.2612$$ ■

The uniform probability density function is used to model experiments in which intervals of equal width have the same probability, as shown in the next example.

Example 3 The train dispatcher at Wahoo Station is asked when the train scheduled to arrive at 10:00 A.M. will actually arrive. His response is "Anytime between 10:15 and 10:30." Find the probability that the train arrives

(a) Between 10:15 and 10:20.

(b) Between 10:20 and 10:25.

(c) During any five-minute interval between 10:15 and 10:30.

Solution If we assume that the dispatcher's response means that the probability density function for the random variable T, the time of arrival, is uniform from $T = 0$ (10:15) to $T = 15$ (10:30), we can write

$$f(t) = k, \qquad 0 \le t \le 15$$

where $k \ge 0$ to satisfy property 1. To satisfy property 2, we write

$$\int_0^{15} k\,dt = 1$$

or

$$k = \frac{1}{15}$$

so

$$f(t) = \frac{1}{15}$$

The graph of the function $f(t) = \dfrac{1}{15}$ is shown in Figure 4.

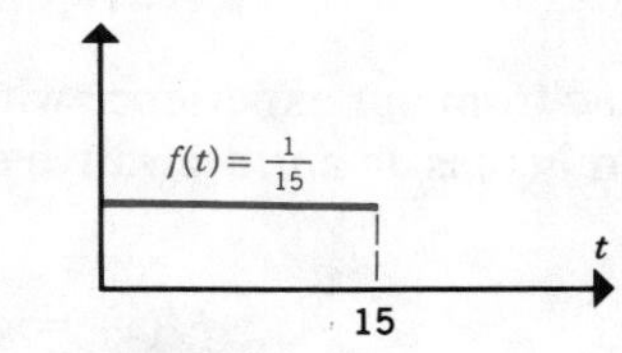

Figure 4

(a) The probability that the train arrives between 10:15 and 10:20 can be written as

$$P(0 \le T \le 5) = \int_0^5 \frac{1}{15}\, dt = \frac{t}{15}\Big|_0^5 = \frac{1}{3}$$

(b) Similarly, the probability that the train arrives between 10:20 and 10:25 becomes

$$P(5 \le T \le 10) = \int_5^{10} \frac{1}{15}\, dt = \frac{1}{3}$$

(c) For any five-minute interval, $a \le T \le a + 5$; where $0 \le a \le 10$, we get

$$P(a \le T \le a + 5) = \int_a^{a+5} \frac{1}{15}\, dt$$

$$= \frac{t}{15}\Big|_a^{a+5} = \frac{a+5}{15} - \frac{a}{15} = \frac{1}{3}$$ ■

Cumulative Probability Distribution Function

In addition to the probability density function $f(x)$, another closely related function called the *cumulative probability distribution* plays an important role in probability. Let $f(x)$ be a probability density function for the random variable X defined on the interval $L \le X \le M$. If x is a number in the interval $[L, M]$, the probability $P(X \le x)$ is a function $F(x)$ known as the **cumulative probability distribution,** defined as

$$F(x) = P(X \le x) = \int_L^x f(t)\, dt$$

Graphically, $F(x)$ can be represented as the area under the graph of the probability density function $y = f(x)$ from $X = L$ to $X = x$, as shown in Figure 5.

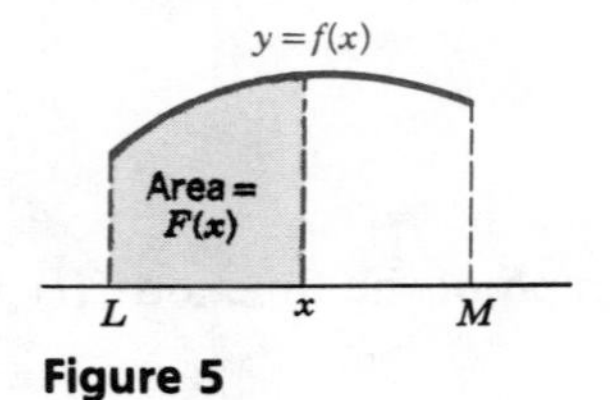

Figure 5

We know from our experience with areas under curves (Chapter 6) that the area function $F(x)$ is an antiderivative of the function $y = f(x)$ that defines the curve, that is,

$$F'(x) = f(x) \quad \text{for } L \leq x \leq M$$

This relationship tells us that the cumulative probability distribution function $F(x)$ is an antiderivative of the probability density function $f(x)$. In addition, $F(x)$ has the following properties:

$$\boxed{\begin{aligned} F(L) &= P(L \leq X \leq L) = 0 \\ F(M) &= P(L \leq X \leq M) = 1 \end{aligned}}$$

Moreover, the probability that the random variable X assumes a value between $X = a$ and $X = b$ can be stated in terms of $F(a)$ and $F(b)$:

$$P(a \leq X \leq b) = \int_a^b f(x)\,dx = F(b) - F(a)$$

Example 4 Suppose that T, the waiting time (in minutes) for a bus, is a random variable whose probability density function is

$$f(t) = \begin{cases} \frac{1}{15}, & 0 \leq t \leq 15 \\ 0, & \text{elsewhere} \end{cases}$$

Find $F(t)$, the cumulative probability distribution function, for this experiment.

Solution If $0 \leq t \leq 15$,

$$\begin{aligned} F(t) = \int_0^t f(x)\,dx &= \int_0^t \frac{1}{15}\,dx \\ &= \frac{t}{15} \end{aligned}$$

If $t > 15$,

$$F(t) = \int_0^{15} f(x)\,dx + \int_{15}^{t} f(x)\,dx$$

$$= 1 + \int_{15}^{t} 0\,dx = 1$$

so we get

$$F(t) = \begin{cases} \dfrac{t}{15}, & 0 \le t \le 15 \\ 1, & t > 15 \end{cases}$$

The graphs of $f(t)$ and $F(t)$ are shown in Figure 6. ■

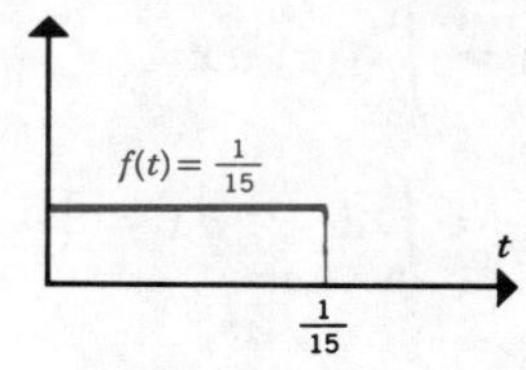

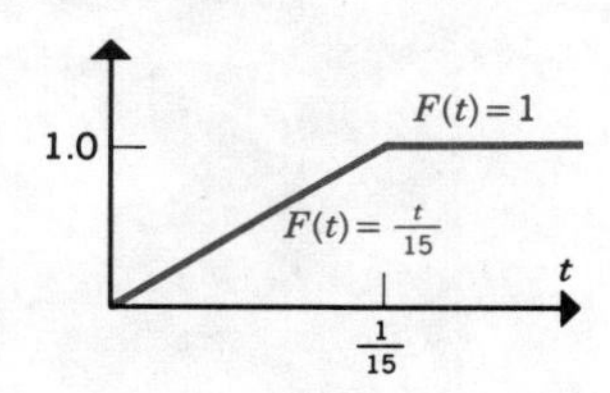

Figure 6

For many experiments the values of L or M or both can be infinite, as the next example illustrates.

Example 5 The time T, in minutes, between incoming calls to a customer service representative is a random variable whose probability density is

$$f(t) = 2e^{-2t}, \qquad t \ge 0$$

(a) Show that $f(t)$ satisfies the criteria for a probability density function.
(b) Find $F(t)$, the cumulative probability distribution.
(c) Find the probability that the time between the next two calls will be between two and five minutes.

Solution (a) For all $t \ge 0$, $f(t) > 0$. In addition, $f(t)$ must satisfy the condition

$$\int_0^{\infty} 2e^{-2t}\,dt = 1$$

Since we have an improper integral, we use the techniques described in Section 7.5:

$$\int_0^{\infty} 2e^{-2t}\,dt = \lim_{b\to\infty} \int_0^{b} 2e^{-2t}\,dt$$

$$= \lim_{b\to\infty} [-e^{-2t}] \Big|_0^b$$

$$= \lim_{b\to\infty} [1 - e^{-2b}] = 1$$

So $f(t) = 2e^{-2t}$ satisfies the criteria for a probability density function.

Note: Unlike a discrete probability function, a probability density function such as $f(t) = 2e^{-2t}$ can take on values that are greater than 1; for example $f(0) = 2$. It is not the individual values of $f(t)$ that are important but the probabilities expressed by the integrals $\int_a^b f(t)\,dt$, which must satisfy the condition

$$0 \le \int_a^b f(t)\,dt \le 1$$

(b) The cumulative probability distribution function $F(t)$ can be found by integrating

$$\begin{aligned} F(t) &= \int_0^t f(x)\,dx \\ &= \int_0^t 2e^{-2x}\,dx = [-e^{-2x}]\Big|_0^t \\ &= 1 - e^{-2t} \end{aligned}$$

The graph of $F(t)$ is shown in Figure 7.

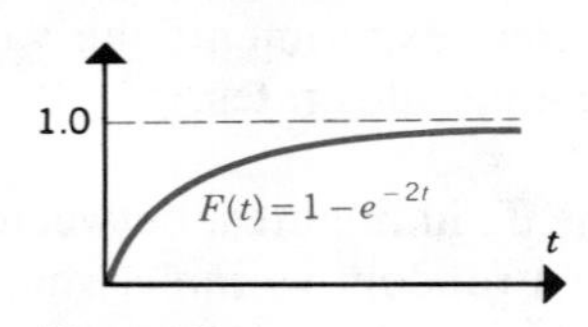

Figure 7

Note: The graph displays an important property of cumulative probability distribution functions. Since $F'(t) = f(t) \ge 0$ for all t, $F(t)$ cannot be decreasing for any t; this means that the graph of $F(t)$ cannot be falling anywhere.

(c) The probability $P(2 \le T \le 5)$ equals $F(5) - F(2)$, so we get

$$\begin{aligned} P(2 \le T \le 5) &= [1 - e^{-2t}]\Big|_2^5 \\ &= e^{-4} - e^{-10} = 0.135 \end{aligned}$$ ■

The cumulative probability distribution function can be an important tool in decision making, as the next example illustrates.

Example 6 Weekly demand for frozen pizzas is a continuous random variable whose cumulative probability distribution function is

$$F(y) = \begin{cases} \dfrac{y^3}{8}, & 0 \le y \le 2 \\ 1, & y > 2 \end{cases}$$

where y is given in hundreds of units. Find the number of units that a retailer should stock each week if he wants the probability of meeting customer demand to be 0.90.

Solution Since $F(y) = P(0 \le Y \le y)$, we let $F(y) = 0.90$ and solve the equation

$$0.90 = \frac{y^3}{8}$$

yielding $y^3 = 7.20$, or $y \approx 1.93$ hundred pizzas. By ordering 193 pizzas each week, the retailer will satisfy customer demand roughly 90 out of every 100 weeks. For the other 10 weeks, there will be one or more unsatisfied customers. ■

10.2 EXERCISES

Determine whether or not each of the following functions qualifies as a probability density function. If any is not, explain why it is not.

1. $f(x) = \begin{cases} \frac{1}{3}, & 2 \le x \le 5 \\ 0, & \text{elsewhere} \end{cases}$

2. $f(x) = \begin{cases} 2x, & 0 \le x \le 1 \\ 0, & \text{elsewhere} \end{cases}$

3. $f(x) = \begin{cases} \dfrac{3}{2x^2}, & 1 \le x \le 3 \\ 0, & \text{elsewhere} \end{cases}$

4. $f(x) = \begin{cases} \frac{1}{3}, & -1 \le x \le 3 \\ 0, & \text{elsewhere} \end{cases}$

5. $f(x) = \begin{cases} \dfrac{x-1}{3}, & 2 \le x \le 3 \\ 0, & \text{elsewhere} \end{cases}$

6. $f(x) = \begin{cases} e^{-x}, & x \ge 0 \\ 0, & \text{elsewhere} \end{cases}$

7. $f(x) = \begin{cases} \dfrac{3x^2 - 1}{2}, & 0 \le x \le 2 \\ 0, & \text{elsewhere} \end{cases}$

8. $f(x) = \begin{cases} \dfrac{3\sqrt{x}}{16}, & 0 \le x \le 4 \\ 0, & \text{elsewhere} \end{cases}$

Determine the value of K for which each of the following functions qualifies as a probability distribution.

9. $f(x) = \begin{cases} Kx, & 0 \le x \le 2 \\ 0, & \text{elsewhere} \end{cases}$

10. $f(x) = \begin{cases} K, & 1 \le x \le 4 \\ 0, & \text{elsewhere} \end{cases}$

11. $f(x) = \begin{cases} Kx(x-1), & 0 \le x \le 1 \\ 0, & \text{elsewhere} \end{cases}$

12. $f(x) = \begin{cases} K\sqrt{x}, & 1 \le x \le 4 \\ 0, & \text{elsewhere} \end{cases}$

13. $f(x) = \begin{cases} \dfrac{K}{x^3}, & x \ge 1 \\ 0, & \text{elsewhere} \end{cases}$

14. $f(x) = \begin{cases} Ke^{-3x}, & x \ge 0 \\ 0, & \text{elsewhere} \end{cases}$

15. $f(x) = \begin{cases} \dfrac{Kx}{x^2 + 1}, & 0 \le x \le 1 \\ 0, & \text{elsewhere} \end{cases}$

16. $f(x) = \begin{cases} \dfrac{K}{\sqrt{x}}, & 1 \le x \le 4 \\ 0, & \text{elsewhere} \end{cases}$

Find the cumulative probability distribution function for each of the following density functions.

17. $f(x) = \begin{cases} \frac{1}{4}, & 2 \le x \le 6 \\ 0, & \text{elsewhere} \end{cases}$

18. $f(x) = \begin{cases} 2 - 2x, & 0 \le x \le 1 \\ 0, & \text{elsewhere} \end{cases}$

19. $f(x) = \begin{cases} e^{-x}, & x \geq 0 \\ 0, & \text{elsewhere} \end{cases}$

20. $f(x) = \begin{cases} \dfrac{3(1 - x^2)}{4}, & -1 \leq x \leq 1 \\ 0, & \text{elsewhere} \end{cases}$

21. If you were in your car waiting in line at a toll booth on the New Jersey Turnpike, which of the probability density functions of Figure 8 would you prefer for the random variable T, the waiting time, in minutes, at the toll booth?

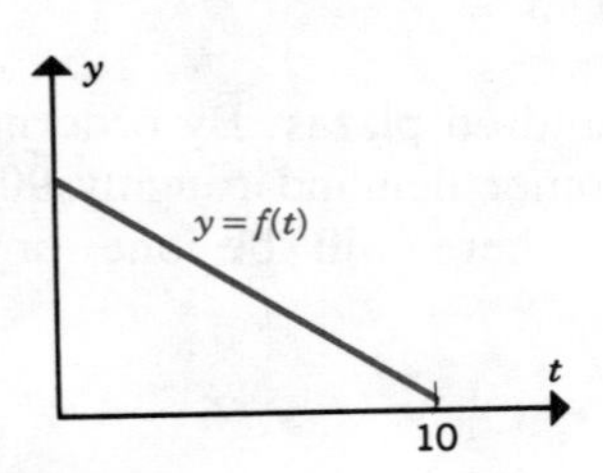

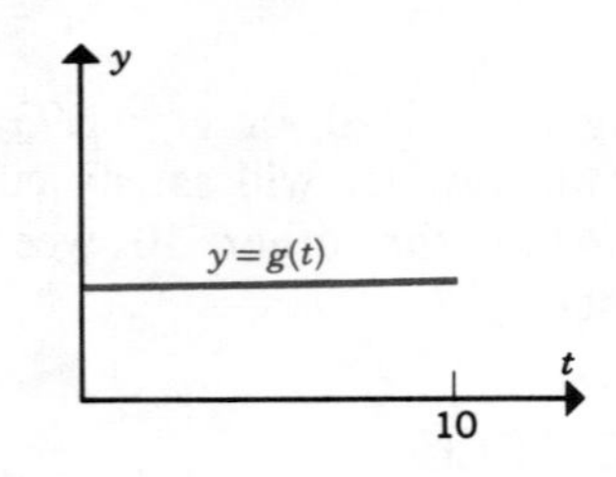

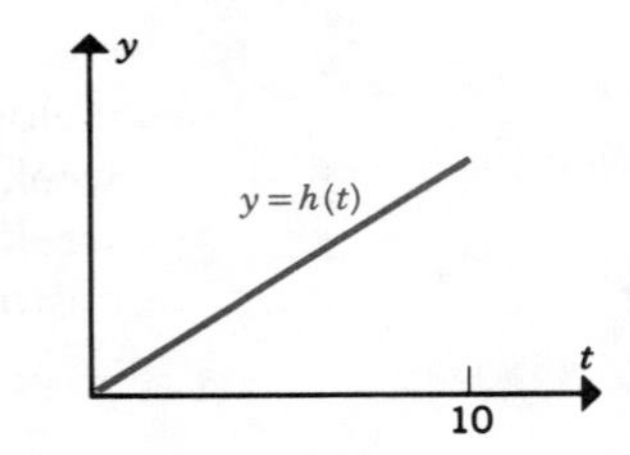

Figure 8

22. The amount of rainfall X, in inches, during the month of July is a continuous random variable whose probability density function is

$$f(x) = \begin{cases} \dfrac{2x(3 - x)}{9}, & 0 \leq x \leq 3 \\ 0, & \text{elsewhere} \end{cases}$$

(a) Find the cumulative probability distribution function $F(x)$.
(b) Find the probability $P(0 \leq X \leq 2)$.
(c) Find the probability $P(X \geq 2)$.

23. The yield Y from a peanut plant, in kilograms, is a continuous random variable whose probability density function is

$$f(y) = \begin{cases} ye^{-y}, & y \geq 0 \\ 0, & \text{elsewhere} \end{cases}$$

(a) Find the probability $P(0 \leq Y \leq 1)$.
(b) Find the cumulative probability distribution function.

Which of the following graphs could represent a cumulative probability distribution function for a probability density function $y = f(x)$ that has the following properties:

$$f(x) \geq 0 \quad (L \leq X \leq M) \quad \text{and} \quad f(x) = 0 \quad (\text{elsewhere})$$

24.

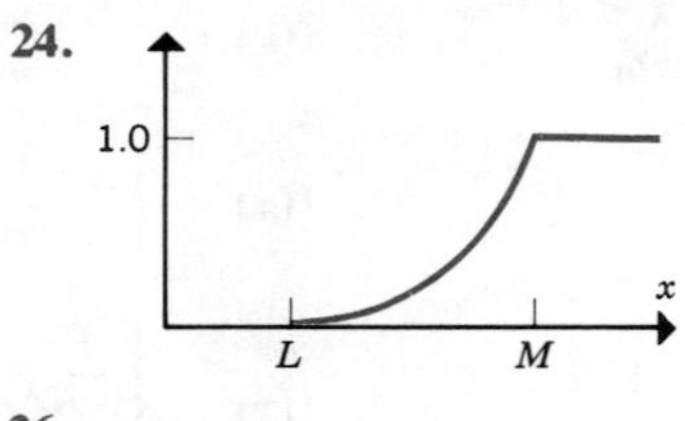

25.

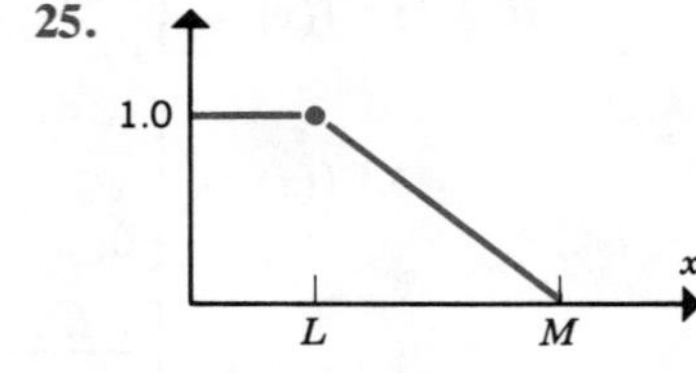

26.

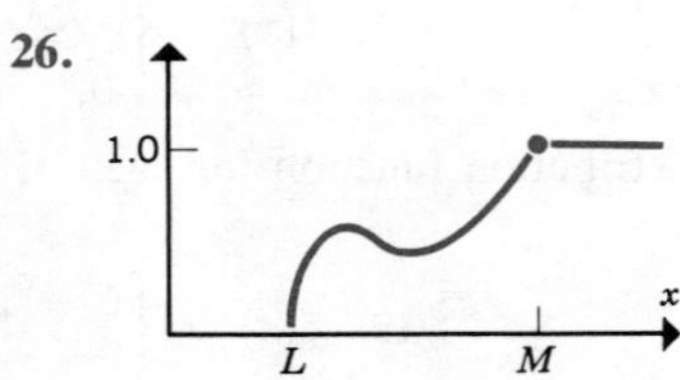

27.

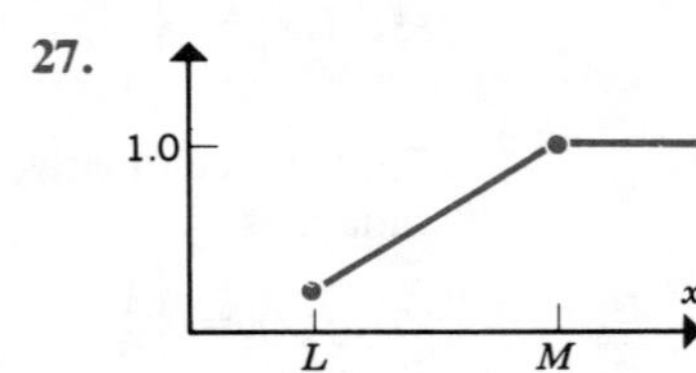

28.

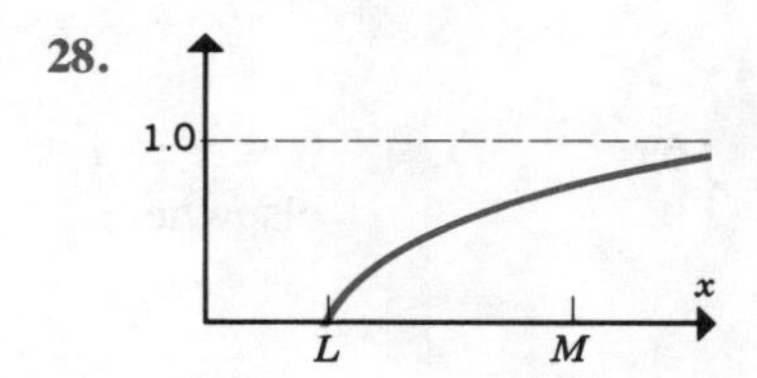

10.3 EXPECTED VALUE AND VARIANCE

In Section 10.1 the expected value (or mean) and variance of a discrete random variable X were defined as

$$E(x) = \sum_{k=1}^{n} x_k P(X = x_k), \qquad k = 1, 2, 3, \ldots, n$$

$$V(x) = \sum_{k=1}^{n} [x_k - E(X)]^2 P(X = x_k), \qquad k = 1, 2, 3, \ldots, n$$

The same approach can be used to define $E(X)$ and $V(X)$ when X is a continuous random variable, with the following variations.

1. The discrete probability distribution $P(X = x_k)$ is replaced by the probability density function $f(x)$.
2. Summation over all values of k is replaced by integration over the interval $L \leq X \leq M$ where $f(x) \geq 0$.

Definitions Let X be a continuous random variable and let $f(x)$ be a probability density function defined on the interval $L \leq X \leq M$. The **expected value,** or **mean,** is defined as

$$E(X) = \int_L^M xf(x)\,dx \tag{1}$$

The **variance** is defined as

$$V(X) = \int_L^M [x - E(X)]^2 f(x)\,dx \tag{2}$$

For a given distribution both $E(X)$ and $V(X)$ are *numbers*. The value of $E(X)$ represents a number around which the probability distribution is balanced, whereas $V(X)$ is a number that describes the spread of the distribution about $E(X)$.

Example 1 Let the continuous random variable X be the proportion of students who successfully complete a demanding premedical program. If the probability density function for X has the form

$$f(x) = \begin{cases} 6x(1 - x), & 0 \le x \le 1 \\ 0, & \text{elsewhere} \end{cases}$$

find $E(X)$ and $V(X)$.

Solution Using Equation 1, we get

$$E(X) = \int_0^1 6x^2(1 - x)\,dx = \int_0^1 (6x^2 - 6x^3)\,dx$$

$$= \left[2x^3 - \frac{3x^4}{2}\right]_0^1 = \frac{1}{2}$$

Similarly, the variance $V(X)$ becomes

$$V(X) = \int_0^1 (x - \tfrac{1}{2})^2 f(x)\,dx$$

$$= \int_0^1 6(x - \tfrac{1}{2})^2 x(1 - x)\,dx$$

$$= \int_0^1 \left(12x^3 - 6x^4 - \frac{15x^2}{2} + \frac{3x}{2}\right) dx$$

$$= \left[3x^4 - \frac{6x^5}{5} - \frac{5x^3}{2} + \frac{3x^2}{4}\right]_0^1 = \frac{1}{20}$$

■

When X is a continuous random variable, $\sigma(X)$, the standard deviation, is obtained from $V(X)$ in the same manner as when X is a discrete random variable, that is,

$$\sigma(X) = \sqrt{V(X)}$$

For the distribution in Example 1,

$$\sigma(X) = \sqrt{\tfrac{1}{20}} \approx 0.2236$$

Figure 1 shows the graph of the function $f(x) = 6x(1 - x)$ together with $E(X)$ and $\sigma(X)$.

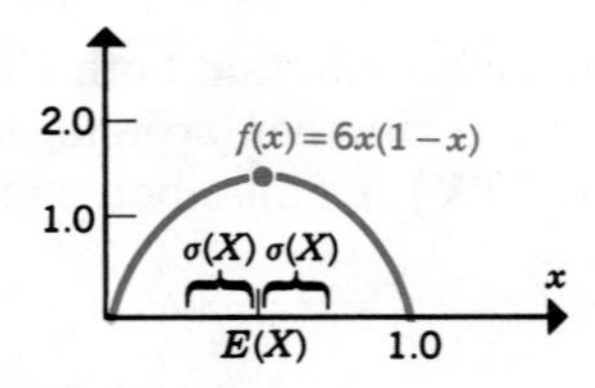

Figure 1

Example 2 The time T, in hours, that a person spends watching television each day is a continuous random variable whose probability density function is

$$f(t) = \begin{cases} \dfrac{4 - t}{8}, & 0 \leq t \leq 4 \\ 0, & \text{elsewhere} \end{cases}$$

Find $E(T)$, $V(T)$, and $\sigma(T)$.

Solution The average, or expected, number of hours spent watching television can be evaluated from Equation 1:

$$E(T) = \frac{1}{8}\int_0^4 t(4 - t)\, dt = \frac{1}{8}\left[\int_0^4 4t\, dt - \int_0^4 t^2\, dt\right]$$

$$= \frac{1}{8}\left[2t^2 - \frac{t^3}{3}\right]\Bigg|_0^4 = \frac{4}{3} \text{ hours}$$

In the same way, $V(T)$ can be found from Equation 2:

$$V(T) = \frac{1}{8}\int_0^4 \left(t - \frac{4}{3}\right)^2 (4 - t)\, dt$$

$$= \frac{1}{8}\int_0^4 \left(\frac{64}{9} - \frac{112t}{9} + \frac{20t^2}{3} - t^3\right) dt$$

$$= \frac{1}{8}\left[\frac{64t}{9} - \frac{56t^2}{9} + \frac{20t^3}{9} - \frac{t^4}{4}\right]\Bigg|_0^4$$

$$= \frac{1}{8}\left(\frac{64}{9}\right) = \frac{8}{9} \text{ (hours)}^2$$

The standard deviation $\sigma(X)$ can be found directly:

$$\sigma(X) = \sqrt{\tfrac{8}{9}} \approx 0.9428 \text{ hour}$$

■

Alternate Formula for $V(X)$

Using Equation 2 to evaluate $V(X)$ in Examples 1 and 2 was somewhat lengthy and tedious because of the algebraic steps required to simplify an integrand. Formula 2 can be converted to a form that is somewhat simpler and therefore easier to use. The shorter version is

$$V(X) = \int_L^M x^2 f(x)\, dx - [E(X)]^2 \tag{3}$$

Example 3 Use formula 3 to find $V(X)$ for the probability density function in Example 1.

Solution We have $f(x) = 6x(1 - x)$ for $0 \le x \le 1$, and we found $E(X) = \frac{1}{2}$. Using Formula 3, we get

$$V(X) = \int_0^1 6x^3(1 - x)\,dx - \left(\frac{1}{2}\right)^2$$
$$= \int_0^1 (6x^3 - 6x^4)\,dx - \frac{1}{4}$$

Integrating term by term gives

$$V(X) = \left[\frac{3x^4}{2} - \frac{6x^5}{5}\right]_0^1 - \frac{1}{4}$$
$$= \frac{3}{2} - \frac{6}{5} - \frac{1}{4} = \frac{1}{20}$$

which is identical to the result obtained in Example 1. ■

Formula 3 can be obtained directly from the definition of $V(X)$ (Equation 2). First, recalling that $E(X)$ is a constant, we let $A = E(X)$. Then we can write Equation 2 as

$$V(X) = \int_L^M (x - A)^2 f(x)\,dx$$

$$= \int_L^M (x^2 - 2xA + A^2) f(x)\,dx$$

Writing the right-hand side as a sum gives

$$V(X) = \int_L^M x^2 f(x)\,dx - \int_L^M 2xAf(x)\,dx + \int_L^M A^2 f(x)\,dx$$

Since A is constant, it can be moved outside the definite integral in the second and third terms, yielding

$$V(X) = \int_L^M x^2 f(x)\,dx - 2A\int_L^M xf(x)\,dx + A^2\int_L^M f(x)\,dx$$

Recalling that $\int_L^M xf(x)\,dx = E(X) = A$ and that $\int_L^M f(x)\,dx = 1$, we get

$$V(X) = \int_L^M x^2 f(x)\,dx - 2A^2 + A^2$$

$$= \int_L^M x^2 f(x)\,dx - A^2 = \int_L^M x^2 f(x)\,dx - [E(X)]^2$$

$$= E(X^2) - [E(X)]^2$$

which is Equation 3.

There are situations in which it is useful to know the probability or likelihood that an observation or value of a random value X will fall within a given number of standard deviations from the mean. This type of calculation is illustrated in the next example.

Example 4 Consider the continuous random variable X defined in Example 1. Find each of the following probabilities.

(a) $P[E(X) - \sigma(X) \leq X \leq E(X) + \sigma(X)]$, the probability that X will assume a value within one standard deviation of the mean, $E(X)$.

(b) $P[E(X) - 2\sigma(X) \leq X \leq E(X) + 2\sigma(X)]$, the probability that X will assume a value within two standard deviations of $E(X)$.

Solution (a) We found that $E(X) = \frac{1}{2}$ and $\sigma(X) \approx 0.2236$; therefore, $E(X) - \sigma \approx 0.2764$ and $E(X) + \sigma \approx 0.7236$. We can determine $P(0.2764 \leq X \leq 0.7236)$ by evaluating the integral

$$\int_{0.2764}^{0.7236} 6x(1 - x)\,dx = [3x^2 - 2x^3]\Big|_{0.2764}^{0.7236} \approx 0.6258$$

This result tells us that roughly 63 percent of the observations are likely to fall within one standard deviation of $E(X)$.

(b) Finding $P[E(X) - 2\sigma(X) \leq X \leq E(X) + 2\sigma(X)]$ is equivalent to finding the following probability:

$$P[\tfrac{1}{2} - 0.4472 \leq X \leq \tfrac{1}{2} + 0.4472] = P[0.0528 \leq X \leq 0.9472]$$

The probability can be written in integral form as

$$P[0.0528 \leq X \leq 0.9472] = \int_{0.5268}^{0.9472} 6x(1 - x)\,dx$$

$$= [3x^2 - 2x^3]\Big|_{0.0528}^{0.9472} \approx 0.9838$$

So there is roughly a 98 percent likelihood that an observed value of X will fall within two standard deviations of $E(X)$. ■

Chebyshev's Theorem

As Example 4 shows, we can determine the probability that a continuous random variable X falls within a specified number of standard deviations of $E(X)$ if we know $f(x)$, the probability density function. For example, the probability that X falls within two standard deviations of $E(X)$ can be written as

$$P[E(X) - 2\sigma(X) \leq X \leq E(X) + 2\sigma(X)] = \int_{E(X)-2\sigma(X)}^{E(X)+2\sigma(X)} f(x)\,dx$$

When $f(x)$, the probability density function, is not known, an important theorem, called *Chebyshev's theorem*, permits us to set lower limits on the probability that the random variable X falls within a given number of standard deviations of $E(X)$.

Chebyshev's Theorem Let X be a continuous random variable for which $E(X)$ and $\sigma(X)$ are known. Then for any $k \geq 1$, the following holds:

$$P[E(X) - k\sigma(X) \leq X \leq E(X) + k\sigma(X)] \geq 1 - \frac{1}{k^2}$$

For example, if $k = 3$, Chebyshev's theorem tells us that the probability that X falls within three standard deviations of the mean is greater than or equal to $1 - 1/3^2 = \frac{8}{9}$. In other words, regardless of the mathematical structure of $f(x)$, the probability that X falls within three standard deviations of $E(X)$ is never less than $\frac{8}{9}$.

Example 5 (a) Use Chebyshev's theorem to find a lower limit on the quantity

$$P[E(X) - 2\sigma(X) \leq X \leq E(X) + 2\sigma(X)]$$

(b) For the probability distribution given in Example 1, find

$$P[E(X) - 2\sigma(X) \leq X \leq E(X) + 2\sigma(X)].$$

Solution (a) According to Chebyshev's theorem, we get

$$P[E(X) - 2\sigma(X) \leq X \leq E(X) + 2\sigma(X)] \geq 1 - \frac{1}{2^2} = \frac{3}{4}$$

In other words, the probability that X falls within two standard deviations of $E(X)$ is always greater than or equal to $\frac{3}{4}$.

(b) In Example 4, for the probability distribution $f(x) = 6x(1 - x)$, we found that

$$P[E(X) - 2\sigma(X) \leq X \leq E(X) + 2\sigma(X)] \approx 0.9838$$

which is greater than the lower limits set by Chebyshev's theorem. ■

10.3 EXERCISES

Find the expected value, variance, and standard deviation for each of the probability density functions given in Exercises 1 through 10.

1. $f(x) = \begin{cases} \frac{1}{5}, & 2 \leq x \leq 7 \\ 0, & \text{elsewhere} \end{cases}$

2. $f(x) = \begin{cases} 2x, & 0 \leq x \leq 1 \\ 0, & \text{elsewhere} \end{cases}$

3. $f(x) = \begin{cases} \dfrac{\sqrt{x}}{18}, & 0 \le x \le 9 \\ 0, & \text{elsewhere} \end{cases}$

4. $f(x) = \begin{cases} 2(x + 1), & -1 \le x \le 0 \\ 0, & \text{elsewhere} \end{cases}$

5. $f(x) = \begin{cases} e^{-x}, & x \ge 0 \\ 0, & \text{elsewhere} \end{cases}$

6. $f(x) = \begin{cases} \dfrac{1}{x}, & 1 \le x \le e \\ 0, & \text{elsewhere} \end{cases}$

7. $f(x) = \begin{cases} 3x^2, & 0 \le x \le 1 \\ 0, & \text{elsewhere} \end{cases}$

8. $f(x) = \begin{cases} \dfrac{2}{x^2}, & 1 \le x \le 2 \\ 0, & \text{elsewhere} \end{cases}$

9. $f(x) = \begin{cases} \dfrac{8}{3x^3}, & 1 \le x \le 2 \\ 0, & \text{elsewhere} \end{cases}$

10. $f(x) = \begin{cases} \dfrac{4x^{1/3}}{3}, & 0 \le x \le 1 \\ 0, & \text{elsewhere} \end{cases}$

11. The time T, in hours, between breakdowns in a flexible manufacturing system is a continuous random variable whose probability density function has the form

$$f(t) = \begin{cases} \dfrac{e^{-t/4}}{4}, & t \ge 0 \\ 0, & \text{elsewhere} \end{cases}$$

(a) Find the expected value and standard deviation of T.
(b) Find the probability that the time between breakdowns will be greater than $E(T)$.
(c) Find the probability that the time between breakdowns will fall within one standard deviation of $E(T)$. In other words, find

$$P[E(T) - \sigma(T) \le T \le E(T) + \sigma(T)].$$

12. The daily high temperature T (in °F) during July is a continuous random variable whose probability density function has the form

$$f(t) = \begin{cases} \frac{1}{10}, & 80 \le t \le 90 \\ 0, & \text{elsewhere} \end{cases}$$

(a) Find the expected value, variance, and standard deviation of T.
(b) Find the probability that the daily high temperature will fall within two standard deviations of $E(T)$. In other words, find

$$P[E(T) - 2\sigma(T) \le T \le E(T) + 2\sigma(T)].$$

13. The time T, (in days) required to dig and pour the foundation for a warehouse is a continuous random variable whose probability density function is

$$f(t) = \begin{cases} \dfrac{(t - 6)(12 - t)}{36}, & 6 \le t \le 12 \\ 0, & \text{elsewhere} \end{cases}$$

(a) Find the expected value, variance, and standard deviation of T.
(b) Find the probability that the random variable T will fall within 1.5 standard deviations of $E(T)$. In other words, find

$$P[E(T) - 1.5\sigma(T) \le T \le E(T) + 1.5\sigma(T)].$$

14. The annual return on investment, R, for a mutual fund is a continuous random variable whose probability density function is

$$f(r) = \begin{cases} \frac{1}{30}, & -10 \le r \le 20 \\ 0, & \text{elsewhere} \end{cases}$$

(a) Find the expected value and standard deviation of R.
(b) Find the probability that the return on investment will be greater than $E(R)$.

(c) Find the probability that the return on investment will fall within one standard deviation of $E(R)$. In other words, find $P[E(R) - \sigma(R) \leq T \leq E(R) + \sigma(R)]$.

15. The length of time T (in hours) that a student spends studying for an examination is a continuous random variable whose probability density function is

$$f(t) = \begin{cases} \dfrac{3\sqrt{t}}{16}, & 0 \leq t \leq 4 \\ 0, & \text{elsewhere} \end{cases}$$

(a) Find the expected value and standard deviation of T.
(b) Find the probability that the time spent studying will be greater than $E(T)$.
(c) Find the probability that the time spent studying will fall within two standard deviations of $E(T)$. In other words, find

$$P[E(T) - 2\sigma(T) \leq T \leq E(T) + 2\sigma(T)].$$

16. The lifetime T (in years) of a new energy-saving light bulb is a continuous random variable whose probability density function is

$$f(t) = \begin{cases} \frac{1}{2}e^{-t/2}, & t \geq 0 \\ 0, & \text{elsewhere} \end{cases}$$

(a) Find the expected value and standard deviation of T.
(b) Find the probability that the lifetime will be greater than $2E(T)$.
(c) Find the probability that the lifetime will fall within one standard deviation of $E(T)$. In other words, find $P[E(T) - \sigma(T) \leq T \leq E(T) + \sigma(T)]$.

17. Find $E(x)$ and $\sigma(X)$ for the random variable X in Exercise 22 of Section 10.2.

18. Find $E(Y)$ and $\sigma(Y)$ for the random variable Y in Exercise 23 of Section 10.2.

19. The distance D (in miles) driven each day by a courier for a delivery service company is a continuous random variable whose expected value $E(D)$ equals 50 miles and whose standard deviation $\sigma(D)$ equals 5 miles. What does Chebyshev's theorem tell us about $P(35 \leq D \leq 65)$?

20. The length of time T (in hours) that it takes to install a vinyl floor in a 12- × 20-ft room is a continuous random variable whose expected value $E(T)$ equals eight hours and whose standard deviation $\sigma(T)$ equals one hour. What does Chebyshev's theorem tell us about $P(6 \leq T \leq 10)$?

If X is a continuous random variable whose probability density function is nonzero over the interval $L \leq x \leq M$, the median m is defined as the value of the random variable that satisfies the equation

$$\int_L^m f(x)\,dx = \frac{1}{2}$$

that is, $P(L \leq X \leq m) = \frac{1}{2}$. In Exercises 21 through 26, find the median for each distribution.

21. $f(x) = \begin{cases} \frac{1}{6}, & 2 \leq x \leq 8, \\ 0, & \text{elsewhere} \end{cases}$

22. $f(x) = \begin{cases} 2x, & 0 \leq x \leq 1 \\ 0, & \text{elsewhere} \end{cases}$

23. $f(x) = \begin{cases} 2e^{-2x}, & x \geq 0 \\ 0, & \text{elsewhere} \end{cases}$

24. $f(x) = \begin{cases} \dfrac{3\sqrt{x}}{14}, & 1 \leq x \leq 4 \\ 0, & \text{elsewhere} \end{cases}$

25. $f(x) = \begin{cases} 3(x-1)^2, & 1 \le x \le 2 \\ 0, & \text{elsewhere} \end{cases}$ **26.** $f(x) = \begin{cases} \dfrac{1}{x}, & 1 \le x \le e \\ 0, & \text{elsewhere} \end{cases}$

KEY TERMS

Probability experiment
Sample space
Rules of probability
Random variable
Probability distribution
Expected value
Variance
Standard deviation
Continuous random variable
Probability density function
Cumulative probability distribution function
Chebyshev's theorem

REVIEW PROBLEMS

1. A fair die is rolled twice and the number of dots on each roll is recorded. A sample space consisting of 36 equally likely outcomes is

$$S = \{(1, 1), (1, 2), (1, 3), \ldots, (4, 6), (5, 6), (6, 6)\}$$

(a) List the elements of the event E: "The number of dots on the first roll equals the number on the second roll."
(b) Find the probability of the event E.
(c) Let the random variable X be defined as the sum of the dots on the two rolls. What values can X assume?
(d) Find the probability distribution for X.
(e) Find $E(X)$, $V(X)$, and $\sigma(X)$.

2. (a) In problem 1, let the random variable Y be defined as the largest number that is rolled. What values can Y assume?
(b) Find the probability distribution for Y.
(c) Find $E(Y)$, $V(Y)$, and $\sigma(Y)$.

3. A number is painted on each of ten marbles in an urn. The number 1 is painted on two marbles, the number 2 on three marbles, and the number 3 on five marbles. One marble is selected at random from the urn. Let the random variable X be the number painted on the marble drawn.
(a) Find the probability distribution for X.
(b) Find $E(X)$, $V(X)$, and $\sigma(X)$.

4. Table 1 shows the probability values for X, the number of patients who are treated daily in the emergency room of a rural hospital.

Table 1

X	16	17	18	19	20	21	22
$P(X)$	0.05	0.20	0.20	0.30	0.10	0.10	0.05

Find $E(X)$, $V(X)$, and $\sigma(X)$.

5. Suppose you are invited to play the following game. A coin is tossed twice, and you receive a dollar amount equal to the number of times the coin shows tails. Let the random variable X equal the dollar amount won. (Assume that you play the game for free.)
   (a) Find the probability distribution for X.
   (b) Find $E(X)$, $V(X)$, and $\sigma(X)$.
   Suppose that you have to pay \$1.50 each time you play the game. Let the random variable G equal your net gain (or loss) on each play.
   (c) Find the probability distribution for G.
   (d) Find $E(G)$, $V(G)$, and $\sigma(G)$.
6. An investor is considering three investments whose annual rates of return are represented by the random variables X, Y, and Z. Table 2 contains the probability values for each of the three variables.

Table 2

Investment 1			Investment 2				Investment 3				
X	0.06	0.08	Y	0.05	0.10	0.15	Z	0.04	0.10	0.14	0.20
$P(X)$	0.50	0.50	$P(Y)$	0.30	0.40	0.30	$P(Z)$	0.20	0.30	0.30	0.20

   (a) Find the expected annual return for each of the three investments.
   (b) Find the variances of the annual returns for each of the three investments.
   (c) Do the results from parts (a) and (b) suggest any relationship between $E(R)$ and $V(R)$ where R is the annual rate of return from an investment?
7. A die is loaded so that each of the even numbers is twice as likely to come up as each of the odd numbers. Let the random variable X represent the number of dots that come up in one roll of the die.
   (a) Find the probability distribution for X.
   (b) Find $E(X)$, $V(X)$, and $\sigma(X)$.
8. A house painter earns \$250 per day when he works, but he loses \$50 per day when he cannot work because of illness or bad weather. If the probability of working on any day is 0.85, find the expected amount that the painter earns each day.

In Exercises 9 through 14, determine whether each function qualifies as a probability density function. If any one does not, explain why it fails to qualify.

9. $f(x) = \begin{cases} 2x - 2, & 1 \le x \le 2 \\ 0, & \text{elsewhere} \end{cases}$

10. $f(x) = \begin{cases} 2e^{-2x}, & x \ge 0 \\ 0, & \text{elsewhere} \end{cases}$

11. $f(x) = \begin{cases} \dfrac{3 - 3x^2}{2}, & 0 \le x \le 1 \\ 0, & \text{elsewhere} \end{cases}$

12. $f(x) = \begin{cases} 1 + x, & -1 \le x \le 0 \\ 1 - x, & 0 \le x \le 1 \\ 0, & \text{elsewhere} \end{cases}$

13. $f(x) = \begin{cases} 1 - x, & -1 \le x \le 1 \\ 0, & \text{elsewhere} \end{cases}$

14. $f(x) = \begin{cases} 1 - x^2, & 0 \le x \le 1 \\ 0, & \text{elsewhere} \end{cases}$

Determine the value of K for which each of the following functions qualifies as a probability density function.

15. $f(x) = \begin{cases} Kx - 1, & 2 \le x \le 3 \\ 0, & \text{elsewhere} \end{cases}$

16. $f(x) = \begin{cases} \dfrac{K}{x}, & 1 \le x \le e \\ 0, & \text{elsewhere} \end{cases}$

17. $f(x) = \begin{cases} Ke^{2x}, & x \leq 0 \\ 0, & \text{elsewhere} \end{cases}$

18. $f(x) = \begin{cases} \dfrac{K}{\sqrt{x}}, & 4 \leq x \leq 9 \\ 0, & \text{elsewhere} \end{cases}$

Find the cumulative probability function for each of the following probability density functions.

19. $f(x) = \begin{cases} \dfrac{x^3}{4}, & 0 \leq x \leq 2 \\ 0, & \text{elsewhere} \end{cases}$

20. $f(x) = \begin{cases} 4e^{-4x}, & x \geq 0 \\ 0, & \text{elsewhere} \end{cases}$

21. $f(x) = \begin{cases} \dfrac{8}{3x^3}, & 1 \leq x \leq 2 \\ 0, & \text{elsewhere} \end{cases}$

22. $f(x) = \begin{cases} \dfrac{\sqrt{x}}{18}, & 0 \leq x \leq 9 \\ 0, & \text{elsewhere} \end{cases}$

23. The time T, in days, to recover from a surgical procedure is a random variable whose probability density function $f(t)$ is

$$f(t) = \begin{cases} e^{-(t-1)}, & t \geq 1 \\ 0, & \text{elsewhere} \end{cases}$$

(a) Find the probability that a patient selected at random recovers within two days.
(b) Find the probability that a patient's recovery period lasts longer than three days.

24. The weight W of a one-pound bag of potato chips is a random variable whose probability density function $f(w)$ is

$$f(w) = \begin{cases} 5, & 0.95 \leq w \leq 1.15 \\ 0, & \text{elsewhere} \end{cases}$$

(a) What is the probability that a bag of potato chips selected at random will weigh less than one pound?
(b) Find $E(W)$ and $V(W)$.

25. The time T, in seconds, between incoming phone calls to a mail order catalog company is a random variable whose probability density function $f(t)$ is

$$f(t) = \begin{cases} \dfrac{e^{-0.20t}}{5}, & t \geq 0 \\ 0, & \text{elsewhere} \end{cases}$$

(a) What is the probability that the time between the next two calls will be greater than five seconds?
(b) What is the expected time between incoming calls?
(c) Find $V(T)$.

26. The amount of snowfall X, in inches, during the month of February in a Canadian province is a random variable whose probability density function $f(x)$ is

$$f(x) = \begin{cases} \dfrac{\sqrt{x}}{144}, & 0 \leq x \leq 36 \\ 0, & \text{elsewhere} \end{cases}$$

(a) Find the probability that snowfall during February will be between 16 and 25 inches.
(b) Find $E(X)$ and $V(X)$.

27. The waiting time T, in minutes, in a doctor's office is a random variable whose expected value $E(T)$ equals 30 minutes and whose standard deviation of $\sigma(T)$ equals 5 minutes. What does Chebyshev's theorem tell us about $P(15 \leq T \leq 45)$?

28. The amount of soda X, in ounces, dispensed by an automatic soft-drink machine at a fast-food restaurant is a continuous random variable whose expected value $E(X)$ equals 12 ounces and whose standard deviation $\sigma(X)$ equals 0.2 ounce. What does Chebyshev's theorem tell us about $P(11.5 \leq X \leq 12.5)$?

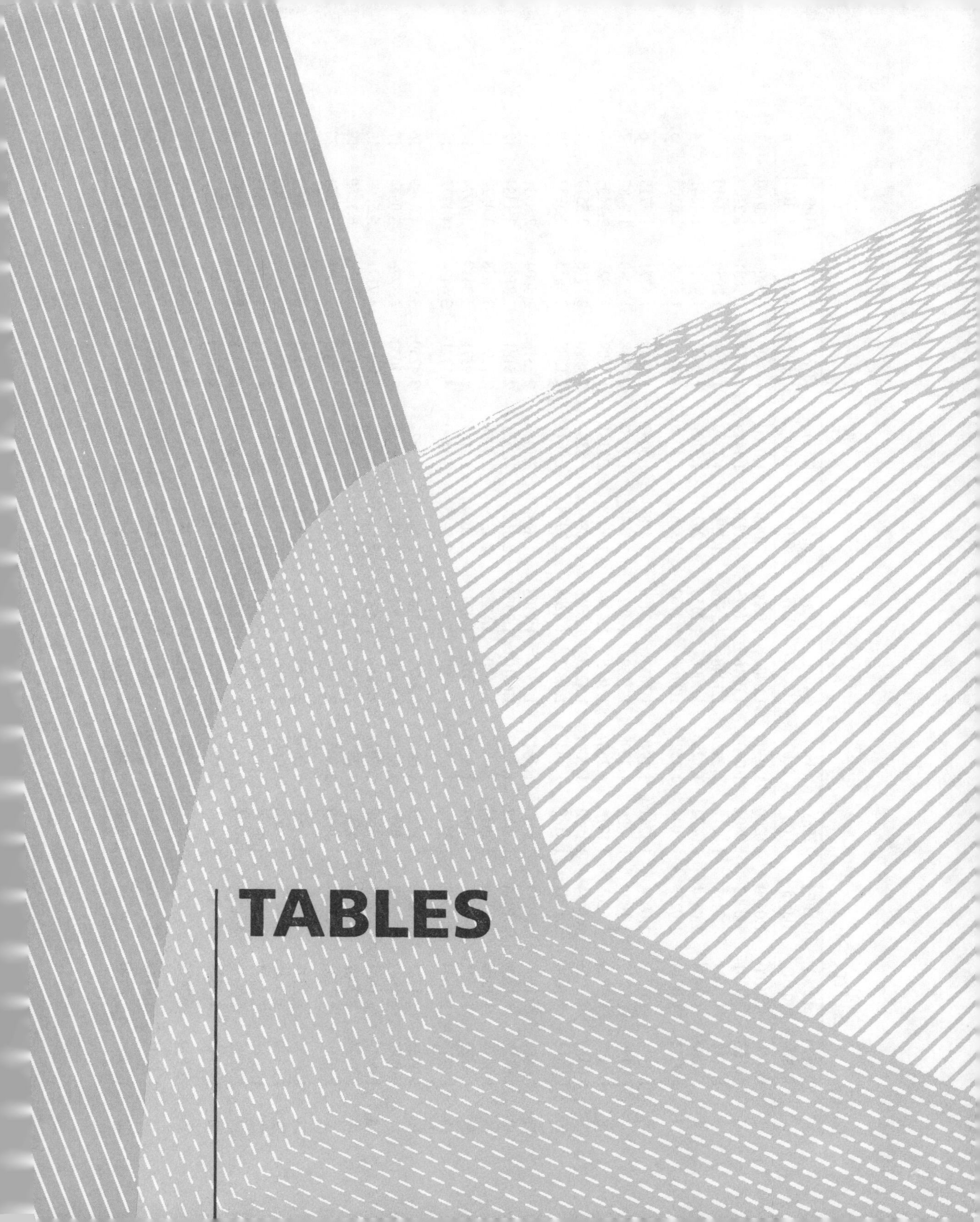

TABLES

Table A Values of $(1 + i)^n$

n \ i	(0.5%) 0.005	(1%) 0.01	(1.5%) 0.015	(2%) 0.02	(3%) 0.03	(4%) 0.04	(5%) 0.05	(6%) 0.06	(7%) 0.07	(8%) 0.08	(9%) 0.09	(10%) 0.10	i / n
1	1.0050	1.0100	1.0150	1.0200	1.0300	1.0400	1.0500	1.0600	1.0700	1.0800	1.0900	1.1000	1
2	1.0100	1.0201	1.0302	1.0404	1.0609	1.0816	1.1025	1.1236	1.1449	1.1664	1.1881	1.2100	2
3	1.0151	1.0303	1.0457	1.0612	1.0927	1.1248	1.1576	1.1910	1.2250	1.2597	1.2950	1.3310	3
4	1.0202	1.0406	1.0614	1.0824	1.1255	1.1698	1.2155	1.2624	1.3107	1.3604	1.4115	1.4641	4
5	1.0253	1.0510	1.0773	1.1040	1.1592	1.2166	1.2762	1.3382	1.4025	1.4693	1.5386	1.6105	5
6	1.0304	1.0615	1.0934	1.1261	1.1940	1.2653	1.3400	1.4185	1.5007	1.5868	1.6771	1.7715	6
7	1.0355	1.0721	1.1098	1.1486	1.2298	1.3159	1.4071	1.5036	1.6057	1.7138	1.8280	1.9487	7
8	1.0407	1.0828	1.1265	1.1716	1.2667	1.3685	1.4774	1.5938	1.7181	1.8509	1.9925	2.1435	8
9	1.0459	1.0936	1.1434	1.1950	1.3047	1.4233	1.5513	1.6894	1.8384	1.9990	2.1718	2.3579	9
10	1.0511	1.1046	1.1605	1.2189	1.3439	1.4802	1.6288	1.7908	1.9671	2.1589	2.3673	2.5937	10
11	1.0564	1.1156	1.1779	1.2433	1.3842	1.5394	1.7103	1.8982	2.1048	2.3316	2.5804	2.8531	11
12	1.0617	1.1268	1.1956	1.2682	1.4257	1.6010	1.7958	2.0121	2.2521	2.5181	2.8126	3.1384	12
13	1.0670	1.1380	1.2136	1.2936	1.4685	1.6650	1.8856	2.1329	2.4098	2.7196	3.0658	3.4522	13
14	1.0723	1.1494	1.2318	1.3194	1.5125	1.7316	1.9799	2.2609	2.5785	2.9371	3.3417	3.7974	14
15	1.0777	1.1609	1.2502	1.3458	1.5579	1.8009	2.0789	2.3965	2.7590	3.1721	3.6424	4.1772	15
16	1.0831	1.1725	1.2690	1.3727	1.6047	1.8729	2.1828	2.5403	2.9521	3.4259	3.9703	4.5949	16
17	1.0885	1.1843	1.2880	1.4002	1.6528	1.9479	2.2920	2.6927	3.1588	3.7000	4.3276	5.0544	17
18	1.0939	1.1961	1.3073	1.4282	1.7024	2.0258	2.4066	2.8543	3.3799	3.9960	4.7171	5.5599	18
19	1.0994	1.2081	1.3270	1.4568	1.7535	2.1068	2.5269	3.0255	3.6165	4.3157	5.1416	6.1159	19
20	1.1049	1.2201	1.3469	1.4859	1.8061	2.1911	2.6532	3.2071	3.8696	4.6609	5.6044	6.7274	20
21	1.1104	1.2323	1.3671	1.5156	1.8602	2.2787	2.7859	3.3995	4.1405	5.0338	6.1088	7.4002	21
22	1.1170	1.2447	1.3876	1.5459	1.9161	2.3699	2.9252	3.6035	4.4304	5.4365	6.6586	8.1402	22
23	1.1216	1.2571	1.4084	1.5768	1.9735	2.4647	3.0715	3.8197	4.7405	5.8714	7.2578	8.9543	23
24	1.1272	1.2697	1.4295	1.6084	2.0327	2.5633	3.2250	4.0489	5.0724	6.3411	7.9110	9.8497	24
25	1.1328	1.2824	1.4509	1.6406	2.0937	2.6658	3.3863	4.2918	5.4274	6.8484	8.6230	10.8347	25
30	1.1614	1.3478	1.5631	1.8114	2.4273	3.2434	4.3219	5.7435	7.6123	10.0627	13.2676	17.4494	30
35	1.1907	1.4165	1.6839	1.9999	2.8138	3.9460	5.5160	7.6861	10.6766	14.7853	20.4140	28.1024	35
40	1.2208	1.4889	1.8140	2.2080	3.2620	4.8010	7.0399	10.2857	14.9744	21.7245	31.4094	45.2592	40

Table B e^x and e^{-x}

x	e^x	e^{-x}	x	e^x	e^{-x}
0.00	1.0000	1.0000	1.5	4.4817	0.2231
0.01	1.0101	0.9901	1.6	4.9530	0.2019
0.02	1.0202	0.9802	1.7	5.4739	0.1827
0.03	1.0305	0.9705	1.8	6.0496	0.1653
0.04	1.0408	0.9608	1.9	6.6859	0.1496
0.05	1.0513	0.9512	2.0	7.3891	0.1353
0.06	1.0618	0.9418	2.1	8.1662	0.1225
0.07	1.0725	0.9324	2.2	9.0250	0.1108
0.08	1.0833	0.9231	2.3	9.9742	0.1003
0.09	1.0942	0.9139	2.4	11.0230	0.0907
0.10	1.1052	0.9048	2.5	12.182	0.0821
0.11	1.1163	0.8958	2.6	13.464	0.0743
0.12	1.1275	0.8869	2.7	14.880	0.0672
0.13	1.1388	0.8781	2.8	16.445	0.0608
0.14	1.1503	0.8694	2.9	18.174	0.0550
0.15	1.1618	0.8607	3.0	20.086	0.0498
0.16	1.1735	0.8521	3.1	22.198	0.0450
0.17	1.1853	0.8437	3.2	24.533	0.0408
0.18	1.1972	0.8353	3.3	27.113	0.0369
0.19	1.2092	0.8270	3.4	29.964	0.0334
0.20	1.2214	0.8187	3.5	33.115	0.0302
0.21	1.2337	0.8106	3.6	36.598	0.0273
0.22	1.2461	0.8025	3.7	40.447	0.0247
0.23	1.2586	0.7945	3.8	44.701	0.0224
0.24	1.2712	0.7866	3.9	49.402	0.0202
0.25	1.2840	0.7788	4.0	54.598	0.0183
0.30	1.3499	0.7408	4.1	60.340	0.0166
0.35	1.4191	0.7047	4.2	66.686	0.0150
0.40	1.4918	0.6703	4.3	73.700	0.0136
0.45	1.5683	0.6376	4.4	81.451	0.0123
0.50	1.6487	0.6065	4.5	90.017	0.0111
0.55	1.7333	0.5769	4.6	99.484	0.0101
0.60	1.8221	0.5488	4.7	109.950	0.0091
0.65	1.9155	0.5220	4.8	121.510	0.0082
0.70	2.0138	0.4966	4.9	134.290	0.0074
0.75	2.1170	0.4724	5.0	148.41	0.0067
0.80	2.2255	0.4493	5.5	244.69	0.0041
0.85	2.3396	0.4274	6.0	403.43	0.0025
0.90	2.4596	0.4066	6.5	665.14	0.0015
0.95	2.5857	0.3867	7.0	1096.60	0.0009
1.0	2.7183	0.3679	7.5	1808.0	0.0006
1.1	3.0042	0.3329	8.0	2981.0	0.0003
1.2	3.3201	0.3012	8.5	4914.8	0.0002
1.3	3.6693	0.2725	9.0	8103.1	0.0001
1.4	4.0552	0.2466	10.0	22026.0	0.00005

Table C Natural Logarithms

x	$\ln x$	x	$\ln x$	x	$\ln x$
		4.5	1.5041	9.0	2.1972
0.1	−2.3026	4.6	1.5261	9.1	2.2083
0.2	−1.6094	4.7	1.5476	9.2	2.2192
0.3	−1.2040	4.8	1.5686	9.3	2.2300
0.4	−0.9163	4.9	1.5892	9.4	2.2407
0.5	−0.6931	5.0	1.6094	9.5	2.2513
0.6	−0.5108	5.1	1.6292	9.6	2.2618
0.7	−0.3567	5.2	1.6487	9.7	2.2721
0.8	−0.2231	5.3	1.6677	9.8	2.2824
0.9	−0.1054	5.4	1.6864	9.9	2.2925
1.0	0.0000	5.5	1.7047	10	2.3026
1.1	0.0953	5.6	1.7228	11	2.3979
1.2	0.1823	5.7	1.7405	12	2.4849
1.3	0.2624	5.8	1.7579	13	2.5649
1.4	0.3365	5.9	1.7750	14	2.6391
1.5	0.4055	6.0	1.7918	15	2.7081
1.6	0.4700	6.1	1.8083	16	2.7726
1.7	0.5306	6.2	1.8245	17	2.8332
1.8	0.5878	6.3	1.8405	18	2.8904
1.9	0.6419	6.4	1.8563	19	2.9444
2.0	0.6931	6.5	1.8718	20	2.9957
2.1	0.7419	6.6	1.8871	25	3.2189
2.2	0.7885	6.7	1.9021	30	3.4012
2.3	0.8329	6.8	1.9169	35	3.5553
2.4	0.8755	6.9	1.9315	40	3.6889
2.5	0.9163	7.0	1.9459	45	3.8067
2.6	0.9555	7.1	1.9601	50	3.9120
2.7	0.9933	7.2	1.9741	55	4.0073
2.8	1.0296	7.3	1.9879	60	4.0943
2.9	1.0647	7.4	2.0015	65	4.1744
3.0	1.0986	7.5	2.0149	70	4.2485
3.1	1.1314	7.6	2.0281	75	4.3175
3.2	1.1632	7.7	2.0412	80	4.3820
3.3	1.1939	7.8	2.0541	85	4.4427
3.4	1.2238	7.9	2.0669	90	4.4998
3.5	1.2528	8.0	2.0794	100	4.6052
3.6	1.2809	8.1	2.0919	110	4.7005
3.7	1.3083	8.2	2.1041	120	4.7875
3.8	1.3350	8.3	2.1163	130	4.8676
3.9	1.3610	8.4	2.1282	140	4.9416
4.0	1.3863	8.5	2.1401	150	5.0106
4.1	1.4110	8.6	2.1518	160	5.0752
4.2	1.4351	8.7	2.1633	170	5.1358
4.3	1.4586	8.8	2.1748	180	5.1930
4.4	1.4816	8.9	2.1861	190	5.2470

Table D Table of Integrals

1. $\int x^n\,dx = \frac{x^{n+1}}{n+1} + C \qquad (n \neq -1)$

2. $\int \frac{1}{x}\,dx = \ln|x| + C$

Forms involving $ax + b$:

3. $\int \frac{x}{ax+b}\,dx = \frac{x}{a} - \frac{b}{a^2}\ln|ax+b| + C$

4. $\int \frac{x}{(ax+b)^2}\,dx = \frac{b}{a^2(ax+b)} + \frac{1}{a^2}\ln|ax+b| + C$

5. $\int \frac{1}{x(ax+b)}\,dx = \frac{1}{b}\ln\left|\frac{x}{ax+b}\right| + C$

6. $\int \frac{1}{x(ax+b)^2}\,dx = \frac{1}{b(ax+b)} + \frac{1}{b^2}\ln\left|\frac{x}{ax+b}\right| + C$

Forms involving $\sqrt{ax+b}$:

7. $\int \frac{x}{\sqrt{ax+b}}\,dx = \frac{2ax-4b}{3a^2}\sqrt{ax+b} + C$

8. $\int \frac{1}{x\sqrt{ax+b}}\,dx = \frac{1}{\sqrt{b}}\ln\left|\frac{\sqrt{ax+b}-\sqrt{b}}{\sqrt{ax+b}+\sqrt{b}}\right| + C, \qquad (b > 0)$

Forms involving $x^2 - a^2$:

9. $\int \frac{1}{x^2-a^2}\,dx = \frac{1}{2a}\ln\left|\frac{x-a}{x+a}\right| + C$

10. $\int \frac{1}{a^2-x^2}\,dx = \frac{1}{2a}\ln\left|\frac{a+x}{a-x}\right| + C$

Forms involving $\sqrt{x^2 \pm a^2}$:

11. $\int \sqrt{x^2 \pm a^2}\,dx = \frac{x}{2}\sqrt{x^2 \pm a^2} + \frac{a^2}{2}\ln\left|x + \sqrt{x^2 \pm a^2}\right| + C$

12. $\int \frac{1}{\sqrt{x^2 \pm a^2}}\,dx = \ln\left|x + \sqrt{x^2 \pm a^2}\right| + C$

Forms involving $\sqrt{a^2 - x^2}$:

13. $\int \frac{1}{x\sqrt{a^2-x^2}}\,dx = -\frac{1}{a}\ln\left|\frac{a+\sqrt{a^2-x^2}}{x}\right| + C$

14. $\int \frac{\sqrt{a^2-x^2}}{x}\,dx = \sqrt{a^2-x^2} - a\ln\left|\frac{a+\sqrt{a^2-x^2}}{x}\right| + C$

Forms involving e^x or $\ln x$:

15. $\int e^{ax}\,dx = \frac{e^{ax}}{a} + C$

16. $\int xe^{ax}\,dx = \frac{e^{ax}}{a^2}(ax-1) + C$

17. $\int x^n e^{ax}\,dx = \frac{x^n e^{ax}}{a} - \frac{n}{a}\int x^{n-1}e^{ax}\,dx$

Table D (*Continued*)

18. $\int \ln x \, dx = x \ln x - x + C$

19. $\int x^n \ln x \, dx = x^{n+1}\left[\frac{\ln x}{n+1} - \frac{1}{(n+1)^2}\right] + C, \qquad (n \neq -1)$

20. $\int (\ln x)^n \, dx = x(\ln x)^n - n \int (\ln x)^{n-1} \, dx$

ANSWERS TO ODD-NUMBERED PROBLEMS

CHAPTER 1

1.1 Exercises

1. (a) 2 (b) 2 (c) 0 (d) 56 **3.** (a) 17 (b) 17 (c) 269 (d) 87

5. (a) 9 (b) 0 (c) $\frac{7}{8}$ (d) $b^3 + 1$ **7.** (a) $d^2 + d + 2$ (b) $d^2 - 3d + 4$ (c) $d^4 - d^2 + 2$ (d) $d^2 + d$

9. (a) 0 (b) $\frac{1}{2}$ (c) $\sqrt{a}$ (d) $\sqrt{a}$ **11.** function **13.** function

15. function **17.** all real numbers **19.** all real numbers **21.** $x \neq -1$

23.

x	$f(x) = x^3$
-2	-8
-1	-1
0	0
1	1
2	8

25.

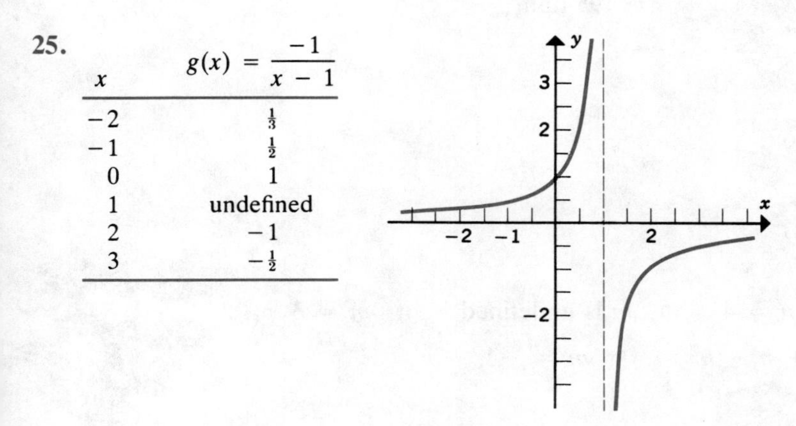

x	$g(x) = \dfrac{-1}{x - 1}$
-2	$\frac{1}{3}$
-1	$\frac{1}{2}$
0	1
1	undefined
2	-1
3	$-\frac{1}{2}$

27.

x	$f(x) = x^2 - x$
-2	6
-1	2
0	0
$\frac{1}{2}$	$-\frac{1}{4}$
1	0
2	2
3	6

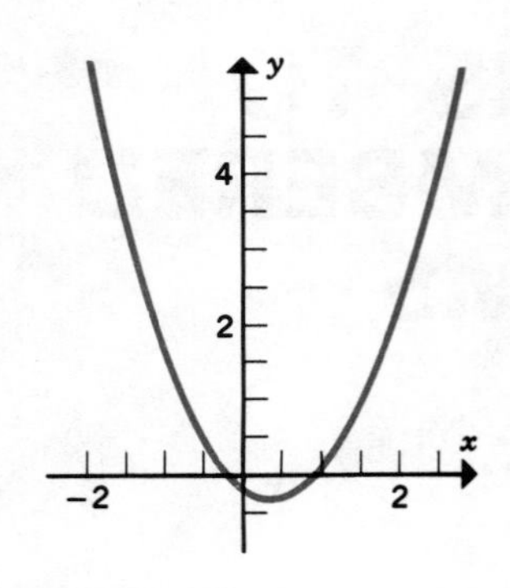

29. (a) $f(-2) = -2$
(b) $f(-1) = -1$
(c) $f(1) = 1$
(d) $f(3) = 9$

31.

33. $S(d) = 400 + 0.03d$

35. x = number of shoes stitched
(a) $S_B = 0.25x$
(b) $S_W = 50$
(c) Worker earns a higher salary at Black Shoe if he or she can stitch more than 200 shoes per day.

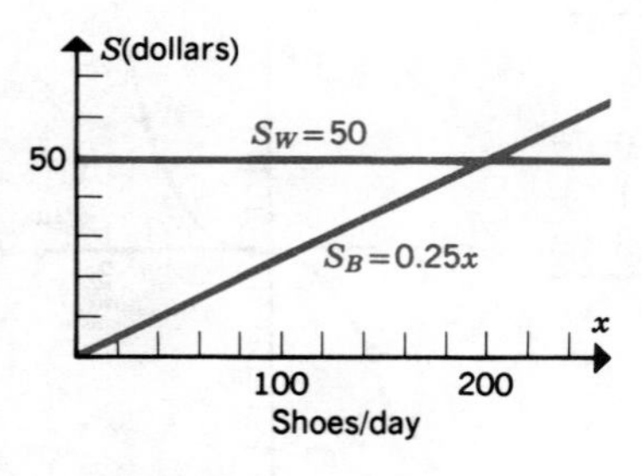

37. (a)

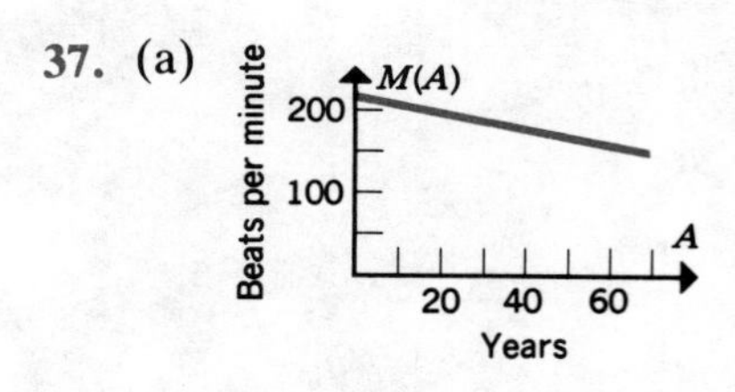

(b) $M(25) = 220 - 25$
$= 195$ bpm

1.2 Exercises

1. $m = 1$ **3.** $m = 0$ **5.** $m = 4$ **7.** m is undefined **9.** $m = 3$

11. $m = 2 + h$ **13.** $m = 2 + h - h^2$ **15.** $m = -1$

17.

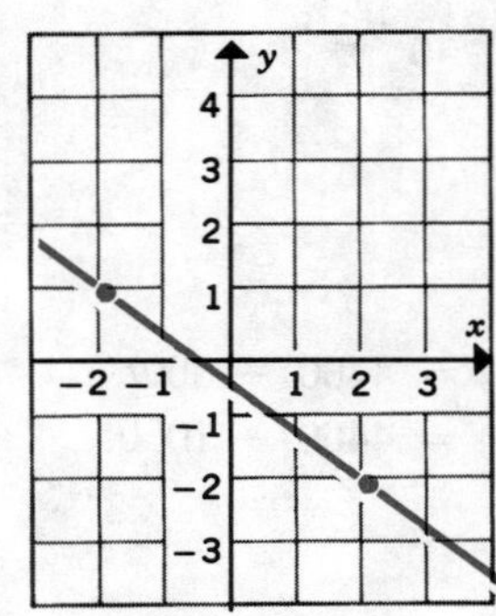

19. 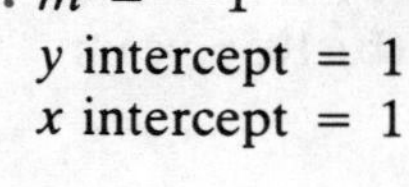
$m = -1$
y intercept $= 1$
x intercept $= 1$

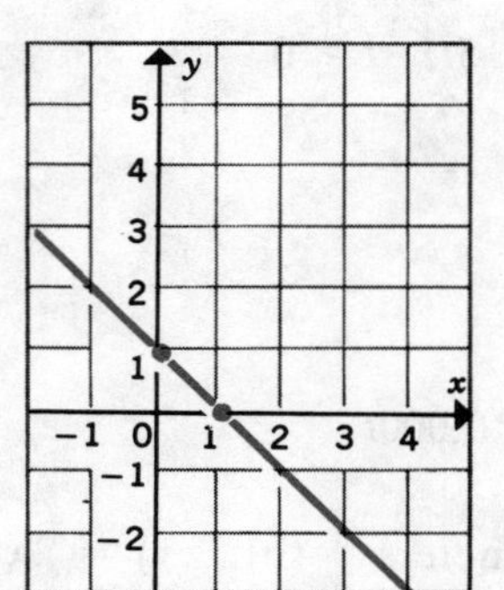

21. $m = \frac{2}{3}$
y intercept $= -2$
x intercept $= 3$

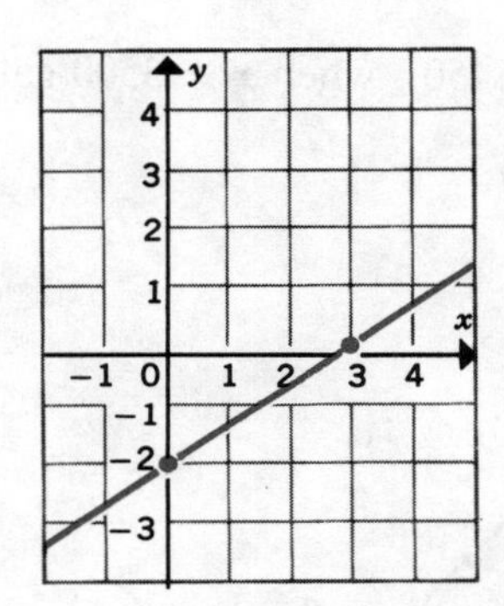

23. $m = 1$
y intercept $= -2$
x intercept $= 2$

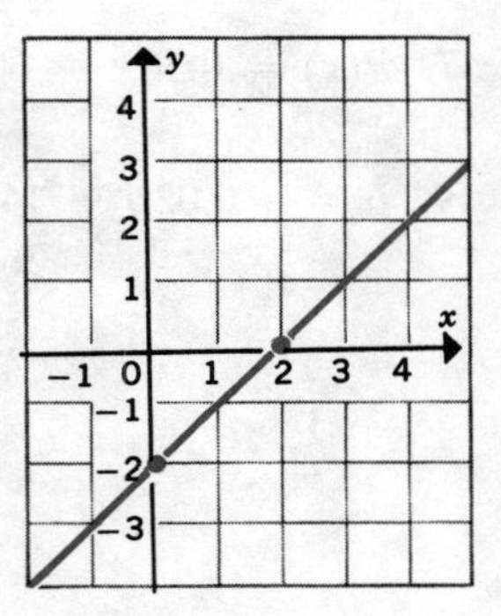

25. $m = -\frac{1}{3}$
y intercept $= 2$
x intercept $= 6$

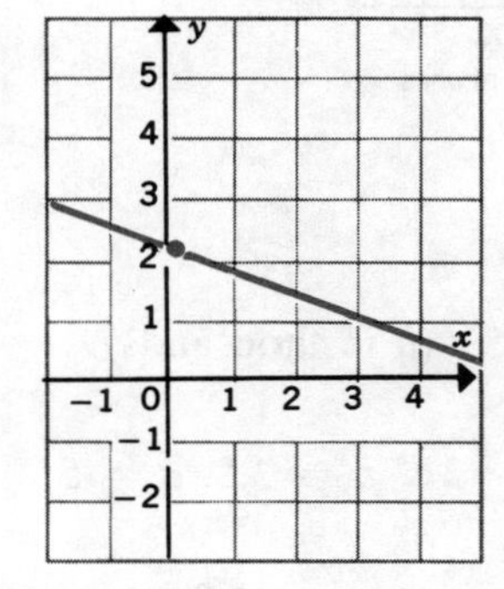

27. m is undefined
x intercept $= 3$

29. $y = x + 5$ **31.** $y = 3x + 7$ **33.** $y = -x + 4$ **35.** $3y + 2x = 12$

37. $y = \sqrt{2}x + 1$ **39.** $y = x + 4$ **41.** $y = x + 3$ **43.** $y = 2x - 5$

45. $x = 3$ **47.** $x = \sqrt{2}$ **49.** $y = \dfrac{2dx}{c}$ **51.** $3y + x = 3$

53. $3y = 2x + 3$ **55.** $2y - 4x + 7 = 0$ **57.** $P = f(p) = -5p + 165$

59. (a) $I(t) = 300 - 15t$
$I(8) = 180$ cars
$I(t) = 0$ when $t = 20$ weeks
(b) $I(t) = 300 - 5t$

61. (a) $A(t) = 1000 + 180t$
(b) $A(t) = 1000 + 15t$

63. (a) $A(t) = 500 - 5t, \quad t \geq 0$
(b) 18 months

65. (a) $W = 5280 - 5t, \quad t \geq 0$
(b) 1056 years

1.3 Exercises

1. $D = \$10{,}000/\text{year}$
$V(t) = 35{,}000 - 10{,}000t$

3. $D_1 = \$800/\text{year}; \quad V_1(t) = 4400 - 800t$
$D_2 = \$1050/\text{year}; \quad V_2(t) = 4400 - 1050t$

5. (a) 12,500 thermometers (b) $P(x) = 2x - 25{,}000$

7. $P(x) = 0.30x - 600; \quad x_{BE} = 2000; \quad P(x) \geq \$450$ when $x \geq 3500$ gal

9. $p = -0.02q + 14$ (q in thousands of units)

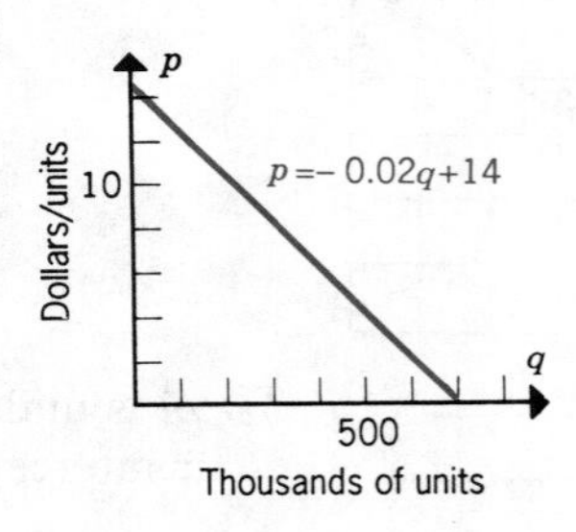

11. $p = -1.25q + 4.25$ (q in millions of units)

13. $p = 0.0025q + 75$ **15.** $p = 0.00625q + 0.75$ (q in thousands)

17. $p = 6, q = 50$ **19.** $p = 4.186, q = 3.143$ **21.** $p = 2.5, q = 5$

23. (a) Plan A: $S(x) = 500 + 0.03x$

Plan B: $S(x) = \begin{cases} 620, & x \leq 20{,}000 \\ 0.05x - 380, & x > 20{,}000 \end{cases}$

(b)

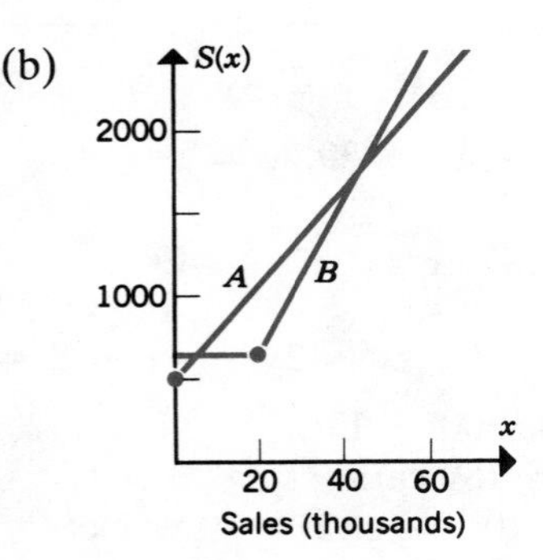

(c) Plan A is preferable to plan B for x satisfying the condition $4000 < x < 44{,}000$.

1.4 Exercises

1. vertex: (0, 1)
no x intercepts
y intercept: (0, 1)

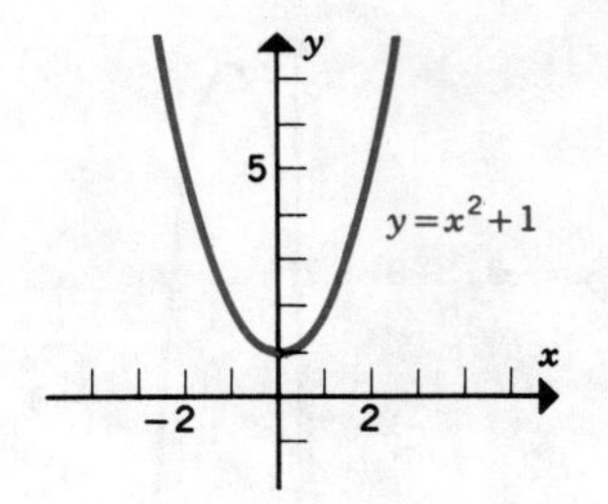

3. vertex: (0, 1)
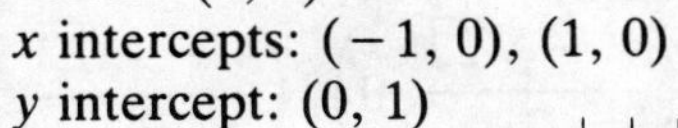
x intercepts: $(-1, 0)$, (1, 0)
y intercept: (0, 1)

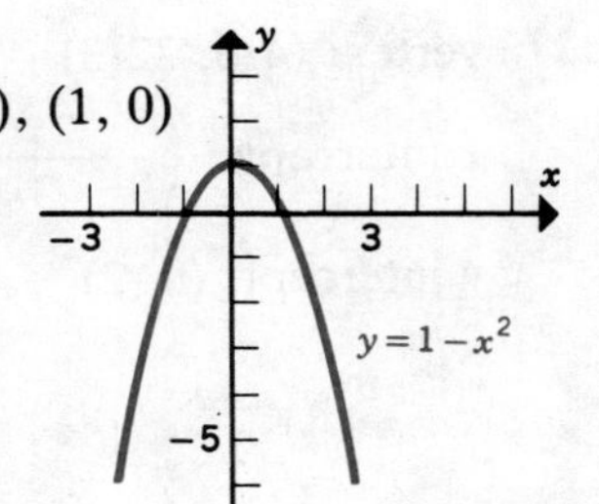

5. vertex: (0, 0)
x intercept: (0, 0)
y intercept: (0, 0)

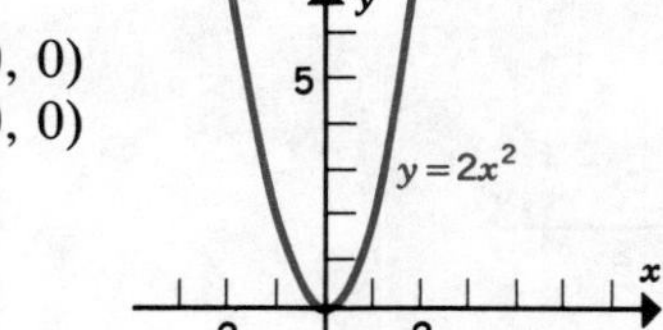

7. vertex: $(-2, 0)$
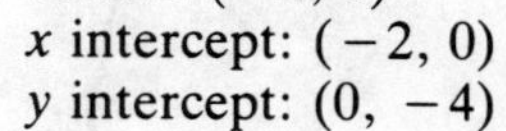
x intercept: $(-2, 0)$
y intercept: $(0, -4)$

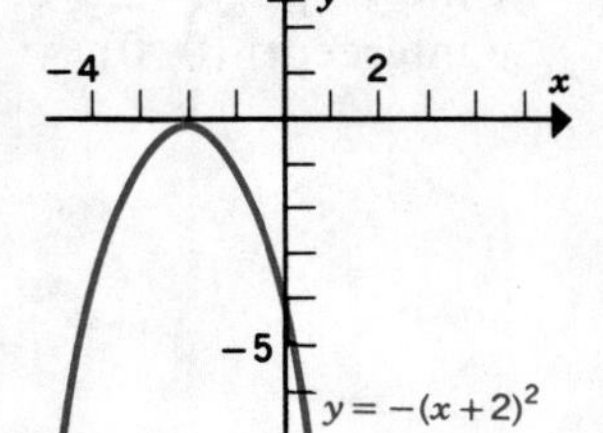

9. vertex: (1, 9)
x intercepts: $(-2, 0)$,
(4, 0)
y intercept: (0, 8)

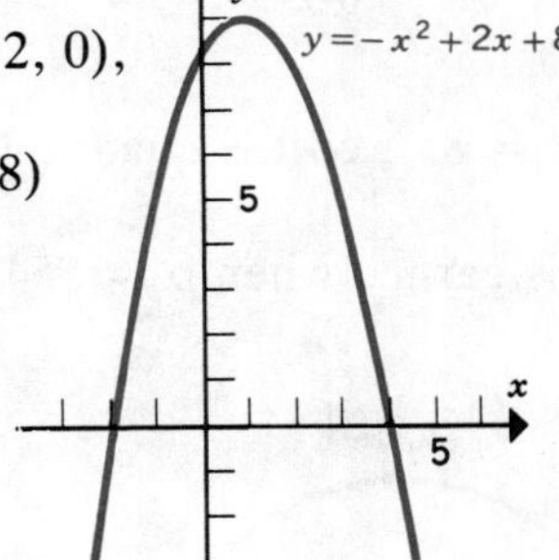

11. vertex: $(-3, 0)$
x intercept: $(-3, 0)$
y intercept: (0, 9)

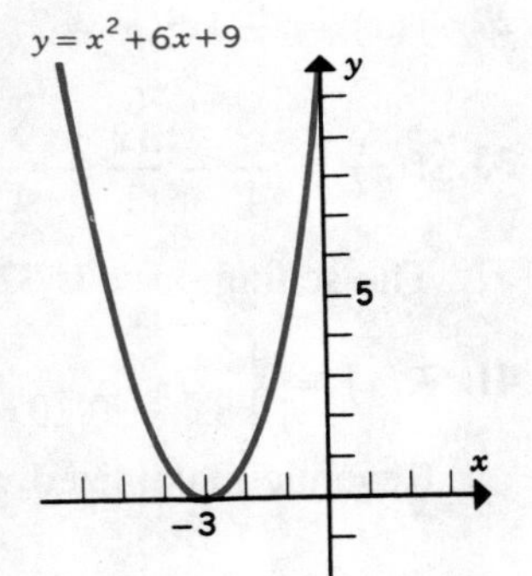

13. vertex: (0, 9)
x intercepts: $(-3, 0)$,
(3, 0)
y intercept: (0, 9)

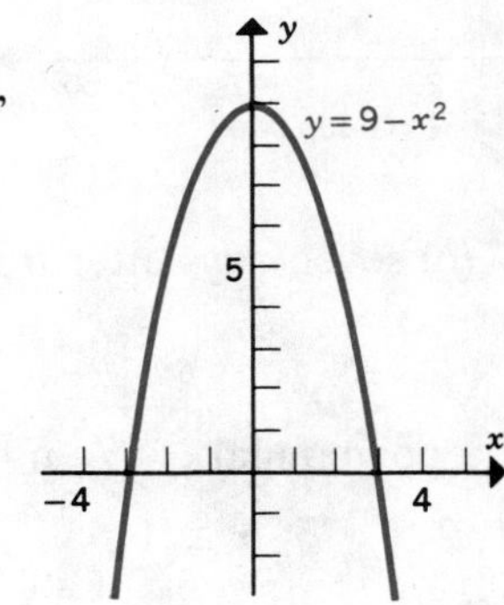

15. vertex: $(-3/2, -9/2)$
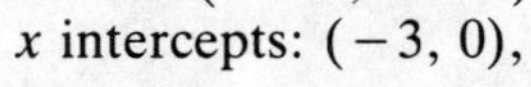
x intercepts: $(-3, 0)$,
(0, 0)
y intercept: (0, 0)

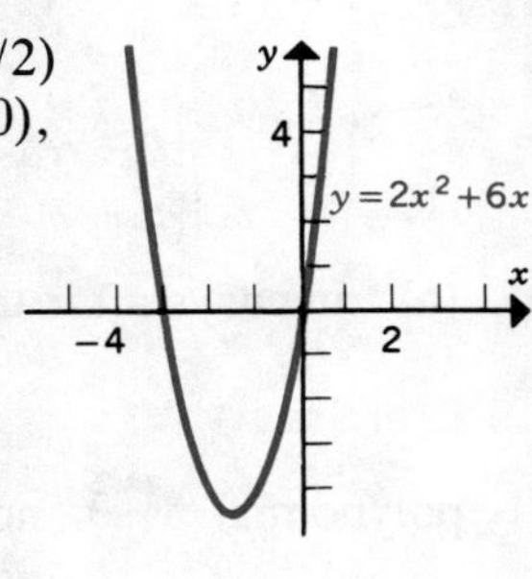

17. vertex: (4/3, 22/3)

x intercepts: $\left(\dfrac{4+\sqrt{22}}{3}, 0\right), \left(\dfrac{4-\sqrt{22}}{3}, 0\right)$

y intercept: (0, 2)

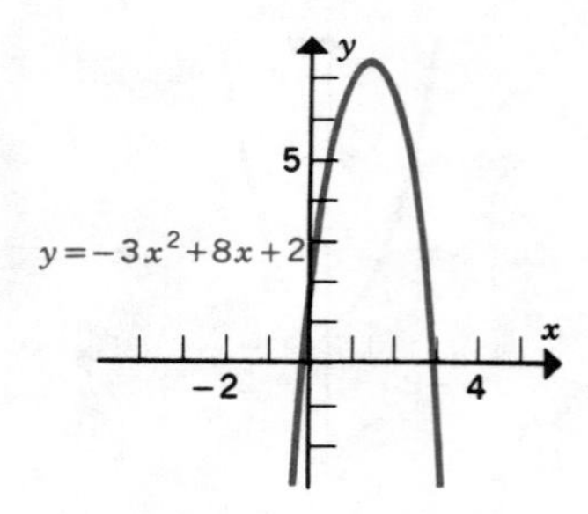

19. vertex: $(-\sqrt{2}, -2\sqrt{2})$
x intercepts: $(-2\sqrt{2}, 0)$, (0, 0)
y intercept: (0, 0)

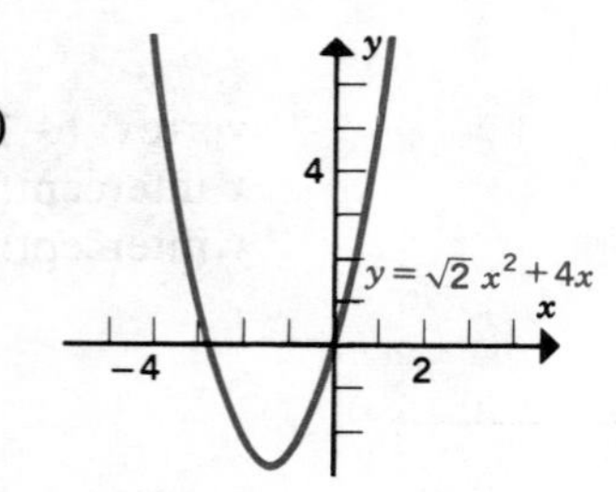

21. $a > 0, b < 0, c > 0$ **23.** $a < 0, b > 0, c < 0$ **25.** $a < 0, b = 0, c > 0$

27. $f(x) = -4x^2 + 4$ **29.** $f(x) = \dfrac{3x^2}{8} - 3x + 6$ **31.** $f(x) = -2x^2 - 8x - 5$

33. $f(x) = \dfrac{x^2}{4} - \dfrac{3x}{2} + \dfrac{9}{4}$ **35.** (a) $C = 2x^2$ (b) $C = 8x^2$; cost is quadrupled (c) $C = 18y^2$

37. The selling price is \$250. **39.** Maximum profit is earned when $p =$ \$54.00.

41. $R(x) = \begin{cases} 6x, & x \le 20 \\ 8x - 0.10x^2, & x > 20 \end{cases}$
Revenue maximized when $x = 40$.

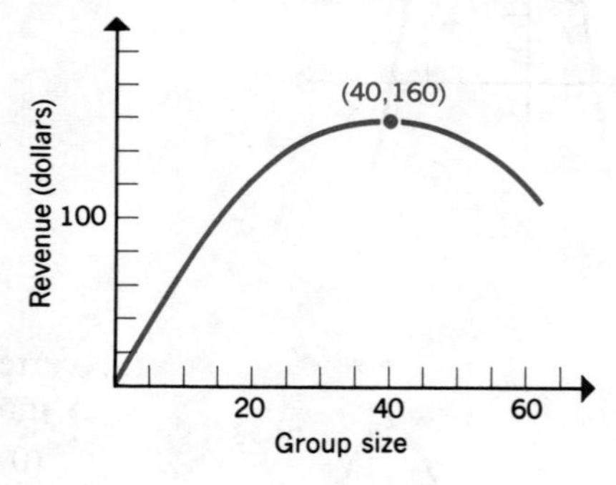

43. (a) four days after treatment (b) one day (c) seven days after treatment

1.5 Exercises

1. polynomial **3.** not a polynomial **5.** not a polynomial **7.** not a polynomial

9. all real numbers except $x = -1$ **11.** $x \neq 1, 2$ **13.** $x \geq 1$

15.

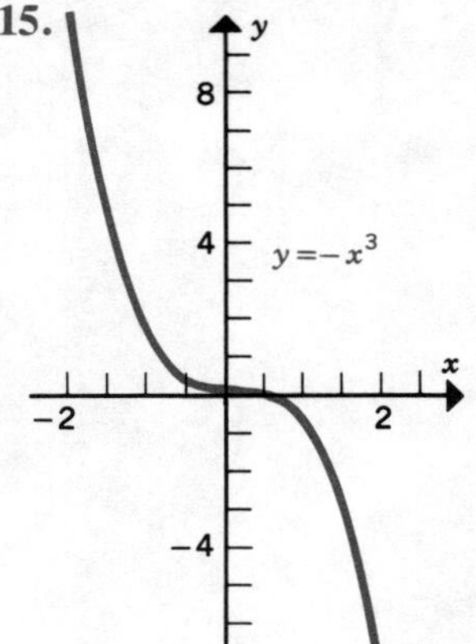

17.

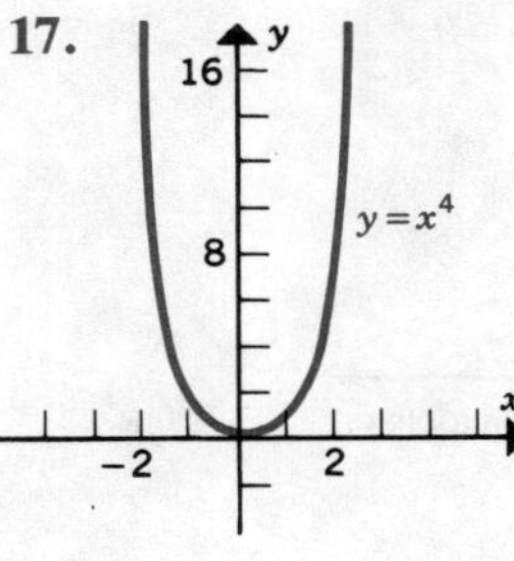

19.

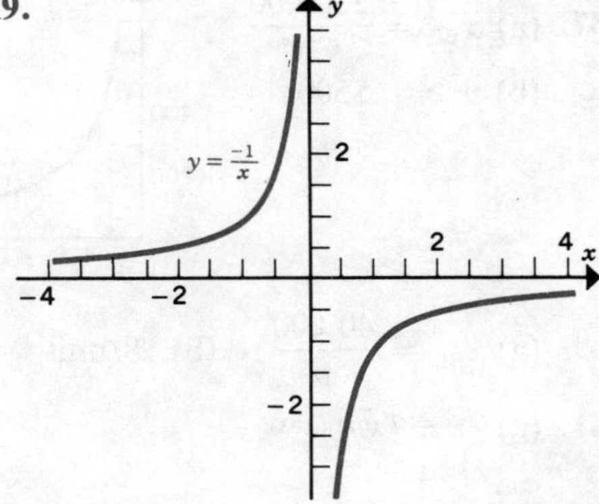

21.

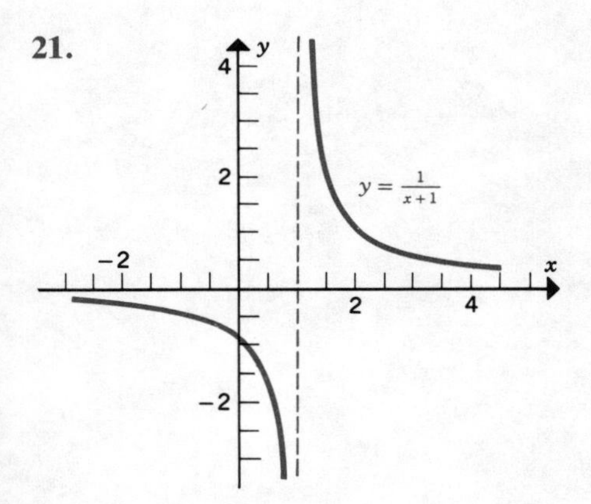

23.

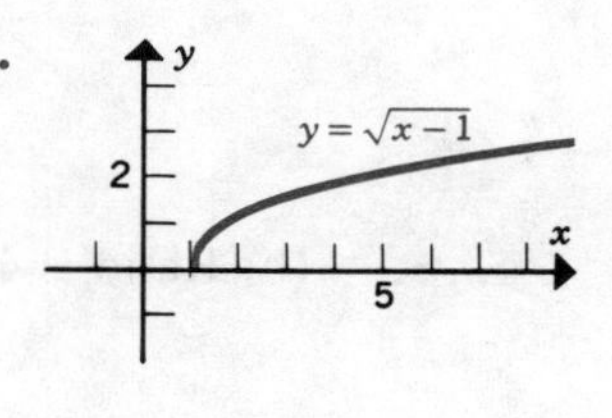

25.

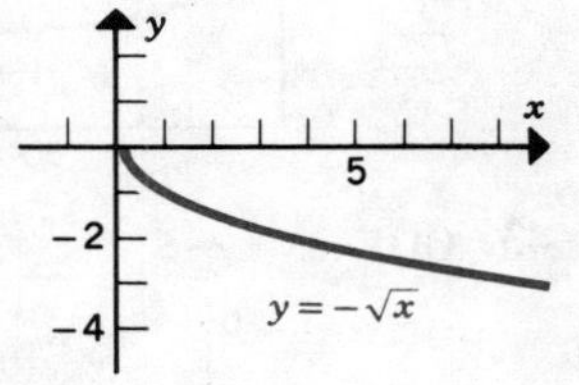

27.

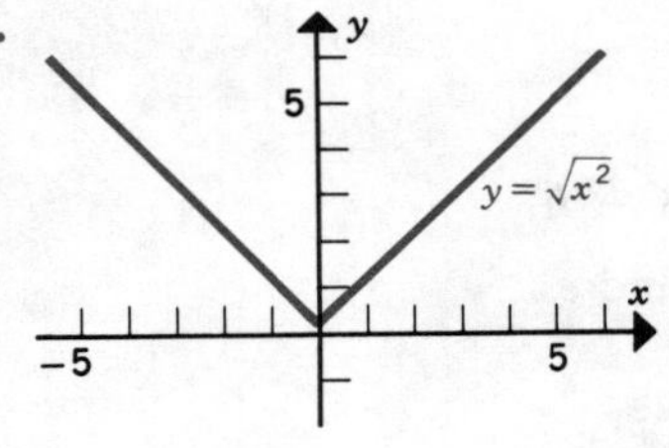

29.

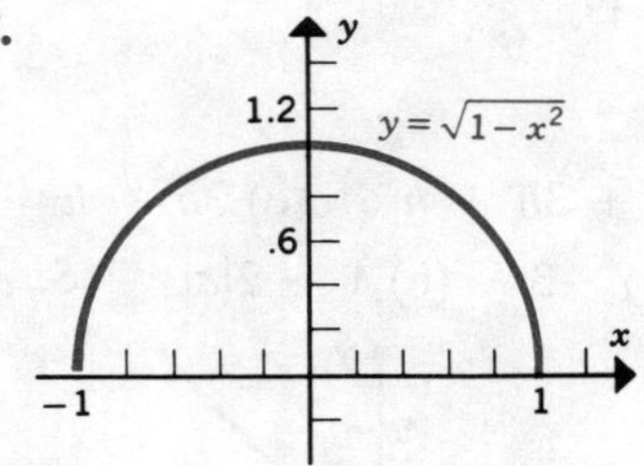

31.

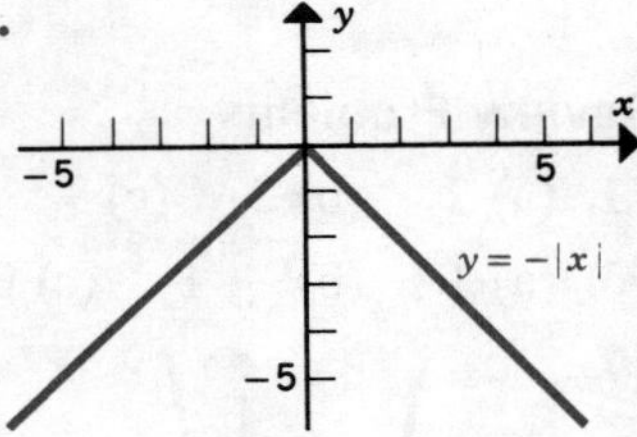

33.

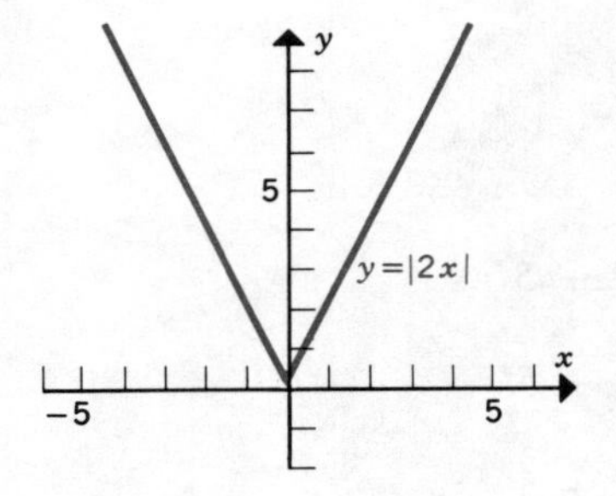

35. (1, 4)

37. (a) $x_{BE} = \dfrac{200{,}000}{CM}$
(b) $p \geq$ \$500

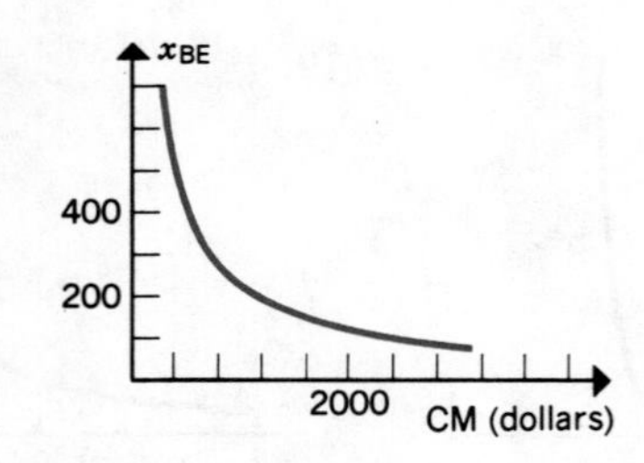

39. (a) $x_{BE} = \dfrac{40{,}000}{v}$ (b) \$8/unit

41. (a)

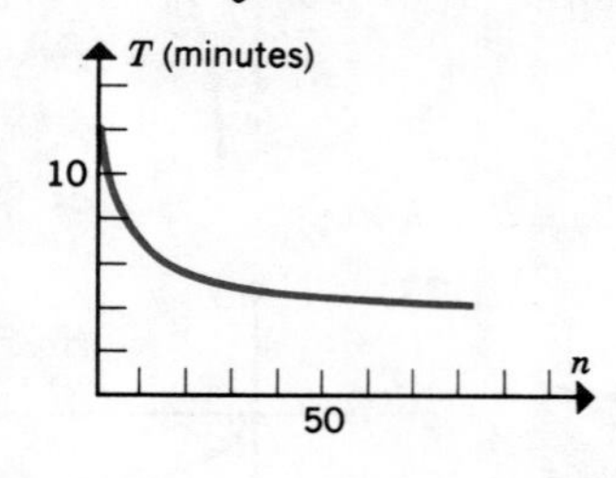

(b) $T(1) = 12$ minutes
$T(16) = 6$ minutes

43. (a)

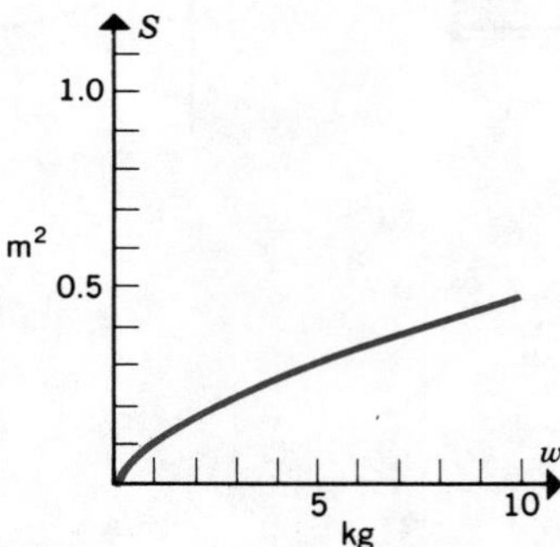

(b) Ratio is 4.

Review Problems

1. (a) 2 (b) 2 (c) $\frac{3}{4}$ (d) $2 + 3h + h^2$ (e) $3h + h^2$ (f) $3 + h$

3. (a) 1 (b) -1 (c) 0 (d) -3 (e) $1 - 2|a|$ **5.** a, d, and e

7.

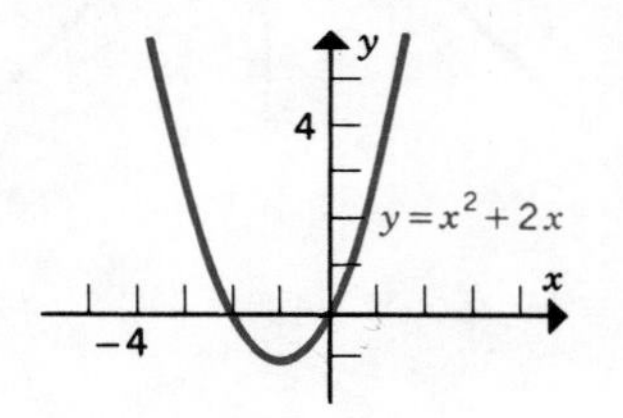

9.

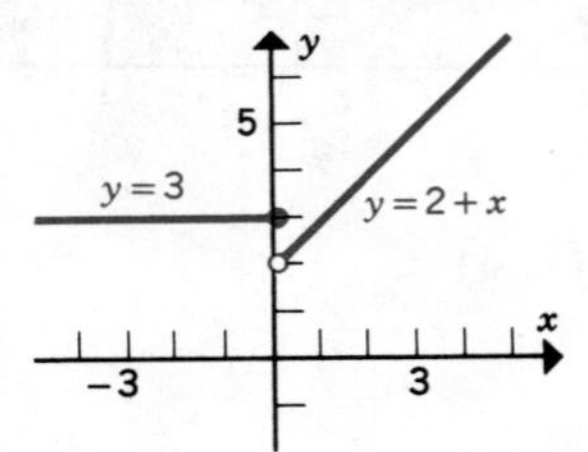

11.

Equation	Line
$3y - 4x = 0$	l_3
$2y + 4x = 8$	l_2
$6y - 2x = 12$	l_4
$3y + 5x = 10$	l_1

13. $y = -2x + 5$

15. $y = -3x + 5k$

17. $y = -3x + 5$

19. vertex: (2, 4)
x intercepts: (0, 0) and (4, 0)
y intercept: (0, 0)

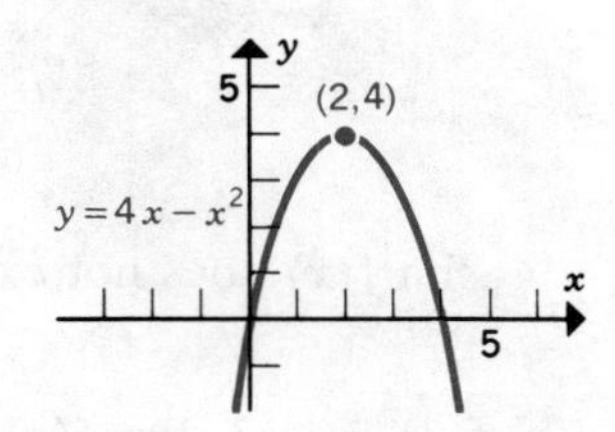

21. vertex: $(-1, 6)$
x intercepts: $(-1 + \sqrt{6}, 0)$ and $(-1 - \sqrt{6}, 0)$
y intercept: (0, 5)

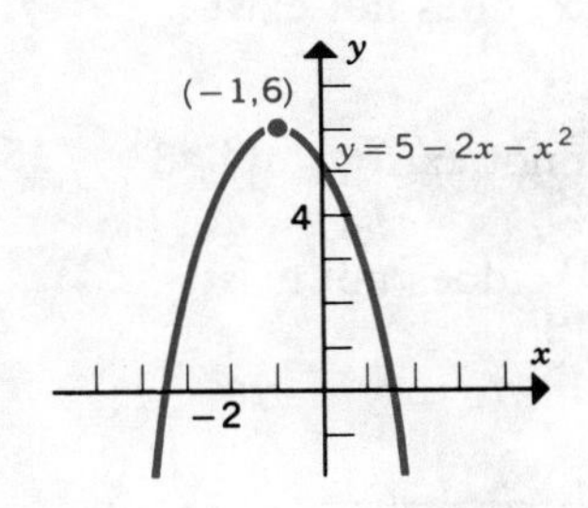

23. vertex: (2, 0)
x intercept: (2, 0)
y intercept: (0, 4)

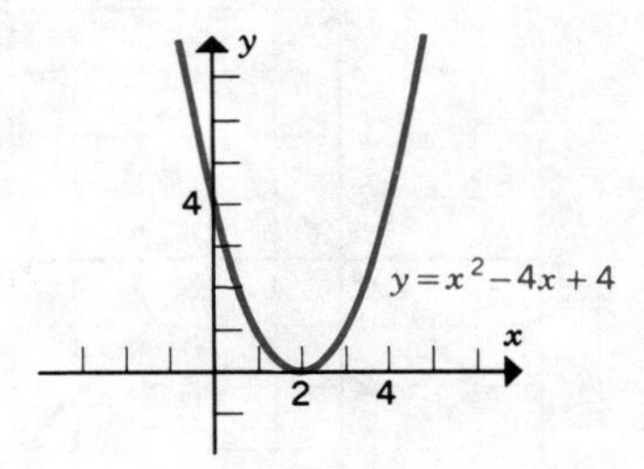

25. $f(x) = -\frac{5}{4}x^2 + 5x - 1$

27. $c_1 > c_2$, $a_2 > a_1$, $b_2 > b_1$

29.

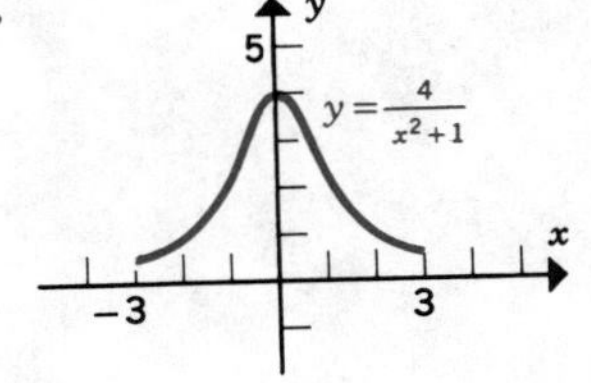

31.

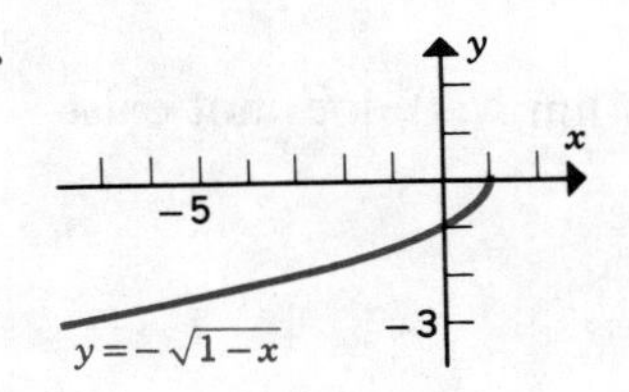

33.

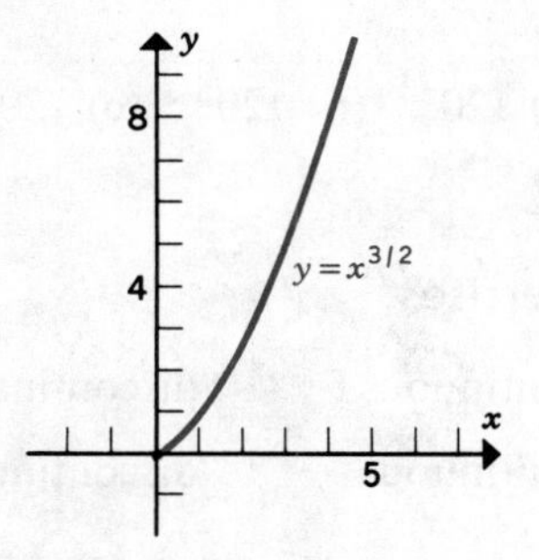

35. $V(t) = 1200 - 240t$

37. (a) $C(x) = \begin{cases} 0.10x, & 0 \le x \le 100 \\ 0.07x, & x > 100 \end{cases}$ (b)

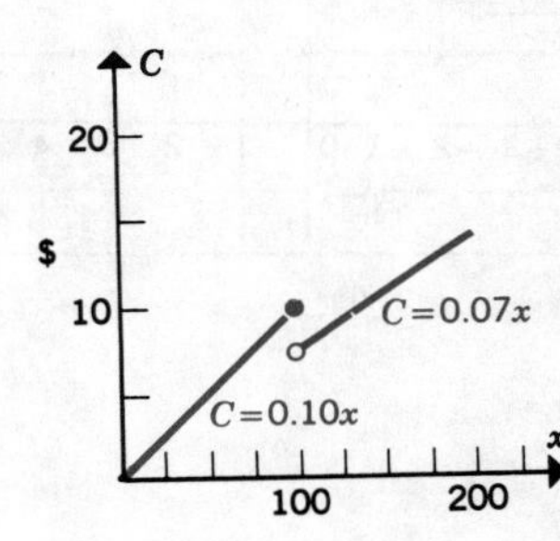

CHAPTER 2

2.1 Exercises

1. $f(2) = 3$; $\lim_{x\to 2} f(x) = 3$ **3.** $f(2) = 4$; $\lim_{x\to 2} f(x)$ does not exist

5. $f(2) = 2$; $\lim_{x\to 2} f(x)$ does not exist **7.** $f(2) = -3$; $\lim_{x\to 2} f(x) = -3$

9. -1 **11.** does not exist **13.** 3 **15.** does not exist

17. 6 **19.** 2 **21.** does not exist **23.** -1

25. does not exist **27.** -2 **29.** 1/4 **31.** 1/12

33. 1

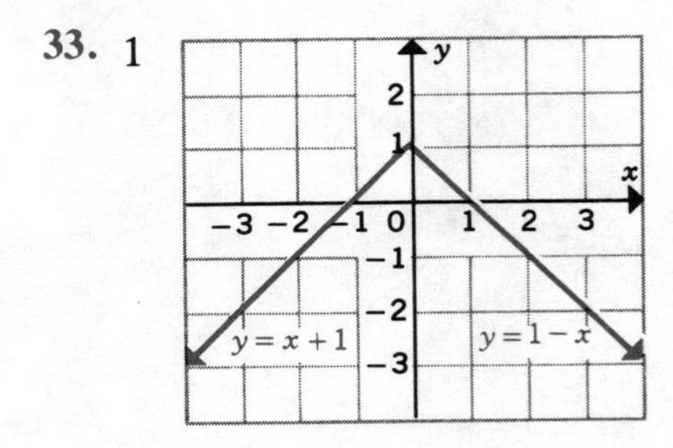

35. does not exist

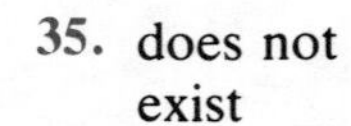

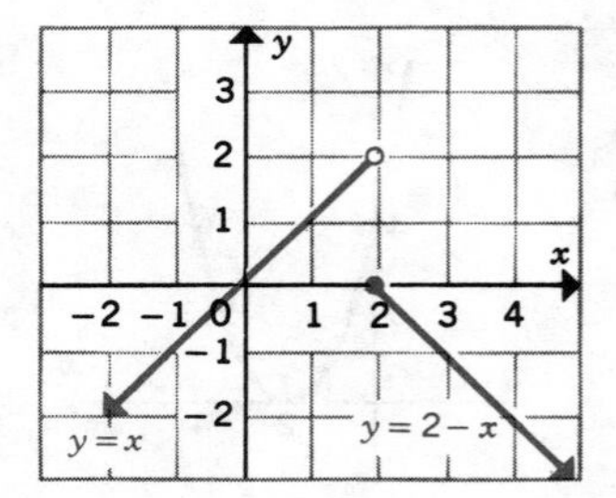

37. 2 **39.** 3 **41.** $\dfrac{a-1}{2a}$

43. (a) 120 (b) 120 (c) 120 (d) $\lim_{t\to 1} N(t)$ does not exist

2.2 Exercises

1. continuous **3.** discontinuous; B **5.** discontinuous; B

7. continuous **9.** discontinuous at $x = 1$; A **11.** discontinuous at $x = 0$; D

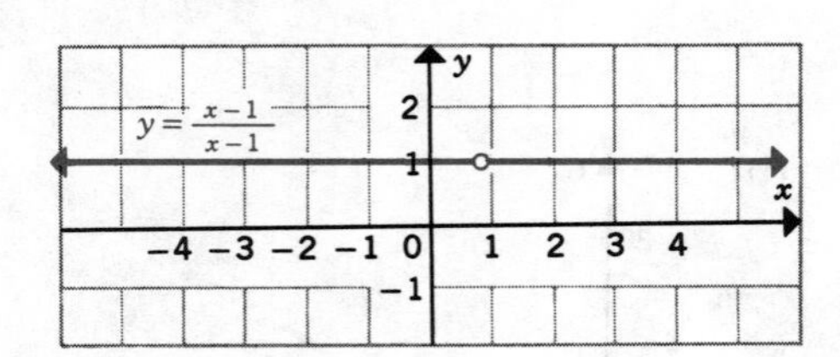

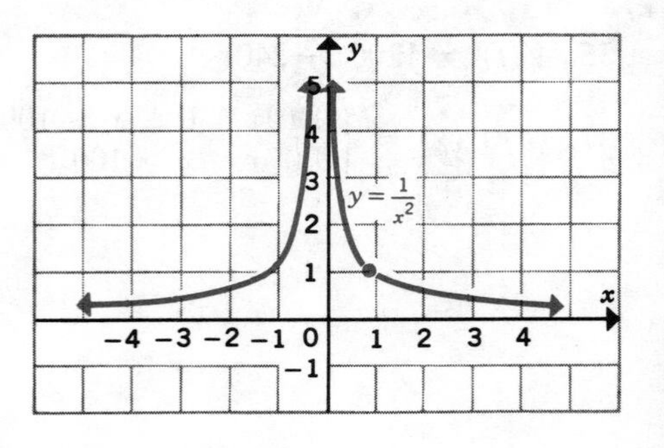

13. discontinuous at $x = 1$; B

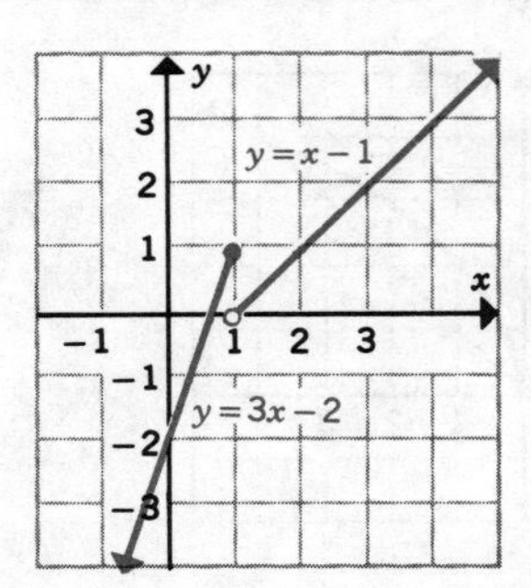

15. discontinuous at $x = 0$; C

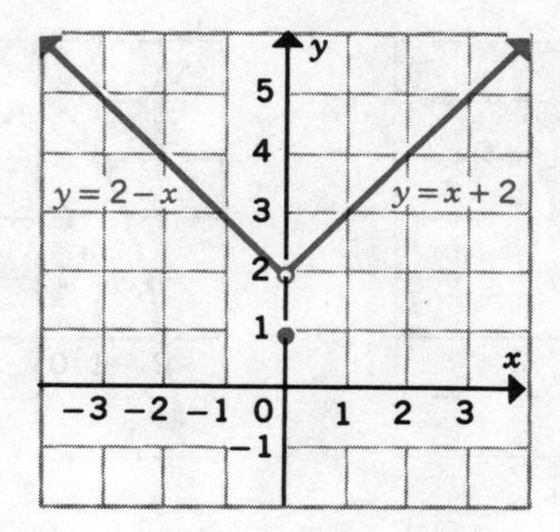

17. discontinuous at $x = 1$; B

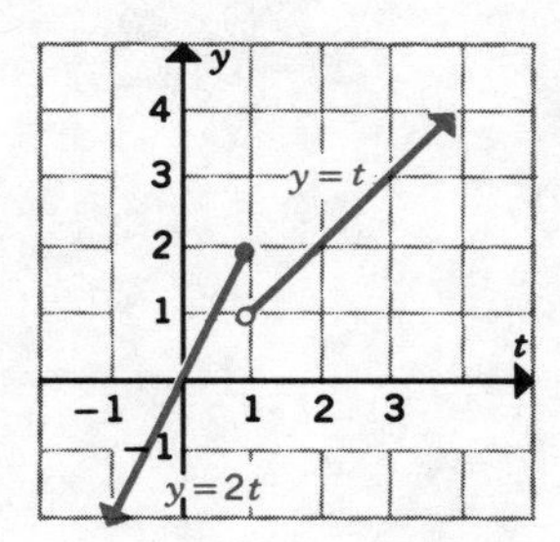

19. discontinuous at $x = 1, -1$; D

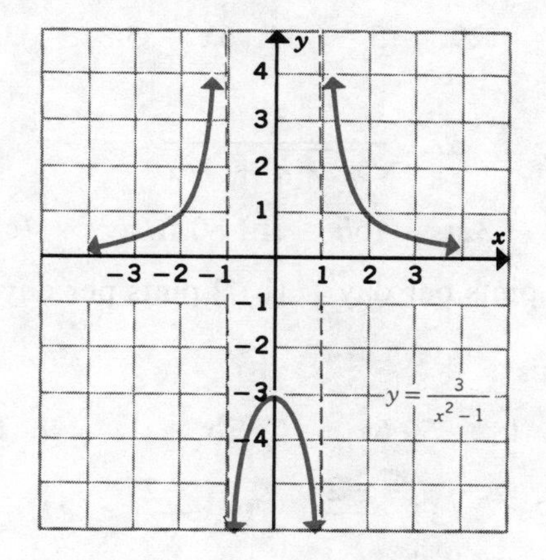

21. The salesman drove the car 100 miles the first day; there was an additional \$20 charge for keeping the car another day.

23. The tire blew out when $t = t_1$.

25. (a) $C(x) = \begin{cases} 4x, & x < 25 \\ 3.6x, & x \geq 25 \end{cases}$ (b) $C(x) = \begin{cases} 5x, & x < 20 \\ 4.5x, & x \geq 20 \end{cases}$

(c) $C(x) = \begin{cases} 4x + 8, & x < 25 \\ 3.6x + 10, & x \geq 25 \end{cases}$; $C(x) = \begin{cases} 5x + 8, & x < 20 \\ 4.5x + 10, & x \geq 20 \end{cases}$

27. $f(2) = 5/2$ **29.** $f(2) = 1/4$

2.3 Exercises

1. (a) \$79.1/year (b) \$105.9/year (c) \$124.6/year

3. 2 **5.** 3 **7.** −8 **9.** 3/4 **11.** −2/3

13. $m = 1$ **15.** $m = 2$ **17.** $m = 1$

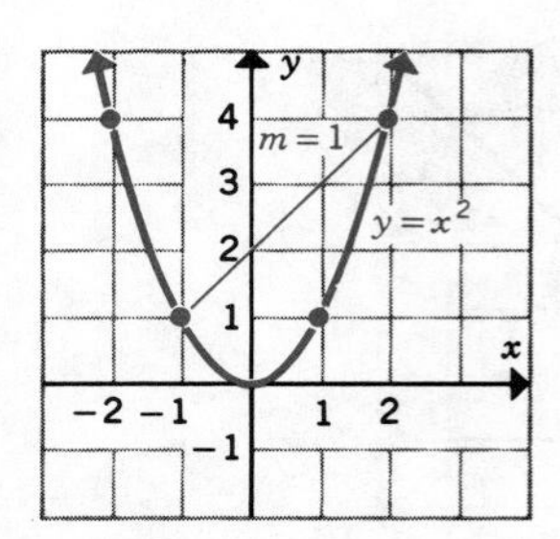

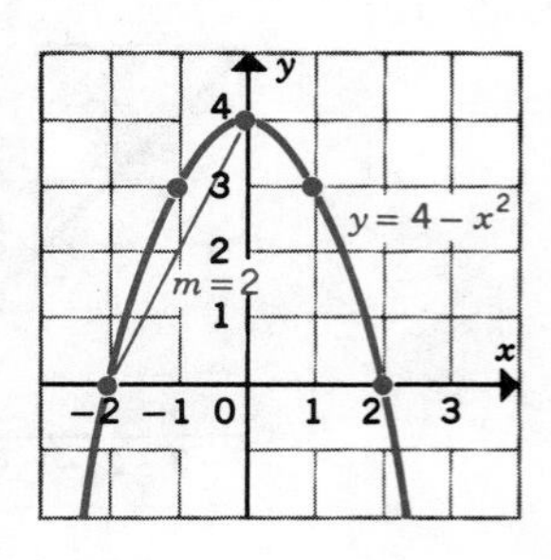

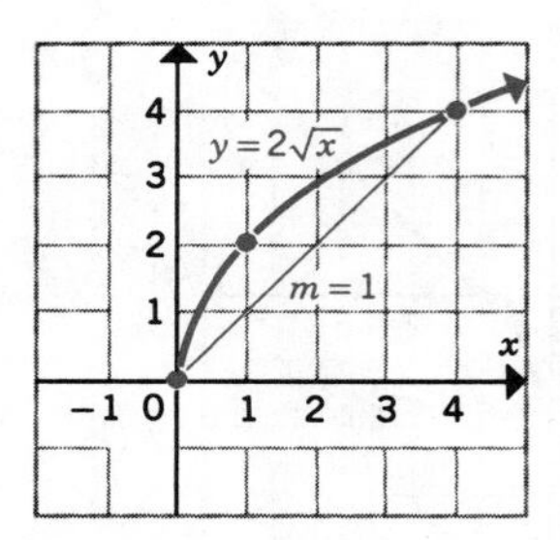

19. $3(h + 2)$ **21.** $h - 5$ **23.** $12 - 3h$ **25.** $\dfrac{2}{\sqrt{1 + h} + 1}$

27. $\dfrac{-3h}{h^2 + 1}$ **29.** -3 **31.** $6x + 3h$ **33.** $2x + 1 + h$

35. $\dfrac{2}{x^2 + xh}$ **37.** $\dfrac{3}{\sqrt{x + h} + \sqrt{x}}$

39. (a) $128 - 32t - 16h$ (b) 80 ft/sec; (c) -32 ft/sec

41. (a) -4 pints per day (b) 2 pints per day (c) $2t + h - 8$

2.4 Exercises

1. 2 **3.** 1 **5.** $6x$ **7.** $2x + 1$ **9.** $2x - 1$ **11.** $3 - 2x$ **13.** $6x^2$

15. $3x^2 - 2x$ **17.** $\dfrac{-2}{x^2}$ **19.** $\dfrac{-2}{x^3}$ **21.** $\dfrac{1}{\sqrt{2x}}$ **23.** $\dfrac{-1}{(x - 1)^2}$

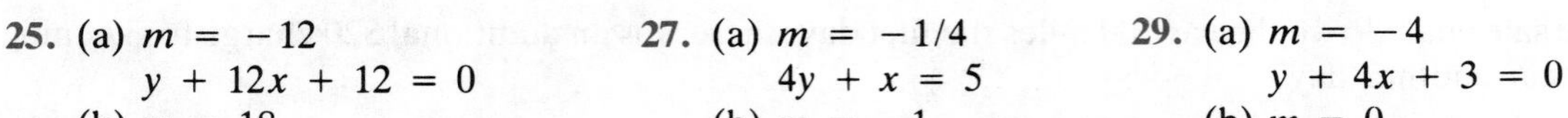

25. (a) $m = -12$
$y + 12x + 12 = 0$
(b) $m = 18$
$y = 18x - 27$

27. (a) $m = -1/4$
$4y + x = 5$
(b) $m = -1$
$y + x + 1 = 0$

29. (a) $m = -4$
$y + 4x + 3 = 0$
(b) $m = 0$
$y = 0$
(c) $m = 32$
$y = 32x - 48$

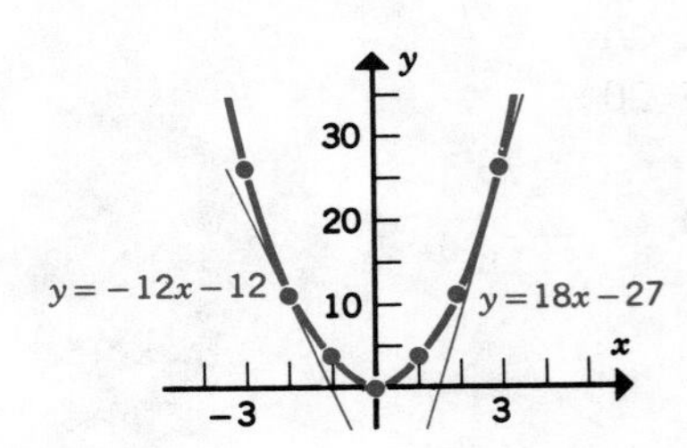

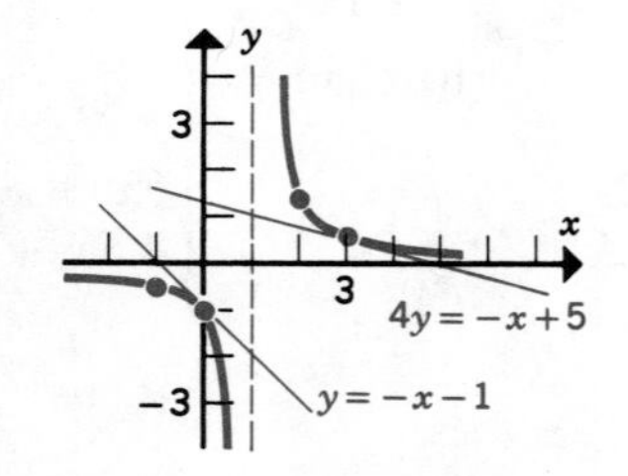

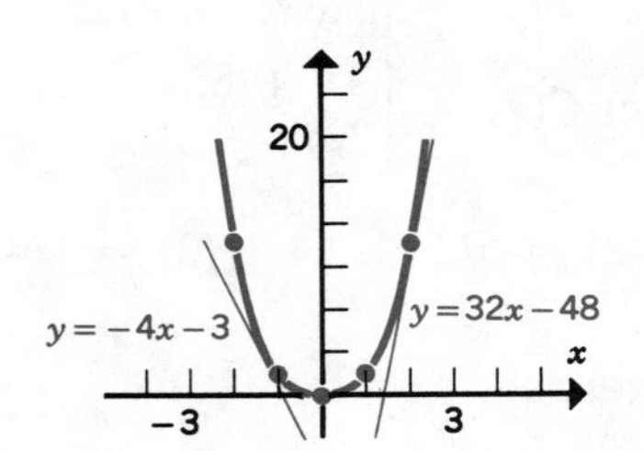

31. $(1, 2)$ **33.** $(3/2, 31/4)$ **35.** $(2, 10), (-2, -10)$ **37.** $f(3) = 5$
$f'(3) = 1$

39. $g(1) = 4$, $g'(1) = -2$ 41. $F(3) = 19/2$, $F'(3) = -5/2$ 43. $\frac{3\sqrt{x}}{2}$ 45. $\frac{a}{2\sqrt{ax}}$

Review Problems

1. $f(1) = 4$; $\lim_{x\to 1} f(x) = 4$ 3. $f(1) = 1$; $\lim_{x\to 1} f(x) = -2$

5. $f(1)$ is not defined; $\lim_{x\to 1} f(x) = 1$ 7. 7 9. 0 11. 2

13. The limit does not exist. 15. 1/2 17. 1 19. 0 21. discontinuous; D

23. continuous 25. discontinuous; B 27. $x = -2$; A 29. $x = 1$; B

31. $x = 0$; D 33. The balloon burst. 35. $f(1) = 1$ 37. 4 39. 2/3

41. $f'(x) = 3$; $m = 3$; $y = 3x$ 43. $f'(x) = 2x + 2$; $m = f'(-1) = 0$; $y = -1$

45. $f'(x) = 3x^2$; $m = f'(1) = 3$; $y = 3x - 1$

47. $f'(x) = \frac{-2}{x^2}$; $m = f'(-1) = -2$; $y = -2x - 4$

49. $f'(x) = 2x + 3$; $m = f'(0) = 3$; $y = 3x - 2$ 51. not differentiable at $x = 1, 3$.

53. not differentiable at $x = 0, 2, 4$. 55. $f(2) = 3$; $f'(2) = 1/2$

CHAPTER 3

3.1 Exercises

1. 0 3. $4x^3 - 6x$ 5. 0 7. $\frac{-10}{x^3}$ 9. $\frac{-8}{3x^3}$

11. $21x^2 + \frac{12}{x^4}$ 13. $\frac{4x^{1/3}}{3}$ 15. $5 + \frac{8}{x^3}$

17. $2x - 1 + \frac{1}{x^2}$ 19. $\frac{x^{3/2} - x^{1/2}}{2x^2}$ 21. $x^n - x^{n-1}$

23. $y + 3x = 3$ 25. $y = 9x - 6$ 27. $2y - 3x = 4$ 29. $4y = 11x - 8$

31. $(0, 1), (-2, 7)$ 33. $(-1, 3)$ 35. $(-1, 0), (1, 0)$ 37. $(1, 4)$

39. $F(3) = 10$, $F'(3) = 2$ 41. $F(1) = 5$, $F'(1) = 0$

43. $F(2) = 10/3$, $F'(2) = 13/3$

45. $F(2) = 0$, $F'(2) = 2$ 47. $v(t) = 2t - 2$, $v(0) = -2$ ft/sec, $v(1) = 0$ ft/sec, $v(2) = 2$ ft/sec; $a(t) = 2$ ft/sec^2

49. $v(t) = 320 - 32t$, $v(5) = 160$ ft/sec, $v(10) = 0$ ft/sec, $v(20) = -320$ ft/sec; $a(t) = -32$ ft/sec^2

51. $v(t) = 48t - 3t^2$, $v(4) = 144$ ft/sec, $v(8) = 192$ ft/sec, $v(12) = 144$ ft/sec; $a(t) = 48 - 6t$

53. (a) $960 - 32t$, $a(t) = -32$ ft/sec^2 (b) 30 sec (c) 60 sec **55.** $N'(4) = 11$

57. (a) 3 in./week (b) 1 in./week

3.2 Exercises

1. $6x + 10$ **3.** $6x^2 + 18x + 8$ **5.** $5x^4 - 4x^3 + 6x^2 - 4x + 3$ **7.** $\dfrac{3x + 1}{2x^{1/2}}$

9. 1 **11.** $2x + 1$ **13.** $\dfrac{-3}{(x - 1)^2}$ **15.** $\dfrac{x^2 + 4x + 1}{(x + 2)^2}$

17. $2nx^{2n-1} + nx^{n-1}$ **19.** $\dfrac{\sqrt{x} - 2}{2(\sqrt{x} - 1)^2}$ **21.** $y = 13x - 10$ **23.** $y = 5x - 8$

25. $y = \dfrac{x}{2}$ **27.** $y = -2$ **29.** $(2, 4)$ **31.** $(1, -1)$, $(-1/9, -13/243)$

33. $(2, 0)$, $(4, 2)$ **35.** $(1, 8)$ **37.** $F(3) = 15$, $F'(3) = 8$

39. $F(1) = 12$, $F'(1) = 2$ **41.** $F(2) = -2/3$, $F'(2) = 7/6$

43. $F(-2) = 1/2$, $F'(-2) = -5/4$ **45.** (a) 12,500 bushels/year (b) 13,500 bushels/year **47.** 800 ounces/year

49. (a) \$20 million/year (b) \$8.92 million/year (c) $-$\$1.38 million/year

3.3 Exercises

1. $8x(x^2 + 3)^3$ **3.** $(36x - 126)(x^2 - 7x + 1)^5$ **5.** $\dfrac{-8(2x + 1)^7}{x^9}$

7. $\dfrac{2t^3 + 1}{(t^4 + 2t - 1)^{1/2}}$ **9.** $\dfrac{-4x}{(x^2 + 1)^3}$ **11.** $\dfrac{-2x}{(x^2 + 1)^{3/2}}$

13. $(x^4 + 5)^5(25x^4 + 48x^3 + 5)$ **15.** $\dfrac{1 - 2x}{(2x + 1)^3}$ **17.** $\dfrac{-6(x + 1)^2}{(x - 1)^4}$

19. $\dfrac{(2x^2 + 3)^3(34x^2 + 3)}{2x^{1/2}}$ **21.** $y = -4x + 5$ **23.** $y = 0$

25. $y = -84x - 163$ **27.** $(-1, 1)$, $(-3, 1)$ **29.** $(-1/8, 3/2)$

31. $g(u(x)) = 9x^2 + 1$, $g(u(1)) = 10$; $u(g(x)) = 3x^2 + 3$, $u(g(1)) = 6$

33. $g(u(x)) = \dfrac{3}{x^2} + 1$, $g(u(1)) = 4$; $u(g(x)) = \dfrac{1}{(3x + 1)^2}$, $u(g(1)) = 1/16$

35. $g(u(x)) = 4x^2 + 20x + 27$, $g(u(1)) = 51$; $u(g(x)) = 2x^2 + 4x + 10$, $u(g(1)) = 16$ 37. $g(u(x)) = \dfrac{1}{x^2 - 2x + 2}$, $g(u(1)) = 1$; $u(g(x)) = -\dfrac{x^2}{x^2 + 1}$, $u(g(1)) = -\dfrac{1}{2}$ 39. $f(2) = 4$, $f(-2) = 1$

41. $(12x - 24)(x^2 - 4x + 5)^5$ 43. $\dfrac{-(3x^2 + 2x + 1)}{(x^3 + x^2 + x)^2}$

45. $\dfrac{-3x}{(1 - 3x^2)^{1/2}}$

3.4 Exercises

1. (a) $200 - 2x$ (b) \$140/unit (c) \$139/unit
3. (a) $16q - 0.001q^2 - 5000$ (b) $16 - 0.002q$ (c) \$14/unit (d) $P'(1000) > 0$, so increasing production will yield higher profits.
5. (a) $500q - 2.01q^2 - 10000$ (b) $500 - 4.02q$ (c) $P'(100) =$ \$98/unit (d) Yes. The company earns \$98 for each additional unit sold.
7. \$6 9. (a) $1 - \dfrac{400}{x^2}$ (b) −\$3/carton (c) −\$3/carton (d) −\$2.64/carton

3.5 Exercises

1. $f'(x) = 4$, $f''(x) = 0$ 3. $f''(x) = 36x - 18$ 5. $f''(x) = \dfrac{48}{x^4}$

7. $f''(x) = \dfrac{-5}{4x^{3/2}}$ 9. $y''(x) = 216(1 - 6x)$

11. $f''(t) = \dfrac{-1}{4(1 - t)^{3/2}}$ 13. $y''(x) = \dfrac{-4}{(1 + x)^3}$

15. $96x^6(x^4 + 1) + 36x^2(x^4 + 1)^2 - \dfrac{2}{x^3}$ 17. $\dfrac{1}{(x^2 + 1)^{3/2}}$ 19. $\dfrac{3}{4x^{5/2}}$

21. 0 23. $uv''(x) + 2u'(x)v'(x) + v(x)u''(x)$

25. $n(u(x))^{n-2}[u(x)u''(x) + (n - 1)(u'(x))^2]$

3.6 Exercises

1. $-5/2$ 3. (a) $\dfrac{7x^2 + 4x^3 - 6x}{(3 + 2x)^2}$ (b) $\dfrac{3x^2 - 2y - 2x}{3 + 2x}$

5. (a) $\dfrac{\pm x}{(4 - x^2)^{1/2}}$ (b) $\dfrac{-x}{y}$ 7. $\dfrac{3y + 2x}{2y - 3x}$

9. $\dfrac{-6xy^2}{5y^4 + 6x^2y}$ 11. $\dfrac{6x - y}{x}$ 13. $\dfrac{6y + 5}{1 - 6x}$

15. $8\sqrt{y}\sqrt{x + \sqrt{y}} - 2\sqrt{y}$ 17. $m = 3/2, 2y = 3x - 5$ 19. $m = -3/4, 4y = -3x + 25$

21. $m = -2, y = -2x + 4$ 23. $\dfrac{dy}{dt} = 3x^2\dfrac{dx}{dt}$ 25. $\dfrac{dy}{dt} = \dfrac{x}{\sqrt{x^2 + 1}}\dfrac{dx}{dt}$

27. $\dfrac{dy}{dt} = -\dfrac{2xy + y^2}{x^2 + 2xy}\dfrac{dx}{dt}$ 29. $\dfrac{dy}{dy} = -\dfrac{4x\sqrt{y} + 2y}{x}\dfrac{dx}{dt}$

31. $\dfrac{dy}{dt} = -\dfrac{y}{x}\dfrac{dx}{dt}$ 33. $\dfrac{dy}{dt} = \dfrac{1 - 6x}{6y}\dfrac{dx}{dt}$ 35. $\dfrac{dR}{dt} = \$1000\text{/week}$

37. $\dfrac{dV}{dt} = -50$ in.^{3}/minute 39. (a) $\dfrac{dH}{dt} = \dfrac{1}{100}$ ft/sec; (b) $\dfrac{dH}{dt} = \dfrac{1}{200}$ ft/sec

41. $\dfrac{dA}{dt} = 80\pi$ cm^2/sec 43. (a) $\dfrac{dI}{dt} = -8$ units/hour; (b) $\dfrac{dI}{dt} = -1$ unit/hour

45. $\dfrac{dC}{dt} = \dfrac{3872}{3}$ dollars/week 47. (a) -0.0034 cm/sec (b) 0.0034 cm/sec

Review Problems

1. 0 3. $\dfrac{-2}{x^2}$ 5. $3x^2 + 2x + 5$ 7. $\dfrac{1}{\sqrt{2x}}$ 9. $\dfrac{3}{2\sqrt{3x}}$

11. $(3x^2 + 2x)(x^4 + 1) + 4x^3(x^3 + x^2 + 1)$ 13. $2x(x^2 + 1)^2(4x^2 + 1)$ 15. $1 - \dfrac{2}{x^3}$

17. $\dfrac{2}{3(2x)^{2/3}} - \dfrac{8}{x^3}$ 19. $x(x + 1)^{1/2}$ 21. $24x - 12$ 23. $\dfrac{6}{t^3} - \dfrac{1}{(8t^3)^{1/2}}$

25. $\dfrac{2}{x^3}$ 27. $m = 21, y = 21x - 39$ 29. $m = 1/4, 4y = x + 16$ 31. $(-1/2, -13/4)$

33. no points 35. $f(1) = 3, f'(1) = -1, g(1) = 4, g'(1) = 2$ 37. $G(1) = 12; G'(1) = 2$

39. $K(1) = 49; K'(1) = 14$ 41. $\dfrac{dy}{dx} = \dfrac{-y^2 - 2xy}{x^2 + 2xy}$ 43. $\dfrac{dy}{dx} = \dfrac{\sqrt{y} - 6\sqrt{xy}}{\sqrt{x}}$

45. (a) $V'(t) = -8000(5 - t)$ (b) $V'(0) = -40{,}000$ dollars/year;
(c) $V'(2) = -24{,}000$ dollars/year

47. (a) $N'(t) = 30 - 20t$ (b) $N'(1) = 10$ hundred or 1000 people per month
(c) $N'(3) = -30$ hundred or -3000 people per month. The negative sign indicates that the number of people who are ill with the flu is decreasing.

49. (a) $R'(q) = \dfrac{40{,}000 - 10{,}000q^2}{(4 + q^2)^2}$ (b) $R'(1) = 1200$ dollars per machine

(c) $P(q) = \dfrac{10{,}000q}{4 + q^2} - 500q$

51. \$374.88/week

CHAPTER 4

4.1 Exercises

1. $(-1, -1)$, relative minimum

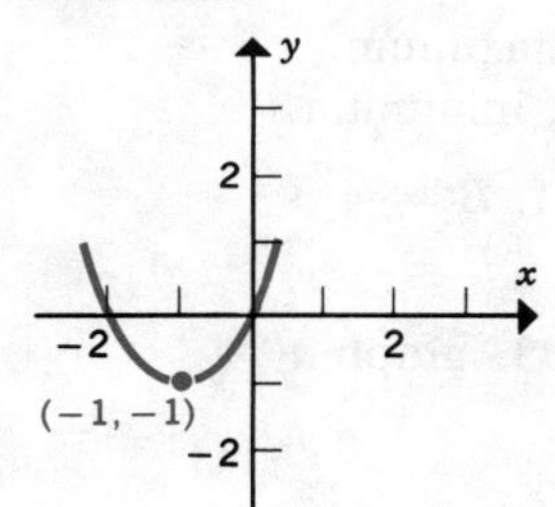

3. $(1, -5)$, relative minimum

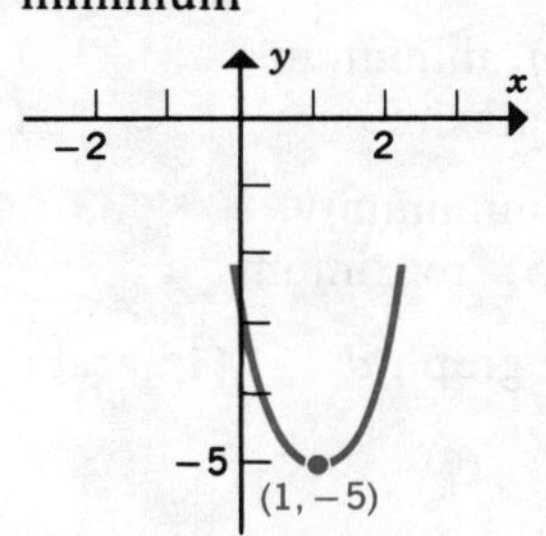

5. $(-5/2, -29/4)$, relative minimum

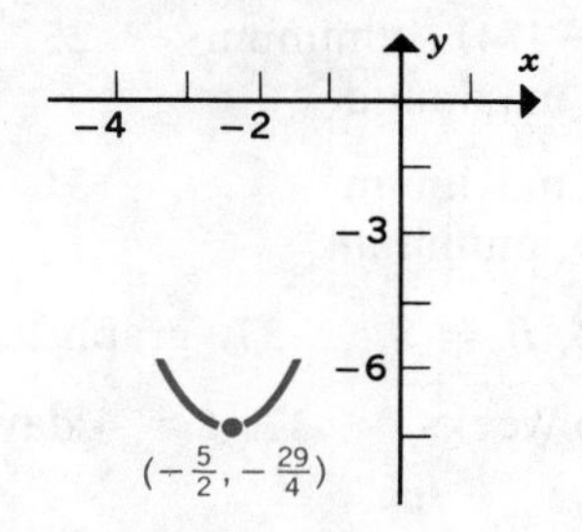

7. $(-2, 3)$, relative maximum
$(0, -1)$, relative minimum

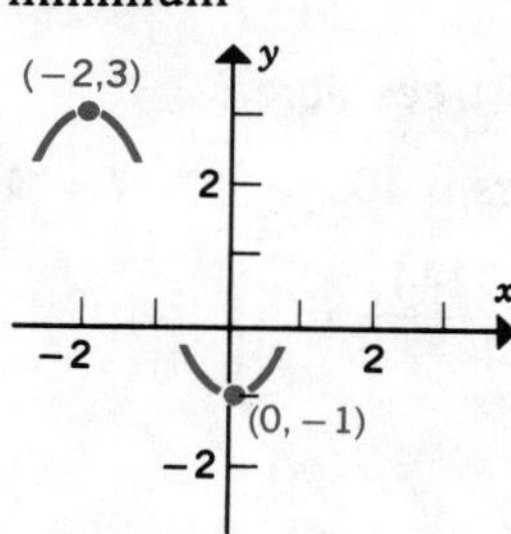

9. $(-2, 94)$, relative maximum
$(3, -31)$, relative minimum

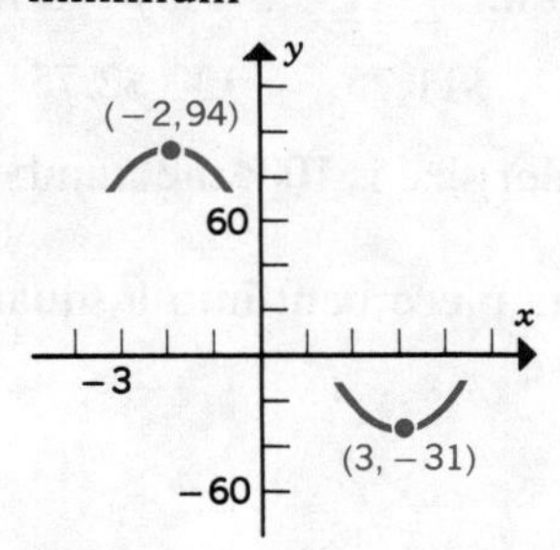

11. $(-1, 0)$, relative minimum
$(0, 1)$, relative maximum
$(1, 0)$, relative minimum

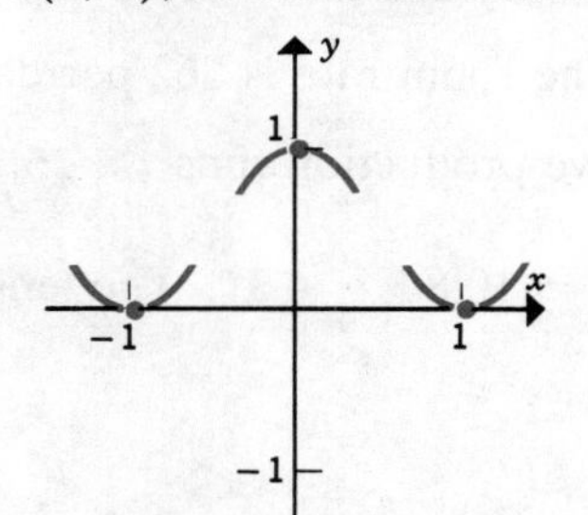

13. $(1, 0)$, relative maximum
$(3, -4)$, relative minimum

15. $(0, 1)$, relative minimum

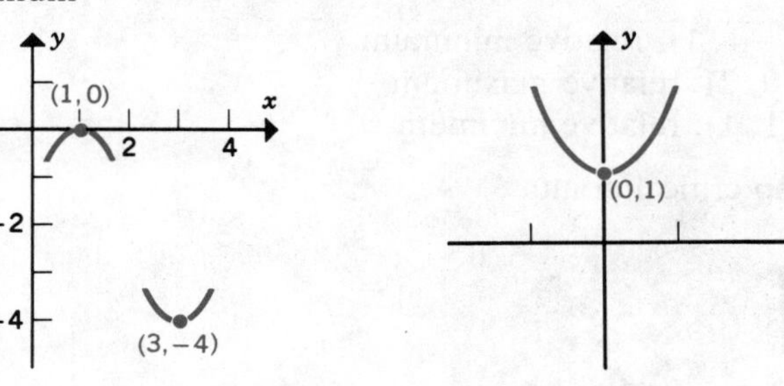

17. $(0, 0)$, relative maximum
$(4, 8)$, relative minimum

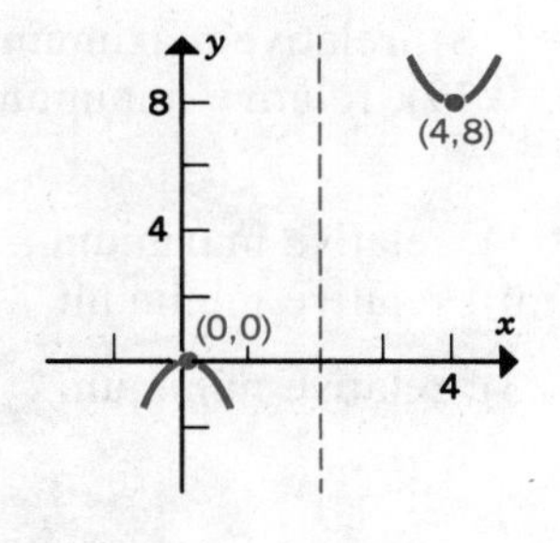

19. $(-1, -1/2)$, relative minimum
$(1, 1/2)$, relative maximum

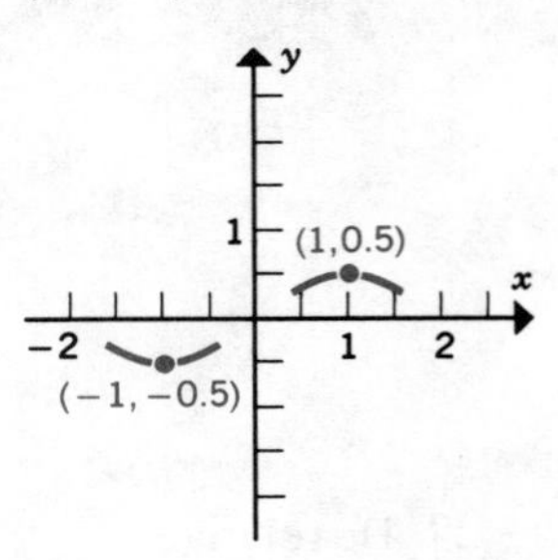

21. $(-1, 3)$, relative maximum
$(0, 1)$, neither
$(1, -1)$, relative minimum

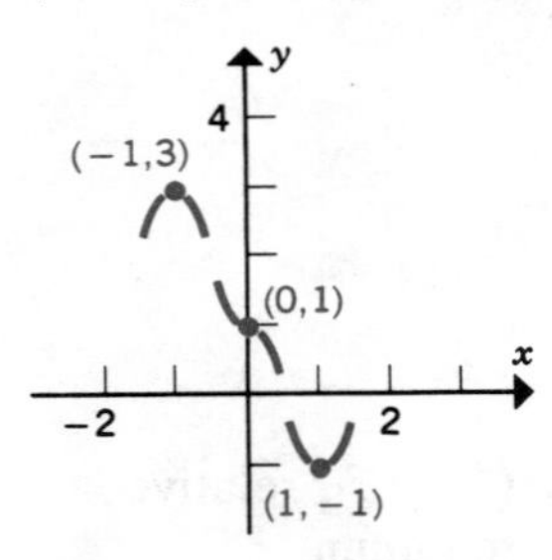

23. $(1/2, -1/4)$, minimum
$(3, 6)$, maximum

25. $(-1, 0)$, minimum
$(1, 2)$, maximum

27. $(1, 1)$, maximum
$(3, 1/3)$, minimum

29. $(4, 4)$, maximum
$(16, 0)$, minimum

31. $(1, 1)$, minimum
$(5, 49/5)$, maximum

33. $A = -1, B = 4$

35. $A = 2, B = 8$ **37.** graph *b* **39.** graph *d* **41.** graph *d* **43.** graph *a*

45. $t = 16$ weeks **47.** $t = 4$ days

4.2 Exercises

1. 3 by 3 ft **3.** 39 and 39 **5.** 16 and 16 **7.** 60 by 30 ft

9. 100 by 200 ft **11.** 20 by 20 by 15 in. **13.** 5/3 by 5/3 in.

15. The room rate is \$62 per day. **17.** \$11.75 **19.** \$2.75 **21.** 65 trees/acre

23. five production runs **25.** The order size is 100; the number of orders is 10. **27.** $I = 12\%$

29. $I = 10.8\%$ **31.** The length of the piece bent into a square equals $\dfrac{144}{4 + \pi}$ in.

4.3 Exercises

1. $(-1, 2)$, relative minimum

3. $(0, 6)$, relative maximum
$(1, 5)$, relative minimum

5. $(-1, 8)$, relative maximum
$(2, -19)$, relative minimum

7. $(-1, 1)$, relative minimum
$(0, 2)$, relative maximum
$(1, 1)$, relative minimum

9. $(0, 1)$, relative maximum
$(1, 0)$, relative minimum

11. no critical points

13. $(4, 3)$, relative minimum

15. (3, −3) is an inflection point

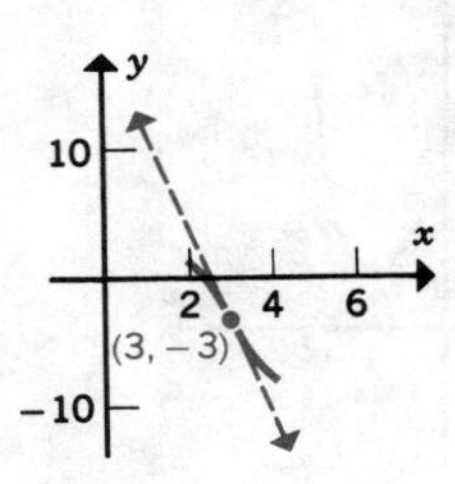

17. (0, 4) is not an inflection point

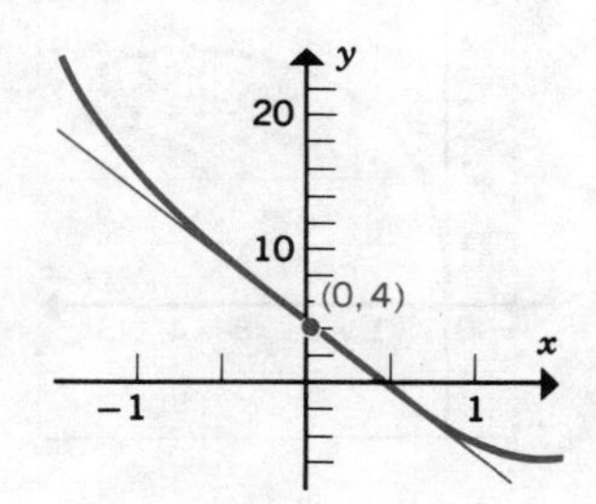

19.

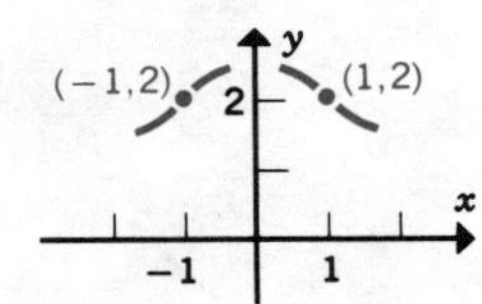

21.

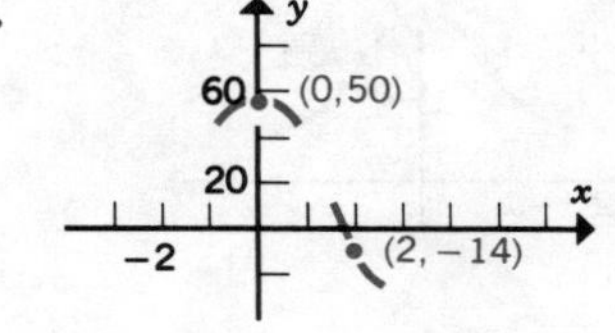

23. graph *e** **25.** graph *c* **27.** graph *c* **29.** graph *d* **31.** graph *a*

33. The headline indicates that $P''(t)$ has become negative, but $P'(t)$ remains positive. So the curve takes on the shape / .

4.4 Exercises

1.

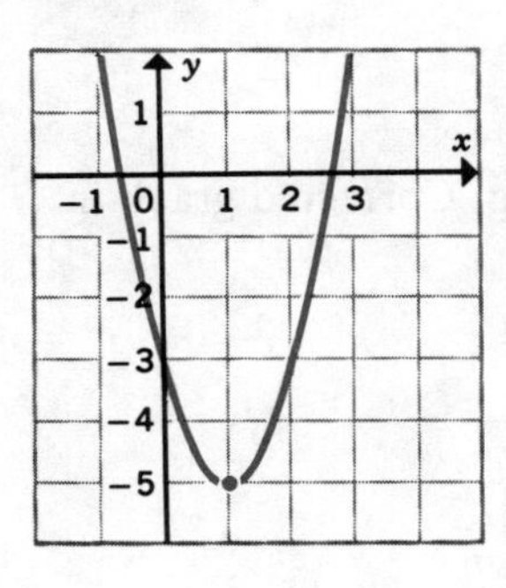

3.

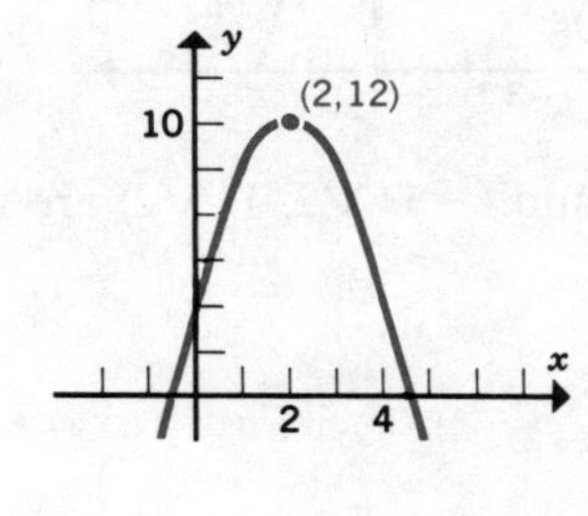

5.

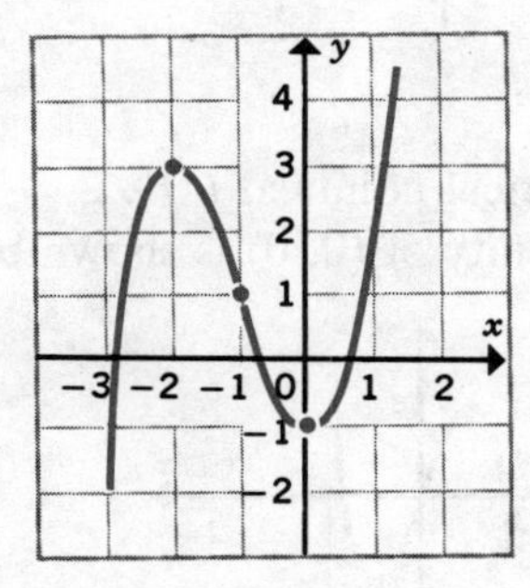

7.

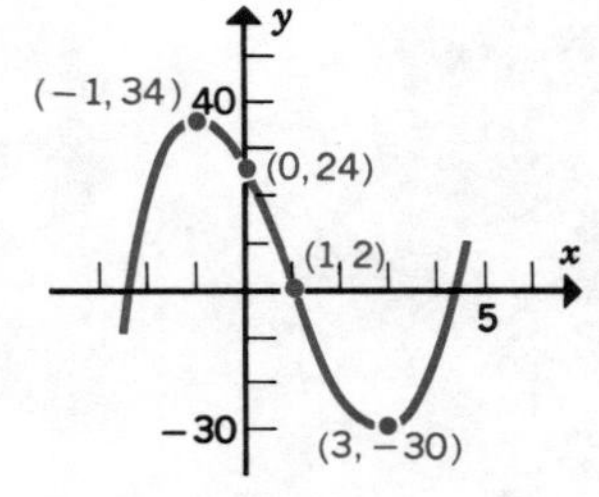

9.

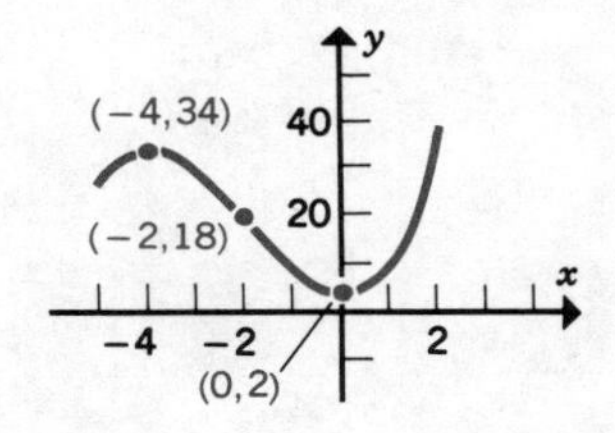

11.

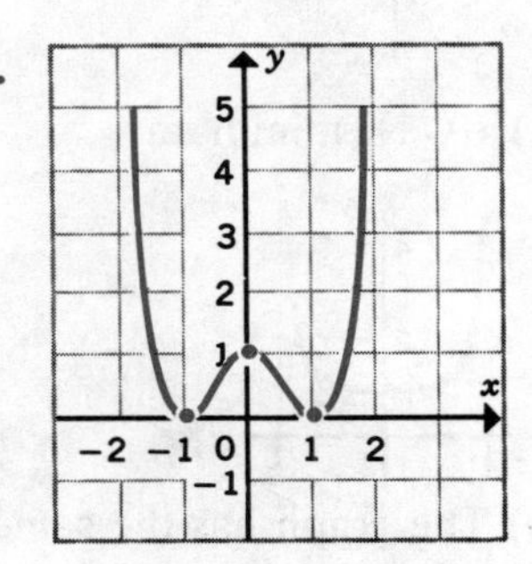

13.

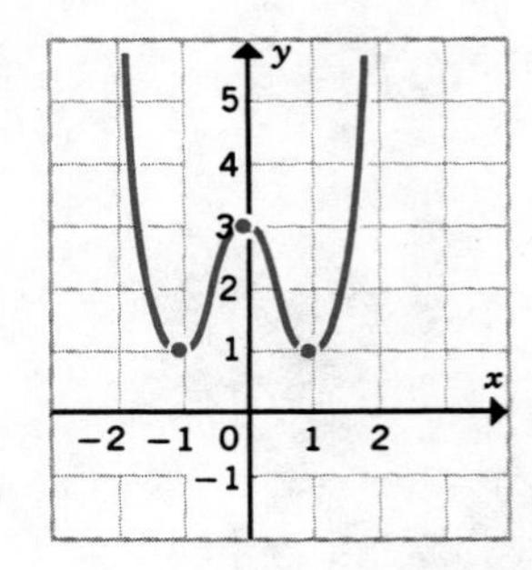

15.

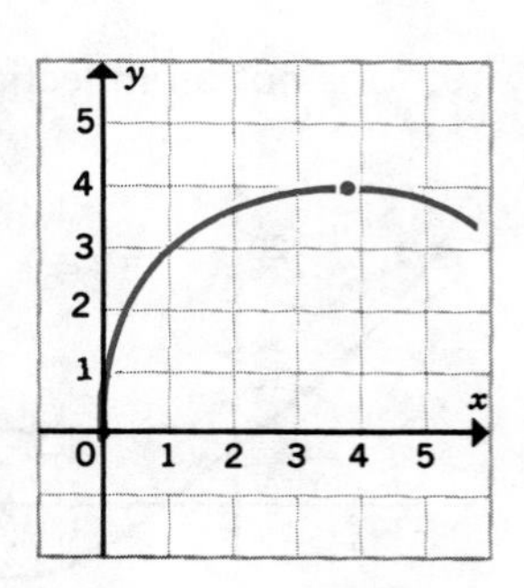

17.

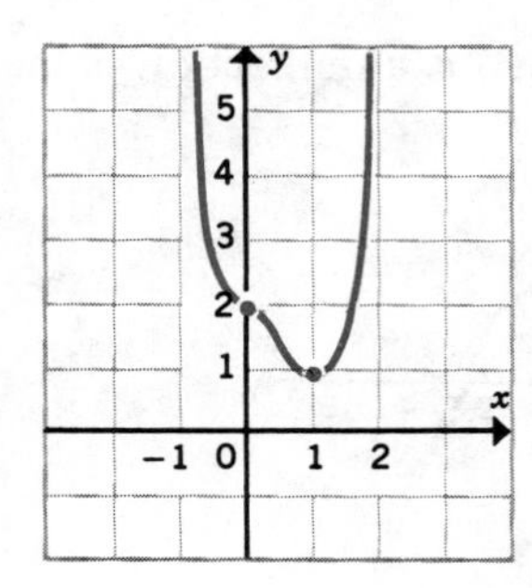

19.

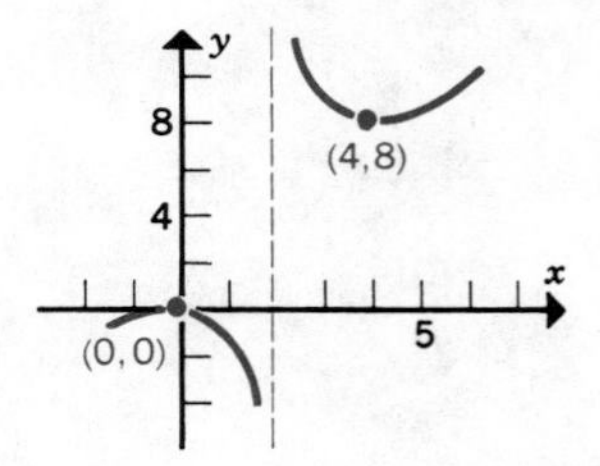

21.

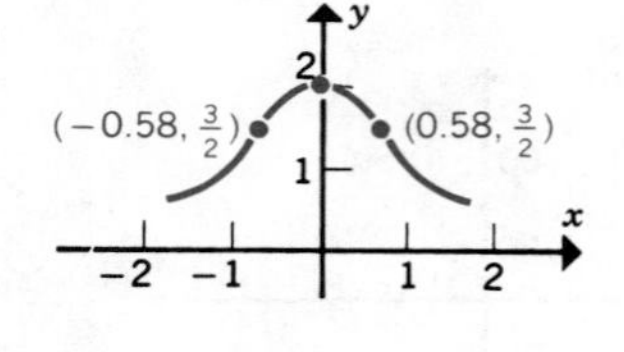

23.

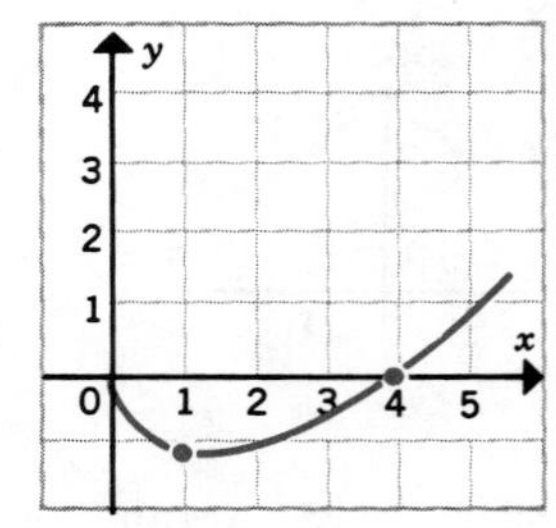

25.

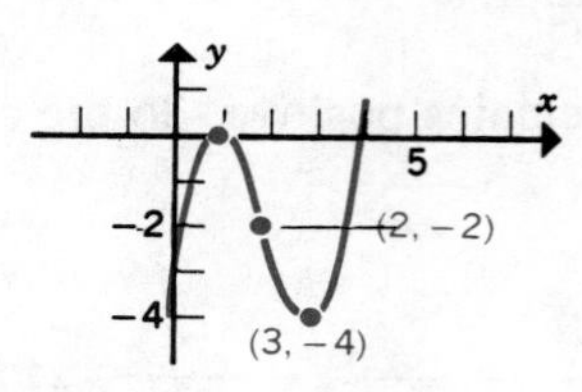

27.

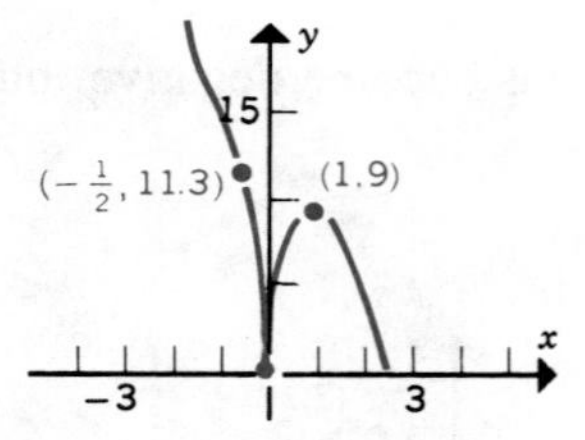

29. Critical points at $(1/\sqrt{2}, -1/\sqrt{2})$ and $(-1/\sqrt{2}, 1/\sqrt{2})$ are missing. Corrected graph in vicinity of (0, 0) is shown below.

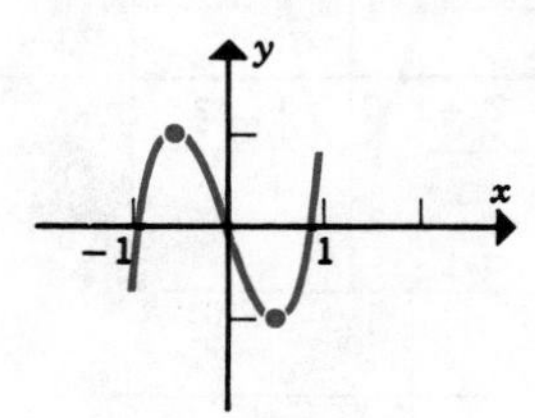

31. (a) (0, 1), minimum

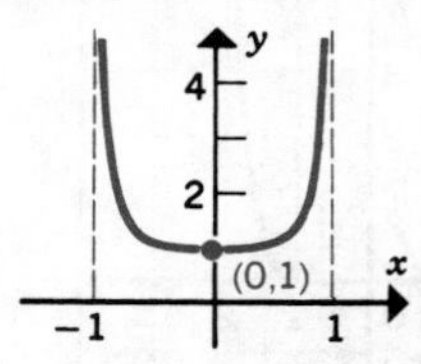

(b) The graph has the same features as the graph in part (a); the minimum is at $(0, m_0)$ and vertical asymptotes are $v/c = \pm 1$

4.5 Exercises

1. $dy = 18x^2\, dx;\ dy = 1.8$ 3. $dy = (2x - 5)\, dx;\ dy = 1$

5. $dy = \dfrac{3}{(3 - 2x)^2}\, dx;\ dy = -1.50$ 7. $dy = \dfrac{-2x}{3(1 - x)^{2/3}}\, dx;\ dy = -0.15$

9. $dy = \dfrac{1}{(x + 1)^2}\, dx;\ dy = 1/9$ 11. $dy = \dfrac{-1}{(2x)^{2/3}}\, dx;\ dy = -1/32$

13. 5.10 15. 3.04 17. 637/320 19. 0.283 21. $\dfrac{dA}{A} = \pm 0.02$

23. $\dfrac{dP}{P} = \pm 0.016$ 25. $\dfrac{dR}{R} = \pm 0.033$ 27. The plant grows approximately $\frac{1}{3}$ in.

29. (a) $dy = (10x - 6x^2)\, dx$ (b) $dy = -4$ (c)

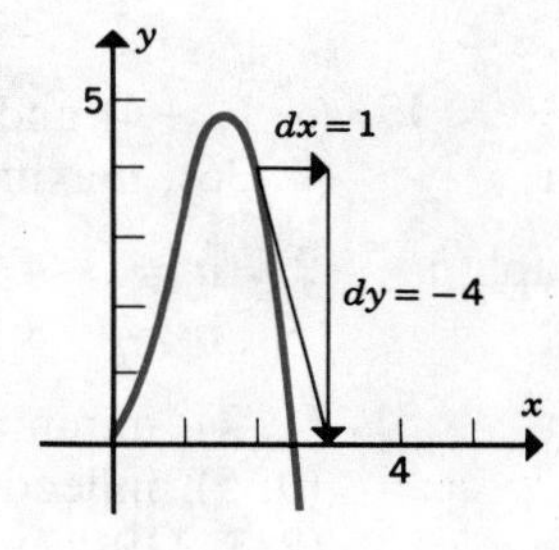

Review Problems

1. (3, 13), relative maximum

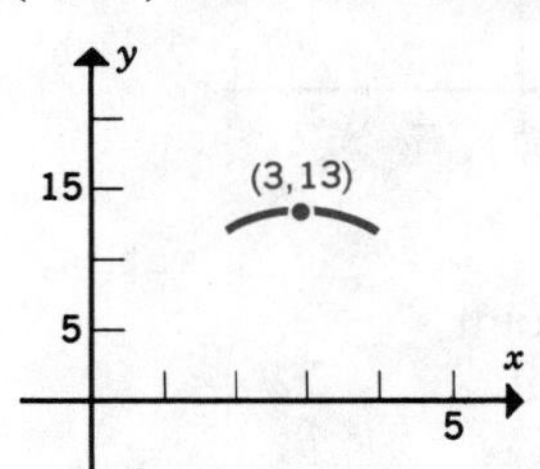

3. (2, 1), relative minimum

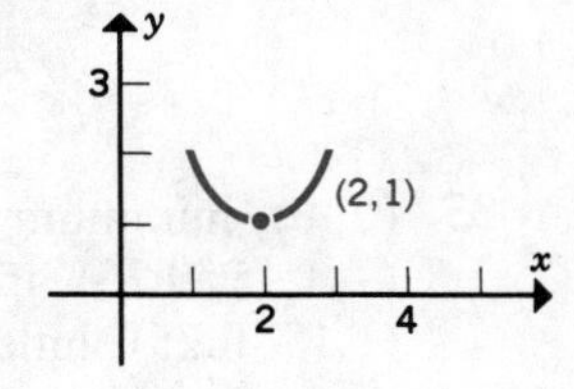

5. (−1, 1), relative minimum
(0, 6), relative maximum
(2, −26), relative minimum

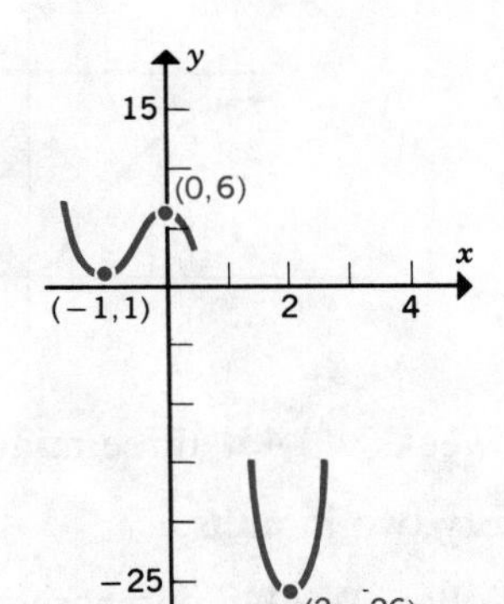

7. (1, 6), relative minimum

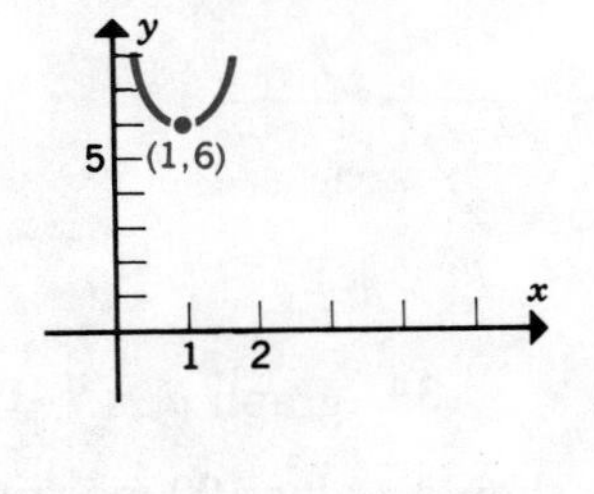

9. $(-2, 166)$, relative maximum
$(0, -10)$, inflection point
$(1, -23)$, relative minimum

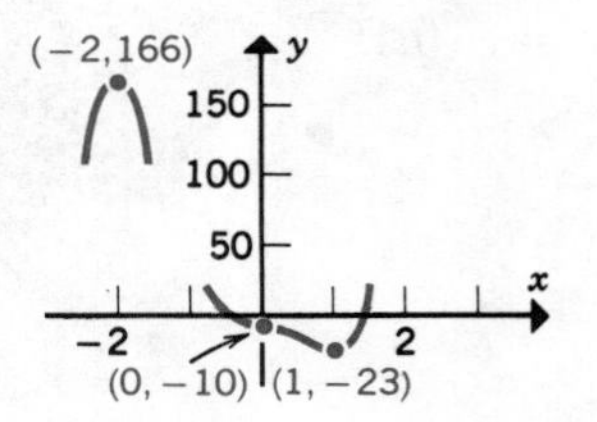

11. $(0, 0)$, relative maximum
$(2, -12{,}500)$, relative minimum
$(7, 0)$, inflection point

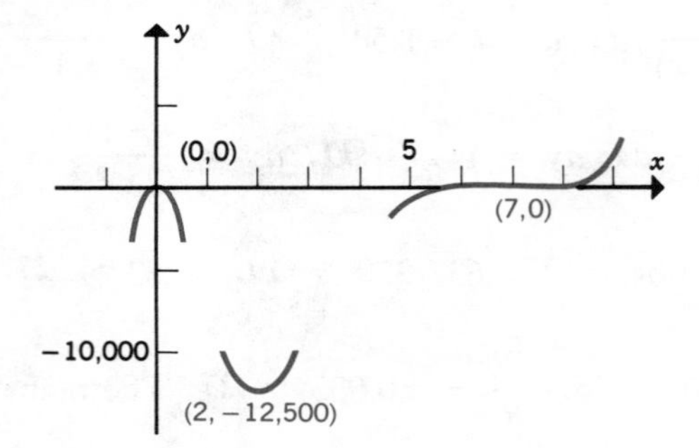

13. $(0, 2)$, maximum
$(3, -7)$, minimum

15. $(-1, -4)$ and $(2, -4)$, minima
$(4, 36)$, maximum

17. graph *c*

19. graph *e*

21. graph *b*

23. $a = -4$
$b = -8$

25. $a = 2$
$b = 8$

27. $a = 2$
$b = 6$

29. $(2, -1)$, minimum

31. $(1, 4)$, minimum
$(0, 5)$, inflection point
$(2/3, 119/27)$, inflection point

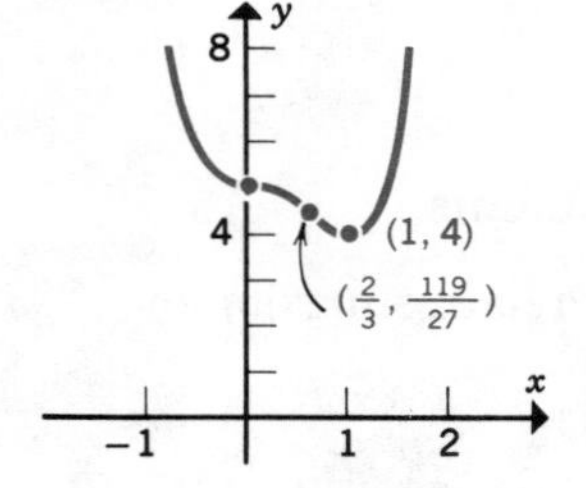

33. $(0, 0)$, relative maximum
$(2, -108)$, relative minimum
$(5, 0)$, inflection point
$(0.78, -45.72)$, inflection point
$(3.22, -58.48)$, inflection point

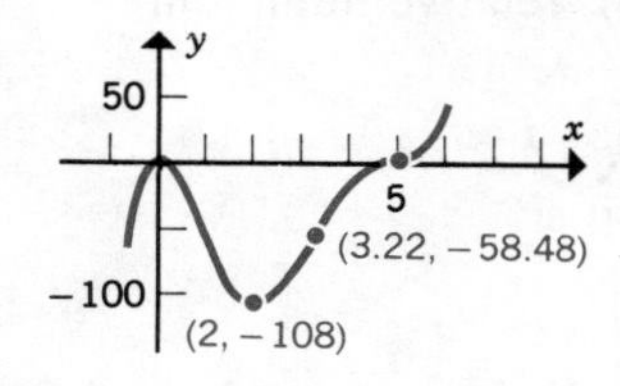

35. $(0, 0)$, minimum
$(-1.15, 0.25)$, inflection point
$(1.15, 0.25)$, inflection point

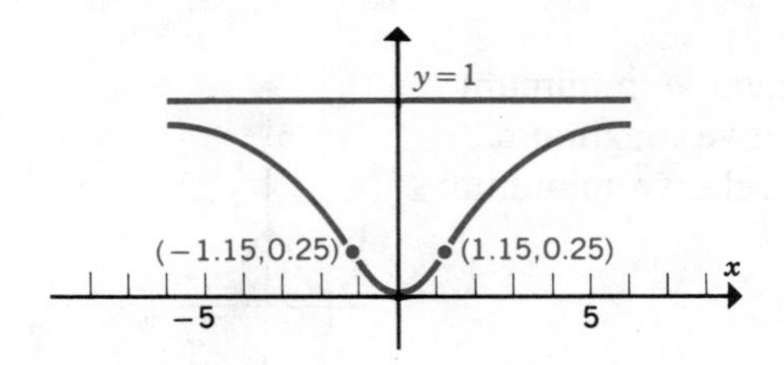

37. graph *e*

39. graph *d*

41. two weeks

43. three minutes

45. Acme should order 100 machines every two months.

47. The sales tax should be 6 cents per dollar. Weekly purchases will decline to \$300 million.

49.

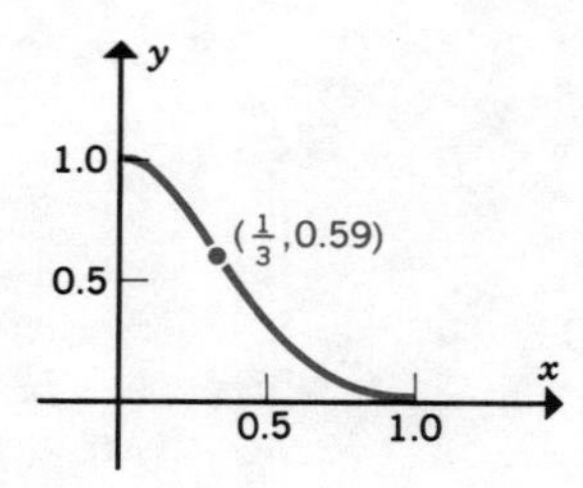

CHAPTER 5

5.1 Exercises

1.

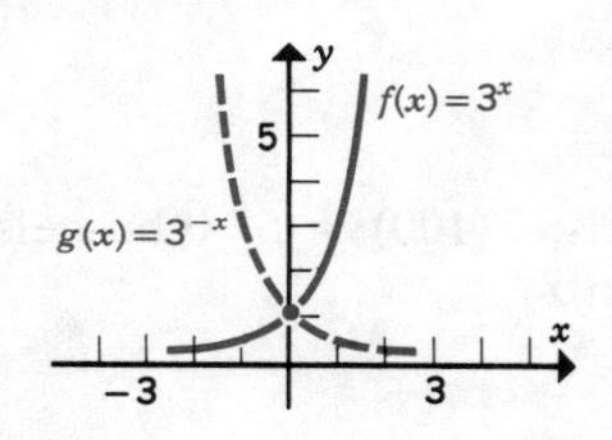

3.

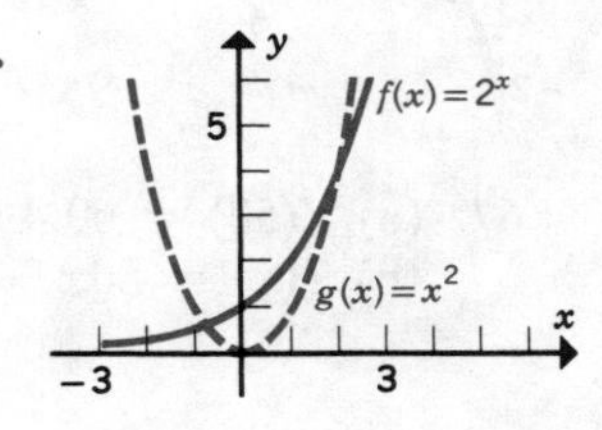

5.

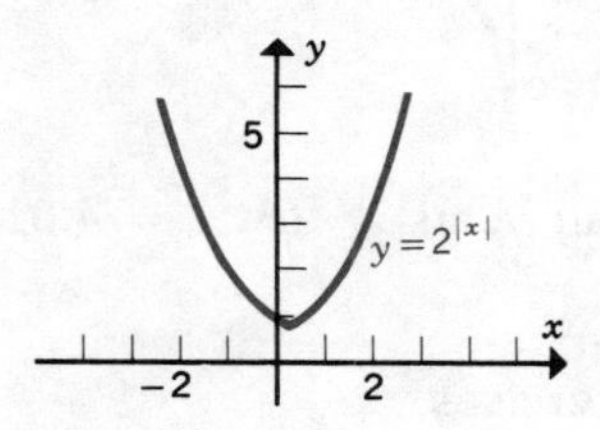

7. $f(x) = 8^x$ **9.** $f(x) = 2^x$ **11.** $f(x) = 2^x$ **13.** $f(x) = 4^x$
15. $f(x) = 625^x$ **17.** $F(x) = 3^x$ **19.** \$3,573.05
21. (a) \$3,173.75 (b) \$3,202.06 (c) \$3,216.87 **23.** \$16,117.63
25. \$186.28 **27.** (a) \$16,288.95 (b) \$17,192.14

29.

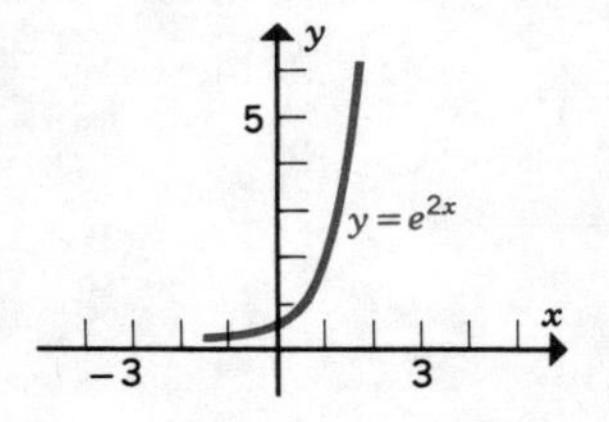

31.

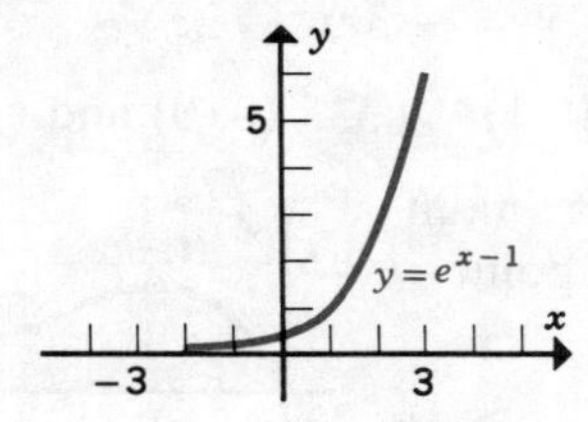

33.

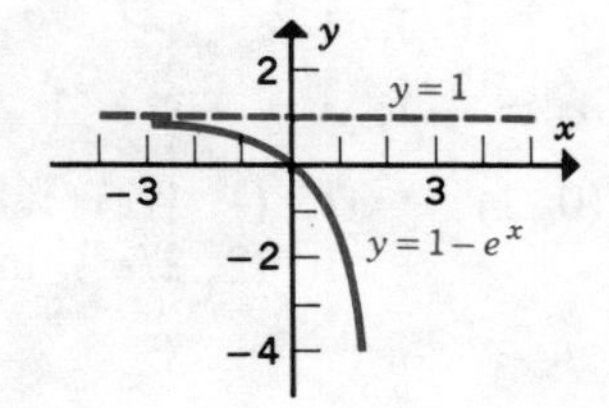

35. \$6,885.64 **37.** \$6,065.31
39. (a) 30 words per minute (b) $N(4) \approx 80$ words per minute; $N(8) \approx 102$ words per minute
41. (a) 8.2 mg/L (b) 4.5 mg/L **43.** $V = 60{,}000(0.8)^t$; value at the end of two years is \$38,400

5.2 Exercises

1. $\log_7 49 = 2$ **3.** $\log_{2/3}(16/81) = 4$ **5.** $\log_4 10 = x$ **7.** $\log_a 12 = 3$
9. $\log_3 2 = x$ **11.** $x = 1/8$ **13.** $x = 5$ **15.** $x = 3/2$ **17.** $x = 1/2$
19. $x = 2$ **21.** $x = 1/2$ **23.** $x = 7$ **25.** $\log_b 12 = 2.1584$
27. $\log_b 16 = 2.4084$ **29.** $\log_b 9 = 1.9084$ **31.** $\log_b 36 = 3.1126$

33. $\log_b(1/27) = -2.8626$ **35.** $x = -2/3$ **37.** $x = 25/8$

39.

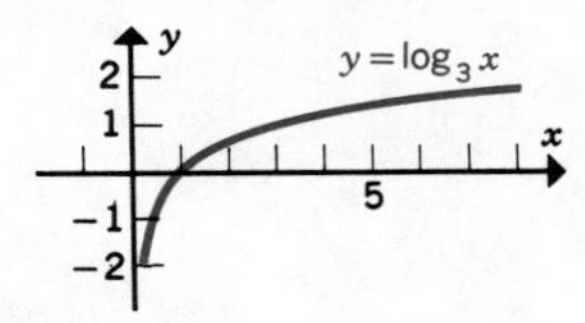

41. 6.1527 **43.** 6.9078

45. -5.3817 **47.** -2.8824

49. $x = 1.5440$ **51.** $x = -0.2619$

53. $x = 1.3023$ **55.** $x = 1.1565$ **57.** $x = \sqrt{2}\,e$

59.

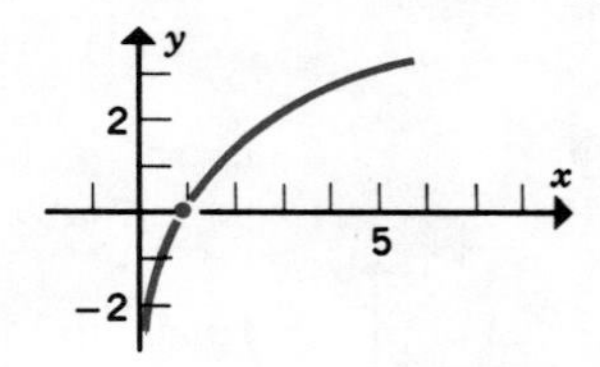

61.

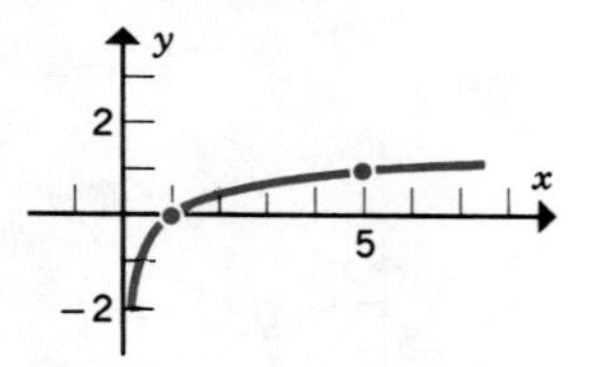

63. four years **65.** $t = 4.02$ hours **67.** (a) $Y(50) = 60.4$ bushels; $Y(100) = 57.0$ bushels
(b) $T(x) = 80x - 5x \ln x$

5.3 Exercises

1. $2e^{2x}$ **3.** $xe^x(x + 2)$ **5.** $4^x \ln 4$ **7.** $2xe^{x^2} - 2xe^{-x^2}$

9. $\dfrac{e^x(x - 1)}{x^2}$ **11.** $2^x(x \ln 2 + 1)$ **13.** xe^x **15.** $\dfrac{e^{x/2}}{2}$

17. $e^{e^x + x}$ **19.** $2^{e^x}e^x \ln 2$ **21.** ex^{e-1} **23.** $m = e$; $y = ex$

25. $m = 0$; $y = 2$ **27.** $m = -3e$; $y = -3ex - 2e$

29. $m = -1$; $y = -x + 1$ **31.** $(0, 1)$ **33.** $(0, 0)$ and $(-2, 4/e^2)$

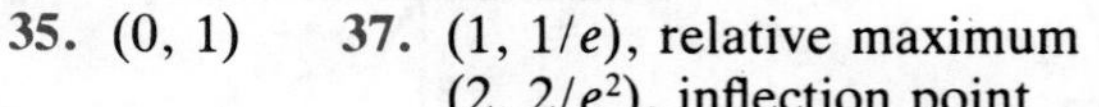

35. $(0, 1)$ **37.** $(1, 1/e)$, relative maximum
$(2, 2/e^2)$, inflection point

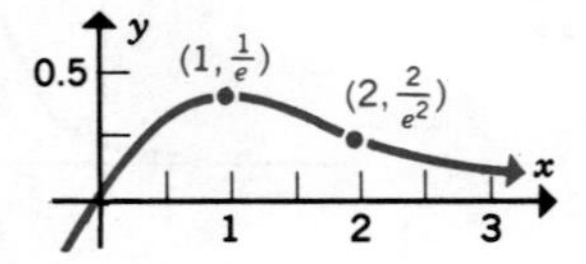

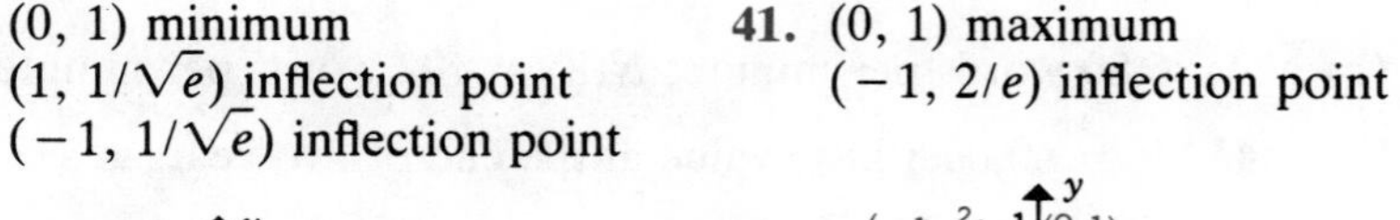

39. $(0, 1)$ minimum
$(1, 1/\sqrt{e})$ inflection point
$(-1, 1/\sqrt{e})$ inflection point

41. $(0, 1)$ maximum
$(-1, 2/e)$ inflection point

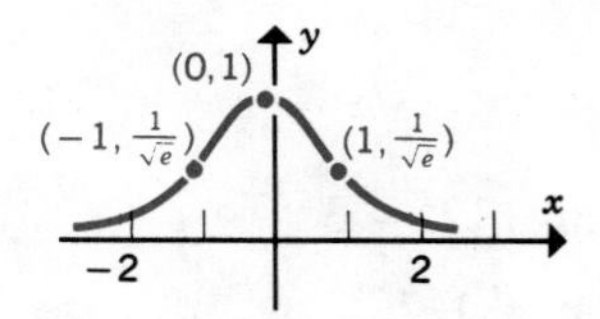

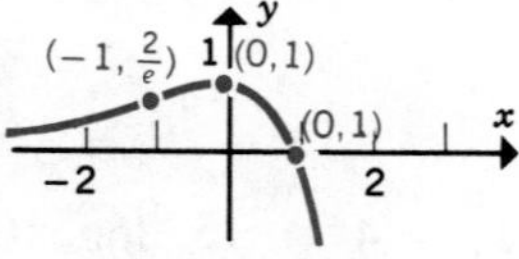

43. $f(1) = e^3$; $f'(1) = 2e^3$ **45.** $(3 - 3x^2)e^{-y}$ **47.** $\dfrac{4e^{-y}}{y + 1}$

49. $\dfrac{y}{2ye^{y^2} - x}$ **51.** (a) $1000e^{0.06t}$ (b) $A'(t) = 60e^{0.06t}$ (i) \$63.71/year (ii) \$81.00/year

53. (a) $C'(0) = 8$ mg/(L · hr); $C'(5) = 0.05$ mg/(L · hr)

5.4 Exercises

1. $\dfrac{x+1}{x}$ **3.** $\dfrac{2}{x}$ **5.** $\dfrac{1 - \ln x}{x^2}$ **7.** $\dfrac{(\ln x)^2 - 1}{x(\ln x)^2}$

9. $\dfrac{2x}{x^2 + 1}$ **11.** $\dfrac{2x}{(x^2 - 1)\ln 4}$ **13.** $2x \ln x + x - 1$

15. $\dfrac{1}{2x\sqrt{\ln x}}$ **17.** $\dfrac{4}{x(x^2 + 2)}$ **19.** $f'(x) = 1$

21. $(\ln 2)x^{(\ln 2 - 1)}$ **23.** $m = \dfrac{1}{e}$; $ey = x$ **25.** $m = 1$; $y = x - 1$

27. $m = 2/3$; $3y - 3 \ln 3 = 2x$ **29.** $m = \dfrac{1}{4 \ln 2}$; $y - 3 = \dfrac{x - 4}{4 \ln 2}$

31. $(2, \ln 2)$ **33.** $(1, 0)$ **35.** $(1, 1)$ **37.** (e, e)

39.

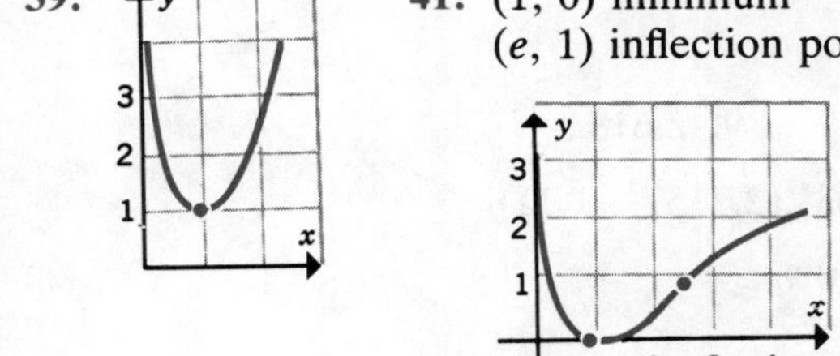

41. $(1, 0)$ minimum
$(e, 1)$ inflection point

43. $(e, 1/e)$ minimum
$(4.48, 0.34)$ inflection point

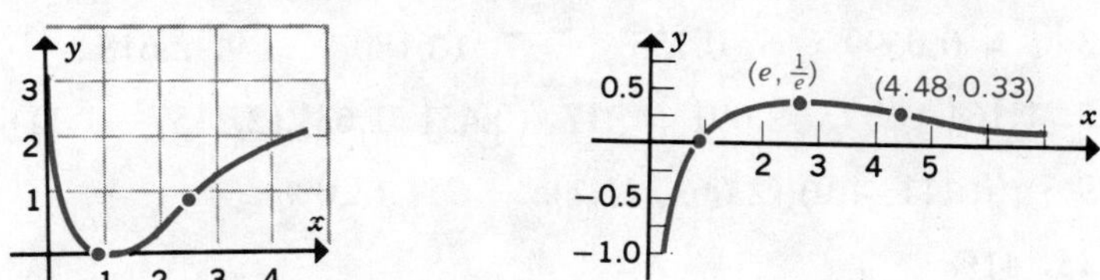

45. $f(1) = 1.10$; $f'(1) = 2/3$ **47.** $\dfrac{y \ln y + y}{2y^2 - x}$ **49.** $\dfrac{2x}{1 + \ln y}$

51. $\dfrac{xye^x - y}{x}$ **53.** $\dfrac{2xy}{e^y + ye^y \ln y}$ **55.** 1097 trees

57. (a) $G(0) = 40$ (b) $G'(t) = \dfrac{45}{1.5t + 1} > 0$ for $t > 0$. This means that $G(t)$ increases as t increases.

(c) $G''(t) = \dfrac{-67.50}{(1.5t + 1)^2} < 0$ for $t > 0$. This means that $G(t)$ increases at a slower rate as t increases.

5.5 Exercises

3. $q = 20$, $p = \dfrac{600}{e} = \$220.73$ **5.** $\dfrac{f'(1)}{f(1)} = 4$; $\dfrac{f'(3)}{f(3)} = \dfrac{4}{3}$

7. $\frac{F'(1)}{F(1)} = -1, \frac{F'(-1)}{F(-1)} = 1$ 9. $\frac{g'(1)}{g(1)} = \frac{1}{\ln 2}; \frac{g'(3)}{g(3)} = \frac{3}{(5 \ln 10)}$

11. (a) \$205 (b) 0.146 or 14.6% (c) 0.114 or 11.4%

13. (a) $\frac{e^{-0.10t}}{10 - 10e^{-0.10t}}$ (b) $\frac{N'(1)}{N(1)} = 0.9508; \frac{N'(10)}{N(10)} = 0.0582$
(c) 2000, the maximum number of homes that will be built

15. (a) $\frac{4}{4 + e}$ (b) $\frac{4}{4 + e^5}$ 17. $E(q) = 1 + \frac{1}{q}$; $E(4) = 1.25$, elastic

19. $E(q) = \frac{25}{q}$; $E(3) = 25/3$, elastic

21. $E(q) = \frac{100 - q}{q}$; $E(40) = 1.50$, elastic 23. $E(q) = \frac{1}{n}$

25. $E(p) = 1$ 27. $E(p) = \frac{p}{p - 1}$ 29. $E(p) = \frac{p}{40 - p}$

31. (a) $S'(1) = -4.76$; $S'(5) = -3.89$ (b) $\frac{S'(1)}{S(1)} = \frac{S'(5)}{S(5)} = -0.05$ (c) 13.9 time periods

5.6 Exercises

1. 1.710 3. $i = 0.0699$ 5. 0.755 7. 1.6180 9. 2.618
11. 2.317 13. 1.164 15. 2.091 17. (0.451, 1.63),(2.215, −1.11)
19. (−2.175, 29.5),(0.111, 4.9),(2.063, 21.1) 21. 13.0%
23. 28.6% 25. 41%

Review Problems

1. 3. 5.

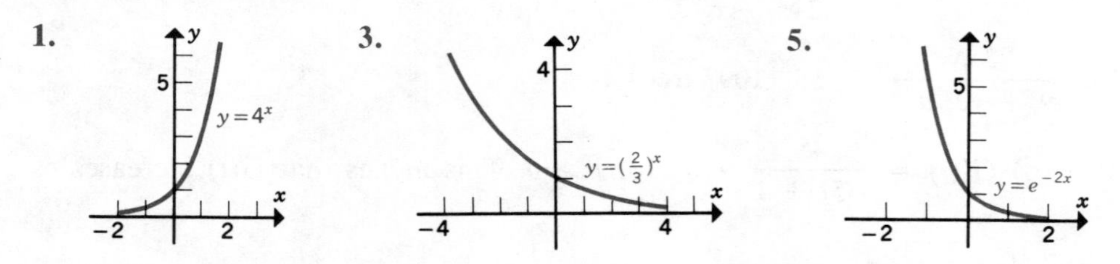

7. $\log_3 81 = 4$ 9. $\log_5 7.477 = 1.25$ 11. $\ln 7.389 = 2$ 13. $2^6 = 64$
15. $e^{2.303} = 10$ 17. $\log_3 81 = 4$ 19. $\log_{13} 13 = 1$ 21. $\log_2 \frac{1}{2} = -1$
23. $\log_7 49 = 2$ 25. $x = 12$ 27. $x = e^{e/2} = 3.89$ 29. 2.32

31. $x = 0.0718$ 33. $2e^x$ 35. $12e^{3x}$ 37. $\frac{2x}{x^2 + 1}$

39. $\frac{1 - \ln x}{x^2}$ 41. $-xe^{-x}$ 43. graph f 45. graph c

47. (0, 0), minimum
(−1, 0.135), relative maximum
(−0.29, 0.048) and (−1.71, 0.095), inflection points

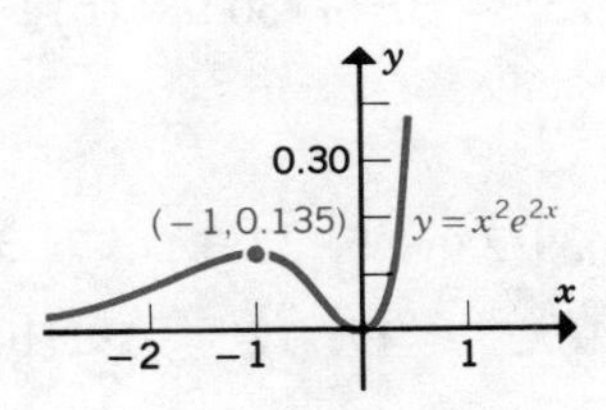

49. (0, 2), relative minimum
(−1, 2.26), inflection point
$y = 3$, horizontal asymptote

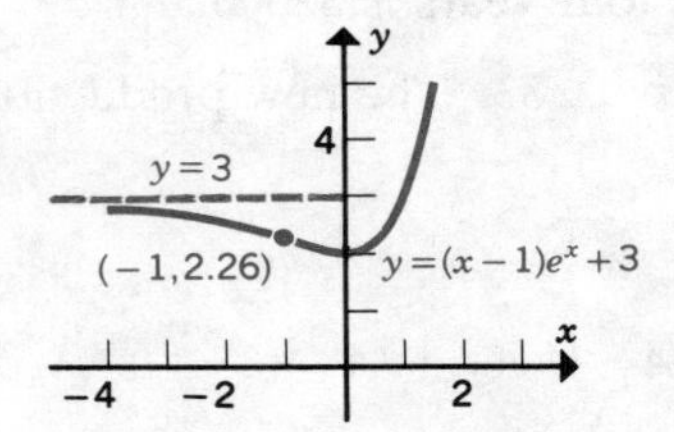

51. (a) \$3147.04 (b) \$3187.70 (c) \$3224.45 (d) \$3232.15

53. (a) 14.78 million (b) 5.44 million cells per day (c) 1, or 100% per day

55. (a) 0.3 (b) −1 per hour

CHAPTER 6

6.1 Exercises

1. $5x + C$ 3. $\frac{9t^{4/3}}{2} = 2 + C$ 5. $2 \ln|x| + 2x^4 + C$ 7. $-2e^{-5x} + C$

9. $4x + C$ 11. $\frac{-2}{x^5} + C$ 13. $x^4 - \ln x + C$ 15. $4x^{5/2} + C$

17. $6x^{3/2} + C$ 19. $\frac{x^7}{7} + \frac{x^4}{2} + x + C$ 21. $x^3 + x^2 + x + C$

23. $\frac{x^2}{2} - 4x + C$ 25. $e^x + x - e^{-x} + C$

27. $x^n + C$ 29. true 31. true 33. false 35. true

37. $F(x) = x^2 - 3x - 1$ 39. $f(t) = 2e^{3t} + 2$ 41. $F(s) = s^3 + \frac{1}{s} + 4$

43. $f(x) = x^3 - 2x^2 + 3$ 45. (a) $s(t) = -16t^2 + 256$ (b) four seconds

47. (a) $R(x) = 120x - x^2$ (b) \$425 49. (a) $N(t) = 100e^{0.10t} + 100$
(b) $t = 10 \ln 5 \approx 16.09$ years

51. (a) $V(t) = 180(10 - t)^2$ (b) $V(6) - V(2) = -\$8640$

53. (a) $N(t) = 25t + 3t^2 + 500$ (b) $t = 10$ days **55.** $F'(x) = e^{-x^2}$

6.2 Exercises

1. 27/4 **3.** 39/2 **5.** 32/3 **7.** 35/6 **9.** ln 4 **11.** 9

13. 16 **15.** 15/2 **17.** 12 **19.** 101/24 **21.** 7 − ln 2 **23.** 93

25. $8000 **27.** The decrease over four years is $2000. **29.** 32/3 in.

31. (a) 9530 gallons (b) 80.5 hours **33.** The new production level is 12 VCRs.

6.3 Exercises

1. Sum of the areas is $\frac{2}{3}$ **3.** (a) $A_1 + A_2 + A_3 + \cdots + A_n = \frac{2(n + 1)}{n}$
(b) area under the curve = 2

5. 7/2 **7.** 6 ln 3 **9.** 662/3 **11.** 1/2 **13.** 45/2 **15.** 124

17. −22/3 **19.** 7/2 **21.** −0.69 **23.** $\frac{e^n - 1}{n}$ **25.** $\frac{4e + 1}{4}$

27. $d = -3$ or $d = 3$ **29.** $d = 2$ **31.** $d = -4$ or $d = 3$ **33.** 1/4 **35.** 8/3

37. 91 **39.** $31,606 **41.** $5,175

43. (a) −40 million gallons (b) nine years (c) The reservoir should contain at least 84 million gallons.

6.4 Exercises

1. 44/3 **3.** $\frac{e^2 - 1}{e}$ **5.** 12 **7.** $\frac{5e - 1}{5}$ **9.** $e^2 - e - \ln 2$

11. 32/3 **13.** 32/3 **15.** 8/3 **17.** 9 **19.** 256/5

21. $6.35 million **23.** $128,960 **25.** 1 **27.** 1000/3

29. 80 **31.** 2.745 **33.** consumers' surplus = 4
producers' surplus = 8

35. consumers' surplus = 3.07
producers' surplus = 1.93 **37.** (a) $61.3 million
(b) $70.3 million **39.** 64 mm²

6.5 Exercises

1. $29,142 **3.** $46,402 **5.** The annual contribution by the employees is $303.57, the weekly contribution is $5.84.

7. $835.31 **9.** $51,184,000 **11.** 5 **13.** 8

15. 12.78 **17.** 0.549 **19.** 2^{n-1}

21. 19,400 tons

Review Problems

1. $8x + C$ **3.** $\sqrt{5}x + C$ **5.** $2x^3 - x^2 + 3x + C$ **7.** $2x^4 - \frac{1}{x} + C$

9. $2e^{2x} + 2\ln|x| + C$ **11.** $\frac{x^{1-e}}{1-e} - e^{-x} + C$ **13.** true **15.** true **17.** 4.35

19. 22/3 **21.** 12 **23.** 8 **25.** 2.44 **27.** −8/3 **29.** $3 - \ln 2$

31. $x^2e^{-x} + \sqrt{xe^{-x}}$ **33.** $3t^2 - 2t + 4$ **35.** 125/6 **37.** 64/15

39.

rank	1	2	3	4
integral	$\int_a^b k(x)\,dx$	$\int_a^b h(x)\,dx$	$\int_a^b f(x)\,dx$	$\int_a^b g(x)\,dx$

41. (a) 40,000 operations (b) \$400 million (c) \$308.3 million

43. (a) \$1.96 million (b) \$2.25 million (c) \$1.21 million

45. (a) \$9553 (b) \$13,206 **47.** 1295 **49.** 56.67 hundred

Chapter 7

7.1 Exercises

1. $\frac{(x-5)^4}{4} + C$ **3.** $e^{x+2} + C$ **5.** $\frac{(2x+1)^6}{6} + C$ **7.** $\ln|4x+3| + C$

9. $\frac{(x^2-1)^5}{10} + C$ **11.** $\frac{-e^{-x^2}}{2} + C$ **13.** $\frac{-5}{6(3x^2+7)} + C$

15. $\frac{(x^6+2)^2}{12} + C$ **17.** $\frac{-e^{1-x^2}}{2} + C$ **19.** $\ln(2+e^x) + C$

21. $\frac{e^{x^2}}{2} - \ln(x^2+1) + C$ **23.** $-e^{1/x} + C$

25. $\frac{2(x+3)^{5/2}}{5} - \frac{4(x+3)^{3/2}}{3} + C$ **27.** $\frac{(\ln x)^3}{3} + C$

29. $\frac{2(x+1)^{7/2}}{7} - \frac{4(x+1)^{5/2}}{5} + \frac{2(x+1)^{3/2}}{3} + C$

31. 0 **33.** 250/3 **35.** 26/3 **37.** $2 + \ln 0.6$

39. 4/15 **41.** \$10.4 thousand

7.2 Exercises

1. $\frac{xe^{2x}}{2} \frac{-e^{2x}}{4} + C$ **3.** $\frac{x^2e^{4x}}{4} - \frac{xe^{4x}}{8} + \frac{e^{4x}}{32} + C$

5. $x \ln(3x) - x + C$ **7.** $\frac{2x^{3/2}\ln x}{3} - \frac{4x^{3/2}}{9} + C$

9. $2x^{1/2} \ln x - 4x^{1/2} + C$ **11.** $\frac{3x(x+2)^{4/3}}{4} - \frac{9(x+2)^{7/3}}{28} + C$

13. $(x+1)\ln(x+1) - x + C$ **15.** $\frac{(\ln x)^2}{2} + C$ **17.** $\frac{-2}{e}$

19. $9 \ln 3 - 4 \ln 2 - 5/2$ **21.** 1 **23.** $\frac{(\ln 2)^2}{2}$ **25.** 1

27. \$304,049 **29.** \$603.52

7.3 Exercises

1. $\frac{x^4 \ln x}{4} - \frac{x^4}{16} + C$ **3.** $\frac{\ln|x + \sqrt{x^2+1}|}{3} + C$

5. $\frac{1}{16(4x+1)} + \frac{\ln|4x+1|}{16} + C$ **7.** $\frac{3x}{5} - \frac{6 \ln|5x+2|}{25} + C$

9. $-\frac{1}{2} \ln \frac{2 + \sqrt{4 - 9x^2}}{x} + C$ **11.** $\frac{1}{8} \ln\left|\frac{4+x}{4-x}\right| + C$ **13.** $\frac{1}{4} \ln\left|\frac{x^2-1}{x^2+1}\right| + C$

15. $\frac{4x-12}{3}\sqrt{2x+3} + C$ **17.** $2 \ln|x + \sqrt{x^2-9}| + C$ **19.** $\frac{1}{4} \ln\left|\frac{x-1}{x+3}\right| + C$

21. $\frac{\ln 3}{4}$ **23.** $\frac{3}{2} - \frac{3 \ln 3}{4}$ **25.** $6 + \frac{7 \ln 7}{4}$

27. $\ln 2 - 0.25$ **29.** $6 - 2e$

7.4 Exercises

FTC = fundamental theorem of calculus; T = trapezoidal rule; S = Simpson's rule

1. 8/3 (FTC)
2.75 (T) **3.** 1.7183 (FTC)
1.7272 (T) **5.** 1.8692 (FTC)
1.8690 (T) **7.** 0.3466 (FTC)
0.3414 (T)

9. 9 (FTC)
9 (S) **11.** 0.3863 (FTC)
0.3863 (S) **13.** 26.799 (FTC)
27.019 **15.** 0.7828 (T)
0.7854 (S)

17. 2.3822 (T)
2.3783 (S) **19.** 1.0825 (T)
1.0815 (S) **21.** 1.4373 (T)
1.4332 (S) **23.** $E_T \leq 0.0625$
$E_S \leq 0.00052$

25. $E_T \le 0.0052$
$E_S \le 1/46{,}080$

27. $E_T \le 1/96$
$E_S \le 1/1920$

29. $n \ge 150$

7.5 Exercises

1. $2/3$ 3. 2 5. 1 7. $1/2$ 9. 6 11. $1/2$ 13. $-1/4$
15. $-1/6$ 17. ∞ 19. $k = 1$ 21. no solution 23. $k = 4$
25. $5000/3$ tons 27. 5000 gallons

7.6 Exercises

1. 0 3. $-\infty$ 5. 0 7. 2 9. 0 11. ∞ 13. 1 15. $\dfrac{1}{e}$
17. ∞ 19. 0 21. 1 23. $-1/8$ 25. 1 27. $-1/4$ 29. ∞
31. \$177,778 33. 20,000 barrels

Review Problems

1. $(x^2 + 1)^5 + C$ 3. $-e^{-x^2} + C$ 5. $3\ln(x^2 + 2x + 2) + C$

7. $\dfrac{(2x + 3)^{5/2}}{10} - \dfrac{(2x + 3)^{3/2}}{2} + C$ 9. $\dfrac{2(x^3 + x + 5)^5}{5} + C$

11. $-\dfrac{xe^{-2x}}{2} - \dfrac{e^{-2x}}{4} + C$ 13. $x \ln 4x - x + C$ 15. $-e^{-2x}(2x^2 + 2x + 1) + C$

17. $-\dfrac{\ln x}{(n - 1)x^{n-1}} - \dfrac{1}{(n - 1)^2 x^{n-1}} + C, \quad n \neq 1; \quad \dfrac{(\ln x)^2}{2} + C, \quad n = 1$

19. $\dfrac{x^2(2x + 3)^{3/2}}{3} - \dfrac{2x(2x + 3)^{5/2}}{15} + \dfrac{2(2x + 3)^{7/2}}{105} + C$ 21. $\dfrac{5}{3} \ln \left| \dfrac{x}{2x + 3} \right| + C$

23. $\ln |2x + \sqrt{4x^2 + 25}| + C$

25. (a) 9.374, trapezoidal rule (b) 9.300, Simpson's rule

27. (a) 1.104, trapezoidal rule
(b) 1.105, Simpson's rule

29. (a) 5.353, trapezoidal rule
(b) 5.379, Simpson's rule

31. (a) 1.542 (b) 1.571 33. 3 35. $\dfrac{9}{2e^2}$ 37. -1

39. The limit does not exist; $\dfrac{e^{2x}}{x} \to \infty$ as $x \to \infty$. 41. 0.354

43. (a) −\$18,964
(b) \$30,000

45. (a) 10 million board-feet
(b) \$0.12 million
(c) \$0.064 million

47. (a) 990.1 gallons
(b) 1000 gallons

CHAPTER 8

8.1 Exercises

1. (a) 7
(b) 10
(c) -3
(d) $5\sqrt{2}$
(e) $8 + 2h$
(f) $2h$

3. (a) 2
(b) 1/2
(c) $2 + k$
(d) k
(e) 1
(f) 1

5. (a) 0
(b) 0
(c) 1
(d) -1
(e) 2
(f) -2

7. (a) 1
(b) 0
(c) e^2
(d) 1
(e) e
(f) 4

9. (a) 36π
(b) 36π
(c) 125π
(d) 1000π

11.

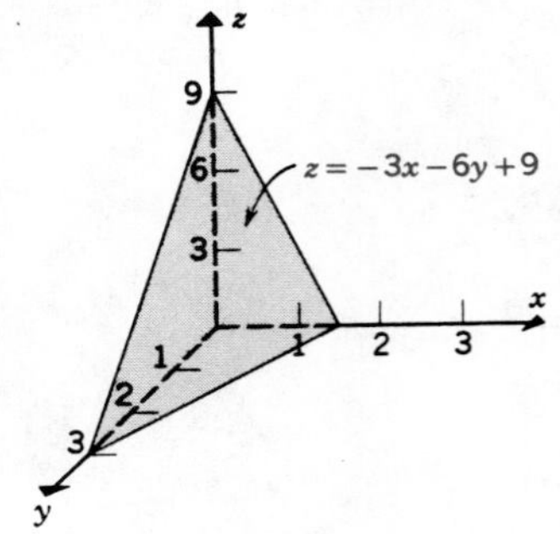

13.

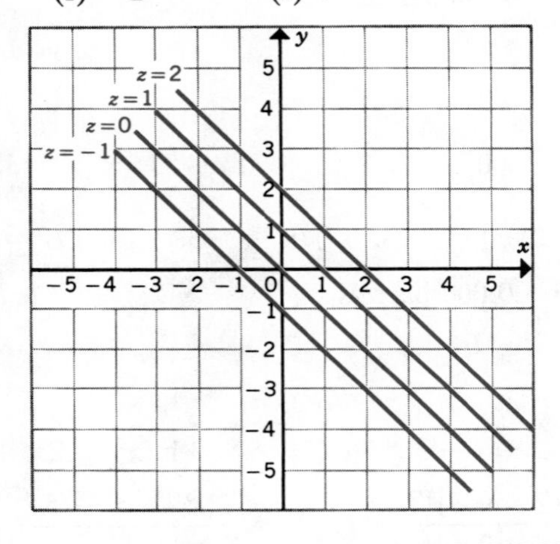

15.

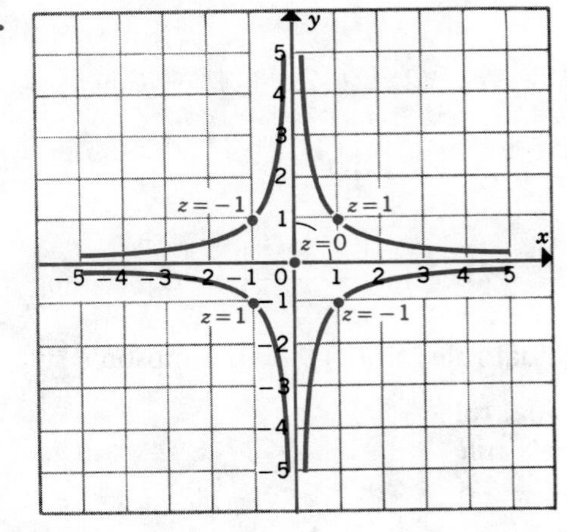

17.

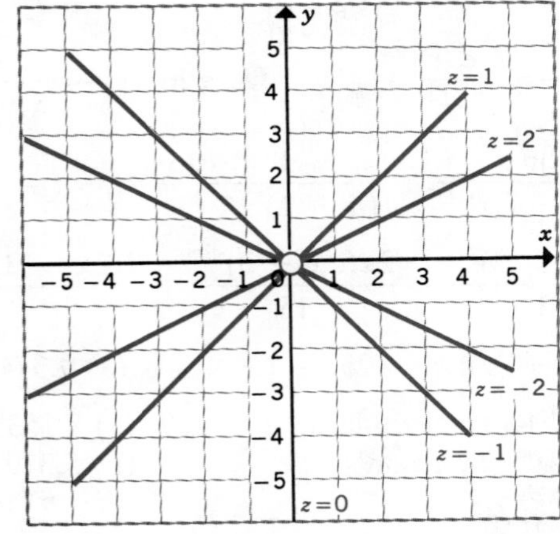

19.

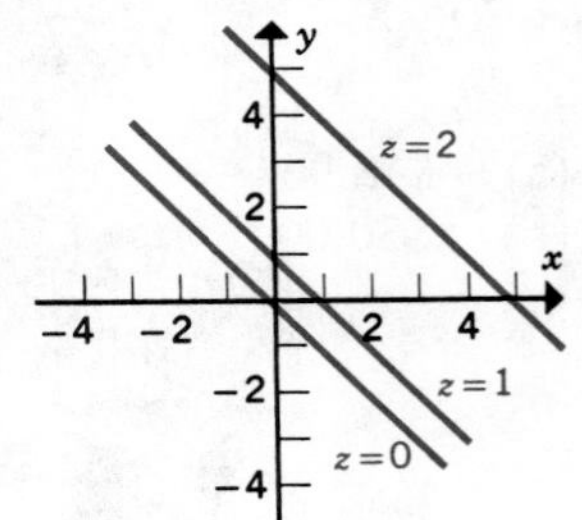

21. $R(x, y) = 0.75x + 0.85y$

23. (a) $R(x, y) = 200x + 300y$
(b) $C(x, y) = 10{,}000 + 100x + 175y$
(c) $P(x, y) = 100x + 125y - 10{,}000$

25. $C(r, h) = 0.02\pi r^2 + 0.08\pi rh$

27. (a) $q = 2{,}400$
(b) $q = 4{,}800$
(c) $q = 2[50L^{0.2}M^{0.8}]$
The output is doubled.

29. $R_1(p_1, p_2) = 10p_1 - (p_1)^2 + 0.5p_1p_2$
$R_2(p_1, p_2) = 8p_2 + 0.4p_1p_2 - (p_2)^2$

31. (a) 120 (b) 100 (c) 15.6 years

8.2 Exercises

1. $f_x(x, y) = 2x$
$f_y(x, y) = -2y$

3. $f_x(x, y) = 2xy^3$
$f_y(x, y) = 3x^2y^2$

5. $f_x(x, y) = \dfrac{1}{y}$
$f_y(x, y) = \dfrac{-x}{y^2}$

7. $f_x(x, y) = \dfrac{\sqrt{y}}{2\sqrt{x}}$
$f_y(x, y) = \dfrac{\sqrt{x}}{2\sqrt{y}}$

9. $f_x(x, y) = e^y$
$f_y(x, y) = xe^y$

11. $f_x(x, y) = ye^{xy}$
$f_y(x, y) = xe^{xy}$

13. $f_x(x, y) = \dfrac{2y}{y^2 - x^2}$
$f_y(x, y) = \dfrac{2x}{y^2 - x^2}$

15. $f_x(x, y) = \dfrac{1}{2x + 2y} = f_y(x, y)$

17. $f_x(x, y) = \dfrac{x}{(x^2 + y^2)^{1/2}}$
$f_y(x, y) = \dfrac{y}{(x^2 + y^2)^{1/2}}$

19. $f_y(x, y) = 2y$
slope $= f_y(1, 1) = 2$

21. $f_x(x, y) = \dfrac{y}{2\sqrt{xy}}$
slope $= f_x(1, 4) = 1$

23. $f_x(x, y) = \dfrac{1}{x}$
slope $= f_x(e, e) = \dfrac{1}{e}$

25. $f_{xx} = 6$
$f_{yy} = -14$
$f_{xy} = f_{yx} = 2$

27. $f_{xx} = e^x - \dfrac{1}{x^2}$
$f_{yy} = -\dfrac{1}{y^2}$
$f_{xy} = f_{yx} = 0$

29. $f_{xx} = \dfrac{-y}{4x^{3/2}}$
$f_{yy} = 0$
$f_{xy} = f_{yx} = \dfrac{1}{2\sqrt{x}}$

31. $f_{xx} = f_{yy} = \dfrac{-1}{(x + y)^2}$
$f_{xy} = f_{yx} = \dfrac{-1}{(x + y)^2}$

33. $f_{xx} = \dfrac{2}{(x - y)^3}$
$f_{yy} = \dfrac{2}{(x - y)^3}$
$f_{xy} = f_{yx} = \dfrac{-2}{(x - y)^3}$

35. (a) $V_r(r, h) = 2\pi rh$
$V_h(r, h) = \pi r^2$
(b) $V_r(8, 10) = 160\pi$
$V_h(8, 10) = 64\pi$

37. (a) $PV_R(R, r) = \dfrac{1}{r}$, $PV_r(R, r) = \dfrac{-R}{r^2}$
(b) $PV_R(500, 0.10) = 10$
$PV_r(500, 0.10) = -50{,}000$

39. (a) $q_L = 1.92$, $q_M = 2.22$ (b) $q_L = 0.93$, $q_M = 3.2$ (c) $q_L = 4/3$, $q_M = 8/3$

41. $A_w = 0.12$ m²/kg, $A_h = 6.22$ meters

8.3 Exercises

1. (0, 0, 1), minimum **3.** (0, 0, 3), saddle point

5. (4, −1, −12), minimum **7.** (1, −4, 22), saddle point

9. (−2, 1, 9), maximum
(0, 1, 5) saddle point

11. (0, 0, −5), saddle point
(1, 1, −6), minimum

13. (−1, 1, 9), saddle point
(2, 1, −18), minimum

15. (0, 0, 4), saddle point
(6, 18, 112), maximum

17. (4/3, 9/2, 18), minimum **19.** (0, −1, 1), saddle point

21. $x = y = z = 15$; $P = 15^3 = 3375$ **23.** $l = 50$, $w = 100$, $h = 75$

25. $w = l = 4$, $h = 2$ **27.** $p_1 = \$21.50$, $p_2 = \$30.00$

8.4 Exercises

1. $x = y = 1$, $f(1, 1) = 1$

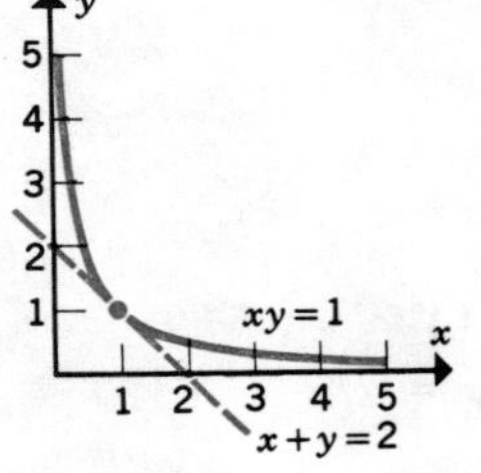

3. $x = 2/3$, $y = 1/3$, $f(2/3, 1/3) = 4/3$

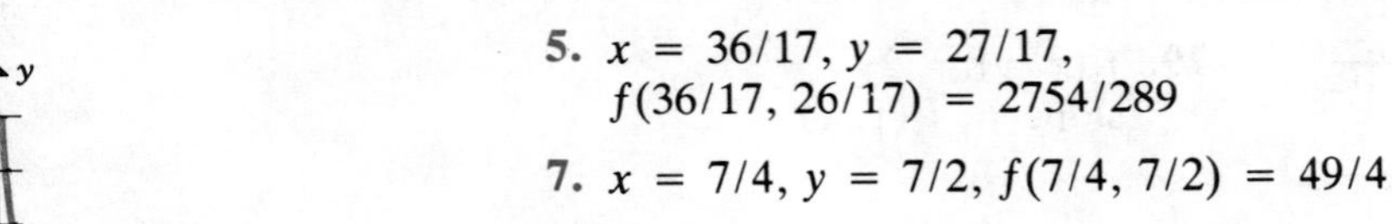

5. $x = 36/17$, $y = 27/17$,
$f(36/17, 26/17) = 2754/289$

7. $x = 7/4$, $y = 7/2$, $f(7/4, 7/2) = 49/4$

9. $x = y = 2$, $f(2, 2) = 2$ **11.** $x = y = 3$, $f(3, 3) = 3\sqrt{2}$

13. length = width = height = $30\sqrt[3]{4}$ ft.

15. L = M = 625, production costs = \$50,000 **17.** 40 and 40

19. larger number = 7
smaller number = −7

21. length = 60 ft
width = 30 ft

23. length = 12 in.
width = 12 in.
height = 10 in.

8.5 Exercises

1. $y = 0.63x + 0.77$

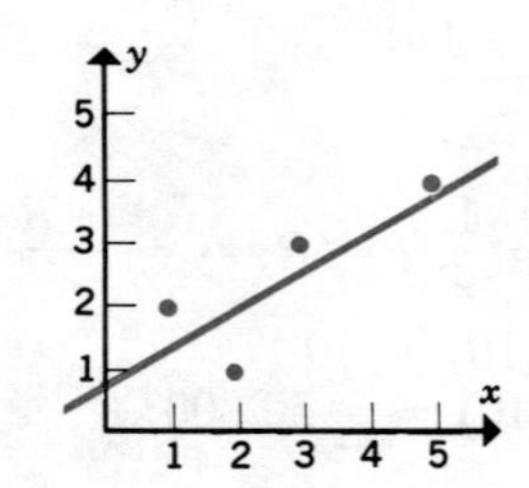

3. $y = 0.62x + 0.96$

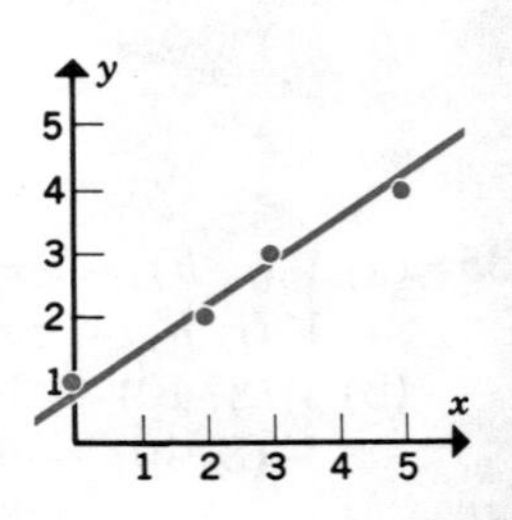

5. $y = -0.58x + 4.95$
7. $y = -1.23x + 5.63$
9. $y = -1.5x + 7.50$
11. $p = -0.34x + 3.38$
13. (a) $y = 0.10x - 0.47$
(b) \$2.03 thousand
15. (a) $y = 5x - 167$
(b) 188 lb

8.6 Exercises

1. $dz = 2xy^4\,dx + 4x^2y^3\,dy$
3. $dz = -e^{y-x} + e^{y-x}\,dy$
5. $dz = \dfrac{3x^2}{2\sqrt{x^3 + y^2}}\,dx + \dfrac{y}{\sqrt{x^3 + y^2}}\,dy$
7. $dz = \dfrac{y}{x}\,dx + (\ln x)\,dy$
9. $dz = \dfrac{-4x}{(x^2 + y^2)^2}\,dx - \dfrac{2}{(x^2 + y^2)^2}\,dy$
11. $dz = \dfrac{e^y}{x}\,dx + (\ln x)e^y\,dy$
13. $dz = \dfrac{2}{3(2x + 5y)^{2/3}}\,dx + \dfrac{5}{3(2x + 5y)^{2/3}}\,dy$
15. $dz = \dfrac{x + y - 1}{(x + y)^2}(dx + dy)$
17. $dz = \dfrac{4}{3}\left(\dfrac{y}{x}\right)^{2/3} dx + \dfrac{8}{3}\left(\dfrac{x}{y}\right)^{1/3} dy$
19. $dz = \dfrac{1}{x(1 + \ln y)}\,dx - \dfrac{1 + \ln x}{y(1 + \ln y)^2}\,dy$
21. $dz = 6.50$
23. $dz = 1$
25. $dz = \dfrac{1}{2}$
27. $dz = \dfrac{e - 4}{4e}$
29. $dz = 0.10$
31. $dV = 15{,}000$ ft^3
33. $dA = 1.7266$ m^2
35. $dV = -2000\pi$ ft^3
37. ± 0.015
39. ± 0.03

8.7 Exercises

1. $2x^5y^2 + C(y)$
3. $\dfrac{x^2 + x}{y} + C(y)$
5. $2x^3 + 3y^2 \ln|x| + C(y)$
7. $e^{xy} + C(y)$
9. $\frac{1}{2}\ln|x^2 - y^2| + C(y)$
11. $\dfrac{2yx^{3/2}}{3(y + 1)} + C(y)$
13. $48y^2$
15. $104y$
17. $y \ln 2 + 3/2$
19. $1 - e^{-y}$
21. $12y(3 - y^2)^{1/2} - 12y(1 - y^2)^{1/2}$
23. $ye^{3-y} - ye^{2-y}$
25. 4
27. 32
29. $\dfrac{11 \ln 11 - 10 \ln 10 - 2 \ln 2}{2}$
31. 10.57
33. $\dfrac{2e + 1 - e^2}{2e}$
35. 18
37. $e^3 - 2e + \dfrac{1}{e}$
39. 216

8.8 Exercises

1. 6

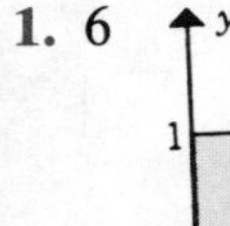

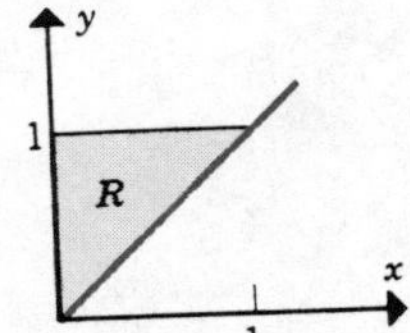

3. 1

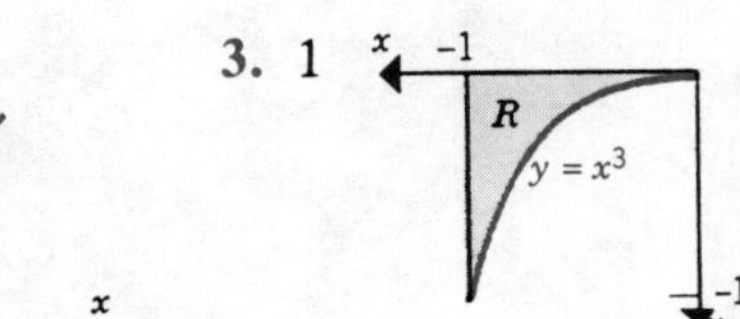

5. 62/15

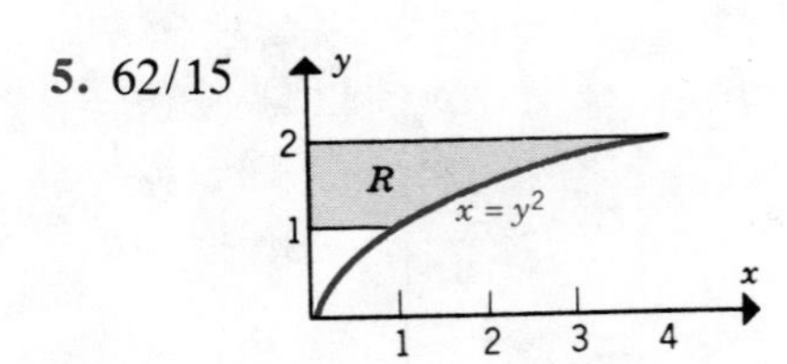

7. 1447/84

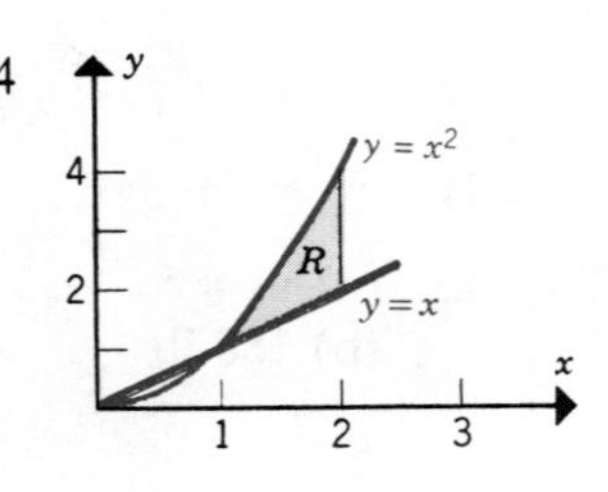

9. 64 11. $2 \ln 2 - 1$ 13. 0 15. $\frac{1}{2e}$ 17. 32/15 19. 0

Review Problems

1. (a) -1 (b) 5 (c) $3h^2 - 2h - 1$ (d) $3h - 2$ (e) -2 (f) -2

3. (a) $\frac{2}{e}$ (b) 1 (c) 1 (d) $\frac{e^2 - 1}{e}$ 5. $f_x = 2x$, $f_y = -2y$

7. $f_x = 6xe^y$, $f_y = 3x^2e^y$ 9. $f_x = \frac{x}{(x^2 - y^3)^{1/2}}$, $f_y = \frac{-3y^2}{2(x^2 - y^3)^{1/2}}$

11. $f_x = 2e^{2x} + 2e^{x-y}$
$f_y = -2e^{x-y} - 2e^{-2y}$
13. $f_{xx} = 2$, $f_{xy} = 4$, $f_{yy} = -6y$

15. $f_{xx} = y^2e^{xy}$, $f_{xy} = e^{xy} + xye^{xy}$, $f_{yy} = x^2e^{xy}$

17. $f_{xx} = 0$, $f_{xy} = -e^{-y} = f_{yx}$, $f_{yy} = xe^{-y}$ 19. $(2, 3, -19)$, relative minimum

21. $(0, 0, 1)$, relative maximum 23. $(0, 2, -19)$, saddle point

25. $(0, 0, 0)$, relative minimum 27. $(3, 3, 18)$ 29. $(24, 12, 624)$

31. $dz = 1.4$ 33. $dz = 0.08$ 35. $5x^3y^4 + g(x)$

37. $\frac{y^2 \ln x}{2} + g(x)$ 39. $126y^2 + 9y$ 41. $ye^2 - ye$ 43. 4 45. $3(e^2 - 1)$

47. 28 49. 1.486 51. 364/3 53. $-1/54$ 55. 2.40

57. (a) \$1221.40 (b) $A_P = e^{0.10t}$, $A_t = 0.10Pe^{0.10t}$ (c) 1.22 (d) \$122/year

59. $x = 27$, $y = 17$ 61. (a) $y = 0.34x + 66$ (b) 134

CHAPTER 9

9.1 Exercises

11. $y = 2x^2 + 3x + 1$ 13. $y = 2x \ln x - 1$ 15. $y = e^{2x}$

17. $y = \frac{1}{x + 4}$ 19. $y = 3e^{x^2}$ 21. $y = x^2 + 3$

23. $y = e^x + e^{-x} + 1$ **25.** $y = \frac{2}{9}x^{3/2} + \frac{8}{3}x - \frac{8}{9}$ **27.** $y = 2e^{-2x} + 16x + 2$

29. $y = (x + 1)^4 - x - 2$ **31.** $s(t) = -16t^2 + 128t + 600$

9.2 Exercises

1. $y = 2x^2 + C$ **3.** $y = Ce^x$ **5.** $y^3 = x^2 + C$

7. $y = Ce^{x^2}$ **9.** $y^2 = x^3 + C$ **11.** $y = 1 + Ce^{x^2}$

13. $y = C(x^2 + 1)^{1/2}$ **15.** $y^2 = x \ln x - x + C$ **17.** $y = -\ln(C - x^2)$

19. $y = \ln(e^x + C)$ **21.** $y^3 = 2x^2 + 1$ **23.** $y = \sqrt{x^3 + 16}$

25. $y = \sqrt{(x^2 + 1)^{1/2} + 1}$ **27.** $y = (x^2 + 8)^{2/3}$ **29.** $y = \frac{-1}{\ln|1 + x| - 1}$

31. $3y = \sqrt{25x^2 + 56}$ **33.** $y^2 + 2y = x^3 + 1$ **35.** $T = 50 + 35e^{-0.028t}$; $t = 44.7$ minutes

9.3 Exercises

1. $xy = C$ **3.** $x^2 - 2xy - 2y = C$ **5.** $x + x^2y - y^3 = C$ **7.** $x^4 + xy - y^3 = C$

9. $x^2y + x = C$ **11.** $ye^x - x^2 + y^2 = C$ **13.** $x^3 - y^3 = 0$

15. $xy - x + y = 2$ **17.** $y^2 - x^2 - 2xy = 2$ **19.** $x^2y^3 - 3x + y = 28$

21. $pq^2 + p = 52$

9.4 Exercises

1. (a) $A = Pe^{0.06t}$ (b) 11.55 years (c) 23.10 = 2(11.55)years

3. 14.65 years **5.** $T = 7$ years **7.** (a) 61.8 days (b) 144 days

9. (a) 586 years (b) 78%

11. $s(t) = s_0 + v_L t + \frac{v_L - v_0}{k}(e^{-kt} - 1)$; $s(t) = 10{,}000 - 200t - 1250(e^{-0.16t} - 1)$

13. (a) $T = 80 + 70e^{-0.056t}$ (b) $t = 6.37$ minutes **15.** 130 minutes

17. $A(t) = 125{,}000 - 25{,}000e^{0.08t}$; 20.1 years **19.** $V = 0.02(1 - e^{-t/200})$

Review Problems

5. $y = \frac{1}{x^2 + 2}$ **7.** $y = e^{2x}$ **9.** $y = 3x^2 - 2x + 4$

11. $y = \frac{x^2 \ln x}{2} - \frac{x^2}{4} + \frac{13}{4}$ **13.** $2y^4 = x^3 + 31$ **15.** $e^y = x \ln x - x + 2$

17. $6y^{3/2} = x^2 + 48$ **19.** $y = \sqrt[4]{x^3 - 7}$ **21.** $x^3 - 2y^2 = C$

23. $ye^x - xe^y = C$ **25.** (a) $\frac{dA}{dt} = 4 - kA$, where k is the constant of proportionality

(b) $A(t) = \frac{4}{k} + Ke^{-kt}$, where K is a constant

$A(t) = 20 - 20e^{-0.20t}$

27. (a) $\frac{dA}{dt} = -kA\ (k > 0)$; $A(t) = 30e^{-0.55t}$; (b) $\frac{dA}{dt} = 20 - 0.55A$; $A(t) = 36.4 - 31.4e^{-0.55t}$

29. $T = 8$ years

CHAPTER 10

10.1 Exercises

1. $S = \{Hd, Hc, Hh, Hs, Td, Tc, Th, Ts\}$, where H is heads, T is tails, d is diamonds, c is clubs, h is hearts, and s is spades. $P(Hd) = P(Hc) = \ldots = P(Ts) = 1/8$

3. $S = \{A, B, C\}$; $P(A) = 6/11$, $P(B) = 3/11$, $P(C) = 2/11$

5. The assignment of probabilities is in accord with the rules of probability.

7. The sum of the probabilities is $17/16 \neq 1$. **9.** The assignment of probabilities is in accord with the rules of probability. **11.** $E(X) = 1.5$; $V(X) = 1.25$; $\sigma(X) = 1.12$

13. It qualifies as a probability distribution. **15.** The assignment $P(x = 3) = -0.01$ violates the nonnegativity requirement. **17.** The sum of the probabilities is $1.01 \neq 1.00$.

19. $E(X) = -\$0.50$; $V(X) = 499.75$; $\sigma(X) = \$22.4$

10.2 Exercises

1. Since $f(x) \geq 0$ for $2 \leq x \leq 5$ and $\int_2^5 \frac{1}{3}\,dx = 1$, the function qualifies as a probability density function.

3. Since $f(x) \geq 0$ for $1 \leq x \leq 3$ and $\int_1^3 \frac{3}{2x^2} = 1$, the function qualifies as a probability density function.

5. Because $\int_2^3 \frac{x-1}{2}\,dx = \frac{1}{2} \neq 1$, the function does not qualify.

7. Because $f(x) < 0$ for all x in the interval $0 < x < 0.577$, the function does not meet the requirements for a probability density function.

9. $K = \frac{1}{2}$ **11.** $K = -6$ **13.** $K = 2$ **15.** $K = \frac{2}{\ln 2}$

17. $F(x) = \begin{cases} 0, & x < 2 \\ \dfrac{x-2}{4} & 2 \le x \le 6 \\ 1, & x > 6 \end{cases}$ 19. $F(x) = \begin{cases} 0, \\ 1 - e^{-x}, & x \geq \end{cases}$

21. Graph (a). If $T = t$, where $0 \le t \le 10$, $P(T \le t)$ is largest for the probabili... function in graph (a).

23. (a) $\dfrac{e-2}{2}$ (b) $F(y) = \begin{cases} 0, & y < 0 \\ 1 - e^{-y} - ye^{-y}, & y \ge 0 \end{cases}$

25. $F'(x) = f(x) < 0$ for all x in the interval $L \le x \le M$, so the nonnegativity condition is violated.

27. $F(L) \neq 0$

10.3 Exercises

1. $E(X) = 4.5$; $V(X) = 2.08$; $\sigma(X) = 1.44$
3. $E(X) = 5.4$; $V(X) = 5.55$; $\sigma(X) = 2.36$
5. $E(X) = 1$; $V(X) = 1$; $\sigma(X) = 1$ 7. $E(X) = 3/4$; $V(X) = 0.0375$; $\sigma(X) = 0.1936$
9. $E(X) = 4/3$; $V(X) = 0.0706$; $\sigma(X) = 0.2657$
11. (a) $E(T) = 4$ hours; $\sigma(T) = 4$ hours (b) $P(T > 4) = \dfrac{1}{e}$ (c) 0.8647
13. (a) $E(T) = 9$ days; $V(T) = 1.8$ days^2; $\sigma(T) = 1.34$ days (b) 0.85
15. (a) $E(T) = 2.4$ hours; $\sigma(T) = 1.047$ hours (b) 0.535 (c) 0.979
17. $E(X) = 1.5$ in.; $\sigma(T) = 0.67$ in. 19. $P(35 \le D \le 65) \ge 8/9$ 21. $m = 5$
23. $m = 0.347$ $m = 1.794$

Review Problems

1. (a) $E = \{(1, 1),(2, 2),(3, 3),(4, 4),(5, 5),(6, 6)\}$ (b) $P(E) = 1/6$ (c) $X = 2,3,4, \ldots ,11,12$ (d) $P(X = 2) = P(X = 12) = 1/36$; $P(X = 3) = P(X = 11) = 2/26$; $P(X = 4) = P(X = 10) = 3/36$ $P(X = 5) = P(X = 9) = 4/36$; $P(X = 6) = P(X = 8) = 5/36$; $P(X = 7) = 6/36$ (e) $E(X) = 7$; $V(X) = 5.83$, $\sigma(X) = 2.42$
3. (a) $P(X = 1) = 2/10$; $P(X = 2) = 3/10$; $P(X = 3) = 5/10$ (b) 2.3 (c) 0.78
5. (a) $P(X = 2) = 1/4$, $P(X = 1) = 1/2$, $P(X = 0) = 1/4$ (b) $E(X) = 1.0$, $V(X) = 1/2$, $\sigma(X) = 0.71$ (c) $P(G = -1.50) = 1/4$, $P(G = -0.50) = 1/2$, $P(G = 0.50) = 1/4$ (d) $E(G) = -0.50$, $V(G) = 0.50$, $\sigma(G) = 0.71$
7. (a) $P(X = 1) = P(X = 3) = P(X = 5) = 1/9$; $P(X = 2) = P(X = 4) = P(X = 6) = 2/9$ (b) $E(X) = 3.67$, $V(X) = 2.89$, $\sigma(X) = 1.70$

$x \le 2$, and $\int_1^2 (2x - 2)\, dx = 1$, the function

9. Since $f(x) \ge 0$ for all x in th... nction.

qualifies as a probabili... in the interval $0 \le x \le 1$ and $\frac{1}{2}\int_0^1 (3 - 3x^2)\, dx = 1$, the function

11. Since $fx \ge$ probability density function.

qualif...
because $\int_{-1}^{1} (1 - x)\, dx = 2 \ne 1$, the function does not qualify. **15.** $K = 4/5$

17. $K = 2$

19. $F(x) = \begin{cases} 0, & x \le 0 \\ \dfrac{x^4}{16}, & 0 \le x \le 2 \\ 1, & x > 2 \end{cases}$ **21.** $F(x) = \begin{cases} 0, & x < 1 \\ \dfrac{4}{3}\left(1 - \dfrac{1}{x^2}\right), & 1 \le x \le 2 \\ 1, & x > 2 \end{cases}$

23. (a) $\dfrac{e - 1}{e}$ (b) $\dfrac{1}{e^2}$ **25.** (a) $\dfrac{1}{e}$ (b) 5 seconds (c) 25 seconds^2

27. $P(15 \le T \le 45) = P[E(T) - 3\sigma(T) \le T \le E(T) + 3\sigma(T)] \ge 8/9$

INDEX

Printed by
Fong and Sons Printers Pte Ltd